Methods in Enzymology

Volume 305

BIOLUMINESCENCE AND CHEMILUMINESCENCE

Part C

METHODS IN ENZYMOLOGY

EDITORS-IN-CHIEF

John N. Abelson Melvin I. Simon

DIVISION OF BIOLOGY
CALIFORNIA INSTITUTE OF TECHNOLOGY
PASADENA, CALIFORNIA

FOUNDING EDITORS

Sidney P. Colowick and Nathan O. Kaplan

Methods in Enzymology

Volume 305

Bioluminescence and Chemiluminescence

Part C

EDITED BY

Miriam M. Ziegler

UNIVERSITY OF ARIZONA
TUCSON, ARIZONA

Thomas O. Baldwin

UNIVERSITY OF ARIZONA
TUCSON, ARIZONA

ACADEMIC PRESS

San Diego London Boston New York Sydney Tokyo Toronto

This book is printed on acid-free paper. ♾

Academic Press
A Harcourt Science and Technology Company
525 B Street, Suite 1900, San Diego, California 92101-4495, USA
http://www.academicpress.com

Academic Press Limited
24-28 Oval Road, London NW1 7DX, UK
http://www.hbuk.co.uk/ap/

International Standard Book Number: 0-12-182206-0

PRINTED IN THE UNITED STATES OF AMERICA
00 01 02 03 04 05 MM 9 8 7 6 5 4 3 2 1

Table of Contents

Section I. Chemiluminescence and Bioluminescence: Overviews

Section II. Instrumentation

Section III. Luciferases, Luminescence Accessory Proteins, and Substrates

Section IV. Bacterial Autoinduction System and Its Applications

Section V. Luminescence-Based Assays *in Vitro*

Section VI. Luminescence Monitoring *in Vivo*

Section VII. Bioluminescence as an Education Tool

Contributors to Volume 305

Article numbers are in parentheses following the names of contributors.
Affiliations listed are current.

ROBERT C. ALLEN (41), *Department of Pathology and Laboratory Medicine, School of Medicine, Emory University, Atlanta, Georgia 30335*

CAROL R. ANDERSSON (37), *CSIRO Plant Industry, Wembley, Australia WA 6014*

JOSEPH D. ANDRADE (43), *Department of Bioengineering and Center for Science Education and Outreach, University of Utah, Salt Lake City, Utah 84112-9202*

MICHAEL N. BADMINTON (34), *Department of Medical Biochemistry, University of Wales College of Medicine, Heath Park, Cardiff CF4 4XN, United Kingdom*

THOMAS O. BALDWIN (9, 11, 13), *Department of Biochemistry, University of Arizona, Tucson, Arizona 85721-0088*

MARIO BARALDINI (8, 40), *Institute of Chemical Sciences, University of Bologna, 40127 Bologna, Italy*

FRITZ BERTHOLD (4), *Berthold Intellectual Properties, D-75175 Pforzheim, Germany*

JOHN R. BLINKS (17), *Friday Harbor Laboratories, University of Washington, Friday Harbor, Washington 98250*

VLADIMIR S. BONDAR (17), *Photobiology Laboratory, Institute of Biophysics, Russian Academy of Sciences, 660036 Krasnoyarsk, Russia*

BRUCE R. BRANCHINI (14), *Department of Chemistry, Connecticut College, New London, Connecticut 06320*

IRENA BRONSTEIN (26, 29, 30, 31, 33), *Tropix, Inc., Bedford, Massachusetts 01730*

VALERIE J. BUSH (28), *Mayo Foundation for Medical Education and Research, Mayo Clinic and Foundation, Rochester, Minnesota 55905*

ANTHONY K. CAMPBELL (34), *Department of Medical Biochemistry, University of Wales College of Medicine, Heath Park, Cardiff CF4 4XN, United Kingdom*

LISA CAMPBELL (35), *Department of Oceanography, Texas A&M University, College Station, Texas 77843-3146*

A. CLAY CLARK (11), *Department of Biochemistry and Molecular Biophysics, North Carolina State University, Raleigh, North Carolina 27695-7622*

JEAN E. CRABTREE (40), *Molecular Medicine Unit, St. James's University Hospital, Leeds LS9 7TRF, England*

DAVID C. DALE (41), *Department of Medicine, University of Washington, Seattle, Washington 98195*

JERRY H. DEVINE (20), *Department of Cell Biology and Biochemistry, Texas Tech University Health Science Center, Lubbock, Texas 79430*

ANATOL EBERHARD (22), *Department of Chemistry, Ithaca College, Ithaca, New York 14850-7279*

THOMAS A. EBERT (28), *Mayo Foundation for Medical Education and Research, Mayo Clinic, Rochester, Minnesota 55905*

W. HOWARD EVANS (34), *Department of Medical Biochemistry, University of Wales College of Medicine, Heath Park, Cardiff CF4 4XN, United Kingdom*

F. FERNÁNDEZ-PIÑAS (36), *Departamento de Biologia, Facultad de Ciencias, Universidad Autùnoma de Madrid, Cantoblanco, E-28049 Madrid, Spain*

LUDMILA A. FRANK (17), *Photobiology Laboratory, Institute of Biophysics, Russian Academy of Sciences, 660036 Krasnoyarsk, Russia*

S. GELMINI (39), *Clinical Biochemistry Unit, University of Florence, 50139 Florence, Italy*

CHRISTOPHER H. GEORGE (34), *Department of Medical Biochemistry, University of Wales College of Medicine, Heath Park, Cardiff CF4 4XN, United Kingdom*

Bruce G. Gibson (12), *Department of Biochemistry and Molecular Biology, University of Georgia, Athens, Georgia 30602*

Susan S. Golden (37), *Department of Biology, Texas A&M University, College Station, Texas 77843-3258*

Daniel G. González (16, 44), *Department of Biochemistry and Microbiology, Cook College, Rutgers University, New Brunswick, New Jersey 08901-8525*

Vicki A. Green (9, 13), *Department of Biochemistry and Biophysics, Texas A&M University, College Station, Texas 77843-2128*

E. Peter Greenberg (21), *Department of Microbiology, University of Iowa, Iowa City, Iowa 52242*

Brenda J. Hallaway (27, 28), *Mayo Foundation for Medical Education and Research, Mayo Clinic, Rochester, Minnesota 55905*

Mara Hammer (43), *Department of Bioengineering and Center for Science Education and Outreach, University of Utah, Salt Lake City, Utah 84112-9202*

Brian L. Hanzelka (21), *Department of Microbiology, University of Iowa, Iowa City, Iowa 52242*

Klaus Herick (4), *Berthold GmbH and Company KG, D-75323 Bad Wildbad, Germany*

James C. Hu (35), *Department of Biochemistry and Biophysics and Center for Macromolecular Design, Texas A&M University, College Station, Texas 77843-2128*

Boris A. Illarionov (17), *Photobiology Laboratory, Institute of Biophysics, Russian Academy of Sciences, 660036 Krasnoyarsk, Russia*

Victoria A. Illarionova (17), *Photobiology Laboratory, Institute of Biophysics, Russian Academy of Sciences, 660036 Krasnoyarsk, Russia*

Helen E. Jones (34), *Department of Medical Biochemistry, University of Wales College of Medicine, Heath Park, Cardiff CF4 4XN, United Kingdom*

Jonathan M. Kendall (34), *Department of Medical Biochemistry, University of Wales College of Medicine, Heath Park, Cardiff CF4 4XN, United Kingdom*

Larry J. Kricka (24, 26), *Department of Pathology and Laboratory Medicine, University of Pennsylvania, Philadelphia, Pennsylvania 19104*

J. Kuhn (38), *Department of Food Engineering and Biotechnology, Technion, Haifa 32000, Israel*

Nadya V. Lebedeva (37), *Timiryazev Institute of Plant Physiology, 127276 Moscow, Russia*

John Lee (5, 12), *Department of Biochemistry and Molecular Biology, University of Georgia, Athens, Georgia 30602*

F. Leganés (36), *MSU-DOE Plant Research Laboratory, Michigan State University, East Lansing, Michigan 48824-1312*

Liming Li (18), *Department of Molecular Genetics and Cell Biology, University of Chicago, Chicago, Illinois 60637*

Betty Liu (31), *Tropix, Inc., Bedford, Massachusetts 01730*

Arne Lundin (25), *BioThema AB, S-130 54 Dalarö, Sweden*

Chris S. Martin (31, 33), *Tropix, Inc., Bedford, Massachusetts 01730*

Frank McCapra (1, 42), *Seaford, East Sussex BN25 4PG, England*

Hongtao Min (37), *Department of Biology, Texas A&M University, College Station, Texas 77843-3258*

Maria Mittag (19), *Botanisches Institut der Ludwig-Maximilians-Universität München, D-80638 München, Germany*

David Morse (19), *Biology Department, University of Montreal, Montreal, Quebec, Canada H1X 2B2*

Jennifer Mosier (29, 30, 31), *The Center for Blood Research, Boston, Massachusetts 02115*

Monica Musiani (8, 40), *Department of Clinical and Experimental Medicine, Division of Microbiology, University of Bologna, 40138 Bologna, Italy*

Brian W. Noland (11), *Department of Biochemistry, University of Arizona, Tucson, Arizona 85721*

DENNIS J. O'KANE (5, 27, 28, 32), *Department of Laboratory Medicine and Pathology, Mayo Clinic and Foundation, Rochester, Minnesota 55905*

CORINNE E. M. OLESEN (29, 30, 31), *Tropix, Inc., Bedford, Massachusetts 01730*

MATTHEW R. PARSEK (21), *Department of Microbiology, University of Iowa, Iowa City, Iowa 52242*

PATRIZIA PASINI (8, 40), *Department of Pharmaceutical Sciences, University of Bologna, 40126 Bologna, Italy*

M. PAZZAGLI (39), *Clinical Biochemistry Unit, University of Florence, 50139 Florence, Italy*

VALENTIN N. PETUSHKOV (12), *Institute of Biophysics, Academy of Sciences of Russia (Siberian Branch), 660036 Krasnoyarsk, Russia*

P. PINZANI (39), *Clinical Biochemistry Unit, University of Florence, 50139 Florence, Italy*

S. JAMES REMINGTON (15), *Department of Physics, Institute of Molecular Biology, University of Oregon, Eugene, Oregon 97403*

ANGELA R. RIBEIRO (34), *Department of Medical Biochemistry, University of Wales College of Medicine, Heath Park, Cardiff CF4 4XN, United Kingdom*

ALDO RODA (8, 40), *Department of Pharmaceutical Sciences, University of Bologna, 40126 Bologna, Italy*

TILL ROENNEBERG (7), *Institute for Medical Psychology, University of Munich, D-80336 Munich, Germany*

GRACIELA B. SALA-NEWBY (34), *Department of Medical Biochemistry, University of Wales College of Medicine, Heath Park, Cardiff CF4 4XN, United Kingdom*

AMY L. SCHAEFER (21), *Department of Microbiology, University of Iowa, Iowa City, Iowa 52242*

JEFFREY B. SCHINELLER (22), *Department of Chemistry, Humboldt State University, Arcata, California 95521*

GERALD S. SHADEL (20), *Department of Biochemistry, Emory University School of Medicine, Atlanta, Georgia 30322*

JEFFREY SHELTON (37), *Timiryazev Institute of Plant Physiology, 127276 Moscow, Russia*

DEBORAH A. SIEGELE (35), *Department of Biology, Texas A&M University, College Station, Texas 77843-3258*

RUTH M. SIEWE (4), *Berthold GmbH and Company KG, D-75323 Bad Wildbad, Germany*

JAMES F. SINCLAIR (10), *Department of Microbiology and Immunology, Uniformed Services University of the Health Sciences, Bethesda, Maryland 20814-4712*

PHILIP E. STANLEY (2, 6), *Cambridge Research and Technology Transfer Limited, Cambridge CB1 2HF, United Kingdom*

GAVIN C. SWIATEK (44), *Department of Biochemistry and Microbiology, Cook College, Rutgers University, New Brunswick, New Jersey 08901-8525*

ANITA VAN TILBURG (23), *Department of Biochemistry, University of Arizona, Tucson, Arizona 85721-0088*

FLETCHER B. TAYLOR, JR. (41), *Oklahoma Medical Research Foundation, Oklahoma City, Oklahoma 73190*

WALTER TAYLOR (7), *Perkin Elmer Applied Biosystems, Foster City, California 94404*

KATHLEEN S. TENNER (32), *Department of Laboratory Medicine and Pathology, Mayo Clinic, Rochester, Minnesota 55905*

MICHAEL D. THOMAS (9, 23), *Department of Biochemistry, University of Arizona, Tucson, Arizona 85721-0088*

NICHOLAS F. TSINOREMAS (37), *Pangea Systems, Inc., Oakland, California 94612*

S. ULITZUR (38), *Department of Food Engineering and Biotechnology, Technion, Haifa 32000, Israel*

ANTONIE J. W. G. VISSER (12), *Department of Biochemistry, Agricultural University, NL-6703 HA Wageningen, The Netherlands*

KENNETH J. VOSS (3), *Physics Department, University of Miami, Coral Gables, Florida 33124*

JOHN C. VOYTA (26, 29, 30, 31, 33), *Tropix, Inc., Bedford, Massachusetts 01730*

EUGENE S. VYSOTSKI (17), *Photobiology Laboratory, Institute of Biophysics, Russian Academy of Sciences, 660036 Krasnoyarsk, Russia*

WILLIAM W. WARD (16, 44), *Department of Biochemistry and Microbiology, Cook College, Rutgers University, New Brunswick, New Jersey 08901-8525*

DAVID M. WILSON (28), *Mayo Foundation for Medical Education and Research, Mayo Clinic and Foundation, Rochester, Minnesota 55905*

C. PETER WOLK (36), *MSU-DOE Plant Research Laboratory, Michigan State University, East Lansing, Michigan 48824-1312*

JUSTIN YARROW (37), *Department of Cell Biology, Harvard Medical School, Boston, Massachusetts 02115*

MIRIAM M. ZIEGLER (9), *Department of Biochemistry, University of Arizona, Tucson, Arizona 85721-0088*

Preface

In recognition of the pioneering efforts of Marlene A. DeLuca and William D. McElroy in the basic science and applications of bioluminescence and chemiluminescence, we wish to dedicate this volume to their memory.

Volumes 57 (edited by DeLuca) and 133 (edited by DeLuca and McElroy) of *Methods in Enzymology,* Bioluminescence and Chemiluminescence Parts A and B, published in 1978 and 1986, respectively, supplied clear templates for the development of this volume. In the field of bioluminescence and chemiluminescence, the late Drs. DeLuca and McElroy are well recognized for their seminal contributions to the basic scientific understanding of many bioluminescent systems; they also demonstrated the value of light, biological or nonbiological, in a vast array of assays. The first meeting of what was to become an approximately biennial series of International Symposia on Bioluminescence and Chemiluminescence was held in Brussels in 1978, the same year that the first *Methods in Enzymology* volume dedicated to the applications of bioluminescence and chemiluminescence was published. DeLuca and McElroy organized the second international symposium in San Diego in 1980, and served as leaders in developing the scientific programs of the international symposia in Birmingham (1984), Freiburg (1986), and Florence (1988).

Marlene DeLuca passed away in 1987 before the symposium in Florence, a meeting she had helped organize. During the Florence symposium, Bill McElroy received an honorary degree from the University of Bologna, in the year of the 900th anniversary of the founding of that distinguished university, in recognition of his numerous contributions to the field of bioluminescence and chemiluminescence and of his long and distinguished career that continued until his death in 1999.

Volume 133 of *Methods in Enzymology* was published shortly after a series of advances derived from the cloning and expression of the bacterial *lux* genes and the subsequent cloning of a large number of other luciferase-encoding sequences. Availability of luminescence-conferring genetic material enabled the development of numerous *in vivo* assays and imaging technologies, some of which are described in this volume. DeLuca and McElroy predicted the vast expansion in applications of luminescence for *in vivo* diagnostics during the 1984 symposium in Birmingham.

In organizing this volume, it was impossible to include all of the exciting and powerful methods currently in use. In selecting topics, our objective was to achieve a mix of chapters dealing with the science behind the methods, chapters that detail specific methods and approaches for working with different luminescent systems, and chapters that describe specific applications of luminescence. An exciting recent development in the field of bioluminescence and chemiluminescence has been the application of luminescence in the teaching of basic science at all grade levels. The final section of the book presents the efforts of three groups to utilize luminescence as a teaching tool.

The efforts of the chapter authors are acknowledged with deep gratitude. The staff of Academic Press was, as usual, excellent. Finally, the value of the efforts of Ms. Shirley Light of Academic Press cannot be overstated. Her attention to detail, and to the calendar, has been an enormous help to us.

Miriam M. Ziegler
Thomas O. Baldwin

METHODS IN ENZYMOLOGY

VOLUME I. Preparation and Assay of Enzymes
Edited by SIDNEY P. COLOWICK AND NATHAN O. KAPLAN

VOLUME II. Preparation and Assay of Enzymes
Edited by SIDNEY P. COLOWICK AND NATHAN O. KAPLAN

VOLUME III. Preparation and Assay of Substrates
Edited by SIDNEY P. COLOWICK AND NATHAN O. KAPLAN

VOLUME IV. Special Techniques for the Enzymologist
Edited by SIDNEY P. COLOWICK AND NATHAN O. KAPLAN

VOLUME V. Preparation and Assay of Enzymes
Edited by SIDNEY P. COLOWICK AND NATHAN O. KAPLAN

VOLUME VI. Preparation and Assay of Enzymes (*Continued*)
Preparation and Assay of Substrates
Special Techniques
Edited by SIDNEY P. COLOWICK AND NATHAN O. KAPLAN

VOLUME VII. Cumulative Subject Index
Edited by SIDNEY P. COLOWICK AND NATHAN O. KAPLAN

VOLUME VIII. Complex Carbohydrates
Edited by ELIZABETH F. NEUFELD AND VICTOR GINSBURG

VOLUME IX. Carbohydrate Metabolism
Edited by WILLIS A. WOOD

VOLUME X. Oxidation and Phosphorylation
Edited by RONALD W. ESTABROOK AND MAYNARD E. PULLMAN

VOLUME XI. Enzyme Structure
Edited by C. H. W. HIRS

VOLUME XII. Nucleic Acids (Parts A and B)
Edited by LAWRENCE GROSSMAN AND KIVIE MOLDAVE

VOLUME XIII. Citric Acid Cycle
Edited by J. M. LOWENSTEIN

VOLUME XIV. Lipids
Edited by J. M. LOWENSTEIN

VOLUME XV. Steroids and Terpenoids
Edited by RAYMOND B. CLAYTON

VOLUME XVI. Fast Reactions
Edited by KENNETH KUSTIN

VOLUME XVII. Metabolism of Amino Acids and Amines (Parts A and B)
Edited by HERBERT TABOR AND CELIA WHITE TABOR

VOLUME XVIII. Vitamins and Coenzymes (Parts A, B, and C)
Edited by DONALD B. MCCORMICK AND LEMUEL D. WRIGHT

VOLUME XIX. Proteolytic Enzymes
Edited by GERTRUDE E. PERLMANN AND LASZLO LORAND

VOLUME XX. Nucleic Acids and Protein Synthesis (Part C)
Edited by KIVIE MOLDAVE AND LAWRENCE GROSSMAN

VOLUME XXI. Nucleic Acids (Part D)
Edited by LAWRENCE GROSSMAN AND KIVIE MOLDAVE

VOLUME XXII. Enzyme Purification and Related Techniques
Edited by WILLIAM B. JAKOBY

VOLUME XXIII. Photosynthesis (Part A)
Edited by ANTHONY SAN PIETRO

VOLUME XXIV. Photosynthesis and Nitrogen Fixation (Part B)
Edited by ANTHONY SAN PIETRO

VOLUME XXV. Enzyme Structure (Part B)
Edited by C. H. W. HIRS AND SERGE N. TIMASHEFF

VOLUME XXVI. Enzyme Structure (Part C)
Edited by C. H. W. HIRS AND SERGE N. TIMASHEFF

VOLUME XXVII. Enzyme Structure (Part D)
Edited by C. H. W. HIRS AND SERGE N. TIMASHEFF

VOLUME XXVIII. Complex Carbohydrates (Part B)
Edited by VICTOR GINSBURG

VOLUME XXIX. Nucleic Acids and Protein Synthesis (Part E)
Edited by LAWRENCE GROSSMAN AND KIVIE MOLDAVE

VOLUME XXX. Nucleic Acids and Protein Synthesis (Part F)
Edited by KIVIE MOLDAVE AND LAWRENCE GROSSMAN

VOLUME XXXI. Biomembranes (Part A)
Edited by SIDNEY FLEISCHER AND LESTER PACKER

VOLUME XXXII. Biomembranes (Part B)
Edited by SIDNEY FLEISCHER AND LESTER PACKER

VOLUME XXXIII. Cumulative Subject Index Volumes I–XXX
Edited by MARTHA G. DENNIS AND EDWARD A. DENNIS

VOLUME XXXIV. Affinity Techniques (Enzyme Purification: Part B)
Edited by WILLIAM B. JAKOBY AND MEIR WILCHEK

VOLUME XXXV. Lipids (Part B)
Edited by JOHN M. LOWENSTEIN

VOLUME XXXVI. Hormone Action (Part A: Steroid Hormones)
Edited by BERT W. O'MALLEY AND JOEL G. HARDMAN

VOLUME XXXVII. Hormone Action (Part B: Peptide Hormones)
Edited by BERT W. O'MALLEY AND JOEL G. HARDMAN

VOLUME XXXVIII. Hormone Action (Part C: Cyclic Nucleotides)
Edited by JOEL G. HARDMAN AND BERT W. O'MALLEY

VOLUME XXXIX. Hormone Action (Part D: Isolated Cells, Tissues, and Organ Systems)
Edited by JOEL G. HARDMAN AND BERT W. O'MALLEY

VOLUME XL. Hormone Action (Part E: Nuclear Structure and Function)
Edited by BERT W. O'MALLEY AND JOEL G. HARDMAN

VOLUME XLI. Carbohydrate Metabolism (Part B)
Edited by W. A. WOOD

VOLUME XLII. Carbohydrate Metabolism (Part C)
Edited by W. A. WOOD

VOLUME XLIII. Antibiotics
Edited by JOHN H. HASH

VOLUME XLIV. Immobilized Enzymes
Edited by KLAUS MOSBACH

VOLUME XLV. Proteolytic Enzymes (Part B)
Edited by LASZLO LORAND

VOLUME XLVI. Affinity Labeling
Edited by WILLIAM B. JAKOBY AND MEIR WILCHEK

VOLUME XLVII. Enzyme Structure (Part E)
Edited by C. H. W. HIRS AND SERGE N. TIMASHEFF

VOLUME XLVIII. Enzyme Structure (Part F)
Edited by C. H. W. HIRS AND SERGE N. TIMASHEFF

VOLUME XLIX. Enzyme Structure (Part G)
Edited by C. H. W. HIRS AND SERGE N. TIMASHEFF

VOLUME L. Complex Carbohydrates (Part C)
Edited by VICTOR GINSBURG

VOLUME LI. Purine and Pyrimidine Nucleotide Metabolism
Edited by PATRICIA A. HOFFEE AND MARY ELLEN JONES

VOLUME LII. Biomembranes (Part C: Biological Oxidations)
Edited by SIDNEY FLEISCHER AND LESTER PACKER

VOLUME LIII. Biomembranes (Part D: Biological Oxidations)
Edited by SIDNEY FLEISCHER AND LESTER PACKER

VOLUME LIV. Biomembranes (Part E: Biological Oxidations)
Edited by SIDNEY FLEISCHER AND LESTER PACKER

VOLUME LV. Biomembranes (Part F: Bioenergetics)
Edited by SIDNEY FLEISCHER AND LESTER PACKER

VOLUME LVI. Biomembranes (Part G: Bioenergetics)
Edited by SIDNEY FLEISCHER AND LESTER PACKER

VOLUME LVII. Bioluminescence and Chemiluminescence
Edited by MARLENE A. DELUCA

VOLUME LVIII. Cell Culture
Edited by WILLIAM B. JAKOBY AND IRA PASTAN

VOLUME LIX. Nucleic Acids and Protein Synthesis (Part G)
Edited by KIVIE MOLDAVE AND LAWRENCE GROSSMAN

VOLUME LX. Nucleic Acids and Protein Synthesis (Part H)
Edited by KIVIE MOLDAVE AND LAWRENCE GROSSMAN

VOLUME 61. Enzyme Structure (Part H)
Edited by C. H. W. HIRS AND SERGE N. TIMASHEFF

VOLUME 62. Vitamins and Coenzymes (Part D)
Edited by DONALD B. MCCORMICK AND LEMUEL D. WRIGHT

VOLUME 63. Enzyme Kinetics and Mechanism (Part A: Initial Rate and Inhibitor Methods)
Edited by DANIEL L. PURICH

VOLUME 64. Enzyme Kinetics and Mechanism (Part B: Isotopic Probes and Complex Enzyme Systems)
Edited by DANIEL L. PURICH

VOLUME 65. Nucleic Acids (Part I)
Edited by LAWRENCE GROSSMAN AND KIVIE MOLDAVE

VOLUME 66. Vitamins and Coenzymes (Part E)
Edited by DONALD B. MCCORMICK AND LEMUEL D. WRIGHT

VOLUME 67. Vitamins and Coenzymes (Part F)
Edited by DONALD B. MCCORMICK AND LEMUEL D. WRIGHT

VOLUME 68. Recombinant DNA
Edited by RAY WU

VOLUME 69. Photosynthesis and Nitrogen Fixation (Part C)
Edited by ANTHONY SAN PIETRO

VOLUME 70. Immunochemical Techniques (Part A)
Edited by HELEN VAN VUNAKIS AND JOHN J. LANGONE

VOLUME 71. Lipids (Part C)
Edited by JOHN M. LOWENSTEIN

VOLUME 72. Lipids (Part D)
Edited by JOHN M. LOWENSTEIN

VOLUME 125. Biomembranes (Part M: Transport in Bacteria, Mitochondria, and Chloroplasts: General Approaches and Transport Systems)
Edited by SIDNEY FLEISCHER AND BECCA FLEISCHER

VOLUME 126. Biomembranes (Part N: Transport in Bacteria, Mitochondria, and Chloroplasts: Protonmotive Force)
Edited by SIDNEY FLEISCHER AND BECCA FLEISCHER

VOLUME 127. Biomembranes (Part O: Protons and Water: Structure and Translocation)
Edited by LESTER PACKER

VOLUME 128. Plasma Lipoproteins (Part A: Preparation, Structure, and Molecular Biology)
Edited by JERE P. SEGREST AND JOHN J. ALBERS

VOLUME 129. Plasma Lipoproteins (Part B: Characterization, Cell Biology, and Metabolism)
Edited by JOHN J. ALBERS AND JERE P. SEGREST

VOLUME 130. Enzyme Structure (Part K)
Edited by C. H. W. HIRS AND SERGE N. TIMASHEFF

VOLUME 131. Enzyme Structure (Part L)
Edited by C. H. W. HIRS AND SERGE N. TIMASHEFF

VOLUME 132. Immunochemical Techniques (Part J: Phagocytosis and Cell-Mediated Cytotoxicity)
Edited by GIOVANNI DI SABATO AND JOHANNES EVERSE

VOLUME 133. Bioluminescence and Chemiluminescence (Part B)
Edited by MARLENE DELUCA AND WILLIAM D. MCELROY

VOLUME 134. Structural and Contractile Proteins (Part C: The Contractile Apparatus and the Cytoskeleton)
Edited by RICHARD B. VALLEE

VOLUME 135. Immobilized Enzymes and Cells (Part B)
Edited by KLAUS MOSBACH

VOLUME 136. Immobilized Enzymes and Cells (Part C)
Edited by KLAUS MOSBACH

VOLUME 137. Immobilized Enzymes and Cells (Part D)
Edited by KLAUS MOSBACH

VOLUME 138. Complex Carbohydrates (Part E)
Edited by VICTOR GINSBURG

VOLUME 139. Cellular Regulators (Part A: Calcium- and Calmodulin-Binding Proteins)
Edited by ANTHONY R. MEANS AND P. MICHAEL CONN

VOLUME 140. Cumulative Subject Index Volumes 102–119, 121–134

VOLUME 141. Cellular Regulators (Part B: Calcium and Lipids)
Edited by P. MICHAEL CONN AND ANTHONY R. MEANS

VOLUME 142. Metabolism of Aromatic Amino Acids and Amines
Edited by SEYMOUR KAUFMAN

VOLUME 143. Sulfur and Sulfur Amino Acids
Edited by WILLIAM B. JAKOBY AND OWEN GRIFFITH

VOLUME 144. Structural and Contractile Proteins (Part D: Extracellular Matrix)
Edited by LEON W. CUNNINGHAM

VOLUME 145. Structural and Contractile Proteins (Part E: Extracellular Matrix)
Edited by LEON W. CUNNINGHAM

VOLUME 146. Peptide Growth Factors (Part A)
Edited by DAVID BARNES AND DAVID A. SIRBASKU

VOLUME 147. Peptide Growth Factors (Part B)
Edited by DAVID BARNES AND DAVID A. SIRBASKU

VOLUME 148. Plant Cell Membranes
Edited by LESTER PACKER AND ROLAND DOUCE

VOLUME 149. Drug and Enzyme Targeting (Part B)
Edited by RALPH GREEN AND KENNETH J. WIDDER

VOLUME 150. Immunochemical Techniques (Part K: *In Vitro* Models of B and T Cell Functions and Lymphoid Cell Receptors)
Edited by GIOVANNI DI SABATO

VOLUME 151. Molecular Genetics of Mammalian Cells
Edited by MICHAEL M. GOTTESMAN

VOLUME 152. Guide to Molecular Cloning Techniques
Edited by SHELBY L. BERGER AND ALAN R. KIMMEL

VOLUME 153. Recombinant DNA (Part D)
Edited by RAY WU AND LAWRENCE GROSSMAN

VOLUME 154. Recombinant DNA (Part E)
Edited by RAY WU AND LAWRENCE GROSSMAN

VOLUME 155. Recombinant DNA (Part F)
Edited by RAY WU

VOLUME 156. Biomembranes (Part P: ATP-Driven Pumps and Related Transport: The Na,K-Pump)
Edited by SIDNEY FLEISCHER AND BECCA FLEISCHER

VOLUME 157. Biomembranes (Part Q: ATP-Driven Pumps and Related Transport: Calcium, Proton, and Potassium Pumps)
Edited by SIDNEY FLEISCHER AND BECCA FLEISCHER

VOLUME 158. Metalloproteins (Part A)
Edited by JAMES F. RIORDAN AND BERT L. VALLEE

VOLUME 159. Initiation and Termination of Cyclic Nucleotide Action
Edited by JACKIE D. CORBIN AND ROGER A. JOHNSON

VOLUME 160. Biomass (Part A: Cellulose and Hemicellulose)
Edited by WILLIS A. WOOD AND SCOTT T. KELLOGG

VOLUME 161. Biomass (Part B: Lignin, Pectin, and Chitin)
Edited by WILLIS A. WOOD AND SCOTT T. KELLOGG

VOLUME 162. Immunochemical Techniques (Part L: Chemotaxis and Inflammation)
Edited by GIOVANNI DI SABATO

VOLUME 163. Immunochemical Techniques (Part M: Chemotaxis and Inflammation)
Edited by GIOVANNI DI SABATO

VOLUME 164. Ribosomes
Edited by HARRY F. NOLLER, JR., AND KIVIE MOLDAVE

VOLUME 165. Microbial Toxins: Tools for Enzymology
Edited by SIDNEY HARSHMAN

VOLUME 166. Branched-Chain Amino Acids
Edited by ROBERT HARRIS AND JOHN R. SOKATCH

VOLUME 167. Cyanobacteria
Edited by LESTER PACKER AND ALEXANDER N. GLAZER

VOLUME 168. Hormone Action (Part K: Neuroendocrine Peptides)
Edited by P. MICHAEL CONN

VOLUME 169. Platelets: Receptors, Adhesion, Secretion (Part A)
Edited by JACEK HAWIGER

VOLUME 170. Nucleosomes
Edited by PAUL M. WASSARMAN AND ROGER D. KORNBERG

VOLUME 171. Biomembranes (Part R: Transport Theory: Cells and Model Membranes)
Edited by SIDNEY FLEISCHER AND BECCA FLEISCHER

VOLUME 172. Biomembranes (Part S: Transport: Membrane Isolation and Characterization)
Edited by SIDNEY FLEISCHER AND BECCA FLEISCHER

VOLUME 173. Biomembranes [Part T: Cellular and Subcellular Transport: Eukaryotic (Nonepithelial) Cells]
Edited by SIDNEY FLEISCHER AND BECCA FLEISCHER

VOLUME 174. Biomembranes [Part U: Cellular and Subcellular Transport: Eukaryotic (Nonepithelial) Cells]
Edited by SIDNEY FLEISCHER AND BECCA FLEISCHER

VOLUME 175. Cumulative Subject Index Volumes 135–139, 141–167

VOLUME 176. Nuclear Magnetic Resonance (Part A: Spectral Techniques and Dynamics)
Edited by NORMAN J. OPPENHEIMER AND THOMAS L. JAMES

VOLUME 177. Nuclear Magnetic Resonance (Part B: Structure and Mechanism)
Edited by NORMAN J. OPPENHEIMER AND THOMAS L. JAMES

VOLUME 178. Antibodies, Antigens, and Molecular Mimicry
Edited by JOHN J. LANGONE

VOLUME 179. Complex Carbohydrates (Part F)
Edited by VICTOR GINSBURG

VOLUME 180. RNA Processing (Part A: General Methods)
Edited by JAMES E. DAHLBERG AND JOHN N. ABELSON

VOLUME 181. RNA Processing (Part B: Specific Methods)
Edited by JAMES E. DAHLBERG AND JOHN N. ABELSON

VOLUME 182. Guide to Protein Purification
Edited by MURRAY P. DEUTSCHER

VOLUME 183. Molecular Evolution: Computer Analysis of Protein and Nucleic Acid Sequences
Edited by RUSSELL F. DOOLITTLE

VOLUME 184. Avidin–Biotin Technology
Edited by MEIR WILCHEK AND EDWARD A. BAYER

VOLUME 185. Gene Expression Technology
Edited by DAVID V. GOEDDEL

VOLUME 186. Oxygen Radicals in Biological Systems (Part B: Oxygen Radicals and Antioxidants)
Edited by LESTER PACKER AND ALEXANDER N. GLAZER

VOLUME 187. Arachidonate Related Lipid Mediators
Edited by ROBERT C. MURPHY AND FRANK A. FITZPATRICK

VOLUME 188. Hydrocarbons and Methylotrophy
Edited by MARY E. LIDSTROM

VOLUME 189. Retinoids (Part A: Molecular and Metabolic Aspects)
Edited by LESTER PACKER

VOLUME 190. Retinoids (Part B: Cell Differentiation and Clinical Applications)
Edited by LESTER PACKER

VOLUME 191. Biomembranes (Part V: Cellular and Subcellular Transport: Epithelial Cells)
Edited by SIDNEY FLEISCHER AND BECCA FLEISCHER

VOLUME 192. Biomembranes (Part W: Cellular and Subcellular Transport: Epithelial Cells)
Edited by SIDNEY FLEISCHER AND BECCA FLEISCHER

VOLUME 193. Mass Spectrometry
Edited by JAMES A. MCCLOSKEY

VOLUME 194. Guide to Yeast Genetics and Molecular Biology
Edited by CHRISTINE GUTHRIE AND GERALD R. FINK

VOLUME 195. Adenylyl Cyclase, G Proteins, and Guanylyl Cyclase
Edited by ROGER A. JOHNSON AND JACKIE D. CORBIN

VOLUME 196. Molecular Motors and the Cytoskeleton
Edited by RICHARD B. VALLEE

VOLUME 197. Phospholipases
Edited by EDWARD A. DENNIS

VOLUME 198. Peptide Growth Factors (Part C)
Edited by DAVID BARNES, J. P. MATHER, AND GORDON H. SATO

VOLUME 199. Cumulative Subject Index Volumes 168–174, 176–194

VOLUME 200. Protein Phosphorylation (Part A: Protein Kinases: Assays, Purification, Antibodies, Functional Analysis, Cloning, and Expression)
Edited by TONY HUNTER AND BARTHOLOMEW M. SEFTON

VOLUME 201. Protein Phosphorylation (Part B: Analysis of Protein Phosphorylation, Protein Kinase Inhibitors, and Protein Phosphatases)
Edited by TONY HUNTER AND BARTHOLOMEW M. SEFTON

VOLUME 202. Molecular Design and Modeling: Concepts and Applications (Part A: Proteins, Peptides, and Enzymes)
Edited by JOHN J. LANGONE

VOLUME 203. Molecular Design and Modeling: Concepts and Applications (Part B: Antibodies and Antigens, Nucleic Acids, Polysaccharides, and Drugs)
Edited by JOHN J. LANGONE

VOLUME 204. Bacterial Genetic Systems
Edited by JEFFREY H. MILLER

VOLUME 205. Metallobiochemistry (Part B: Metallothionein and Related Molecules)
Edited by JAMES F. RIORDAN AND BERT L. VALLEE

VOLUME 206. Cytochrome P450
Edited by MICHAEL R. WATERMAN AND ERIC F. JOHNSON

VOLUME 207. Ion Channels
Edited by BERNARDO RUDY AND LINDA E. IVERSON

VOLUME 208. Protein–DNA Interactions
Edited by ROBERT T. SAUER

VOLUME 209. Phospholipid Biosynthesis
Edited by EDWARD A. DENNIS AND DENNIS E. VANCE

VOLUME 210. Numerical Computer Methods
Edited by LUDWIG BRAND AND MICHAEL L. JOHNSON

VOLUME 227. Metallobiochemistry (Part D: Physical and Spectroscopic Methods for Probing Metal Ion Environments in Metalloproteins)
Edited by JAMES F. RIORDAN AND BERT L. VALLEE

VOLUME 228. Aqueous Two-Phase Systems
Edited by HARRY WALTER AND GÖTE JOHANSSON

VOLUME 229. Cumulative Subject Index Volumes 195–198, 200–227

VOLUME 230. Guide to Techniques in Glycobiology
Edited by WILLIAM J. LENNARZ AND GERALD W. HART

VOLUME 231. Hemoglobins (Part B: Biochemical and Analytical Methods)
Edited by JOHANNES EVERSE, KIM D. VANDEGRIFF, AND ROBERT M. WINSLOW

VOLUME 232. Hemoglobins (Part C: Biophysical Methods)
Edited by JOHANNES EVERSE, KIM D. VANDEGRIFF, AND ROBERT M. WINSLOW

VOLUME 233. Oxygen Radicals in Biological Systems (Part C)
Edited by LESTER PACKER

VOLUME 234. Oxygen Radicals in Biological Systems (Part D)
Edited by LESTER PACKER

VOLUME 235. Bacterial Pathogenesis (Part A: Identification and Regulation of Virulence Factors)
Edited by VIRGINIA L. CLARK AND PATRIK M. BAVOIL

VOLUME 236. Bacterial Pathogenesis (Part B: Integration of Pathogenic Bacteria with Host Cells)
Edited by VIRGINIA L. CLARK AND PATRIK M. BAVOIL

VOLUME 237. Heterotrimeric G Proteins
Edited by RAVI IYENGAR

VOLUME 238. Heterotrimeric G-Protein Effectors
Edited by RAVI IYENGAR

VOLUME 239. Nuclear Magnetic Resonance (Part C)
Edited by THOMAS L. JAMES AND NORMAN J. OPPENHEIMER

VOLUME 240. Numerical Computer Methods (Part B)
Edited by MICHAEL L. JOHNSON AND LUDWIG BRAND

VOLUME 241. Retroviral Proteases
Edited by LAWRENCE C. KUO AND JULES A. SHAFER

VOLUME 242. Neoglycoconjugates (Part A)
Edited by Y. C. LEE AND REIKO T. LEE

VOLUME 243. Inorganic Microbial Sulfur Metabolism
Edited by HARRY D. PECK, JR., AND JEAN LEGALL

VOLUME 244. Proteolytic Enzymes: Serine and Cysteine Peptidases
Edited by ALAN J. BARRETT

VOLUME 302. Green Fluorescent Protein
Edited by P. MICHAEL CONN

VOLUME 303. cDNA Preparation and Display
Edited by SHERMAN M. WEISSMAN

VOLUME 304. Chromatin
Edited by PAUL M. WASSARMAN AND ALAN P. WOLFFE

VOLUME 305. Bioluminescence and Chemiluminescence (Part C)
Edited by MIRIAM M. ZIEGLER AND THOMAS O. BALDWIN

VOLUME 306. Expression of Recombinant Genes in Eukaryotic Systems
Edited by JOSEPH C. GLORIOSO AND MARTIN C. SCHMIDT

VOLUME 307. Confocal Microscopy
Edited by P. MICHAEL CONN

VOLUME 308. Enzyme Kinetics and Mechanism (Part E: Energetics of Enzyme Catalysis)
Edited by VERN L. SCHRAMM AND DANIEL L. PURICH

VOLUME 309. Amyloid, Prions, and Other Protein Aggregates
Edited by RONALD WETZEL

VOLUME 310. Biofilms
Edited by RON J. DOYLE

VOLUME 311. Sphingolipid Metabolism and Cell Signaling (Part A)
Edited by ALFRED H. MERRILL, JR., AND Y. A. HANNUN

VOLUME 312. Sphingolipid Metabolism and Cell Signaling (Part B) (in preparation)
Edited by ALFRED H. MERRILL, JR., AND Y. A. HANNUN

VOLUME 313. Antisense Technology (Part A: General Methods, Methods of Delivery and RNA Studies)
Edited by M. IAN PHILLIPS

VOLUME 314. Antisense Technology (Part B: Applications)
Edited by M. IAN PHILLIPS

VOLUME 315. Vertebrate Phototransduction and the Visual Cycle (Part A)
Edited by KRZYSZTOF PALCZEWSKI

VOLUME 316. Vertebrate Phototransduction and the Visual Cycle (Part B)
Edited by KRZYSZTOF PALCZEWSKI

VOLUME 317. RNA-Ligand Interactions (Part A: Structural Biology Methods)
Edited by DANIEL W. CELANDER AND JOHN N. ABELSON

VOLUME 318. RNA-Ligand Interactions (Part B: Molecular Biology Methods) (in preparation)
Edited by DANIEL W. CELANDER AND JOHN N. ABELSON

VOLUME 319. Singlet Oxygen, UV-A, and Ozone (in preparation)
Edited by LESTER PACKER AND HELMUT SIES

VOLUME 320. Cumulative Subject Index Volumes 290–319 (in preparation)

VOLUME 321. Numerical Computer Methods (Part C) (in preparation)
Edited by MICHAEL L. JOHNSON AND LUDWIG BRAND

VOLUME 322. Apoptosis (in preparation)
Edited by JOHN C. REED

VOLUME 323. Energetics of Biological Macromolecules (Part C) (in preparation)
Edited by MICHAEL L. JOHNSON AND GARY K. ACKERS

VOLUME 324. Branched-Chain Amino Acids (Part B) (in preparation)
Edited by ROBERT A. HARRIS AND JOHN R. SOKATCH

Section I

Chemiluminescence and Bioluminescence: Overviews

[1] Chemical Generation of Excited States: The Basis of Chemiluminescence and Bioluminescence

By FRANK MCCAPRA

Introduction

The purpose of this review is to summarize current understanding of the mechanisms of efficient chemiluminescence, primarily in relation to bioluminescence. Certain aspects of chemiluminescence, remote from this intention, will therefore not be treated. The references have been chosen to round out what is specifically a discussion of mechanism. It is in effect a progress report on current problems, some of which are stubbornly unresolved; in the last decade new experimental evidence and theories have been in short supply. It would seem that we are currently on a plateau on the journey to a full understanding of the mechanism of light generation in bioluminescence and that there are still several outstanding peaks to be conquered. There have been several excellent reviews[1–6] of all the topics of interest over the years, and rather than repeat the conclusions and information contained in them, a summary of the unchanged position will be provided and attention will be focused on the unresolved problems and more controversial interpretations. A brief review lecture[6] was entitled "Mechanisms in chemiluminescence and bioluminescence: unfinished business" that had the intention of drawing attention to inadequacies in the current explanations. However, the situation is not totally fluid, and there is much common ground between the investigators of the mechanisms of excited state formation, giving comfort to the nonspecialist.

The most successful generalization in the field is that provided by dioxetanes, but there is little doubt that there are efficient reactions not covered by this rationalization. It remains to be seen whether an equally inclusive and satisfying additional scheme can be discovered or whether the answer

[1] K. D. Gundermann and F. McCapra, "Chemiluminescence in Organic Chemistry." Springer-Verlag, Berlin, 1987.

[2] A. K. Campbell, "Chemiluminescence Principles and Applications in Biology and Medicine." Ellis Horwood, Chichester, 1988.

[3] A. Mayer and S. Neuenhoffer, *Angew. Chem. Int. Ed. Engl.* **33,** 1044 (1994).

[4] "Chemi- and Bioluminescence" (J. G. Burr, ed.). Dekker, New York, 1985.

[5] T. Wilson, *in* "Singlet Oxygen" (A. A. Frimer, ed.), Vol. II, 37. CRC Press, Boca Raton, FL, 1985.

[6] F. McCapra, *in* "Bioluminescence and Chemiluminescence: Reporting with Photons" (J. W. Hastings, L. J. Kricka, and P. E. Stanley, eds.), p. 7. Wiley, Chichester, 1997.

0076-6879/00 $30.00

to this particular difficulty is to be found in a reworking of the problematic chemically initiated electron exchange luminescence (CIEEL) mechanism.[7,8] The involvement of dioxetanes in the chemiluminescence of organic compounds was first explicitly proposed[9] as long ago as 1964, and although the principle has been consolidated[10–12] steadily since that time, no equally successful generalization has yet been applied to those reactions that cannot be included under that umbrella.

Oxygen is an indispensable reactant in all cases of bioluminescence studied and almost so in the chemiluminescence of organic compounds. The intermediacy of organic peroxides follows immediately from this, and at present, there is no simpler starting point for the discussion of mechanism. The overall reactions are shown in the following simple schemes:

$$\text{Reactant} + \text{Oxygen} \rightarrow \text{Peroxide; Peroxide} \rightarrow \text{Products}^*$$
$$\text{Products}^* \rightarrow \text{Products} + \text{Light}$$

and

$$\text{Luciferin} + \text{Luciferase} + \text{Oxygen} \rightarrow \text{Oxyluciferin}^*$$
$$\text{Oxyluciferin}^* \rightarrow \text{Oxyluciferin} + \text{Light}$$

The efficiency of the reaction is described in terms of the quantum yield, the number of einsteins/molecule reacting, with a maximum of 1.0 or 100%. In order to focus on the central event, the step that populates the excited state (excitation yield Φ_e or Φ_{es}), the overall quantum yield Φ_{cl} is factored to include chemical yield (Φ_c) of the excited product and its fluorescence quantum yield (Φ_f). Thus $\Phi_{cl} = \Phi_e \times \Phi_c \times \Phi_f$.

Electron Transfer Chemiluminescence

This mechanism is phenomenologically closest to the formation of an excited state and thus has obvious attractions as a basis on which to build a general mechanism of chemiluminescence. In addition, its fundamentals are examined more easily than those of any other due to its relative simplicity. As a result, there can be no doubt as to its validity, especially in electrochemically generated luminescence. The importation of this conviction into the domain of molecular fragmentation (specifically peroxide decomposition) seemingly provided a bridge to the general world of organic

[7] G. B. Schuster, *Acct. Chem. Res.* **12,** 366 (1979).
[8] B. G. Dixon and G. B. Schuster, *J. Am. Chem. Soc.* **103,** 3068 (1981).
[9] F. McCapra and D. G. Richardson, *Tetrahedron Lett.* **3147** (1964).
[10] W. Adam, *in* "Small Ring Heterocycles" (A. Hassner, ed.), Part 3, p. 232. Wiley, New York, 1983.
[11] W. Adam, M. Heil, T. Mosandl, and C. R. Saha-Möller, *in* "Organic Peroxides" (W. Ando, ed.), p. 251. Wiley, New York, 1992.
[12] T. Wilson, *Int. Rev. Sci. Phys. Chem. Ser.* [2] **9,** 265 (1976).

compound chemiluminescence and bioluminescence. The bridge acquired the name chemically initiated electron exchange luminescence.[7] However, as will be argued later, this coherence is probably illusory, and many of the certainties derived from it must be reexamined.

The theory of electron transfer chemiluminescence[13] has a firm base in general electron transfer theory, particularly as defined by the work of Marcus. An obvious feature is the fact that the rate of electron transfer is greater than that of molecular motion. Because the reactions are of necessity highly exergonic (otherwise the high energy of the quantized photon would not be in evidence), this discriminates against the formation of an energy-rich ground state that has all the energy in vibrational modes.

The early work in this area was entirely chemical in concept. A polynuclear hydrocarbon, such as 9,10-diphenylanthracene (DPA), is reduced to the radical anion by dissolving alkali metal, and on addition of any one of a large variety of oxidants such as 9,10-dichloro-9,10-dihydro-9,10-diphenylanthracene, benzoyl peroxide (I), oxalyl chloride, and aluminum chloride produces light identical to the fluorescence of the hydrocarbon.

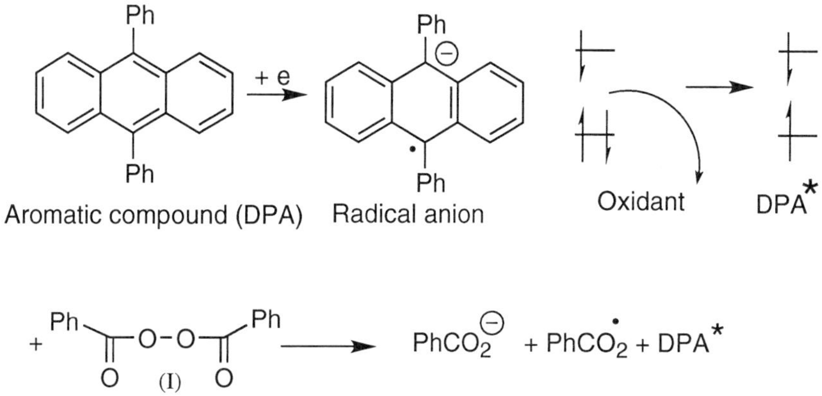

SCHEME 1

This is a very general reaction and lends itself to one of the simplest explanations of excited state generation. The diagram in Scheme 1 is self-explanatory. A refinement of experiments of this type is the use of solvated electrons as the reducing agent. Alkali metals such as potassium, when dissolved in dipolar aprotic solvents such as hexamethylphosphoramide, give the characteristic blue color of solvated electrons. They may also be generated electrochemically in the same solvent, and reducible compounds such as tosylates, halides, and aromatic amines often give a chemilumines-

[13] L. R. Faulkner, *Methods Enzymol.* **57,** 494 (1978).

cent reaction. One of the brightest reactions in the field of electron transfer is obtained in this way, the direct reduction of *N*-*p*-toluenesulfonylcarbazole (II) producing a quantum yield, ϕ_{ecl}, of 3.3%.

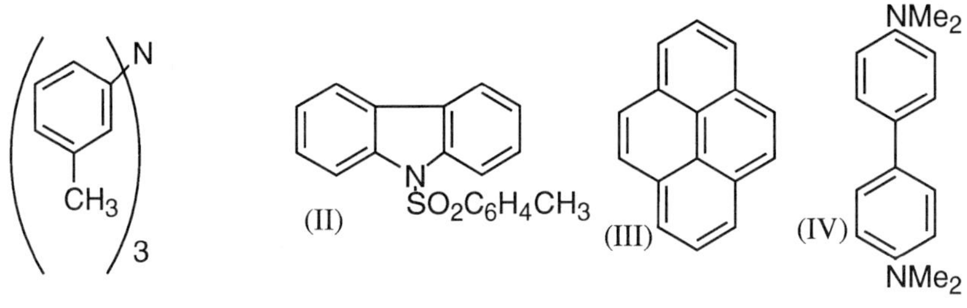

SCHEME 2

More information is obtainable if the anion radical and the oxidant are formed at an electrode in a fully monitored electrochemical experiment using cyclic voltammetry. A particular advantage of this method, often using effective devices such as the rotating disk electrode, is that precise knowledge of the energetics of the reaction is obtainable. Knowledge of the redox potentials of the participants in the reactions has identified two major routes to excitation. In the "energy sufficient route," a radical anion and a radical cation (usually of the same hydrocarbon or heterocycle) are formed by an alternating current in the diffusion layer around the electrode.

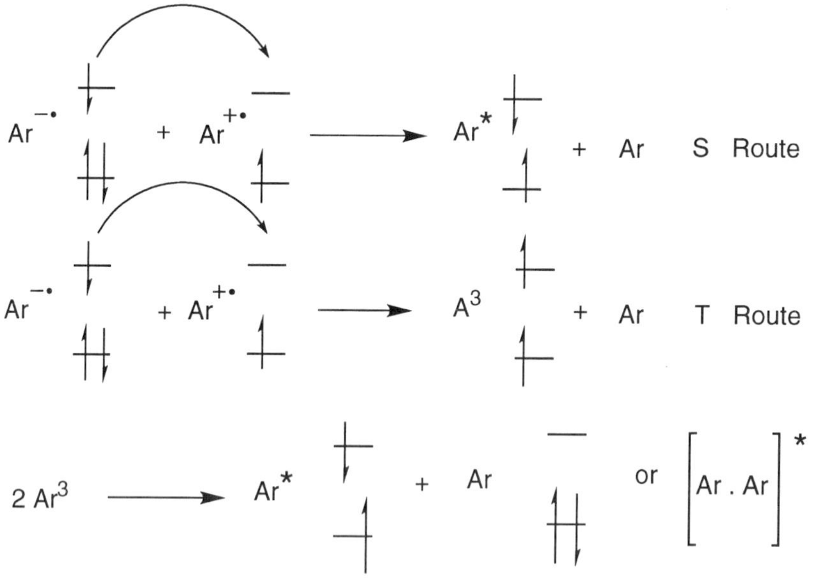

SCHEME 3

This is a convenient result, as such species are very unstable and must be within a short distance of each other when formed for an efficient annihilation reaction to take place. There is sufficient energy in this annihilation reaction for direct formation of the excited singlet state in the case of, say, diphenylanthracene. However, it is important to note that in this best of all cases for efficient excited state formation by electron transfer, the yield is not as high as that of the best peroxide chemiluminescence. A high ECL yield for excited state formation (ϕ_{es}) is 1.5% as against 60% for the active oxalate esters. This should be kept in mind when considering the claim that CIEEL, which is expected to deliver the electron in the excitation step only at the end of a sequence of bond breaking and forming steps, can greatly exceed the efficiency of electron transfer excitation taking place under the best possible circumstances. It would seem an unlikely result, particularly for intermolecular reactions.

There is a large range of aromatic hydrocarbons and heterocycles that react by the energy sufficient route (the so-called S route), and there is no need to detail the essentially similar reactions here. However, light can still be observed when the annihilation of the radical anion and radical cation (however produced) does not yield sufficient energy to populate the first excited singlet state of one of the partners. A great deal of circumstantial evidence shows that radiation from the lower excited state, the triplet, is never the source of even moderately strong light emission. Excited triplet molecules are quenched easily under the conditions of most chemiluminescent experiments by solvent and oxygen so that the expected phosphorescence is rarely seen. Visible light can nevertheless be obtained from energy-deficient systems, with only enough energy to form the triplet state of one of the partners (the T route). An energy pooling process, first discovered in fluorescence and known as slow fluorescence, is the source of the light. In electrochemiluminescent reactions, oxygen is usually excluded, as the radical anions react rapidly with it so that the triplet is less likely to be quenched. The reaction of two triplet molecules then leads to the population of the singlet state as shown.

An interesting consequence of the need for the radical anion and cation to meet in order for the electron transfer to occur is that the excited and ground states are present in the same solvent cage. The contact allows the formation of an excimer, a singly excited dimer. This event is readily detectable, as excimer emission is of lower energy than the isolated singlet and occurs at a longer wavelength, with a loss of fine structure and a broadening of the emission band. Pyrene (III, Scheme 2) is the classical example, and although excimer emission was first seen in fluorescence experiments, the concentration required for the same excimer intensity is higher. The annihilation of two triplets is particularly likely to produce emission from an excimer or exciplex (the term used when the partners are not identical). Such a reaction is that between pyrene and *N,N,N′,N′*-tetramethyl-*p*-phenylenediamine (IV, Scheme 2). The excimer emission from the pyrene dimer is stronger than that in fluorescence experiments on pyrene. Both monomer and excimer emission is seen, a direct result of triplet annihilation.

The very interesting reaction between benzophenone radical anion and the cation radical of tri-*p*-tolylamine is also energy deficient. Here there is insufficient energy to populate *any* of the local excited states of the partners, not even the relatively low energy triplets. The only emission possible is that from the exciplex, a species of uniquely low energy in this system, giving a broad emission band. Exciplexes can serve as precursors of the locally excited states, by dissociation before radiation, and as expected the relative intensities of the bands are dependent on temperature.

A related and consistently efficient series of electron transfer reactions demonstrates that chemically, rather than electrochemically, derived electron transfer reactions can be high yielding. The relatively stable radical cations derived from substituted triphenylamines on reaction with a series of aromatic hydrocarbon radical anions gave quantum yields as high as 8%. The system is energy sufficient, and in this case the locally excited state—the singlet of the hydrocarbon—emits on dissociation of the also emissive exciplex.

These observations certainly act as no barrier to adopting electron transfer as the ultimate step in the formation of excited products from chemiluminescent reactions in general. Thus in extending this concept to peroxide chemiluminescence, the problem is not one of feasibility, but of demonstrating that such a process actually occurs within the very different circumstances of multiple bond reorganization.

Peroxide Chemiluminescence

Simple alkyl peroxides are always weakly chemiluminescent on decomposition, the more so when associated with fluorescent compounds. However, few can be considered efficient, with quantum yields very rarely even approaching 1.0×10^{-4}. More typically the yields are as low as 10^{-8}. A frequently asked question is "Why are peroxides (and hence oxygen) the invariable reactants—why not the nitrogen or sulfur analogs?" This search for alternatives is most often encountered in discussions of dioxetane chemistry, but applies to peroxides in general. Such analogs undoubtedly exist, but the simplest and most compelling answer is that the population of electronically excited states requires a minimum energy. This absolute thermodynamic requirement is not met by the equations for the sulfur and nitrogen analogs, as the O—O bond is weaker than S—S or N—N bonds, and on the other side of the equation, the corresponding C=N and C=S bonds are considerably less strong than the C=O bond found in the respective products. The resulting relatively large shortfall in enthalpy (both differences working in the same direction) is fatal to excited state formation in nonoxygen cases. There are "selection rules" as in the dioxetanes, of course, but simple thermodynamics have a bigger part to play than is commonly realized.

Simple peroxides, such as hydroperoxides and dialkylperoxides, have not yet been shown to be the immediate precursors of excitation sequences with efficiencies high enough to yield visible light emission at any reasonable concentration. They are, of course, usually necessary intermediates on the way to more specific structures that do give highly visible light, but of themselves, they are only relevant to low-level luminescence. A possible exception is bacterial bioluminescence, certainly involving the intermediacy of a flavin hydroperoxide. The exception of this "simple" peroxide (it is actually a dialkyl peroxide) and the respectably efficient bioluminescent reaction from the general rule is one reason for treating the current mechanism with some suspicion.

The presence of a fluorescent acceptor in the reaction does increase the yield, particularly so as simple peroxides produce mainly excited triplet states on fragmentation. These are quenched easily in most circumstances, particularly by oxygen, and the resulting emission of light is very weak. Even with an acceptor of the triplet state energy, the yield is still low. One reason for this is that light is normally easily visible only as fluorescence, with the singlet state being relatively immune to quenching, due to its short lifetime, but there are spin prohibitions acting against the triplet to singlet excited state transfer required. An early appreciation of this was apparent in the work of Vassil'ev,[14] in which the heavy atom effect, found in an acceptor such as 9,10-dibromoanthracene, promotes spin orbit coupling and partially removes the restriction on triplet–singlet energy transfer. This was used to demonstrate the presence of excited triplets in simple peroxide fragmentation.

Most peroxides are formed and decomposed in radical chain reactions, and their formation is promoted by radical initiators in the presence of oxygen. Preformed peroxides are very often sufficiently stable to be studied as pure compounds, with a notable example being tetralin peroxide. There are several good reviews of simple peroxide chemiluminescence, and with the addition of work on the chemiluminescence of polymers, the information on simple peroxide chemiluminescence is considerable. A summary is therefore all that is required in this category, especially since it seems very

[14] R. F. Vassil'ev, *Prog. React. Kinet.* **4,** 305 (1967).

clear from the literature that the direct decomposition of such peroxides has never featured in bright chemiluminescence, far less the very efficient reactions of bioluminescence. They are, however, of some importance in the weak chemiluminescence that accompanies polymer degradation.[15]

The emission spectra accompanying hydrocarbon oxidation are often broad, with many peaks. The strongest peak is usually to be found in the region of 420–70 nm, but several others are seen between this and 670 nm. The weak red light is often ascribed to emission from the singlet oxygen dimol, and the blue-green light to emission from the triplet state of the carbonyl groups formed from the peroxide decomposition. The primary mechanism is usually taken to be the Russell mechanism (Scheme 4), which can generate both excited species. The decomposition path shown requires, for a concerted reaction, that electron spin is conserved, with the result that *either* oxygen should appear in the singlet state or the carbonyl in the triplet state. Both are necessarily excited states.

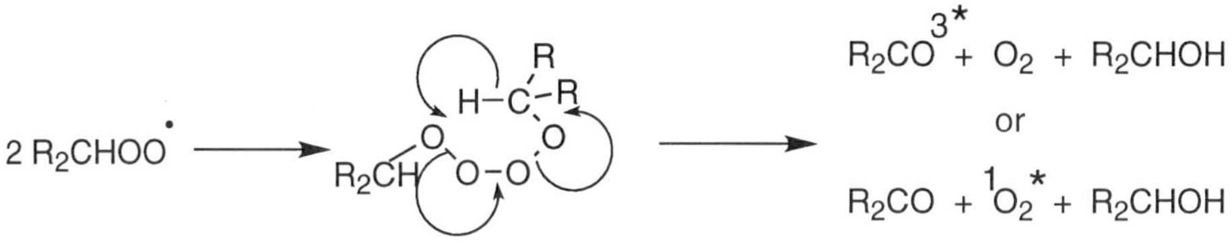

SCHEME 4

We do not have any understanding of the mechanisms of energy distribution within this extended transition state, but it is highly exothermic. Because the oxygen molecule is formed in the same solvent cage as the triplet carbonyl (by far the major contributor to the light intensity), it is possible that the reaction is intrinsically more efficient than is observed, with most of the excited triplet being quenched. However, this does not appear to be the case. Although quenching by oxygen undoubtedly occurs, it has been shown that for the oxidation of dimedone, where the excited product is not very susceptible to oxygen quenching, the quantum yield is still in the region of 10^{-8}.

Structural changes have very little influence on the efficiency of peroxide chemiluminescence,[16] although the oxidation of ketones is slightly more efficient than that of hydrocarbons, typically with a quantum yield of 10^{-8} as against a yield of 10^{-9} for simple hydrocarbons. Note that even when the products are α-diketones, known to have an accessible and fairly emis-

[15] G. A. George, *in* "Developments in Polymer Degradation," Vol. 3, p. 173. Applied Science, New York, 1983.

[16] H. J. Kellog, *J. Am. Chem. Soc.* **91,** 5433 (1969).

sive singlet excited state, the efficiency of this route is not significantly higher. The reactions are never clean, and in the case of polymer oxidation, it can be shown[17] that further oxidation of by-products is detectable spectroscopically. Surprisingly, although these by-products are fluorescent, in this case the longer wavelength emission observed is not the result of energy transfer to these fluorescers, but is the result of a secondary peroxidation. The by-products are less saturated and thus emit at a longer wavelength on further oxidation. Therefore, it must be concluded that fragmentation of simple peroxides is a detectably luminescent reaction, but of no consequence for the understanding of *bright* chemiluminescence.

Peroxides and Electron Transfer

The fundamental position of electron transfer among the various basic mechanisms of excited state population is seductive. It addresses many of the doubts that arise when one is confronted with the question "why should any exothermic reaction choose to generate an electronically excited state when the usual pathway is to the ground state?" First, electron transfer occurs on the same time scale as electronic excitation. This is usually compared to the situation prevailing during the absorption of a photon in photochemistry. The Franck-Condon principle governs the latter, and in its simplest form implies that during the promotion of an electron, the nuclear coordinates of the molecule do not change. Electronic excitation is then said to be a "vertical" transition. Second, unlike the complicated and ultimately unexaminable transition state in peroxide decomposition, the actual excitation step is visualized easily and, at the risk of burying much of the subtlety, uncontroversial and very well understood. Application of these ideas to a suitably exergonic electron transfer transition state allows the conclusion that the activation energy for entrance into the excited state is lower than that for formation of a ground state molecule. Even so, this is only a guiding principle and many other factors must be included to explain why excitation yields are normally rather lower than this "principle" would lead us to expect. However, it is not our purpose to examine the details of electron transfer reactions per se, but to describe the attempt to incorporate their advantages into a scheme for peroxide decomposition.

The first convincing description of what was later called the CIEEL mechanism was provided by Linschitz[18] in 1961, in an investigation of a reaction discovered by Helberger. Tetralin hydroperoxide reacts with zinc tetraphenylporphin (ZnTPP) on heating (148°) to give the bright red light

[17] D. J. Lacey and V. Dudler, *Polym. Degrad. Stab.* **51,** 109 (1966).

[18] H. Linschitz, *in* "Light and Life" (W. D. McElroy and B. Glass, eds.), p. 173. Johns Hopkins Press, Baltimore, 1961.

characteristic of the fluorescence of the porphin. The mechanism is depicted as the formation of a charge transfer complex, formed by transfer of an electron from the electron-rich porphin to the peroxide, with resulting scission of the peroxide bond and the formation of a charge transfer complex. Removal of a proton from the organic radical allows back transfer to the ZnTPP radical cation forming the first excited singlet of the porphin (Scheme 5).

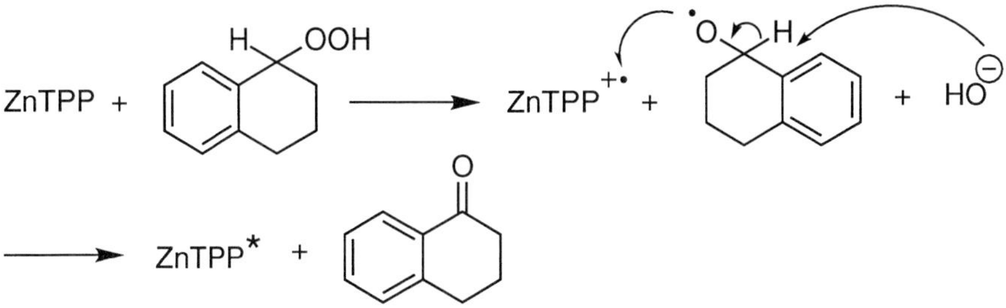

SCHEME 5

However, it is possible that this is actually a very rare example, as the metal clearly is essential (the reaction is general for *metallo*porphyrins, including chlorophyll), and none of the other claimed examples of CIEEL involve metals. Linschitz[18] showed that ZnTPP was a catalyst in the decomposition of the peroxide. The metal-free porphin has no such catalytic effect, nor does it elicit luminescence. A slight inhibitory effect on the rate of decomposition was actually observed. A chain reaction in the decomposition of the peroxide was considered likely.[19] The point concerning the coupling of electron transfer to the movement of the heavy nuclei in the porphin, favoring excitation, was also made. Indeed it is worth quoting a statement covering the generality of the mechanism. "The frequent association of peroxide–breakdown reactions with chemiluminescence of complex organic molecules suggests that the mechanism in all such cases may be essentially the same as that proposed here for the porphyrin reaction, i.e., the molecule being excited, or its precursor, functions always as a catalyst for peroxide decomposition, the excited state being reached following charge–transfer processes involved in this catalysis"—in all, taken with a statement of the concept of "back transfer," a remarkably complete view of CIEEL almost 20 years before the use of the acronym!

CIEEL has the attractive mechanistic property of allowing a back transfer of an electron sufficiently exothermic to populate the singlet excited state of a fluorescer. This exothermicity is derived from the fact that bond

[19] The tendency of peroxides to undergo chain decomposition reactions in the presence of amines and other electron donors is well known.

rearrangement (e.g., the loss of the proton in the example given earlier) allows the formation of a stable molecule. One of the barriers to formulating potential mechanisms for chemiluminescence from *simple* peroxides is that this attractive feature is absent.

Notwithstanding the apparent prescience of this often ignored but significant paper,[18] there are many reasons for doubting the completeness of the rationalization by CIEEL as the study of chemiluminescence and bioluminescence has progressed. In the course of a reexamination of the chemiluminescence associated with the rearrangement of the endoperoxide (V), Catalani and Wilson,[20] recognizing that the reaction conformed to the principles of CIEEL, noted that it was not efficient, with $\phi_{es} = 2 \times 10^{-5}$. This prompted a reevaluation of the principal example of the CIEEL mechanism, the reaction of diphenoyl peroxide (VI) with fluorescent activators.[21] With a quantum yield of 10% and conformity with all the tenets of the mechanism, it stood at the center of the proposal.

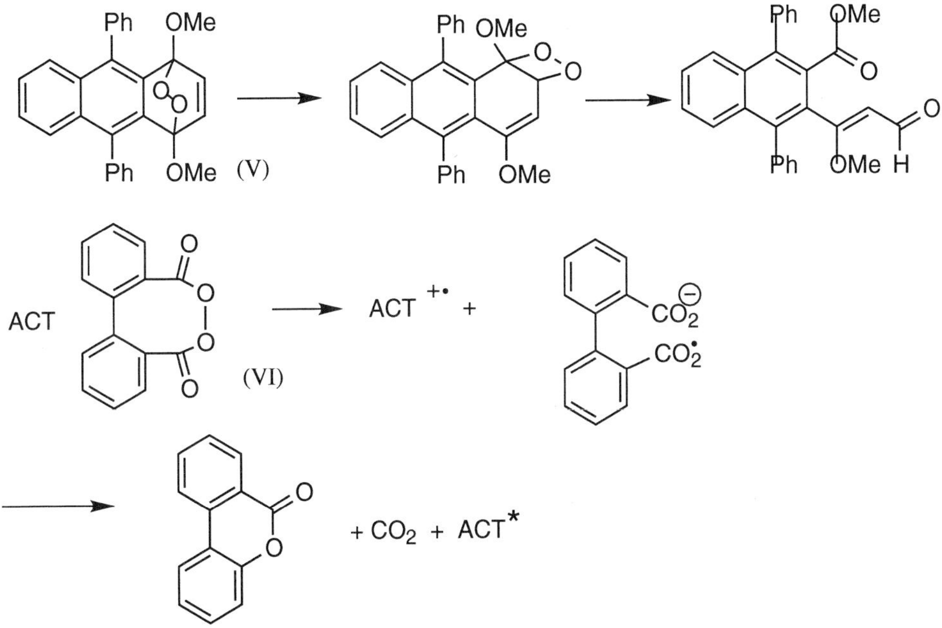

SCHEME 6

[20] L. H. Catalani and T. Wilson, *J. Am. Chem. Soc.* **111,** 2633 (1989).

[21] The chemiluminescent decomposition of simple dioxetanes such as tetramethyldioxetane is not usually catalyzed by fluorescent activators. In the present case, the aromatic rings may be responsible for complex formation with the added aromatic hydrocarbon, leading to unexpected catalytic decomposition.

The endoperoxide (V) rearranges under very mild acid conditions to the dioxetane, which although unstable, is detectable by low temperature nuclear magnetic resonance (NMR). The rate of reaction is dependent on both the concentration and the ionization potential of the fluorescent activators. The similarity between the two reactions prompted the measurement of the quantum yield for excited state formation in the case of diphenoyl peroxide. The same value ($2 \pm 1 \times 10^{-5}$) was found. This is a very low value for any reaction claimed as the basic mechanism for bioluminescent and model chemiluminescent reactions with ϕ_{es} in the range of 0.1 to 0.9. This discovery led Catalani and Wilson[20] to one of the most compelling and thoughtful discussions of the fundamentals of organic chemiluminescence ever published. Before summarizing their conclusions, which lead in the direction of a charge transfer transition state as the immediate precursor of excitation, it is helpful to recall comments by Faulkner[13] concerning the properties of the entities in the solvent cage during the established electron transfer of electrochemiluminescence. He pointed out that, according to Park and Bard,[22] the radiationless processes occurring within the exciplexes found in electrochemiluminescence will be unique to the exciplex and consequently hard to predict. The same can undoubtedly be said of a charge transfer transition state. This uncertainty is particularly important when considering the degree of intersystem crossing to the unproductive triplet state.

The conclusions of Catalani and Wilson,[20] the result of a closely argued case based on experimental observations, reiterate the central observation that the plot of log k vs the oxidation potential of the fluorescent activator is linear. The slope of the line is significant, and from it one can obtain α, an indication of the degree of electron transfer in the transition state. There has been some dispute[23] about the meaning of α in reactions in which electron transfer is a consequence of extensive bond reorganization. However, in any event, the values found in putative CIEEL reactions do not *unambiguously* support full electron transfer in the transition state.

Triplet state products have been sought, without success, but the hydrocarbons most effective in the reaction are always likely to form mainly excited singlets, with the low lying triplet states being undetectable. If these triplets are the most likely products as a result of the special nature of the peroxide–activator complex, then paradoxically, the catalytic properties of the activator merely serve to hasten the decomposition of the peroxide by a dark route! Interestingly, Catalani and Wilson[20] note that this point may be examined by observing the effect of a magnetic field on the light level.

[22] S. M. Park and A. J. Bard, *J. Am. Chem. Soc.* **97,** 2978 (1975).
[23] C. Walling, *J. Am. Chem. Soc.* **102,** 6854 (1980).

The paramagnetism of the triplet state within the radical ion pair renders it susceptible to a magnetic field, with a reduction in intersystem crossing and an increased light emission. This effect has been used to demonstrate the involvement of the T route in electrochemiluminescence.[24] At least one attempt has been made to investigate this possibility by an extremely sensitive method. Instead of simply examining the effect of an applied magnetic field, the field was modulated in an ESR experiment, greatly increasing the sensitivity of the experiment.[25] Oxalate esters, acridinium esters, and hydrazides were examined, with a complete absence of effect. This very important experiment must be repeated by others, especially with the examples of CIEEL discussed in this section, and not least because negative results must always be viewed with caution.

Another of the conclusions is that, given the large range of fluorescent hydrocarbons available, energy deficiency in the back transfer of the electron, known for obvious thermodynamic reasons to be a dark route, is not the problem here. Side reactions of the very reactive oxygen radicals or reactions of the unstable radical ions generally, with the solvent, seem to represent a minor pathway. If they were a substantial cause of the inefficiency, with such an extremely small light yield, the amount of products of such reactions would be correspondingly high. This is not the case.

Catalani and Wilson[20] conclude that the CIEEL mechanism is open to question. The source of the inefficiency remains unclear. An alternative, which we have also favored for some time, although apparently less precise in its formulation, actually has much circumstantial evidence in its support. Full electron transfer is not involved and the charge transfer develops during the transition state. Some direct evidence supports the idea of a charge transfer transition state. Examples are the small effect of solvent polarity and the trend to higher efficiency as the enthalpy of the reaction increases. The small slope of the linear free energy line (or small α) is also suggestive. These ideas are very difficult to pursue, as there are assumed to be no easily examinable intermediates between the encounter complex and the excited state that results.

Other examples of CIEEL, supported by similar evidence, have even lower quantum yields, where the data are available. For example, pyrone endoperoxides[26] such as (VII), apparently form the *ortho*-dioxins, and their rate of decomposition is proportional to the concentration and oxidation potential of the fluorescent activator. Decomposition in the absence of an activator gave a quantum yield of 2.0×10^{-11} einstein/mol. The enhanced

[24] F. Scandola and V. Balzani, *J. Am. Chem. Soc.* **103,** 2519 (1981).

[25] F. McCapra, K. Perring, R. J. Hart, and R. A. Hann, *Tetrahedron Lett.* **5087** (1981)

[26] W. Adam and I. Erden, *J. Am. Chem. Soc.* **101,** 5692 (1979).

(and presumed CIEEL) reaction in the presence of fluorescent activators had a maximum quantum yield of 2.85×10^{-10}. No quantum yield measurements for another similar fully investigated system, *o*-xylylene peroxide[27] (VII), have been reported, but it is likely that the values are in the same range. This last experiment has a bearing on a suggested mechanism for luminol chemiluminescence in which an analog of xylylene peroxide is proposed. The details will be discussed later. Such extremely low quantum yields are even below those of the simplest peroxide, whose mechanism of decomposition gives no prior expectation of chemiluminescence. It is incumbent on anyone wishing to promote the validity of the CIEEL mechanism to explain its manifest failure in these cases. Other examples, such as malonyl peroxides[28] and secondary peroxyesters,[29] may be somewhat more efficient (quantum yields are not always reported), but still fall short of the range of efficiencies required for an explanation of bright chemiluminescence and bioluminescence.

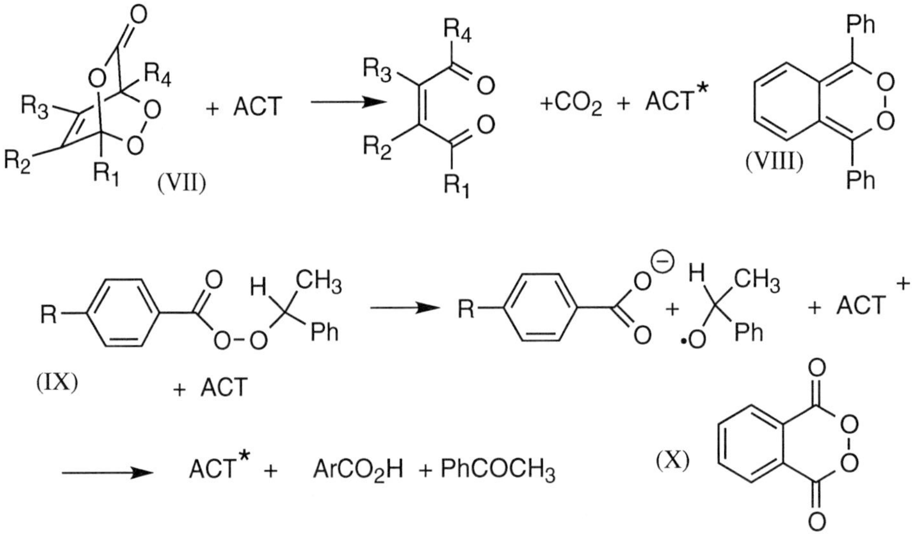

SCHEME 7

The peroxyesters are unusual examples because the direct chemiluminescence in certain cases may well be brighter than the activated chemiluminescence. The peresters [(IX), R = H, MeO, NO_2] show almost undetectable direct chemiluminescence, but (IX) (R = NMe_2) apparently generates

[27] J. P. Smith, A. K. Schrock, and G. B. Schuster, *J. Am. Chem. Soc.* **104,** 1041 (1982).
[28] M. J. Darmon and G. B. Schuster, *J. Org. Chem.* **47,** 4658 (1982).
[29] B. G. Dixon and G. B. Schuster, *J. Am. Chem. Soc.* **103,** 3068 (1981).

triplet excited *p*-dimethylaminobenzoic acid in 3.8% yield, but with no significant visible luminescence. This perester also generates excited singlet acid, as indicated by the chemiluminescence spectrum obtained during the decomposition, with a yield of 0.24%. In the activated chemiluminescence, the *p*-dimethylaminoperester shows no catalytic effect of the activator, and therefore an unchanged efficiency, but the other peresters show enhanced chemiluminescence with the activators, giving light intensities of 0.2–0.4%. The interpretation should be viewed with caution, however, in that the equivalent of the Russell mechanism has not been unequivocally ruled out. If it were to operate, the activation mechanism (and hence CIEEL) could not be maintained.

These reasonably efficient peresters are apparently successful examples of the CIEEL mechanism, and another apparent example had been reported earlier[30] in connection with the mechanism of bacterial bioluminescence (see later). The intramolecular removal of the proton may account for the slightly higher ϕ_{es} of 0.8% found.[30]

The most efficient of the CIEEL examples is that of the reaction of dimethyldioxetanone,[31] with a reported value for ϕ_{es} of 10%. Another dioxetanone, *tert*-butyldioxetanone,[32] however, has a very low excitation yield of 3.3×10^{-4} for both the uncatalyzed and the activated luminescence, using DPA as the activator, at 1000 times the concentration used for the activators for dimethyldioxetanone. Extrapolation to infinite activator concentrations was also made. Schmidt and Schuster[31] indicate in their paper that both of these procedures should have led to "substantial" chemiluminescence. It is not known why there should be this difference between the virtually identical dioxetanones. A further point requiring investigation is that in experiments with dimethyldioxetanone, it is mainly those activators with amino groups or that decompose during the reaction, that have a measurable kinetic effect at the concentrations used to determine the quantum yield. It is possible that the bulk of the reaction in those cases is a ground state chain reaction of the sort usually encountered in reactions between peresters and amines. The same doubt applies to the reactions of the peresters discussed earlier.

Last, the reaction of phthaloyl peroxide (X) with fluorescent hydrocarbons, studied[33] at the same time as diphenoyl peroxide, was shown to have a quantum yield 20 times greater. The reaction is not clean, and the mechanism is quite obscure. It has never been included in the CIEEL canon

[30] F. McCapra, P. D. Leeson, V. Donovan, and G. J. Perry, *Tetrahedron* **42,** 3223 (1986).
[31] S. P. Schmidt and G. B. Schuster, *J. Am. Chem. Soc.* **102,** 306 (1980).
[32] W. Adam and F. Yani, *Photochem. Photobiol.* **31,** 267 (1980).
[33] K. D. Gundermann, M. Steinfatt, and H. Fiege, *Angew. Chem. Int. Ed. Engl.* **83,** 43 (1971).

because it has not proved possible to describe in mechanistic terms any means for facile back transfer of the electron. Thus, not only do predictions of strong emission from convincing electron transfer mechanisms fail, but reactions that apparently do not possess any of the necessary features can be many times more efficient.

To summarize, the theoretical attractions of the CIEEL mechanism are many, and the experimental evidence for the relationship between fluorescer oxidation potential and catalysis compelling. However, it is doubtful whether the reaction is sufficiently efficient or predictable enough to serve as a foundation for the light-generating mechanisms of bioluminescence. There does not seem to be any explanations for the enormous range of efficiencies, from $\phi = 10^{-10}$ to $\phi = 0.1$. Several of the examples of CIEEL have quantum yields so low as to be excluded from any consideration as serious chemiluminescent contenders. The remainder urgently require repetition of the quantum yield measurements. Assuming that these values are actually as reported, then a search for the reasons for the difference becomes worthwhile, either confirming CIEEL as a major route to chemiluminescence or perhaps uncovering a new mechanism.

Second, some important chemiluminescent reactions have been assigned to the CIEEL category largely by analogy, and alternatives have thus been excluded. The possibility of discovering a significant new mechanism is thus diminished. For example, the seductiveness of CIEEL has distorted the perception of the mechanism for bacterial bioluminescence. In this case the prospects for efficient CIEEL are particularly poor, based on the structure of the currently accepted intermediate, as will be discussed in detail later. The alternative view, denying discrete electron transfer in favor of a charge transfer transition state, may seem to some to be an unnecessary refinement.[34] It is reasonable to ask whether the distinction matters, even if the extremely difficult task of devising meaningful distinguishing experiments can be accomplished. The answer will lie in the structural requirements that lead one to favor one or the other, thus guiding both interpretation and experiment.

Almost all mechanisms for organic chemiluminescence will show the same dependence on the oxidation (or ionization) potential of the fluorescent moiety, whether it is a part of the reacting molecule or an additive. This is because all mechanisms have in common electron promotion, charge transfer, and the need to lower the activation energy for excited state formation. These and other requirements respond to the properties of the

[34] In a further refinement, E. H. White, D. F. Roswell, A. C. Dupont, and A. A. Wilson, *J. Am. Chem. Soc.* **109,** 5189 (1987) interpret the reactions of acridine esters with hydrogen peroxide as forming the excited state within a transition state, without the full formation of the dioxetanone intermediate.

fluorescer in the same way, independently of the actual mechanism. The structural requirements for discrete electron transfer and reaction via a charge transfer transition state are not identical, and it is clear from the evidence that structures that have all the properties leading to this sort of electron transfer frequently are virtual dark reactions.

Dioxetanes

The original suggestion[9] that dioxetanes would feature largely in bright chemiluminescence and bioluminescence was made in 1964. It was based on an analysis of the structures of the then known efficient examples of the phenomena. It was also supported by a successful prediction: that if the transient dioxetane was involved, acridinium and acridan derivatives likely to form dioxetanes would be chemiluminescent. A later attempt to provide theoretical support for the experimental success of the concept was instrumental in causing a virtual explosion of interest in dioxetanes, made practical by the first synthesis of the until then assumed structure by Kopecky and Mumford.[35]

The original suggestion was intended to explain the very bright emission from what was clearly the first excited singlet state of acridinium esters, lophine[36] and firefly luciferin. The synthetic access to what have been classified as "simple" dioxetanes immediately produced a surprise. Although a thermodynamic treatment in the 1930s had indicated[37] that the triplet potential energy surface was accessed easily, it was not envisaged that the triplet carbonyl product would be formed in yields up to 100 times greater than the singlet. This finding is general,[38] although the ratios do vary depending on the structure of the dioxetane. A great deal of effort, both experimental and theoretical, has been expended in attempting to explain both the very high absolute quantum yields (as high as 50% for tetramethyldioxetane) and the preponderance of triplet.

There are several excellent, detailed, and exhaustive reviews[5,10–12,39] of this class of dioxetane, and although there are recent calculations of the potential energy surfaces taking advantage of improved methods of calculation, the essential details have not changed much. The experimental work[40]

[35] K. R. Kopecky and C. Mumford, *Can. J. Chem,* **47,** 709 (1969).

[36] E. H. White and M. J. C. Harding, *Photochem. Photobiol.* **4,** 1129 (1965). These authors concluded independently that a dioxetane provided the best explanation for the chemiluminescence of the compounds they were studying.

[37] M. G. Evans, H. Eyring, and J. F. Kincaid, *J. Chem. Phys.* **6,** 349 (1938).

[38] W. Adam, *in* "Chemical and Biological Generation of Excited States" (W. Adam and G. Cilento, eds.), p. 115. Academic Press, New York, 1982.

[39] W. Adam, *Adv. Heterocyclic. Chem.* **21,** 437 (1977).

[40] W. H. Richardson and D. L. Stigall-Estberg, *J. Am. Chem. Soc.* **104,** 4173 (1982).

established that the reaction had a transition state that, from the experimental point of view, certainly had diradical character. Evidence from substitution by aryl groups and deuterium substitution is strongly in favor of the formation of an intermediate oxygen diradical. In simple terms, the cleavage of the very weak O—O bond is to be expected, with the C—C bond breakage following almost immediately after. Spectroscopic evidence suggests that the diradical has an extremely short lifetime (less than 10 psec). Obtaining direct evidence for its existence would seem an extremely difficult, if not impossible, task. Nevertheless, provided the substitution does not supply electrons to the developing carbonyl groups with the ease of, say, an amine, this view of the decomposition seems secure (Scheme 8).

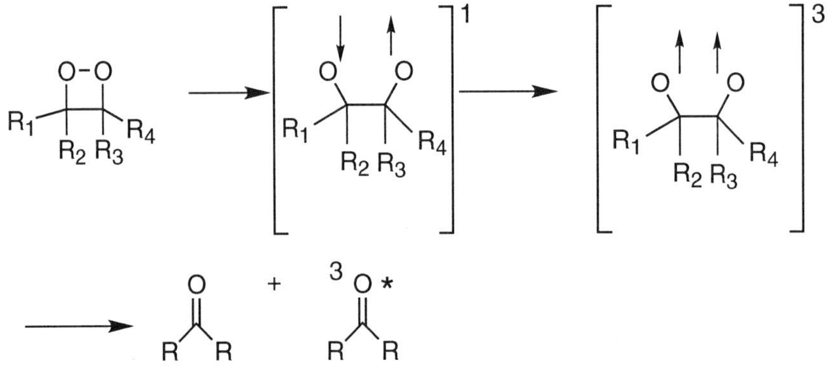

SCHEME 8

If the decomposition of a dioxetane proceeded in the same way as other strained ring compounds, to produce exclusively ground state products, the investigation of the mechanism would probably end at this point. However, the unique and very high yield of excited states requires a more extended analysis.

A large number of molecular orbital calculations have been made,[41–44] sometimes with conflicting results. The simplest "calculation" is provided by a correlation diagram, where orbital symmetry determines the approximate shape of all important potential energy surfaces for the ground and excited states. All of the diagrams reflect the essential feature that first led to the expectation that dioxetanes had theoretical significance in excited state generation. That is, a molecule whose bond rearrangement is ostensibly facile, and whose decomposition will release an enormous amount of free

[41] E. M. Evleth and G. Feler, *Chem. Phys. Lett.* **22,** 499 (1973).
[42] D. R. Roberts, *J. C. S. Chem. Comm.* **683** (1974).
[43] M. J. S. Dewar and S. I. Kirschner, *J. Am. Chem. Soc.* **96,** 7578 (1974).
[44] N. J. Turro and A. J. Devaquet, *J. Am. Chem. Soc.* **97,** 3859 (1975).

energy, is isolable. In other words, the ground state path is discriminated against. This and the high exothermicity are the central facts in dioxetane chemistry. The molecular orbital (MO) calculations are intended to chart the progress of the dioxetane on a ground state potential energy surface into regions where crossing to an excited state surface can occur. Many of these calculations agree with the fact that a triplet state carbonyl fragment is the predominant result of this crossing. In spectroscopic terms, this is intersystem crossing and is normally considered to be too slow to occur within the lifetime of a transition state. Few calculations address this point, but an attractive suggestion[44] is that the rotation about the C—O bond, during the decomposition, mixes the singlet and triplet states by spin orbit coupling, facilitating the entry into the triplet state. This has been questioned by Barnett[45] as being too small (a second order perturbation and by an ineffective light atom only) to achieve the desired effect.

Although these calculations are now more often in agreement with experimental observation, they seem to be wise after the event and do not add any new practical information. When applied to dioxetanes, they differ in their conclusions. For example, one very sophisticated calculation[46] indicates that the rate-determining step occurs on the triplet surface. All the others either imply, or state, that the highest activation energy barrier is to be found on the ground state surface. Another[47] even has the C—C bond breakage as the leading event, justified by the postulate that, unlike O—O bond breakage, the cleavage would be driven by the gain in energy by developing partial double bond character in the C—O bond. Both of these findings are intuitively unlikely. Not surprisingly, a later recalculation[48] apparently confirms the diradical formed by early O—O bond breakage. Where would these calculations take us if they were not guided by existing experimental evidence?! The most recent semiempirical calculation[49] seems more acceptable. Using a PM3 Hamiltonian with configuration interaction, it has excellent agreement with the experimentally determined activation energy, and finds that the ground state PE surface is degenerate with the excited triplet surface for a substantial portion of the reaction coordinate. This would give an opportunity for crossing to the triplet from the ground state singlet. This is consistent with the diradical pathway, forming the excited triplet, depicted by other, more direct methods.

[45] G. Barnett, *Can. J. Chem.* **52,** 3837 (1974).

[46] M. Reguero, F. Bernardi, A. Bottoni, M. Olivucci, and M. A. Robb, *J. Am. Chem. Soc.* **113,** 1566 (1991).

[47] C. W. Eaker and J. Hinze, *Theoret. Chim. Acta* (*Berl.*) **40,** 113 (1975).

[48] S. Wilsey, F. Bernardi, M. Olivucci, S. Murphy, and W. Adam, *J. Phys. Chem.* **103,** 1669 (1999).

[49] T. Wilson and A. M. Halpern, *J. Phys. Org. Chem.* **8,** 359 (1995).

Other discrepancies appear when studies of the gas phase decomposition of dioxetanes are made. Thermochemical calculations[50] suggest that there is insufficient energy in the unsubstituted dioxetane to form even triplet formaldehyde. A similar deficit was seen in MO calculations.[47] However, Bogan *et al.*[51] have observed emission from *singlet* excited formaldehyde during the oxidation of ethylene in the gas phase by singlet oxygen. The presence of the dioxetane was not confirmed directly, however. No triplet was observed in this case. Another, infrared laser-induced, gas phase decomposition of tetramethyldioxetane[52] produced emission from the triplet state, and although the authors doubted their own original conclusion that an emission at 420 nm rather than the expected acetone fluorescence (405 nm) or phosphorescence (460 nm) was that from an excimer, this still seems probable. A suggestion that it was a mixture of two emissions does not seem particularly likely from the shape of the published curves. The very strong 420-nm emission is red shifted from the acetone fluorescence, but if an excimer is involved, this is as expected and therefore indicates largely excited singlet formation. Another example of exciplex emission from dioxetane decomposition that is only seen when the intimate geometry of the two carbonyl products is derived immediately from the geometry of the precursor dioxetane will be discussed later.

The failure to identify triplet emission unequivocally from these two gas phase experiments can have simple experimental reasons such as quenching of the triplet under the experimental conditions, in the case of formaldehyde. It is therefore still probable that the gas phase results mirror the case in a solvent, where there is absolutely no doubt that triplet excited states predominate. However, an explanation worth considering, with important implications if true, is that the triplet-excited carbonyl is derived simply from a triplet diradical, formed as a result of intersystem crossing in the ground state diradical. Although some of the calculations, especially those of Wilson and Halpern,[49] are consistent with this, a simpler description may be adequate.

For example, it could be that the O—O bond in a vibrationally excited dioxetane in the gas phase breaks to give, as it must, an initial singlet diradical. With fewer collisions than in a solvent, the C—C bond immediately breaks to give the singlet carbonyl product, and on the expected short

[50] H. E. O'Neal and W. H. Richardson, *J. Am. Chem. Soc.* **92,** 6553 (1970). Errata *J. Am. Chem. Soc.* **93,** 1828 (1971).

[51] D. J. Bogan, J. L. Durant, R. S. Sheinson, and F. W. Williams, *Photochem. Photobiol.* **30,** 3 (1979).

[52] Y. Haas and G. Yahav, *J. Am. Chem. Soc.* **100,** 4885 (1978).

time scale, retaining the geometry of the precursor dioxetane, could well form an excimer. In a solvent, the lifetime of the diradical will be longer, as excess energy is removed by collision, and essentially irreversible intersystem crossing to form the triplet diradical will eventually lead to the formation of a ground state singlet and excited triplet carbonyl products. Structural influences on the strength of the C—C bond and consequent differences in the lifetime of the first formed singlet diradical could account for the observed variations in the triplet/singlet ratio. The triplet would most often be in excess, however. Note that Bogan *et al.*[51] have provided evidence that the parent dioxetane is vibrationally excited in a way not seen in the condensed phase.

The idea that the triplet excited state is simply derived from a triplet diradical formed by intersystem crossing in the ground state is of course not new. It is essentially that of Richardson, who supported it by very reliable experiments. Nevertheless, it will be helpful to focus on it because it avoids the doubts attending the use of not always convincing calculations. In summary, the excitation arises from three factors. (1) The orbital symmetry (or antiaromaticity) prohibitions preserve a molecule with sufficient energy to form excited products that would otherwise decompose exothermically to ground state products. These prohibitions need not have anything to do with the actual formation of the excited state. (2) The O—O bond breaks with the expected activation energy [other routes being barred by (1)] in the singlet state. Depending on the energy stabilization available (the solvent is an excellent energy "sink"), this may immediately form an excited and a ground state singlet or, by vibrational stabilization by a solvent, find time to intersystem cross to the triplet diradical. (3) If the triplet forms, then given the exothermicity of the decomposition, an excited state (triplet) is formed adiabatically. There is thus no arcane surface crossing to invoke to explain excess triplets, nor any concern about the time available for such an event to occur within the lifetime of a transition state. In addition, the configuration of the oxygen centered radical is virtually that of the observed n,π^* excited state.

Paradoxically, by simplifying the account of the formation of triplets from simply substituted dioxetanes, the principles governing the formation of excited *singlet* states from the luciferins and their models are made much clearer. In effect, it is a return to the original concept of orbital symmetry guiding the formation of large amounts of excited singlets. One must never forget that firefly luciferin, and its models, can deliver excited singlet states in astonishingly high yields (88% in the case of the luciferin). This is not a chance event, but must have a proper explanation. In this context the simple dioxetanes are but a trivial subset whose decomposition takes place with a change in mechanism.

Thus, the relationship between simple dioxetanes and electron-rich dioxetanes can be described as follows. The same prohibitions to a concerted *ground* state formation of products exist. It may be that the O—O bond breakage leads the transition state. However, the structure of the aromatic substituent in electron-rich dioxetanes is such that a dissociation of the C—C bond is made much more likely in these cases. One has only to think of the difference in dissociation energies between ethane (104 kcal/mol) and diphenylethane (85 kcal/mol). This difference is of course not expressed fully in a transition state, and Richardson, who did indeed synthesize simple dioxetanes with just these substitutions, did not reach the point at which C—C dissociation was concerted with O—O bond breakage. Luciferins and their models are nearly all excellent stabilizer of radicals, if we are to think of the dissociation as homolytic, and this substitution effectively destabilizes the C—C bond to give a concerted decomposition. There is another factor at work, in that the electron-rich dioxetanes can deliver charge to the developing carbonyl. A simple outline of how this may lead to excited state formation, in a particular case, has been published.[53] Some form of exciplex between the excited carbonyl and the departing ground state second carbonyl group is likely depending on conditions. Usually the exciplex will dissociate before radiation and only the unperturbed emission from, for example, oxyluciferin is seen. Note that the emission bands are very broad, and small perturbations could pass unnoticed. An example in which the exciplex is readily detectable will be discussed later. Because the decomposition of the dioxetane, or dioxetanone, is concerted, singlet states are predominant, with no opportunity for electron spin decoupling being given.

It has to be said that an earlier proposal,[54] invoking discrete electron transfer, has to be disowned if the explanation described previously is to survive unambiguously. Although the distinction between charge transfer and discrete electron transfer will always remain a fine one, rethinking of the events leading to excitation as described earlier suggests other objections to the older formulation. For example, for an electron to be transferred to the antibonding orbital or lowest unoccupied molecular orbital (LUMO) of the peroxide, there would have to be significant overlap of the orbitals of the aromatic portion of the dioxetane and the LUMO of the peroxide bond. From models, this overlap does not appear to be great. If the back transfer of an electron, in the light of the reappraisal of the CIEEL mechanism is not found to be an efficient process, then it is not a likely constituent of the near unit efficiency of the firefly reaction.

[53] F. McCapra, *Tetrahedron Lett.* **34,** 6941 (1993).

[54] F. McCapra, *J. C. S. Chem. Comm.* 946 (1977).

Last in this section, a dioxetane (XI) of more than average interest, discovered by Nakamura and Goto,[55] epitomizes the foregoing discussion. This dioxetane is significant for several reasons. In the first place, it is remarkably efficient, almost certainly the most efficient[56] light generator of isolated and proven structure in organic chemiluminescence. The light emitted, of unusually short wavelength at 320 nm, has a quantum yield (Φ_e) of more than 50%. The wavelength of this light shifts to 400 nm when the solvent is changed from hexane to the more polar methylene chloride. This change is interpreted as emission from an intramolecular exciplex, involving the excited indole ester and its other carbonyl partner the benzoate (see Scheme 9). This is confirmed[57] by the synthesis of the cis-diester (XIII) on a cyclohexane template, which has the geometry postulated for the transient exciplex. The exciplex was difficult to detect in the fluorescence of the product of the chemiluminescent reaction and is only apparent when the exact geometry derived from the precursor dioxetane is duplicated. The attempt to duplicate the exciplex emission from the fluorescence of the product molecule failed.

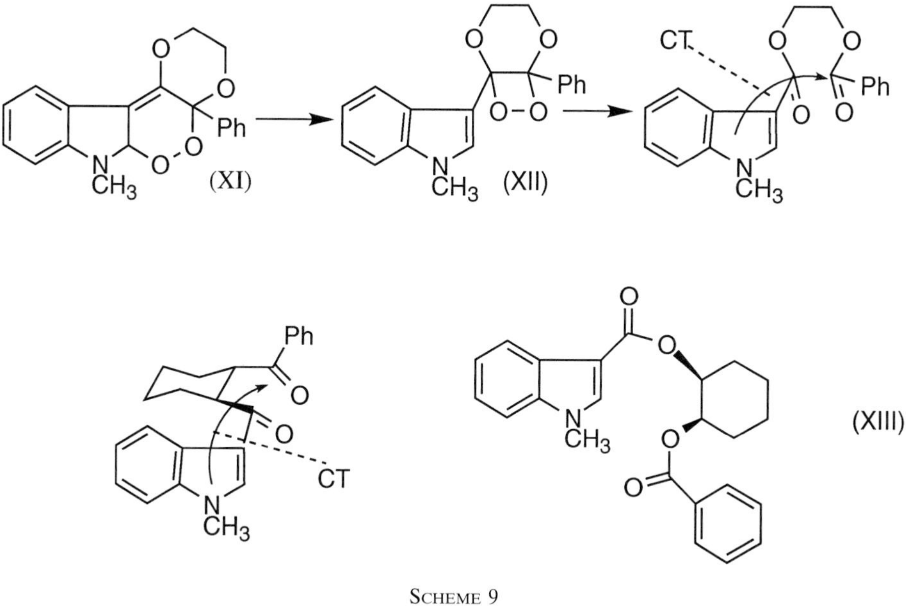

SCHEME 9

[55] H. Nakamura and T. Goto, *Photochem. Photobiol.* **30,** 27 (1979).

[56] However, note that the scintillation standard used in the determination of this quantum yield gives values up to twice those obtained using an alternative, the chemiluminescence of luminol.

[57] H. Nakamura and T. Goto, *Chem. Lett.* 1231 (1979).

It is tempting to try to use the extra detail in the chemistry of this particular dioxetane to distinguish between charge transfer and electron transfer models. Although it is in excellent agreement with the mechanism preferred in the preceeding paragraphs, a case could also be made out for the electron transfer mechanism. Nevertheless, the reaction sequence shown in Scheme 9 serves as a summary of the recommended view of bright chemiluminescence from electron-rich dioxetanes and is also a model for the reactions of firefly luciferin, *Vargula* luciferin, and coelenterazine.

In the chemiluminescence of lophine, there is a mystery that has been explained along similar lines.[36] The emission is visibly yellow, yet no product of the reaction is fluorescent in this region. Although the chromophores are not entirely separate, a charge transfer complex between the two carbonyls could give an excited state of a geometry not found in the spent reaction mixture.

Active Oxalates and Hydrogen Peroxide

This important chemiluminescent system is typical of the brighter examples in that there is agreement on the basic outlines of the mechanism, but a disquieting lack of knowledge of the actual final light-generating event. It is the most highly developed of any practical example, thanks to the thorough and inventive work of Rauhut and his team at American Cyanamid.[58,59] Despite successive investigations by several groups since its discovery in the early sixties, we are still in roughly the position achieved by the first investigators.[58,59] The importance of this chemiluminescence lies in several areas: it is easily the most efficient synthetic system; it uses inexpensive, accessible materials; and it is a uniquely efficient example of an *intermolecular* electron or charge transfer chemiluminescent reaction.

Structural requirements for the oxalate are discerned easily. The carbonyls must be attached to what is loosely called a "good leaving group." This can range from the obvious, such as chloride, to the complex, such as a sulfonamide. A small selection of the structures giving efficient chemiluminescence is shown in Scheme 10.

[58] M. M. Rauhut, *Acc. Chem. Res.* **2,** 80 (1969).

[59] A. G. Mohan, *in* "Chemi- and Bioluminescence" (J. G. Burr, ed.), p. 245. Dekker, New York, 1985.

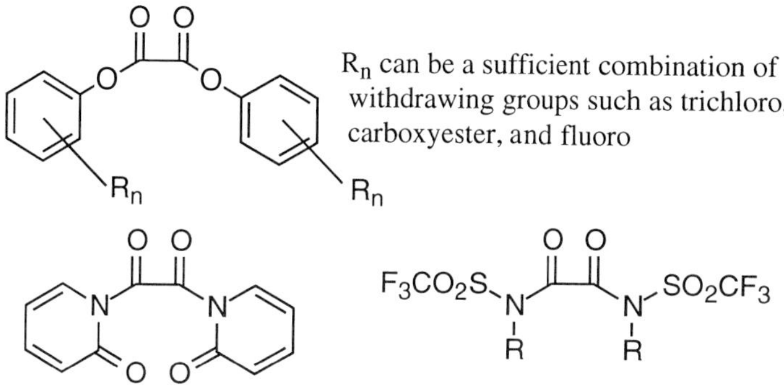

SCHEME 10

A comprehensive study[60] of the reactions of oxalyl chloride has not been bettered in the years since its publication. The conclusion was that monoperoxyoxalic acid was the key intermediate that by reaction in a charge transfer complex gave rise to excited diphenylanthracene. The reaction is probably more complex than the reactions of the more efficient active oxalate esters, and chain reactions play a part. A variety of oxalyl peroxides were also examined for chemiluminescence with aromatic hydrocarbon fluorescers and some were found to be effective, whereas others were not. Because various structures (to be discussed later) have been postulated for the intermediate peroxides, the comparison between successes and failures could be considered valuable. However, some caution is required. The comparison is not quantitative and the difficulty of avoiding traces of hydrogen peroxide in the final peroxide product was not fully appreciated. Some of the active peroxides may react with H_2O_2 to give other reactants. Nevertheless, in view of the doubt still attached to the mechanism of active esters, the implications of this work have been neglected, and a carefully controlled repetition is overdue.

Very many peroxyoxalate structures have been proposed in an attempt to pin down the ultimate generator of excitation. In the foreground is the dioxetanedione (XIV), an early candidate with a chequered history. Suffice it to say that it remains the most intellectually attractive peroxide, but that all attempts to produce direct evidence for its existence have failed. Some 11 different peroxides have been suggested[3] for the key role, but an investigation of the kinetics of the reaction[61] narrows the choice. The scheme

[60] M. M. Rauhut, D. Sheehan, R. A. Clarke, and A. M. Semsel, *Photochem. Photobiol.* **4,** 1097 (1965).

[61] F. L. Alvarez, N. J. Parekh, B. Matuszewski, R. S. Givens, T. Higuchi, and R. L. Schowen, *J. Am. Chem. Soc.* **108,** 6435 (1986).

arising from this study has, as a hypothesis, the peroxides (XIV) to (XIX). A complex kinetic scheme has been simulated by computation and includes the conclusion that there is a storage intermediate that gives rise to the active peroxide or peroxides. It does seem probable that there is more than one light-generating peroxide in the sequence, and the various suggested intermediates are all highly plausible. The ease of retention of nucleophiles on the α-dicarbonyl structure of the oxalate and on the strained four-membered ring suggests such a series of intermediates.

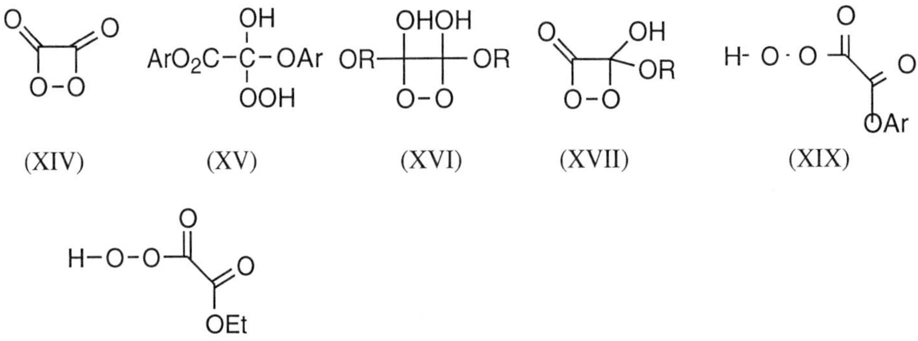

(XIV) (XV) (XVI) (XVII) (XIX)

(XIX)

SCHEME 11

If the dioxetanone and dioxetanedione structures are considered to be especially important (as indeed they might be on evidence from other systems discussed previously), then one can assess the worth of (XIV) easily as the dimethyl analog has been synthesized and has appeared in discussion in another context. The repetition of the quantum yield measurements that gave a value of 10% in the latter case is thus made doubly desirable. However, note that all of the structures (XIV) to (XIX), other than the dimethyldioxetanone, can give rise very easily to the dioxetanedione. It therefore seems possible that a series of ground state complexes of any of these electronegative carbonyl-bearing candidates, with the low ionization potential hydrocarbons, could be the actual precursors. The light-emitting decomposition may not occur until a further reaction produces the (expectedly) very unstable and transient dioxetanedione within the complex.

A significant failure among the peroxyoxalates is the ethyl ester (XIX). In this case, further investigation might serve to establish whether such compounds form ground state complexes with the hydrocarbon, and why, when the peroxyoxalic acid is *apparently* effective, no light is emitted. It is a possibility that the reluctance of the ethoxy group to act as leaving group blocks the path to the dioxetanes and the dioxetanedione. Notwithstanding one's predilection for the dioxetanedione, a general principle may be the basis for progress. Rauhut[58] was aware that charge transfer complexes were likely first steps on the light-emitting path and that these could form reversibly in the ground state so that the *complex* is the species to be examined for reasons for light production. This will not be an easy task, as the precise properties, both experimental and theoretical, may be hard to obtain. The suggestion of an electron transfer mechanism can be transmuted easily into one of charge transfer and, if experimentation justifies it, can be made to include other peroxide intermediates. A reliable finding that a linear peroxide of whatever structure gave quantum yields around 50% would transform our view of the mechanism of bright chemiluminescence.

Hydrazides

Although hydrazides in general are chemiluminescent, only the cyclic hydrazides related to luminol, 5-amino-2,3-dihydrophthalazine-1,4-dione, are efficient. There are some excellent and comprehensive reviews[62,63] of all aspects of the mechanistic studies, and only the difficult matter of the excitation step will be discussed here. The mechanism of chemiluminescence of the cyclic hydrazides appears, at present, to bear no relationship to mechanisms of interest in bioluminescence. However, it is one of the oldest, and certainly the most studied, of the bright organic compounds, and its mechanism of light generation, if properly understood, would be expected to add useful general information.

No linear hydrazide can match the efficiency of luminol, usually being about 100 times less efficient. The most effective actually reacts via a known dioxetane and is thus not actually of the type of reaction discussed here. One might expect that the study of the mechanisms in these hydrazides would assist in the discovery of the more significant luminol mechanism, but this has not proved to be the case so far. The major difference is that in luminol, the diazaquinone (XXI) plays a major role. Examples of the latter can be isolated in certain cases, and their participation, even when they cannot, is strongly indicated by the fact that alkyl substitution, which should prevent their formation, destroys chemiluminescence. Linear hydrazides are not affected greatly by alkyl substition. Many hypotheses have been made concerning the bond rearrangements during the excitation step. They are all sensible and have analogies in ground state chemistry in other systems, but none really carries conviction. All reasonable possibilities have

[62] E. H. White and D. F. Roswell, *in* "Chemi- and Bioluminescence" (J. G. Burr, ed.), p. 215. Dekker, New York, 1985.

[63] D. F. Roswell and E. H. White, *Methods Enzymol.* **57,** 409 (1978).

been discussed in a very clear and open-minded fashion by White and Roswell,[62,63] and these reviews should be read by anyone seeking a menu from which to choose. The missing conviction will only appear when intermediates predicted by any of the mechanisms are synthesized, and these are brightly chemiluminescent with the required characteristics. Athough some attempts have been made in this direction, there is still great scope for further work.

Although there is a surfeit of speculation, some further comments on the mechanisms are appropriate. Among the suggestions made by White and Roswell,[62,63] the dioxetane (XXI) is included. There is often too great a reliance on the value of dioxetanes, and their involvement should not be accepted uncritically. However, the extremely facile rearrangement of endoperoxides such as (V) (Scheme 6) and (XI) (Scheme 9), among others, to dioxetanes is suggestive. As White and Roswell point out, the further reaction required after dioxetane decomposition to reach the accepted emitter phthalate dicarboxylate is most unlikely to compete on the same time scale with emission from the newly formed excited state. If this mechanism were to be considered, the intermediate (XXII) would have to have the same fluorescence spectrum as the dicarboxylate. It is possible that fluorescence spectra of other carboxyl derivatives of the 3-aminophthalic acid, e.g., the monoamide and its dianion, do not differ sufficiently from that of the dicarboxylate, and small differences could be obscured by the broad emission. It might be worth checking this point. Note the similarity between this formulation and the structure of the dioxetane (XXIII).

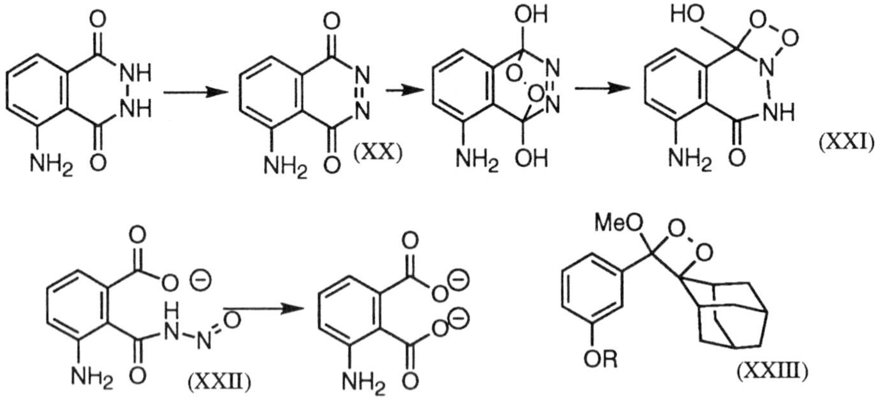

SCHEME 12

A mechanism that has achieved wide acceptance was proposed on theoretical terms by Michl[64] and supported by a wealth of detailed theoretical discussion and experimental evidence by Merenyi and Lind.[65–67] The first difficulty in assessing this work is dealing with the extended array of speculation, forming a long, difficult to grasp, chain of reasoning. Many of the assumptions that have a marked impact on the conclusions can be questioned, but to argue through an alternative case is inappropriate in this review. However, two cardinal assumptions can be addressed. In the Michl formulation[64] of what is, at first sight, an analog of a simple Cope rearrangement, the presence of an O—O bond renders it a special case. It is necessary to assume that the simple [3,3] sigmatropic ground state path occurring in (XXIV) is avoided, and although Michl does give reasons for this, it is hard to avoid the conclusion that strong experimental evidence would be needed to support this rejection of a well-documented path.

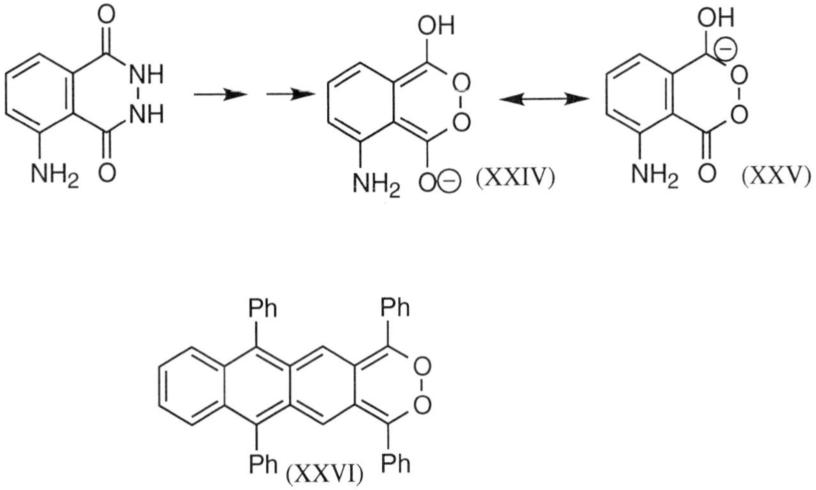

SCHEME 13

The second difficulty arises from the way in which the cyclic peroxide is arrived at by the elimination of nitrogen. The key structure is shown as an analog of xylylene peroxide, yet it is written in an arbitrary fashion. Because of the dominance of benzene aromaticity, (XXV) is a better repre-

[64] J. Michl, *Photochem. Photobiol.* **25,** 141 (1977).
[65] G. Merenyi, J. Lind, and T. E. Eriksen, *J. Am. Chem. Soc.* **108,** 7716 (1986).
[66] J. Lind, G. Merenyi, and T. E. Eriksen, *J. Am. Chem. Soc.* **105,** 7655 (1983).
[67] S. Ljunggren, G. Merenyi, and J. Lind, *J. Am. Chem. Soc.* **105,** 7662 (1983).

sentation of the extended enolate (the reaction would actually be expected to give the protonated form). Even if there is dispute as to the *exact* electron distribution, this structure is no basis for the arguments used by Michl, depending as they do on the ordering of the orbital energies, when the orbitals themselves are not as written in the argument. The same applies to MO calculations, as they apply to an idealized structure that may have no significance in the actual reaction. Last, as was pointed out earlier, peroxides of the structure required as a test of the Michl argument have already been made (see VIII, Scheme 7). The vanishingly small chemiluminescence efficiencies do not increase confidence in the interpretation. It could be argued that these peroxides need to give rise to substantially fluorescent products for a true test. The synthesis of a compound such as (XXVI), or any other easily arrived at variant, is not likely to be difficult, and would be a major contribution to the assessment of this suggestion. In addition, an analog of the suspected endoperoxide itself has also been synthesised,[68] and although it forms the diester by elimination of nitrogen and cleavage of the peroxide, no chemiluminescence results.

Acridinium Esters and Acridans

These effective compounds have been a mainstay of the dioxetane and dioxetanone hypothesis for 30 years and have a well established mechanism.[69,70] There are many reactions and properties of chemiluminescent acridinium compounds that have been studied, but only those studies that relate specifically to the central event of excitation will be discussed. Compounds related to the esters such as the acid chloride and sulfonamides are equally chemiluminescent, reacting by means of the common feature of possessing a good leaving group. The acridine nucleus is an excellent basis for a variety of chemiluminescent reactions, being stable, almost invariably producing strongly fluorescent products when chemiluminescence is involved, and allowing easy substituent changes that facilitate mechanistic studies. A comparison[71] of the acridine, acridinium, acridine N-oxide, and acridan nuclei, with appropriate changes in substitution, gave the results shown in Scheme 14.

[68] T. Goto, M. Isobe, and K. Ienaga, *in* "Chemiluminescence and Bioluminescence" (M. J. Cormier, D. M. Hercules, and J. Lee, eds.), p. 492. Plenum Press, New York, 1973.

[69] F. McCapra, *Acc. Chem. Res.* **9,** 201 (1976).

[70] K. A. Zaklika, D.Phil. Thesis, University of Sussex, 1976.

[71] M. Taheri-Kadkhoda, D.Phil. Thesis, University of Sussex, 1986.

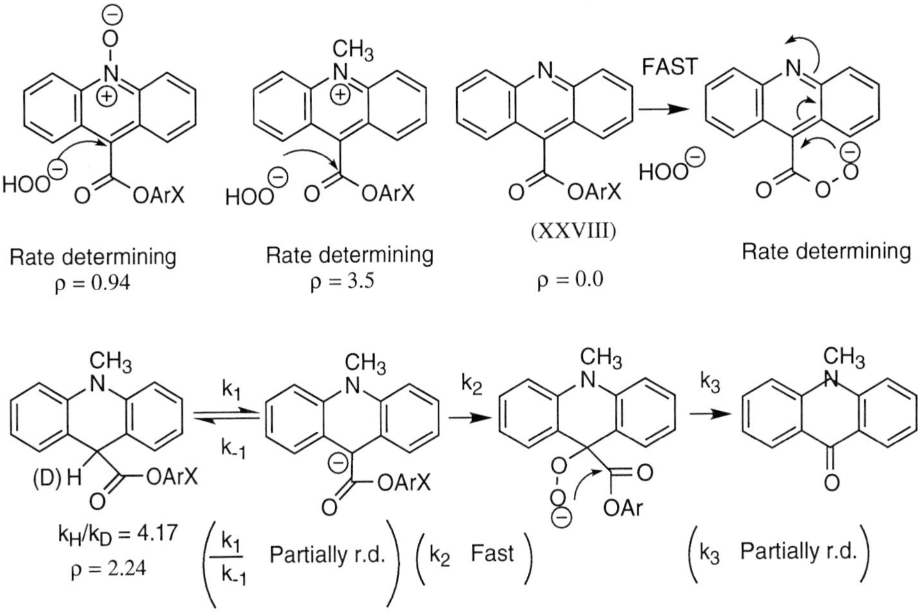

SCHEME 14

A particularly interesting detail has emerged from an intensive study[72] of the acridine ester (XXVII), referred to earlier in connection with the distinction between charge transfer and electron transfer excitation mechanisms. Because the emission from the ester is that of the acridinone *anion,* even when the pH of the solution is too low to generate the anion by an acid–base reaction, the conclusion is that the formation of the excited state occurs before full charge is developed on the N-atom, and that it probably is formed in the transition state itself, rather than at the later stage of full dioxetanone formation.

Although it is possible to envisage electron transfer occurring on this time scale, a more satisfactory picture is that the transition state is itself the charge transfer state that leads to excitation. A similar observation had been made earlier[73] in connection with the chemiluminescence of models for coelenterazine. In this case the amide anion (XXXI) is the emitter, even when the pH of the solution is too low to form it in either the ground or the excited state. A similar conclusion can be drawn, namely that the

[72] E. H. White, D. F. Roswell, A. C. Dupont, and A. A. Wilson, *J. Am. Chem. Soc.* **109,** 5189 (1987).

[73] F. McCapra and M. J. Manning, *J.C.S. Chem. Commun.* 467 (1973).

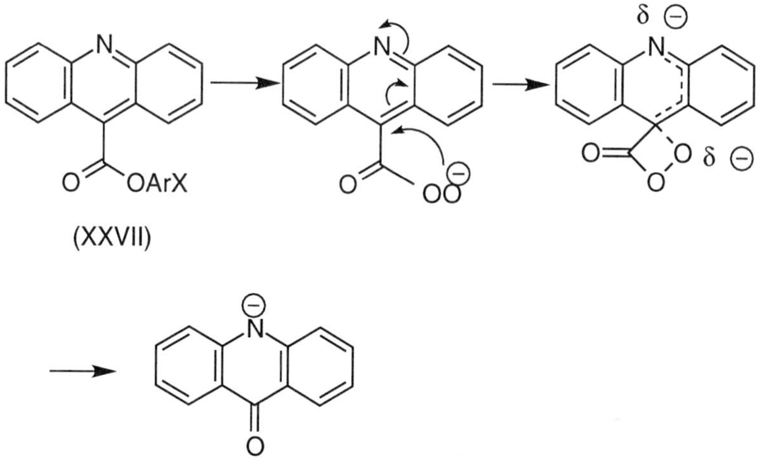

(XXVII)

SCHEME 15

excited state is created on a time scale consistent with a transition state rather than discreet intermediate formation.

The subtleties of this mechanism are more accessible than those of the typical transition state, involving as it does an event (the formation of an easily detected electronically excited state) whose key properties are on the shortest time scale in chemistry. There are nevertheless problems with the interpretation given, as one has to ask how the other bonds of the dioxetanone (or, if preferred, the peracid anion) are broken and how a bond, not yet fully formed, becomes a double bond that is part of the acridone excited state, yet there is no spectroscopic "record" of the other bonds of the peracid that are fully formed. There may be answers to these questions, and if so, much will have been learned concerning excited states in general. Simpler explanations, seemingly excluded by White's work, cannot be dismissed entirely. For example, the concentration of the N-anion formed by full C-O bond formation at C-9 may be exceedingly small, but this species may decompose so very much more quickly than the N-protonated form that this becomes the only route observed. It should be kept in mind that we are discussing whether a minimum exists on the potential energy curve of the reaction and that it could be real, but only the energy of a few vibrations deep. Excitation could occur so quickly that the concentration would need to be infinitely low to avoid this route.

The chemiluminesence of the acridans includes the luminescent decomposition of a dioxetanone, and the similarity to firefly and coelenterate luciferins makes them highly significant compounds. This has been pointed out before; the biggest difference betweeeen them and other acridines is the prior formation of a 9-hydroperoxide by autoxidation. This has been isolated, using the addition of hydrogen peroxide to the acridinium nucleus, and there is no doubt as to its existence nor the mechanism of formation of the dioxetanone. In this case, and in the case of the luciferins, the four-membered peroxide ring arises by the formation of the tetrahedral intermediate common to all the reactions of esters. There is insufficient information available to allow speculation as to the timing of the various bond reorganizations, but it is fairly safe to assume that a dioxetanone rather than dioxetane is formed in view of the undetectably short lifetime of this intermediate. Related dioxetanes have a much longer and easily detectable existence.[74] From the point of view of the generation of excited states, these compounds behave exactly like the luciferins. The similarities are striking, but although the oxidation of the C-H group in the nine position is well understood in the acridans, details of the enzymic counterpart remain to be discovered. It is worth pointing out that this type of reaction is very general, and many other ring systems could be discovered without much thought.

The Reaction of Imines and Enamines with Oxygen

Reactions of this type encompass many apparently distinct structures, but they all conform to the same general principle. Some examples are shown, but the details of their chemistry will not be repeated, as there are few significant developments since their chemistry was last reviewed. The presence of an intermediate dioxetane has been demonstrated in two very different cases by isotopic labeling.[75,76] Perhaps the most fascinating of this class is tetrakisdimethylaminoethylene,[77,78] in that it is spontaneously chemiluminescent in air. A dioxetane is suggested here too, but the evidence is less secure than usual due to the complexity of the reaction. If one were involved, a ground state charge transfer complex with a dioxetane (rather than the more frequently suggested complexes with dioxetanones) would be quite believable because the electron-rich olefin has a uniquely low ionization potential, compensating for the usual inability of the dioxetane to act as an acceptor. Emission is from the olefin itself. A reexamination of this type of reaction is overdue.

[74] F. McCapra, I. Beheshti, A. Burford, R. A. Hann, and K. A. Zaklika, *J. C. S. Chem. Commun.* 944 (1977).

[75] F. McCapra and A. Burford, *J.C.S. Chem. Commun.* 874 (1977).

[76] F. McCapra and Y. C. Chang, *Chem. Comm.* 522 (1966).

[77] W. H. Urry and J. Sheeto, *Photochem. Photobiol.* **4,** 1067 (1965).

[78] F. Roeterdink, J. W. Scheeren, and W. H. Laarhoven, *Tetrahedron Lett.* **24,** 2307 (1983).

Reactions of Schiff bases, peroxides derived from them and enamines, provide some of the more efficient examples of organic chemiluminescence, with quantum yields as high as 18%. A few examples are shown in Scheme 16 and all react via dioxetanes on oxidation or rearrangement.

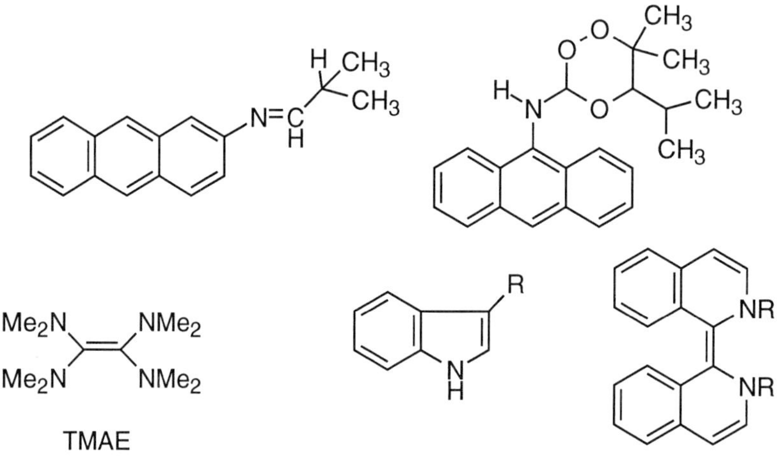

SCHEME 16

Most of these compounds are only chemiluminescent in dipolar aprotic solvents such as dimethyl sulfoxide with a very strong base. The reason for this has not been investigated, but possible answers include the need to desolvate the peroxy anion so that addition to the imine to form the strained four-membered ring can proceed and/or the need to prevent the N-anion from protonating. Circumstantial evidence shows that this highly basic anion is essential for an efficient reaction, perhaps mirroring the behavior of (XXVII) (Scheme 15). In contrast, the peroxide here is formed from an enamine in which the N atom is substituted, with the result that the very electrophilic, positive iminium ion allows formation of the dioxetane in neutral conditions and in most solvents.

Firefly Luciferin

This compound,[79] the first luciferin to have its structure determined and be synthesized, is the most efficient of all organic chemiluminescent compounds, if the enzymic reaction is assumed to reflect the underlying

[79] F. McCapra and K. D. Perring, *in* "Chemi- and Bioluminescence" (J. G. Burr, ed.), p. 359. Dekker, New York, 1985.

organic reaction ($\phi = 88\%$). Even the nonenzymic reaction almost matches the very best results obtained from simple organic reactions ($\phi_{es} = 53\%$). It thus plays a pivotal role in determining mechanisms. For example, the yields of singlet excited states are far in excess of the statistical 3 : 1 triplet-to-singlet ratio. Thus, any mechanism that allows the attainment of a statistical spin distribution must be excluded. The mechanism shown in Scheme 17 is supported by isotopic labeling in the enzymic reaction[80] and by analogy

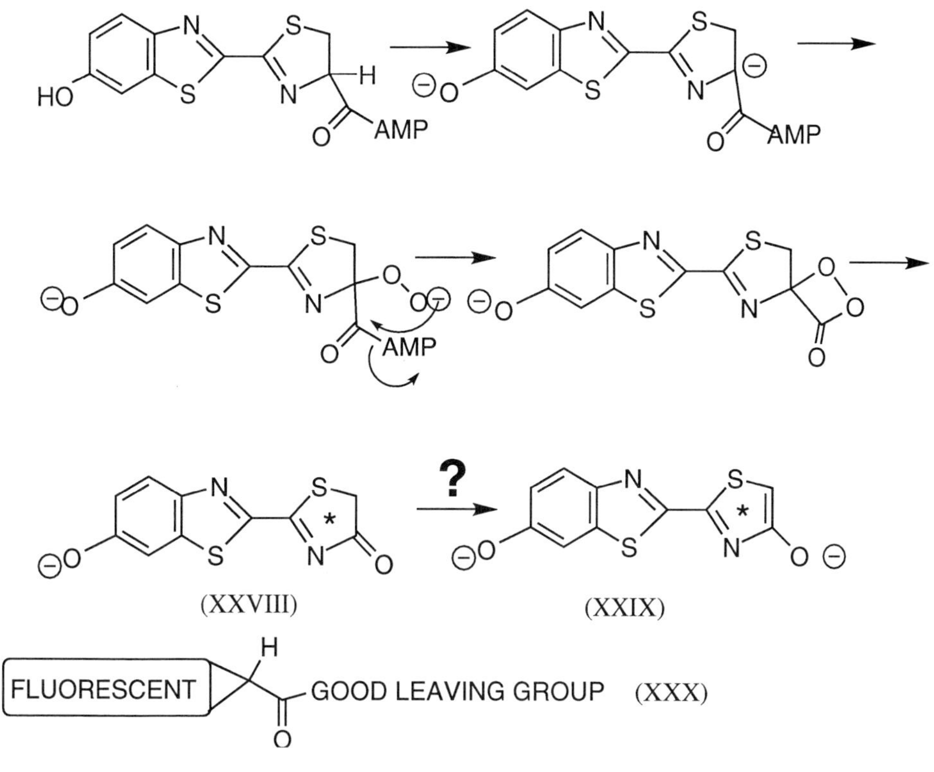

SCHEME 17

with well-studied model compounds. The literature on firefly luciferin is extensive, and only the less well-understood details relating to excited state generation will be discussed.

The first of these is the reaction between luciferin and oxygen. Although the anion is the logical first step, neither it nor the peroxide has been discovered directly. A mention[81] of unpublished work suggests that the

[80] O. Shimomura, T. Goto, and F. H. Johnson, *Proc. Natl. Acad. Sci. U.S.A.* **74,** 2799 (1977).
[81] W. D. McElroy, H. H. Seliger, and E. H. White, *Photochem. Photobiol.* **10,** 153 (1969).

replacement of the 4-H atom by 4-D results in a slower reaction, but because the method of introducing it was by exchange with D_2O, the inevitable presence of 50% L-luciferin would result in considerable inhibition and a marked decrease in rate in any event. Deuterium labeling at this position has been carried out properly[82] with chirally pure D-luciferin, and an isotope effect k_H/k_D of 2.3 obtained. In addition, $^{18}O_2$ labeling work requires an intermediate dioxetanone, and with the similarity of behavior of the model compounds, the case for the whole mechanism is compelling. In fact, it is more than likely that *any* reaction including part reactions of the type shown in (XXX) will be strongly chemiluminescent.

The full details of the reaction after the formation of the peroxide are not known, and various possibilities, most very difficult to prove, exist. It is unlikely that decomposition and hence light emission occur at the dioxetane (tetrahedral intermediate) stage in view of the fast rise time. Similar dioxetanes (e.g., XXIII), while decomposing more quickly when the phenolic oxyanion is generated, do not do so at a suitably fast rate.[83,84] A direct comparison of the different rates of reaction between dioxetane and dioxetanone intermediates in the acridine series reinforces this view. The most likely event catalyzing the decomposition to excited oxyluciferin and CO_2 is the expulsion of the good leaving group, AMP, to give the very unstable dioxetanone.

One of the most intriguing aspects of luminescence in the *Coleoptera* is the variation in color of the light. This has ecological significance. The variation occurs in nature[85] and can be produced genetically *in vitro*, and the mechanism is intimately bound up with that of excitation in general. For many years, an early observation by White and co-workers[86] explained, in a most satisfactory way, the behavior of the *in vitro* luciferase and luciferin reaction. A variety of mild denaturing agents (metals, heat, pH, etc.) alter the prevailing yellow (ca. 560 nm) light to red (ca. 620 nm). This change is clearly biphasic, with one or the other band appearing as a shoulder on the other. It thus is explained easily by the notion outlined earlier where the first formed red-emitting keto excited state (XXVIII) loses a proton before emitting to form the yellow-emitting dianionic species.

[82] F. McCapra, D. J. Gilfoyle, D. W. Young, N. J. Church, and P. Spencer, "Bioluminescence and Chemiluminescence—Fundamentals and Applied Aspects" (A. K. Campbell, L. J. Kricka, and P. E. Stanley, eds.), p. 387. Wiley, Chichester, 1994.

[83] A. P. Schaap, T. S. Chen, R. S. Handley, R. DeSylva, and B. P. Giri, *Tetrahedron Lett.* 1155 (1987).

[84] A. P. Schaap, M. D. Sandison, and R. S. Handley, *Tetrahedron Lett.* 1159 (1987).

[85] H. H. Seliger and W. D. McElroy, *Proc. Natl. Acad. Sci. U.S.A.* **52,** 75 (1964).

[86] E. H. White, E. Rapoport, T. A. Hopkins, and H. H. Seliger, *J. Am. Chem. Soc.* **9,** 2178 (1969).

According to several investigations in the field by Seliger and McElroy,[85] all of the maxima, from 543 to 574 nm, are characteristic of spectra derived from one molecular species.[87] It is not at all reasonable to assume that the two maxima found by White could bridge the range found by Seliger and McElroy[85] without showing shoulders or other evidence, such as very different bandwidths, of additive spectra. Other influences have to be added to the description. There are three ways of looking at the causes of the wide (>50 nm) range of wavelengths found. The simplest is that the local "solvent" effect of the active site, together with as yet undetermined bonding, such as hydrogen bonding to the N-atoms of the oxyluciferin, is different in each of the luciferases. The oxyluciferin-excited state is undoubtedly a linear charge transfer state and, as such, is liable to be affected markedly by solvent polarity. Unfortunately, oxyluciferin is so unstable that any experiments devised to check this are extremely difficult and liable to mislead, as fluorescent impurities are invariably present. It is known that the active site is very hydrophobic, and Wood and co-workers[88] have demonstrated that short wavelength maxima are associated with tighter, more specific active sites in line with this view. Access to water would be denied at these sites, giving a shift toward the blue, and partial denaturation would expose the oxyluciferin to a more polar, aqueous environment, causing a red shift. It is an attractive possibility in its simplicity and ample precedent, with the main worry being that it has not yet been demonstrated that such relatively minor perturbations can give rise to the wide range of colors seen. The position has been made more difficult by the cloning of the beetle luciferases, as the increase in hydrophobicity now has to transform 620 to 538 nm.

A hybrid proposition is that the range derives not solely from the proven red, 620-nm emitter, but from either it or a yellow (ca. 560 nm) emitter. This is an important modification in that the clones of beetle luciferases do occasionally show shoulders on the main emission band, and obviously it will be much easier to cover the range of colors if the starting points are close to either extreme wavelength observed. The difficulty with this explanation (probably the simplest and most acceptable) is that White and Roswell[89] were unable to repeat the observation that started thinking in this way. An exceedingly strong base failed to produce even a trace of the previously observed yellow emission. There is thus no consistent evidence

[87] The molecular species in this case includes, of course, the concept of an enzyme-bound emitter.

[88] G. D. Kutuzova, R. R. Hannah, and K. V. Wood, *in* "Bioluminescence and Chemiluminescence—Reporting with Photons" (J. W. Hastings, L. J. Kricka, and P. E. Stanley, eds.), p. 248. Wiley, Chicester, 1997.

[89] E. H. White and D. F. Roswell, *Photochem Photobiol.* **53,** 131 (1991).

that the dianion (XXIX) exists, however likely it may seem, and none that it has a yellow emission. Other yellow-emitting tautomers and ionized forms were proposed[89] as replacements and may well be involved, thus rescuing this interpretation.

In response to the observations made on living fireflies and the conclusion that a single molecular species was responsible for the whole range of color, a suggestion was made[82] that by twisting the thiazolinone and benzothiazole portions of the molecule about the connecting single bond, a more than sufficient energy range could be obtained from the excited state to provide the colors. A surprising feature revealed by calculations made to test this hypothesis, was that the energy differences were in an excited state of the twisted charge transfer variety and not the ground state conformations. Either explanation is acceptable. It is an attractive idea because it provides a continuum of color changes that can be imposed easily merely by differences in the binding energy between the luciferin (and by extension the oxyluciferin) and the enzyme. All of these mechanisms will respond to changes at remote sites of the enzyme, produced either naturally or by genetic engineering. In the first case, the site will change in hydrophobicity, as a result of changes in conformation, and in the second case, the base at the active site will either be within the appropriate distance for proton removal or not. In the third case, the site will impose variable restrictions on rotamers, depending on overall enzyme conformation. More work needs to be done on both the chemistry and the enzymology to distinguish between these mechanisms.

Imidazolopyrazine Luciferins

Only two distinct members of this class have been discovered so far, but given the origins of both from modifications of ubiquitous peptide structures, it is possible that more will be found, with different amino acid-derived side chains. The first to have its structure determined[90] was that of *Cypridina*, now *Vargula*, an ostracod crustacean. Although it is associated with other organisms, notably certain fish, it has a very restricted occurrence when compared to the other related luciferin, coelenterazine. This is extraordinarily widely spread, even where bioluminescence is not a feature of the organism. Coelenterazine was the first luciferin for which a dioxetanone was proposed,[91] although firefly luciferin was also considered as reacting in this way in the same paper. Many analogs have been prepared

[90] Y. Kishi, T. Goto, S. Inoue, S. Sugiura, and H. Kishimoto, *Tetrahedron Lett.* 3445 (1966).
[91] F. McCapra and Y. C. Chang, *Chem. Commun.* 1011 (1967).

and, like firefly luciferin, it has a rich literature as a chemiluminescent compound.[79] The mechanism shown in Scheme 18 has been established by isotopic labeling among other techniques and is unchanged over many years. A variety of side reactions has been discovered, but none changes the basic principle by which the luciferin generates light. Studies of the oxygenation mechanism agree with the view that all the luciferins discussed earlier and chemiluminescent compounds such as acridans and their relatives', form anions that react with oxygen to form superoxide ion and the luciferin radical in a solvent cage before going on to yield the peroxide.

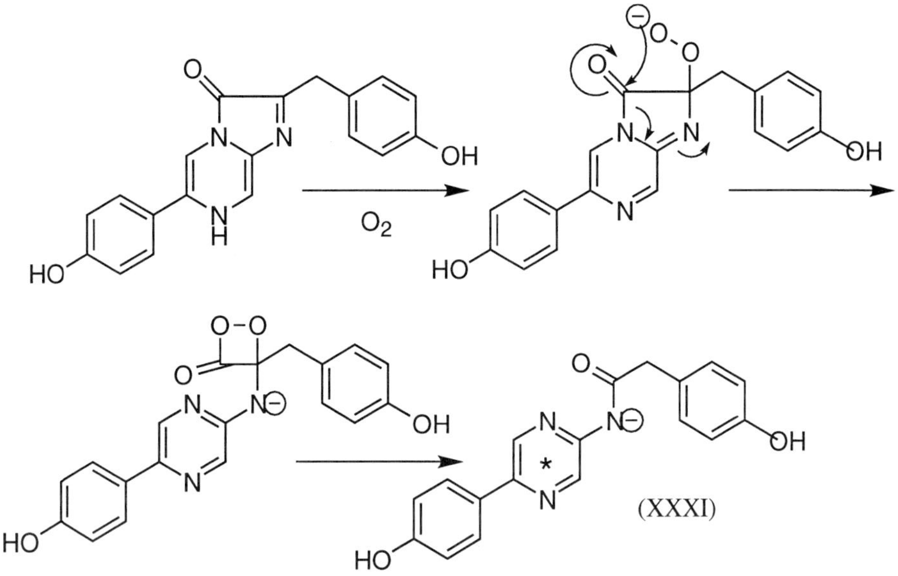

SCHEME 18

An aspect of the charge transfer nature of the decomposition referred to earlier is particularly interesting. The phenolic hydroxyl remains unionized under certain basic conditions, serving as an indicator of the special nature of the formation of the dioxetanone. As the dioxetanone forms, the negative charge on the N-atom is so short-lived that it does not protonate before the dioxetanone decomposes. This means that the emission is from the unprotonated amide of the oxyluciferin, with the phenol still unionized. However, when the base strength is lowered, emission is seen from the protonated amide. This is in marked contrast to the situation with the acridine (XXVII, Scheme 17), in which emission from the anion is found under all conditions. Incidentally, because energy transfer to the green fluorescent protein (GFP) only occurs from the anion,[92] this effectively rules out an electron exchange mechanism with GFP as "activator," for the biological reaction at least. Resonance energy transfer is the most likely alternative, i.e., the neutral dioxetane should have a higher electron affinity than the anion and so should be more effective in a CIEEL mechanism. It is clearly not. However, only the emission spectrum of the amide anion overlaps effectively with the absorption spectrum of GFP, confirming that the energy transfer is of the resonance or Förster type. There are several very interesting studies of this system that add considerably to our knowledge in this area, but cannot be discussed in the space available.

Bacterial Bioluminescence

This is the most studied[93] of all the examples of bioluminescence, and the benefits are many. Due largely to the work of Hastings and co-workers, a large body of evidence concerning a host of topics relevant to general biochemistry has been examined. In addition, we owe to that group the basis on which the mechanism of light generation can be placed. There have been many hypotheses about the mechanism, but all now take as their starting point the conclusions arrived at in recent years by the very effective use of NMR spectroscopy.[94] Formation of the 4a-hydroperoxydihydroflavin (XXXIII, Scheme 20) and its reaction with aldehyde to form the peroxyhemiacetal (XXXIV, Scheme 20) seem a secure base. The structure of the latter has not been demonstrated experimentally, but it is a conclusion that is hard to avoid. It is impossible to review all of the mechanisms proposed over the years in the space available, as each would require an inordinate amount of comment, and it is probable that none is consonant with all the evidence. In addition to the biochemical experiments, many chemical model reactions have been carried out. Few are compelling because none has suggested a general mechanism that can be used to construct other examples of chemiluminescence of the same or similar efficiency. Until this is achieved, support for any one proposal is difficult. That being said, because of the (misplaced) confidence in the generality of the CIEEL mechanism, many experiments appear to support an electron transfer mechanism and clearly require comment.

Studies of the enzymic reaction have to assume the peroxyhemiacetal (XXXIV) as the ultimate precursor of the light reaction. It is a reasonable

[92] R. C. Hart, J. C. Matthews, K. Hori, and M. J. Cormier, *Biochemistry* **18,** 2204 (1979).

[93] J. Lee, *in* "Chemi- and Bioluminescence" (J. G. Burr, ed.), p. 401. Dekker, New York, 1985.

[94] J. Vervoort, F. Muller, J. Lee, W. A. M. van den Berg, and C. T. W. Moonen, *Biochemistry* **25,** 8062 (1986).

assumption, and two hypotheses are based on it. In the first, a Baeyer-Villiger reaction is postulated,[95] supported by the observation[96–98] of a deuterium isotope effect, varying in size between a k_H/k_D of about 1.5 to 5.4 Other work[99] using deuterated aldehydes did not confirm these values, finding only a very small effect. Nevertheless, the weight of evidence, including that obtained in the most recent investigation,[100] is in favor of a small but significant isotope effect of 1.5–1.9. In the second hypothesis,[101] supported by the equivalent of a Hammett linear-free energy relationship, the Baeyer-Villiger reaction was discounted, and evidence was obtained for the possibility of electron donation by a fluorescent activator (in this case 4a-hydroxyflavin) in an intramolecular CIEEL reaction of the peroxyhemiacetal. It has, however, been pointed out[6] that the rate data may in fact be *compatible* with a Baeyer-Villiger pathway, although this is not necessarily support for that particular mechanism. Rather, one should simply conclude that kinetic evidence does not demonstrate unequivocally that electron donation from the 4a-hydroxyflavin is a rate-determining event.

The electron transfer construction rests, not on the biochemical experiments, but on confidence in CIEEL and the use of model compounds. Because no fluorescence can be detected from the 4a-hydroxyflavin in solution, all claims for meaningful *chemiluminescence* from this system (and there are several) cannot be taken seriously. Electrochemical studies are similarly beside the point.[102,102a] In any event, the quantum yields are extremely low (typically less than 10^{-5}). Another objection is that adducts of alkyl peroxides such as *n*-butyl peroxide and flavinium salts are moderately stable, surviving for many hours. One must include the enzyme-bound peroxide in this group. There is thus no evidence at all that the simple peroxide O—O bond in the peroxyhemiacetal (XXXIV) is capable of accepting an electron from the flavin, whether the latter is thought of as a

[95] A. Eberhard and J. W. Hastings, *Biochem. Biophys. Res. Commun.* **47,** 348 (1972).
[96] R. B. Presswood and J. W. Hastings, *Photochem. Photobiol.* **30,** 93 (1977).
[97] R. P. Shannon, R. B. Presswood, R. Spencer, J. E. Becvar, J. W. Hastings and C. Walsh, *in* "Mechanism of Oxidising Enzymes" (T. P. Singer and R. N. Ondarza, eds.), p. 69. Elsevier, Amsterdam, 1978.
[98] S.-C. Tu, L.-H. Wang, and Y. Yu, *in* "Flavins and Flavoproteins" (D. E. Edmondson and D. B. McCormick, eds.), p. 539. De Gruyter, New York, 1987.
[99] P. Macheroux, S. Ghisla, M. Kurfürst, and J. W. Hastings, *in* "Flavins and Flavoproteins" (R. C. Bray, P. C. Engel, and S. G. Mayhew, eds.), p. 669. De Gruyter, Berlin, 1984.
[100] W. A. Francisco, H. M. Abu-Soud, A. J. DelMonte, D. A. Singleton, T. O. Baldwin, and F. M. Raushel, *Biochemistry* **37,** 2596 (1998).
[101] J. W. Eckstein, J. W. Hastings, and S. Ghisla, *Biochemistry* **32,** 404 (1993).
[102] H. I. X. Mager, D. Sazou, Y. H. Liu, S.-C. Tu, and K. M. Kadish, *J. Am. Chem. Soc.* **110,** 3759 (1988).
[102a] T. W. Kaaret and T. C. Bruice, *Photochem. Photobiol.* **51,** 629 (1990).

good reductant or not. This is supported amply by evidence from the failed attempts to force still more electronegative peroxides, such as the simple dioxetanes, into acting as electron acceptors in putative CIEEL reactions.

A very good case has been made for bacterial bioluminescence as a sensitized chemiluminescence, and neither of the two mechanisms discussed earlier can accommodate this. The most recent work[100] satisfies this requirement and interprets the isotope effect in terms of a previously suggested mechanism for excitation requiring the formation of a dioxirane.[103] The isotope effect on the rate of reaction is seen as a partially rate-determining rearrangement of the radical anion (XXXII) to give the species that back donates the electron in the final step of the CIEEL mechanism.

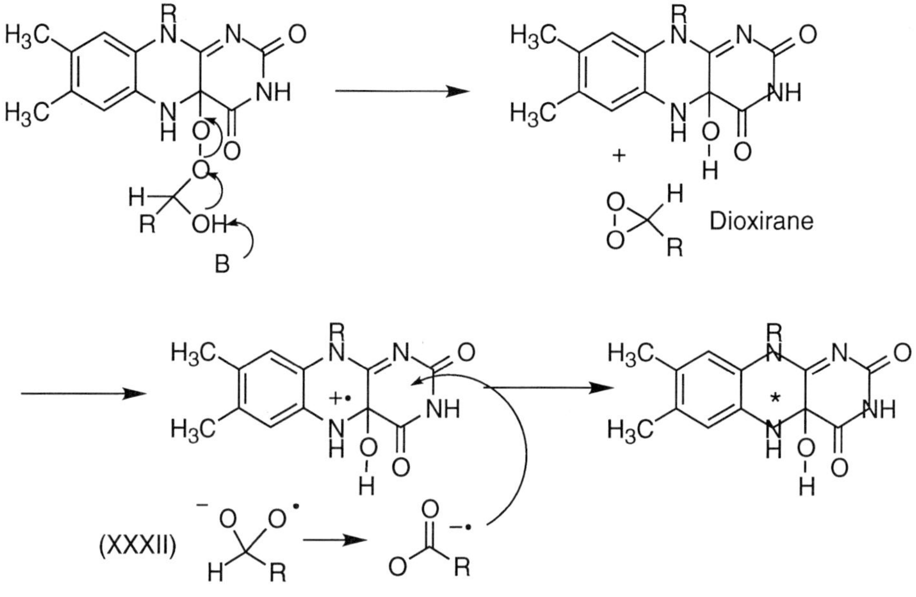

SCHEME 19

Despite the very solid experimental observations, the proposal involving dioxirane has many flaws. First, formation of a dioxirane requires a leaving group of considerable power, most commonly the sulfate ion formed as the result of the decomposition of the adduct of persulfate and the carbonyl group. The 4a-oxyflavin is commonly accepted as being a very poor leaving group. Second, no dioxirane has ever been formed from an aldehyde. It is virtually certain that elimination of a proton always occurs to give a carbox-

[103] F. M. Raushel and T. O. Baldwin, *Biochem. Biophys. Res. Commun.* **164,** 1137 (1989).

ylic acid. Third, dioxiranes are exceedingly reactive species and would tend to react by transfer of an oxygen atom to the surrounding amino acids of the enzyme in a fast reaction. None of the many examples of flavin hydroperoxide-catalyzed oxidations incorporates this inherently unlikely entity.

An alternative mechanism,[104] shown later, derived from studies of fluorescent 10a-adducts[105] and quinoxalinium salts has the following merits, with all components and steps having ample precedents. It is the only suggestion with a *chemiluminescent* counterpart and is consistent with the isotope evidence and, depending on the rate-determining step in the enzymic reaction, with substitution evidence.

As the experiment with added hydrogen peroxide[106] shows, the latter is a limiting factor in the quantum yield. It is postulated that there is an equilibrium among the peroxyflavin, aldehyde peroxyhemiacetal, and hydrogen peroxide such that the adduct (XXXV) with a second aldehyde is formed. Reaction then proceeds as shown in Scheme 20.

The perester is an excellent candidate for the sensitized mechanism[93] and is expected to form a complex with any low ionization potential sensitizer (such as lumazine and flavins) by virtue of the carbonyl group. In its undeveloped form, it gave a quantum yield of 0.8%. Enzyme binding would be of considerable assistance in enhancing this yield by improving contact between the perester and flavin. In addition, it explains extremely well the observation[106] that added hydrogen peroxide is a potent enhancer of the bioluminescence, with a change in kinetics. Hydrogen peroxide and aldehydes are known to form adducts spontaneously, including two aldehydes.[106] Note that the actual *order* of the bond reorganization steps is not yet known.

Finally, it would also be very satisfactory if it were possible to include the enzyme from luminescent bacteria among those that perform Baeyer-Villiger reactions such as cyclohexanone oxygenase,[107] with the obvious and acceptable difference being that an *elimination* reaction (or hydride migration followed by proton loss) is possible from the aldehyde–peroxide adduct.

[104] F. McCapra, P. D. Leeson, V. Donovan, and G. J. Perry, *Tetrahedron* **42,** 3223 (1986).

[105] It may be objected that we know that the addition of peroxides in the presence of the luciferase is to the 4a and not the 10a position. However, inspection of the reactivity at either site shows that they are equivalent from the point of view of peroxide adducts. The 10a system was chosen for the chemiluminescence model reaction because only these adducts are fluorescent in free solution.

[106] H. Watanabe and J. W. Hastings, *J. Biochem.* **101,** 279 (1987).

[107] C. C. Ryerson, D. P. Ballou, and C. Walsh, *Biochemistry* **21,** 2644 (1982).

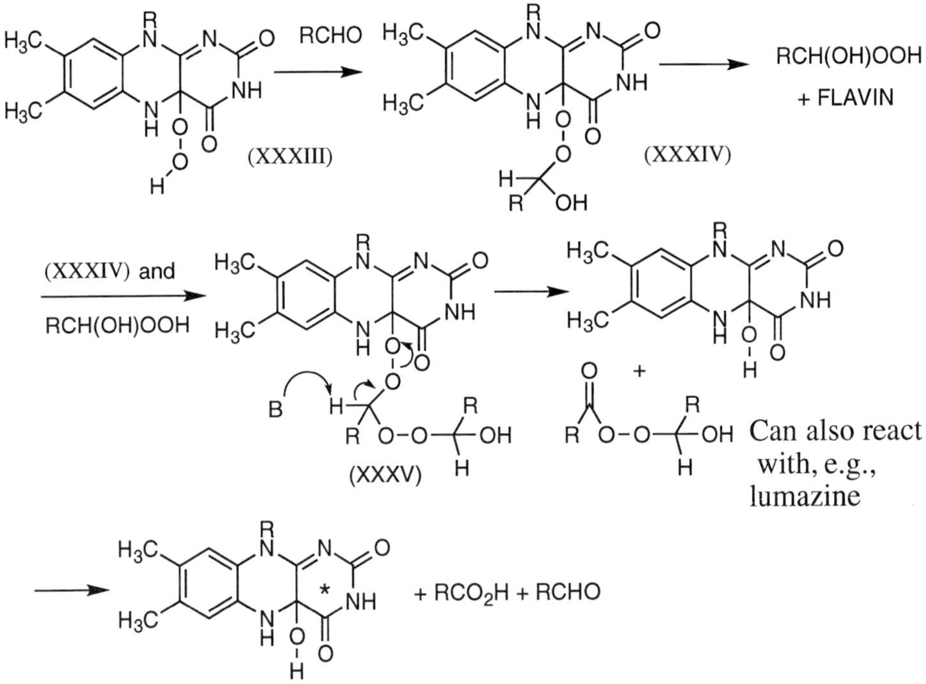

SCHEME 20

Last, the mechanism does have analogy with other reactions of peresters with fluorescent compounds, but this takes us full circle in that there is a serious gap in our understanding of nondioxetane/dioxetanone peroxide chemiluminescence that must be filled by further experimentation before bacterial bioluminescence can be considered to be understood.

Other Luciferins

There are many chemiluminescent reactions not dealt with in this review, but almost all can be considered under one of the types discussed. The situation in bioluminescence is different. Apart from the many organisms for which a luciferin structure is unavailable, yet show all the signs of major differences between their mechanisms of light generation and, say, firefly and coelenterate bioluminescence, there are some bright examples for which the luciferin structure is known. Three in particular, dinoflagellate, *Latia*, and *Diplocardia* luciferins, cannot at present be assigned to any category of chemiluminescence mechanism. The largest part of the difficulty is a lack of detailed information concerning the biochemical reaction, mainly as a result of lack of material. However, it should be possible to devise chemiluminescence mechanisms independently, with a little work with model compounds. It is the author's contention that every bioluminescent reaction will have a counterpart in chemiluminescence and that the light-generating step in any example of bioluminescence is not fully determined until that counterpart is found.

The high point of the studies in mechanism may have passed, but new experiments and a more determined search for better rationlizations for the reactions deemed to have unsatisfactory explanations are both urgently required.

[2] Some Brief Notes on Nomenclature and Units and Standards Used in Bioluminescence and Chemiluminescence

By PHILIP E. STANLEY

Nomenclature

A hundred years ago Wiedemann[1,2] proposed that a luminescent substance "becomes luminous by the action of an external agency which does not involve an appropriate rise in temperature." He defined six kinds of luminescence, including chemiluminescence that is produced as a result of a chemical reaction and this definition still holds today. Bioluminescence is now generally taken to be a special form of chemiluminescence in which light production occurs by the action of a biologically catalyzed chemical reaction.

O'Kane[3] has pointed out that scientists working in these fields have a common language and common vocabulary but that the words used do not always have a common definition, which has led to needless arguments. Words such as luciferase, luciferin, and photoprotein, for example, still need to have a consensus definition. This article does not intend to propose definitions, merely to concur with the need for such an agreement. This

[1] E. Wiedemann, *Wied. Ann. Phy. Chem.* **34,** 446 (1888).

[2] E. Wiedemann, *Wied. Ann. Phy. Chem.* **54,** 201 and 604 (1895).

[3] D. J. O'Kane, *in* "Bioluminescence and Chemiluminescence: Fundamental and Applied Aspects" (A. K. Campbell, L. J. Kricka, and P. E. Stanley, eds.), p. 651. Wiley, Chichester, 1994.

0076-6879/00 $30.00

would be a suitable topic for a specialized session at an international symposium when many points of view can be distilled.

Words used in the study of bioluminescence and chemiluminescence are also important for another reason. When using them to search the Internet or electronic databases, mishits are sometimes found. Thus when searching for the bacterial *lux* gene one also retrieves references, including the light unit "lux," and with the firefly luciferase gene *luc* one will also find not only authors with the last name Luc but also hospitals named Saint Luc. The latter two examples are detected readily but the first mentioned will require some detailed reading. Search terms using acronyms such as GFP and ATP should be used with great care because they will also find some unexpected items, including glomerular filtration protein and Association of Tennis Professionals. Wild card characters should be used carefully. For example, the search term lucifer*, which one would use to collectively find luciferase(s), luciferin(s), and luciferyl, will also find lucifer yellow and lucifer.

Units and Standards

New workers in the field soon become aware of the lack of a common unit for reporting light measurements in bioluminescence and chemiluminescence studies. This is particularly noticed by workers using these techniques in analytical procedures, particularly in the clinical laboratory.[3] One would consider that the best unit to use is the photon, but at present there is no universally convenient technique for the calibration of a luminometer based on photons. So-called photon-counting luminometers do indeed count photons but they do not count *all* the photons produced in the reaction vessel; instead, they count the number of photoelectrons generated at the photocathode of the photomultiplier. A very large proportion of photons generated in the reaction get absorbed, reflected, or lost before leaving the reaction vessel. Some are lost elsewhere or dispersed en route to the photomultiplier, and the photocathode will then only generate a photoelectron with a probability given by the photocathode quantum efficiency at the wavelength of the impinging photon. For many bioluminescent and chemiluminescent sources, this quantum efficiency is usually less than 20% and often less than 8%. A Monte Carlo computer model has been used to study the generation and loss of photons in glass vessels,[4–7] and in

[4] P. J. Malcolm and P. E. Stanley, *in* "Liquid Scintillation Counting" (M. A. Crook and P. Johnson, eds.), Vol. 4, p. 16. Heyden & Son, London, 1977.

[5] P. E. Stanley and P. J. Malcolm, *in* "Liquid Scintillation Counting" (M. A. Crook and P. Johnson, eds.), Vol. 4, p. 45. Heyden & Son, London, 1977.

[6] P. J. Malcolm and P. E. Stanley, *Int. J. Appl. Radiat, Isotopes* **27,** 397 (1976).

[7] P. J. Malcolm and P. E. Stanley, *Int. J. Appl. Radiat. Isotopes* **27,** 415 (1976).

most luminometers the number of photons counted is only a small fraction of 1% of those generated in the reaction vessel. Luminometers fitted with photomultipliers operated in the current measuring mode suffer the same kinds of losses.

Most workers routinely give their results in "relative light units" (RLU) as reported by the luminometer. Any day-to-day variation in instrument performance is generally ignored. There is a further problem in that different luminometers can give vastly different RLU for the same sample. Even luminometers of the same model can give results that differ by factors of 10 or more. How then can the worker relate his results to those reported by others using different luminometers? A formal group study showed how this can be attained but it was not a trivial exercise.[8] Ten evaluators used 10 different luminometers to measure ATP using a standard firefly luciferase-based procedure. The largest conversion factor required to normalize results between two of the luminometers was around 10^4. An 800-fold difference in sensitivity was shown between the most and the least sensitive instrument. This clearly demonstrated the vast range of instrument capability and the difficulty in comparing results produced by different models.

The user should be aware that the result units displayed by a luminometer may well have been generated by software manipulation of the true number of photons counted (or photomultiplier anode current). Thus, for example, if the background is really 1000 events (photoelectrons) and a sample gives a result showing 2000 events, a division by 100 by the software will produce a background of 10 and a sample result of 20. This manipulation is considered to produce a "better" result, as a background of 1000 is generally not viewed by the user to be as acceptable as 1 of 10.

The availability of calibrated standards has been reported.[9] Methods for standards based on luminol have been published.[10] Other workers have used quenched scintillation solutions spiked with ^{14}C[11] or tritium,[12] whereas solid phosphors spiked with radionuclides (such as ^{63}Ni)[13] have also been

[8] P. H. Jago, W. J. Simpson, S. P. Denyer, A. W. Evans, M. W. Griffiths, J. R. Hammond, T. P. Ingram, R. F. Lacey, N. W. Macey, B. J. McCarthy, T. T. Salusbury, P. S. Senior, S. Sidorowicz, R. Smither, G. Stanfield, and P. Stanley, *J. Biolumin. Chemilumin.* **3,** 131 (1989).

[9] D. H. Leaback, *in* "Bioluminescence and Chemiluminescence: Status Report" (A. A. Szalay, L. J. Kricka, and P. E. Stanley, eds.), p. 33. Wiley, Chichester, 1993.

[10] J. Lee, A. S. Wesley, J. F. Ferguson III, and H. H. Seliger, *in* "Bioluminescence in Progress" (F. H. Johnson and Y. Haneda, eds.), p. 35. Princeton Univ. Press, Princeton, 1966.

[11] J. W. Hastings and G. Weber, *J. Opt. Soc. Am.* **53,** 1410 (1963).

[12] E. Schram, H. Van Esbroeck, and H. Roosens, *in* "Proceedings of the Symposium on Analytical Applications of Bioluminescence and Chemiluminescence" (E. Schram and P. E. Stanley, eds.), p. 687. State Printing & Publishing, Westlake Village, CA, 1979.

[13] D. J. O'Kane and J. Lee, *Photochem. Photobiol.* **52,** 723 (1990).

used, as have ^{14}C-spiked aminofluoranthracene and lanthanide chelates[14] and electronically controlled light-emitting diodes.[15] Probably only the luminol standards provide true single photon events suitable for the current purpose as radionuclide-based ones will produce scintillations. Each scintillation is made up of many photons that are produced so close together in time that the photomultiplier "sees" the scintillation as one event and does not resolve the individual photons. Standards made using radionuclides will decay according to the half-life of the nuclide (12 years for tritium). The long-term chemical stability, and thus light output, of such standards has yet to be established.

[14] T. E. T. Oikari, I. A. Hemmila, and E. J. Soini, *in* "Analytical Applications of Bioluminescence and Chemiluminescence" (L. J. Kricka, P. E. Stanley, G. H. G. Thorpe, and T. P. Whitehead, eds.), p. 475. Academic Press, Orlando, 1984.

[15] R. Bräuer, B. Lübbe, and R. Ochs, *In* "Bioluminescence and Chemiluminescence: Status Report" (A. A. Szalay, L. J. Kricka, and P. E. Stanley, eds.), p. 13. Wiley, Chichester, 1993.

Section II

Instrumentation

[3] Physics of Low Light Level Detectors

By KENNETH J. VOSS

The measurement of light by most detectors involves the conversion of light, or photons, into an electrical signal. Photons carry an increment of energy, E_{ph}, which depends inversely on the wavelength (λ) of the photon, $E_{ph} \alpha h/\lambda$ (h is Planck's constant = 6.63E-34 J-s). When a photon is absorbed by a material, the energy is used to excite the atoms or molecules in the material. If the photon carries energy greater than, or equal to, the "work function," E_w, of the molecule, an electron will be ejected. The energy above the work function ($E_{ph} - E_w$) is carried by the electron as kinetic energy. This process is called the photoelectric effect.[1]

Selection of a device to perform this conversion to an electrical signal depends on the type of measurement required. For low light level single point measurements the most commonly used device is the photomultiplier tube (PMT). This device has the advantage of being reasonably efficient, with high gain in a linear mode (on the order of 10^6) and a relatively low cost. A solid-state device, the avalanche photodiode (APD), has been introduced in many applications. The APD requires a lower supply voltage (100 V vs 500 V or greater) but has lower gain (approximately 100) when operating in a linear mode. When operating an APD in a nonlinear mode, gain can be increased greatly, thus the APD is more suited to counting single events (photon counting).

For two-dimensional (2D) measurements, array detectors are required. In this case, charge-coupled device (CCD) arrays dominate the market. Two methods exist for making a CCD into a low light detector. When longer integration times (0.1 sec to minutes) can be utilized, cooled CCD arrays allow the noise intrinsic to the device to be lowered, thus lower signals can be detected. This method of increasing the sensitivity of the CCD allows the device to maintain a large intrascene dynamic range (up to 16,000–60,000 to 1).

For some measurements, longer integration times are not appropriate, such as looking at the time development of a process. In this case, another device is added to the CCD array to preamplify the optical signal. This device is called an image intensifier and is much like adding a PMT for each point in the 2D CCD array. Image intensifiers increase the signal

[1] A. Einstein, *in* "Great Experiments in Physics" (M. H. Shamosed, ed.), p. 232. Dover, New York, 1987.

METHODS IN ENZYMOLOGY, VOL. 305
0076-6879/00 $30.00

obtained for each incident photon, but tend to reduce the dynamic range of the array by increasing the noise. The technology of these image intensifiers is developing rapidly and the costs are decreasing as they move into consumer markets (low light vision systems for sailors, etc.). The following discussion provides a simple description of how each type of detector works, but there are many similarities between the devices.

Photomultiplier Tubes

The oldest technique and most commonly commercially available low light level detector is the photomultiplier tube. Figure 1 shows a diagram of a side-illuminated PMT. The PMT consists of a photocathode, electron multiplier plates, and anode all contained in a vacuum tube. The photocathode is the site where the photoelectric effect takes place. Photoelectrons generated at the photocathode are accelerated in a large electric field, gaining energy, to an electron multiplier plate, or dynode. The collision of

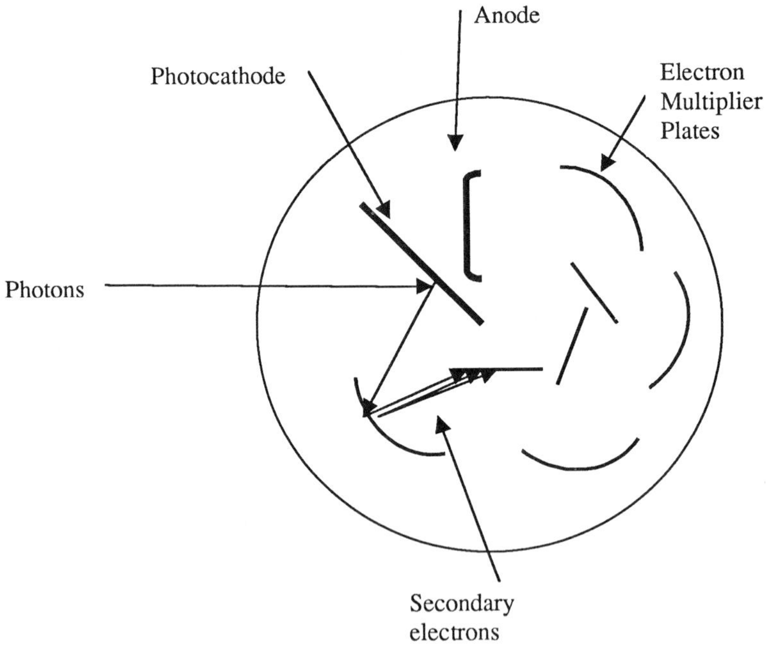

FIG. 1. Top view of a side-illuminated photomultiplier tube. Photons strike the photocathode, generating a photoelectron. These photoelectrons are accelerated into the electron multiplier plates, or dynodes, to produce multiple electrons ("secondary emission"), which are eventually collected by the anode.

these high-energy electrons with the plate causes the dynode to emit multiple electrons for each incident electron in a process called secondary emission. By having a series of dynodes, the initial photoelectron can be amplified greatly. At the end of a series of dynodes, the electrons are captured at the anode and are read as a current between the photocathode and the anode. The signal from the PMT depends on two factors: the quantum efficiency of the photocathode and the efficiency of the amplification process. Obviously the more efficient the photoelectric effect at the photocathode, the greater the output from the PMT. Similarly, the amplification efficiency of the PMT depends on the collection efficiency of the dynodes and the acceleration of the electrons between plates by the applied electric field. Selection of a specific PMT depends on the wavelength of the light to be measured, sensitivity required, and time response desired. These factors are used to select the photocathode material, window material, and tube design.

Photomultiplier Tube Design

Photomultiplier tubes come in two basic types: side illuminated and front end illuminated. The first commercially available PMT was a side-illuminated device, the 931 by RCA.[2] Side-illuminated PMTs are generally low priced and used in many commercial devices. In a side-illuminated PMT, the photocathode is opaque and electron multiplier plates are arranged in a circular cage structure, as shown in Fig. 1. These tubes have good sensitivity and amplification at relatively low supply voltage (typically $\leq$1000 V). Front end-illuminated PMTs use a semitransparent photocathode (diagram in Fig. 2). These PMTs typically have better spatial uniformity because the photoelectrons can be focused onto the electron multiplier plates more efficiently. The photocathode can have photosensitive areas ranging from tens of square millimeters to hundreds of square centimeters.[3] There are several designs for electron multiplier plate arrangements. Figure 2 shows a box and grid type of arrangement. Time response, spatial uniformity of the response, sensitivity to magnetic fields, and cost are all factors in choosing the design of electron multiplier plates. These factors tend to trade off so that selection is made on which factors are important for the particular application.

Photocathodes

The photocathode is the stage where the photoelectric process takes place. The conversion of photons to electrons is described by the quantum

[2] R. W. Engstrom, "Photomultiplier Handbook." RCA Solid State Division, Lancaster, 1980.
[3] "Photomultiplier Tubes." Hamamatsu, Shimokanzo, Japan, 1990.

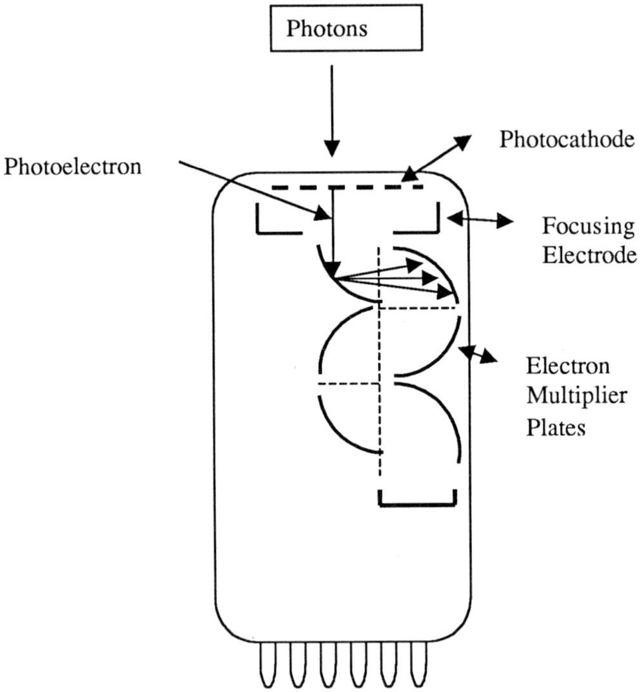

FIG. 2. Side view of an end-on illuminated photomultiplier tube. The operating structure is similar to the side-illuminated PMT shown in Fig. 1.

efficiency (ratio of electrons generated per incident photon) of the photocathode. The quantum efficiency varies with wavelength of the incident photon as it depends on the incident photon having enough energy to ionize the photocathode and cause the ejection of the electron. Because the energy of the photon varies inversely with wavelength, photocathodes for red to infrared radiation must have lower work functions then those used for violet or ultraviolet radiation. With the lower work functions, intrinsic thermal noise can eject electrons easier. Thus red sensitive photocathodes tend to have higher noise. The dark signal is exponentially proportional to temperature. Thus by cooling these PMTs, the thermal noise can be reduced. Cooling is generally required for infrared detectors.

The photocathode material used depends on the application and the desired wavelength sensitivity required. Typically the spectral limit imposed by the photocathode occurs on the red end of the spectrum, as the photons on the blue end increase in energy. The blue end cutoff of a PMT is typically due to the window material used for the PMT envelope. Optical glasses start

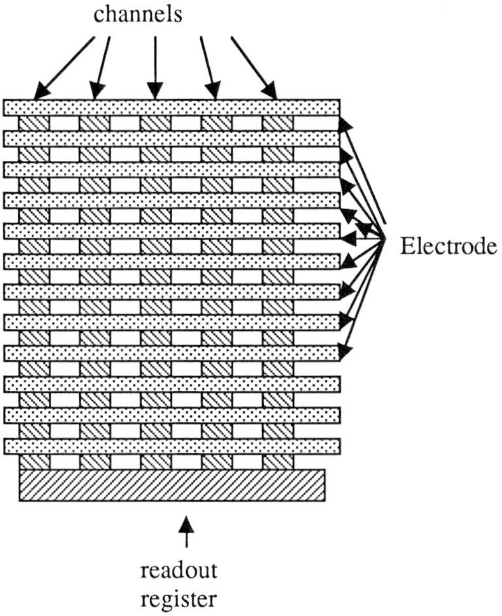

FIG. 3. Top view of a simplified structure for CCD array. Conduction channels are separated by insulating barriers. Electrodes are used to control charge location along the conduction channels. Photoelectrons are transported along the conduction channels to the readout register, where they are moved off of the side of the array to be read electronically.

to absorb light at around 350–400 nm. For ultraviolet light measurement, materials such as quartz must be used for the optical window.

Charge-Coupled Device Arrays

Charge-coupled device arrays are one- to two-dimensional arrays of "potential wells," pixels, which hold the electrons generated by the photoelectric process.[4] Figures 3 and 4 illustrate the top and side view of a simplified structure, whereas Fig. 5 shows the CCD operation. In the CCD array structure, channels of p-type silicon are separated by insulating strips. The whole structure is overlaid with an insulating layer of SiO_2, onto which transparent conducting electrodes are deposited. In a structure such as this, three electrodes and a conducting channel define a single pixel. For example, Fig. 3 would illustrate a five column by four row imager, thus 20 pixels. Looking along one of the columns, Fig. 4 shows a side view of the device.

[4] G. F. Amelio, *Sci. Am.* **230,** 23 (1974).

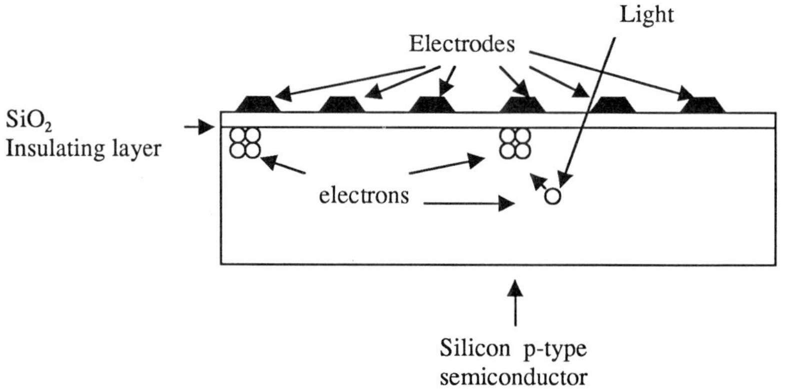

FIG. 4. Side view, along a conduction channel, of a CCD array. Each pixel is defined in this structure by the conduction channel and three electrodes. Photoelectrons are captured by setting one electrode in each pixel to a positive voltage when compared to the other two electrodes. This creates a potential well, which holds the photoelectrons in place. An insulating layer exists between the electrodes and the bulk semiconductor.

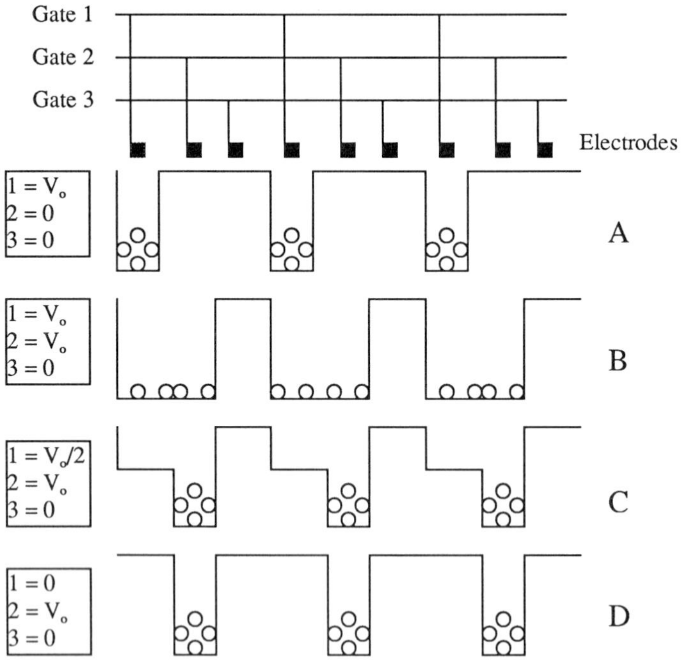

FIG. 5. Schematic of the three-phase system to transport photoelectrons along the conduction channels in the array. By varying the voltage to the three electrodes in each pixel, the photoelectrons can be transported to the readout register and then off of the array.

Light incident on the array generates a photoelectron in the bulk material and in each pixel one electrode is charged to a positive voltage attracting the photoelectron. To measure the amount of charge (electrons) in each pixel, the pixels are shifted down along the conduction column to the readout register. The readout register is then shifted to the left or right sequentially by the same process and the charge in each pixel is measured.

Figure 5 illustrates how the electrons contained in one pixel can be shifted into a neighboring pixel in a three-step process. This design is referred to as a three-phase system because there are three electrodes per pixel. While the array is being exposed to light, a positive voltage is applied to gate 1. This captures the electrons as shown in diagram A. After the light is blocked, gate 2 is brought to voltage V_o, allowing the electrons to spread into the area under this electrode. Gate 1 is then brought to $V_o/2$, which pushes the electrons into the area under gate 2. When the electrons are all under gate 2, gate 1 is brought to 0 V, which results in all the electrons being confined under gate 2. Electrons can be moved sequentially in this manner from the array onto the readout register, and then off the array to be counted. The total charge measured for each pixel is proportional to the incident flux, thus giving a measure of the incident light. This process must be very efficient at transporting all the photoelectrons for the array to work. Transfer efficiencies limit the speed an array may be emptied (read out) and the size of an array. These transfer efficiencies are greater than 0.9999/transfer for modern arrays.[4]

The process just described is for a three-phase transfer in a surface channel device (photoelectrons carried near the insulator–silicon interface). Other designs exist for the transferring mechanism, such as two-phase systems, which reduce the number of conduction gates required. Buried channel devices use ion implantation to keep the potential wells from being near the surface. These are more complicated to build, but have lower noise and better transfer efficiencies.[5] The light incident on the array must go through the electrodes and insulating layer before interacting with the doped silicon to produce photoelectrons. Because these electrodes are only semitransparent, there are losses associated with this. For low light applications, backside-illuminated CCDs can be used to avoid this problem. Photoelectrons can also recombine with the silicon substrate on the way to the pixel potential well. "Thinned" arrays can be used to reduce this effect by reducing the pathlength the photoelectron must travel in the material.

Variables for CCDs include quantum efficiency, dark noise, electronic noise (from amplification and reading out of the array), and array size and

[5] A. F. Milton, *in* "Topics in Applied Physics: Optical and Infrared Detectors" (R. J. Keyes, ed.), p. 197. Springer-Verlag, Berlin New York, 1977.

pixel dimensions. One important variable for a CCD array is the pixel well capacity. In other words, how many electrons can be contained within a pixel before overflowing the potential well into neighboring pixels. This process is called "blooming" and can be recognized by bright pixels extending along columns in the array or image. The electron capacity of each well is dependent on the size of the pixel and the design of the potential well, but in general the more pixels (higher resolution) in the array the lower the electron capacity.

Low light levels can be investigated with CCD arrays in two ways. By cooling the array first, thermally generated dark noise can be reduced significantly. This allows the light falling on the array to be integrated for longer times, thus increasing the signal-to-noise ratio (more photoelectrons than noise electrons from dark noise and fixed readout and electronic noise). Cooled arrays are commonly used where large dynamic range and spectral range are required and the integration times can be extended.

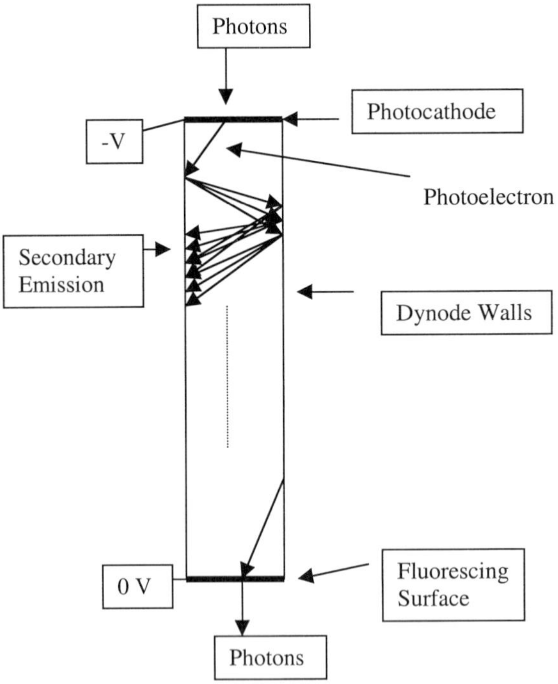

FIG. 6. Schematic of a single image intensifier tube. Operation is much like a PMT tube, but in the end the electrons are turned into photons again by the fluorescing surface. These amplified photons are collected by the CCD array. An image intensifier will typically have 1.5 million of these tubes for the array.

The other method of increasing the signal from a CCD array is by adding an intensifier stage to the device. This is somewhat like adding a photomultiplier tube to each pixel in an array. A diagram of a single intensifier tube is shown in Fig. 6. An intensifier stage would typically contain an array of 1.5 million of these tubes. Here a photocathode converts the incident photon into an electron. The electron is accelerated down a dynode "tube" interacting with the walls of the tube and causing secondary emission from the walls. At the end of the tube the electrons hit a fluorescing surface that converts these electrons into photons, which are then incident on the CCD. Thus one incident photon can be amplified into many photons by the intensifier and collected on the CCD. An important issue is, as with a PMT, determining the proper photocathode for the intensifier. Intensifiers tend to reduce dynamic range and increase noise, as opposed to cooled CCDs. However, they can help when gain is required, a large dynamic range is not necessary, and long integration times are not available.

Avalanche Photodiodes

Avalanche photodiodes are modern solid-state devices that convert a detected photon into a large number of electrons. They operate much like a PMT, except the electrons are transported in a doped silicon medium rather than a vacuum. The basic device contains two regions, an absorption region and a gain region. The absorption region is a relatively thick region where the photons are converted to photoelectrons. A small electric field is applied in this region to sweep the photoelectrons to the gain region. In the gain region a large electric field is applied. This causes the photoelectrons to accelerate inside the material and gain energy. The high energy electrons interact with the substrate and, through the process of impact ionization, cause additional free electrons. As these electrons increase in energy they can cause additional electrons to be created and a cascade of electrons is formed from the single initial photoelectron. When operated in a linear mode, APDs can have gains of 100 or more with applied voltages of 100 V. For more gain the APD can be operated in "Geiger"mode.[6] In this mode a voltage greater than the breakdown voltage is applied to the APD (approximately 400 V or more). When either a photoelectron or an electron generated by noise (typically thermal) is produced, large gains can occur (10^4–10^8). Low noise APDs can be selected and the devices can be cooled to reduce the number of thermally generated electrons. In the Geiger mode, single photon detection efficiencies of close to 70% can be achieved.

[6] H. Dautet, P. Deschamps, B. Dion, A. D. MacGregor, Darleene MacSween, Robert J. McIntyre, Claude Trottier, and P. P. Webb, *Appl. Opt.* **32,** 3894 (1993).

[4] Luminometer Design and Low Light Detection

By FRITZ BERTHOLD, KLAUS HERICK, and RUTH M. SIEWE

Introduction

This article describes the methods of low level light detection that are used in bioluminescence and chemiluminescence and the application of these principles in instrumentation, both luminometers and imaging systems.

The availability, quality, and decreasing cost of commercial luminometers have reached a level that makes the use of "home-built" instruments a rare exception. Accordingly, we will focus on those physical and engineering principles that are, or could be, employed in commercial instrumentation and describe and discuss specific features in a critical fashion.

For low level light imaging, progress has been driven by dramatic advances in solid-state technology, particularly charge-coupled devices (CCDs). Tube and microplate luminometers, however, have been improved in more of an incremental fashion.

It is amazing how well the photomultiplier has held up as the detector of choice in luminometry. In view of the enormous progress in solid-state optoelectronics, it had seemed to be only a question of time until the photomultiplier, a vacuum tube, would be replaced. However, photomultipliers have improved considerably in recent years, and especially when operated as photon counters, they are the most sensitive light detectors presently available at a reasonable cost.

Following the wide popularity of microplates, many plate luminometers are now available. Luminometers for 384-well plates have also become available, and as the trend goes to an even higher sample density, the development of instrumentation can be expected to follow.

Portable, battery-driven luminometers have been developed in recent years, mostly in the use of rapid ATP measurement in industrial hygiene monitoring and other microbiological applications.

First, basic principles of luminometry and instrument design will be explained, in particular the function of the photomultiplier, the measuring chamber, and reagent injectors. Furthermore, we will discuss quality control, calibration, the definition and determination of sensitivity, dynamic range, and units used in luminometry. This should enable the reader to judge the performance of his luminometer critically and to optimize his experimental procedures to obtain correct results.

METHODS IN ENZYMOLOGY, VOL. 305
0076-6879/00 $30.00

The features needed in a luminometer depend both on the applications and on the type of luminescent chemistry used. For example, an injector is needed for flash-type kinetics, but is not necessary for glow kinetics. In some applications, such as reporter gene assays using a luciferase gene, high sensitivity may be the first consideration; in other cases, a high dynamic range is more important. The requirement for sample throughput will determine the level of automation. We will then describe luminometers by categories rather than individual instruments. Commercial suppliers are listed in the overview by Phil Stanley in this volume.

Basics of Luminometry

Instrumentation

The basic design, including measuring chamber, detector, and reagent injectors, will be discussed using a semiautomatic instrument as an example (Fig. 1).

Detector. The standard detector used in almost all luminometers is still the photomultiplier tube. So far, it has held up well against semiconductor technology, as it is hard to match the signal-to-noise ratios of photomultipliers at comparable cost. However, in the future, avalanche diodes may be developed into an alternative to multipliers, especially for portable luminometers with space constraints. Different types of low-noise silicon

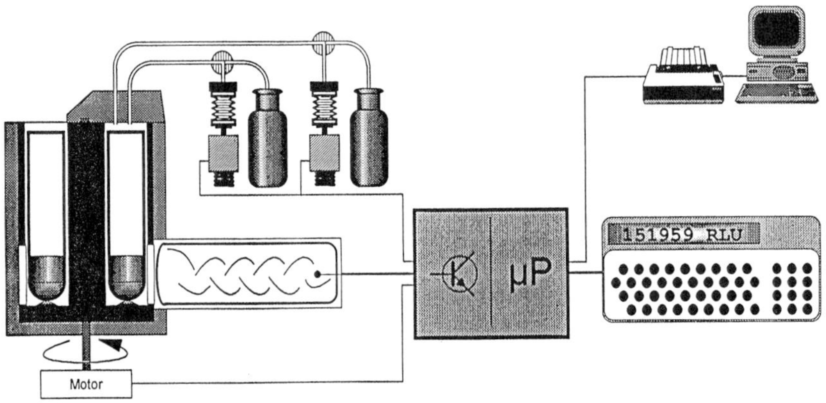

FIG. 1. Schematics of a semiautomatic tube luminometer. Hardware components include a motor-driven turret with a light-tight measuring chamber, two automatic injectors, a photomultiplier unit, an electronic board with microprocessor, and an alphanumeric keyboard and display. Note that connection of an external PC is optional.

detectors have been used in luminometry, but their sensitivity is several orders of magnitude lower than for photomultipliers.

At the cathode of a photomultiplier, photons release electrons by the photoelectric effect. These electrons are accelerated electrostatically and focused onto the first dynode, where they release secondary electrons, typically four to six. These electrons subsequently pass through a chain of further dyodes, with the multiplication process recurring every time. In total, up to 14 dynodes are used in commercial photomultipliers. Finally, the electron bunch is collected on an anode.

The amplitude–time function of anode current signals at low light levels is shown in Fig. 2. Individual events have full width at half maximum of about 2–6 nsec in relatively fast multipliers. The pulse amplitudes vary considerably and can be classified into three categories.

Only pulses with amplitudes between A and B correspond to electrons released at the cathode, representing the photons to be measured, plus noise electrons released spontaneously from the cathode.

Pulses smaller than A are due to noise events from dynodes or other low amplitude noise sources, whereas pulses greater than B originate from radioactivity in the glass and cosmic radiation.

Two methods to process these signals are available.

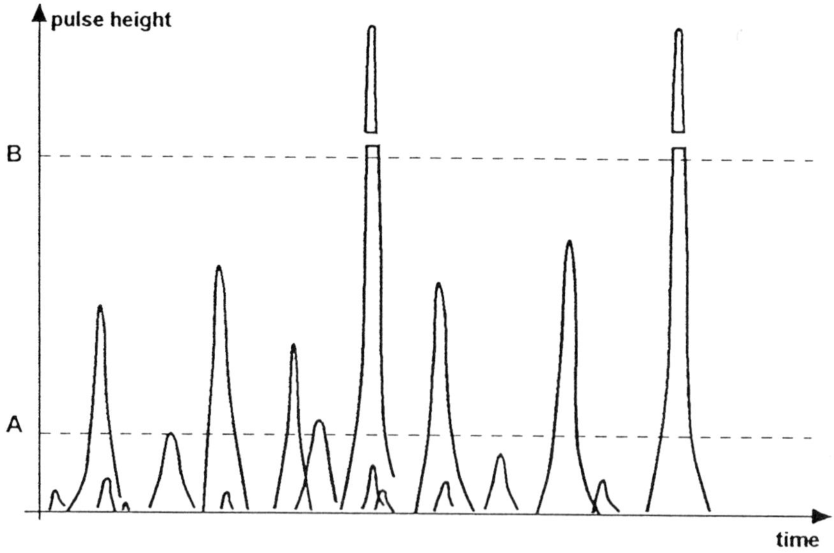

FIG. 2. Output pulses from a photomultiplier falling into three pulse height ranges: below A, dynode noise, etc.; between A and B, signal plus cathode noise; and above B, high energy events.

i. Current measurement by integrating the charge signals with a direct current (DC) amplifier.
ii. Amplifying the signals with a fast pulse amplifier and digitally counting only those pulses with an amplitude greater than A using a threshold discriminator circuit that is set accordingly. This is called photon counting. To exclude the contribution from radioactivity, one could use an upper level discriminator set at level B that would exclude all pulses greater than B. However, this is rarely done because the contribution from this noise source is small.

Photon counting leads to higher sensitivity than current measurement, mainly because the noise contribution from small pulses is excluded. Another advantage is much better stability, making frequent readjustments unnecessary.[1] The only disadvantage of photon counting is that saturation at high light levels occurs at pulse rates of around 2.5×10^7 counts per second, whereas current measurement extends to light levels higher by about one or two orders of magnitude.

To avoid the effect of saturation, it is normally possible to reduce light levels, e.g., by placing a neutral-density light filter between the sample and the photo cathode. It should also be noted that after exposure to very high light levels, most photomultipliers have elevated background levels for a certain time. Some luminometers are equipped with overload warning and protection.

The spectral sensitivity of a photomultiplier is determined by the type of cathode used. For bioluminescence and chemiluminescence measurements, bialkali or rubidium-bialkali photocathodes are normally used, covering a spectral range from about 350 to 700 nm. For special applications, like the measurement of singlet oxygen, a sensitivity extension into the red range is desirable. This would require the use of S20 or extended S20 cathodes covering the spectrum up to about 900 nm or S1 cathodes that extend even to 1100 nm. As a general rule, more red sensitivity also means more dark current so that cooling becomes almost mandatory. Bialkali tubes, however, can be operated at ambient temperature with tolerable dark count rates: about 10–50 counts per second. This value can be reduced by a factor of four by cooling to 6°.

Measuring Chamber. The measuring chamber and sample-loading mechanism have the following functions.

[1] F. Berthold, *in* "Advances in Scintillation Counting" (S. A. McQuarrie, C. Ediss, and L. I. Wiebe, eds.), p. 230. Faculty of Pharmacy and Pharmaceutical Sciences, University of Alberta, Edmonton, Canada, 1983.

i. The sample must be transported from outside the instrument into the measuring position in a reliable way, and the sample-loading mechanism must also be well designed ergonomically.
ii. The measuring chamber has to be absolutely light tight in its closed position. When it is open, the detector should be protected against excessive light levels that could cause memory effects in the detector (see section on detectors).
iii. Because light emission occurs practically isotropically, only a fraction of the light will reach the detector directly. Therefore, reflecting surfaces must be used, and a well-designed geometry employed. The efficiency ε of a measuring chamber is defined as the ratio of the number of photons reaching the photocathode divided by the number of photons emitted from the sample. A value for ε of 40% can be obtained (F. Berthold, unpublished results), but it is normally considerably lower. This number is dependent on sample format and wavelength.
iv. According to its design, the measuring chamber may allow quite different sample formats.

Reagent Injectors. The injection of starter reagent into the sample while the sample is in the measuring chamber in front of the detector allows tracing of the light emission instantaneously from the beginning. This is imperative for luminescence systems with so-called flash kinetics such as acridinium esters or aequorin, where light emission reaches its peak after 1 sec or less, and 90% of the light is emitted within 2 to 3 sec. However, in the case of glow kinetics, the reagent(s) starting the reaction may be added outside the luminometer. The definition of "glow" versus "flash" is somewhat arbitrary. For measurement with a microplate, the light output should not vary by more than about 1% per minute to allow operation without an on-board reagent injector. However, for a manual luminometer, it may be possible to start measurement within 1 to 2 sec after manual reagent addition. Therefore, the requirements regarding constant light output may be relaxed greatly, and a signal decay of 10–20% per minute might not cause a problem.

Injection of liquid volumes, such as 100–300 μl, may take several hundred milliseconds. Note that this is about the time for a flash kinetics signal to reach maximum, which has two implications: (i) the intrinsic rise time of the luminescence flash is slowed down and (ii) until injection is completed, the reagent mixture is in a transient state of rapidly changing concentrations. Good precision can only be obtained when not only the volume, but also the time and pressure profile of injection (velocity over time) are highly reproducible. In many applications this profile must also ensure rapid mixing of reagents. Rotation or shaking devices normally do not allow mixing fast enough for a flash kinetics system, but are useful in maintaining the long-term homogeneity of samples.

The liquid injector may have the additional function to resuspend magnetic or other microparticles, which can be achieved with a "jet"-type injection. In practice, a 1–2% precision of relative light unit (RLU) values can be achieved and maintained, which is generally considered to be sufficient.

Automatic reagent injection may also be used to initiate sequential luminescent reactions, as in dual luciferase assays. Furthermore, injection and measurement of an internal standard following measurement of the analyte can be advantageous because it can make the establishment of a standard curve unnecessary. Such protocols can be used for ATP determination alone or for the simultaneous measurement of ATP, ADP, and AMP when only very low amounts of ATP are present.[2]

Two types of injector design are used, based on either pistons or bellows. Total dead volume, from reagent container to injector nozzle, should be minimized. A useful feature is the possibility of pumping unused reagents back into their original container.

The pressure profile has to be adapted to the actual measuring problem. In the case of sequential reactions such as dual reporter gene assays or internal calibrations, the injection process must avoid squirting droplets to the upper part of the tube as these cannot be reached by subsequent injections. Furthermore, the integrity of cells might be impaired by too high injection speed.

In conclusion, an ideal injector allows setting the volume over a wide range with high precision and has an adjustable pressure profile.

Special Considerations for Microplate Luminometers. For fast flash kinetics, the reagent triggering the light emission has to be injected into the sample while it is in the measurement position. For opaque microplates, which are used in most luminometers, both the nozzle of the reagent injector and the light detector have to be positioned above the microplate, therefore competing for space. An effective and elegant solution is the use of an optical fiber transporting the light from the microplate to the photomultipliers, while an optically transparent reagent duct practically replaces one of the fibers in the bundle.[3]

Commercial microplate luminometers are equipped with up to four reagent injectors. Two is the number used mostly, with the one injecting into the measuring position the most critical and important one.

[2] F. Berthold, S. Kolehmainen, and V. Tarkkanen, U.S. Patent 4,390,274 (1983).
[3] F. Berthold and W. Lohr, FRG Patent 41 23 818 (1994).

Units and Performance

Units

Raw data from luminometers are generally presented as relative light units. RLU is not a scientifically defined unit, but only a relative measure for the light output from the sample. A priori one cannot expect identical RLU readings from different luminometers when measuring the same sample. Each luminometer has to be calibrated individually using known analyte concentrations or special calibrators (see section on calibration and quality control). The situation differs somewhat between photon counting and current measuring systems.

In a current measuring device using a photomultiplier, the output signal not only depends on light intensity, but also on parameters such as applied high voltage, amplifier gain, and the digitizing factor converting current (or voltage) into RLUs.

Let us now look at a photon counter. As explained in the section on detectors, practically each photoelectron produces a logic pulse (except very few that do not reach the first dynode or do not produce secondary electrons). The number of photoelectrons released is equal to the number of photons hitting the photocathode multiplied by the quantum efficiency. Therefore, the counting rate registered by a photon counter, expressed as counts per second (cps), is a meaningful unit. It is generally used by the manufacturers of photomultiplier tubes to specify their product for their intended use as photon counters.

The situation becomes somewhat blurred because most luminometers convert the count rate into RLUs internally, using different conversion factors that are frequently not known or accessible to the user.

Some instruments allow the user to enter so-called RLU factors on top of the internally set conversion factors. These can be set so that different instruments of the same type show identical RLU readings for samples with the same photon emission, compensating for inevitable variations in quantum efficiency between individual photomultiplier tubes.

We think that an option to present results as counts per second, in addition to RLUs, is desirable. Basic performance criteria of photon counters, such as dead time and background that determine dynamic range, are reported as counts per second.

Furthermore, it is recommended to always report measuring results normalized to time as RLU per second. In the case of glow kinetics producing a practically constant signal during measurement, results reported as RLU per second are preferable over RLU because the results are independent of measuring time.

Dynamic Range and Sensitivity

Both dynamic range and sensitivity are important specifications for a luminescence system and will be discussed along a calibration curve for ATP measurements (Fig. 3). This example was chosen because this chemistry exhibits both a high dynamic range and a low reagent background.

Three regions can be distinguished clearly. The horizontal part in region 1 is background, normally the addition of instrument and reagent background. It should be noted that the term "relative light unit" might be a misleading unit for instrument background because this is (normally) due to noise electrons emitted from a photocathode, without any contribution from photons. Region 2 is the linear range that is best suited for quantitative work. Region 3 is characterized by increasing saturation effects of the photon counter, leading to a flat part of the curve without any further increase of output vs concentration. With very high light levels the output will decrease eventually.

In the case of a photon counter, the saturation effect is due to its dead time, determining the maximum intensity that can be registered (I_{max}).

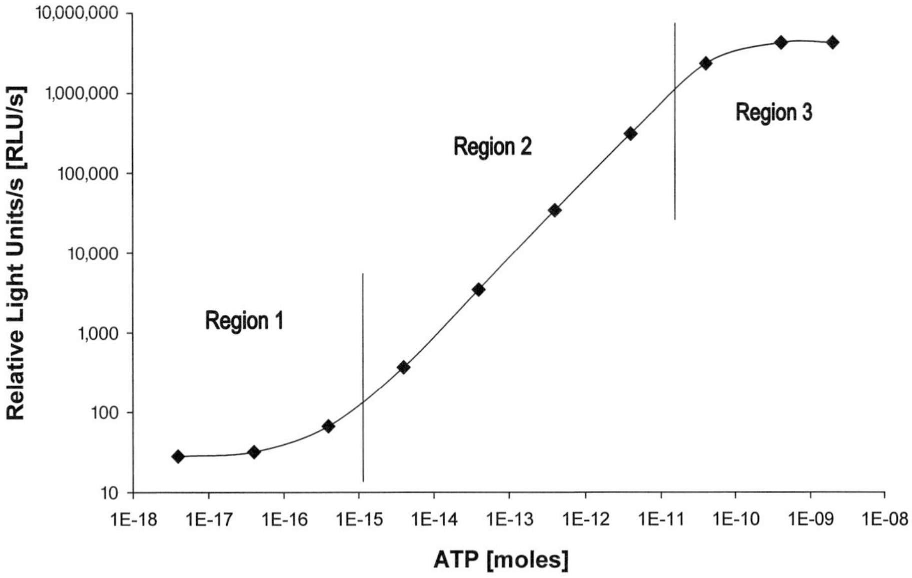

FIG. 3. ATP standard calibration curve measured on a EG&G Berthold (Bad Wildbad, FRG) Lumat LB 9507. ATP was diluted in distilled water to given quantities. Measurement was performed using the ATP detection kit from Analytical Luminescence Laboratory (Ann Arbor, MI) according to manufacturer's instructions. Different regions indicated are described in the text.

Over a limited range, a dead time correction is possible using, in good approximation, the following formula:

$$I_{true} = \frac{I_m}{1 - \dfrac{I_m}{I_{max}}}$$

where I_{true} is the corrected true intensity, I_m is the measured intensity, and I_{max} is the maximum intensity that can be detected. It may be useful to remember that if I_m is 10% of I_{max}, the result is already 10% too low due to counting losses. This formula is a good approximation up to $I_m/I_{max} = 0.5$; above this value errors become too great to be corrected properly.

For photomultipliers in the current measuring mode, the linear range is extended by typically two orders of magnitude before saturation occurs. A good way to increase the linear range at the high end is to use an optical filter, say OD 2 absorbing 99% of the light.

Calibration and Quality Control

The standard calibration procedure is to establish a calibration curve as shown in Fig. 3. We will now discuss additional methods for calibration and quality control and their usefulness.

Absolute Calibration. This means calibrating a luminometer in terms of the absolute number of photons emitted from a sample (see by O'Kane and Lee, this volume). Ideally, the calibrator source should have the same optical spectrum as the samples containing unknown analyte quantities and should also have the same sample format. Such calibrations are necessary in basic luminescence research, but would rarely be required by those using luminescence as a tool, as in biochemistry and molecular biology.

System Performance Check. This test allows checking not only of the function of the light detector, but also of the entire system, including reagent quality and injector function. It is performed by measuring the signal obtained for a standardized amount of a suitable luminophor. The result is then compared to a reference value and deviations are indicated. Such performance tests are particularly useful in clinical applications to indicate system failure, thereby avoiding false results for patient samples.

Light Calibrators. Their purpose is to calibrate or to just check the function of the light detector in a luminometer. These light sources are normally not themselves calibrated in terms of the number of photons emitted, but must have excellent long-term stability. Radioactive light sources using scintillation as the light generation process are very stable, and any decrease due to radioactive decay is exactly calculable. A common approach is the combination of tritium or another low energy β emitter, with a suitable scintillator.[4,5] Because luminescence is a random emission of single photons, scintillation test sources should be sufficiently quenched to exhibit single-photon characteristics.[6]

The use of radioactivity is considered undesirable, as one of the driving forces of luminescence technology is to avoid the use of radioactivity. An interesting approach is the use of naturally occurring ^{238}U, minimizing regulatory concerns.[7] Autoradioluminescence in luminescent solid uranyl salts is generated from the α particles emitted.

Light-emitting diodes (LEDs) are convenient light sources and can be stabilized to obtain low drift. Green LEDs emit around 560 nm, matching the spectrum of firefly bioluminescence quite well. Some luminometers are equipped with built-in LEDs. For instruments using DC measurement, it is advantageous to use such light sources for automatic gain control. Good photon counters, however, are so stable that recalibrations are normally not required, and they are more stable than most nonisotopic light calibrators. Nevertheless, light calibration can be very useful in indicating any major problem in the luminometer before actual samples are measured.

Internal Calibration. The purpose is to correct inhibition of the light-generating chemistry (chemical quench) or absorption of light within the sample (optical quench). It is applied to glow-type reagent systems. The procedure is to add a calibrated quantity of analyte to the sample, comparing the signal before and after calibrator addition, and correcting the result using straightforward calculations. Internal calibration may be performed with manual calibrator addition, but automatic injection is more convenient.[2] The ratio of calibrator to sample volume must be sufficiently small so that the conditions to be tested are not influenced by the calibrator addition. This might necessitate sample rotation or shaking to affect mixing.

Sensitivity

Determining sensitivity as the minimum detectable number of photons would be an interesting specification. However, this is only possible when the luminometer is first calibrated in terms of photons emitted from the sample. As discussed in the preceding section, this is somewhat complicated

[4] T. ET. Oikari, I. A. Hemmil, and E. J. Soini, *in* "Analytical Applications of Bioluminescence and Chemiluminescence" (L. J. Kricka, P. E. Stanley, G. H. G. Thorpe, and T. P. Whitehead, eds.), p. 475. Academic Press, New York, 1984.

[5] D. J. O'Kane and J. Lee, *Photochem. Photobiol.* **52,** 723 (1990).

[6] E. Schram, H. Van Esbroeck, and H. Roosens, *in* "Proc. 1978 Int. Symp. on Anal. Appl. of Biolumin. Chemilumin" (E. Schram and P. E. Stanley, eds.). State Printing and Publishing, Westake Village, CA, 1979.

[7] D. J. O'Kane and D. F. Smith *in* "Bioluminescence and Chemiluminescence: Current Status" (P. E. Stanley and L. Kricka, eds.), p. 38. Wiley, Chicester, 1993.

to perform and may also be of limited interest to the user employing luminescence as a tool wishing to know the minimum measurable quantity of a particular analyte.

Most data sheets for commercial instrumentation therefore state sensitivity in terms of the minimum amount of analyte detectable, such as ATP or firefly luciferase. However, it must be understood that this number is not a characteristic of the luminometer alone, but also of reagent quality. As shown in the section on microplate luminometers, one can get almost one order of magnitude difference in signal levels between black and white microplates for the same luminometer and analyte concentration. An exact definition of sensitivity is therefore only possible for one particular system configuration, comprising luminescence reagent, sample cuvette or microplate, and sample volume. Furthermore, it would be useful to adhere to certain definitions and procedures when defining sensitivity. The work of Jago *et al.*[8] determining the sensitivity of 10 different luminometers is still exemplary because of its methodological approach.

We recommend the following simple method for determining the sensitivity of a luminometer for a specific reagent system.

i. Take a statistically significant number ($\geq$10) of background measurements (with all reagents, but zero analyte concentration).
ii. Measure a known amount of analyte in the midrange, e.g., 100 to 1000 times above background.
iii. Calculate sensitivity using the formula

$$S = (A * 3\sigma_{BG})/(S_A - BG)$$

where S is the sensitivity (same unit as A), A is the amount of analyte used for calibration (in mol, g, etc.), S_A is the result for measurement of A (in counts, cps, relative light units, or any other unit), BG is the mean value for background measurement (same unit as S_A), and σ_{BG} is the standard deviation of background measurement.

Example: In Fig. 3, the amount of 3.94×10^{-13} mol ATP (A) results in 34,140 RLU/sec (S_A). The mean value with a standard deviation of background measurement is 25 ± 4 RLU/sec (BG and σ_{BG}). Therefore, a sensitivity of the ATP detection kit used, measured on the indicated instrument, can be calculated to be 1.39×10^{-16}.

[8] P. H. Jago, W. J. Simpson, S. P. Denyer, A. W. Evans, M. W. Griffiths, J. R. M. Hammond, T. P. Ingram, R. F. Lacey, N. W. Macey, B. J. McCarthy, T. T. Salusbury, P. S. Senior, S. Sidorowicz, R. Smither, G. Stanfield, and P. E. Stanley, *J. Biolumin. Chemilumin.* **3,** 131 (1989).

Practical Hints

Because luminometers are about the most sensitive instruments to measure light, they should be placed where they are not exposed to very bright light, particularly sunlight.

Many reagent tubes exhibit strong phosphorescence after being exposed to light so it is good practice to always store tubes in the dark (in a drawer, not on a table).

Most instruments have no cooling for their detector (and normally do not need it). However, it should be remembered that background increases considerably with temperature; an increase of a factor of two for a temperature increase of 7° might be typical.

Factors Determining the Choice of Instrumentation

In almost all analytical fields of chemistry, biochemistry and medicine, bioluminescence and chemiluminescence techniques are used for the measurement of analytes in the attomole and femtomole range. Luminescent reactions are also widely used in food technology (for review, see Stanley *et al.*[9] and Navas and Jimenez[10]). The exploding market is reflected in the increasing number of luminescence-based reagents and the variety of luminometers currently available (for a review, see Stanley[11]).

The application determines the choice of a suitable luminometer because special features, e.g., injectors or temperature control, are required for some applications. All assays can be performed either in a microplate or in a tube luminometer. Additionally, for spatial resolution, imaging systems are used (see later).

Luminophors and Light-Generating Systems

A variety of bioluminescent and chemiluminescent luminophors or light-generating systems are described. Table I shows some common luminophors and their characteristics. The given values summarize characteristics under particular *in vitro* conditions. Kits and reagents available on the market often show different behavior of the same luminophor.

In principle, light emission kinetics are distinguished functionally into two groups: the flash type, when light is emitted within a few seconds, and

[9] P. E. Stanley, R. Smither, and W. J. Simpson (eds.), *in* "A Practical Guide to Industrial Uses of ATP-Luminescence in Rapid Microbiology." Cara Technology Limited, Lingfield, UK, 1997.

[10] M. J. Navas and A. M. Jimenez, *Food Chem.* **55,** 7 (1996).

[11] P. E. Stanley, *J. Biolumin. Chemilumin.* **12,** 61 (1997).

TABLE I
CHARACTERISTICS OF COMMON LIGHT-GENERATING SYSTEMS

Light-generating system	λ_{max}(nm)[a]	Emission kinetics
Acridiniume esters (e.g., Lucigenin)	530	Flash
Dioxetane derivatives	450–477	Glow
Aequorin	469	Flash
Luciferin/luciferases (Vibrio)	490	Flash
Luciferin/luciferase (firefly),	562 (615)	Flash
e.g., addition of coenzyme A		Glow
Luminol/isoluminol,	425	Flash
e.g., with enhancers		Glow

[a] Values are dependent on reaction conditions.

the glow type, when light emission is extended for a longer time period, e.g., up to several hours. The time scale of this classification is arbitrary. It should be used to distinguish between reactions where, on the one hand, the manual pipetting of reagents is possible (glow) and, on the other hand, when automatic injection to allow immediate measurement of the samples is essential (flash). Even very similar proteins, e.g., luciferases of the glowworm *Lampyris noctiluca* and the firefly *Photinus pyralis,* can show different emission kinetics.[12] The firefly emits light within seconds (flash type) whereas light emission of the glowworm is much more continuous (glow type).

It is important to note that for some applications, the luminophor itself, e.g., luminol, is used as the label, whereas in other applications the label is an enzyme catalyzing a luminescent reaction. The quantum efficiency reflects the number of photons produced per reacting molecule of luminophor. When an enzyme such as alkaline phosphatase is used as a label, one should consider that light output can be high despite a low quantum efficiency, as it can be overcompensated by a high number of substrate molecules turned over per molecule enzyme.

As shown in Table I, the label affects not only the emission kinetics, but also the spectrum of light emission. The wavelength of emitted light depends on the conditions used. The light output of firefly luciferase reaches a maximum at 562 nm. In the presence of certain divalent metal ions, such as Cd^{2+} or Zn^{2+}, or at low pH, light output is shifted into the red region of the spectrum (λ_{max} 610–615 nm). This affects light measurement because the quantum efficiency of photomultipliers depends on the wavelength.

[12] G. B. Sala-Newby, C. M. Thomson, and A. K. Campbell, *Biochem. J.* **313,** 761 (1996).

Some kits available on the market include enhancers. These compounds also often lead to a shift in wavelength.

Applications and Demands for Instrumentation

This section discusses common bioluminescent and chemiluminescent applications in the context of the required luminometer features. Depending on the other reagents, the same luminophor may show flash or glow characteristics. The handling of glow-type luminescence not requiring an automatic injection of reagents is more convenient, but luminol (without enhancer)-, acridinium ester-, and aequorin-based assays with flash kinetics maintain their ground because of certain advantages.

Reporter Gene Assays

In most of the reporter gene assays currently on the market (for review, see Bronstein *et al.*[13]), light emission is quite stable (glow type), which allows, in general, measurement without using injectors. Common reporter genes are luciferases, β-galactosidase, β-glucuronidase, secreted alkaline phosphatase, and the photoprotein aequorin. For the use of aequorin, injectors are required due to the flash kinetics of light emission. In general, temperature control is not necessary for reporter gene assays.

For the use of dual-label assays,[14,15] injectors might be recommended. Some of the available luminometers even provide software that allows the automatic injection of reagents, measurement, and evaluation (ratio calculation) of measured data.

For spatial resolution of gene expression, imaging systems can be used (see section on imaging).

ATP Measurements

In principle, the firefly luciferase leads to a flash-type luminescence, requiring reagent injectors, but a glow-type light emission can be obtained, e.g., when coenzyme A is present.[16] In this case, injectors are not needed. However, it should be mentioned that injectors can also be used for the addition of different reagents, e.g., for lysis buffer and luciferin/luciferase, leading to high sample throughput.

[13] I. Bronstein, J. Fortin, P. E. Stanley, G. S. A. B. Stewart, and L. J. Kricka, Anal. *Biochem.* **219,** 169 (1994).

[14] B. A. Sherf, S. L. Navarro, R. R. Hannah, and K. V. Wood, *J. NIH Res.* **9,** 56 (1997).

[15] C. S. Martin, P. A. Wight, A. Dobretsova, and I. Bronstein, *Biotechniques* **21,** 520 (1996).

[16] K. V. Wood, *in* "Bioluminescence and Chemiluminescence: Current Status" (P. E. Stanley and L. J. Kricka, eds.), p. 543. Wiley, New York, 1991.

Glow-type luminescence is applied in industrial hygiene applications, where small portable luminometers without injectors are used.

Measurement of Enzymes and Metabolites

Many enzymes and metabolites can be coupled to the production or utilization of ATP, NAD(P)H, or H_2O_2, and luminescent detection of these compounds is possible. Luminescent detection methods have been described for more than 100 enzymes and metabolites (for review, see Campbell[17]). General statements on instrumentation requirements are difficult, as, for example, enzymes are temperature dependent and reaction kinetics of light emission depend on the particular assay.

Room temperature to 25° is usual. Because the temperature should be maintained constant, especially for continuous measurements, temperature control might be useful. Luminescent systems for the measurement of ATP, NAD(P)H, or H_2O_2 in principle show flash kinetics, but, as described earlier, glow luminescence can be obtained by the addition of other compounds, such as enhancers.

Cellular Chemiluminescence (Phagocytosis)

Because the determination of cellular chemiluminescence uses living cells, adjustable temperature control (e.g., from 25 to 41°) is necessary. The kinetic of light emission is measured over long time periods of up to several hours. Frequently a group of samples derived from a primary sample must be measured simultaneously. This necessitates parallel measurement, which can be affected in an automatic sample changer (tube or microplate), taking cyclic measurements from a multitude of samples, or a multidetector instrument.

Immunoassays

From a commercial point of view, immunoassays are certainly the most important application of bioluminescence and chemiluminescence. In clinical diagnostics, fully automated systems, which include pipetting, incubation, separation, and measurement, are available, but on a smaller scale, tube and microplate luminometers are widely used.

In clinical diagnostics, acridinium esters and dioxetane derivatives are the most common luminophors; however, other sensitive luminescent systems, based for example on acetate kinase or aequorin, can be attractive.[18]

[17] A. K. Campbell, "Chemiluminescence: Principles and Applications in Biology and Medicine," Chapter 5. Ellis Horwood, Chichester, UK, 1988.

[18] L. J. Kricka, *Pure Appl. Chem.* **68,** 1825 (1996).

Temperature control is not necessary for this application. Referring to Table I, the number of required injectors depends on the light emission kinetics of the luminescence system and also on the degree of automatic sample preparation.

DNA Probe Techniques

Nonisotopic DNA probe techniques are becoming more and more important in the field of nucleic acids research (for review, see Kricka[19]). For many luminescent applications, the use of photographic films is still sufficient for light detection. If quantification of light emission is required, luminometers or imaging systems are recommended. Using filter membrane adapters, dot blots in a 96-well format can be measured in microplate luminometers.[20] For a detailed analysis of gels and blots of different formats, low light imaging systems are best suited.

From Single Sample to Fully Automated Luminometers

Single Sample Tube Luminometers

We will start with the simplest (and least expensive) type of luminometer, which is best suited for low or medium throughput. Each sample has to be loaded into the instrument manually, is usually measured promptly, then unloaded, and the sequence can be repeated.

Single sample instruments can be equipped with automatic or manual reagent injectors so that every assay using flash or glow kinetics can be performed. Their only limitation is lack of automatic sample changing.

In certain respects, single sample luminometers can even be superior to automatic instruments. Some models can accommodate a wide range of sample formats, including different tubes with diameters of 8–16 mm and a height of 40–100 mm, a design with cap, liquid scintillation vials, and also flat sample formats such as culture tubes and filters with up to a 35 mm diameter.

Because a single sample instrument may have less space constraint for the design of an optimized measuring chamber (see measuring chamber section), one finds instruments with the highest sensitivity in this category.

Portable Luminometers. Portable, battery-driven luminometers have been developed for industrial microbiology applications.[9] A typical protocol for rapid hygiene monitoring may be the following: wettened swab samples

[19] L. J. Kricka (ed.), *in* "Nonisotopic DNA Probe Techniques." Academic Press, San Diego, 1992.

[20] N. Walter and C. Steiner, *Biotechniques* **15,** 926 (1993).

are taken from the surface to be tested. The ATP content is then determined using the firefly reaction. Some systems use extraction reagents, extracting ATP from somatic cells or bacteria, whereas others measure free ATP only. Light-generating reagents (luciferin/luciferase) are frequently either freeze-dried or immobilized for convenience and stability.

Reagents and instrumentation are both designed for error-free operation under adverse conditions and must observe sterility requirements in food-processing plants.

In advanced commercial products, all consumables come in single-dose packages, avoiding any pipetting steps. This type of packaging frequently results in special sample formats so that most portable luminometers are part of a proprietary reagent/instrument system. For instance, it may be necessary to measure a swab of 10 cm length directly, which is only possible in specially designed luminometers. Weight is an important consideration; it is typically less than 1500 g.

Portable luminometers may allow storage of the results of 1000 or more samples that have been measured in the field, together with sample identification entered via keyboard or downloaded from a computer. These results can be downloaded to a PC for further analysis.

Portable luminometers for hygiene applications are usually less sensitive than those for scientific use. However, portable instrumentation has become available that is sensitive enough for scientific or industrial use beyond rapid hygiene monitoring. Such applications include the distinction of bacterial from somatic ATP, identification of coliform bacteria by specific enzymatic reactions, resulting in a luminescence signal, and immunoassay and DNA probe assays also using luminescent labels.

Automatic Tube Luminometers

Automatic tube luminometers are available in two configurations, either with a chain or with racks for sample holding and transport. A popular rack-type luminometer can load and measure up to 400 samples in 40 racks with 10 samples each.

Some automatic tube luminometers are equipped with temperature stabilization and other features, making them particularly useful for cellular chemiluminescence studies.

Microplate Luminometer

Because of the popularity of microplates, automatic plate luminometers have found a considerable market. Some not only accommodate 96-well plates, but also plates with 384 wells.

In a microplate luminometer the measuring chamber consists of both the instrument and the microplate. Therefore, microplates have to be chosen carefully in terms of detection performance parameters such as cross-talk (see later).

Measuring luminescence from microplates presents some unique design challenges, which are reflected in considerable performance differences among such instruments, particularly cross-talk and reagent injection.

Cross-talk. Because of the close proximity of neighboring sample wells, it is difficult to avoid spillover of light between wells, called cross-talk. A practical method for cross-talk determination is the following: A quasi-constant signal (S_m) is generated in well m, and the apparent signals in neighboring empty wells S_d (diagonal position) and S_a (adjacent position) are measured. For neighboring wells in the diagonal position, the cross-talk is defined as $CT_d = S_d/S_m$; for neighboring wells in adjacent positions, the cross-talk is defined as $CT_a = S_a/S_m$ (Fig 4). Ideally, cross-talk would be zero. In reality, values of 10^{-4} are normally acceptable. Because adjacent wells are closer than diagonal ones, CT_a is higher. Of course, cross-talk is only an issue for glow luminescence because light emission for flash-type luminescence will occur only from the well in the measurement position. Cross-talk and sensitivity are characteristics of the combined system (luminometer and microplate), a fact that is frequently not reflected in the manufacturers' documentation.

Table II shows diagonal and adjacent cross-talk values for different commercially available 96-well microplates, i.e., white, black, clear, clear bottom, and strips. In addition, the relative efficiency of S_m is given, which is related to a white plate with maximum reflection.

Clear microplates are used if, in addition to luminescence, absorbance is also of interest, e.g., in cell growth determination. Using clear plates, or

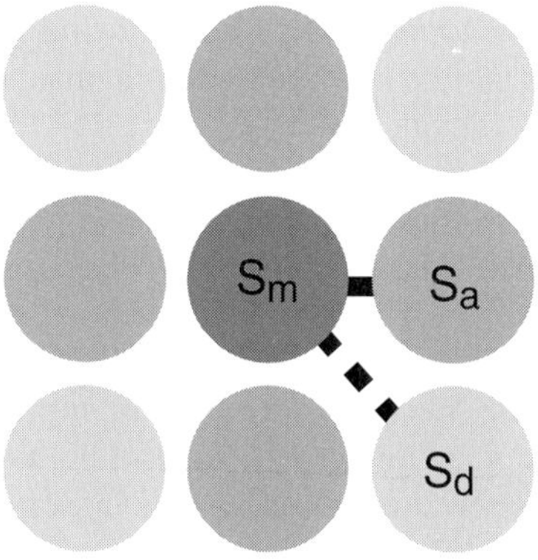

FIG. 4. Determining cross-talk.

TABLE II
CROSS-TALK VALUES AND RELATIVE SIGNAL EFFICIENCIES FOR DIFFERENT 96-WELL MICROPLATES

Type	Relative signal efficiency (%)	Cross-talk CT_d	CT_a
White, opaque	81–100	$3.55–5.94 \times 10^{-5}$	$1.55 \times 10^{-5}–1.15 \times 10^{-4}$
White strips	50–80	$1.70 \times 10^{-6}–3.37 \times 10^{-5}$	$4.56 \times 10^{-6}–6.26 \times 10^{-5}$
White, clear bottom	34–37	$1.83–3.37 \times 10^{-5}$	$3.02–6.27 \times 10^{-5}$
Clear	12–14	$3.45–3.73 \times 10^{-2}$	$8.43–8.93 \times 10^{-2}$
Clear strips	12–15	$2.01–3.37 \times 10^{-2}$	$6.30–7.08 \times 10^{-2}$
Black	6–7	$<10^{-6}–1.09 \times 10^{-5}$	$6.31 \times 10^{-6}–3.42 \times 10^{-5}$

[a] Diagonal and adjacent cross-talk values (CT_d and CT_a, respectively) were determined using 18 different commercially available 96-well microplates, the ATP bioluminescence assay kit CLS II from Boehringer Mannheim (FRG) according to the instructions, and the luminometer MicroLumat-Plus LB 96V from EG&G Berthold (FRG). The same quantity of ATP was pipetted into each plate; the relative signal efficiency of 100% corresponds to 7×10^6 RLUs per 10 sec. The cross-talk, e.g., from well B2 into A1, A3, C1, and C3, was averaged as CT_d and from well B2 (S_m) to B1, B3, A2, and C2 as CT_a ($n = 3$). The averaged background of wells containing buffer was subtracted before calculation. Standard deviations were in a range of 5–30%.

strips inserted into plate holders, the highest cross-talk values were obtained in a range of 2 to 9% of S_m. Adjacent cross-talk could be measured even in wells two (CT_{a2}) and three positions (CT_{a3}) away from the well generating the signal S_m: $CT_{a2} = 0.63–2.33 \times 10^{-2}$ and $CT_{a3} = 1.05–4.02 \times 10^{-3}$. The relative efficiency of S_m shows only small values of about 12 to 15% because the isotropic light is not reflected inside the well. Because light shines through the clear plate material, resulting in enormous cross-talk, clear microplates are not suitable for measuring luminescence.

To reduce this high cross-talk, white microplates with clear bottoms, which are optically not connected to neighboring wells, are used. These plates show CT values up to three orders of magnitude less than completely clear plates. Also, the efficiency is increased to 34–37%, due to more reflection from the white walls of the wells.

Maximum signal efficiency was obtained using white opaque microplates as well as white opaque strips inserted into holders, as here the optical reflection reaches the highest values. Furthermore, these plates show very small cross-talk values that are in the same range as clear bottom plates. White microplates have become standard in luminescent measurement because they combine very low cross-talk and high signal efficiency.

Black microplates show the best cross-talk values—about 10 times lower than for white ones—but because of their low efficiency, they are recommended only when enough light is available.

Furthermore, there are differences in cross-talk and efficiency between plates of the same type from different manufacturers; e.g., S_m of white strip plates differs from 50 to 80%.

Reagent Injection. The limited space for both detector and injector(s) above a well is an engineering challenge that can be solved quite elegantly by using light conductors to the detector, which are partly substituted by optically transparent injection nozzles.

Additional design criteria for plate luminometers are their interfaces to robot and stacker systems. For high throughput screening, either the integration into a fully automated robot system or the use of a stacker system, where a series of microplates is measured automatically, is a prerequisite.

The trend to higher sample numbers and smaller volumes has led to the design of microplates with 384 wells and beyond. Correspondingly, several luminometers accommodating 384-well plates have been introduced. The measuring time for a plate with 384 wells is roughly four times that for 96 wells and can therefore reach 10 min or more.

Throughput can be increased with multidetector systems. A system with eight detectors operated in parallel reduces the measuring time by a factor of 8, compared to only one detector.

The problem of cross-talk is even more critical for 384-well plates than for 96-well plates because of the greater proximity of neighboring wells. Cross-talk values between 10^{-3} and 10^{-4} may be considered acceptable. The authors are not aware of a 384-well instrument allowing reagent injection into the measuring position, which would be difficult but not impossible to realize. Consequently, 384-well luminometers are presently used for glow kinetics.

For even higher sample density, the conventional luminometer design with the sequential or parallel measurement of single wells may no longer be suitable; the use of imaging systems might be the better choice, albeit associated with a different set of limitations.

Multi-functional Instruments

Some instruments have been designed to measure not only luminescence, but also fluorescence, light absorption, or radioactivity. A combination of different instruments in one has the advantage of lower costs and less use of space. However, there are also some disadvantages. Just one type of measurement at one time can be done. Therefore, the use of the instrument by many people is difficult. The major problem is that a combined instrument cannot be optimized for the measurement of each label. For example, luminescent and fluorescent labels differ in emission spectrum.

Whereas luminescent labels emit their light mainly in the range of 400 to 600 nm (Table I), the emission of light of some common fluorescent labels is at a much higher wavelength, e.g., Cy5.5™ 694 nm, Cy7™ 767 nm. Because detectors with sensitivity in the red area show a higher background noise, these detectors are less suitable for low light luminescence applications.

In a microplate scintillation counter, cross-talk values for luminescence measurements are higher than for scintillation applications. This is due to the fact that coincidence and other discrimination methods used in scintillation counting not only suppress noise, but cross-talk as well, and corresponding methods are not available in luminescence measurement.

Luminescence Imaging

For particular applications, besides the amount of luminescence, information about spatial resolution is required. Imaging is the point of interest in this recently established field of luminescence measurements. The sample size ranges from a single cell, using microscopes, to complete animals, such as a mouse, using light-tight housings.[21] Examples involving spatial resolution include luminescent blot detection and analysis of reporter gene expression.[22,23]

The spatial measurement of bio- and chemiluminescent signals demands a high-density array of highly sensitive detectors. The photon multiplier technique as described earlier is not suitable for this purpose, So imaging detectors are used. The most important detector technologies for luminescence imaging are (i) slow-scan cooled CCD cameras, (ii) intensified CCD cameras, and (iii) video cameras with an intensifier.

Detector Techniques

The CCD is a semiconductor detector containing an array of picture elements called pixels. These arrays show various formats and consist of up to about 9×10^6 pixels. Because the pixel size is in the range of a few micrometers, i.e., less than 10 μm to about 25 μm, a high-density array performance leads to high optical resolutions. Inside the pixels, electron-hole pairs are generated by the impinging photons, and the electrons are accumulated. The number of generated electrons is proportional to the number of incoming photons. At the end of the exposure, the charges are

[21] G. Taubes, *Science* **276,** 1993 (1997).

[22] J. C. Nicolas, *J. Biolumin. Chemilumin.* **9,** 139 (1994).

[23] B. Kost, M. Schnorf, I. Potrykus, and G. Neuhaus, *Plant J.* **8,** 155 (1995).

read out pixel by pixel and are stored. The resulting pattern represents the image.

In principle, CCDs are divided into standard (front)- and back-illuminated chips according to their installed orientation. The spectral sensitivity of a front-illuminated CCD differs from that of a bialkali photocathode and ranges from 400 to 1100 nm; it can be extended down to 200 nm by a UV-to-VIS converter. At 650 nm a maximum quantum efficiency of about 40% is reached. Spectral sensitivity and quantum efficiency are limited by absorption of the incoming light through the silicon dioxide layer before generating charges inside the pixel. By exposing the chip from the other side (back-illuminated CCD), quantum efficiency can be increased to a maximum of about 85% at 650 nm, as there is no light absorption by the silicon dioxide layer.

Low light imaging is restricted by two sources of background noise: thermal noise (dark current) and readout noise. Thermal noise is generated due to statistical fluctuations of thermal electrons within the pixel and can be reduced by cooling either cryogenically or by Peltier elements. Because the use of liquid nitrogen or water cooling may not be convenient in many laboratories, only multiple-stage Peltier cooling will be considered here. At 200 K (four-stage Peltier/air cooling of the CCD chip) the thermal noise of a state-of-the-art detector is less than four electrons/pixel/hour. This means the CCD chip is almost free of thermal noise even for exposure times up to hours.

The readout noise is generated during the readout process. It increases with scan speed. Because the readout noise becomes significant at low signal levels, such as luminescence imaging, CCD detectors usually operate in a slow-scan mode of less than 100 kHz, e.g., a chip of 512 × 512 pixels can be read out at a rate of 50 kHz within about 5 sec. The readout noise, given in statistical terms, is called root mean square (RMS) and can be reduced to less than six electrons RMS.[24]

In the case of intensified CCD systems, where an image intensifier is placed in front of a CCD, the intensified signal-to-background noise ratio is influenced only marginally by the noise level. Therefore, intensified CCD's usually operate at higher readout rates of up to 5 MHz. Further cooling is not strictly required.[25,26] Yet another approach is the combination of a very

[24] C. D. Mackay, *Photonics Spectra* **2,** 113 (1992).

[25] P. L. Jones, I. M. Greenfield, C. E. Hooper, R. Ansorge, and M. A. Stanley, *in* "Bioluminescence and Chemiluminescence: Current Status" (P. E. Stanley and L. J. Kricka, eds.), p. 175. Wiley, New York, 1991.

[26] C. E. Hooper and R. E. Ansorge, *in* "Bioluminescence and Chemiluminescence: Current Status" (P. E. Stanley and L. J. Kricka, eds.), p. 337. Wiley, New York, 1991.

low noise saticon tube, which is a special type of high dynamic, low noise vidicon tube, with an age intensifier.[27]

The relative advantages of different detector types have been discussed repeatedly.[28,29] In our view, it has not been proved that any particular type of detector is superior to others, but there are considerable performance differences within each class of detectors. The benefits of cooled slow-scan CCDs are image quality, sensitivity, and a broad dynamic range (1>10,000), whereas the advantage of an intensified detector is fast image acquisition, e.g., in the case of calcium ion imaging. All detector principles are used in high-quality commercial instruments.

Technical Construction

Sample formats vary from microscopic, such as single cells or tissue sections, to macroscopic, such as culture dishes, blots, thin-layer chromatograms, or gels. Therefore, the detector can be adapted to a microscope for microscopic imaging or, for imaging of a macroscopic-sized sample, the detector can be mounted inside a light-tight dark housing containing a suitable sample support.

Figure 5 presents a schematic of a macroscopic imaging system. The detector is mounted mechanically onto a vertical drive inside the dark cabinet to allow adjustment of the magnification. The lens is mounted independently on a second vertical precision drive for focus adjustment. By combining both positioning capabilities under computer control, it is possible to bring images from about a 3- to 30-cm edge length into full view of the detector, with the system autofocused continually. The sensitivity of microscopic imaging systems is mainly influenced by the quality of the microscope.

Imaging Software

The software of an imaging system controls three main functions: (i) image acquisition, (ii) image correction, and (iii) image processing.

[27] R. Bräuer, B. Lübbe, R. Ochs, H. Helma, and J. Hoffmann, *in* "Bioluminescence and Chemiluminescence: Current Status" (P. E. Stanley and L. J. Kricka, eds.), p. 13. Wiley, New York, 1991.

[28] R. E. Ansorge, C. E. Hooper, W. W. Neale, and J. G. Rushbrooke, *in* "Bioluminescence and Chemiluminescence: Current Status" (P. E. Stanley and L. J. Kricka, eds.), p. 349. Wiley, New York, 1991.

[29] R. A. Wick, G. Iglesias, and M. Oshiro, *in* "Bioluminescence and Chemiluminescence: Current Status" (P. E. Stanley and L. J. Kricka, eds.), p. 47. Wiley, New York, 1991.

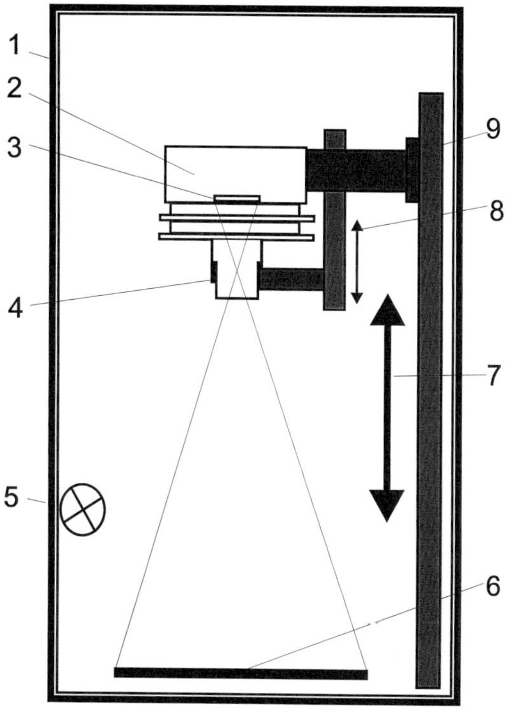

FIG. 5. Schematics of a low level imager. The light-tight housing (1) of a macroscopic imaging system is presented containing the detector (2) with a motor-driven vertical adjustment of magnification (7, 9), the sensor element (3), the lens (4) with a second vertical precision drive for focus adjustment (8), a weak and fluorescence light source with interchangeable filters (5), and the sample table (6).

i. The acquisition software has to control camera parameters such as the exposure time, optical parameters such as the focus adjustment/magnification, and the display/storage of raw data.
ii. The image correction software includes functions to control background subtraction, sensitivity correction to compensate for variations in pixel sensitivity, and geometrical correction, which is necessary because of "cushion" or "fish-eye" distortion associated with large solid angle imaging. Additionally, the image brightness and contrast have to be controlled because the 14- to 16-bit gray levels of the CCD camera have to be converted into 8-bit gray levels that can be recognized by the human eye.
iii. Finally, the image processing software performs the addition and subtraction of images and the conversion of black and white images

of the CCD into pseudocolor images. Furthermore, this software controls pattern recognition to allow counting and classifying objects such as cells or colonies, as well as quantitative reading of the microplate. Blot and gel documentation function is also included in imaging processing software. For these purposes, many analysis programs are available from other types of imaging, including fluorescence or radioactivity methods.

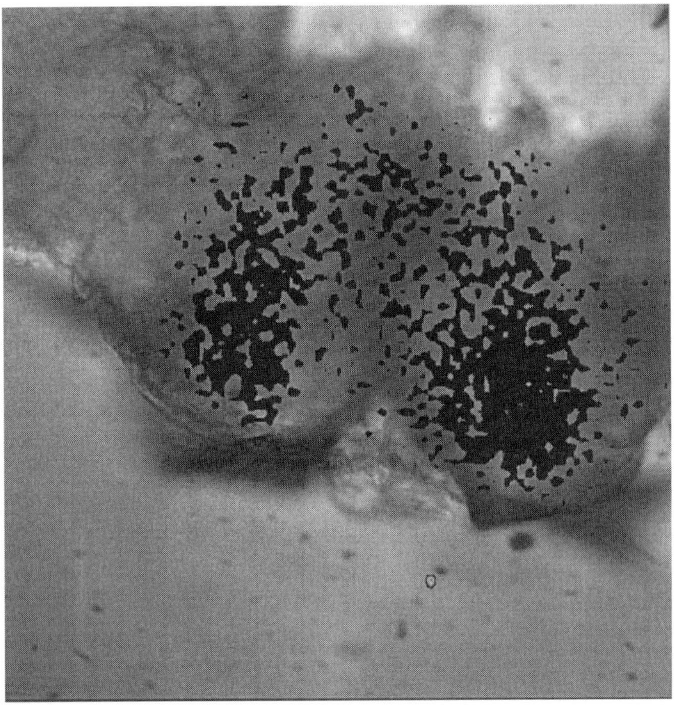

FIG. 6. Detection of luminescence emitted from the jelly fish photoprotein aequorin expressed in a brain section of the fruit fly *Drosophila melanogaster* (line 103Y) using a back-illuminated slow-scan CCD camera coupled to an inverse microscope (courtesy of D. J. Amstrong, Glasgow). Luminescence, shown as rough black areas and spots, appeared from the Calyx region (approximately 30–40 μm across) of the mushroom bodies of the brain of an adult fruit fly. The imaging system NightOWL from EG&G Berthold (Bad Wildbad, FRG) was connected to the inverse microscope Axiovert TV 100 from Zeiss (Jena, FRG) equipped with a Neofluar 20× lens (0.50 NA). A 20-fold magnification was used. An overlay of a transilluminated and a luminescent image is presented. Two copies of aequorin (cDNA clone pAQ2) have been targeted into the mushroom bodies using the P[GAL4] expression system as described in Rosay *et al.*,[31] Tettamanti *et al.*,[32] and Yang *et al.*[33]; coelenterazine was purchased from Molecular Probes (Eugene, OR).

Example of Application

Luminescence imaging is only possible for glow-type luminescence because it would be almost impossible to dispense reagents into (intact) samples and take an image at the same time. Even so, the weak and chemical flash-type luminescence of the jellyfish photoprotein aequorin can be detected. Figure 6 shows aequorin expression in the brain of a transgenic fruit fly *Drosophila melanogaster* as an actual example for reporter gene analysis.

The macroscopic and microscopic imaging system NightOWL from EG&G Berthold[30] was used for this spatial detection of the luminescent label aequorin. Figure 6 presents an overlay of two images. First, a transilluminated image was acquired using a 20-msec exposure time. For this image, less sensitivity was required due to the illumination, but high resolution was required for maintaining good spatial resolution. Therefore, each pixel was read out separately. Second, a luminescent image was acquired. Here, camera parameters were set mainly for high sensitivity but less for spatial resolution: An exposure time of 3 min was chosen and groups of 5 × 5 pixels were connected together (pixel binning) to increase sensitivity. For comparison, the emitted aequorin luminescence was quantified to about 200 to 500 RLU per 15 sec using a LB 9507 luminometer from EG&G Berthold (Bad Wildbad, FRG) (J. D. Amstrong, personal communication).

[30] B. Möckel, J. Grand, and R. Ochs, *in* "Bioluminescence and Chemiluminescence" (J. W. Hastings, L. J. Kricka, and P. E. Stanley, eds.), p. 539. Wiley, Chicester, 1996.

[31] P. Rosay, S. A. Davies, Y. Yu, M. A. Sozen, K. Kaiser, and J. A. T. Dow, *J. Cell Sci.* **110,** 1683 (1997).

[32] M. Tettamanti, J. D. Amstrong, K. Endo, M. Y. Yang, K. Furukubo Tokunaga, K. Kaiser, and H. Reichert, *Dev. Genes Evol.* **207,** 242 (1997).

[33] M. Y. Yang, J. D. Amstrong, I. Vilinsky, N. J. Strausfeld, and K. Kaiser, *Neuron* **15,** 45 (1995).

[5] Absolute Calibration of Luminometers with Low-Level Light Standards

By DENNIS J. O'KANE and JOHN LEE

Relative light units are commonly reported in the measurement of bioluminescent and chemiluminescent reactions. The reason is that the calibration of the absolute photon sensitivity of the detector system entails overcoming severe technical problems. Absolute photon calibration is essential if one is to determine the absolute quantum yields of these reactions, important information required for elucidating the mechanisms of

0076-6879/00 $30.00

these reactions. At the present time, the exploding applications of luminescence measurements would seem to render a knowledge of the absolute values of the photon fluxes rather desirable, for example, for the comparison of methodologies or to quantify the performance of instrumentation.

The standard of photon flux is a blackbody radiation source which are maintained in many national standards laboratories. The National Institute of Standards and Technology in Gaithersburg, Maryland, for example, issues a standard tungsten lamp of radiance calibrated with reference to the standard blackbody, and such a lamp is employed in many photon calibration applications. A difficulty is that this lamp is many orders of magnitude too bright to be used directly for the calibration of a photomultiplier as normally employed in a luminometer. Calibration of the absolute photon sensitivity of a photomultiplier was achieved by precise attenuation of the output of a standard lamp, and the bioluminescence quantum yield of the firefly reaction and the chemiluminescence quantum yield of the luminol reaction were determined to satisfactory accuracy, about 25%.[1,2] The value of this experimental approach is that the quantum yields are directly traceable to the standard blackbody.

Another technical problem in luminometer calibration is that of detector geometry.[2] The preferred geometrical arrangement for a quantum yield measurement is that of point source geometry, i.e., the luminescing solution should be, say, in a 1-ml volume, at about 25 cm from the face of the detector, and in a suitably baffled holder. This last baffling is so that only light truly emanating in the direction of the detector is allowed to pass, i.e., no reflections from the cuvette walls or light piping down the sides. These conditions are important on two counts. The first is that the size of light source viewed from the detector depends on the square of the refractive index of the reaction medium. Whereas bioluminescence reactions are carried out in aqueous medium, many chemiluminescence reactions are in organic solution. The detector as calibrated with the standard lamp sits in air, of course, so the refractive index correction is considerable. This correction has been verified experimentally.[2] Unfortunately, this simple refractive index-squared relationship only holds for the point source, and most luminometers have the reaction chamber quite close to the detector in order to maximize sensitivity. Under such a condition the correction becomes impossible to estimate directly. Therefore, the secondary calibration with reference to the luminol chemiluminescence reaction has a distinct advantage under this circumstance.

[1] H. H. Seliger and W. D. McElroy, *Arch. Biochem. Biophys.* **88,** 136 (1960).

[2] J. Lee and H. H. Seliger, *Photochem. Photobiol.* **4,** 1015 (1965).

The second problem with the nonpoint source and "close-up" arrangement is that detection is prone to variations in mixing. Many reactions are initiated by injection and, depending on the rate of mixing, the light emission may be uniform throughout the reaction volume, giving rise to considerable variation in distance from the detector face.

It is relevant to mention in this introductory section that the employment of a silicon photodiode as a detector obviates the requirement for absolute photon calibration as these photodiodes are "self-calibrating."[3] Photodiodes have the advantage of much more uniform wavelength dependence of the sensitivity than a photomultiplier and they are also unaffected by exposure to ambient light, in contrast to photomultipliers.[4] A photomultiplier-based luminometer must always be designed to strictly avoid exposing the detector to room light. The disadvantage of the photodiode is that the visible photon sensitivity is about 100 times less than a photomultiplier, although if the photodiode has a "close-up" geometry to the reaction chamber, the sensitivity is still enough for many luminescence applications. However, this geometry again entails calibration with reference to a standard chemiluminescence reaction.[5]

Luminol Chemiluminescence Standard

The absolute quantum yield from the luminol (3-aminophthalhydrazide) chemiluminescence reaction is 0.0124. This means that for each 100 molecules of luminol reacting, 1.24 photons on average are emitted within the spectral envelope, depending on the solvent. The same quantum yield is obtained for reactions in aqueous solution or in dimethyl sulfoxide (DMSO), although the conditions are very different.[2] The quantum yield translates to a value of 9.75×10^{14} photons from 1 ml of a luminol solution having an absorbance of unity at the maximum wavelength, 347 nm for aqueous solution at pH 11.5, and 359 nm in DMSO.[5]

The photon sensitivity of photomultiplier detectors used in most luminometers depends strongly on wavelength. Therefore, the detector signal from the luminol or any other chemiluminescence reaction will be a result of the overlap between the spectral emission distribution and the spectral response curve of the detector. For high accuracy, this response needs to be determined directly for the detector being employed, but this is not

[3] E. F. Zalewski, *in* "Optical Radiation Measurements" (K. D. Mielenz, ed.), Vol. 3, p. 89. Academic Press, New York, 1982.

[4] J. Lee, A. S. Wesley, J. F. Ferguson III, and H. H. Seliger, *in* "Bioluminescence in Progress" (F. H. Johnson and Y. Haneda, eds.), p. 35. Princeton Univ. Press, Princeton, NJ, 1966.

[5] I. B. C. Matheson, J. Lee, and E. F. Zalewski, *Proc. Soc. Photo-Opt. Instrum. Eng.* **489,** 380 (1984).

practical unless the laboratory is set up for photometric measurements. It is usually sufficient to accept the spectral response curve provided by the manufacturer, e.g., for a commonly used photomultiplier with S11 response, which has a maximum around 420 nm and falls to 50% by 550 nm. An advantage of the photodiode is that its spectral response is practically constant over the visible wavelength region.[6]

The emission maximum for the aqueous reaction is at 428 nm and for DMSO is at 486 nm. If the geometry of the emitter is a point source with respect to the detector, then the detector will view the DMSO reaction $(1.33/1.49)^2 = 80\%$ less efficiently than from the aqueous reaction, where the ratio is of the solvent refractive indices. For proximate geometry, this refractive index correction is closer to unity. For the detector arrangement commonly employed with luminometers, this correction is hard to calculate accurately for the reasons discussed above but can be estimated to be in the range 90–95%. It is the experience of this laboratory that the integrated signal measured from the DMSO reaction is about half that from the aqueous reaction.

The important property of the luminol calibration method, therefore, is that both detector geometry and spectral response for the reactions in the two solvents are different but the final result after taking these factors into account must be the same. This condition provides a vital cross-check on the reaction integrity. The accuracy of the spectral overlap and geometrical corrections also contribute, but to a lesser extent.[6]

Aqueous Luminol Chemiluminescence

High-purity 3-aminophthalhydrazide can be purchased from a number of commercial sources but nevertheless should be assayed for purity by its absorption spectrum: $\varepsilon(347\ \text{nm}) = 7640\ M^{-1}\ \text{cm}^{-1}$. A solution containing 0.1 *M* potassium carbonate to maintain pH 11.5 is used for all reactants. It is recommended to make up a stock solution of the luminol to absorbance around 20, measured accurately by dilution.

The other reagents are H_2O_2 (30%, about 12 *M*) and a catalyst; heme proteins are usually employed, hemoglobin[5] or peroxidase[6] is recommended (indeed, fresh blood is quite superior). The protein concentration in the carbonate buffer should be such as to give a just perceptible brown color due to the heme absorption. If the catalyst concentration is kept dilute, then on addition to the reaction, the absorbance in the 420-nm region will not exceed, say, $0.02\ \text{cm}^{-1}$ and the chemiluminescence emission intensity will not be attenuated.

[6] D. J. O'Kane, M. Ahmad, I. B. C. Matheson, and J. Lee, *Methods Enzymol.* **133B,** 109 (1986).

The amount of luminol used in the reaction should be adjusted to the response range of the luminometer. Typically one would take 0.9 ml of a luminol solution with A(347 nm) = 0.0001 and add 10 μl of freshly 1:1000 diluted H_2O_2. The amount of H_2O_2 used has to be determined by trial and error. The 30% concentration of the stock can be quite uncertain and a highly diluted solution of peroxide is unstable.

After taking appropriate precautions to ensure light tightness, chemiluminescence is initiated by rapid injection of the catalyst solution (0.05 ml) into the luminol sample. The emitted light intensity will rise quickly to a maximum, then should decay slowly over about 100 sec. If the light decays rapidly over only a few seconds, the amount of H_2O_2 added should be reduced (usually) to achieve the more slowly decaying condition. After the slow decay is nearly complete, several minutes of time usually, a second addition of catalyst should be made. This second addition will increase the total light measured typically by another 25%. A third addition may also be necessary, also with more H_2O_2 addition, until the operator is convinced that the luminol is completely exhausted. The maximum integrated intensity should be attained after only two or three additions. It should also be cautioned that a few percent of the total light may be generated on the first addition of the H_2O_2 before the addition of catalyst.

It is important to practice the reaction technique until the maximum total signal response is attained. It is our experience that in laboratories that have not had previous experience with this reaction, the operator may need many repetitions to achieve confidence in the procedure.

Dimethyl Sulfoxide Luminol Chemiluminescence

In this method it is important to keep the reaction mixture relatively anhydrous. High-purity and water-free DMSO is inexpensive, so we have always found it convenient to use a new unopened bottle of DMSO on each occasion the reaction is performed. The luminol stock solution can be made up and, for the reaction, diluted to the appropriate A(359) necessary to calibrate the luminometer on the desired scale. The only other reagent needed is a saturated solution of potassium *tert*-butoxide in dry *tert*-butanol. Because this solution freezes around 20°, it needs to be kept slightly above this temperature. It may have to be prepared every few hours due to the instability of the *tert*-butoxide.

To 1.0 ml of the luminol solution in DMSO is rapidly injected 0.02 ml of the potassium *tert*-butoxide/*tert*-butanol. The emission intensity should rise to a maximum in about 1 min, then decay to a negligible level in 10–15 min. A second injection could be tried but should not be necessary. Sometimes the basicity is not sufficient with only one injection. Too much butoxide, however, is equivalent to the water contamination problem, slowing the reaction and decreasing the total light yield. If the reaction time is much slower than about 45 min, contamination by traces of water can be suspected and the total light emitted will be too low.

As pointed out earlier, the integrated signal from the DMSO reaction in a luminometer employing a photomultiplier should be about one-half that for the aqueous reaction for the same amount of luminol employed. After corrections, therefore, the quantum yield will be the same for aqueous and DMSO reactions. This provides a critical cross-check on the operator and reaction procedures.

Day-to-Day Calibrators

Although absolute photometer calibration is essential for research applications and for standardizing routine measurements, this approach is too tedious to implement on a daily basis. Instead, a secondary calibrator is used to monitor variation both in instrument response and in the chemiluminescence or bioluminescence system. Several different day-to-day calibrators have been utilized for these purposes. The most commonly used calibrators include radioactive liquid standards,[7] encapsulated phosphors with adsorbed radioactive compounds,[8,9] uranyl salts,[10,11] light-emitting diodes, and commercially available acridinium chemiluminescence calibrators (Quest Diagnostic Laboratories, San Juan Capistrano, CA). Each calibrator has advantages and disadvantages.

Radioactive Liquid Calibrators

These calibrators use soft β disintegration to excite fluorescence emission from scintillators in organic solvents. The major optical advantage of these calibrators is that they produce isotropic emissions. A disadvantage, however, is that the emission is structured and may not overlap significantly the bioluminescence or chemiluminescence emission of interest.[9] Two different approaches have been reported for developing liquid radioactive standards: the use of tritiated[7] and quenched ^{14}C-labeled compounds.[12]

[7] J. W. Hastings and G. Weber, *J. Opt. Soc. Am.* **53,** 1410 (1963).

[8] W. H. Biggley, E. Swift, R. J. Buchanan, and H. H. Seliger, *J. Gen. Physiol.* **54,** 95 (1969).

[9] D. J. O'Kane and J. Lee, *Photochem. Photobiol.* **52,** 723 (1990).

[10] G. D. Mendenhall and X. Hu, *J. Photochem. Photobiol A* **52,** 285 (1990).

[11] D. J. O'Kane and D. F. Smith, *in* "Bioluminescence and Chemiluminescence: Status Report" (A. Szalay, L. Kricka, and P. Stanley, eds.), p. 38. Wiley, Chichester, 1993.

[12] T. E. T. Oikari, I. A. Hemmil, and E. J. Soini, *in* "Analytical Applications of Bioluminescence and Chemiluminescence" (L. J. Kricka, P. E. Stanley, G. H. G. Thorpe, and T. P. Whitehead, eds.), p. 475. Academic Press, New York, 1984.

Tritiated calibrators have the advantage of producing single-photon emissions due to the low energy β^- particles. Because of the higher energy β^- particles, ^{14}C-liquid calibrators will generate multiphoton emissions that will not register correctly in all luminometers. Consequently, these calibrators should be quenched. The main disadvantage encountered with these standards is the decrease in emission output over time. The emission fall off from tritiated standards is predominantly attributable to the half-life of tritium (12.3 years; Table I). This decrease in emission can be corrected using standard tables and the time between initial and subsequent measurements. However, radiation damage appears to contribute to the decrease in the emission level from ^{14}C calibrators and this is more difficult to predict and correct over time.

Encapsulated Radioactive Phosphors Calibrators

Different radionuclides can be deposited on the surface of phosphor powders to generate radiophosphorescence emissions.[8,9] ^{14}C-bicarbonate is precipitated in a slurry of inorganic phosphor powder by $BaCl_2$ addition. The phosphor is then embedded in epoxy resin.[8] This procedure produces calibrators in a tube format that are stable for many years and that are useful in the range of intensities found in most chemiluminescence-emitting systems. The disadvantage is that it is difficult to prepare calibrators that emit at sufficiently high intensities to permit calibrating luminometers for use with bioluminescent organisms. Because the amount of light emitted by

TABLE I
CHARACTERISTICS OF RADIONUCLIDES POTENTIALLY SUITABLE FOR LIGHT STANDARDS

Radionuclide[a]	$t_{1/2}$ (years)	Mode of emission (Mev)		Specific activity (Ci/g)	
				Theoretical[b]	Commercial
^{3}H	12.3	β^-	0.019	9688	9688
^{121m}Sn	55	β^- and γ	0.357	54	0.02
^{232}U	73.6	α and γ	5.41	21	NCA[c]
^{151}Sm	93 ± 8	β^- and γ	0.076	25	NCA
^{63}Ni	96	β^-	0.066	59	12
^{32}Si	104	β^-	0.227	107	NCA
^{14}C	5730	β^-	0.156	4.5	4.5

[a] Radionuclides arranged in increasing order of half-life. Half-life and maximum particle energy taken from NCRP Report 58 or from "The Handbook of Chemistry and Physics."
[b] Theoretical specific activity (Ci/g) = $kn/(3.7 \times 10^{10}$ dps), where k is the first-order radioactive decay constant (sec^{-1}) and n is the number of atoms per gram of radionuclide.
[c] Not available commercially.

these standards is proportional to the number of radioactive disintegrations, very large amounts of ^{14}C must be used to prepare high-intensity standards. Alternatively, radionuclides with shorter half-lives, a greater number of disintegrations per unit time, and potentially greater levels of emission may be useful for this purpose (Table I). The use of ^{63}Ni ($t_{1/2}$ = 96 years) is advantageous because it is available commercially at high specific activity, precipitates quantitatively as the insoluble alkaline carbonate, emits soft β radiation that does not produce measurable damage to the phosphors, and is inexpensive compared to ^{14}C-bicarbonate. Calibrators can be prepared in several formats using different phosphors that span wide spectral emission ranges that correspond closely to many bioluminescence and chemiluminescence emissions.[9] The radioactive phosphors may be embedded in plastics either by simply mixing the phosphor with the plastic or by a combined vacuum infiltration and centrifugal infiltration of a phosphor layer. The second procedure produces calibrators with greater emission intensities than those produced by mixing phosphor with plastic. The major advantage of these calibrators is that they are robust physically and the emission levels do not decay quickly over time (~1% per year initial loss). However, the calibrators are not necessarily isotropic, and this must be tested.

Uranyl Salt Calibrators

Uranyl salts emit low-intensity radioluminescence, as first noted by Becquerel.[13] In solution, emission is attributed predominantly to the ^{232}U daughter radionuclide generated in the ^{238}U decay series.[10] However, radioluminescence from solid uranyl salts appears to result from the decay of ^{238}U directly. Different uranyl salt forms emit different levels of radioluminescence, with acetate and nitrate salts producing the highest levels of radioluminescence. The degree of hydration of the uranyl salt affects the level of radioluminescence. Consequently, calibrators should be sealed in order to prevent dehydration and changes in environmental humidity from affecting the level of radioluminescence. In contrast to the radioactive phosphor calibrators described earlier, uranyl salt calibrators should preferably not be embedded in plastic. Over time, radiation damage from the high energy α-particles emitted from ^{238}U and radioactive daughter nuclides darkens the plastic, resulting in decreased radioluminescence over time. Calibration of a luminometer with a liquid uranyl nitrate standard and a

[13] H. Becqueral, *C.R. Acad. Sci.* **122,** 689 (1896).

solid uranyl acetate standard is shown in Fig. 1. The luminometer and calibrators were moved several times and resited in different laboratories during the period indicated. Caution is required when using uranyl salt calibrators with new luminometers. In certain applications, optical filters may be placed in the photomultiplier housing to reduce background chemiluminescence noise. These filters may absorb the radioluminescence as well and must be removed in order to calibrate with uranyl salts. Unlike the components of the other calibrators listed earlier, the amount of uranyl salt used in a calibrator (as much as 10 g) may pose a health hazard to personnel through dermal exposure or by inhalation if the calibrator is broken. A second disadvantage is that the disposal of uranyl salts is regulated by statute and must be disposed of properly.

Discussion

Little attention has as yet been given to absolute standardization of the various luminometers and bioluminescence and chemiluminescence assay systems in use commercially and for diagnostic procedures. The National Institute of Standards and Technology has initiated a program in the United States to develop an absolute, traceable calibration standard for luminometers. Unfortunately, this may not be entirely applicable for calibrating all bioluminescence and chemiluminescence reaction systems such as the firefly luciferase reaction. This is attributed to photocathode spectral sensitivity differences between photomultiplier tubes and dissimi-

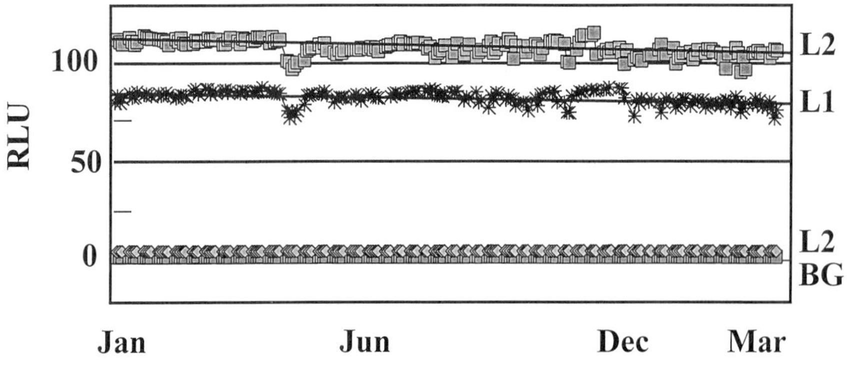

FIG. 1. Calibration of a luminometer with a liquid uranyl nitrate standard. A solid uranyl acetate standard is also shown.

larities between emission spectra of potential chemiluminescence calibrators and the oxyluciferin. Another area where much needs to be accomplished is in the area of low-level light imaging where calibration is still primitive.[14-16]

[14] D. J. O'Kane and J. E. Wampler, *Proc. Soc. Photo-Opt. Instr. Eng.* **1161,** 205 (1989).

[15] F. Berthold, K. Hafner, and D. DiPrato, *in* "Bioluminescence and Chemiluminescence: Fundamental and Applied Aspects" (A. K. Campbell, L. J. Kricka, and P. E. Stanley, eds.), p. 617. Wiley, Chichester, 1994.

[16] D. H. Leaback, *in* "Bioluminescence and Chemiluminescence: Molecular Reporting with Photons" (J. W. Hastings, L. J. Kricka, and P. E. Stanley, eds.), p. 545. Wiley, Chichester, 1997.

[6] Commercially Available Luminometers and Low-Level Light Imaging Devices

By PHILIP E. STANLEY

The study of bioluminescence (BL) and chemiluminescence (CL) and assays and techniques based on them usually requires the measurement of light output. Today, the life scientist uses a commercially available instrument because there are many from which to choose in the marketplace. This article discusses briefly some aspects of the various types of instruments and provides a list of many of the suppliers.

In Volume 133 of "Methods in Enzymology" the specifications of various types of luminometers are described in some detail.[1] In this volume, Berthold *et al.*[2] review this topic in light of today's technology, whereas Voss[3] describes the mode of action of detectors used in this instrumentation. Technical specifications of many other luminometers are given elsewhere.[4-12]

[1] P. E. Stanley, *Methods Enzymol.* **133,** 587.

[2] F. Berthold, K. Herick, and R. M. Siewe, *Methods Enzymol.* **305** [4] 2000 (this volume).

[3] K. J. Voss, *Methods Enzymol.* **305** [3] 2000 (this volume).

[4] P. E. Stanley, *in* "Clinical and Biochemical Luminescence" (L. J. Kricka and T. J. N. Carter, eds.), p. 219. Dekker, New York, 1982.

[5] P. E. Stanley, *J. Biolumin. Chemilumin.* **7,** 77 (1992).

[6] P. E. Stanley, *J. Biolumin. Chemilumin.* **7,** 157 (1992).

[7] P. E. Stanley, *J. Biolumin. Chemilumin.* **8,** 237 (1993).

[8] P. E. Stanley, *J. Biolumin. Chemilumin.* **9,** 51 (1994).

[9] P. E. Stanley, *J. Biolumin. Chemilumin.* **9,** 123 (1994).

[10] P. E. Stanley, *J. Biolumin. Chemilumin.* **11,** 175 (1996).

[11] P. E. Stanley, *J. Biolumin. Chemilumin.* **12,** 61 (1997).

[12] P. E. Stanley, *Luminescence* **14,** 215 (1999).

0076-6879/00 $30.00

Luminometers typically comprise a photomultiplier that is operated either as a current measuring device or, more commonly now, as a photon counting device, a sample chamber to hold a cuvette or other container, perhaps a device(s) for injecting reagents automatically into the cuvette either before placing it in the sample chamber or while in the darkened sample chamber itself. Automatic sampling handling may also be featured.

In the last decade or so there has been a move toward multisample processing and the use of a personal computer to control the instrument and also software to deal with the data produced. There has also been an increase in sensitivity of the instruments by virtue of improved optics as well as photomultiplier performance.

Some luminometers are now available that are designed for a specific task or range of tasks; for example, hygiene testing in the food processing industry in which microbial adenosine triphosphate is measured with firefly luciferase built into a disposable unit within the luminometer. Another example would be luminometers designed for toxicity assays using luminous bacteria or for testing water for the tracing of pollutants.

A major change that has occurred in the last decade is the ever increasing number of microtiter plate luminometers coming to the marketplace. Some are designed to measure BL or CL which exhibit glow luminescence whereas others will measure BL and CL with either glow or flash luminescence. Some manual or semiautomatic cuvette luminometers will measure glow or both glow and flash luminescence.

So-called glow luminescence occurs when the light output from the reaction remains more or less constant over many tens of seconds (often minutes or even hours). In this situation, light measurement can be made comfortably at any reasonable time during the glow period. In contrast, flash luminescence occurs when the light output rises to a maximum and then falls rather quickly; the whole process may take only a very few seconds. In this situation, the time when the light measurement is made is of critical importance. In measurements involving flash luminescence there must be consistency in timing and reagent mixing, and injector design plays an important role. Luminometers only able to measure glow reactions may not be equipped with automatic injectors. At present, luminometers are available for 96- to 384-well plates and beyond. A feature that should be addressed in choosing a microtiter plate luminometer is its ability to discriminate the signal produced by adjacent wells, so-called cross-talk. It may be that one well is producing many photons while the adjacent one is producing none at all. A unit with good cross-talk characteristics will discriminate the light generated by the two wells and show minimum spill of light from the bright well into the dark well. Another important feature is the ability of the injectors to ensure good mixing of reagents in the well while avoiding splashing. Some workers require the microtiter plate to be temperature controlled and this feature is not always available as an option. A number of manufacturers of microtiter plate luminometers provide models that may be used in conjunction with robotic systems to enable large numbers of samples to be processed, which are finding considerable usage by those workers engaged in high throughput screening in the pharmaceutical discovery arena.

Automatic injection of reagents is often considered a very desirable feature. Care should be taken in choosing a system with the following points in mind: ease of cleaning; sterilizability, if needed; ease of disassembly; and ease of replacement of disposable parts. Some injectors have software that will permit control of the injected volume as well as variation of the injection rate. Others have a simple single injection volume that is not variable by the user.

There is a very wide range of software available, either built-in or available as an option. Sometimes the software is available on a PC or a Mac computer that is connected to, and may also operate, the luminometer.

A number of low-level light imaging systems are now available. These are used for the two-dimensional imaging of, for example, chemiluminescent gels, and some have software to assay the bands. Some systems can be connected to a microscope so that bio- or chemiluminescence produced by cells can be assessed. In the list that follows are a number of companies that producc high-sensitivity charge-coupled device (CCD) cameras that may be built into custom designed instruments.

New models of luminometers are now being introduced into the marketplace at a steady rate and some older models are inevitably withdrawn from time to time. Consequently, a contribution such as this becomes outdated rather quickly. The following list of manufacturers was compiled in August 1998 and adjusted in January 2000 to take account of some changes to company names and contact details. See Reference 12 for more detailed information about some current instruments. Contact details are given together with a brief mention of the types of luminometers they currently provide.

When choosing a luminometer or CCD imaging system the purchaser would be advised to test the competing instruments in his laboratory with his own reagents and ancillary equipment.

Manufacturers of Luminometers

Although the list is extensive, the author does not guarantee that it is comprehensive, complete, or free from errors. Instruments not included in this list are (1) large automatic dedicated luminometers used in clinical laboratories for immunoassays and (2) imaging systems based on phosphor screen detectors.

Anthos Labtec Instruments GmbH, Lagerhausstrasse 507, A-5071 Wals/Salzburg, Austria
Tel: +43 662 857-220; Fax: +43 662 857-223

Microtiter plate luminometers

ATTO Corporation, 2-3, Hongo 7-chome, Bunkyo-Ku, Tokyo 113, Japan
Tel: +81 3 3814 4861; Fax: +81 3 3814 4868

Microtiter plate luminometer

Azur Environmental, 2232 Rutherford Road, Carlsbad, CA 92008
Tel: +1 619 438-8282; Fax: +1 619 438-2980

Specialized luminometers for toxicity testing

BBI-Source Scientific, Inc., 7390 Lincoln Way, Garden Grove, CA 92841
Tel: +1 714 898-9001; Fax: +1 714 891-1229

Luminometers

Berthold Detection Systems GmbH, Bleichstrasse 3A, D-75173 Pforzheim, Germany
Tel: +49 7231 92060; Fax: +49 7231 920650

High sensitivity luminometer, microtiter plate luminometer

Bio-Orbit Oy, P.O. Box 36, FIN-20521 Turku, Finland
Tel: +358 2 410-1100; Fax: +358 2 410-1123

Luminometers, manual and automatic, microtiter plate

Bio-Rad, Inc., 1000 Alfred Nobel Drive, Hercules, CA 94547
Tel: +1 510 724-7000; Fax: +1 510 741-5817

CCD imaging system and microtiter plate system

Bioscan, Inc., 4590 MacArthur Bvld. NW, Washington, DC 20007
Tel: +1 202 338-0974; Fax: +1 202 338-8514

Luminometer

BioThema AB, Strandvägen 36, S-130 54 Dalarö, Sweden
Tel: +46 8501 51641; Fax: +46 8501 50952

Portable luminometer

Biotrace Ltd., The Science Park, Bridgend, CF31 3NA, UK
Tel: +44 1656 768844; Fax: +44 1656 768835

Specialized luminometers for hygiene and similar assays in food and allied industries

BMG-LabTechnologies GmbH, Hanns-Martin-Schleyer-Strasse 10, D-77656 Offenburg, Germany
Tel: +49 781 969-680; Fax: +49 781 969-6867

Microtiter plate luminometer

Boehringer-Mannheim GmbH, Sandhoffer Strasse 116, D-68305 Mannheim, Germany
Tel: +49 621 7598545; Fax: +49 621 7598830

Imaging device for analysis of blots and microtiter plates

Camspec Ltd., 11 High Street, Sawston, Cambridge CB2 4BG, UK
Tel: +44 1223-8326971; Fax: +44 1223-836414

Chemiluminescence detector for HPLC

Capitol Controls Ltd., 8 Hawksworth, Southmead Industrial Park, Didcot, Oxfordshire OX11 7HR, UK
Tel: +44 1235-512000; Fax: +44 1235-512020

Specialized luminometer for water pollution assays

Cardinal Associates Inc., P.O. Box 5220, Santa Fe, NM 87502
Tel: +1 505 473-0033; Fax: +1 505 473-5637

Microtiter plate luminometer

Celsis · Lumac, Cambridge Science Park, Milton Road, Cambridge CB4 4FX, UK
Tel: +44 1223 426008; Fax: +44 1223 426003

A range of specialized luminometers for hygiene assays in food and allied industries

Charm Sciences Inc., 36 Franklin Street, Malden, MA 02148
Tel: +1 617 322-1523; Fax: +1 617 322-3141

A range of specialized luminometers for hygiene assays in food and allied industries

CLIMBI Ltd., P.O. Box 20, Moscow 125422, Russia
Tel: +7 095-976-4055; Fax: +7 095-976-7586

Portable and manual luminometer

ConCell BV, Wevelinghoven 26, D-41334 Nettetal, Germany
Tel: +49 2153 911-833; Fax: +49 2153 911-835

Specialized luminometer for hygiene and similar assays in food and allied industries

Cypress Systems, Inc., 2500 West 31st Street, Suite D, Lawrence, KS 66047
Tel: +1 913 842-2511; Fax: +1 913 832-0406

Luminometer for hygiene and environmental monitoring

Dr Bruno Lange GmbH Berlin, Willstättersreasse 11, D-40549 Düsseldorf, Germany
Tel: +49 211-52-880; Fax: +49 211-52-88175

Specialized luminometers for toxicity testing

Dynex Technologies Inc., 14340 Sullyfield Circle, Chantilly, VA 20151
Tel: +1 703 631-7800; Fax: +1 703 631-7816

A range of microtiter plate luminometers

Electron Tubes Ltd., Bury Street, Ruislip, Middlesex HA4 7TA, UK
Tel: +44 1895-630771; Fax: +44 1895-635953

Manufacturer of photomultipliers and associated equipment

F.A.T. Research Institute for Antioxidant Therapy Ltd., Chausseestrasse 8, D-10115 Berlin, Germany
Tel: +49-30-39789376; Fax: +49-30-39789386

Specialized luminometers for assay of antioxidants.

GEM Biomedical, MGM Instruments, 925 Sherman Avenue, Hamden, CT 06514
Tel: +1 203 248-4008; Fax: +1 203 288-2621

A range of luminometers (manual, automatic, and microtiter plate) and hygiene and reporter genes

Hamamatsu Photonics K.K., 314-5 Shimokanzo, Toyooka-village, Iwata-gun, 436-01 Japan
Tel: +81 539-62-5248; Fax: +81 539-62-2205

Manufacturers of photomultipliers and associated equipment and of CCD imaging systems

Hughes Whitlock Ltd., Wonostow Road, Singleton Court, Monmouth, Gwent NP5 3AH, UK
Tel: +44 1600-715632; Fax: +44 1600-715674

Specialized luminometer for hygiene and similar assays in food and allied industries

IDEXX Laboratories, Inc., One IDEXX Drive, Westbrook, ME 04092
Tel: +1 207 856-0300; Fax: +1 207 856-0630

Specialized luminometer for hygiene and similar assays in food and allied industries

Imaging Associates, 8 Thame Park Business Centre, Wenman Road, Thame, Oxfordshire OX9 3XA, UK
Tel: +44 1844-213790; Fax: +44 01844-213644

CCD cameras and imaging systems

Kikkoman Corporation, 399 Noda, Noda City, Chiba Pref, 278 Japan
Tel: +81 471 23 5517; Fax: +81 471 23 5959

Luminometer for hygiene testing

Labsystems Oy, P.O. Box 8, FIN-00881 Helsinki, Finland
Tel: +358 9 329 100; Fax: +358 9 3291 0312

Microtiter plate luminometers and tube luminometer

Labtech International Ltd., 1 Acorn House, The Broyle, Ringmer, East Sussex BN8 5NW, UK
Tel: +44 1273 814-888; Fax: +44 1273 814-999

Manual luminometer, microtiter plate luminometer; hygiene monitoring

Mediators Diagnosticka GmbH, Simmeringer Haupstrasse 24, A-1100 Vienna, Austria
Tel: +43 1 740-40-350; Fax: +43 1 740-40-359

Microtiter plate luminometer

Merck KgaA, Frankfurterstrasse 250, 64271 Darmstadt, Germany
Tel: +49 6151 725410; Fax: +49 6151 723380

Specialized luminometer for hygiene testing of foods and similar products

Millipore Corporation, 80 Ashby Road, Bedford, MA 01730
Tel: +1 617 275-9200; Fax: +1 617 533-2889

Specialized luminometers for microbiological testing of foods and similar products; also CCD camera for sterility testing using filters

Molecular Devices Corporation, 1311 Orleans Drive, Sunnyvale, CA 94089
Tel: +1 408 747-1700; Fax: +1 408 747-3602

Specialized CCD imaging microtiter plate reader

New Horizons Diagnostics Corporation, 9110 Red Branch Road, Columbia, MD 21045
Tel: +1 410 992-9357; Fax: +1 410 992-0328

Specialized luminometers mainly for food safety and sanitation

Packard Instrument Co., 800 Research Parkway, Meriden, CT 06450
Tel: +1 203 238-2351; Fax: +1 203 639-2172

Microtiter plate luminometers

PerkinElmer Berthold, P.O. Box 100163, D-75312 Bad Wildbad, Germany
Tel: +49 7081 1770; Fax: +49 7081 177166

A wide range of manual and automatic luminometers, microtiter plate luminometers, CCD imaging unit for blots, microtiter plates, etc.

PerkinElmer Wallac, P.O. Box 10, FIN-20101 Turku, Finland
Tel: +358 2-2678-111; Fax: +358 2-267-8357

A range of microtiter plate luminometers and CCD imaging units for blots, microtiter plates, etc.

Photek Ltd., 26 Castleham Road, St. Leonards-on-Sea, East Sussex TN38 9NS, UK
Tel: +44 1424-850555; Fax: +44 1424-850051

Manufacturer of photomultipliers, image intensifiers, and custom-built CCD cameras

Photonic Science Ltd., Millham, Mountfield, Robertsbridge, East Sussex TN32 5LA, UK
Tel: +44 1580 881199; Fax: +44 1580 880910

Manufacturer of specialized CCD cameras and imaging systems

Randox Laboratories Ltd., Diamond Road, Crumlin, Co. Antrim, UK
Tel: +44 1849-422413; Fax: +44 1849-422413

Specialized luminometer for water quality assays

Roper Scientific, Tucson Office: 3440 East Britannia Drive, Tucson, AZ 85706
Tel: +1 602 889-9933; Fax: +1 602 573-1944
Trenton Office: 3660 Quakerbridge Road, Trenton, NJ 08619
Tel: +1 609 587-9797; Fax: +1 609 587-1970

Manufacturer of specialized CCD cameras and imaging systems

St. John Associates, Inc., 4806 Price George's Avenue, Beltsville, MD 20705
Tel: +1 301 595-5605; Fax: +1 301 595-2738

Manual luminometer for cuvettes and flow

STRATEC Electronic GmbH, Gewerbestrasse 11, D-75217 Birkenfeld 2, Germany
Tel: +49 7082-79160; Fax: +49 7082-20559

Manual luminometer and microtiter plate luminometer

TECAN Austria GmbH, Untersbergstrasse 1A, 5082 Grödig/Salzburg, Austria
Tel: +43 6246 8933; Fax: +43 6246 72770

Microtiter plate instruments

Tropix, Inc., 47 Wiggins Avenue, Bedford, MA 01730
Tel: +1 617 271-0045; Fax: +1 617 275-8581

Microtiter plate luminometer

Turner Designs, 845 W. Maude Avenue, Sunnyvale, CA 94086
Tel: +1 408 749-0994; Fax: +1 408 749-0998

Manual luminometer

Valbiotech, 57 bvd de la Villette, F-75010 Paris, France
Tel: +33 1 40 03 89 14; Fax: +33 1 44 52 92 69

Microtiter plate luminometer

Wright Instruments Ltd., Unit 10, 26 Queensway, Enfield, Middlesex EN3 4SA, UK
Tel: +44 181-4433339; Fax: +44 181-4433638

CCD cameras

Zylux Corporation, 1742 Henry G. Lane Street, Maryville, TN 37801
Tel: +1 423 379-6016; Fax: +1 423 379-6018

Manual luminometer and microtiter plate luminometer

[7] Automated Recordings of Bioluminescence with Special Reference to the Analysis of Circadian Rhythms

By TILL ROENNEBERG and WALTER TAYLOR

Introduction

To the day-active human, light provides the main source of information about the structure of the environment. Our visual sense is so dominant that we often use machines to "translate" nonphotic physical signals, such as heat or ultrasound, into images. It is, therefore, not surprising that our photic window is also used extensively to probe the world with scientific experiments.

Unlike sound, which relies on mechanical "devices" for both its production and perception, chemical, thermal, and photic information can be directly produced and perceived by molecules. Although thermal conversions are mostly unspecific in their molecular interactions, both chemical and photic information depend on specific qualities. Specialized molecules absorb light or produce photons, be it as bioluminescence, fluorescence, or phosphorescence. Due to our visual preference, we have always been especially fascinated by organisms that can produce light by biochemical reactions, such as fireflies, luminescent bacteria, or dinoflagellates.

J. W. Hastings has been a pioneer in research on the biochemistry of bioluminescence in microorganisms and insects.[1] Together with Beatrice Sweeney, he also was one of the first who used light as reporter for a biological function, namely circadian rhythms.[2] Sweeney and Hastings took advantage of the natural light production of the marine dinoflagellate *Gonyaulax polyedra,* which is under the strict control of the circadian system in this unicellular alga. Approximately 40 years after introduction of this natural light reporter, circadian biologists are now using molecular constructs,[3] fusing the promoters of rhythmically expressed genes to the genes of luciferases or green fluorescent protein (GFP) with the same aim: to "see" the circadian clock. The method of constructing "reporter genes" is, of course, not restricted to the circadian field, but is becoming increasingly important in the investigation of many different biological questions, such as cell cycle, development, or sensory responses.

[1] J. W. Hastings, *J. Cell Comp. Physiol.* **40,** 1 (1952).
[2] B. M. Sweeney and J. W. Hastings, *J. Cell Comp. Physiol.* **49,** 115 (1957).
[3] J. D. Plautz, M. Kaneko, J. C. Hall, and S. A. Kay, *Science* **278,** 1632 (1997).

0076-6879/00 $30.00

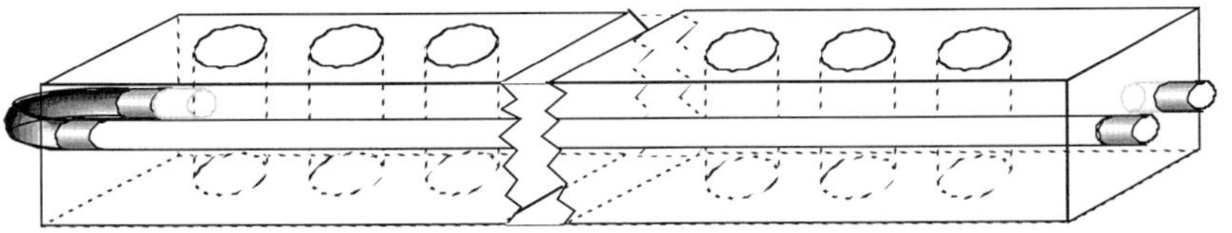

FIG. 1. Basic structural element of the recording apparatus, consisting of a modified aluminum bar (6 × 12 × 200 cm) holding up to 30 culture vessels (vials) and allowing even temperature control.

One of the pioneers in circadian biology, Colin Pittendrigh, is often quoted as having said "if you can't automate it, don't measure it." Adhering to this rule, Woody Hastings modified a scintillation counter so that it could automatically measure the bioluminescence rhythm in many *Gonyaulax* cultures at the same time for many days.[4,5] Since then, the advent of inexpensive and ever more powerful personal computers and the availability of affordable data acquisition and control hardware have made possible the development of successive versions of the machine with improved reliability and performance.[6–8] The latest of these versions (1989) was developed in collaboration between the two *Gonyaulax* laboratories at Harvard and in Munich. The details of this instrumentation will be described here, together with the evaluation software, which allows sophisticated analysis of circadian rhythms.

Recording Apparatus

The central unit of the recording setup consists of a hollow aluminum bar (approximately 2 m long, Fig. 1) serving as a temperature-controlled vial holder across which a "photomultiplier cart" is moved with the help of chains driven by a stepper motor. At regular intervals, 30 aluminum tubes have been welded into this bar to hold cylindrical vials containing algal cultures (the vials' commercial height of 6.3 cm was cut down to 4.7 cm, Fischerbrand No. 03-339-26E, 2.3 cm diameter).

The ends of the bar are closed with water-tight plates. Inlets allow water to be circulated through a metal tubing in one direction and through the

[4] J. W. Hastings, *Cold Spring Harbor Symp, Quant. Biol.* **25,** 131 (1960).

[5] J. W. Hastings and V. C. Bode, *Ann. N.Y. Acad. Sci.* **98,** 876 (1962).

[6] R. Krasnow, J. C. Dunlap, W. Taylor, J. W. Hastings, W. Vetterling, and V. D. Gooch, *J. Comp. Physiol.* **138,** 19 (1980).

[7] W. Taylor, S. Wilson, R. P. Presswood, and J. W. Hastings, *J. Interdiscipl. Cycle Res.* **13,** 71 (1982).

[8] H. Broda, V. D. Gooch, W. R. Taylor, N. Aiuto, and J. W. Hastings, *J. Biol. Rhythms* **1,** 251 (1986).

inner volume in the other. This counterflow arrangement ensures an even temperature across the length of the bar. For the long-term survival of the photosynthetic alga, light can be provided either from the top or from the bottom of the bar. The Harvard laboratory uses two white fluorescent light tubes along the entire length of the bar, whereas the Munich laboratory uses single halogen lamps for each vial, which can be adjusted to either the same intensity for all vials or varied in intensity depending on experimental needs.

The wider top of each vial holder (see details in Fig. 2) allows placements of circular glass covers as well as neutral density or spectral filters (1-in. diameter), and a small ridge at the bottom serves to retain the vials in place. Bioluminescence produced in each channel is detected by a photomultiplier housed in the light-tight chamber of a movable cart. A Macintosh computer controls cart movements by sending appropriate digital pulses through a National Instruments data acquisition board (Lab NB, National Instruments) to a stepper motor control unit (Berger Lahr, Lahr, Germany), which in turn controls the stepper motor that moves the cart (see wiring diagram in Fig. 3). After amplification via a Keithley picoammeter (Model 485), the bioluminescent signal is digitized by the National Instruments data acquisition board and is processed in real time by an algorithm described later. Software for the system was written in C using the Metrowerks development environment. During measurements, the housing of the cart covers three vials while measuring from the middle one, where a glass

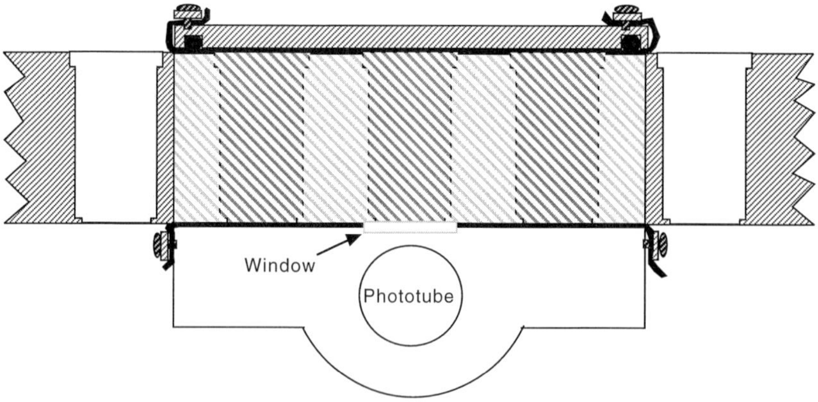

FIG. 2. Detailed side view of the vial holders and the cart housing of the photomultiplier tube. The lower end of the vial slots is narrower to hold the vials; the top end is wider to hold circular filters for intensity reduction or for spectral restrictions. All sides of the cart holding the photomultiplier are lined with black velvet to prevent light leaks. During each measurement, three of the vials are always in darkness for approximately 3 to 5 min.

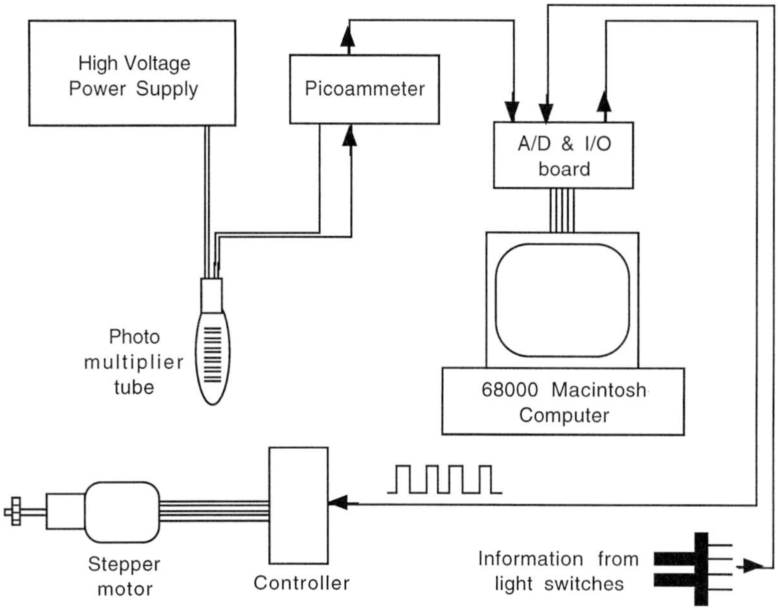

FIG. 3. Wiring diagram of data acquisition and cart control. The photomultiplier unit consists of a high voltage power supply (Kepko) providing voltage for the phototube, which is housed in a light-tight cart (see Fig. 2). The light-dependent current is amplified by a picoammeter (Keithley). The signal is digitized (Lab-NB card by National Instruments, Austin, TX) and stored in memory of a Macintosh computer, which also controls the movements of the cart with the help of a stepper motor. Light switches ensure the positioning of the cart and signal the end points of the bar to the computer program (see text for details).

window allows photons to reach the photomultiplier. The positioning of the cart over a specific vial is ensured with the help of an infrared light switch, mounted on the cart, sliding along a fixed rail containing an aperture at the appropriate positions. Similar optical switches are mounted at the ends of the aluminum bar, informing the computer program when the cart has reached an end.

At the end of an experiment, data are analyzed by software (see program description later) that assumes a constant time interval from one round to the next. Because the exact amount of time that a round consumes is subject to mechanical, electrical, and programmatic variation, the program adds a variable-length buffer period at the beginning of each round to ensure that all rounds take the same amount of time. Once this buffer time has elapsed, the cart is moved to a position just before the first vial, where the phototube is exposed to darkness, enabling the measurement of the photomultiplier dark current, which is subtracted from all subsequent vials.

The photomultiplier cart is stopped at each vial position for a length of time determined by the experimenter, but in most cases this is 40 sec, a time that is a compromise between the time required to complete a round and the length of time needed to obtain an accurate estimation of the frequency of bioluminescent flashes. Times longer than 40 sec provide better estimation of the frequency of flashes per unit of time, but result in longer times to complete a round, and thus fewer points to define the circadian waveform over a 24-hr period. The 40-sec period results in a bioluminescent glow determination approximately every 20 min, which allows a precise glow waveform measurement, but a somewhat less precise flash frequency measurement. This bias in favor of precision of glow waveform measurement was adopted because it has been determined in prior continuous measurements of both glow and flashing rhythms of *Gonyaulax* that the glow rhythm is much more precise and noise free than the flashing rhythm and thus presumably is a more precise indicator of the position of the circadian oscillator in its daily cycle.

Rather than containing cells, the first vial position houses a light standard (a sealed vial containing a quantity of ^{14}C in scintillation fluid, producing a known number of light quanta per second). Because the signal from the photomultiplier depends on many factors, including the high voltage applied to the tube and its age, the light standard is useful for the comparison of light levels determined in differing experiments. The first experimental measurement is performed when the cart moves to position 2, containing the first culture. During the rest of the round, the cart moves from position to position, measuring the amount of light from each vial and storing data in memory.

For each vial, the computer measures the bioluminescent glow emitted by all of the cells, which is essentially continuous in intensity over the 40-sec measurement period, and the number of flashes produced by all of the cells over the same 40-sec time period. Because the light intensity of a flash from a single cell is similar to the flow from 10,000 cells, the computer program uses an algorithm described previously to quantify glow level and number of flashes.[6]

After reading the last vial (No. 30), the cart continues to move in the same direction until the end switch is reached. The program then saves data accumulated in memory during the entire experiment to disk, and the direction of the stepper motor is reversed. The cart returns (without measurements) to the start position (also checked by a light switch) and a new round begins.

In our systems, the computer controls and collects data from two identical units (aluminum bar with cart, stepper motor and its controller, separate voltage supply as well as picoammeter), each of which contains a separate light standard in the first position, providing 58 channels for each experiment.

Evaluation Methods

For the evaluation of bioluminescence data collected, a program was written allowing a detailed mathematical analysis of the time series' circadian properties, i.e., period (τ), phase (ϕ), amplitude (CHRONO II, written in Metrowerks Pascal for the Macintosh). The program can directly use the binary data files created by the data collection program described earlier. There are few commercial programs available for the special needs of circadian analysis, and CHRONO II has been developed since 1990 into one of the most comprehensive circadian analysis programs. Its multiple data import routines allow the analysis of many different file types, generated in experiments with different organisms measuring different variables (activity, drinking, CO_2, and many more).

Figure 4 shows an example of the glow and flashing rhythms recorded from a single *Gonyaulax* culture in constant light conditions (19° and 30 $\mu E\ m^{-2}\ sec^{-1}$, white incandescent halogen light). Under entrained conditions, i.e., when the circadian system is synchronized to an external zeitgeber cycle (e.g., an alternation of 12 hr light and 12 hr dark, LD 12:12), both forms of bioluminescence are produced during the dark phase. The frequency of flashing peaks in the middle of the dark phase, whereas the intensity of the glow reaches its maximum around the end of the dark phase and is inhibited abruptly by the onset of light. Under constant conditions (free run), the endogenous day and night phases are referred to as subjective day and subjective night. During free run, the two bioluminescence rhythms still peak during the subjective night with a similar phase relationship to each other, but the frequency of flashes is reduced greatly and the glow is distributed in a bell shape.

The rhythms' period or phase can be calculated with several different methods. One way is to determine the daily times of the respective peak values (either by eye or by a peak picking algorithm of the program). These peak times can be presented as so-called phase plots (Fig. 5A) with the days of the experiment on the vertical axis (the actual abscissa) and the local time on the horizontal axis (the actual ordinate). The period of the rhythm can be calculated by a linear regression through the daily peak times. The peak times shown in Fig. 5A correspond to the original data shown in Fig. 4. As can be seen, flash and glow rhythms exhibit the same circadian period (24.3 hr), the former leading the latter by approximately 7.5 hr.

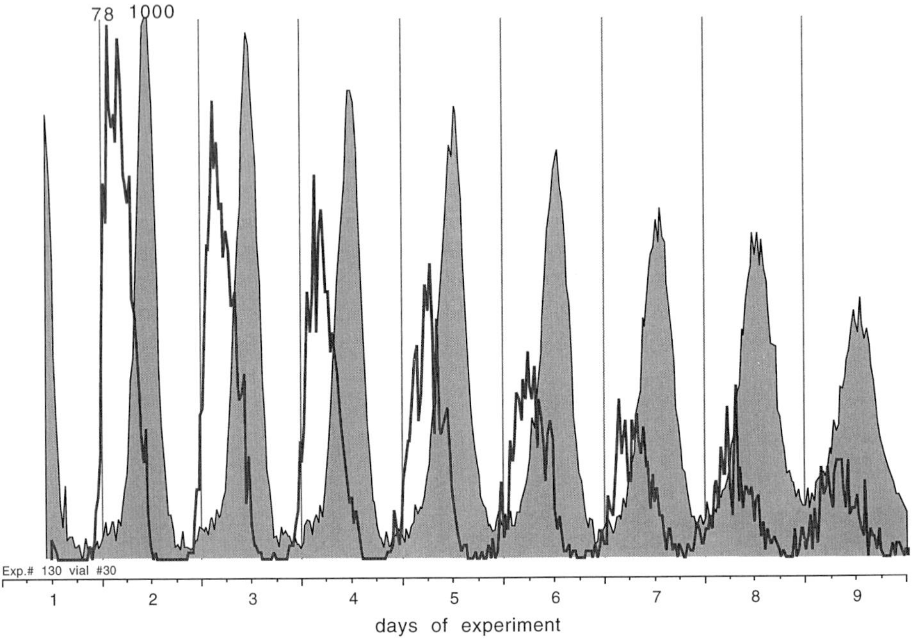

FIG. 4. Original data collected for glow intensity (shaded curve) and the number of flashes measured within 40 sec approximately every 20 min over the course of 9 days. Maximum glow intensity of the time series (in arbitrary units) was 1000, whereas the maximum number of flashes within 40 sec was 78. The *Gonyaulax* culture was kept in constant temperature (19°) and light (white incandescent halogen lamp, 30 μE m^{-2} sec^{-1}).

The period length can also be determined by different time series analysis methods (e.g., cosine fits, periodogram, spectral analysis, or autocorrelation). The results of the periodogram, based on the algorithms developed originally by Enright[9] and modified later by Sokolove and Bushell,[10] are shown in Fig. 5B. This method is based on a transformation of data into a matrix, the lines of which consist of sequential "chunks" of the time series (their length is defined as "test period," drawn on the abscissa of Fig. 5B). An average value is then calculated for each column of the resulting matrix, giving rise to an average distribution of the values within the selected period, and the variance of this average distribution (or "curve") is used to quantify the inherent rhythmicity (ordinate in Fig. 5B). If the chosen test period is not the circadian period inherent in the time series, then the average curve will be flat, i.e., will have a low variance. The highest variance

[9] J. T. Enright, *J. Theoret. Biol.* **8,** 426 (1965).

[10] P. G. Sokolove and W. N. Bushell, *J. Theoret. Biol.* **72,** 131 (1978).

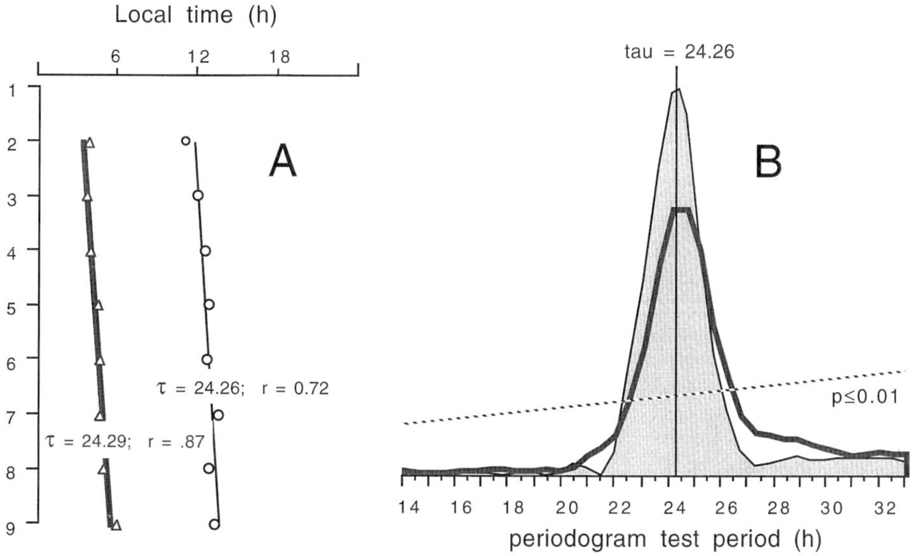

FIG. 5. Different methods of determining the circadian period, either by linear regression through the daily peak times (A) or by the periodogram method (B, see text for details).

results if the optimal period is chosen to transform the data series into the described matrix.

A similar method can also be used to actually graph the average curve of a rhythm, as long as the circadian period is stable throughout the experiment or for an investigated subset. The average curve of the rhythm shown in Fig. 4, calculated on the basis of an inherent circadian period of 24.3 hr, is shown in Fig. 6.

Experimental Possibilities

A frequent experimental procedure is to compare cultures that have been kept in either experimental or control conditions. The measurement system described here has been designed to enable as many of these comparisons as possible. As a prerequisite, recordings of each vial within the aluminum bar should have a high degree of reproducibility. The bioluminescent glow rhythms from two separate cultures (two control cultures placed in positions 7 and 25 of the aluminum bar) are shown in Fig. 7. The amplitude of these rhythms may decrease over the course of longer experiments. This amplitude reduction is due, at least in part, to a desynchronization within

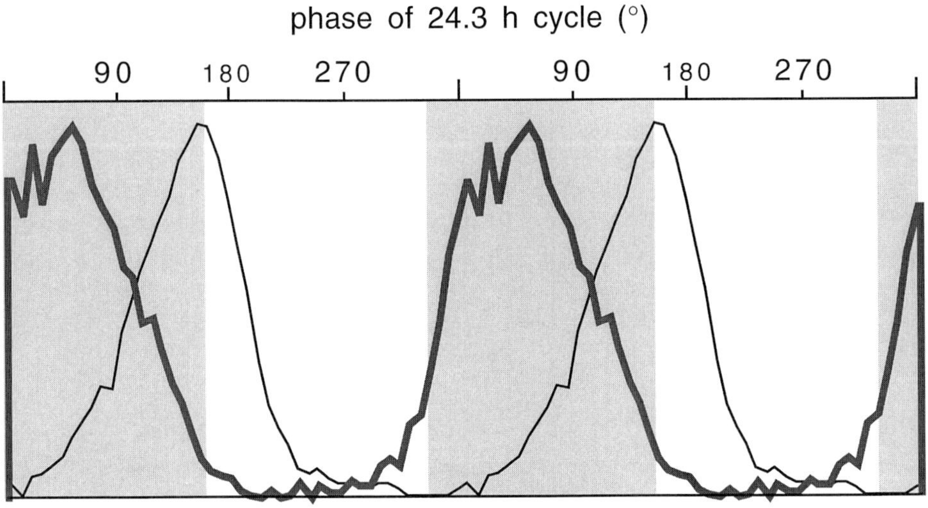

FIG. 6. Average curves of the time series for the flash (line) and the glow (shaded curve) rhythm shown in Figs. 4 and 5 based on the inherent period of 24.3 hr (see text for details). The subjective night is indicated by the gray areas. Data for an average cycle are plotted twice.

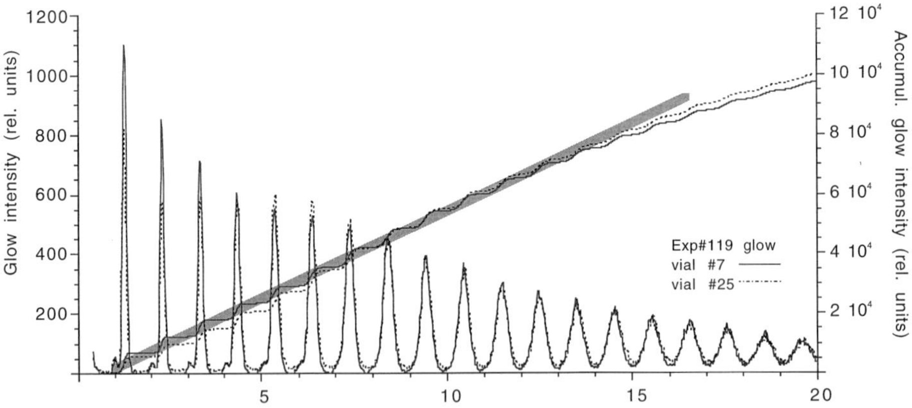

FIG. 7. Reproducibility of circadian rhythmicity among cultures kept under identical conditions. Note that the decrease of the rhythms' amplitudes is mainly due to a desynchronization between individual cells of the culture leading to a broadening of the peak and to increased trough levels. The amount of daily rhythmicity remains constant for the experiment (see straight line of the integrated data, right ordinate).

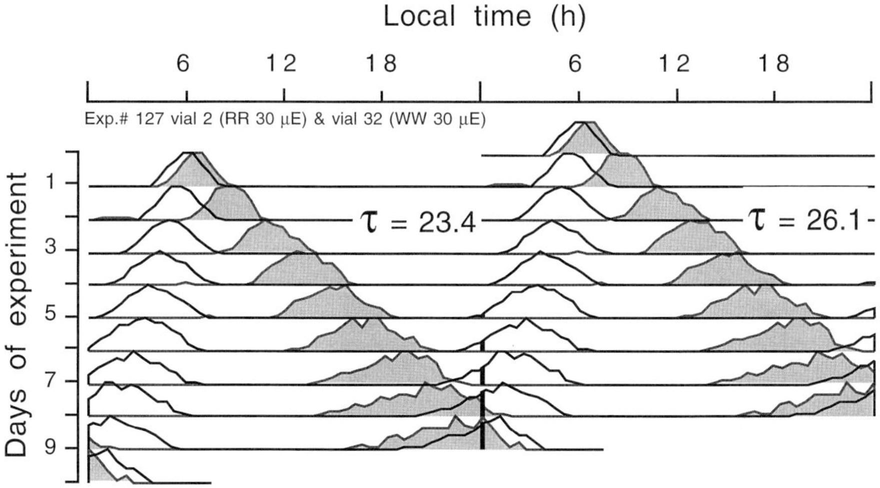

FIG. 8. Double plots of two cultures kept in white (open curve) and in red (shaded curve) light. Note the large difference in the circadian period.

the population.[11] Although the rhythms of individual cells are probably very stable, individual periods may differ slightly so that, over the course of the experiment, each cell will peak increasingly at different local times. As can be seen in the integrated glow values (right ordinate of Fig. 7), the total amount of glow produced by the culture is linear for almost 2 weeks (fat gray line), whereas the daily distribution of the glow becomes progressively wider, and the basic level between peaks gradually increases. This is consistent with the hypothesis of a gradual desynchronization within the population of cells.

The high reproducibility between different cultures in the recording apparatus allows direct comparisons between experimental cultures and controls or between cultures kept in different experimental conditions, e.g., in constant light of different intensities or spectral quality. In the experiment shown in Fig. 8, cells grown in artificial seawater (ASW)[12] were exposed in the measurement apparatus either to constant red or to constant white light [lights were adjusted to 30 $\mu E\ m^{-2}\ sec^{-1}$ either by neutral density filters or by a spectral cutoff filter (600 nm), Rosco, Port Chester, NY]. Data in Fig. 8 are "double plotted"; as in the phase plot of Fig. 5A, the horizontal axis represents local time, but here it spans 2 days. Sequential days of the experiment are represented on the vertical axis. Data shown

[11] D. Njus, V. D. Gooch, and J. W. Hastings, *Cell Biophys.* **3,** 223 (1981).

[12] T. Roenneberg and J. Rehman, *FASEB J.* **10,** 1443 (1996).

in the left and right half of the graph are identical; the first line shows days 1 and 2, the next line days 2 and 3, and so on so that the course of a non-24-hr rhythm can be followed across midnight. The results of this experiment show the drastic effects of spectral quality of light on the circadian period in *G. polyedra*.[12] Because the period in red light is almost 3 hr longer than the one in white light, the two cultures peak at approximately the same local time on day 9, despite the fact that the white light sample has gone through 10 cycles and the red light sample only through 9 cycles. The drastic effects of spectral light quality on period, as shown here for ASW cultures, also occur in cells grown in supplemented culture medium, F/2.[13]

Although the period can vary greatly, depending on the quality of the constant conditions, the effect of interventions on the phase of the circadian system is investigated under a different experimental protocol. In such experiments, single pulses of light, temperature, or chemicals are given to one culture and the resulting circadian rhythm is compared to that of a control culture. In Fig. 9, glow rhythms, produced in constant red light (30 μE m^{-2} sec^{-1}), are presented as phase plots similar to the one shown in Fig. 5A. In this case, however, the horizontal axis does not represent local time but the length of the circadian cycle, i.e., 26.3 hr expressed in 360° ("modulo τ"). In these plots, the phases of the subsequent peaks fall on a vertical line (see control culture in Fig. 9). At the beginning of the third cycle, one of the cultures received 500 μM nitrate, an important nutrient for marine algae. As a consequence, the subsequent bioluminescent glow peak appears approximately 6 hr (or 82°) later compared to the control, and the subsequent rhythm of the experimental culture shows a steady-state phase delay of approximately 5 hr (69°).

The bioluminescence recording apparatus, described here, as well as its previous versions have mainly been used to investigate the circadian system of *G. polyedra,* but also of other dinoflagellates.[14,15] Among many other questions, the experiments investigated the system's reactions to light[16–19] and temperature,[20–22] to inhibitors of protein syn-

[13] R. R. L. Guillard and J. H. Ryther, *Can. J. Microbiol.* **8,** 229 (1962).

[14] P. Colepicolo, T. Roenneberg, D. Morse, W. R. Taylor, and J. W. Hastings, *J. Phycol.* **29,** 173 (1993).

[15] R. Knaust, T. Urbig, L. M. Li, W. Taylor, and J. W. Hastings, *J. Phycol.* **34,** 167 (1998).

[16] B. M. Sweeney, F. T. Haxo, and J. W. Hastings, *J. Gen. Physiol.* **43,** 285 (1959).

[17] J. W. Hastings, *in* "Photophysiology" (A. C. Giese, ed.), p. 333. Academic Press, New York, 1964.

[18] C. H. Johnson and J. W. Hastings, *J. Biol. Rhythms* **4,** 417 (1989).

[19] T. Roenneberg and J. W. Hastings, *Photochem. Photobiol.* **53,** 525 (1991).

[20] J. W. Hastings and B. M. Sweeney, *Proc. Natl. Acad. Sci. U.S.A.* **43,** 804 (1957).

[21] B. M. Sweeney and J. W. Hastings, *Cold Spring Harbor Symp. Quant. Biol.* **25,** 87 (1960).

[22] D. Njus, L. McMurry, and J. W. Hastings, *J. Comp. Physiol. B* **117,** 335 (1977).

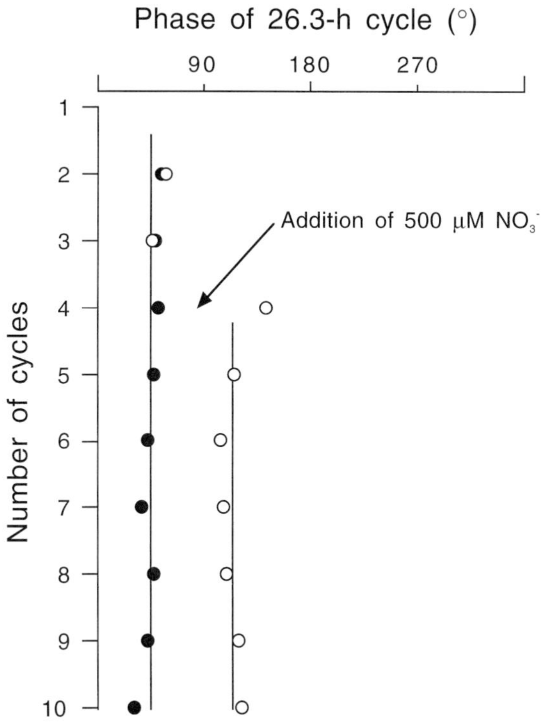

FIG. 9. Bioluminescent glow rhythms of two ASW cultures (see text) represented as phase plots (phases of the respective glow peaks) on the basis of the circadian period (horizontal axis: "modulo τ" = 26.3 hr/360°). The experimental culture (open symbols) received 500 μM NO_3^- during the third cycle, which resulted in a phase delay of approximately 5 hr compared to the control culture (filled symbols).

thesis,[23,24] those affecting protein phosphorylation,[25–27] or inhibitors of photosynthesis[4,28] and nitrogen assimilation.[12] In addition, the involvement of general metabolism[29] or membrane composition[30–33] in the generation of

[23] J. C. Dunlap, W. R. Taylor, and J. W. Hastings, *J. Comp. Physiol.* **138,** 1 (1980).
[24] W. Taylor, J. C. Dunlap, and J. W. Hastings, *J. Exp. Biol.* **97,** 121 (1982).
[25] J. Comolli, W. Taylor, and J. W. Hastings, *J. Biol. Rhythms* **9,** 13 (1994).
[26] J. Comolli, W. Taylor, J. Rehman, and J. W. Hastings, *Plant Physiol.* **111,** 285 (1996).
[27] T. Roenneberg, H. Nakamura, and J. W. Hastings, *Nature* **334,** 432 (1988).
[28] T. Roenneberg, *in* "Evolution of Circadian Clock" (T. Hiroshige and K.-I. Honma, eds.), p. 3. Hokkaido Univ. Press, Sapporo, 1994.
[29] M. McDaniel, F. M. Sulzman, and J. W. Hastings, *Proc. Natl. Acad. Sci. U.S.A.* **71,** 4389 (1974).
[30] D. Njus, F. M. Sulzman, and J. W. Hastings, *Nature* **248,** 116 (1974).
[31] D. Njus, V. D. Gooch, D. Mergenhagen, F. Sulzman, and J. W. Hastings, *Fed. Proc.* **35,** 2353 (1976).

circadian rhythmicity as well as the interactions between the circadian system and the cell cycle[34–37] were subjects of extensive research. Other experiments addressed the question of whether the circadian clocks of individual cells communicate within a population.[38,39] Finally, more recent work has concentrated on the interactions between bioluminescence rhythms and other circadian phenotypes, which will be described in more detail in the following section.

Interactions between the Bioluminescence Rhythm and Other Rhythms in *Gonyaulax*

The circadian system of the marine dinoflagellate *G. polyedra* has been studied extensively since the late 1950s.[39–41] One of its great advantages over other systems is the large number of known circadian output rhythms, ranging from protein synthesis and enzyme activity to behavior.[42] The circadian control of behavior is expressed in the daily vertical migration and in the formation of surface-near aggregations during the subjective day.[43,44] In order to record these behavioral rhythms together with the bioluminescent rhythms, flashing and glow, we attached a digital camera to the photomultiplier cart, recording either from the bottom or the top (depending where the lights are located). Its plane of focus is adjusted just beneath the water surface so that the dense daytime aggregations can be seen as a black spot contrasting the loose nightly carpet of cells (Fig. 10). Due to the extension of the light-tight cart (see Fig. 2), pictures are taken of the vial two positions in advance of the one that is measured for its bioluminescence. The pixel density within a user-controllable analysis rect-

[32] W. Taylor, V. D. Gooch, and J. W. Hastings, *J. Comp. Physiol. B* **130,** 355 (1979).
[33] W. Taylor and J. W. Hastings, *J. Comp. Physiol.* **130,** 359 (1979).
[34] K. Homma and J. W. Hastings, *J. Biol. Rhythms* **3,** 49 (1988).
[35] K. Homma and J. W. Hastings, *Exp. Cell Res.* **182,** 635 (1989).
[36] K. Homma and J. W. Hastings, *J. Cell Sci.* **92,** 303 (1989).
[37] K. Homma, E. Haas, and J. W. Hastings, *Cell Biophys.* **16,** 85 (1990).
[38] H. Broda, D. Brugge, K. Homma, and J. W. Hastings, *Cell Biophys.* **8,** 47 (1985).
[39] T. Roenneberg and M. Mittag, *Semin. Cell Dev. Biol.* **7,** 753 (1996).
[40] J. W. Hastings, B. Rusak, and Z. Boulos, *in* "Neural and Integrative Animal Physiology" (C. L. Prosser, ed.), p. 435. Wiley-Liss, Cambridge, 1991.
[41] T. Roenneberg and J. Rehman, *in* "Microbial Responses to Light and Time" (M. X. Chaddick, S. Baumberg, D. A. Hodgson, and M. K. Phillips-Jones, eds.), p. 237. University Press, Cambridge, 1998.
[42] T. Roenneberg, M. Merrow, and B. Eisensamer, *Zoology* **100,** 273 (1998).
[43] T. Roenneberg, G. N. Colfax, and J. W. Hastings, *J. Biol. Rhythms* **4,** 201 (1989).
[44] T. Roenneberg and J. W. Hastings, *in* "Oscillations and Morphogenesis" (L. Rensing, ed.), p. 399. Dekker, New York, 1993.

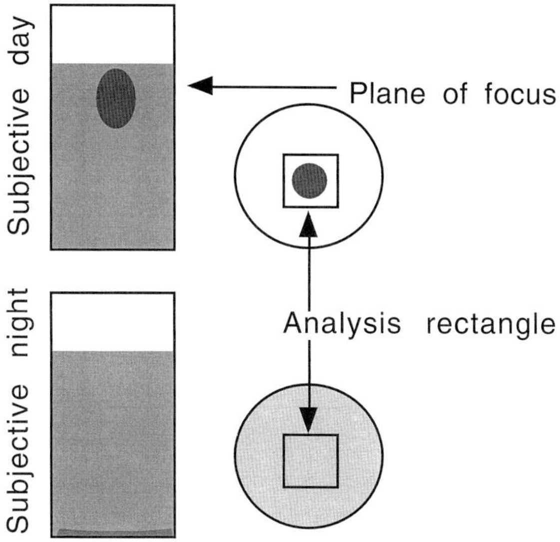

FIG. 10. Recording the circadian rhythms of vertical migration and aggregation behavior of *Gonyaulax* in vials of the bioluminescence recording apparatus. The digitized picture of a video camera looking through the openings in the aluminum bar into the vials is analyzed by the computer program. The pixel density within a controllable analysis rectangle correlates with the density of the surface-near aggregation of the cells during the subjective night. The nightly carpet appears less dense because the cells spread out on the bottom of the vial in the subjective night and the camera's plane of focus is adjusted below the water surface.

angle, covering the center of the vial, is calculated in the digitized picture (using a video card for Macintosh by Scion, Frederick, MD) and is stored together with the bioluminescence measurement of the same vial.

Surprisingly, these simultaneous recordings showed that the behavioral rhythm can be desynchronized from the bioluminescence rhythms, indicating that these rhythms are controlled by two separate circadian oscillators (A and B oscillator).[45] Further experiments revealed that the bioluminescence rhythms and the behavioral rhythm respond differently to light[46] and nutrients.[12,42]

The behavioral rhythm consists of two separate components, the vertical migration and the horizontal movement of the daytime aggregations along the surface selecting for different light intensities over the course of a circadian cycle. Detailed experiments in petri dishes show that it is the

[45] T. Roenneberg and D. Morse, *Nature* **362,** 362 (1993).
[46] D. Morse, J. W. Hastings, and T. Roenneberg, *J. Biol. Rhythms* **9,** 263 (1994).

horizontal component that is controlled by the A oscillator, whereas the vertical migration is controlled together with the bioluminescence rhythms by the B oscillator (Eisensamer and Roenneberg, unpublished results). An example of this "internal desynchronization" is shown in Fig. 11.

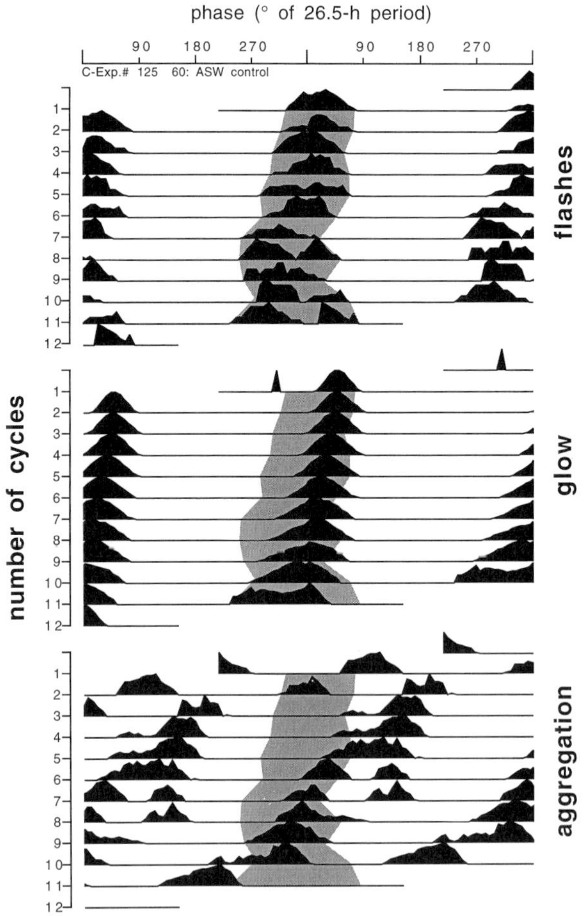

FIG. 11. Simultaneous recording of the bioluminescent rhythms (flash and glow) and the aggregation rhythm of a culture kept in constant red light. The horizontal axis of this double plot is adjusted to the circadian period of the flashing rhythm. Note that the glow rhythm (center) maintains a constant phase relationship with the flashing rhythm (top) similar to the one shown in Figs. 4 and 6. In contrast, the aggregation rhythm (bottom) displays a much shorter period and therefore "moves" through the bioluminescence rhythms over the course of the experiment. The flashing rhythm at the top of the graph is traced by a shaded area, which is repeated in the two other graphs to facilitate comparisons.

Outlook

The system described here was designed originally for circadian experiments using bioluminescence of the marine alga *G. polyedra* as the output rhythm. However, both the recording system and the evaluation program are useful for a variety of different experiments using various organisms. The recording system can be modified easily to record the expression of light produced by reporter gene constructs (lux genes) in microorganisms, plants, and animals (which are not only used to probe circadian outputs). The advantage of our system compared to other automated photon counters lies in the wide ranges of volumes and organisms that can be placed in the vials. These range from microorganisms plated on agar covering the vials' bottom to whole organisms. The addition of video recordings would permit looking at two rhythms at the same time. Transformed flies, for example, can be recorded for their rhythmic expression of a luciferase reporter gene, while their activity can be monitored simultaneously (our image analysis allows recording of the position of an object as well as its relative changes over time). The possibility of applying different filters for each vial allows experiments under light of different intensity as well as spectral composition. The temperature control of the aluminum bar permits experiments probing circadian rhythms or the kinetics of the expression of lux genes at different temperatures.

The evaluation program (CHRONO II) provides an excellent platform for the analysis of circadian data, not only for those accumulated in our recording setup. A variety of different import modules allow the analysis of any given set of time series (presently the program can handle up to 120 channels, including a maximum of three different variables per channel, e.g., glow, flashing, and aggregation, or activity, drinking, and feeding). CHRONO II is currently used for the circadian analysis of circadian rhythms in fungi, higher plants, frogs, fish, rats, and hamsters.

Acknowledgments

The article is dedicated to Woody Hastings, whose pioneering spirit has stimulated so many important developments in the biology of bioluminescence and circadian systems. Work was supported by the Deutsche Forschungsgemeinschaft, the Friedrich-Baur-Stiftung, the Meyer-Struckmann-Stiftung, and the NIMH.

[8] Chemiluminescence Imaging Systems for the Analysis of Macrosamples: Microtiter Format, Blot Membrane, and Whole Organs

By ALDO RODA, PATRIZIA PASINI, MONICA MUSIANI, and MARIO BARALDINI

Introduction

Chemiluminescence (CL) is a versatile, ultrasensitive analytical tool with a wide range of applications in fields such as biotechnology, pharmacology, molecular biology, and clinical chemistry.[1,2]

By coupling enzymatic reactions to light-emitting systems, CL can be used to detect enzyme activity or substrates or to study the effect of enzyme inhibitors by the resultant modifications of the light emission kinetic profile.[3–5] The involvement of radical species in the mechanisms of CL reactions such as luminol/peroxidase/H_2O_2 or luminol/FeEDTA/xanthine/xanthine oxidase makes it possible to determine the antioxidant capacity of biologically active compounds in simple aqueous solution or in biological fluids such as serum.[6,7] Chemiluminescence also represents a sensitive detection system for filter membrane biospecific reactions such as Southern, Northern, or Western blot tests and dot blot hybridization reactions that avoids the use of radioisotopes and long exposure times.[8,9] Finally, ultraweak and enhanced CL have been widely employed to study oxygen-free radical (OFR) formation in different biological systems, including whole organs, tissue homogenates, isolated cells, and subcellular organelles, made possible

[1] L. J. Kricka, P. E. Stanley, G. H. G. Thorpe, and T. P. Whitehead, "Analytical Applications of Bioluminescence and Chemiluminescence." Academic Press, London/New York, 1984.

[2] A. K. Campbell, "Chemiluminescence: Principles and Applications in Biology and Medicine." Ellis Horwood, Chichester, 1988.

[3] M. DeLuca and W. McElroy, *Methods Enzymol.* **133,** 331 (1986).

[4] D. C. Vellom and L. J. Kricka, *Methods Enzymol.* **133,** 229 (1986).

[5] A. Roda, P. Rauch, E. Ferri, S. Girotti, S. Ghini, G. Carrea, and R. Bovara, *Anal. Chim. Acta* **294,** 35 (1994).

[6] T. P. Whitehead, G. H. G. Thorpe, and S. R. J. Maxwell, *Anal. Chim. Acta* **266,** 265 (1992).

[7] C. Pascual and C. Romay, *J. Biolumin. Chemilumin.* **7,** 123 (1992).

[8] S. Beck and H. Köster, *Anal. Chem.* **62,** 2258 (1990).

[9] M. Musiani, M. Zerbini, M. Gibellini, D. Gentilomi, S. Girotti, and E. Ferri, *Anal. Biochem.* **194,** 394 (1991).

0076-6879/00 $30.00

by a weak spontaneous photon emission associated with OFR-related oxidative processes.[10]

The light output generated by biospecific reactions is usually detected by photomultiplier- or photodiode-based luminometers or by instant photographic film. Luminographs, recently developed low-light imaging devices based on Vidicon or CCD video cameras, can be utilized to analyze macrosamples such as gel, membrane, microtiter plates, or petri dishes or, in conjunction with optical microscopy, microsamples such as tissue sections, single cells, or microchip devices.[11–14]

These imaging systems offer some advantages over conventional instrumentation. For example, when a microtiter format is used for enzymatic or antioxidant activity studies, a video camera-based imaging system allows the simultaneous measurement of 96 to 384 samples, whereas microtiter plate luminometers read microtiter wells one by one or strip by strip. This feature and the availability of CL reagents that reach steady-state light emission quickly (i.e., in a few minutes) make CL imaging particularly suitable for the development of high throughput screening (HTS) assays. As far as blot applications are concerned, the main advantage of CL imaging with respect to photographic film is the possibility of performing direct semiquantitative analysis. For the study of oxygen-free radical formation in whole organs, imaging devices allow not only quantification of the emitted photons, but also visualization of the spatial distribution of the CL signal corresponding to radical production in a specific area of the organ.

This article describes some systems for CL imaging of macrosamples developed in our laboratory, lists some applications, and critically evaluates their analytical performance in comparison with other systems.

Instrumentation

The basic instrumentation for CL imaging includes an ultrasensitive video camera, an optical system, and appropriate software for image analysis. The video camera has to be designed specifically to pick up the static CL signal and must be characterized by very low instrumental noise. The

[10] E. Cadenas, A. Boveris, and B. Chance, *in* "Free Radicals in Biology" (W. A. Pryor, ed.), p. 211. Academic Press, New York, 1984.

[11] Y. Hiraoka, J. W. Sedat, and D. A. Agard, *Science* **238,** 36 (1987).

[12] R. A. Wick, *BioTechniques* **7,** 262 (1989).

[13] A. Roda, P. Pasini, M. Musiani, S. Girotti, M. Baraldini, G. Carrea, and A. Suozzi, *Anal. Chem.* **68,** 1073 (1996).

[14] A. Roda, P. Pasini, M. Baraldini, M. Musiani, G. Gentilomi, and C. Robert, *Anal. Biochem.* **257,** 53 (1998).

instrumentation thus differs from that used to evaluate the fast transient excited state utilized in fluorescence analysis.

Detection and analysis of CL signals were performed using a high-performance, low-light level imaging apparatus (Luminograph LB 980, EG&G Berthold, Bad Wilbad, Germany), which permits the measurement of emitted light at the single-photon level. The video system consists of a 1-in. Saticon high dynamic range pick-up tube (a Vidicon-type tube with Se-As-Tl light target photoconductor) linked to an image intensifier, by high transmission lenses, and also to a video amplifier. This system is connected to a PC for quantitative image analysis, and a sample dark box is provided to prevent contact with external light. The system operates in the following consecutive steps: (i) the illuminated sample images are recorded; (ii) the luminescent signal is measured with optimized photon accumulation lasting 1–5 min, with 2-sec interval integration; and (iii) after computer elaboration of the luminescent signal with pseudocolors corresponding to the light intensity, an overlay of the sample photograph with the luminescent signal allows visualization of the spatial distribution of the target analytes. Light emission from the samples is quantified by delineating a given area and counting the number of photon fluxes from within it.

Alternatively, when higher detectability is required, a Luminograph LB 981 (EG&G Berthold, Bad Wilbad, Germany) based on a back-illuminated cooled CCD is used; thanks to its high sensitivity, it does not require the addition of an image intensifier. The instrument setup and CL imaging processing are quite similar to those of the LB 980 luminograph.

The analytical signal is usually expressed as relative light units for qualitative analysis or as photons/second/surface unit (pixel or mm^2) for quantitative data.

Materials and Methods

Reagents

Peroxidase (type VI-A, from horseradish, 1100 U/mg), acetylcholinesterase (AChE) (type V-S, from electric eel, 1000–2000 U/mg), choline oxidase (from *Alcaligenes* genus, approximately 10 U/mg), acetylcholine iodide, 5-amino-2,3-dihydro-1,4-phthalazinedone (luminol sodium salt), bis-*N*-methylacridinium nitrate (lucigenin), 9-amino-1,2,3,4-tetrahydroacridine (tacrine hydrochloride), and ascorbic acid are from Sigma (St. Louis, MO). Trolox, a water-soluble analog of α-tocopherol, and 4-iodophenol are from Aldrich (Milwaukee, WI). The ECL-enhanced chemiluminescent reagent (luminol/enhancer/H_2O_2) for horseradish peroxidase (HRP) detection is from Amersham (Amersham, UK).

Principle of the Methods

Microtiter Format Chemiluminescent Assays

Evaluation of Antioxidant Activity. The luminol/enhancer/peroxidase/H_2O_2 reaction produces relatively steady-state light output for at least 5–40 min in the presence of an excess of peroxidase substrate. The addition of antioxidant compounds to the CL cocktail results in an activity- and concentration-dependent reduction in photon emission, followed by light resumption after a certain time period. The kinetic profile of the light output allows calculation of the relative antioxidant potency of a given molecule with respect to reference substances.

Evaluation of Acetylcholinesterase Inhibitors. The CL detection of AChE is based on coupled enzymatic reactions that involve choline oxidase and horseradish peroxidase as the indicator enzymes and uses luminol/enhancer to detect the H_2O_2 formed, thus leading to stable light emission. The addition of AChE inhibitors to the CL cocktail reduces the light emission as a function of inhibitor activity and concentration. AChE inhibitors include drugs commonly used in the treatment of Alzheimer's disease, such as tacrine, and pesticides like Aldicarb and Paraoxon.

Dot Blot Hybridization for the Detection of Human B19 Parvovirus DNA

Nucleic acid hybridization techniques are able to detect viral genomes directly in clinical samples, allowing rapid and sensitive diagnosis of viral infections, especially for viruses that do not grow in cell cultures or have a long replication cycle. In dot blot hybridization reactions the specimens are dotted on membrane and are hybridized with a specific gene probe labeled with a hapten and then the hybrid is detected using an enzyme-conjugated antihapten antibody and CL substrate such as peroxidase/luminol/H_2O_2 or alkaline phosphatase/dioxetane.

Chemiluminescent Imaging of Oxygen-Free Radical Formation in Isolated and Perfused Rat Liver

Oxygen-free radicals are known to be involved in the ischemia–reperfusion injury of organs.[15,16] OFR-related oxidative processes are char-

[15] J. M. McCord, *N. Engl. J. Med.* **312,** 159 (1985).
[16] H. Jaeschke, *Chem. Biol. Interact.* **79,** 115 (1991).

acterized by a weak spontaneous photon emission. Chemical light enhancers are employed to increase luminescent output by interaction with OFR. In particular, the acridinium salt lucigenin has been shown to react specifically with superoxide radical $\cdot O_2^-$ to form *N*-methylacridone in the excited state, which reverts to its ground state with the emission of light.[17]

Procedures

Microtiter Format Chemiluminescent Assays

Evaluation of Antioxidant Activity. Prepare aqueous serial dilutions (two to three decades) of the compounds to be examined for antioxidant activity and of ascorbic acid and Trolox as positive controls. In this system, ascorbic acid and Trolox are generally used at final concentrations ranging from 0.3 to 30 μM. Prepare a CL cocktail by adding 100 μl of a 100-ng/ml HRP solution in 0.1 *M* Tris buffer, pH 8.6, to 5 ml of ECL-enhanced chemiluminescent reagent containing luminol/H_2O_2/enhancer. Add 10 μl of antioxidant solution and 20 μl of CL cocktail to each well of a black 384-well microtiter plate (Corning Costar, Acton, MA). Acquire the CL signal for 1 min and monitor the reaction kinetics at 2-min intervals for 20–30 min.

Evaluation of Acetylcholinesterase Inhibitors. Prepare aqueous serial dilutions (two to three decades) of the compounds to be examined for AChE inhibition activity and of tacrine hydrochloride as a positive control. In this system, tacrine chlorohydrate is generally used at final concentrations ranging from 0.1 to 10 n*M*. Prepare a CL cocktail with the following: 0.1 m*M* acetylcholine iodide, 0.1 U/ml AChE, 0.3 U/ml choline oxidase, 6 μM luminol, 2 μM 4-iodophenol, and 0.22 U/ml HRP in 0.5 *M* potassium phosphate buffer, pH 8.0. Add 10 μl of inhibitor solution and 20 μl of CL cocktail to each well of a black 384-well microtiter plate (Corning Costar). Acquire the CL signal for 1 min and monitor the reaction kinetics at 2-min intervals for 20–30 min.

Dot Blot Hybridization for the Detection of Human B19 Parvovirus DNA

Specimen Preparation. Add 5-μl aliquots of serum (with B19 parvovirus present during active infection) to 195 μl of distilled water and filter the mixture by the Bio Dot apparatus on a nylon membrane (Amersham Hybond N, Amersham, UK) equilibrated in distilled water. Air dry and treat

[17] S. J. Rembish and M. A. Trush, *Free Radic. Biol. Med.* **17,** 117 (1994).

the membrane with UV for 5 min. Alkali denature the specimens on the membrane by 0.1 *M* NaOH–1 *M* NaCl and neutralize by 0.1 *M* Tris–HCl, pH 7.4. Soak membranes in 50 m*M* Tris–HCl, pH 7.6, 5 m*M* EDTA containing 1 mg/ml pronase and incubate with gentle shaking at 37° for 1 hr. After incubation, remove the pronase by three 5-min rinses in 2× SSC (1× SSC is 0.15 *M* NaCl, 0.015 *M* sodium citrate, pH 7).

As a control of sensitivity, prepare serial dilutions (100-0.001 pg) of B19 parvovirus DNA in 200 μl of distilled water, filter on nylon membrane, and then treat as described earlier.

Hybridization Reaction. Seal the nylon membrane with dotted specimens in a polypropylene bag with 200 $\mu l/cm^2$ of prehybridization mixture containing 4× SET (20× SET is 3 *M* NaCl, 0.4 *M* Tris–HCl, pH 7.8, 20 m*M* EDTA), 5× Denhardt solution (50× Denhardt is 1% Ficoll, 1% polyvinylpyrrolidone, 1% bovine serum albumin in distilled water), 100 μg/ml of denaturated calf thymus DNA, and 0.5% sodium dodecyl sulfate (SDS). Incubate the membranes at 65° for 1 hr. Remove the prehybridization mixture and replace it with 100 $\mu l/cm^2$ of hybridization mixture containing 4× SET, 5× Denhardt solution, 50% formamide, 0.5% SDS, 100 μg/ml denaturated calf thymus DNA, and 50 ng/ml of denaturated parvovirus B19 DNA probe labeled with digoxigenin. Incubate the membrane in a shaking water bath at 42° for 16–18 hr. After hybridization, wash the membrane in (i) 2× SSC, 0.5% SDS at room temperature for 5 min, (ii) 2× SSC, 0.5% SDS at room temperature for 15 min, (iii) 0.1× SSC, 0.1% SDS at 65° for 1 hr, and (iv) 0.1× SSC, 0.1% SDS at room temperature for 5 min.

Chemiluminescent Immunoenzymatic Detection. Wash briefly in 100 m*M* Tris–HCl buffer, pH 7.5, with 150 m*M* NaCl and then apply the blocking reagent (Boehringer, Mannheim, Germany) to the nylon membrane at room temperature for 30 min. Incubate the membrane at room temperature for 30 min with sheep antidigoxigenin Fab fragments conjugated to alkaline phosphatase (Boehringer, Mannheim, Germany) diluted 1:5000. Equilibrate it for 2 min with equilibration buffer (100 m*M* Tris–HCl, 100 m*M* NaCl, 50 m*M* $MgCl_2$, pH 9.5). Seal the dried nylon membrane in a polypropylene bag with 100 $\mu l/cm^2$ of a freshly prepared solution of Lumi-Phos Plus (Lumigen, Southfield, MI) chemiluminescent substrate. After a 45-min incubation, acquire the luminescent signal from the hybrid formation for 1 min.

Chemiluminescent Imaging of Oxygen-Free Radical Formation in Isolated and Perfused Rat Liver

Isolate livers from anesthetized Wistar male fed rats weighing 200–250 g. Cannulate the portal vein with an 18-gauge catheter, wash the organ with

Ringer's lactate solution, and expose to 1 hr ischemia at 37°. Place the liver in the luminograph chamber and perfuse for 1 hr at 37° through the portal catheter with oxygenated Krebs Henseleit buffer, containing 10 μM lucigenin. Drive the perfusion medium (maintained at 37°) by a Model 312 Minipuls 3 peristaltic pump (Gilson, Villiers le Bel, France) at a flow rate of 18 ml/min. Monitor the light emission from the organ surface for 50–60 min by acquiring the chemiluminescent signal for 1 min at 5-min intervals and record live images at the beginning and at the end of the experiment. Acquire chemiluminescent signals during ischemia and also during the reperfusion of nonischemic livers as negative controls.

Results and Discussion

Microtiter Format Chemiluminescent Assays

Evaluation of Antioxidant Activity. The light output from the luminol/enhancer/peroxidase/H_2O_2 reaction reached a maximum in a few minutes and maintained a steady state for at least 20–30 min. Figure 1a shows a 384-well microtiter plate used for determining activities of antioxidant compounds and AChE inhibitors at different concentrations. The addition of antioxidant compounds to the corresponding CL cocktail resulted in a concentration-dependent reduction in light output, as shown in Fig. 1b for ascorbic acid. After initial light suppression, photon emission resumed, starting at lower concentrations, then increasing to values similar to those of the control wells, followed by a subsequent decrease in light output in all wells. This pattern is consistent with CL reaction kinetics, indicating that the antioxidant activity should be evaluated during steady-state emission.

Evaluation of Acetylcholinesterase Inhibitors. The detection of AChE with a microtiter format was optimized to permit simultaneous analysis of the 384 wells of a microtiter plate with a luminograph device that uses a cooled ultrasensitive CCD camera. Light emission reached a steady state after 10–15 min and was stable for at least 10 min. Figure 1a shows a 384-well microtiter plate used for determining activities of AChE inhibitors and antioxidant compounds at different concentrations. The addition of AChE inhibitors to the corresponding CL cocktail resulted in a concentration-dependent reduction in light output, as shown in Fig. 1c for tacrine. Photon emission did not resume after this reduction and it never reached values similar to those of the control wells, even though tacrine is a reversible AChE inhibitor. The relatively rapid decrease in light output observed in all wells indicates that inhibition activity should be evaluated during steady-state emission.

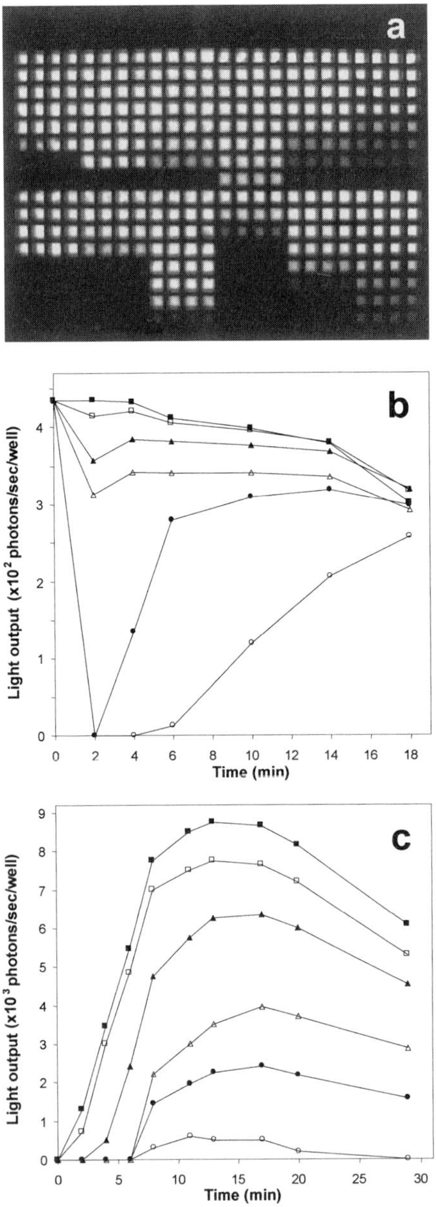

FIG. 1. Chemiluminescent signal detected at 15 min in a 384-well microtiter plate used for determining activities of antioxidant compounds and AChE inhibitors at different concentrations (a). Kinetic profile of the luminol/enhancer/peroxidase/H_2O_2 reaction and concentration-dependent reduction in light output due to the addition of ascorbic acid: ■, control; □, 0.3; ▲, 1.0; △, 3.0; ●, 10; and ○, 30 μM (b). Kinetic profile of the CL reaction for the detection of AChE and concentration-dependent reduction in light output due to the addition of tacrine: ■, control; □, 0.1; ▲, 0.25; △, 0.5; ●, 1.0; and ○, 10 nM (c).

For both of these microtiter format assays, with the luminograph software, it was possible to calculate the integral of the light emitted by each well or to directly visualize the dose-dependent inhibitory effect of the drugs. The methods fulfill the requirements of precision as the coefficients of variation calculated over different sets of experiments never exceeded 5%. When compounds poorly soluble in water have to be analyzed, a mixed solvent containing not more than 10–20% of dimethyl sulfoxide (DMSO) or dimethylformamide (DMF) should be used, and determination of the drug antioxidant activity or enzyme inhibitory effect must take into consideration the light-quenching effects of DMSO or DMF.

Chemiluminescence imaging has proved to be a suitable tool for the development of high throughput screening assays of drugs. Using a CCD camera, 384-well black polystyrene microtiter plates, and CL reagents that quickly (in a few minutes) reach a steady state of light output, an automated system can be set up easily to screen more than 5000 samples per hour. The high detectability of CL labels makes the technique very sensitive, thus allowing analysis of small sample amounts and very diluted specimens. For quantitative measurements, optical distortion and geometry must be considered; thus, image corrections, which can be included easily in the software for signal processing, are required.

Dot Blot Hybridization for the Detection of Human B19 Parvovirus DNA

It has been demonstrated that chemiluminescent enzyme substrates are the most sensitive tools available for detecting enzyme conjugates in a wide range of enzyme-based applications.[18] In order to assess the sensitivity of the developed assay using the imaging luminograph and Lumi-Phos Plus chemiluminescent substrate for alkaline phosphatase, different dilutions of B19 parvovirus DNA (ranging from 100 pg to 1 fg) were analyzed, and it was found that 20-10 fg of homologous DNA could be detected at end point dilution (Fig. 2a). This detection limit is similar to that achieved with photographic film, but lower than that obtained with colorimetric detection that has a detection limit of about 100 fg (Fig. 2b). The advantage of luminographic over photographic detection is that the amount of photons emitted by each dot can be quantified simultaneously, thus rendering the data analysis easy and rapid. Moreover, with luminographic detection, chemiluminescent reactions can be recorded permanently, as all sample images are stored in a computer; they can also be printed or sent to other laboratories for evaluation using floppy disks or computer network.

[18] L. J. Kricka, *in* "Nonisotopic Probing, Blotting, and Sequencing" (L. J. Kricka, ed.), p. 3. Academic Press, New York, 1995.

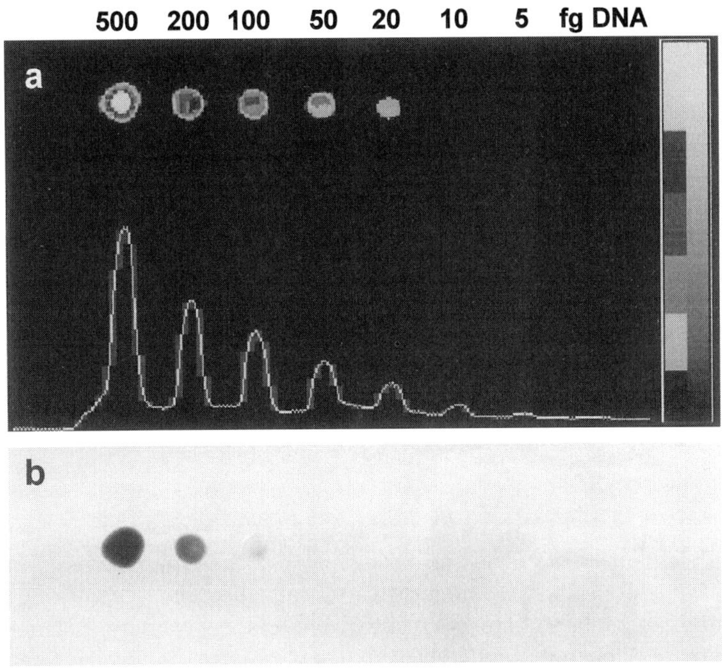

FIG. 2. Dot blot hybridization for the detection of serial dilutions (500-5 fg) of human B19 parvovirus DNA using chemiluminescent (a) or colorimetric (b) substrates. A bidimensional profile of light output showing a linear decrease in photon emission is reported in a.

The specificity of the chemiluminescent assay was ascertained on the basis of the following experiments: (a) chemiluminescent signal was detected when B19 parvovirus-positive reference sera were hybridized with the labeled DNA probe; (b) no chemiluminescent signal was detected when B19 parvovirus-negative reference sera were hybridized with the labeled DNA probe; (c) no chemiluminescent signal was observed when B19 parvovirus-positive reference sera were hybridized with the unlabeled DNA probe and treated with antidigoxigenin Fab fragments conjugated with alkaline phosphatase, followed by chemiluminescent detection; and (d) no chemiluminescent signal was shown after the hybridization of B19 parvovirus-positive reference sera with digoxigenin-labeled probe when incubation with antidigoxigenin Fab fragments was either omitted or replaced by incubation with sheep nonimmune serum.

The chemiluminescent dot blot hybridization system with luminographic detection can be a useful tool for the detection, quantitation, and study of specific viral genetic sequences inside clinical specimens and, thus, for sensitive, specific, and rapid diagnosis of viral infections.

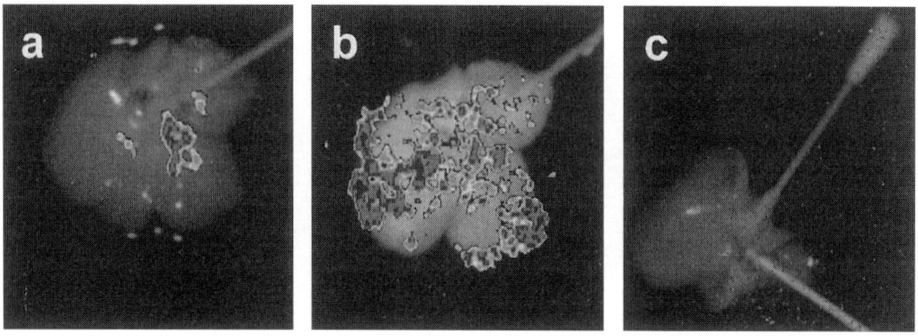

FIG. 3. Chemiluminescent imaging of oxygen-free radical formation in isolated and perfused rat liver. Spatial distribution of the luminescent signal in ischemic liver at 5 (a) and 40 (b) min after reperfusion. Luminescent emission is not detected in liver during ischemia (c).

Chemiluminescent Imaging of Oxygen-Free Radical Formation in Isolated and Perfused Rat Liver

The aim of our study was to develop a chemiluminescent method for *real time* quantitative detection and localization of oxygen-free radical formation in isolated and perfused rat livers exposed to ischemia–reperfusion.[19] A chemiluminescent signal was observed a few minutes after reperfusion. Superimposition of the chemiluminescent and live images acquired successively during reperfusion allowed localization of the luminescent emission, showing that it started around the hepatic vascular pedicle (Fig. 3a) and then spread progressively over the whole organ surface (Fig. 3b). Light output kinetics were evaluated by measuring the photon flux/second/organ surface at consecutive 5-min intervals: the signal increased, reaching the maximum value after 15–20 min, and then it decayed slowly, disappearing after 50–60 min. The supposition that the luminescent peak could correspond to a burst of superoxide radical production is supported by the following findings. The addition of 30 U/ml superoxide dismutase, a specific scavenger of $\cdot O_2^-$, to the perfusion medium caused a rapid decrease in light output, showing that the observed signal was truly dependent on superoxide radical formation. On the contrary, luminescent emission was not observed during ischemia, when the O_2 required for oxygen radical formation was, of course, lacking (Fig. 3c). Furthermore, light production was not detected during the reperfusion of nonischemic livers, probably because the postulated conversion of xanthine dehydrogenase to its oxygen radical-producing form, xanthine oxidase, during hypoxia did

[19] A. Gasbarrini, P. Pasini, B. Nardo, S. De Notariis, M. Simoncini, A. Cavallari, E. Roda, M. Bernardi, and A. Roda, *Free Radic. Biol. Med.* **24,** 211 (1998).

not occur.[20] Although the light emission may well correspond to superoxide radical formation on the liver surface, in previous studies it was postulated that it reflects metabolism of the whole organ and correlates with more global parameters such as lipid peroxidation.[10]

We have shown the possibility of assessing the rate and spatial distribution of oxygen-free radical formation on the surface of intact organs *in real time*. This system opens new prospects for the study of the physiopathogenesis of oxidative stress and provides a suitable model for screening substances with antioxidant activity. In addition, it could represent a useful tool for testing the effectiveness of storage conditions and media utilized in the preservation of organs for transplantation.

Concluding Remarks

The applications of CL imaging described earlier demonstrate the high versatility of this system. In the chapter "*In situ* Hybridization and Immunohistochemistry with Enzyme-Triggered Chemiluminescent Probes" by Roda *et al.* in this volume, we report the use of CL imaging coupled with optical microscopy, showing additional applications in immunohistochemistry and *in situ* hybridization.

These results have demonstrated that CL imaging is a valuable bioanalytical tool suitable for a wide range of applications. When used to measure CL reactions occurring in microtiter plates, the main advantage is that it permits simultaneous monitoring of up to 4×384 microtiter wells, thus facilitating kinetic studies and the measurement of time-dependent CL signals. However, the main limitation of CCD-based imaging is the narrow dynamic range with respect to photomultiplier detection (two to three decades lower). In addition, its sensitivity is 5–10 times lower than those of conventional CL microtiter readers. When imaging is used to measure CL reactions detecting blot format biospecific probes, its analytical performance is superior to those of systems such as photographic film or other systems using different principles to detect immunological or genetic reactions such as radioisotopes or color-producing substrates.[21,22] The main advantages are higher detectability as well as the possibility of performing quantitative analysis. CL imaging of whole organs to detect oxygen-free radicals or, in principle, any kind of CL reaction is a promising tool. Because

[20] Y. Amaya, K. I. Yamazaki, M. Sato, K. Noda, and T. Nishino, *J. Biol. Chem.* **265,** 14170 (1990).

[21] I. Bronstein, J. C. Voyta, K. G. Lazzari, O. Murphy, B. Edwards, and L. J. Kricka, *BioTechniques* **8,** 310 (1990).

[22] S. Girotti, M. Musiani, P. Pasini, E. Ferri, G. Gallinella, M. L. Zerbini, A. Roda, G. Gentilomi, and S. Venturoli, *Clin. Chem.* **41,** 1693 (1995).

the analytical information is derived from an intact organ or even an intact live laboratory animal, it offers an enormous advantage over *in vitro* systems using cultured isolated cells. The system must be standardized carefully in terms of sample geometry, and the permeability, diffusion, and partition of the CL substrate in a given organ, cellular, and subcellular compartment have to be considered.

In conclusion, CL imaging is an important tool in the modern bioanalytical laboratory, at times complementary to other systems but, in many cases, unique.

Section III

Luciferases, Luminescence Accessory Proteins, and Substrates

[9] Overexpression of Bacterial Luciferase and Purification from Recombinant Sources

By THOMAS O. BALDWIN, MIRIAM M. ZIEGLER, VICKI A. GREEN, and MICHAEL D. THOMAS

Several methods for the purification of luciferase from bioluminescent marine bacteria have been published.[1,2] All take advantage of two general characteristics of the protein. First, the subunits of bacterial luciferase are rather acidic polypeptides with isoelectric points around pH 4.[3,4] This characteristic causes the enzyme to bind tightly to anion-exchange resins under conditions that cause most proteins to bind weakly or not at all. Second, because the catalytic turnover rate of bacterial luciferase is very slow, high levels of light emission require high levels of the enzyme. Consequently, luciferase is an abundant protein, comprising on the order of 5% of the soluble protein in *Vibrio harveyi*.[5] These two features of the system account for the ease of purification of bacterial luciferases. With the cloning of *lux* genes and overexpression of the enzymes in *Escherichia coli*, the purification process has become even simpler. We have found conditions that allow bacterial luciferase to accumulate in *E. coli* to levels of well over 75% of the soluble protein. Furthermore, the analysis of luciferase subunits and contaminating cytosolic proteins from *V. harveyi* and from *E. coli* by two dimensional gel electrophoresis[6] revealed that the contaminating *E. coli* proteins, on average, are more basic than the contaminating *V. harveyi* proteins from which the luciferase must be resolved in the course of purification. The relative acidity of the luciferase is therefore much greater in a background of contaminating *E. coli* protein than in a background of contaminating *V. harveyi* protein. As a consequence of these differences, it is even easier to purify bacterial luciferase from *E. coli* than from the native host organism. The discussion that follows will describe our approach to purification of the *V. harveyi* luciferase from *E. coli* cells carrying the

[1] J. W. Hastings, T. O. Baldwin, and M. Z. Nicoli, *Methods Enzymol.* **57,** 135 (1978).
[2] T. O. Baldwin, T. F. Holzman, R. B. Holzman, and V. W. Riddle, *Methods Enzymol.* **133,** 98 (1986).
[3] D. H. Cohn, A. J. Mileham, M. I. Simon, K. H. Nealson, S. K. Rausch, D. Bonam, and T. O. Baldwin, *J. Biol. Chem.* **260,** 6139 (1985).
[4] T. C. Johnston, R. B. Thompson, and T. O. Baldwin, *J. Biol. Chem.* **261,** 4805 (1986).
[5] A. Gunsalus-Miguel, E. A. Meighen, M. Z. Nicoli, K. H. Nealson, and J. W. Hastings, *J. Biol. Chem.* **247,** 398 (1972).
[6] J. J. Waddle, T. C. Johnston, and T. O. Baldwin, *Biochemistry* **26,** 4917 (1987).

0076-6879/00 $30.00

plasmid pJHD500.[7,8] We present our preferred methods for cell growth and purification of the enzyme. One of the most important developments since the last published update in luciferase purification methods has been the availability of various fast-flowing anion-exchange chromatographic media. Use of these newer ion-exchange resins allows for much faster separations, and in all purification schemes, time is perhaps the most important consideration in final yield and ultimate quality of the purified enzyme.

Maintenance of Plasmid-Carrying Cells

We have employed a variety of plasmids to overexpress bacterial luciferase and have found pJHD500[7,8] to give the highest yield. All the plasmids that we have used are derived from pBR322 and are maintained in *E. coli* by selection for resistance to ampicillin or the ampicillin analog carbenicillin. *Escherichia coli* cells in stationary-phase liquid cultures or on agar plates have a tendency to lose these plasmids, so it is essential that inocula be taken from freshly grown cells on agar plates and that liquid cultures not be allowed to grow to high cell density during the process of preparation for the inoculation of the large volumes of media from which cells will be harvested for enzyme purification because the antibiotic ampicillin prevents cell division, but it does not kill cells. To grow in the presence of ampicillin, the cells produce the enzyme β-lactamase, which is secreted into the medium where it degrades the antibiotic. Any antibiotic-sensitive cells in the culture, such as those that have lost the recombinant plasmid, will be able to grow in the medium after the antibiotic has been degraded, and because these cells do not carry the additional metabolic load of the plasmid, they will grow faster and potentially overgrow the population of plasmid-carrying cells, thereby greatly reducing the yield of recombinant protein. To minimize the potential for growing cells that do not contain the plasmid, and therefore the desired protein, two precautions are taken. First, cells are streaked from glycerol stocks, described later, onto agar plates and single colonies are picked for inoculation from fresh overnight growth on the solid media. Second, cells are collected by centrifugation from small volumes of media and these cells are used to inoculate larger volumes. This process removes cells needed for inoculation from β-lactamase-containing medium, which would hasten the degradation of the antibiotic in the larger volume of medium, thereby allowing antibiotic-sensitive cells to begin growth sooner. Careful attention to these details has resulted in routine high yield of the enzyme.

[7] J. H. Devine, G. S. Shadel, and T. O. Baldwin, *Proc. Natl. Acad. Sci. U.S.A.* **86,** 5688 (1989).
[8] M. D. Thomas and A. B. van Tilburg. *Methods Enzymol.* **305** [23], 315 (2000) (this volume).

Preparation of Glycerol Stocks

We store glycerol stocks at −80° (25–50% glycerol) for the long-term maintenance of bacterial strains. To prepare tubes for glycerol stocks, add 0.5 ml glycerol to each of a series of screw-top 2-ml microfuge tubes and screw the lids on tightly prior to autoclaving. These tubes may be stored indefinitely at room temperature. Mix 0.5–1.5 ml from a fresh overnight liquid bacterial culture with the glycerol in a tube, reseal the tube, and store it at −80°. There is no need to quick-freeze the cells in liquid nitrogen prior to storage. For maximum viability, the stocks should not be allowed to thaw. Note that stock tubes will thaw more quickly on ice than on the laboratory bench due to the higher thermal conductivity of water. To inoculate an agar plate from a glycerol stock, remove the tube from the freezer, scrape a minute portion of the frozen bacterial suspension from the surface of the frozen suspension with a sterile loop or toothpick, and streak the cells across the surface of the solid growth medium. Return the stock to the −80° freezer expeditiously.

Bacterial Cell Growth Conditions

This protocol is for the overexpression of proteins encoded by genes under regulatory control of the *Vibrio fischeri* autoinducer system in plasmids derived from pJHD500[7,8] in common laboratory strains of *E. coli.* The *V. fischeri* bioluminescence system is a regulon consisting of two divergently transcribed operons.[9–11] One operon, the leftward operon, consists of a single gene, *luxR,* which encodes a transcriptional activator, LuxR, required for expression of the luminescence functions that are encoded on the rightward operon.[9] Expression of *luxR* is under control of a CRP/cAMP-dependent promoter.[12,13] The rightward operon encodes *luxICDABE* genes.[9] The α and β subunits of luciferase are encoded by *luxAB,* and the enzymes required to supply the aldehyde substrate *in vivo* are encoded by *luxCDE.* The *luxI* gene encodes a synthase that produces the autoinducer,[9,14] *N*-(3-oxo) hexanoyl-DL- homoserine lactone,[15] from cytoplasmic

[9] J. Engebrecht, K. Nealson, and M. Silverman, *Cell* **32,** 773 (1983).

[10] J. H. Devine, C. Countryman, and T. O. Baldwin, *Biochemistry* **27,** 837 (1988)

[11] T. O. Baldwin, J. H. Devine, R. C. Heckel, J.-W. Lin, and G. S. Shadel, *J. Biolumin. Chemilumin.* **4,** 326 (1989).

[12] P. V. Dunlap and E. P. Greenberg, *J. Bacteriol.* **164,** 45 (1985).

[13] P. V. Dunlap and E. P. Greenberg, *J. Bacteriol.* **170,** 4040 (1988).

[14] J. Engebrecht and M. Silverman, *Proc. Natl. Acad. Sci. U.S.A.* **81,** 4154 (1984).

[15] A. Eberhard, A. L. Burlingame, C. Eberhard, G. L. Kenyon, K. H. Nealson, and N. J. Oppenheimer, *Biochemistry* **20,** 2444 (1981).

precursors.[16] The LuxR transcriptional activator requires the autoinducer to stimulate transcription of the *lux* genes. The LuxR-binding site is between the two operons.[7,10] Binding of LuxR stimulates transcription of both operons, causing transcriptional positive feedback.[17–19] We have developed an expression vector based on this system that employs the *luxR* gene, the control region between the two operons, and either an either an intact *luxI* gene or a truncated *luxI* gene. With the intact *luxI* gene, most strains of *E. coli* can produce autoinducer, but with the truncated *luxI,* the expression of genes downstream of the LuxR-activated promoter requires exogenous autoinducer. Using this system, cells are allowed to grow to modest density without the addition of autoinducer and then, upon the addition of autoinducer, excellent expression of recombinant protein is obtained. We have used the *V. fischeri lux* gene expression system to overexpress not only the luciferase from *V. harveyi,* but also a number of other proteins.[8]

To attain high yields of recombinant protein, it is necessary to obtain a high yield of cells and to maximize the level of the target protein within those cells. Terrific Broth[20] is used to support enhanced cell densities. The high cell densities attained with this protocol require vigorous aeration of the cultures. Media should be prepared immediately before use as the short autoclave time (20 min for 1.5 liters medium) employed to minimize caramelization may only pasteurize the broth and may allow microbial growth upon extended storage. As cells can enter the stationary phase before they have exhausted the medium nutrients, perhaps due to the accumulation of secondary metabolites, the protocol calls for washing the cells by centrifugation and increasing their number in stages. A second advantage of the washing procedure described here is that it removes the (plasmid-encoded) β-lactamase secreted by the cells to degrade the antibiotic. Inoculation with large volumes of spent medium will introduce large amounts of β-lactamase and hasten degradation of the antibiotic, potentially allowing cells that do not carry the plasmid to overgrow the culture, thereby reducing the yield of the recombinant protein. At higher cell densities, aeration is very important: covering the Fernbachs with foil or cotton plugs or shaking them at slower speeds causes reduced growth rates. Alternation between aerobic and anaerobic growth is likely to reduce expression of the foreign protein and will increase the number of contami-

[16] M. I. Moré, L. D. Finger, J. L. Stryker, C. Fuqua, A. Eberhard, and S. C. Winans, *Science* **272,** 1655 (1996). This paper demonstrated the enzymatic activity of TraI, the LuxI homolog from *Agrobacterium tumifaciens.*

[17] G. S. Shadel and T. O. Baldwin, *J. Bacteriol.* **173,** 568 (1991).

[18] S. H. Choi and E. P. Greenberg, *J. Bacteriol.* **174,** 4064 (1992).

[19] G. S. Shadel and T. O. Baldwin, *J. Biol. Chem.* **267,** 7696 (1992).

[20] K. D. Tartof and C. A. Hobbs, *Bethesda Res. Lab.* **9,** 12 (1987).

nating proteins present in the cell, complicating purification of the protein of interest. An OD_{600} of 15 to 20 is usually reached using this protocol. At an OD of 18, there is normally about 200 g cell paste for 9 liters. This amount of cell paste (200 g) yields approximately 30 g of total protein, or roughly 15 g of luciferase or other recombinant protein for which expression with this system has been optimized.

Materials for Cell Growth

Materials required to purify luciferase are readily available in virtually any biochemistry laboratory. The following list of materials will allow growth of the recombinant cells and purification of the enzyme by any of several methods. There are several suppliers of these materials. We have not used materials from all suppliers, but it is our belief that all supply good products.

Growth Medium Components

1. Glycerol (IMScience)
2. Peptone, tryptone, or casein hydrolysate (GIBCO/BRL; use interchangeably)
3. Yeast extract (GIBCO/BRL)
4. Mono- and dibasic potassium phosphate (ICN)
5. Carbenicillin, disodium salt (Sigma)
6. Agar (GIBCO/BRL)
7. *Vibrio fischeri* autoinducer [Sigma; *N*-(3-oxo)hexanoyl-DL-homoserine lactone]

Equipment and Glassware

1. Autoclave for sterilization of glassware, medium, and medium components.
2. Static incubator (37°) for growth of cells on agar plates.
3. Air bath shaker for 37° growth (New Brunswick Model G-25 or equivalent) with clamps for 250-ml flasks.
4. Air bath shaker for subambient (18–25°) growth (New Brunswick PsycroTherm or equivalent) with large clamps for Fernbach flasks.
5. Refrigerated clinical centrifuge (Beckman Model J2-21 or equivalent) with either a JA10 or a JA14 rotor (for harvesting cells from large volumes of growth medium) and a JA20 rotor.
6. Oak Ridge centrifuge tubes (50 ml) to fit the JA20 rotor and centrifuge bottles to fit either a JA10 or a JA14 rotor.
7. A top-loading balance (Sauter Model RL4 or equivalent).
8. French press for cell disruption (SLM/Aminco).

9. Spectrophotometer for measuring the absorbance of cultures at 600 nm (Milton Roy Spectronic 601 or equivalent).

Cell Growth Procedure

Note that growth temperature is critically important. The normal biological habitat of marine bacteria is at temperatures well below the 37° commonly used to grow laboratory strains of *E. coli.* While bacterial luciferase is stable at 37°, the enzyme does not fold properly and assemble efficiently at temperatures above ca. 30°, and yield is compromised above ca. 27°. At 25°, the growth rate of *E. coli* is reduced, but the yield of the correctly folded protein is enhanced greatly. For some luciferases, and some mutant forms of the *V. harveyi* enzyme, growth at 18° is required for a high yield of enzyme.

1. Day 1. Inoculate a fresh culture from a −80° stock of the plasmid-containing strain on solid medium containing the selective antibiotic at the appropriate concentration, e.g., ampicillin or carbenicillin at 50 μg/ml. The seed culture should be used within 48 hr.
2. Day 2. Make up 3 liters of 3× Terrific Broth[20] as follows (note that the final volume will be closer to 3.2 liters): 50 g glycerol, 114 g Bacto-tryptone (peptone or enzymatic casein hydrolysate works well also), 228 g Bacto or other yeast extract, and 3 liters H_2O.

   Dilute 150 ml of the 3× Terrific Broth with 300 ml H_2O; dispense 45 ml into each of ten 250-ml culture flasks. Dispense 500 ml of the 3× Terrific Broth into each of six Fernbach flasks. Add 850 ml of H_2O to each Fernbach.

   Prepare 1 liter of potassium phosphate buffer solution in a 2-liter flask: 23.1 g KH_2PO_4, 125 g K_2HPO_4, and 950 ml H_2O.

   Cover the 250-ml culture flasks, the Fernbach flasks and the potassium phosphate buffer flask with foil, autoclave for 20 min on the liquid cycle and remove the medium promptly to avoid caramelization. *Do not autoclave longer.* The medium should be used within 24 hr as it may only be pasteurized.
3. Late on Day 2. Finish preparing the medium in four of the 250-ml culture flasks by adding the appropriate antibiotic and 5 ml potassium phosphate buffer to each. Inoculate each flask with a single colony from the seed culture made in step 1 and allow cells to grow at 37° with the shaker set to 350 rpm.
4. Early on Day 3. Measure and record the OD_{600} of the overnight cultures. Collect cells by centrifugation for 5 min at 5000 rpm in 50-ml centrifuge tubes.

5. Add the selection antibiotic to the remaining potassium phosphate buffer and dispense 5 ml into each of the six remaining 250-ml culture flasks and 150 ml into each Fernbach flask.
6. Resuspend the cell pellets in a small volume of Terrific Broth and divide equally among the 250-ml culture flasks. Allow cells to grow at 37° with high aeration (350 rpm) for 8 hr. Measure and record the OD_{600} of the cultures.
7. Collect the cells in separate (six total) 50-ml tubes as in step 4.
8. Resuspend the cells in each centrifuge tube in a few milliliters of fresh broth and pour all of the cells from one centrifuge tube into each Fernbach.
9. Allow cells to grow for 5 hr with high aeration (350 rpm); the incubation temperature may need to be changed if the protein of interest is not stable or does not fold properly at 37°. For example, cells producing luciferase encoded by *V. harveyi luxAB* must be grown at 25° or lower for optimal expression. Do not cover the flasks tightly while they are incubating. Optimally, the flasks should remain uncovered; however, phage contamination is a potential problem, and some strains produce aromatic secondary products that some may find offensive.
10. After the 5 hr, check the OD_{600}; if it is not at least 4, grow 1–2 hr more. Add 15 μl of 40 m*M* *V. fischeri* autoinducer to each 1.5-liter culture, which will yield a final concentration of 400 n*M* *N*-(3-oxo)hexanoyl-DL-homoserine lactone.
11. Allow the cultures to continue to grow at the appropriate temperature with shaking at 350 rpm for 12 to 24 hr, depending on the growth temperature. It is prudent to perform a trial growth in which samples are withdrawn at various times to determine the window of time during which accumulation of the protein of interest is maximal. Measure and record the OD_{600}, and harvest the cells by centrifugation for 15 min at 5000 rpm in tared 1-liter centrifuge bottles; measure and record the yield of cell paste.

Lysis of Bacterial Cells. Disruption of *E. coli* is significantly more difficult than lysis of marine bacteria, which can be accomplished by osmotic shock.[1] The disruption of *E. coli* may be achieved by sonication or with a French press. It is also possible to disrupt *E. coli* by digestion of the outer cell wall with lysozyme. We do not recommend this method because many commercial sources of lysozyme are contaminated with proteases, and bacterial luciferase is exquisitely sensitive to protease attack.[21,22] Regardless

[21] D. Njus, T. O. Baldwin, and J. W. Hastings, *Anal. Biochem.* **61,** 280 (1974).
[22] T. O. Baldwin, J. W. Hastings, and P. L. Riley, *J. Biol. Chem.* **253,** 5551 (1978).

of the method, disruption may be facilitated by a slow freeze-thaw cycle. Cells collected from the growth medium by centrifugation may be frozen following decantation of the growth medium by placing the centrifuge bottles in a $-20°$ freezer for a few hours or more. The cells may then be thawed on ice prior to lysis. Cells treated in this way are much more fragile and lyse more completely than cells taken directly following harvest from the growth medium.

1. Resuspend the cells in an appropriate lysis buffer at a ratio of 1.25 ml buffer/g cells. Cells grown as described here tend to be somewhat fragile, so to avoid premature cell lysis, the buffer should be supplemented with NaCl to 100 m*M*.
2. Although cells often lyse better if they are subjected to a slow freeze-thaw cycle, we have found that cells grown by the protocol outlined earlier will lyse very efficiently without the freeze-thaw cycle. The efficient disruption of cells produced by this method is accomplished with a French press set to a cell pressure of 16,000 psi.
3. Centrifuge the crude cell lysate for 45 min at 10,000 rpm (JA10 rotor) or at 14,000 rpm (JA14 rotor). For quality control purposes, take clarified lysate samples before proceeding with purification. A clarified lysate sample of 0.5 μl prepared by this procedure should provide a good starting point for the amount of sample to load on a Coomassie blue-stained polyacrylamide gel. Proceed with the purification procedure.

Purification Procedure

This section gives protocols for the preparation of buffers and other materials that can be used for several different purification methods. Luciferase from *V. harveyi* is stable and rather forgiving. However, it is very sensitive to inactivation by proteases, a consequence of a disordered loop within the α subunit.[23–25] Therefore, it is especially prudent when working with bacterial luciferases to maintain sterile conditions to the extent possible. For example, centrifuging samples of enzyme will remove any bacterial cells that may be in the solutions. We also supplement all buffers with 1 m*M* EDTA to retard bacterial growth and with 0.5 m*M* dithiothreitol (DTT) to prevent oxidative damage.

[23] T. F. Holzman and T. O. Baldwin, *Proc. Natl. Acad. Sci. U.S.A.* **77,** 6363 (1980).
[24] A. J. Fisher, F. M. Raushel, T. O. Baldwin, and I. Rayment, *Biochemistry* **34,** 6581 (1995).
[25] A. J. Fisher, T. B. Thompson, J. B. Thoden, T. O. Baldwin, and I. Rayment, *J. Biol. Chem.* **271,** 21956 (1996).

Materials for Purification

Buffers. Bacterial luciferase has a phosphate-binding site[24,26,27] and is stabilized by phosphate-containing buffers. Because of its protective effects, we routinely use phosphate buffers for all steps of purification. We prepare concentrated buffer stocks and store them at 4°. Working buffers are then prepared by dilution of the stock solutions.

Stock Solutions

0.2 *M* EDTA: 14.88 g EDTA (disodium salt, dihydrate) per 200 ml, adjusted to pH 7.0

2 *M* K_2HPO_4: 348.36 g K_2HPO_4 (anhydrous) per 1 liter

2 *M* NaH_2PO_4: 276.02 g $NaH_2PO_4 \cdot H_2O$ per 1 liter

To prepare 1 *M* phosphate buffer, pH 7.0, mix 307 ml 2 *M* K_2HPO_4 stock, 193 ml 2 *M* NaH_2PO_4 stock, and 500 ml H_2O, and adjust the pH to 7.0.

Chromatographic Media. Because of its acidic character, bacterial luciferase is purified most easily using anion-exchange resins. Earlier purification procedures relied primarily on DEAE-Sephadex A50.[1,5] The only disadvantage of the A50 resin relative to the newer anion exchangers is the slower column buffer flow rates that must be employed. Even so, when we wish to purify very large amounts of luciferase, we still use DEAE-Sephadex A50 because of its lower cost. Using a 5 × 40-cm column with DEAE-Sephadex A50, we can isolate multiple grams of the enzyme in a single chromatographic step.

When our needs for enzyme are more modest, in the range of 1 g, we use a 5 × 30-cm column of DEAE-Sepharose Fastflow. This single column yields protein of adequate purity for most experiments, but for crystal growth and certain other uses, we employ a second column (2.6 × 18 cm) of Q-Sepharose.

Purification Protocol

On the basis of Coomassie blue staining of protein in SDS polyacrylamide gels, bacterial luciferase overexpressed in *E. coli* as described here comprises at least 50% of the soluble protein in a clarified crude lysate (Fig. 1, lane 2). As a result of the high initial concentration of luciferase, we have modified the previous[1] purification scheme. We now employ essentially two steps, an ammonium sulfate fractionation and a single anion-exchange column, and the entire process can be accomplished in approximately 4 hr.

[26] E. A. Meighen and R. E. MacKenzie, *Biochemistry* **12,** 1482 (1973).

[27] T. F. Holzman and T. O. Baldwin, *Biochem. Biophys. Res. Commun.* **94,** 1199 (1980).

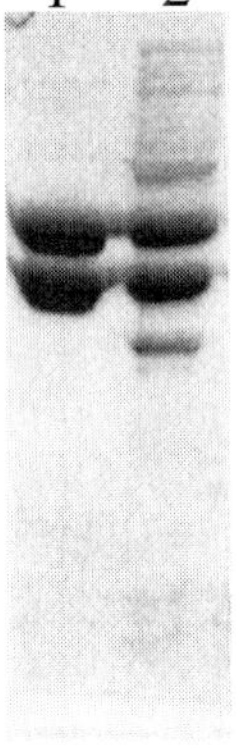

FIG. 1. Purification of *Vibrio harveyi* luciferase. Final product (lane 1) after one column chromatographic step and starting material (lane 2) are shown on a Coomassie blue-stained sodium dodecyl sulfate polyacrylamide gel (12%, w/v) following electrophoresis. Lane 1, pooled product following column chromatography on DEAE-Sepharose Fastflow; lane 2, clarified crude lysate.

Ammonium Sulfate Fractionation and Dialysis. It is well known that the fractional saturation of ammonium sulfate at which a specific protein will precipitate is strongly dependent on the concentration of that protein, in addition to other factors, especially temperature and pH.[28] It is therefore not surprising that the extraordinarily high concentrations of luciferase in lysates of *E. coli* require changes in the ammonium sulfate fractionation protocol. Published methods for the purification of bacterial luciferase call for the collection of protein precipitating between 40 and 75% saturation of ammonium sulfate.[1,5] Using *E. coli* carrying pJHD500, grown as described here, we are now using a fractionation range of 30–55% saturation ammonium sulfate at 0° and pH 7.0, as described later.

To 100 ml of clarified crude lysate on ice, slowly add 16.4 g ammonium sulfate. The lysate should be stirred slowly but continuously during the salt addition to allow the salt to dissolve and for 10 min following the addition of the last of the salt. Remove the precipitated protein by centrifugation for 10 min at 15,000 rpm in a JA20 rotor at 4°. Collect the supernatant in a beaker, place it on ice, and slowly add an additional 13.2 g ammonium sulfate to the solution while stirring slowly, as before. After ca. 10 min of stirring, collect the precipitated protein by centrifugation for 10 min at 15,000 rpm at 4° in a JA20 rotor. Decant the supernatant and dissolve the pellet in a minimal volume of 100 m*M* phosphate, pH 7.0, 0.5 m*M* DTT,

[28] M. Dixon and E. C. Webb, *Adv. Protein Chem.* **16,** 197 (1961).

1 m*M* EDTA. Place the solution in dialysis tubing and dialyze against 1 liter of the same buffer at room temperature for 20 min. Change the buffer twice, giving a total dialysis time of 1 hr. The sample is now ready for centrifugation in preparation for application to the anion-exchange column.

Anion-Exchange Chromatography. Following the final step of dialysis, centrifuge the sample for 10 min at 15,000 rpm. Carefully decant the soluble material from any precipitate, save ca. 100 μl for analysis, and apply the remainder to a column (5 $\times$ 30 cm) of DEAE-Sepharose Fastflow that has been equilibrated in the dialysis buffer. Wash the column with 50 ml of 200 m*M* phosphate, pH 7.0, 0.5 m*M* DTT, 1 m*M* EDTA at 2 ml/min and then elute the column at the same flow rate with a 300-ml linear gradient from 400 to 700 m*M* phosphate, both at pH 7.0 with 0.5 m*M* DTT and 1 m*M* EDTA. Collect 5-ml fractions and determine the absorbance at 280 nm and the bioluminescence activity of each fraction. Pool fractions with constant specific activity. A Coomassie blue-stained gel of luciferase purified by this protocol is shown in Fig. 1 (lane 1).

If the protein from the DEAE-Sepharose Fastflow is not sufficiently pure, we use Q-Sepharose for a final purification step. This column is 2.6 $\times$ 18 cm and is eluted at 2 ml/min. The protein is applied in 100 m*M* phosphate and is eluted with a 300-ml linear gradient from 100 to 400 m*M* phosphate. All buffers are pH 7.0 with 0.5 m*M* DTT and 1 m*M* EDTA.

Alternatively, we have successfully employed molecular sieve chromatography for a final step of purification. The choice of chromatographic method depends on the nature of the contaminating species, which vary depending on the host *E. coli* strain that is used for the overexpression of the enzyme.

Methods for Luciferase Enzyme Assay

The following assay procedures are those used currently in our laboratory and represent modified versions of the assay methods described previously in this series.[1]

Materials for Assay

Luciferase assays are usually performed in buffers containing bovine serum albumin (BSA). The aliphatic aldehyde substrate presents some difficulty because it is not water soluble and must either be solubilized with a detergent or organic solvent, which may cause problems for the luciferase, or it must be dispersed in water as an emulsion by sonication. Because either method leads to an increased rate of autoxidation of the aldehyde to the carboxylic acid, it is necessary to prepare fresh aldehyde solutions at ca. 3- to-4 hr intervals. The FMN substrate is photosensitive, so to prevent photochemical degradation, it is necessary to keep the flavin solutions in darkened containers. We routinely prepare the assay buffer and FMN solutions as concentrated stocks and prepare working solutions by dilution of the stocks.

Assay Buffer. The stock assay buffer is prepared by dissolving 20 g BSA in 1.0 *M* phosphate buffer, pH 7.0, and adjusting the final volume to 500 ml with the buffer, yielding a 4% BSA solution in 1 *M* phosphate, pH 7.0. The stock assay buffer is stored in 5-ml aliquots at −20°. The working assay buffer is prepared by diluting 5 ml of the stock to 100 ml with water to yield 0.2% BSA, 50 m*M* phosphate, pH 7.0.

FMN Solution. The stock FMN solution is prepared by dissolving 239 mg of the monosodium salt in water and adjusting the volume to 100 ml, yielding a 5 m*M* stock solution in water. After checking the actual concentration by visible absorbance at 450 nm ($\varepsilon_{450\,nm} = 12{,}200\ M^{-1}\ cm^{-1}$),[29] the stock is stored frozen at −20° in 1.0-ml aliquots in dark brown snap-cap microfuge tubes. The working FMN solution is prepared by diluting a 1.0-ml aliquot of the stock with 5 ml of 1.0 *M* phosphate, pH 7.0, and adjusting to a final volume of 100 ml with distilled water. If the flavin is to be photoreduced, 1.0 ml of 0.2 *M* EDTA is added and the amount of water is reduced accordingly. The resulting working FMN solution is 50 μ*M* FMN, 2 m*M* EDTA in 50 m*M* phosphate, pH 7.0.

Photoreduction of FMN

The flavin is photoreduced[30,31] by drawing ca. 1 ml of the FMN/EDTA solution into a 1.0-ml tuberculin syringe fitted with a 3-in.-long 19-gauge hypodermic needle. We use a circular fluorescent tube of the type commonly used in kitchens as the source of illumination for the syringes. The fluorescent tube is mounted in a wooden box together with a small fan that circulates air to prevent heat buildup near the light source. The needles are inserted into a block of foam insulation in the bottom of the wooden box so that the syringes are held vertically a few inches from the fluorescent tube. Reduction of the flavin is apparent from the loss of yellow color and requires 2–3 min. We routinely photoreduce flavin in 10–20 syringes at a time and use these for assays within ca. 10–15 min.

Catalytic Reduction of FMN

For catalytic reduction, it is not necessary to add EDTA to the working solution. Catalytic reduction is effected by bubbling hydrogen gas through

[29] L. G. Whitby, *Biochem. J.* **54,** 437 (1953).
[30] Q. H. Gibson and J. W. Hastings, *Biochem. J.* **83,** 368 (1962).
[31] J. W. Hastings and Q. H. Gibson, *J. Biol. Chem.* **238,** 2537 (1963).

the flavin solution in the presence of a small (ca. 5–10 mg on the tip of a microspatula) amount of platinum-covered activated charcoal (Fluka). Reduction is carried out with flavin in a 100-ml dark round-bottom flask fitted with a vaccine bottle stopper. The hydrogen gas is introduced through a 3-in.-long hypodermic needle and is vented through a second needle and tubing into a good chemical fume hood. *Extreme* caution should be exercised in using the hydrogen gas. Our laboratory has completely switched to using photoreduction in preparing the substrate for the bacterial luciferase-catalyzed reaction.

Instrumentation and Enzyme Assay

It is most convenient, although not essential, to monitor purification of the enzyme by assay of the bioluminescence activity using any of a variety of luminometers.[32] If a luminometer is not available, light emission from reactions described later may be observed by eye and/or the protein may be monitored by gel electrophoresis of small aliquots of material from individual steps. Quantitative assays of the enzyme activity require the measurement of the intensity of light emitted from the enzyme-catalyzed reaction. The various formats for the assays reflect the chemical reactivity of the reduced flavin mononucleotide with molecular oxygen in the absence of enzyme, necessitating segregation of these ingredients until immediately prior to initiation of the reaction. The overall reaction[31] is given in Scheme I.

$$FMNH_2 + O_2 + RCHO \xrightarrow{\text{Luciferase}} FMN + RCOOH + H_2O + \text{light}$$

SCHEME I

The enzyme has three substrates, $FMNH_2$, O_2, and the aldehyde, and three products, FMN, the carboxylic acid, and water, in addition to the blue-green light. The order of addition of the substrates can have a strong impact on the apparent quantum efficiency of the reaction catalyzed by the enzyme. The kinetic complexity of the reaction is apparent from the kinetic mechanism shown in Fig. 2.[33–37]

In Fig. 2, X denotes an inhibitor, such as an aliphatic alcohol or acid

[32] P. E. Stanley, *Methods Enzymol,* **305** [6], 96 (2000) (this volume).
[33] H. Abu-Soud, L. S. Mullins, T. O. Baldwin, and F. M. Raushel, *Biochemistry* **31,** 3807 (1992).
[34] H. Abu-Soud, A. C. Clark, W. A. Francisco, T. O. Baldwin, and F. M. Raushel, *J. Biol. Chem.* **268,** 7699 (1993).
[35] W. A. Francisco, H. M. Abu-Soud, T. O. Baldwin, and F. M. Raushel, *J. Biol. Chem.* **268,** 24734 (1993).
[36] W. A. Francisco, H. M. Abu-Soud, R. Topgi, T. O. Baldwin, and F. M. Raushel, *J. Biol. Chem.* **271,** 104 (1996).
[37] W. A. Francisco, H. M. Abu-Soud, A. J. DelMonte, D. A. Singleton, T. O. Baldwin, and F. M. Raushel, *Biochemistry* **37,** 2596 (1998).

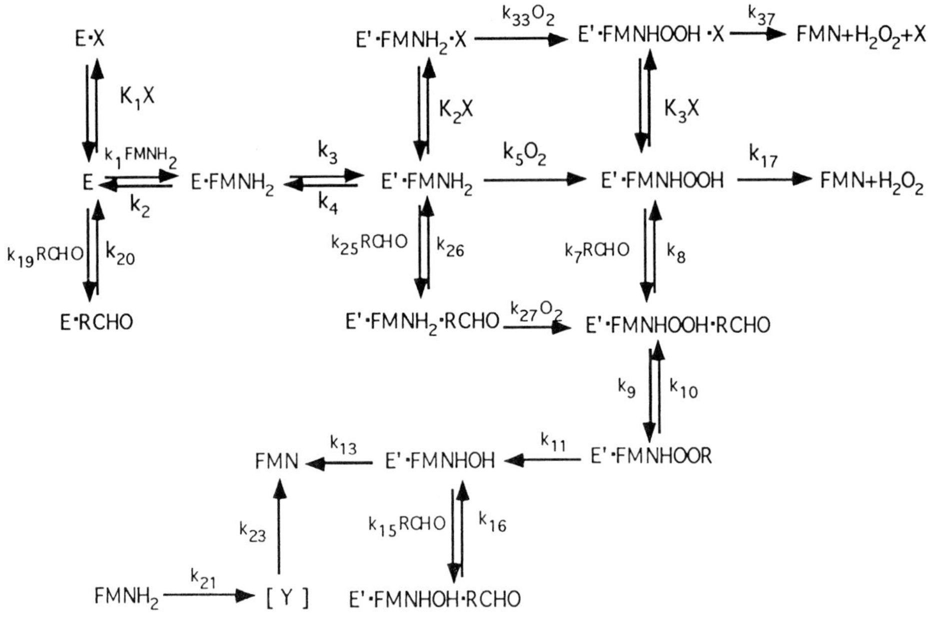

FIG. 2. Kinetic mechanism of the bacterial luciferase-catalyzed reaction. X, an inhibitor, such as an aliphatic alcohol or acid, that binds competitively with the aldehyde substrate; E′, an altered conformational state of the enzyme that forms following binding of the reduced flavin but prior to reaction with molecular oxygen; E′-FMNHOOH, the C4a peroxydihydroflavin intermediate; E′,-FMNHOH, the proposed light-emitting species (the enzyme-bound flavin pseudobase). Rate constants determined for the *V. harveyi* enzyme at 25° and pH 7.0 are given in Table I.

that binds competitively with the aldehyde substrate.[35] E′ denotes an altered conformational state of the enzyme that forms following binding of the reduced flavin but prior to the reaction with molecular oxygen.[33,38] The C4a peroxydihydroflavin intermediate is E′-FMNHOOH, and the proposed light-emitting species is the enzyme-bound flavin pseudobase, E′-FMNHOH.[39] Following the injection of reduced flavin into an enzyme solution containing the aldehyde substrate and oxygen, light emission increases to a maximum that is proportional to the concentration of enzyme and then decays in an apparent first-order fashion.[31,40] As can be seen from the scheme, however, the reaction is far from first order, and the fact that the

[38] N. K. AbouKhair, M. M. Ziegler, and T. O. Baldwin, *Biochemistry* **24,** 3942 (1985).
[39] M. Kurfuerst, S. Ghisla, and J. W. Hastings, *Proc. Natl. Acad. Sci. U.S.A.* **81,** 2990 (1984).
[40] J. W. Hastings, K. Weber, J. Friedland, A. Eberhard, G. W. Mitchell, and A. Gunsalus, *Biochemistry* **8,** 4681 (1969).

TABLE I
RATE CONSTANTS AND EQUILIBRIUM CONSTANTS FOR THE MECHANISM IN FIG. 2[a]

k_1	$1.7 \times 10^7\ M^{-1}\ sec^{-1}$
k_2	$1200\ sec^{-1}$
k_3	$200\ sec^{-1}$
k_4	$14\ sec^{-1}$
k_5	$2.4 \times 10^6\ M^{-1}\ sec^{-1}$
k_7[b]	$1.9 \times 10^7\ M^{-1}\ sec^{-1}$
k_8[b]	$120\ sec^{-1}$
k_9[b]	$1.6\ sec^{-1}$
k_{10}[b]	$1.2\ sec^{-1}$
k_{11}[b]	$1.1\ sec^{-1}$
k_{13}	$0.60\ sec^{-1}$
k_{15}[b]	$3.0 \times 10^3\ M^{-1}\ sec^{-1}$
k_{16}[b]	$0.06\ sec^{-1}$
k_{17}	$0.10\ sec^{-1}$
k_{19}[b]	$9.1 \times 10^5\ M^{-1}\ sec^{-1}$
k_{20}[b]	$5.8\ sec^{-1}$
k_{21}	$4.7\ sec^{-1}$
k_{23}	$11.5\ sec^{-1}$
k_{25}[b]	$1.2 \times 10^6\ M^{-1}\ sec^{-1}$
k_{26}[b]	$37\ sec^{-1}$
k_{27}[b]	$5.1 \times 10^4\ M^{-1}\ sec^{-1}$
k_{33}[c]	$7.7 \times 10^4\ M^{-1}\ sec^{-1}$
k_{37}[c]	$0.004\ sec^{-1}$
K_1[d]	$3.9 \times 10^3\ M^{-1}$
K_2[d]	$6.1 \times 10^3\ M^{-1}$
K_3[d]	$3.5 \times 10^4\ M^{-1}$

[a] Determined at pH 7.0, 25°.
[b] Determined with *n*-decanal.
[c] Determined with *n*-decanol.
[d] Equilibrium constants determined with *n*-decanol.

decay is apparent first order is completely fortuitous. The bioluminescence decay rate does not (necessarily) report the rate of the slowest step in the overall reaction, but it is a complex mixture of many rate constants. The rate constants determined for the *V. harveyi* enzyme at 25° and pH 7.0 are given in Table I.[33–37]

Bacterial luciferase from *V. harveyi* is sensitive to aldehyde substrate inhibition[35,40,41] due to the reduction or loss of binding of $FMNH_2$ to the

[41] T. F. Holzman and T. O. Baldwin, *Biochemistry* **22,** 2838 (1983).

enzyme : aldehyde complex.[35,42] Assays that involve mixing of enzyme with buffer containing aldehyde and oxygen result in lower levels of light emission, because in this format, some fraction of the enzyme will have aldehyde bound at the moment of injection of the reduced flavin. That fraction will not bind reduced flavin and will therefore not participate in the reaction. An additional opportunity for aldehyde interference with the reaction occurs with the enzyme : reduced flavin complex. If aldehyde binds to the enzyme : reduced flavin complex, the rate of reaction of oxygen with the flavin is reduced substantially,[35] thereby reducing the yield of luminescence. The highest levels of luminescence are obtained with assay formats that allow equilibration of the enzyme with $FMNH_2$ in the absence of aldehyde and oxygen. The luminescence reaction may be initiated by the rapid injection of air-equilibrated buffer containing the aldehyde substrate. If the concentrations are correct, the rate of oxygen reaction with the enzyme-bound flavin will be greater than the rate of binding of aldehyde so that little, if any, inhibition will occur.

Flavin Injection Method. In the flavin injection assay, the luciferase is first mixed with an air-equilibrated buffer solution of the aldehyde substrate and dissolved O_2. The bioluminescence reaction is initiated by the rapid injection of 1 ml of $FMNH_2$, reduced either catalytically or photochemically, into 1 ml of the enzyme solution. Light emission from the reaction mixture rises rapidly to a maximum within about 1 sec and then decays with a rate that is dependent both on the species source[40] of the enzyme and on the chain length of the aldehyde substrate.[40,43] The peak light emission intensity is directly proportional to the concentration of luciferase in the reaction mixture over many orders of magnitude.[43]

For photoreduction of the flavin, a solution of 50 μM FMN is prepared in 50 mM phosphate buffer with 2 mM EDTA, pH 7.0. The EDTA serves as a source of electrons for the photoreduction process. If it is not included, the FMN will suffer some photodegradation. Each of a series of 1-ml tuberculin syringes is filled with the flavin solution and placed in front of a fluorescent light bulb. A small fan is used to circulate room air around the syringes to minimize heating. After the dissolved oxygen is consumed by the reaction with reduced flavin, the yellow color of the flavin will be lost as the flavin becomes fully reduced. Reduction times depend on the intensity of the fluorescent light bulb, but the process is usually complete in 2–3 min. When performing assays, we usually keep about 10 syringes of

[42] B. Lei, K. W. Cho, and S.-C. Tu, *J. Biol. Chem.* **269,** 5612 (1994).

[43] J. W. Hastings, Q. H. Gibson, J. Friedland, and J. Spudich, *in* "Bioluminescence in Progress" (F. H. Johnson and Y. Haneda, eds.), p. 151. Princeton Univ. Press, Princeton, NJ, 1966.

flavin in the reduction queue to avoid the wait between assays for reduction of the substrate.

Catalytic reduction is also very effective, but extremely hazardous. Catalytic reduction is accomplished by bubbling H_2 gas through a 100-ml solution of 50 μM flavin in 50 mM phosphate, pH 7.0, containing a small amount (5–10 mg) of platinum-covered charcoal. The flavin solution is maintained in the dark in a 100-ml round-bottom flask (covered with black masking tape) fitted with a rubber vaccine bottle stopper. The gas is introduced and vented by tygon tubing and 3-in.-long hypodermic needles. Aliquots of 1 ml are withdrawn from the round-bottom flask and are injected rapidly into the luciferase enzyme solution, as with the photoreduced flavin. Catalytic reduction requires about 5 min when the hydrogen gas is supplied as a slow, steady stream of bubbles. It is essential with this method to vent the hydrogen gas into a good chemical fume hood, as even a very small accumulation of hydrogen gas can cause an extremely powerful explosion. As a result of the danger of working with hydrogen gas, our laboratory now uses exclusively the photoreduction approach for routine assays.

Dithionite Method. The dithionite method[1,44] of assay of bacterial luciferase differs from the flavin injection method in the order of addition of substrates. In this method, assay buffer containing 50 μM FMN is prepared by mixing 5 ml of the assay buffer stock solution with 1 ml of the FMN stock solution and diluting to 100 ml with water. The FMN-containing assay mix is introduced (1-ml aliquots) into 20-ml scintillation vials and is allowed to temperature equilibrate before the enzyme to be assayed is introduced. The flavin is reduced and molecular oxygen is removed by the addition of a small excess of sodium dithionite powder. Reduction of the flavin is apparent when the yellow color of the oxidized flavin is lost from the solution. The bioluminescence reaction is then initiated by rapid injection of 1 ml of an air-equilibrated suspension of the aldehyde substrate [0.01% (v/v) of *n*-decanal in water]. This order of addition allows reaction of the reduced flavin bound to the enzyme with the oxygen followed by binding of the aldehyde substrate. The dithionite assay routinely yields higher levels of bioluminescence activity than flavin injection methods. Aldehyde is not inhibitory in this assay, even at high concentrations, because the enzyme : flavin complex is formed prior to introduction of the aldehyde.

Dithionite reduction may be performed in a more reproducible fashion by using a solution of dithionite.[44] This is accomplished by dissolving 150 mg $Na_2S_2O_4$ in 10 ml O_2-free water. The dithionite solution should be protected from air by constant flushing of the container with argon or by overlaying the solution with mineral oil. The minimum volume of the

[44] S.-C. Tu and J. W. Hastings, *Biochemistry* **14,** 4310 (1975).

dithionite solution required to reduce the flavin in the assay vial must be determined experimentally, but should be in the range of 10–20 μl.

Oxidoreductase-Coupled Assay. Vibrio harveyi possesses two enzymes capable of supporting the flavin-dependent bioluminescence reaction of the luciferase by supplying reducing equivalents from reduced pyridine nucleotide to the flavin. These enzymes have been studied most extensively by Tu and colleagues[45]; the high-resolution crystal structure of one of these oxidoreductases has been reported.[46]

The coupled oxidoreductase assay, performed in 1 ml assay buffer containing 4 μM FMN, decanal (10 μl of 0.1%, v/v), the luciferase, and the flavin oxidoreductase, is initiated by the addition of 0.1 ml of 0.1 mM NADH. The light emission rises to a peak and continues at a relatively constant level until the NADH is depleted. To use this assay, it is essential to establish conditions, dependent on the oxidoreductase, under which the light intensity is linearly dependent on the concentration of the luciferase. That is, there should be sufficient oxidoreductase to supply an excess of the reduced flavin substrate.

[45] B. Lei, M. Liu, S. Huang, and S.-C. Tu, *J. Bacteriol.* **176,** 3552 (1994).
[46] J. J. Tanner, B. Lei, S.-C. Tu, and K. L. Krause, *Biochemistry* **35,** 13531 (1996).

[10] Purification of Luciferase Subunits from Recombinant Sources

By JAMES F. SINCLAIR

Bacterial luciferase, the enzyme responsible for light production in luminescent bacteria, is a heterodimer[1] composed of two homologous subunits,[2] designated α and β. While only the heterodimer has biologically significant activity,[3] a study of the individual subunits is of interest both for understanding their structure in the absence of the other subunit[4,5] and for elucidating

[1] J. W. Hastings, K. Weber, J. Friedland, A. Eberhard, G. W. Mitchell, and A. Gunsalus, *Biochemistry* **8,** 4681 (1969).
[2] T. O. Baldwin, M. M. Ziegler, and D. A. Powers, *Proc. Natl. Acad. Sci. U.S.A.* **76,** 4887 (1979).
[3] T. O. Baldwin and M. M. Ziegler, *in* "Chemistry and Biochemistry of Flavoenzymes" (F. Müller, ed.), Vol. III, p. 467. CRC Press, Boca Raton, FL, 1992.
[4] E. G. Ruby and J. W. Hastings, *Biochemistry* **19,** 4989 (1980).
[5] J. F. Sinclair, M. M. Ziegler, and T. O. Baldwin, *Nature Struct. Biol.* **1,** 320 (1994).

METHODS IN ENZYMOLOGY, VOL. 305
0076-6879/00 $30.00

the mechanism by which they associate to form the active enzyme.[6] A procedure is described here for the expression and purification of the individual subunits. Using this method, 15–20 mg of the purified subunit was obtained per liter of cell culture. Although the protocol given was developed using luciferase from a strain of *Vibrio harveyi*, it should be readily adaptable to the isolation of subunits from any of the other bacterial luciferases.

Plasmid Construction

The α and β subunits are encoded by the *luxA* and *luxB* genes, respectively. These genes have been cloned[7] and sequenced.[8,9] For expression of the α subunit, a DNA fragment containing the *luxA* and *luxB* genes was isolated from the plasmid pTB7[7] using the enzymes *Sal*I and *Bam*HI and then cloned into the pUC9 plasmid. Subsequently, the *luxB* gene was removed from this construct by removal of the *Eco*RI fragment, leaving only the *luxA* gene intact.[10] *Escherichia coli* strain LE392, which had been shown to be suitable for the overexpression of the heterodimer,[11] was then transformed to ampicillin resistance with this plasmid.

For expression of the β subunit, the *Pst*I to *Sac*I fragment containing the *luxB* gene was isolated from the plasmid pTB7[7] and cloned into the plasmid pT7-6 behind the T7 polymerase promoter.[12] This plasmid was transformed into the *E. coli* strain BL21 from Novagen (Milwaukee, WI), which contains the gene for T7 polymerase under control of the *lac* promoter. Both the plasmids containing the *luxA* and *luxB* genes conferred ampicillin resistance upon the host cells, allowing for efficient selection of the transformed cells.

Cell Growth and Protein Expression

Cells are grown in LB media supplemented with 100 mg per liter carbenicillin. The subunits are produced in the soluble cell fraction only when cells

[6] J. F. Sinclair, J. J. Waddle, E. F. Waddill, and T. O. Baldwin, *Biochemistry* **32,** 5036 (1993).

[7] T. O. Baldwin, T. Berends, T. A. Bunch, T. F. Holzman, S. K. Rausch, L. Shamansky, M. L. Treat, and M. M. Ziegler, *Biochemistry* **23,** 3663 (1984).

[8] D. H. Cohn, A. J. Mileham, M. I. Simon, K. H. Nealson, S. K. Rausch, D. Bonam, and T. O. Baldwin, *J. Biol. Chem.* **260,** 6139 (1985).

[9] T. C. Johnston, R. B. Thompson, and T. O. Baldwin, *J. Biol. Chem.* **261,** 4805 (1986).

[10] J. J. Waddle, T. C. Johnston, and T. O. Baldwin, *Biochemistry* **26,** 4917 (1987).

[11] T. O. Baldwin, L. H. Chen, L. J. Chlumsky, J. H. Devine, and M. M. Ziegler, *J. Biolumin. Chemilumin.* **4,** 40 (1989).

[12] S. Tabor, *in* "Current Protocols in Molecular Biology" (F. Ausubel *et al.,* eds.), 16.2.1. Green Publishing and Wiley Interscience, New York, 1990.

are grown at temperatures below 30°.[13] Therefore, all biological materials are grown at 25° to maximize the yield of soluble protein. Cultures of 5 ml are inoculated with single colonies picked from overnight agar plates and allowed to grow at 37° for 6 hr. These cultures are used to inoculate 70 ml of LB media, which is incubated overnight at 25°. Finally, 15 ml of the overnight culture is added to 1.5 liters of LB media and grown at 25° for 24 hr. We found that aeration is crucial for optimal expression. A total of 6 liters of cell culture are used in a typical preparation. The α subunit is expressed constitutively throughout the growth period. Expression of the β subunit is induced by the addition of 5 m*M* lactose when the culture reaches an optical density at 600 nm of 0.8.

Cell Lysis

Cells are collected by centrifugation at 6370*g* for 15 min at 10°. All subsequent purification steps are performed in a sodium potassium phosphate buffer system containing 0.5 m*M* dithiothreitol (DTT) and 1 m*M* disodium ethylenediaminetetraacetate (EDTA). The cell pellet from 6 liters of culture is resuspended in 70 ml of cold 200 m*M* phosphate buffer. Purification of the α subunit is carried out at pH 7.0, whereas that of the β subunit is performed a pH 6.2, conditions that promote maximal resolution of the subunit during chromatography. The resuspended cells are lysed by high pressure in an SLM-Aminco French pressure cell. A pressure of 1000 psi is applied to the bacterial cells; cells are allowed to drain slowly from the pressure cell at a rate of not more than 6 ml min^{-1}. The lysed cells are collected in a beaker placed on wet ice. The insoluble portion of the cell lysate is removed by centrifugation at 27200*g* for 20 min at 10°; the supernatant is retained. For the β subunit, a selective precipitation step is utilized. Ammonium sulfate is added to the cell lysate to 40% of saturation, a concentration at which the β subunit remains soluble. The solution is centrifuged and the supernatant is adjusted to 75% of the saturation concentration with ammonium sulfate. The precipitated β subunit is then collected by centrifugation at 27,200*g* for 15 min. The lysate containing the α subunit is subjected only to a 75% ammonium sulfate treatment because this subunit precipitates over a much wider range of salt concentration and selective precipitation cannot be achieved. The precipitated protein is resuspended in 15 ml of 200 m*M* phosphate buffer, pH 7.0, and dialyzed extensively against the same buffer at 4°.

[13] J. J. Waddle and T. O. Baldwin, *Biochem. Biophys. Res. Commun.* **178,** 1188 (1991).

Column Chromatography

All chromatography steps are carried out using a Pharmacia FPLC system (Piscataway, NJ) maintained at a constant temperature of 4°. A Pharmacia DEAE Sepharose Fast Flow column is equilibrated by washing with at least 3 column volumes of 200 m*M* phosphate buffer. The buffers used are pH 7.0 for the α subunit and 6.2 for the β subunit. Under these conditions, the subunits show the greatest resolution from contaminating proteins on elution from the ion-exchange column. The dialyzed cell lysate is injected onto the column and allowed to equilibrate for 30 min. The column is then washed with 1 column volume of the 200 m*M* phosphate buffer. Protein is eluted from the column by 3 column volumes of a linear gradient from 200 to 700 m*M* phosphate at a flow rate of 1 ml min^{-1}. Fractions are analyzed by SDS–polyacrylamide gel electrophoresis, and those that show the highest concentration of the required purity of subunit are retained. The pooled fractions are concentrated to a volume of approximately 10 ml in an Amicon ultrafiltration cell (Danvers, MA) fitted with a PM30 membrane.

The dialyzed protein is injected onto a second anion-exchange column, a Pharmacia HiLoad 26/10 Q Sepharose column. The protein is allowed to equilibrate with the column matrix for 20 min and is then eluted by 3 column volumes of a linear gradient from 100 to 400 m*M* phosphate at a flow rate of 1 ml min^{-1}. As described earlier, column fractions are analyzed by SDS–polyacrylamide gel electrophoresis, and those fractions that show the highest concentration of subunit are retained. The selected fractions are concentrated to a volume of approximately 5 ml. The concentrated subunit is then dialyzed against 200 m*M* phosphate buffer at 4°.

At this point, the individual subunits are judged to constitute greater than 95% of the total protein as estimated from Coomassie blue staining after SDS–polyacrylamide gel electrophoresis. The protein is used without further purification for spectroscopic characterization. For crystal trials, the subunits are chromatographed on an additional size-exclusion column to ensure even greater purity. An aliquot (0.2 ml) of the concentrated protein is injected onto a Pharmacia Sepharose 12 HR 10/20 column and eluted at a flow rate of 0.1 ml min^{-1} with 100 m*M* phosphate buffer. Fractions of 0.5 ml are collected, and only fractions with the highest absorbance at 280 nm are retained.

Protein Storage

Concentrated protein is stored at −20° in small (0.1 ml) aliquots. Repeated freezing and thawing causes precipitation of the subunits. It is recommended that once an aliquot has been thawed, any unused protein be discarded. Further, long-term storage appears to cause aggregation of the subunit. For this reason, it appears preferable to use protein that has been purified recently. Crystals of the β subunit can only be grown from freshly prepared protein. Stored protein produces primarily aggregated material under conditions where fresh β subunit yields large crystals.

Physical Properties of Subunits

Extinction coefficients have been shown to be 1.41 and 0.71 mg ml^{-1} cm^{-1} for the α and β subunits, respectively.[6] The large difference in extinction coefficients results from the fact that the α subunit has six tryptophan residues,[8] whereas the β subunit has only two tryptophans.[9] The molecular weight of the α subunit, calculated from the amino acid composition,[8] is 40,000, whereas the molecular weight of the β subunit, also calculated from its sequence,[9] is 36,000. The purified β subunit is a homodimer as demonstrated by analytical ultracentrifugation.[5] This technique also shows that the α subunit is monomeric, but begins to dimerize at higher concentrations with a dissociation constant of approximately 2 μM.[14] Both subunits have a weak bioluminescence activity that is approximately four to five orders of magnitude less than that of the heterodimer.[6]

The stability and folding properties of the β_2 homodimer are worthy of special mention. The association reaction is very slow: the estimated bimolecular rate constant is approximately 150 M^{-1} sec^{-1} at 18° in pH 7.0 phosphate buffer.[5] The estimated dissociation rate constant for the homodimer is 10^{-14} sec^{-1} under the same conditions.[5] The half-time for dissociation of the complex, about 10^6 years, is clearly too long to be biologically significant. Therefore, from the biological perspective, formation of the homodimer may be regarded as irreversible. These slow processes are evident in the hysteresis observed in urea-induced unfolding transitions. The β_2 homodimer is fully native in 5 M urea, pH 7.0,[5] conditions that render the $\alpha\beta$ heterodimer fully unfolded.[15] This is clearly an example of kinetic control of a protein-folding reaction. Careful analysis of the crystal structure of the β_2 homodimer has yielded no clear suggestions of the structural basis for the remarkable kinetic stability of the β_2 homodimer.[16]

[14] J. F. Sinclair, Ph.D. thesis, Texas A&M University, College Station, TX, 1995.

[15] M. M. Ziegler, M. E. Goldberg, A. F. Chaffotte, and T. O. Baldwin, *J. Biol. Chem.* **268,** 10760 (1993).

[16] J. B. Thoden, H. M. Holden, A. J. Fisher, J. F. Sinclair, G. Wesenberg, T. O. Baldwin, and I. Rayment, *Protein Sci.* **6,** 13 (1997).

Formation of the Heterodimer from Purified Subunits

The active, heterodimeric luciferase may be produced by the association of the purified subunits.[6] Because the β subunit is a homodimer, it must be unfolded before it is able to combine with the α subunit. The β homodimer was denatured by incubation in 5 *M* guanidine hydrochloride, then dialyzed against 5 *M* urea, a urea concentration that will not unfold the β dimer but will prevent the monomer from refolding. Folding of the β monomer was initiated by rapid dilution into pH 7.0 phosphate buffer at 18° in the presence of α subunit in an equimolar concentration with the folding β subunit. After 6 to 8 hr, the folding was essentially complete, as determined by monitoring luciferase activity.

[11] A Rapid Chromatographic Method to Separate the Subunits of Bacterial Luciferase in Urea-Containing Buffer

By A. CLAY CLARK, BRIAN W. NOLAND, and THOMAS O. BALDWIN

Introduction

Bacterial luciferase from *Vibrio harveyi* is a flavin monooxygenase of M_r 76,457 that is heterodimeric in structure.[1] The nonidentical but homologous α and β subunits are composed of 355 and 324 amino acid residues, respectively, and the genes encoding both subunits have been cloned.[2,3] The recombinant protein can be purified in large quantities when expressed in *Escherichia coli*.[4,4a]

The protein-folding properties of the individual subunits and their assembly into active heterodimer have been studied as independent, experimentally distinguishable steps.[5,6] These studies demonstrate that when the

[1] T. O. Baldwin and M. M. Ziegler, *in* "Chemistry and Biochemistry of Flavoenzymes" (F. Muller, ed.), p. 467. CRC Press, Boca Raton, FL, 1992.

[2] D. H. Cohn, R. C. Ogden, J. N. Abelson, T. O. Baldwin, K. H. Nealson, M. I. Simon, and A. J. Mileham, *Proc. Natl. Acad. Sci. U.S.A.* **80,** 120 (1983).

[3] T. O. Baldwin, T. Berends, T. A. Bunch, T. E. Holzman, S. K. Rausch, L. Shamansky, M. L. Treat, and M. M. Ziegler, *Biochemistry,* **23,** 3663 (1984).

[4] T. O. Baldwin, L. H. Chen, L. J. Chlumsky, J. H. Devine, and M. M. Ziegler, *J. Biolumin. Chemilumin.* **4,** 40 (1989).

[4a] T. O. Baldwin, M. M. Ziegler, V. A. Green, and M. D. Thomas, *Methods Enzymol.* **305,** [9], 135 (2000) (this volume).

[5] J. J. Waddle, T. C. Johnston, and T. O. Baldwin, *Biochemistry* **26,** 4917 (1987).

[6] J. F. Sinclair, J. J. Waddle, E. F. Waddill, and T. O. Baldwin, *Biochemistry* **32,** 5036 (1993).

0076-6879/00 $30.00

subunits are folded independently, i.e., in the absence of the other subunit, they fold into stable structures that do not interact to form active luciferase. The formation of native luciferase is a kinetically controlled process that proceeds through the formation of several folding intermediates.[7,8] If the subunits do not associate during folding, they ultimately form structures that are incapable of assembly. The β subunit, for example, forms a β_2 homodimer, which is a kinetically trapped structure with a very slow rate of dissociation (half-time of $>10^6$ years).[9] In addition, when refolded at low protein concentrations, the β subunit forms a stable monomeric species that is incapable of assembly with the α subunit.[10] The crystal structures of the native, heterodimeric luciferase[11] and of the β_2 homodimer[12] have been reported and reveal that the structure of the β subunit is the same in both $\alpha\beta$ and β_2. Therefore, based on the crystal structures, there are no obvious reasons for these differences.

Studies of luciferase folding and assembly have been aided by the separation of the individual subunits in urea-containing buffer, allowing for studies in which the concentration of one subunit can be varied relative to the other subunit. In 1967, Friedland and Hastings[13] demonstrated that the individual luciferase subunits could be separated using ion-exchange chromatography (DEAE-cellulose) in 8 *M* urea-containing buffer. Other methods[8,10,14] that describe the separation of the luciferase subunits in urea-containing buffer rely, to a large extent, on the original methods of Hastings.

This article describes a rapid protocol to separate luciferase subunits in 5 *M* urea-containing buffer using ion-exchange chromatography on Q-Sepharose. We show by the absorbance properties and by SDS–PAGE analysis of the eluant that this method gives baseline resolution between the subunits. In addition, we demonstrate that the refolded subunits have fluorescence emission properties that are similar to those of subunits that were purified individually following overexpression in *E. coli.*[6]

[7] M. M. Ziegler, M. E. Goldberg, A. F. Chaffotte, and T. O. Baldwin, *J. Biol. Chem.* **268,** 10760 (1993).

[8] T. O. Baldwin, M. M. Ziegler, A. F. Chaffotte, and M. E. Goldberg, *J. Biol. Chem.* **268,** 10766 (1993).

[9] J. F. Sinclair, M. M. Ziegler, and T. O. Baldwin, *Nature Struct. Biol.* **1,** 320 (1994).

[10] A. C. Clark, S. W. Raso., J. F. Sinclair, M. M. Ziegler, A. F. Chaffotte, and T. O. Baldwin, *Biochemistry* **36,** 1891 (1997).

[11] A. J. Fisher, F. M. Raushel, T. O. Baldwin, and I. Rayment, *Biochemistry* **34,** 6581 (1995).

[12] J. B. Thoden, H. M. Holden, A. J. Fisher, J. F. Sinclair, G. Wesenberg, T. O. Baldwin, and I. Rayment, *Protein Sci.* **6,** 13 (1997).

[13] J. Friedland and J. W. Hastings, *Proc. Natl. Acad. Sci. U.S.A.* **58,** 2336 (1967).

[14] S.-C. Tu, *Methods Enzymol.* **57,** 171 (1978).

Overexpression and Purification of Luciferase

The method described here for the separation of luciferase subunits requires purified luciferase. *Vibrio harveyi* luciferase can be overexpressed in *E. coli* and purified as described previously.[4,4a] This expression system and purification protocol result in the accumulation of luciferase to levels of over 50% of the soluble protein in *E. coli.*

Separation of Luciferase α and β Subunits by Ion-Exchange Chromatography on Q-Sepharose.

An overview of this protocol is given in Table I. The following extinction coefficients[6] were used to determine the concentrations of the luciferase heterodimer and of the individual α and β subunits: luciferase, 1.13 $(mg/ml)^{-1}\ cm^{-1}$ (or $8.69 \times 10^4\ M^{-1}\ cm^{-1}$); α subunit, 1.41 $(mg/ml)^{-1}\ cm^{-1}$ (or $5.64 \times 10^4\ M^{-1}\ cm^{-1}$); and β subunit, 0.71 $(mg/ml)^{-1}\ cm^{-1}$ (or $2.59 \times 10^4\ M^{-1}\ cm^{-1}$).

1. Preparation of the column. An ion-exchange chromatography column of Q-Sepharose (2.6 × 10 cm; Pharmacia) is equilibrated at 4° with 200 ml of Q-Sepharose low salt buffer (Table II). The flow rate is 2 ml per minute.
2. Preparation of the protein. The protein is prepared by dialyzing 50 ml of purified luciferase (approximately 7 mg/ml) against 500 ml of Q-Sepharose low salt buffer (Table II). A dialysis membrane with a molecular weight cutoff (MWCO) of 12–14 kDa is used. The dialysis is carried out at 4°, and the buffer is changed three times (total

TABLE I
OVERVIEW OF COLUMN CHROMATOGRAPHY

Q-Sepharose ion-exchange column (2.6 × 10 cm; run at 4°)
Column is preequilibrated with 200 ml of Q-Sepharose low salt buffer (Table II) at a flow rate of 2 ml per minute
Protein is prepared by dialyzing (12–14 kDa MWCO membrane) 50 ml of purified luciferase (approximately 7 mg/ml) against 500 ml of Q-Sepharose low salt buffer (Table II) with three changes of buffer
Following dialysis, the protein is filtered through a 2.2-μm pore size syringe filter
Dialyzed protein (40 ml) is loaded onto the column at a flow rate of 2 ml per minute
Column is washed with 100 ml of Q-Sepharose low salt buffer (Table II)
Subunits are eluted with a linear gradient from 20 mM phosphate (Q-Sepharose low salt buffer; Table II) to 120 mM phosphate (Q-Sepharose high salt buffer; Table II) at a flow rate of 2 ml per minute (total volume of 1100 ml); 15-ml fractions are collected
Column fractions are monitored by absorbance at 280 nm and by SDS–PAGE

TABLE II
BUFFERS FOR LUCIFERASE SUBUNIT SEPARATION

1. Q-Sepharose low salt buffer (20 m*M* Na/K-phosphate, pH 7.0, 1 m*M* EDTA,[a] 1 m*M* DTT, 5 *M* urea)
   - K_2HPO_4, 8.8 g
   - NaH_2PO_4-H_2O, 3.97 g
   - EDTA, 1.49 g
   - DTT, 617 mg
   - Ultrapure urea, 1201.2 g
   - ddH_2O to 4 liters
   - Adjust to pH 7.0 with 12 *N* HCl
2. Q-Sepharose high salt buffer (120 m*M* Na/K-phosphate, pH 7.0, 1 m*M* EDTA, 1 m*M* DTT, 5 *M* urea)
   - K_2HPO_4, 26.64 g
   - NaG_2PO_4-H_2O, 11.92 g
   - EDTA, 745 mg
   - DTT, 308 mg
   - Ultrapure urea, 600.6 g
   - ddH_2O to 2 liters
   - Adjust to pH 7.0 with 12 *N* HCl
3. Refolding buffer (50 m*M* Na/K-phosphate, pH 7.0, 1 m*M* EDTA, 1 m*M* DTT)
   - K_2HPO_4, 22.2 g
   - NaH_2PO_4-H_2O, 9.93 g
   - EDTA, 1.49 g
   - DTT, 617 mg
   - ddH_2O to 4 liters
   - Adjusted to pH 7.0 with 12 *N* HCl

[a] Abbreviations used: EDTA, ethylenediaminetetracetate; DTT, dithiothreitol; ddH_2O, doubly distilled water; K_2HPO_4, dibasic potassium phosphate; NaH_2PO_4-H_2O, monobasic sodium phosphate monohydrate.

volume of 1500 ml). Following dialysis, the protein is filtered through a 2.2-μm pore-size syringe filter.

3. Column loading and elution. The protein (40 ml) in Q-Sepharose low salt buffer, described earlier, is added to the column at a flow rate of 2 ml per minute, and the column is washed with 100 ml of Q-Sepharose low salt buffer (Table II). The subunits are eluted with a linear gradient from 20 m*M* phosphate (Q-Sepharose low salt buffer; Table II) to 120 m*M* phosphate (Q-Sepharose high salt buffer; Table II). The gradient (total volume of 1100 ml) is run at a flow rate of 2 ml per minute, and 15-ml fractions are collected.

In order to determine the fractions to pool for each subunit, the column fractions are monitored by absorbance at 280 nm (Fig. 1) and by polyacryl-

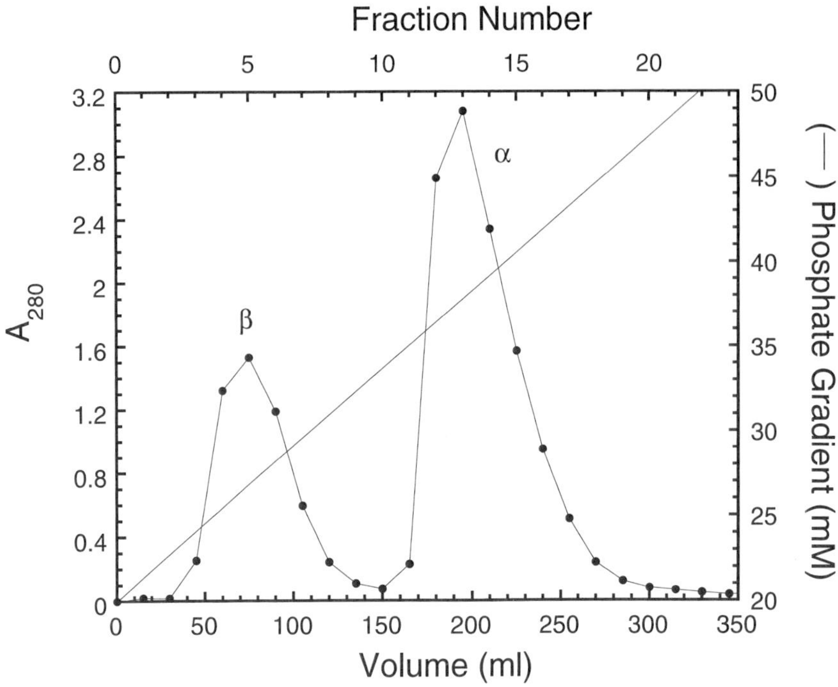

FIG. 1. Absorbance profile at 280 nm of Q-Sepharose ion-exchange column fractions. The symbols correspond to the absorbance at 280 nm for each column fraction, and the diagonal line represents the phosphate gradient during the elution.

amide gel electrophoresis in sodium dodecyl sulfate (SDS–PAGE) (Fig. 2). A typical column profile is shown in Fig. 1. These data demonstrate that the β subunit elutes prior to the α subunit and has a maximum protein concentration at an elution volume of approximately 75 ml. As shown by the diagonal line in Fig. 1, this elution profile corresponds to phosphate

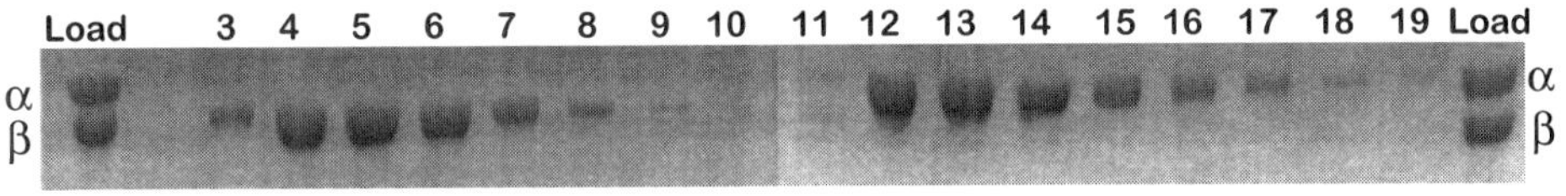

FIG. 2. SDS–PAGE of Q-Sepharose ion-exchange column fractions. Lane 1 (load), 10 μl of a 1:10 dilution of sample loaded onto the column (approximately 7 μg of luciferase); lanes 3–19, column fractions 3–19, respectively (5 μl of each column fraction was used); lane 20 (load), 10 μl of a 1:10 dilution of sample loaded onto the column (same as lane 1). The gel was a 12% Tris–glycine acrylamide gel with a 4.8% acrylamide stacking gel and was run at 150 V for approximately 45 min. The gel was stained with Coomassie brilliant blue R-250.

concentrations between approximately 23 and 33 m*M*. The α subunit elutes between approximately 150 and 300 ml of elution buffer (Fig. 1), with a maximum protein concentration at approximately 200 ml of buffer. As shown by the diagonal line in Fig. 1, this corresponds to phosphate concentrations between approximately 33 and 50 m*M*.

The fractions were also examined by SDS–PAGE, as shown in Fig. 2. These data demonstrate that the β subunit elutes in fractions 3–9 (see Fig. 1), whereas the α subunit elutes in fractions 11–19 (see Fig. 1). These data also demonstrate that the protein in each peak of the elution profile (Fig. 1) contains only the individual subunit with little or no contamination from the other subunit. Based on these data, fractions 3–9 were pooled for the β subunit, and fractions 12–19 were pooled for the α subunit. The final volume of pooled α subunit (1.15 mg/ml) was 118 ml, and the final volume of pooled β subunit (1.07 mg/ml) was 105 ml.

Refolding of Luciferase α and β Subunits

The isolated α subunit was refolded by adding 6.23 ml of pooled α subunit to 472 ml of refolding buffer (50 m*M* Na/K-phosphate, pH 7.0, Table II) at 4° using a flow rate of 13.3 ml per minute for 28 sec. This procedure is done using a multirate infusion pump (Sage Pump, ATI Orion). Following the initial dilution, the flow rate is adjusted to 1.17 ml per minute until the entire volume is added.

The isolated β subunit is refolded by a similar procedure. In this case, 6.00 ml of the pooled β subunit is added to 420 ml of refolding buffer (50 m*M* Na/K-phosphate, pH 7.0, Table II) at 4° using a flow rate of 13.3 ml per minute for 27 sec. The flow rate is then adjusted to 1.67 ml per minute until the entire volume is added. For both proteins, the initial, rapid dilution allows a protein concentration of 15 μg/ml to be reached quickly, which is then followed by a slower increase in the protein concentration.

For both refolded subunits, the solutions are centrifuged to remove precipitate (8000 rpm at 4° for 25 min). At this point, the concentration of the α subunit is approximately 0.18 mg/ml, and the concentration of the β subunit is approximately 0.21 mg/ml. The subunits are concentrated approximately 10-fold using an Amicon concentrator (PM10 membrane) at 4°. Following concentration, the refolded subunits are dialyzed against refolding buffer (Table II) to remove the residual urea. A dialysis membrane with a molecular weight cutoff (MWCO) of 12–14 kDa is used. The dialysis is carried out at 4°, and the buffer (500 ml) is changed three times (total volume of 1500 ml). Following dialysis, the subunits are frozen in liquid nitrogen and stored at −20°. This refolding procedure results in >85% yield of refolded subunit. In addition, refolding the subunits on elution from the Q-Sepharose column prevents modification of the proteins, such as carbamylation, which may occur due to prolonged exposure to the urea.

Fluorescence Properties of Separated Luciferase Subunits

The fluorescence emission of the subunits isolated from the Q-Sepharose ion-exchange chromatography column demonstrates a maximum at approximately 350 nm, indicating that the proteins are largely unfolded (Fig. 3). Following refolding (see earlier), the α subunit has a fluorescence emission maximum at approximately 335 nm, whereas the emission maximum of the refolded β subunit is approximately 325 nm. These fluorescence properties

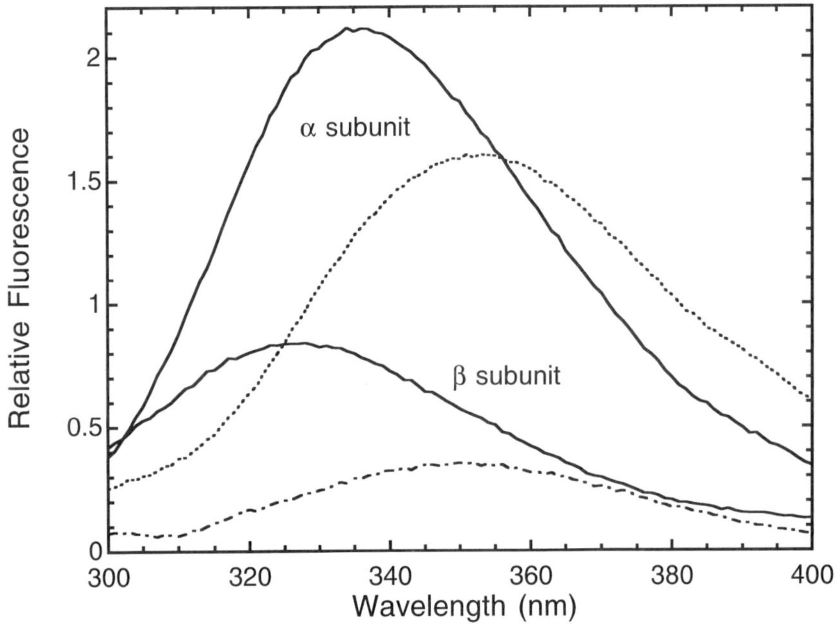

FIG. 3. Fluorescence emission of luciferase α and β subunits. The samples were excited at 280 nm and fluorescence emission was monitored from 300 to 400 nm. The concentration of each protein was 30 μg/ml. The refolded subunits were in a buffer of 50 m*M* Na/K-phosphate (Table II), pH 7.0, and the unfolded subunits were in the same buffer except that urea was added to a final concentration of 5 *M*. The temperature was maintained at 18° using a circulating water bath, and data were collected on an SLM 8000C spectrofluorometer. Fluorescence emission of the refolded subunits is represented by solid lines and is labeled in the figure. ---, α subunit from the Q-Sepharose column; –·–, β subunit from the Q-Sepharose column.

are similar to those reported previously for the individual subunits that were overexpressed independently in *E. coli.*[6] In that study, the α subunit was also shown to have a fluorescence emission maximum that was red shifted relative to that of the β subunit. Data demonstrate that the luciferase subunits isolated by the protocol described here can be refolded with high efficiency.

[12] Purification and Ligand Exchange Protocols for Antenna Proteins from Bioluminescent Bacteria

By VALENTIN N. PETUSHKOV, BRUCE G. GIBSON, ANTONIE J. W. G. VISSER, and JOHN LEE

The emission spectrum from bioluminescent bacteria has been observed to vary depending on the type of bacteria. *Photobacterium phosphoreum* species usually show bioluminescence maxima around 472 nm and *Photobacterium leiognathi* species emit at a slightly longer wavelength. A certain strain (Y1) of *Vibrio fischeri* has a yellow bioluminescence with maximum emission at 542 nm. These differences are due to the bioluminescence originating from the fluorescence transition of an "antenna" protein that participates in the bioluminescence reaction along with the enzyme bacterial luciferase. The bioluminescence from a number of coelenterates involves a similar participation of an antenna protein, with the famous "green-fluorescent protein" being the origin of the bioluminescence emission from these organisms.

Two types of bacterial antenna protein have received detailed study. The first are the lumazine proteins (LumP) from *Photobacterium,* named for the ligand, 6,7-dimethyl-8-(1′-D-ribityl)lumazine (Lum).[1] The second is the "yellow fluorescence protein" (YFP) from the Y1 strain, of similar primary sequence to LumP but with a flavin as a ligand.[2–6] It has also been

[1] P. Koka and J. Lee, *Proc. Natl. Acad. Sci. U.S.A.* **76,** 3068 (1979).
[2] S. C. Daubner, A. M. Astorga, G. B. Leisman, and T. O. Baldwin, *Proc. Natl. Acad. Sci. U.S.A.* **84,** 8912 (1987).
[3] P. Macheroux, K. U. Schmidt, P. Steinerstauch, S. Ghisla, P. Colepicolo, R. Buntic, and J. W. Hastings, *Biochem. Biophys. Res. Commun.* **146,** 101 (1987).
[4] V. N. Petushkov, B. G. Gibson, and J. Lee, *Biochem. Biophys. Res. Commun.* **211,** 774 (1995).
[5] T. O. Baldwin, M. L. Treat, and S. C. Daubner, *Biochemistry* **29,** 5509 (1990).
[6] D. J. O'Kane, B. Woodward, J. Lee, and D. C. Prasher, *Proc. Natl. Acad. Sci. U.S.A.* **88,** 1100 (1991).

0076-6879/00 $30.00

found that these proteins exist in bacteria with different noncovalently bound ligands, the lumazine derivative, FMN, and riboflavin (Rf), and that their bioluminescence activity depends on the particular ligand as well as the species origin of the protein and the luciferase.[7]

From the Y1 strain another blue fluorescence protein (BFP) was discovered and purified.[4,7–9] The N-terminal amino acid sequence showed BFP to be related to the antenna proteins[7] and a member of the expanding class of "lumazine-binding" proteins along with the riboflavin synthases from Eubacteria.[7,10] However, BFP shows neither bioluminescence effects nor synthase activity. Antenna proteins and BFP are monomers with one binding site compared with two binding sites per monomer for riboflavin synthase, which is active as an asymmetrical trimer.[11] The monomer mass of all these proteins is in the 20- to 26-kDa range.

In the course of the bioluminescence reaction of luciferase with its substrates, $FMNH_2$, O_2, and a fatty aldehyde such as tetradecanal, a metastable luciferase peroxyflavin intermediate is formed. Complexation of this intermediate with the antenna protein is the necessary step for the antenna function and the ligand appears to exert critical influence in this process because, for example, the apo-LumP recharged with Rf fails to complex and to exhibit the bioluminescence spectral shift.[7,9,12–15] It is proposed that in this protein–protein complex, the antenna ligand and the peroxyflavin exert a weak dipole–dipole interaction in the excited state. As a result, excitation released at the luciferase site is emitted as radiation from the antenna ligand. Applying Förster's equation, a separation of the ligand from the peroxyflavin of 15 Å would account for the fluorescence dynamics observations. The formation of this protein complex also explains the effects on the bioluminescence kinetics.[14]

The development of efficient methods for the exchange of ligands on these proteins is of both practical and fundamental interest. Many diagnostic applications of bioluminescence are enhanced by the ability to shift the color of the emission, which could be achieved by the use of these different

[7] V. N. Petushkov and J. Lee, *Eur. J. Biochem.* **245,** 790 (1997).
[8] H. Karatani, T. Wilson, and J. W. Hastings, *Photochem. Photobiol.* **55,** 293 (1992).
[9] H. Karatani, T. Wilson, and J. W. Hastings, *J. Photochem. Photobiol.* **B18,** 227 (1993).
[10] K. Schott, J. Kellermann, F. Lottspeich, and A. Bacher, *J. Biol. Chem.* **265,** 4204 (1990).
[11] J. Scheuring, M. Fischer, M. Cushman, J. Lee, A. Bacher, and H. Oschkinat, *Biochemistry* **35,** 9637 (1996).
[12] A. J. W. G. Visser, A. van Hoek, N. V. Visser, Y. Lee, and S. Ghisla, *Photochem. Photobiol.* **65,** 570 (1997).
[13] V. N. Petushkov, B. G. Gibson, and J. Lee, *Biochemistry* **34,** 3300 (1995).
[14] V. N. Petushkov, M. Ketelaars, B. G. Gibson, and J. Lee, *Biochemistry* **35,** 12086 (1996).
[15] V. N. Petushkov, B. G. Gibson, and J. Lee, *Biochemistry* **35,** 8413 (1996).

ligands. For example, mutants of the green-fluorescent protein have been produced with different fluorescence spectra for such applications. Variation of the spectroscopic properties of the bacterial protein–protein complex is also a valuable tool for obtaining experimental support for the electronic coupling model by such methods as fluorescence dynamics. Indeed it has already been shown that the donor–acceptor spectral overlap in the complex with YFP is an order of magnitude greater than with LumP, corresponding to the ratio of energy transfer rates measured by fluorescence dynamics.[7] The most direct test of the dipole–dipole interaction model, however, is a determination of the three-dimensional structure of the protein complex. A "structure-based mechanism" is a popular goal in the field of biochemistry. An X-ray crystal study is presently not feasible, but as a first step, the solution structure of LumP is being attempted by nuclear magnetic resonance (NMR) techniques. For these purposes, the protocols being used for easy ligand exchange among the antenna proteins, and the method for producing isotope-enriched protein for the NMR measurements, will be detailed here.

Materials

Reagents used are of the highest purity available commercially. The lumazine derivative was a gift of Professor A. Bacher, Munich Technical University, Garching, Germany.

Standard buffer has the composition: 50 m*M* Na/K phosphate with 1 m*M* DTT, pH 7.0.

Spectroscopy

Absorption spectra were measured at 2° with an SLM Aminco UV-vis spectrophotometer (SLM Instruments Inc., Model DW 2000) with a resolution of 1 nm. Steady-state fluorescence emission spectra were measured with an SLM Aminco spectrofluorimeter (SLM Instruments Inc., Model SPF 500C). No correction was made for instrument response. The emission and excitation resolution was 1 nm. Concentrations of free Lum and Rf were measured from absorbance at 408 and 545 nm using molar extinction coefficients, 10,300 and 12,500 M^{-1} cm^{-1}, respectively. For proteins with bound Lum or Rf, the molar extinction coefficients used were 10,100 M^{-1} cm^{-1} for LumP at 420 nm and Lum-BFP at 417 nm, and 12,500 M^{-1} cm^{-1} for Rf-LumP at 463 nm and Rf-BFP at 457 nm.

Growth Media

Yeast-tryptone (YT) growth medium: In 1 liter of standard buffer without dithiothreitol (DTT), dissolve 10 g tryptone, 5 g yeast extract, and 5 g sodium chloride; adjust pH to 7.4 and autoclave.

Defined growth medium: In 1 liter of standard buffer without DTT, dissolve 2 g glycerol and 0.5 g NaCl; adjust pH to 7.4 and autoclave. The following addition is made by filter sterilization to prevent precipitation of the metal constituents and because of the instability of the ampicillin and thiamine at autoclave temperatures. The addition is 100 ml standard buffer without DTT containing 1 g NH_4Cl, 0.25 g $MgSO_4 \cdot 7H_2O$, 0.14 g $CaCl_2 \cdot 2H_2O$, 0.005 g $FeCl_2 \cdot 6H_2O$, 3 g sodium acetate, 0.1 g ampicillin, 0.05 g thiamine.

Bioluminescent bacteria medium: In 1 liter of standard buffer without DTT, dissolve 30 g NaCl, 0.5 g NH_4Cl, 0.1 g $MgSO_4 \cdot 7H_2O$, 10 g bactopeptone (Difco), 3 g yeast extract, 3 ml glycerol; adjust pH to 7.4 and then autoclave.

Solid medium for both *Escherichia coli* or bioluminescent bacteria is the respective complete medium with 15 g/liter bactoagar.

Recombinant Lumazine Protein

The recombinant protein was produced by the procedures described by Illarionov *et al.*[16] *Escherichia coli* BL21 (DE3) cells containing the pPHL36 plasmid inserted downstream of the T7 gene 10 promoter were first isolated as a single colony grown on ampicillin plates, inoculated, and grown in complete medium to verify the efficient production of the desired r-LumP. The procedures are as described in detail elsewhere.[16] The protein is first prepared with Rf as the ligand, and replacement by the natural lumazine derivative or other ligands is accomplished by mass action.

Selection

From the stock of frozen BL21 cells described ealier, a drop is streaked onto solid medium containing ampicillin (0.1 g/liter), kept overnight at 37°, and then at 4–8 hr at room temperature until single colonies appear. A single colony is selected and inoculated to 20 ml YT medium and shaken overnight. The next day, the culture is centrifuged, and the precipitate is used to inoculate the larger 1-liter volume. Ampicillin (0.1 g/liter) is included in the YT medium by filter sterilization.

[16] B. Illarionov, V. Illarionova, J. Lee, W. van Dongen, and J. Vervoort, *Biochim. Biophys. Acta* **1201,** 251 (1994).

Growth and Extraction from Complete Medium

These centrifuged cells can be used to inoculate 10 liters of YT medium and the growth made at 37° with a shaking speed to achieve vigorous aeration, for about 16 hr until midexponential cell density, corresponding to OD(550 nm) = 1.5. Induction is made by the addition of IPTG (100 mg/liter), sucrose (50 m*M*), and riboflavin (0.5 m*M*) at final concentrations indicated in parentheses. After a further 3 hr of growth the cells are harvested by centrifugation (10,000*g*, 30 min). The yield of wet cells is about 2 g/liter.

The cell cake is extracted by suspending it in standard buffer (3 ml/g wet cells) with 5 m*M* DTT, 6 *M* urea (Sigma HPLC grade), benzamidine (10 mg/liter), RNase (10 mg/liter), and DNase (10 mg/liter), and either passing three times through a French press or sonicating three times at 1-min bursts. These and the following procedures are carried out at 2–4° and in low light. The suspension is kept for 4 hr, and then dialyzed (12 hr) against standard buffer with riboflavin (0.5 m*M*), DTT (1 m*M*), and urea (2 *M*). The dialysis buffer is changed to one without the urea and dialysis is continued for a further 12 hr. The material is centrifuged (27,000*g*, 1 hr) and the pellet discarded. The supernatant contains the solubilized protein having bound riboflavin (Rf-LumP).

The yield of protein is more than 30 mg/liter and it is 95% pure by SDS–PAGE.[16] The solution can be concentrated by ultrafiltration (Amicon microflow, YM-10 membrane) to the desired volume.

Growth and Extraction from Defined Medium

The molecular mass of the antenna proteins is in the range that a three-dimensional structure solution is feasible by NMR methods. However, this technique requires that the proteins be enriched by NMR-active nuclei; in addition to the protons, these are ^{15}N and ^{13}C. This is only achievable by growth in a medium where the sole carbon and nitrogen sources are enriched with these isotopes.

The pellet from the centrifuged 2-ml growth is used to inoculate 1 liter of the defined medium made from uniformly ^{13}C-labeled glycerol (Cambridge Isotopes Ltd., Cambridge, MA; CIL-1524), sodium [^{13}C]acetate (CLM-440), and [^{15}N]ammonium chloride (NLM-467). The growth temperature is 28° and induction is performed when the cell density is in the region of OD(550 nm) = 0.5–0.8. Growth is continued for a further 3–4 hr after which the cells are harvested and the protein extracted as described earlier.

The yield of the Rf-LumP from a 1-liter growth in this defined medium is up to 20 mg. As the cost of the labeled chemicals is about $2500 and the conversion to desired product is very inefficient, methods of recycling the growth medium were investigated. The supernatant from the cell harvest is reacted with 1 ml 30% H_2O_2 at room temperature overnight. Catalase (10 mg) is then added to remove all H_2O_2 detected by the luminol test. The medium is then autoclaved and centrifuged to remove precipitate. Ampicillin is added, the medium is reinoculated, and the cells are grown as described earlier. The yield from the recycled growth medium was about 8 mg/liter. This recycling method merits further study for improved product yield.

Purification of Labeled Rf-LumP

SDS–PAGE of the labeled Rf-LumP showed major impurities. Therefore the product was subjected to column chromatography steps using a Pharmacia FPLC system.

The centrifuged cell extract (5–10 ml) is applied to a 2.5 × 14-cm column of Q-Sepharose equilibrated with buffer A (standard buffer containing 10 m*M* 2-mercaptoethanol and 1 m*M* EDTA). The column is then washed with 1 column volume of buffer A and eluted by a linear gradient of 10 column volumes going from buffer A to buffer B (buffer A with 1.0 *M* NaCl). The elution pattern is shown in Fig. 1. All fractions with A(280)/A(462) < 8 are pooled (270–330 ml) and concentrated for application to the size-exclusion column. However, the product resulting from growth in

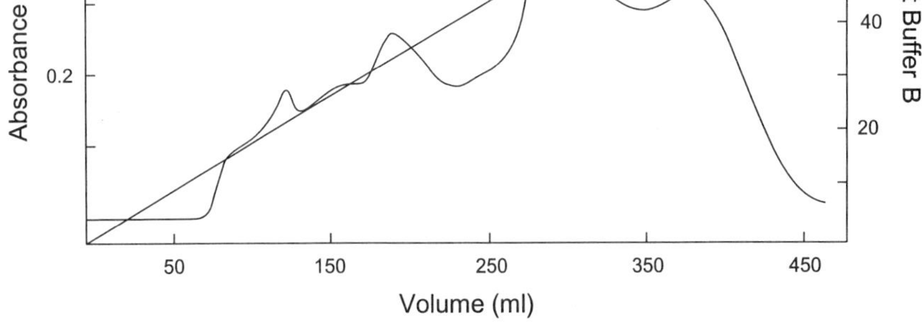

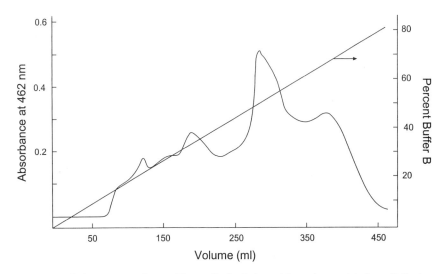

FIG. 1. Elution pattern of recombinant riboflavin-bound lumazine protein from Q-Sepharose at 2°. Buffer A is 50 m*M* Pi, 10 m*M* 2-mercaptoethanol, 1 m*M* EDTA, pH 7.0. Buffer B is buffer A with 1 *M* NaCl.

the recycled medium is rechromatographed on Q-Sepharose by combining the main protein fractions (e.g., 260–450 ml, in Fig. 2) and dialyzing for 12 hr against two changes of buffer A. The dialysate is concentrated, loaded, and eluted by the gradient as before.

Size-exclusion chromatography is on SR-100 HR resin, 2.5 × 70 cm, equilibrated with buffer C (buffer A with 0.1 *M* NaCl). The pooled and concentrated protein from the Q-Sepharose is applied to this column and eluted with buffer C at 2.0 ml/min. Fractions with $A(280)/A(462) < 6$ (Fig. 3) are pooled (#35-50) and desalted by dialysis as described earlier.

The desalted pool from size-exclusion chromatography is applied and eluted again on the same Q-Sepharose column as described previously (Fig. 4). All fractions with $A(280)/A(462) < 4.2$ are pooled and are 95–99% homogeneous on SDS–PAGE. Side fractions with $A(280)/A(462)$ in the range 4.2–8, although containing only about 10% of the total protein, can be pooled, desalted by dialysis, and reapplied to the Q-Sepharose column.

Ligand Exchange

Conditions for the growth of different types of bioluminescent bacteria and purification of the proteins have been described elsewhere.[7,13] In the procedures described here, the "exchange buffer" used is 25 m*M* NaPi, 1

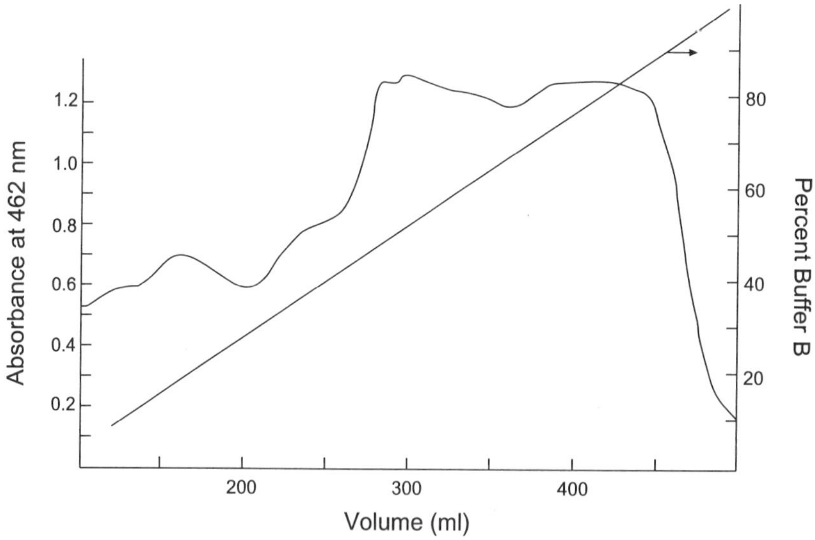

FIG. 2. Recombinant protein product from recycled medium, eluted from Q-Sepharose as in Fig. 1.

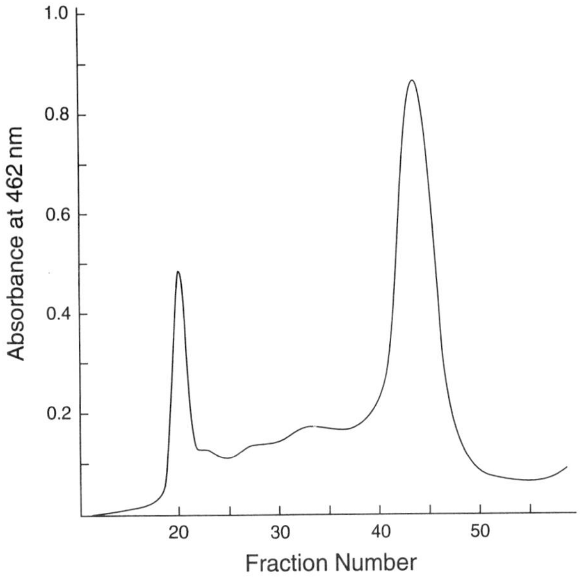

FIG. 3. Elution of the recombinant riboflavin protein from size-exclusion SR-100 with buffer C (buffer A with 0.1 *M* NaCl, 2°).

m*M* EDTA, 10 m*M* 2-mercaptoethanol, pH 7. All experiments are conducted at 2°, unless otherwise indicated.

Riboflavin or Lum ligands on LumP, BFP and YFP from *V. fischeri* Y1, are exchanged by two methods. The first is equilibrium displacement by mass action. A threefold or greater excess of the ligand to be exchanged is included with the protein and the sample is concentrated repeatedly.[7] The second method (GuHCl) affects a more complete and efficient replacement. To a 1-ml solution of protein or a protein precipitate suspended in 1 ml of exchange buffer, 10 ml 6 *M* GuHCl is added and then concentrated to 1 ml using an Amicon membrane-type YM-10. This addition and concentration is repeated three times. The apoprotein in 1 ml 6 *M* GuHCl is then taken in a syringe and is injected rapidly into 50 ml of exchange buffer containing fivefold excess of free ligand. A similar method of fast renaturation (with the exception that urea is used for denaturation) was suggested by Liu *et al.*[17] for another flavoprotein: NADPH : FMN oxidoreductase from *V. harveyi.* After concentration again on the Amicon membrane

[17] M. Y. Liu, B. F. Lei, Z. H. Ding, J. C. Lee, and S.-C. Tu, *Arch. Biochem. Biophys.* **337,** 89 (1997).

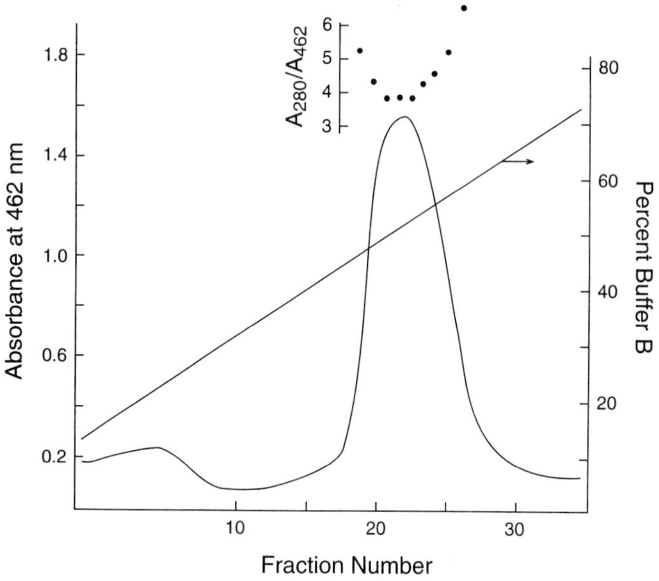

FIG. 4. Final Q-Sepharose elution, 2°. Selected fractions from size exclusion (Fig. 3) are pooled, desalted, and reapplied to ion exchange (Fig. 1).

to 1 ml, the protein is applied to a gel-filtration column (1 × 5-cm Sephadex G25) to remove any excess free ligand.

Chromatography of YFPs and BFP

A detailed account of the growth and extraction of the crude proteins from *V. fischeri* Y1 is available elsewhere.[7,13] Here we will detail only from the first column step. Figure 5 shows the chromatographic separation of the various proteins, including the two forms of YFP, Rf-YFP and FMN-YFP, and the BFP. Accurate absorbance measurements of the YFPs as they come off this column were frustrated by contamination with cytochromes. After the denaturation–exchange procedure with 6 *M* GuHCl, the cytochromes were no longer seen, possibly being removed by aggregation. Also, although the BFP comes mainly with Lum as a ligand, Lum-BFP, it was contaminated with Rf-BFP on the shoulder (Fig. 5, 380–430 ml). To get an accurate spectrum of the BFP, this fraction was exchanged with either the Lum or Rf ligands as desired. Absorption spectra of the BFP fraction before and after these ligand exchanges are shown in Fig. 6.

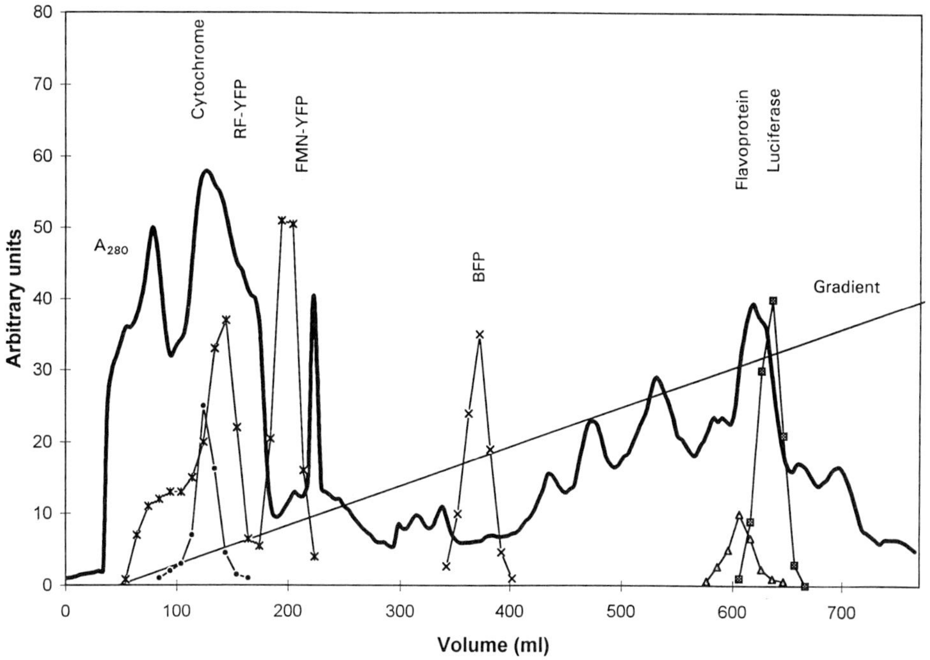

FIG. 5. Elution profile (A_{280}, heavy line; 100 = 2 absorbance units) from the DEAE Q-Sepharose column of *V. fischeri* Y1 cell lysate. Cytochrome: a mixture of cytochromes c4 and c5 (filled circles) assayed by A(418 nm) (100 = 0.5 absorbance units). Luciferase is assayed by bioluminescence activity. The other proteins were assayed by fluorescence (excitation, emission maximum, nm): Rf-YFP and FMN-YFP (crosses, 380, 540). Flavoprotein: thioredoxin reductase (triangles, 380, 520); BFP: blue fluorescence protein (crosses, 420, 462). The fluorescence units are arbitrary. For the gradient, 40 units = 0.4 M NaCl. From V. N. Petushkov, B. G. Gibson, and J. Lee, *Biochem. Biophys. Res. Commun.* **211,** 774 (1995).

Stability

The proteins from Y1, especially the YFP, were found to be much less stable than LumP. After incubation at 100° for a few minutes, LumP from *P. leiognathi* was still 50% in the native form,[13] whereas the Y1 proteins precipitated at temperatures above 30°. It is surprising and not understood that the Y1 proteins cannot be stored frozen for long and under many variations of solution conditions without precipitation and release of the ligand on thawing. Precipitation of YFP was also observed if 6 M urea was used for the exchange and refolding. The only reliable method found to recover the Y1 proteins was by complete unfolding and refolding in the exchange buffer with 6 M GuHCl.

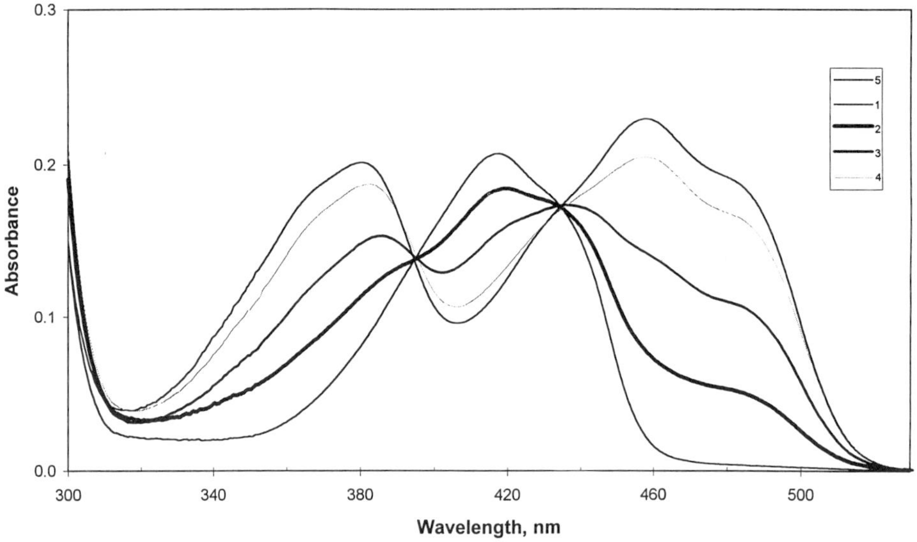

FIG. 6. Samples of BFP protein from *V. fischeri* Y1. Sample 1: Pure Lum-BFP (417 nm max) made by titration by free Lum into the BFP fraction of Fig. 5. Sample 2: This BFP fraction is a mixture of Lum-BFP and Rf-BFP and is used for the recharging experiments and for the samples shown here. After two chromatography purification steps of the Y1 cell extract, samples 3 and 4 were made from sample 2 by simple equilibrium with an excess of riboflavin. Sample 5: Pure Rf-BFP made from the apoprotein by the addition of free riboflavin. Apoprotein was made after removing all Lum by the GuHCl method. There are two isobestic points located at 396 and 436 nm.

Absorption Spectra

The absorption spectra of Lum and Rf, free and ligated to the different proteins, are compared in Figs. 7 and 8. The spectral parameters are found in Table I. In every case the absorbance of the bound ligand is shifted to a longer wavelength relative to the free ligand. Also, the protein spectra, although similar, can be distinguished from each other. Spectra in Fig. 8 are typical of flavoproteins with two bands in the near UV-visible region. Rf-YFP (Fig. 8, filled squares) and Rf (Fig. 8, open circles) have relatively smooth bands, whereas Rf-BFP and Rf-LumP show a vibronic structure, a shoulder around 480 nm. The presence of this shoulder indicates that the flavin-binding environment in these last two proteins is not the same as in YFP. Absorption maxima are compared in Table I.

The Lum-YFP dissociates during anion-exchange chromatography; the association is too weak to allow determination of absorption spectrum of the complex. The fluorescence spectrum of a solution is also dominated by that of the dissociated ligand.[7]

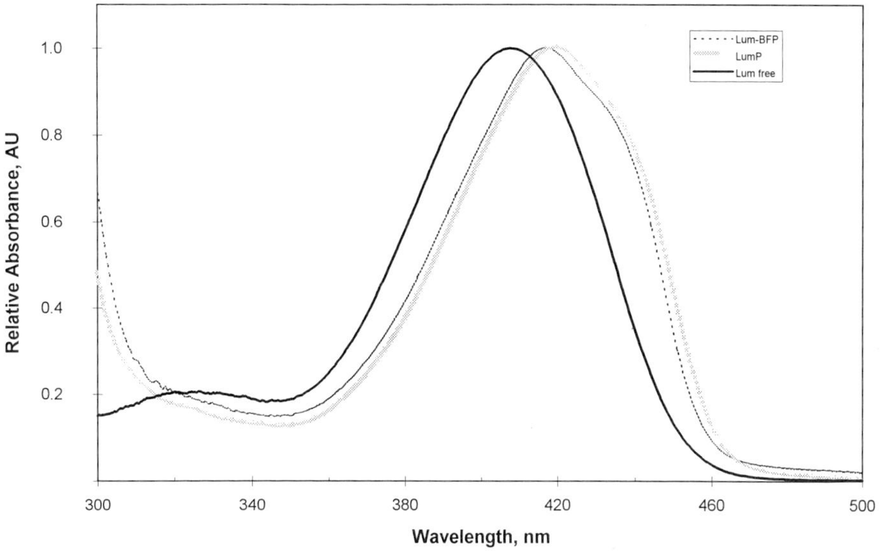

FIG. 7. Comparative normalized absorption spectra of Lum-BFP (max 417 nm) from *V. fischeri* Y1, LumP (max 420 nm) from *P. leiognathi,* and free Lum (max 408 nm) (25 m*M* NaPi buffer, pH 7, 1 m*M* EDTA, 10 m*M* 2-mercaptoethanol, with 10–20 μM protein or free ligand; room temperature).

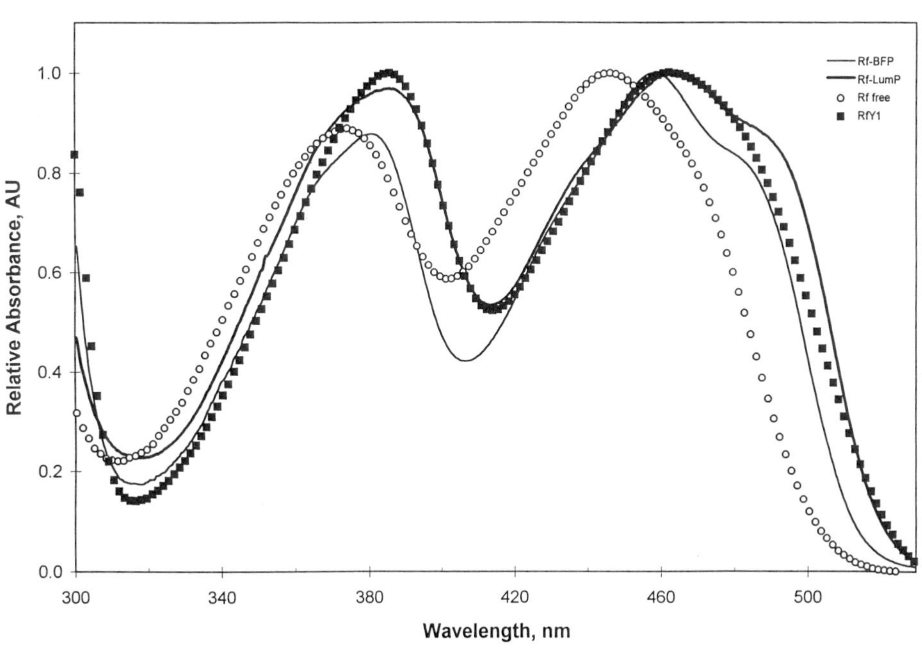

FIG. 8. Comparison of normalized absorption spectra of riboflavin proteins. Rf-YFP (yellow fluorescence protein from *V. fischeri* Y1); Rf-LumP (LumP from *P. leiognathi*); Rf-BFP (riboflavin form of BFP from *V. fischeri* Y1); and free riboflavin. Conditions as in Fig. 7.

TABLE I
SPECTROSCOPIC PROPERTIES OF FREE AND PROTEIN-BOUND RIBOFLAVIN (Rf) AND 6,7-DIMETHYL-8-RIBITYLLUMAZINE (Lum)[a]

	Absorbance max (nm)		Fluorescence max[b] (nm)		Fluorescence lifetime (nsec)	
Sample	Rf	Lum	Rf	Lum	Rf	Lum
BFP	381, 457	417	520	463	5.7	13.4
YFP	387, 464	nd	544[c]	480[d]	7.5	13.4
LumP	386, 463	420	529	469	6.0	15.3
Rf	373, 445		526		4.9	
Lum		408		481		10.0

[a] BFP, blue fluorescence protein with from *V. fischeri* Y1: LumP, lumazine protein from *P. leiognathi.* Rf-BFP and Rf-LumP, BFP and LumP recharged by riboflavin (25 m*M* NaPi buffer, pH 7.0, 1 m*M* EDTA, 10-m*M* 2-mercaptoethanol, 22°); nd, not determined. Fluorescence lifetimes are measured at 0° and are from V. N. Petushkov and J. Lee, *Eur. J. Biochem.* **245,** 790 (1997).

[b] Fluorescence spectra are uncorrected.

[c] From V. N. Petushkov, B. G. Gibson, and J. Lee, *Biochem. Biophys. Res. Commun.* **211,** 774 (1995).

[d] From V. N. Petushkov and J. Lee, *Eur. J. Biochem.* **245,** 790 (1997).

Fluorescence Spectra

Fluorescence spectra in Fig. 9, along with absorption spectra (Figs. 7 and 8), clearly indicate a different environment of the ligands on the proteins as compared with water. For bound Lum the fluorescence maxima are shifted to shorter wavelength, more in the case of BFP, but for Rf-LumP and Rf-YFP (Table I), fluorescence maxima are at a longer wavelength, and Rf-BFP at a shorter wavelength than for free Rf. The Rf proteins all show a long wavelength shoulder (Rf-YFP not shown).

Stoichiometric Replacement of Rf by Lum on BFP

Figure 10 shows the near-UV visible absorption spectrum of Rf-BFP (22.4 μM) upon the addition of Lum. After the first few additions, an isobestic point at 457 nm can be seen and, by comparison with Fig. 8, this is simply explained as conversion of bound Rf to free. On more additions, as the Lum concentration exceeds 20 μM, spectra become more complicated, resulting from contributions from Rf-BFP, Lum-BFP, and Lum. Clearer evidence for the displacement of Rf from the protein is obtained by plotting the absorbance changes at 489.7 nm, where the contribution of Lum (free and bound) was very low (Fig. 7). At this wavelength the absorbance is almost entirely from the Rf, free or bound (compare Fig. 8). The break

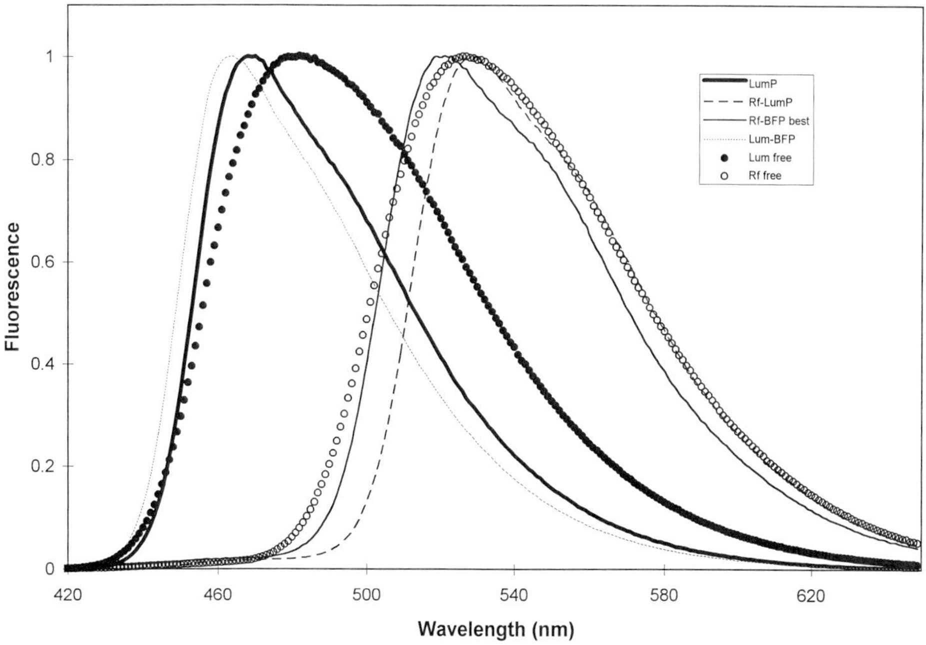

FIG. 9. Normalized fluorescence spectra of free and bound ligands. Fluorescence maxima from left to right are 463, 469, 481, 520, 526, and 529 nm. Thin line, left is Lum-BFP, right is Rf-BFP. The heavy line, left is LumP, right is Rf-LumP (recharged by riboflavin). The closed circles are for free Lum and the open circles for free riboflavin. All spectra were measured at room temperature (25 m*M* NaPi buffer, pH 7, 1 m*M* EDTA, 10 m*M* 2-mercaptoethanol, with 10–20 μM protein or free ligand). For the Lum free and bound, excitation was at 420 nm, and for riboflavin and riboflavin proteins, excitation was at 380 nm. Spectra are not corrected for instrument response. Excitation and emission resolution was 1 nm.

point at a 1:1 ratio of Lum to Rf protein indicates that the stoichiometry of binding of both ligands is the same (Fig. 11). We conclude that on BFP both Lum and Rf have the same binding site.

A similar experiment was made with the Rf-LumP, and this isobestic point was at 461 nm (data not shown). It is also concluded that for both BFP and LumP proteins, the Lum binds more strongly than Rf. The spectroscopic parameters for these proteins are shown in Table I.

The interpretation of the titration curve is supported by spectra of the purified proteins (Fig. 6). These spectra are taken after the removal of excess ligand by gel filtration. After normalization on protein concentration, the two isobestic points located at 396 and 436 nm are evidence of competition between Lum and Rf for one binding site. For the best Rf-BFP sample, the ratio 457/405 nm = 2.38. Other samples give higher ratios due to

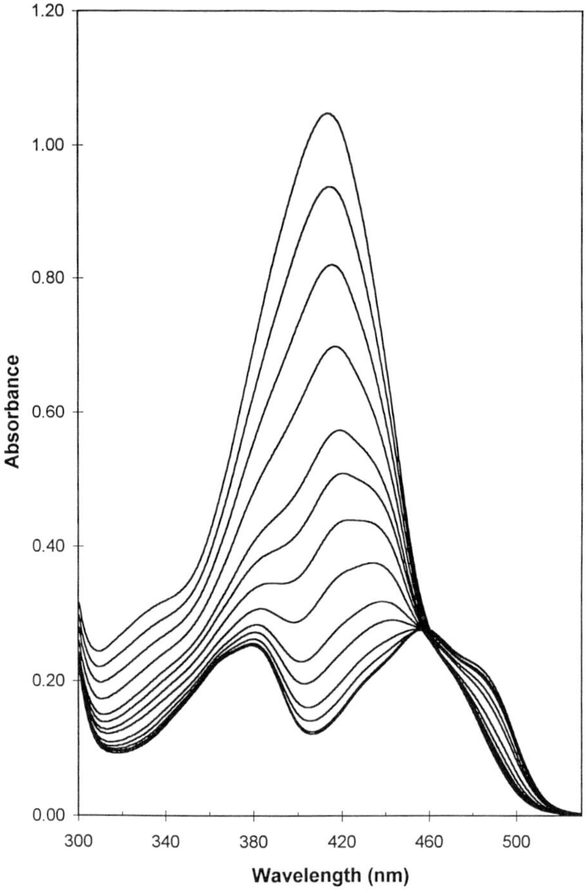

FIG. 10. Titration of Rf-BFP by free Lum. Initially, BFP was recharged by riboflavin to a final concentration of 22.4 μM for protein. Then the first spectrum was taken (heavy line). At 1-min intervals, microvolumes of a 3 mM solution of Lum were added to the sample in the spectrophotometer cell. Spectra were scanned after each addition. The final concentration of Lum in the cell for the family of curves from lowest to the highest was 0 μM (heavy line) and 1.5, 3, 6, 9, 15, 21, 27, 33, 45, 57, 69, and 91 μM (20°, in NaPi buffer, pH 7.0).

contamination by bound Lum. Complete replacement of Lum by Rf was possible only by using the GuHCl method.

Discussion

Bioluminescent organisms are a rich source of highly fluorescent proteins. The fluorescence property has an obvious biological function for

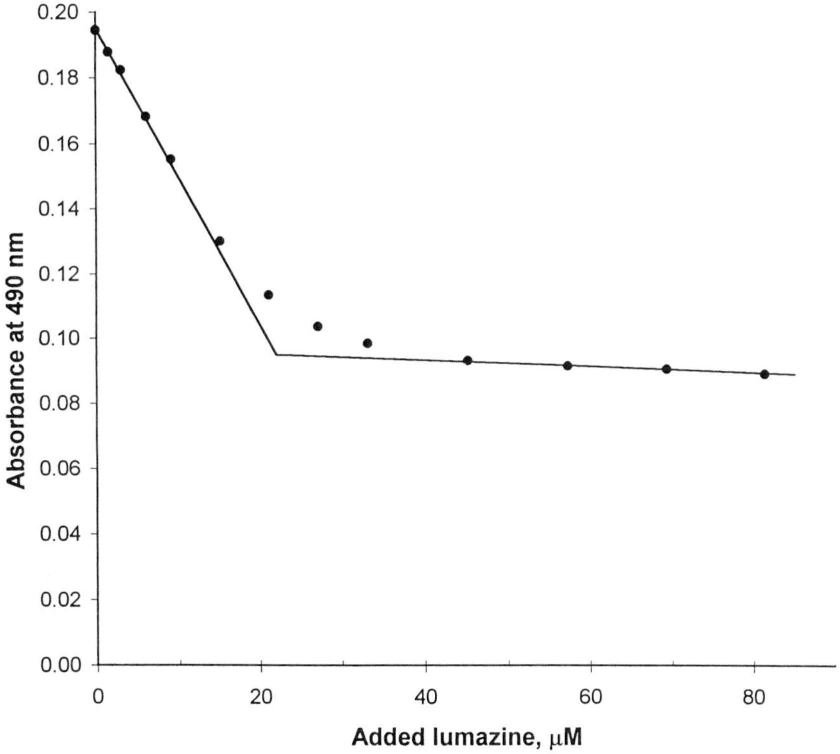

FIG. 11. Each point is the absorbance at 490 nm from the family of spectra in Fig. 10, plotted against the final concentration of added free Lum. The titration break point indicates the amount of free Lum needed to replace the riboflavin on the protein is close to 1:1.

those proteins that participate in the bioluminescence reaction. It is remarkable that for proteins having bound flavin their fluorescence is unusually high, as most flavoproteins exhibit only weak fluorescence, being quenched by interactions with aromatic residues near the binding site, e.g., flavodoxin. As most flavoproteins play a role in electron transfer reactions, the quenching of their fluorescence may be significant in this respect. For the present cases, however, it is concluded that the binding site environment is distinct from that of the major classes of flavoproteins in having no local quenching residues. The location of this binding site for the subject proteins is unknown, as are other details of their structural features. Only LumP has received extensive spectral and protein characterization, and it is found that the ligand is rigidly bound yet solvent exposed. The binding environment also appears to be highly polar, as suggested by the fluorescence spectral positions and from NMR results.[18]

The ligands examined here are natural metabolites in procaryotes. The binding strength depends on the ligand structure and, as determined for the *Photobacterium* case, the stereochemistry of the side group.[19] In *Photobacterium,* only the LumP species has bioluminescence, yet the cells contain major amounts of the Rf-LumP without evident function.[13] For *V. fischeri* (Y1), the YFP with either FMN or Rf has bioluminescence activity, but if Lum is the ligand this is less certain, and there is an additional related protein BFP, also binding these same ligands, at a single site and in an analogous fashion. The variety of related proteins now appearing in the lumazine-binding family implies that they may have additional functions yet to be discovered.

Acknowledgments

V. N. Petushkov was a recipient of a fellowship from The Netherlands Organization for Scientific Research (NWO). Work was supported in part by NWO and 7F0107 from the Krasnoyarsk Regional Science Foundation.

[18] J. Lee, *Biophys. Chem.* **48,** 149 (1993).
[19] J. Lee, B. G. Gibson, D. J. O'Kane, A. Kohnle, and A. Bacher, *Eur. J. Biochem.* **210,** 711 (1992).

[13] Purification of Firefly Luciferase from Recombinant Sources

By THOMAS O. BALDWIN and VICKI A. GREEN

Firefly luciferase has been used for several decades for the measurement of ATP levels in biological samples. Until relatively recently, the source of this enzyme was the lantern organs of fireflies, and during the 1950s the late Professor W. D. McElroy paid school children a penny per insect to collect fireflies in the Baltimore area.[1] During the 1970s, Sigma Fine Chemicals sponsored the Firefly Club and supplied firefly collection kits and financial rewards to children to collect fireflies as a source of this valuable enzyme.

[1] J. W. Hastings, *J. Biolumin. Chemlumin.* **4,** 29 (1989).

0076-6879/00 $30.00

It has been well appreciated by biochemists for many years that it is critical to choose a biological source for a specific enzyme that is an abundant source of that enzyme. Obviously, for firefly luciferase, it was necessary to use the firefly as the source of the enzyme, whereas bovine trypsin is the most commonly used trypsin for obvious reasons. An abundant source of the desired enzyme was a necessary prerequisite for obtaining highly purified enzyme. However, with the development of recombinant DNA technology, the preferred source of enzymes has become the host in which the highest levels of overexpression of the recombinant protein can be obtained. In the case of firefly luciferase, the highest reported levels of expression of the enzyme have been from bacteria in which, interestingly, expression is under control of the genetic system that is responsible for expression of the bacterial bioluminescence genes.[2] The bacterial *lux* gene system has been demonstrated to be one of the best systems currently available for the overexpression of foreign proteins in *Escherichia coli*,[3] and firefly luciferase is a prime example. As in days long past, it is still true that one of the most important steps in the purification of an enzyme is in choosing the best biological source for the enzyme. However, today, the admonition to find the best source should be directed to the host/vector system for overexpression of the recombinant protein.

Description of the *lux* Vector System and the Autoinduction Mechanism

The *Vibrio fischeri lux* vector system has three primary control elements: the *luxR* gene and the *luxI* gene, which are transcribed in opposite directions, and the control region between *luxR* and *luxI* (Fig. 1).[4–7] The *luxR* gene encodes a transcriptional activator protein, LuxR,[4] and *luxI* encodes a synthase, LuxI,[4,8] which produces a small molecule "autoinducer", *N*-(3-oxo)hexanoyl-DL-homoserine lactone.[9] To obtain high level transcription of genes downstream of the *luxI* gene requires the LuxR protein, the

[2] J. H. Devine, G. D. Kutuzova, V. A. Green, N. N. Ugarova, and T. O. Baldwin, *Biochim. Biophys. Acta* **1173,** 121 (1993).

[3] M. D. Thomas and A. van Tilburg, *Methods Enzymol.* **305** [23], 315 (2000) (this volume).

[4] J. Engebrecht, K. Nealson, and M. Silverman, *Cell* **32,** 773 (1983).

[5] J. H. Devine, C. Countryman, and T. O. Baldwin, *Biochemistry* **27,** 837 (1988).

[6] T. O. Baldwin, J. H. Devine, R. C. Heckel, J.-W. Lin, and G. S. Shadel, *J. Biolumin. Chemilumin.* **4,** 326 (1989).

[7] J. H. Devine, G. S. Shadel, and T. O. Baldwin, *Proc. Natl. Acad. Sci. U.S.A.* **86,** 5688 (1989).

[8] J. Engebrecht and M. Silverman, *Proc. Natl. Acad. Sci. U.S.A.* **81,** 4154 (1984).

[9] A. Eberhard, A. L. Burlingame, C. Eberhard, G. L. Kenyon, K. H. Nealson, and N. J. Oppenheimer, *Biochemistry* **20,** 2444 (1981).

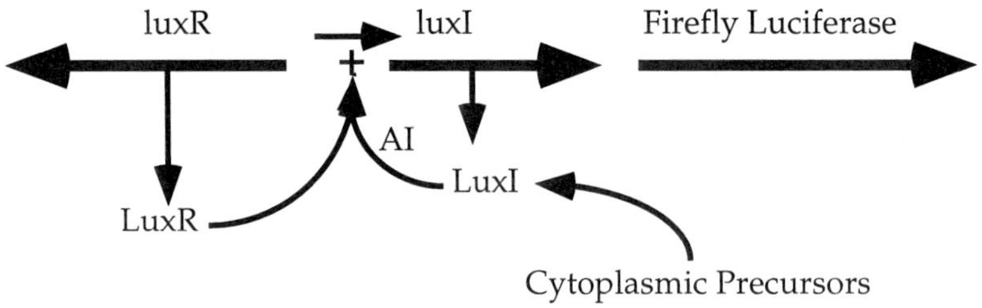

FIG. 1. Schematic representation of the bacterial *lux* expression system as designed for overexpression of the luciferase from the European firefly, *Luciola mingrelica*.

autoinducer, and a binding site for the LuxR protein in the control region that is upstream of the transcription start for the *luxI* gene and the downstream region. In the natural setting in the bioluminescent bacterium *V. fischeri,* the genes downstream of *luxI* encode the α and β subunits of the bacterial luciferase as well as the genes that encode the enzymes that supply the aldehyde substrate for the bacterial bioluminescence reaction.[4] When bioluminescent bacteria are grown in liquid culture, an intriguing phenomenon called autoinduction is observed.[10] From a dilute inoculum, cells undergo multiple divisions without any increase in the expression of bioluminescence. Following this lag, once cells reach late exponential phase growth, expression of the luminescence function is induced as a consequence of accumulation of the autoinducer in the growth medium and expression of the LuxR protein that is regulated by the CRP/cAMP system.[11,12] If the *luxI* gene is defective, or if it has been deleted, it is necessary to add exogenous autoinducer to obtain expression of the protein of interest. However, in most laboratory strains of *E. coli,* the LuxI protein is able to supply the required levels of autoinducer.

By replacing bioluminescence-related genes with genes that encode other proteins of interest, we have been able to develop a procaryotic expression system that is capable of yielding both high bacterial cell densities and high levels of protein of interest per cell.[3,13] The discussion that follows describes how to grow cells containing the *lux* expression vector with the coding sequence for the luciferase from the European firefly *Luciola mingrelica* and how to extract the protein from the cells and to purify the protein to homogeneity in high yield.[2] For this enzyme, we used a vector

[10] K. H. Nealson, T. Platt, and J. W. Hastings, *J. Bacteriol.* **104,** 313 (1970).

[11] P. V. Dunlap and E. P. Greenberg, *J. Bacteriol.* **164,** 45 (1985).

[12] P. V. Dunlap and E. P. Greenberg, *J. Bacteriol.* **170,** 4040 (1988).

[13] T. O. Baldwin, L. H. Chen, L. J. Chlumsky, J. H. Devine, and M. M. Ziegler, *J. Biolumin. Chemilumin.* **4,** 40 (1989).

with an intact *luxI* gene, so cells carrying this plasmid produce autoinducer and do not require an exogenous autoinducer to direct the overexpression of the firefly luciferase. This is a distinct difference from the format for this vector that we have employed in the overexpression of bacterial luciferase in *E. coli.*[14]

A schematic representation of the expression vector is shown in Fig. 1. This plasmid, designated pJGR,[2] has three main components: the *luxR* and *luxI* genes, which control expression of the luciferase, and the firefly luciferase coding sequence. Plasmids of this type are discussed in this volume by Thomas and van Tilburg,[3] and the quorum-sensing system of luminous bacteria is discussed in this volume by Greenberg and co-workers.[15] In pJGR, the *luxI* gene is intact, so cells carrying this plasmid vector produce the autoinducer from endogenous substrates. The LuxR protein is expressed in late log-phase growth. Together with the autoinducer, which accumulates as the cell density increases, the LuxR protein enhances the transcription of *luxI* and other genes downstream by binding to a site, the "lux box," upstream of *luxI.*[7]

Growth of Bacterial Cells Expressing Firefly Luciferase

Considerations in Choosing a Host Strain

In choosing a host strain of *E. coli* for overexpression of firefly luciferase using the bacterial *lux* expression system, there are several considerations. First, it is essential that the strain support the biosynthesis of autoinducer. Currently, we know of no strains of *E. coli* that do not, but it is possible that strains with lower than normal levels of *S*-adenosyl methionine or 3-oxohexanoyl acyl carrier protein, the proposed substrates for the LuxI protein,[16] could cause difficulty using the vector construct with the intact *luxI*. Second, a strain of *E. coli* that grows to high cell densities is essential. Third, in some cases, elevated levels of cellular chaperones could be advantageous.[17] Finally, the strain of *E. coli* should be compatible with the antibiotic selection used on the plasmid vector. In the present case, it is necessary that the host be sensitive to ampicillin.

[14] T. O. Baldwin, M. M. Ziegler, V. A. Green, and M. D. Thomas, *Methods Enzymol.* **305** [9], 135 (2000) (this volume).

[15] A. L. Schaefer, B. L. Hanzelka, M. R. Parsek, and E. P. Greenberg, *Methods Enzymol.* **305** [21], 288 (2000) (this volume)

[16] M. I. Moré, L. D. Finger, J. L. Stryker, C. Fuqua, A. Eberhard, and S. C. Winans, *Science* **272,** 1655 (1996). This paper demonstrated the enzymatic activity of TraI, the LuxI homolog from *Agrobacterium tumifaciens.*

[17] K. M. Dolan and E. P. Greenberg, *J. Bacteriol.* **174,** 5132 (1992).

Growth Medium

For overexpression of recombinant proteins in *E. coli,* we use exclusively the rich growth medium known as Terrific Broth.[18] Because the medium is so rich, it is prone to caramelization if it is autoclaved too long. Therefore, one should pay strict attention to the following instructions and use the medium within a short time after autoclaving as it is prone to becoming contaminated due, in part, to the brief sterilization times employed.

The following recipe yields 3 liters of 3× Terrific Broth (the actual yield is closer to 3.2 liters), sufficient for six Fernbach flasks, each with 1.5 liters of medium, which should yield ca. 200–250 g of packed cells: 50 g of glycerol, 114 g of Bacto-tryptone (peptone or enzymatic casein hydrolysate works well also), 228 g of Bacto or other yeast extract, and 3 liters of H_2O.

For medium to grow the inocula for the larger flasks, take 150 ml of the 3× Terrific Broth, dilute with 300 ml H_2O, and dispense 45 ml into each of ten 250-ml culture flasks. For the main growth, dispense 500 ml of the 3× Terrific Broth into each of six Fernbach flasks and add 850 ml of H_2O to each.

The antibiotic is added to the medium in a sterile phosphate buffer solution immediately prior to inoculation. The following recipe is for 1 liter of the phosphate buffer, sufficient for the 9-liter growth described here.

In a 2-liter flask, combine the following ingredients: 23.1 g of KH_2PO_4, 125 g of K_2HPO_4, and 950 ml of H_2O.

Cover the 250-ml culture flasks, the Fernbach flasks, and the phosphate buffer flask with foil and autoclave for 20 min on the liquid cycle and remove the medium promptly to avoid caramelization. *Do not autoclave longer.* The medium should be used within 24 hr as it may not be fully sterilized.

Detailed Cell Growth Procedure

This procedure starts with single cells carrying the firefly luciferase-encoding plasmid growing on a fresh overnight agar plate prepared with ampicillin (or carbenicillin) at 50 μg/ml. To begin the cell growth procedure, finish preparing the medium in four of the 250-ml culture flasks by adding the antibiotic in 5 ml phosphate buffer to each. Inoculate each flask with a single colony from the seed culture agar plate and allow the cells to grow at 37° with the shaker set to 350 rpm.

The next morning, measure and record the OD_{600} of the overnight cultures. Collect the cells by centrifugation for 5 min at 5000 rpm in 50-ml

[18] K. D. Tartof and C. A. Hobbs, *Bethesda Res. Lab.* **9,** 12 (1987)

conical centrifuge tubes. While the cells are being collected, add the selection antibiotic to the remaining buffer salt solution and dispense 5 ml into each of the six remaining 250-ml culture flasks and 150 ml into each Fernbach flask. Resuspend the cell pellets in a small volume of Terrific Broth and divide equally among the 250-ml culture flasks. Allow cells to grow at 37° with high aeration (350 rpm) for 6 hr. Measure and record the OD_{600} of the cultures.

Again, collect the cells in separate (6 total) 50-ml tubes, resuspend the cells in each centrifuge tube in a few milliliters of fresh broth, and pour all of the cells from one conical tube into each Fernbach.

Allow the cells to grow 24–36 hr with high aeration (350 rpm). The incubation temperature should be reduced to 18° for optimal expression. At 25° or higher, the levels of accumulation of firefly luciferase were reduced significantly. *Do not cover the flasks tightly while they are incubating.* The cell density obtained in this method is so high that oxygen will become limiting. Cell yield will be reduced by roughly 50% if the flasks are covered with foil. Optimally, the flasks should remain uncovered. However, phage contamination is a potential problem, and some strains produce aromatic secondary products that some may find offensive.

It is a good idea to do a trial growth in which samples are withdrawn at various times to determine the window of time during which accumulation of the protein of interest is maximal. Measure and record the OD_{600}, and harvest the cells by centrifugation for 15 min at 5000 rpm in tared 1-liter centrifuge bottles. Measure and record the yield of cell paste. This method normally yields 200–250 g of cell paste.

Resuspend the cells in 1.25 ml lysis buffer (0.1 *M* Tris–acetate, 10 m*M* $MgSO_4$, 2 m*M* EDTA, pH 7.8) per gram cell paste. Because cells grown as described here tend to be somewhat fragile and are prone to premature lysis we supplement the buffer with 100 m*M* NaCl. Lysis is accomplished most readily with a French press using a cell pressure of 16,000 psi to disrupt the cells. Clarify the resulting cell lysate by centrifugation for 30 min at 15,000 rpm in a Beckman JA-20 rotor, or equivalent. Collect the supernatant and proceed with the purification.

Purification Protocol for *Luciola mingrelica* Luciferase

The purification method described here has been reported elsewhere,[2] as has the assay for firefly luciferase.[19] While methods for purification have improved since the original publication, the methods described are efficient and result in an overall yield >50% of nearly homogeneous protein. The

[19] K. V. Wood and M. A. DeLuca, *Anal. Biochem.* **161,** 501 (1987).

methods described should be translated readily to more modern chromatographic media to shorten the time required for purification.

Ammonium Sulfate Fractionation

Add solid ammonium sulfate to the clarified crude lysate on ice to 35% saturation. Allow the solution to stir slowly for ca. 15 min and then pellet precipitated protein by centrifugation for 15 min at 15,000 rpm in a Beckman JA-20 rotor, or equivalent. Collect the supernatant and add additional ammonium sulfate to bring the concentration to 75% saturation. Again, allow ca. 15 min for protein precipitation and then pellet the precipitated protein by centrifugation as before. Discard the supernatant and save the pellet. Dissolve the protein that precipitated in 75% saturation ammonium sulfate in a minimum amount of assay buffer (0.1 *M* Tris–acetate, 10 m*M* $MgSO_4$, 2 m*M* EDTA, pH 7.8). The crude protein may be stored at $-20°$ following flash freezing with liquid N_2 at this stage.

Chromatographic Purification

To remove ammonium sulfate, dialyze the protein exhaustively against 0.05 *M* Tris–acetate, 2 m*M* EDTA, 35 m*M* $MgSO_4$, and 12% glycerol. Centrifuge to remove any precipitated protein and apply to a DEAE-Sephadex A-50 column (5×20 cm, 300 ml bed volume) equilibrated in the same buffer. After loading, wash the column with 500 ml of the loading buffer and then elute the column at 15 ml/hr with a linear gradient formed from 1 liter each of the starting buffer and the same buffer with 80 m*M* $MgSO_4$. Monitor the luciferase activity[19] and protein concentration of each fraction, and pool and concentrate those fractions having the highest specific activity. A representative elution profile is shown in Fig. 2A.

Partially purified luciferase (see Fig. 2D, lane 4) is purified further using a column identical to the first but with a smaller bed volume (5×15 cm). Elute the column with the same gradient with the same volumes and flow rate as the first column. Pool the fractions with the highest specific activity, concentrate and dialyze against the same buffer with 60 m*M* $MgSO_4$, and apply to a 1.5×19-cm column of Blue Sepharose CL-6B equilibrated in the same buffer. Elute the column isocratically; the luciferase is not bound tightly by the Blue Sepharose and elutes shortly after the breakthrough volume. Again, monitor fractions for protein concentration and luciferase activity, and pool those fractions with a constant specific activity.

Elution profiles for the three chromatography steps are shown in Fig. 2, together with a Coomassie blue-stained SDS polyacrylamide gel showing the purity of the luciferase following each step. As can be seen from the gel, this method produces highly pure luciferase in good overall yield.

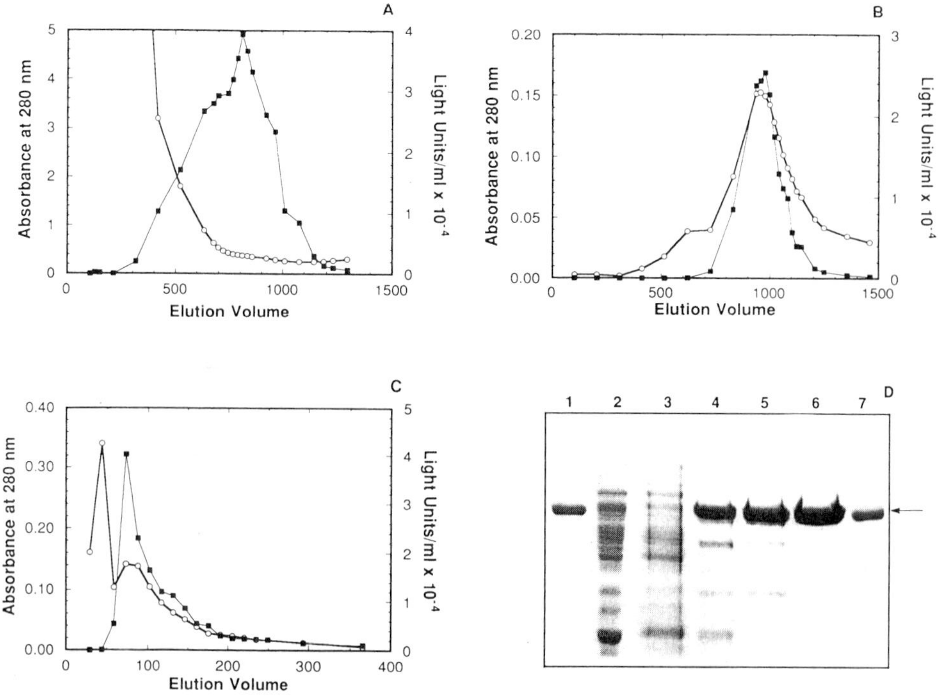

FIG. 2. Purification of *Luciola mingrelica* luciferase from *E. coli* carrying the plasmid pJGR (Fig. 1). (A) The elution profile of cellular protein, including luciferase, from the first DEAE Sephadex A-50. (B) The elution profile obtained by chromatography of the pooled luciferase containing fractions from A on a second, smaller DEAE Sephadex columnn using an identical gradient. (C) The elution profile of proteins contained in luciferase-containing fractions from B. (D) A Coomassie blue-stained polyacrylamide gel in which proteins were resolved by electrophoresis in SDS-containing buffers. Lanes 1 and 7 represent commercial *Photinus pyralis* firefly luciferase purchased from Boehringer Mannheim. Lane 2 contains proteins from the clarified crude lysate, and lane 3 contains proteins precipitating between 35 and 75% saturation ammonium sulfate. Lane 4 contains pooled protein from the chromatogram shown in A, lane 5 contains pooled protein from B, and lane 6 contains pooled protein from C. Reproduced from J. H. Devine, G. D. Kutuzova, V. A. Green, N. N. Ugarova, and T. O. Baldwin, *Biochim. Biophys. Acta* **1173,** 121 (1993), with permission.

Summary Comments

Our experience with purification of the firefly luciferase provides another example of the principle that one of the most important steps in modern protein purification is optimization of expression of the recombinant protein. Overexpression of the protein makes purification much faster and simpler, with much higher yields. It should be stressed that in purification of firefly luciferase, as with all proteins, time is a critical parameter. To optimize yields and minimize chemical heterogeneity in the final product, it is essential to execute all steps as quickly as possible and to avoid as much as possible maintaining the protein under dilute conditions. We should point out that use of DEAE-Sepharose Fastflow, rather than DEAE A-50, would allow flow rates of 5 ml/min, rather than the 15 ml/hr described here. Using DEAE-Sepharose Fastflow, the first two columns should be completed in ca. 3 hr each, dramatically shortening the time for purification. Ideally, using the newer column materials, this purification protocol should be completed within a total of 4 hr.

[14] Chemical Synthesis of Firefly Luciferin Analogs and Inhibitors

By BRUCE R. BRANCHINI

Introduction

Synthetic structural analogs and inhibitors of natural enzyme substrates have proved very useful in the investigation of structure–function relationships of a wide variety of enzymes, including luciferases. Beetle luciferases use a common substrate firefly luciferin, D-(-)-2-(6′-hydroxy-2′-benzothiazolyl)-Δ^2-thiazoline-4-carboxylic acid, first isolated from the North American firefly *Photinus pyralis* by Bitler and McElroy.[1] Firefly luciferin was first chemically synthesized by White *et al.*,[2,3] and the structure, including stereochemistry, was confirmed by X-ray crystallography[4] and is shown in Fig. 1. Previously, Bowie[5] has reviewed the synthetic pathways to firefly luciferin and several structurally related compounds. Additionally, he has described a detailed procedure for the preparation of firefly luciferin. This article updates Bowie's contribution,[5] focusing on improved synthetic methods and novel luciferase substrates and inhibitors. New compounds, such as those discussed here, may be useful for future luciferase structure–function studies, especially since the crystal structure of *P. pyralis* luciferase (without bound substrates) has been reported.[6]

[1] B. Bitler and W. D. McElroy, *Arch. Biochem. Biophys.* **72,** 358 (1957).
[2] E. H. White, F. McCapra, G. F. Field, and W. D. McElroy, *J. Am. Chem. Soc.* **83,** 2402 (1961).
[3] E. H. White, F. McCapra, and G. F. Field, *J. Am. Chem. Soc.* **85,** 337 (1963).
[4] G. E. Blank, J. Pletcher, and M. Sax, *Biochem. Biophys. Res. Commun.* **42,** 583 (1971).
[5] L. J. Bowie, *Methods Enzymol.* **57,** 15 (1978).
[6] E. Conti, N. P. Franks, and P. Brick, *Structure* (*Lond.*) **4,** 287 (1996).

METHODS IN ENZYMOLOGY, VOL. 305 0076-6879/00 $30.00

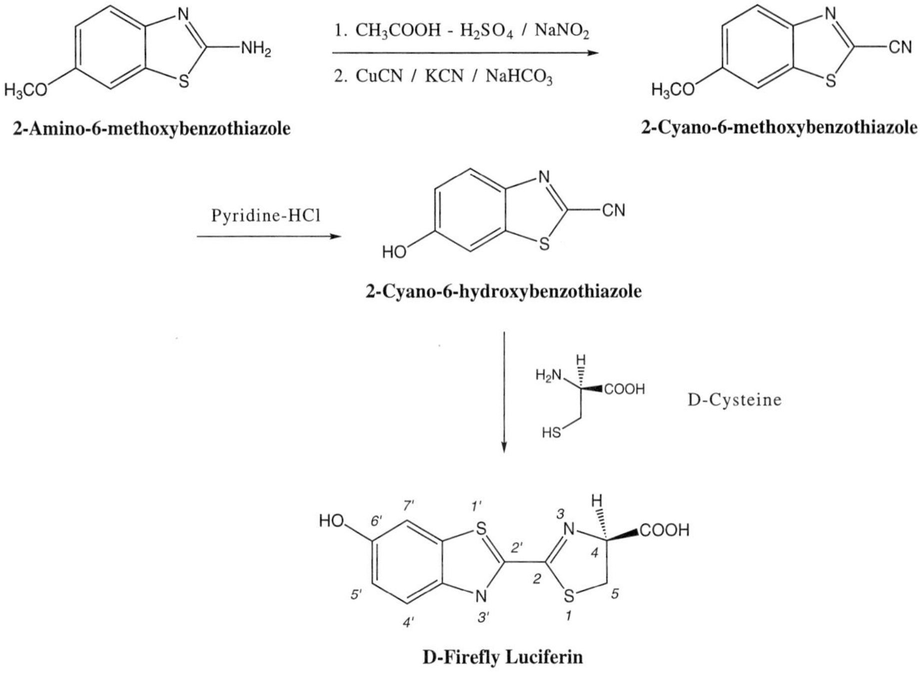

FIG. 1. Synthetic route to the preparation of firefly luciferin. From S. Seto, K. Ogura, and Y. Nishiyama, *Bull. Chem. Soc. Jpn.* **36,** 332 (1963); E. H. White, H. Wörther, G. F. Field, and W. D. McElroy, *J. Org. Chem.* **30,** 2344 (1965); and Y. Toya, M. Takagi, H. Nakata, N. Suzuki, M. Isobe, and T. Goto, *Bull. Chem. Soc. Jpn.* **65,** 392 (1992).

Luciferin Synthesis

Bowie[5] has provided a detailed description of the chemical synthesis of firefly luciferin, which is based on the method of Seto *et al.*[7] The key intermediate in this pathway (and in White's original synthesis[2,3]) is 2-cyano-6-methoxybenzothiazole, which is prepared in three steps from commercially available *p*-anisidine. White *et al.*[8] later reported an improvement in the preparation of the key compound that required only two steps, starting with commercially available 2-amino-6-methoxybenzothiazole. White's method was refined further by Goto *et al.*,[9] who used a one-step Sandmeyer cyanation reaction to prepare 2-cyano-6-methoxybenzothiazole

[7] S. Seto, K. Ogura, and Y. Nishiyama, *Bull. Chem. Soc. Jpn.* **36,** 332 (1963).

[8] E. H. White, H. Wörther, G. F. Field, and W. D. McElroy, *J. Org. Chem.* **30,** 2344 (1965).

[9] Y. Toya, M. Takagi, H. Nakata, N. Suzuki, M. Isobe, and T. Goto, *Bull. Chem. Soc. Jpn.* **65,** 392 (1992).

from the same commercial starting material. The synthetic pathway shown in Fig. 1 reportedly[9] provides a large-scale (tens of grams) route to the key synthetic intermediate. The method employed by Toya *et al.*[9] to remove the methoxy-protecting group of 2-cyano-6-methoxybenzothiazole (see Fig. 1) is similar to the one described in this article and is much more convenient than Bowie's procedure.[5]

Firefly luciferin is also available now as a "caged" substrate,[10] which is actually the 1-(4,5-dimethoxy-2-nitrophenyl) ester of firefly luciferin. This luciferin derivative is cell and organelle permeable and can release luciferin on photolysis or as a result of endogenous esterolytic activity.[10]

Luciferyl and Dehydroluciferyl Adenylate

The adenylates of luciferin and dehydroluciferin, the former a substrate for luciferase-catalyzed bioluminescence in the presence of oxygen, can be synthesized chemically according to the original method of Morton *et al.*[11] Preparations of the adenylates have also been described in detail by Bowie.[5] Subsequent improvements in the synthesis and characterization of these compounds have been reported.[12,13] Significantly, the stability of these compounds on storage can be improved dramatically by adjusting the pH of solutions to approximately 3.5 and keeping them at −20 or −70°.

Structural Analogs of Luciferin

Many structural analogs of firefly luciferin have been synthesized and these are also discussed in Bowie's[5] review and the references provided therein. All of the reported analogs contain the benzothiazole ring system, and those few that function as light-emitting substrates with luciferase display considerably red-shifted bioluminescence emission spectra. Our laboratory[14] has prepared and characterized two novel luciferin analogs, D-naphthylluciferin and D-quinolylluciferin. The structures of these substrates are shown in Fig. 2, and procedures for their preparation are described later. Interestingly, the naphthyl-containing analog is the only

[10] J. Yang and D. B. Thomason, *Biotechniques* **15,** 848 (1993).
[11] R. A. Morton, T. A. Hopkins, and H. H. Seliger, *Biochemistry* **8,** 1598 (1969).
[12] K. Imai and T. Goto, *Agric. Biol. Chem.* **52,** 2803 (1988).
[13] A. Kukhovich, A. Sillero, and M. A. Günther Sillero, *FEBS Lett.* **395,** 188 (1996).
[14] B. R. Branchini, M. H. Hayward, S. Bamford, P. M. Brennan, and E. J. Lajiness, *Photochem. Photobiol.* **49,** 689 (1989).

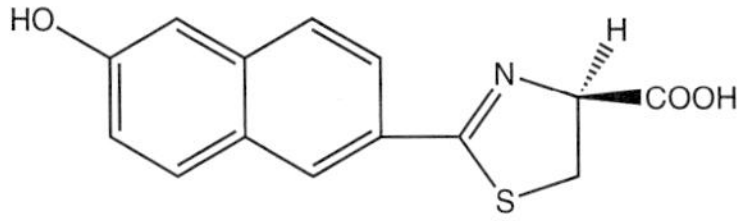

D-Naphthylluciferin ($NpLH_2$)

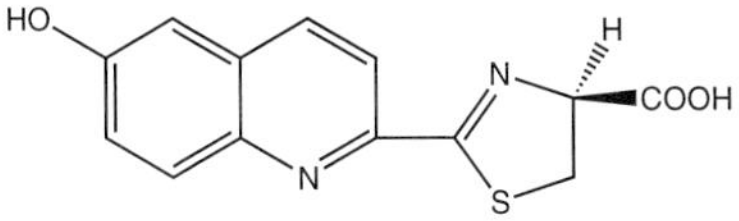

D-Quinolylluciferin (QLH_2)

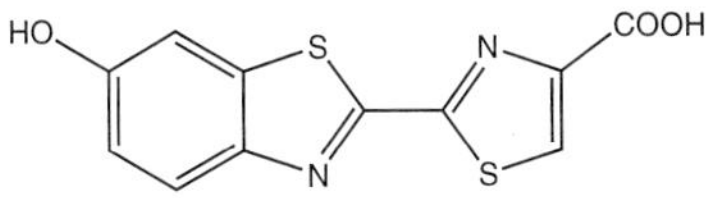

Dehydroluciferin

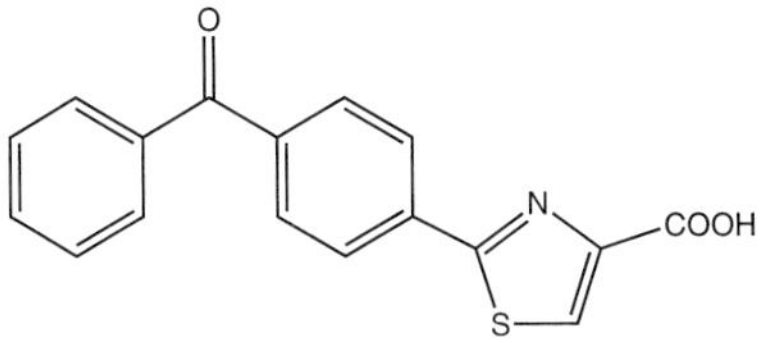

BPTC

FIG. 2. Chemical structures of firefly luciferin structural analogs and inhibitors.

example of a *P. pyralis* luciferase substrate whose emission spectrum is blue-shifted (λ_{max} = 524 nm) compared to that of the native substrate (λ_{max} = 560 nm). In contrast, D-quinolylluciferin is an orange-red light (λ_{max} = 608 nm)-emitting substrate for luciferase. A comparison of the properties of the substrate analogs to those of firefly luciferin are presented in Table I.

Procedure

The substrate analogs D-naphthylluciferin and D-quinolylluciferin are prepared by the general reaction sequence shown in Fig. 1 for the preparation of firefly luciferin, starting with the syntheses of the 2-cyano-6-methoxy-substituted naphthalene and quinoline compounds. These intermediates are deprotected with pyridine hydrochloride, which must be dried over

TABLE I
COMPARISON OF SUBSTRATE PROPERTIES OF FIREFLY LUCIFERIN AND ANALOGS WITH FIREFLY LUCIFERASE

Property	pH	$NpLH_2$	QLH_2	Luciferin
Bioluminescence				
pH optima	—	9.3	8.6	8.0
λ_{max} (nm)	6.0	—	608	611
	7.8	524	—	560
	8.6	524	608	560
	9.3	524	608	560
Relative light yield[a]	7.8	—	—	100.0
	8.6	—	7.0	83.0
	9.3	1.5	—	24.0
Light emission kinetics				
Time (sec) to maximum intensity	6.0	—	—	0.86
	7.8	—	—	0.55
	8.6	—	3.1	0.58
	9.3	9.0	—	0.60
Decay time (min to 10% of maximum intensity)	6.0	—	—	5.2
	7.8	—	—	6.2
	8.6	—	11.5	2.0
	9.3	28.4	—	1.3
K_M (μM)	7.8	—	—	15 ± 4
	8.6	—	31 ± 5	22 ± 5
	9.3	24 ± 7	—	42 ± 9

[a] All light yields are reported relative to the value for luciferin at pH 8.0, which was set arbitrarily to 100.

phosphorus pentoxide overnight *in vacuo* prior to use. The 2-cyano-6-hydroxy-substituted compounds are then condensed with D-cysteine to yield the respective substrates.

Preparation of 2-Cyano-6-hydroxynaphthalene

6-Methoxy-2-naphthonitrile (1.0 g, 5.5 mmol) and pyridine hydrochloride (13 g, 0.12 mol) are stirred together at 200° (bath temperature) for 1 hr in a stoppered flask. The mixture solidifies on cooling to room temperature, and 10% sodium carbonate is added until the resulting suspension reaches pH 8 (litmus). A light brown solid is isolated by filtration, dried *in vacuo*, and recrystallized from toluene to yield 0.77 g (4.6 mmol, 83%)

of product with m.p. 164.5–165.5° (m.p. 164–168°, Curd and Raison[15]). IR (KBr, cm^{-1}) 3400, 2220, 1605, 1485, 1390, and 1175.

Preparation of D-(-)-*2-(6′-Hydroxy-2′-naphthyl)-Δ^2-thiazoline-4-carboxylic acid (*D-*naphthylluciferin)*

The following procedures are based on the method of White *et al.*[3] Nitrogen is bubbled for 15 min through all solutions and, under a nitrogen atmosphere, a solution of 2-cyano-6-hydroxynaphthalene (266 mg, 1.6 mmol) in 20 ml methanol is added dropwise to a freshly prepared aqueous solution (18 ml) of D-cysteine (568 mg, 4.7 mmol) that has been adjusted to pH 8.0. Methanol (16 ml) is added and stirring continued for 18 hr. The work up of the reaction mixture is as described by White *et al.*[3] for the preparation of luciferin. The crude material is recrystallized from methanol, yielding 257 mg (0.93 mmol, 60%) of a yellow crystalline product with m.p. 201.5–203.5° (decomp). IR (KBr, cm^{-1}) 2500(b), 1575(b), and 1260(b); UV (25 m*M* Tris buffer, pH 9.3) λ_{max} 251 nm (log ε = 4.63) and 315 nm (log ε = 4.10); optical rotation $[\alpha]_D^{18} = -74.15°$ (c = 0.0126 DMSO); ^{1}H-NMR (DMSO-d_6) δ 3.61 (*dd*, 1H,H_{5a}), 3.73 (*dd*, 1H,H_{5b}), 5.30(*t*, 1H,H_4), 7.15 (*m*, 2H,$H_{5'}$,$H_{7'}$), 7.74 (*d*, 1H,$H_{4'}$), 7.83(*d*, 1H,$H_{8'}$), 7.90 (*d*, 1H,$H_{3'}$), and 8.17(*s*, 1H,$H_{1'}$); ^{13}C-NMR (DMSO-d_6) δ 34.8(C-5), 78.2(C-4), 108.6(C-5′), 119.35(C-7′), 124.4(C-3′), 126.3(C-4′), 126.6(6-2′), 126.6(C-8a′), 129.0(C-1′), 130.5(C-8′), 136.1(C-4a′), 157.0(C-6′), 168.0(C-2), and 171.6(COOH). Analysis, $C_{14}H_{11}NO_3S$; Calc. %: C, 61.52; H, 4.06; N, 5.125; S, 11.73; Found: C, 61.35; H, 4.18; N, 5.11; S, 11.63.

Preparation of 2-Cyano-6-hydroxyquinoline

According to the method described earlier for 2-cyano-6-hydroxynaphthalene, crude 2-cyano-6-hydroxyquinoline is prepared from dry pyridine hydrochloride (24 g, 0.20 mol) and 2-cyano-6-methoxyquinoline (2.0 g, 11 mmol). A dark-green solid is obtained that is extracted into ethyl acetate (100 ml) for 30 hr using a Sohxlet extractor. The solution is boiled down to 20 ml, 20 ml of toluene is added, and the volume of the solution is reduced to half. On standing, a white crystalline product forms, which is isolated by filtration and dried *in vacuo* to yield 0.70 g (4.1 mmol, 38%) with m.p. 222–224.5° (m.p. 218–219°, Kaneko *et al.*[16]) IR (KBr, cm^{-1}) 3000(b), 2217, 1550(b), 1375, 1325, 1200, and 1110.

[15] F. H. S. Curd and C. G. Raison, *J. Chem. Soc.* 160 (1947).
[16] C. Kaneko, H. Hasegawa, S. Tanaka, K. Sunayashiki, and S. Yamada, *Chem. Lett.* **2,** 133 (1974).

Preparation of D-(-)-*2-(6'-Hydroxy-2'-quinolyl)-Δ^2-thiazoline-4-carboxylic acid* (D-*quinolylluciferin*)

Nitrogen is bubbled for 15 min through all solutions used in the following procedure, which is based on Seto *et al.*[7] Under a nitrogen atmosphere, a solution of 2-cyano-6-hydroxyquinoline (133 mg, 0.78 mmol) in methanol (7 ml) is added slowly to an aqueous solution (7 ml) of D-cysteine · HCl (283 mg, 2.0 mmol) and potassium carbonate (0.276 g, 2.0 mmol). The mixture is stirred for 18 hr, the methanol is removed by rotary evaporation, and the pH is adjusted to 2 (litmus) with 1 *N* HCl. The yellow precipitate, which forms on cooling to 5°, is collected by filtration and dried *in vacuo* to yield 148 mg (0.54 mmol, 69%) of a solid with m.p. 147–148° (decomp). IR (KBr, cm^{-1}) 3000, 1710, 1570, 1460, 1375, 1210; UV (50 m*M* ammonium acetate buffer, pH 8.0) λ_{max} 217 nm ($\log \varepsilon = 4.27$), 259 nm ($\log \varepsilon = 4.36$), 329 nm ($\log \varepsilon = 3.86$), and 342 nm ($\log \varepsilon = 3.87$); optical rotation $[\alpha]_D^{18} = -139.72°$ ($c = 0.0025$, DMSO); ^{1}H-NMR (DMSO-d_6) δ 3.61(*dd,* 1H,H_{5a}), 3.73 (*dd,* 1H,H_{5b}), 5.40(*t,* 1H,H_4), 7.18(*d,* 1H,$H_{5'}$), 7.34(*dd,* 1H,$H_{7'}$), 7.88(*d,* 1H,$H_{8'}$), 7.98(*d,* 1H,$H_{3'}$), and 8.20(*d,* 1H,$H_{4'}$); ^{13}C-NMR (DMSO-d_6) δ 33.5(C-5), 78.8(C-4), 108.4(C-5'), 118.6(C-7') 122.9(C-3'), 130.3(C-4a'), 130.8(C-8') 135.0(C-4'), 141.8(C-8a'), 147.1(C-2'), 157.0(C-6'), 171.7(C-2), and 171.8(COOH). Analysis (high resolution mass spectrum); Calculated for $C_{13}H_{10}N_2O_3S$: 274.0412: Found: 274.0408.

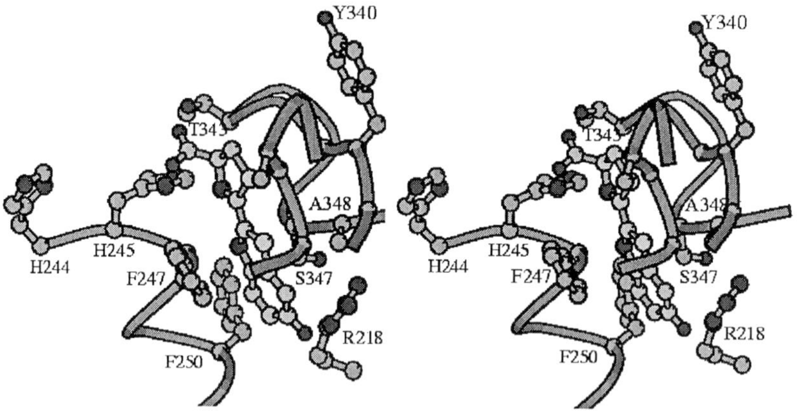

FIG. 3. Stereo diagram showing interactions predicted by molecular modeling[21] of the *Photinus pyralis* luciferase X-ray crystal structure[6] with substrate firefly luciferin added. Peptide backbone traces through the α carbons of regions 244–252, 314–319, and 338–351 are shown as coils. The α carbons of glycine 246, glycine 315, glycine 316, alanine 317 (and side chain methyl), glycine 339, and glycine 341 are shown but are not labeled. H, histidine; F, phenylalanine; T, threonine; S, serine; A, alanine; Y, tyrosine; and R, arginine. This diagram was generated using the program MOLSCRIPT.[23]

Inhibitors Related to Firefly Luciferin

Several inhibitors of firefly luciferase structurally similar to firefly luciferin have been prepared and studied.[17–19] Although these compounds effectively modified or abolished enzyme activity, no specific luciferase amino acid residues involved in the chemical modification processes were ever identified. Our laboratory[20] has reported the preparation and characterization of a benzophenone-containing substrate analog BPTC, 2-(4-benzoylphenyl)thiazole-4-carboxylic acid, and this structure is shown in Fig. 2. BPTC mediated the efficient photooxidation of *P. pyralis* luciferase very likely by catalyzing the formation of singlet oxygen at the active site of the enzyme. The photooxidation of histidine 245 has been suggested[21] to account for the observed loss of luciferase activity. According to our proposed model[21] of the luciferase substrate-binding pocket for luciferin shown in Fig. 3, histidine 245, as well as arginine 218, is within the hydrogen-bonding distance of the luciferin carboxylate and phenolate ions, respectively. Our model[21] of the luciferase active site was produced by molecular modeling methods and is based on the X-ray structures of *P. pyralis* luciferase[6] and the functionally related phenylalanine-activating subunit of gramicidin synthetase 1 in a complex with its substrate phenylalanine and AMP.[22] Perhaps our model[21] and the available X-ray structure[6] together will be of use in the design of new substrates and inhibitors for firefly luciferase.

[17] M. DeLuca, *Adv. Enzymol.* **44,** 37 (1976).

[18] E. H. White and B. R. Branchini, *J. Am. Chem. Soc.* **97,** 1243 (1975).

[19] E. H. White and B. R. Branchini, *Methods Enzymol.* **46,** 537 (1977).

[20] B. R. Branchini, R. A. Magyar, K. M. Marcantonio, K. J. Newberry, J. G. Stroh, L. K. Hinz, and M. H. Murtiashaw, *J. Biol. Chem.* **272,** 19359 (1997).

[21] B. R. Branchini, R. A. Magyar, M. H. Murtiashaw, S. M. Anderson, and M. Zimmer, *Biochemistry* **37,** 15311 (1998).

[22] E. Conti, T. Stachelhaus, M. A. Marahiel, and P. Brick, *EMBO J.* **16,** 4174 (1997).

[23] P. J. Kraulis, *J. Appl. Crystallogr.* **24,** 946 (1991).

[15] Structural Basis for Understanding Spectral Variations in Green Fluorescent Protein

By S. JAMES REMINGTON

Introduction

Few molecules have captured the imagination of cell and molecular biologists in the way that the green fluorescent protein (GFP) from the Pacific Northwest jellyfish *Aequorea victoria* has done. In just a few years GFP has gone from nearly complete obscurity to become one of the most widely studied and valuable probes of subcellular structure and gene expression, and new uses and more interesting properties of this unusual protein are being discovered on nearly a daily basis. The ability of the protein to spontaneously generate a visually appealing, nontoxic and efficient fluorophore in a host-independent process, combined with ease of genetic manipulation and new techniques in video microscopy, has opened the door to the study of cellular processes in ways that could hardly have been imagined just a few years ago. The engineering of GFP to generate chimeric proteins can give us, in a living cell, X, Y, Z, *and* T for nearly any process involving proteins. Finally, high-resolution crystal structures are permitting the rational design of differently colored GFPs that will provide simultaneous probes of two or more cellular processes. GFP has been firmly established as the probe of choice for environmental monitoring, protein targeting and gene expression in living cells.

Short History of Green Fluorescent Protein

GFP was discovered by Shimomura *et al.*[1] in the early 1960s while isolating the luciferase (aequorin) of the luminescent jellyfish *A. victoria.* They described a protein that was weakly greenish in sunlight but was intensely fluorescent in the ultraviolet of a Mineralite, which already hinted at the complex optical properties of the molecule. *In vitro,* aequorin emits blue light; however, the jellyfish emits green light, suggesting resonant energy transfer from aequorin to GFP *in vivo.* Morin and Hastings[2] later described a similar protein isolated from the sea pansy *Renilla* and pointed out that some luminescent coelenterates appeared to lack GFP, i.e., the emitted light is blue both *in vivo* and *in vitro.* Why the jellyfish uses GFP to emit green light and what the advantage of doing so, remain a mystery.

[1] O. Shimomura, F. H. Johnson, and Y. Saiga, *J. Cell. Comp. Physiol.* **59,** 223 (1962).
[2] J. G. Morin and J. W. Hastings, *J. Cell. Physiol.* **77,** 313 (1971).

0076-6879/00 $30.00

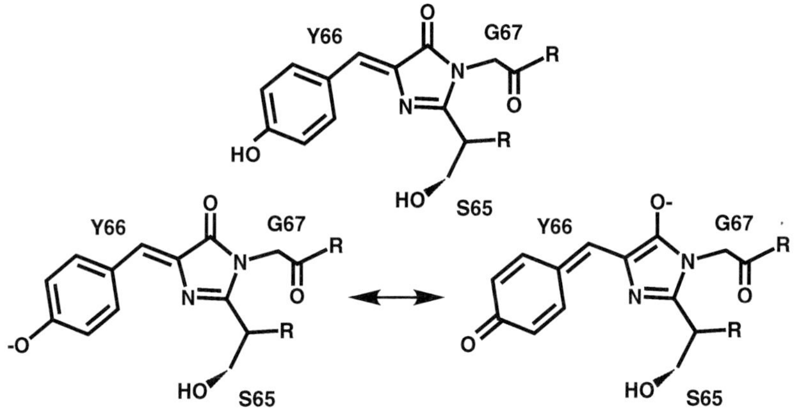

FIG. 1. (Top) Chemical structure of the chromophore in green fluorescent protein in its neutral form. (Bottom) Proposed resonant species of the anionic form of the chromophore. The "R" groups represent continuation of the polypeptide backbone. Attack of the amide nitrogen of Gly67 on the carbonyl carbon of Ser65 forms the heterocycle, whereas subsequent oxidation of the α-β bond of Tyr66 conjugates the phenol ring with the heterocycle.

GFP studies continued in relative obscurity for the next two decades. Its molecular mass was established by Prendergast and Mann[3] to be about 30 kDa. Shimomura[4] concluded that the chromopeptide derived by proteolysis contained 4-(*p*-hydroxybenzylidine)imidazolidin-5-one, covalently attached to the backbone of the protein (Fig. 1), later determined to be derived from the internal protein sequence -Ser-dehydroTyr-Gly-.[5] The isolated peptide containing the chromophore has a similar absorption spectrum as the intact protein, but is not fluorescent. However, when model compounds and the proteolytically derived peptide are frozen in ethanol glass, the resulting material is highly fluorescent,[6] suggesting that the intact protein prevents dissipation of absorbed light energy by nonfluorescent processes, most likely *cis-trans* isomerization of the $C\alpha$-$C\beta$ bond of the chromophore.[6] Furthermore, the protein must be folded properly to be fluorescent. Denaturation and renaturation behavior were characterized by Bokman and Ward,[7] and the renatured protein has a spectrum identical to the native

[3] F. G. Prendergast and K. G. Mann, *Biochemistry* **17,** 3448 (1978).

[4] O. Shimomura, *FEBS Lett.* **104,** 220 (1979).

[5] C. W. Cody, D. C. Prasher, W. M. Westler, F. G. Prendergast, and W. W. Ward, *Biochemistry* **32,** 1212 (1993).

[6] G. Niwa, S. Inouye, T. Hirano, T. Matsuno, S. Kojima, K. Masayuki, M. Ohashi, and F. I. Tsuji, *Biochemistry* **93,** 13617 (1996).

[7] S. H. Bokman and W. W. Ward, *Biochem. Biophys. Res. Commun.* **101,** 1372 (1981).

protein.[8] The protein is remarkably stable and, over the range of pH ~3 to 12, it is stably folded, even in 8 *M* urea. It is also quite thermostable and resists the attack of most proteases for many hours.

One of the most significant steps toward the understanding of GFP biology was the cloning of the gene and determination of the primary structure by Prasher *et al.*[9] in 1992. The internal peptide that forms the chromophore was determined to be derived from the sequence -Ser65-Tyr66-Gly67-. The protein consists of 238 amino acids, and subsequent deletion analysis determined that residues 2–232 are required for the formation of fluorescence.[10] However, at that time it was not known whether the chromophore required jellyfish enzymes for formation. Thus, it was not until Chalfie *et al.*[11] demonstrated the heterologous expression of GFP cDNA in both eukaryotes and prokaryotes in 1994 that it became clear to the scientific community that chromophore formation was almost certainly autocatalytic and thus could in principle lead to the formation of visible fluorescence in virtually any living cell. The autocatalytic formation of the chromophore and the requirement for molecular oxygen for maturation[12] have been confirmed by Reid and Flynn.[13] Their detailed kinetic analysis showed that folding to the native, nonfluorescent conformation is fairly slow and is followed by the relatively rapid cyclization of the imidazolidinone, whereas the final maturation of the chromophore, the oxidation of the $C\alpha$-$C\beta$ bond of Tyr 66, is very slow (in wild type on the order of 4 hr) and requires molecular oxygen. These workers also showed that the chromophore could be reduced in the unfolded protein and that reacquisition of fluorescence follows a similar time course, as does the maturation of the native protein.

Spectral Properties

Light absorption by wild-type *Aequorea* GFP is characterized by a broad absorption spectrum with two peaks; the major peak is in the long-wave UV region at about 395 nm and the minor peak at about 475 nm (Fig. 2). The intensity ratios of these two peaks are typically about 4 : 1, but variations in protein concentration, salt, and pH influence this ratio.[14] A considerable

[8] W. W. Ward and S. H. Bokman, *Biochemistry* **21,** 4535 (1982).

[9] D. C. Prasher, V. K. Eckenrode, W. W. Ward, F. G. Prendergast, and M. J. Cormier, *Gene* **11,** 229 (1992).

[10] J. Dopf and T. M. Horiagon, *Gene* **173,** 39 (1996).

[11] M. Chalfie, Y. Tu, G. Euskirchen, W. W. Ward, and D. C. Prasher, *Science* **263,** 802 (1994).

[12] R. Heim, D. C. Prasher, and R. Y. Tsien, *Proc. Natl. Acad. Sci. U.S.A.* **91,** 12501 (1994).

[13] B. G. Reid and G. C. Flynn, *Biochemistry* **36,** 6786 (1997).

[14] W. W. Ward, H. J. Prentice, A. F. Roth, C. W. Cody, and S. C. Reeves, *Photochem. Photobiol.* **35,** 803 (1982).

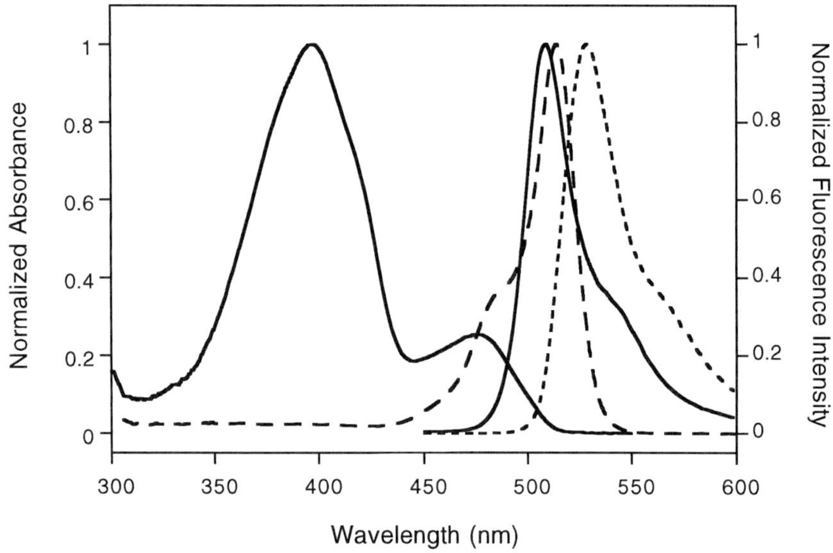

FIG. 2. Absorption and emission spectra of wild-type GFP (solid lines) superimposed on absorption and emission spectra of a rationally designed yellow mutant T203Y (dashed lines). For wild type, the emission spectrum was recorded with irradiation at 395 nm; for the mutant at 510 nm.

amount of evidence favors the interpretation that the chromophore species that absorbs at 395 nm in the protonated phenol form, whereas the longer wavelength peak may be attributed to the phenolate anion (see later section). In solution, the chromophore has a pK_a of 8.1,[14] but interestingly, in the intact protein the chromophore begins to titrate only at a pH where the protein is beginning to unfold (at about pH 12[14]), therefore the chromophore seems to be completely isolated from bulk solvent at physiological pH. In contrast to wild-type *Aequorea* GFP, the *Renilla* GFP has a single absorption maximum at about 495 nm, but the chromophore has been shown spectrophotometrically to be identical to the *Aequorea* chromophore.[15] This suggests that details of the interactions between the chromophore and the rest of the protein can have profound effects on the spectral properties, and indeed this has been firmly established. Fluorescence emission is from the phenolate form of the chromophore and peaks at 508 nm regardless of which absorption band is excited.[6]

[15] W. W. Ward, C. W. Cody, R. C. Hart, and M. J. Cormier, *Photochem. Photobiol.* **31,** 611 (1980).

Spectral Variations by Random Mutagenesis

The availability of GFP cDNA quickly led several groups to search for point mutations that altered the absorption spectrum, the emission spectrum, or both. Three major classes of mutations were discovered, some independently by more than one group. Ehrig *et al.*[16] discovered two mutations that effectively eliminated one or the other of the two major absorption bands of wild-type protein. The mutation T203I eliminated the minor absorption band at 475 nm, whereas E222G eliminated the 395-nm absorption band, presumably by shifting the equilibrium between charged and uncharged chromophore species. Similar results were obtained with mutations at residue Ile167 (Val, Thr) by Heim *et al.*[12] These results can be rationalized easily on the basis of the three-dimensional structure (discussed later).

It is genuinely surprising that the tripeptide that becomes the chromophore is very tolerant of substitutions. The central tyrosine and the adjacent serine have been mutated to a variety of other amino acids and fluorescent proteins are still obtained. Mutations at Ser65 are all very similar to the E222G mutation and suppress the 395-nm absorption with a concomitant increase in the 475-nm peak. Heim *et al.*[12] mutated Ser65 to Ala, Leu, Cys, and Thr, and all of these showed a single absorption band peaking at 470–490 nm. S65T has particularly desirable properties. It has an extinction coefficient that is six times higher than wild type, and red shifts in the absorption to 489 nm and in the emission to 510 nm. The protein also matures approximately four times faster than the wild type with a $t_{1/2}$ of about 90 min. Delagrave *et al.*[17] mutagenized the hexapeptide 64–69 and recovered similar mutations at residue 65: Leu, Cys, Gly, and Ala. One of these, RSGFP4 (RS for "red shifted"), a triple mutant F64M/S65G/Q69L, has been used by several groups for a variety of purposes. It has absorption and emission maxima of 490 and 505 nm, thus the emission is actually slightly blue shifted relative to wild type.

Finally, Tyr66 has been mutated to all other aromatic possibilities: His, Trp, and Phe. All are fluorescent (Phe only weakly so) but have radically altered absorption *and emission* spectra.[12] Of particular interest is Y66H, which has absorption and emission maxima of 382 and 448 nm (blue fluorescent protein, BFP). The emission maximum overlaps the absorption maximum of S65T or RSGFP4, leading to the possibility of resonance energy transfer between the two molecules. In the Applications for Biotechnology section, some uses of this effect are discussed. Finally, Y66W has

[16] T. Ehrig, D. J. O'Kane, and F. G. Prendergast, *FEBS Lett.* **367,** 163 (1995).

[17] S. Delagrave, R. E. Hawtin, C. M. Silva, M. M. Yang, and D. C. Youvan, *Bio/Technology* **13,** 151 (1995).

absorption and emission maxima of 458 and 480 nm, intermediate between Y66H and S65T (cyan fluorescent protein).

Structure of GFP and Chromophore Environment

Although diffraction quality crystals of GFP were reported as early as 1988,[18] the crystal structures of dimeric wild-type GFP[19] and monomeric S65T[20] were determined independently by Yang *et al.* and us in 1996 using multiple isomorphous replacement and anomalous dispersion techniques. Other groups subsequently determined the structures of monomeric wild type[21] and the mutant Y66H[22,23] by molecular replacement from the S65T structure. Coordinates for the protein are available from the Protein Data Bank (PDB). The two original structures have the PDB identification codes 1GFL[19] and 1EMA.[20] They are all very similar and reveal the molecule to be an extremely unusual 11-stranded β-barrel surrounding a rather distorted helix. Short, distorted α-helices cap each end of the barrel, which forms a nearly perfect cylinder that has been described as a "β-can."[19] The chromophore is located close to the exact center of the molecule (Fig. 3) and is formed from the central helix. The presumably rigid barrel and α-helical end caps undoubtedly have at least two functions: (1) to prevent access of bulk solvent to the chromophore (reducing quenching by molecular oxygen) and (2) to prevent the chromophore from dissipating energy by isomerization or vibrational phenomena. The structure is also an important key to understanding how the chromophore is generated. The initial, presumably α-helical, conformation of the central helix brings the amide nitrogen of Gly67 close to the carbonyl carbon of residue 65, which would promote nucleophilic attack.

Why then don't helices in other proteins form similar imidazolinone structures? The answer appears to be in several lines of evidence. GFP does not fold easily at temperatures above about 25°, indicating a fairly large energy barrier in the folding process, and tends to form inclusion

[18] M. A. Perozzo, K. B. Ward, R. B. Thompson, and W. W. Ward, *J. Biol. Chem.* **263,** 7713 (1988).

[19] F. Yang, L. G. Moss, and G. N. Phillips, Jr., *Nature Biotechnol.* **14,** 1246 (1996).

[20] M. Ormö, A. B. Cubitt, K. Kallio, L. A. Gross, R. Y. Tsien, and S. J. Remington, *Science* **273,** 1392 (1996).

[21] K. Brejc, T. K. Sixma, P. A. Kitts, S. R. Kain, R. Y. Tsien, M. Ormö, and S. J. Remington, *Proc. Natl. Acad. Sci. U.S.A.* **94,** 2306 (1997).

[22] G. J. Palm, A. Zdanov, G. A. Gaitanaris, R. Stauber, G. N. Pavlakis, and A. Wlodawer, *Nature Struct. Biol.* **4,** 361 (1997).

[23] R. Wachter, B. A. King, R. Heim, K. Kallio, R. Y. Tsien, S. G. Boxer, and S. J. Remington, *Biochemistry* **36,** 9759 (1997).

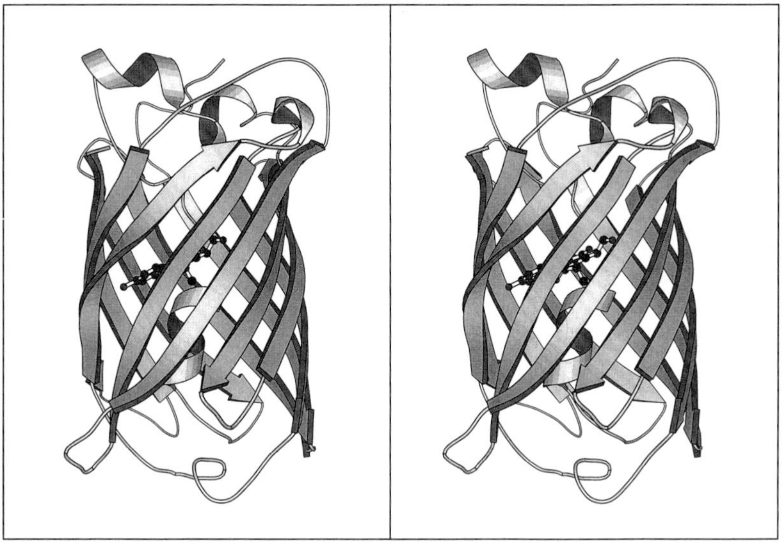

FIG. 3. Stereo ribbon diagram of the fold of GFP, with the chromophore shown as a ball and stick representation. The figure was produced using the program MOLSCRIPT.[37]

bodies when expressed in *Escherichia coli* at temperatures much above this.[13] If one attempts to model an α-helix at the location of the chromophore in the context of the final structure, there are rather severe steric clashes (S.J.R., unpublished observations; Palm *et al.*[22]), suggesting that the interior of the protein is overcrowded. A number of mutants have been discovered that help alleviate the folding problem (discussed, for example, in Palm *et al.*[22]). Many of these mutations are internal to the β-barrel and result in a smaller side chain,[22] perhaps reducing internal crowding. Finally, neighboring side chains make presumably strong interactions with the chromophore (Fig. 4) and may actually force a conformation favoring ring closure.[20]

Thus, one can imagine that on folding, the apoprotein β-barrel exerts pressure on the central helix. Once the chromophore is formed, the internal pressure is partly relieved and the molecule assumes its unusually stable and rigid structure. Whether certain residues are catalytic in this process has not been determined, probably because there is little interest in dark GFP mutants. Although Arg96 was proposed by Ormö *et al.*[20] to be catalytic in this process due to its intimate interaction with the exocyclic oxygen of the imidazolinone ring (Fig. 4), mutation of Arg96 to Cys does not prevent chromophore formation (R. Ranganathan, personal communication). In fact, the emission from this mutant is slightly blue shifted (by 7 nm),

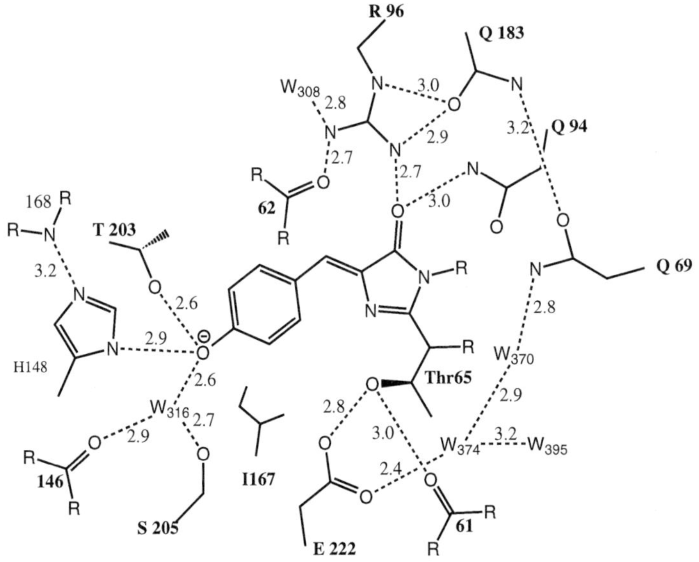

FIG. 4. Details of interactions of side chains with the chromophore in the S65T mutant of green fluorescent protein. Hydrogen bonds are indicated by dashed lines and the donor–acceptor distance is given in angstroms. The diagram was based on the 1EMA structure deposited in the Protein Data Bank. The "R" groups represent continuation of the polypeptide backbone.

suggesting that the purpose of Arg96 is to alter the electron density distribution of the chromophore to achieve the required emission maximum.

The chromophore environment is unusually polar and for the S65T structure is shown in detail in Fig. 4. It is immediately obvious that several amino acids found by mutagenesis to affect the absorption spectrum are in contact with the chromophore (e.g., Thr203, Ile167, Glu222), and from the hydrogen bonding one can deduce that in this mutant, the chromophore is in the anionic phenolate state. The phenolate oxygen accepts three hydrogen bonds and must be charged, whereas Glu222 is neutral because it donates a hydrogen bond to Thr65, which in turn donates a hydrogen bond to a backbone carbonyl. Arg96 would stabilize a negative charge on the oxygen of the heterocyclic ring, which would tend to favor that resonance form of the chromophore (Fig. 1).

Clearly, the T203I replacement would disfavor a negative charge on the chromophore by eliminating the potential hydrogen bond, and the I167V,T replacements would reduce the hydrophobicity near the phenolic oxygen, favoring the charge. The E222G and S65T mutations are harder to explain and require one to take the wild-type structure into account.

However, it is clear that since these residues interact with each other, one could imagine that stabilizing a negative charge to reside on Glu222 would disfavor a negative charge on the chromophore. This in fact helps one to also at least partially understand the complex photodynamic behavior of GFP, which is described and accounted for in a following section. Overall, these structural features are perfectly in accord with the dual band structure of the absorption spectrum of wild-type GFP, with the 395-nm absorbance corresponding to the neutral phenol and the 475-nm absorbance corresponding to the phenolate forms of the chromophore. Rather small effects are apparently sufficient to push the equilibrium between the two forms one way or the other.

Palm *et al.*[22] and Wachter *et al.*[23] independently reported structures of GFP mutants containing the Y66H substitution (blue fluorescent protein, BFP). The chromophore is fully formed, indicating that the machinery required to generate the chromophore is insensitive to the nature of the side chain at position 66. It has also been shown by picosecond time-resolved fluorescence that the BFP chromophore is neutral and does not ionize in the excited state.[23] Thus, the blue emission can be rationalized on the basis of the chemical differences between the BFP chromophore and the wild-type chromophore, and not to structural perturbations in the protein.

Rationally Designed Emission Mutants

The S65T structure revealed a cluster of water molecules in a hydrogen-bonding network adjacent to the chromophore and close to the side chain of Thr203. Model building studies suggested that Thr203 could be replaced by aromatic groups that would presumably displace some of these water molecules, and the corresponding conformation would allow a ring-stacking interaction with Tyr66. We reasoned[20] that the polarizability of the π electrons of the introduced aromatic group would lower the energy of the ground and/or excited state of the chromophore and would thus red shift the spectrum. The mutant 10C was constructed to test this hypothesis (T203Y/S65G/V68L/S72A, the latter three mutations affect folding efficiency but do not affect the spectral properties)[20] and it was found that both absorption and emission spectra were substantially red shifted to 513 and 527 nm, respectively. This mutant is distinctly yellow to the eye (yellow fluorescent protein, YFP) and one does not even require a side-by-side comparison with S65T or wild type to see the difference. Filter sets are available for fluorescence microscopy that clearly distinguish S65T and YFP, so YFP should be useful in double (or with BFP, in triple) label experiments. Spectra of wild type and 10C are presented in Fig. 2. It

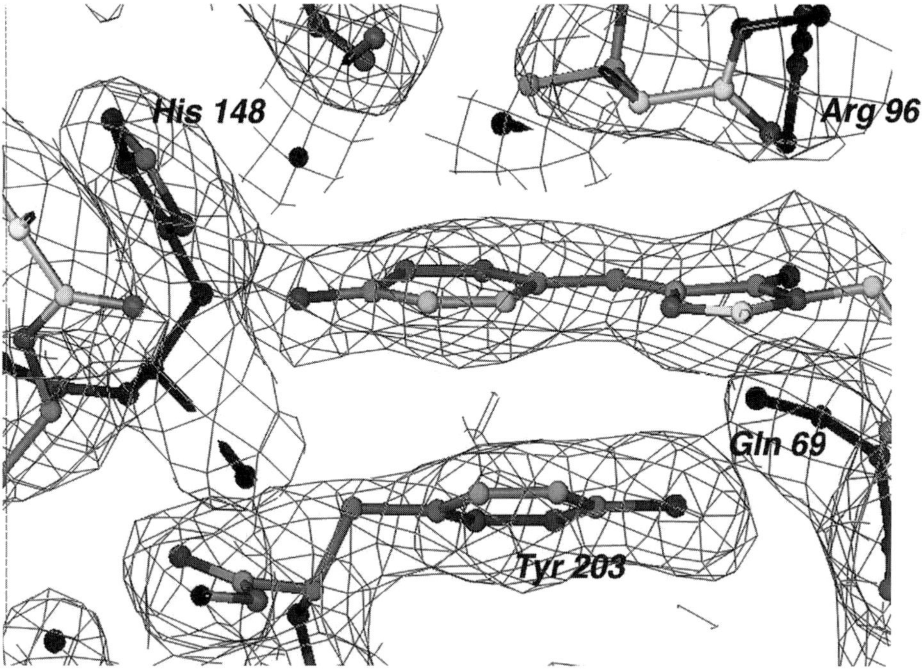

FIG. 5. Portion of the $2F_o$-F_c electron density map at 3.0-Å resolution, contoured at 1 SD, showing a portion of the chromophore and the side chain of Y203 for the T203Y yellow variant of GFP. The mutant was rationally designed based on the three-dimensional structure and it was expected that the polarizable Y203 side chain would stack against the chromophore, lowering the energy of the ground and excited states.

is interesting that the Stokes shift (i.e., wavelength difference between absorption and emission maxima) is so small. Perhaps the internal packing of YFP is optimal for minimal energy dissipation by vibrational states. It is important to note that substitution of Tyr for Thr requires *all three bases* of the codon to be replaced, thus this mutant would have been nearly impossible to discover by random mutagenesis.

The crystal structure of YFP has been solved in two different crystal forms at 2.6- and 2.5-Å resolution[23a] and reveals that the aromatic group of Tyr203 does indeed stack against the phenolate chromophore as it was designed to do (Fig. 5). However, to justify the spectral shifts on a quantitative basis by theoretical calculations does not seem to be feasible at present. Along these lines, it will be interesting to determine if combinations of

[23a] R. M. Wachter, M.-A. Elsliger, K. Kallio, G. T. Hanson, and S. J. Remington, *Structure* **6,** 1269 (1998).

changes such as those discussed here are additive, e.g., would a Y66H/T203Y mutant compared to Y66H be as red shifted in absorption and emission as T203Y compared to S65T?

Structural Basis for Photodynamic Behavior of GFP

Both wild-type GFP and T203Y YFP show remarkably rich photodynamic behavior. It has been known for some time[24,25] that GFP undergoes a photoisomerization process. On exposure of dark-adapted wild-type GFP to long-wavelength UV light, absorption of the 395-nm band decreases with time with a concomitant enhancement of the 475-nm absorption band. When stored in the dark, the effect slowly reverses and is not complete after 24 hr.[25] Chattoraj *et al.*[25] applied extremely fast fluorescence upconversion techniques to study the time dependence of the fluorescence and were able to resolve subpicosecond behavior of wild type and partially deuterated GFP. They discovered that on excitation at 398 nm, fluorescence was observed at 460 nm that decayed rapidly (lifetime 2–3 psec), whereas fluorescence at 508 nm showed a rise time that matched the decay of the 460-nm fluorescence. These lifetimes are increased dramatically to about 40 psec by partial deuteration of the sample. This very large isotope effect strongly implies that proton motion is coupled to fluorescence emission at 508 nm, which ties in nicely with the idea that the phenolate anion is the emitter at 508 nm. The photoisomerization behavior and dual absorption maxima of wild-type GFP were proposed by Chattoraj *et al.*[25] to result from excitation of two different ground states (A and B) of the chromophore that can interconvert in either the ground or the excited state (Fig. 6a).

Brejc *et al.*[21] compared the wild type and S65T hydrogen-bonding patterns around the chromophores and proposed an atomic model for the A and B ground states in the equilibrium diagram of Fig. 6. They suggested that in wild type there is a proton relay connecting the chromophore with Glu222, as shown in Fig. 6b. The phenol oxygen is connected to Glu222 via a hydrogen bond network that includes a water molecule and Ser205. Upon excitation, the pK_a of the phenol oxygen drops from its uncharged value, and a concerted series of proton shifts allows picosecond transfer of the proton to the carboxylate of Glu222 in the unfavorable *anti* configuration (Fig. 6b). After fluorescence emission, the pK_a of the chromophore increases and normally the proton is rapidly transferred back from the carboxyl to the phenolate. However, in S65T, this hydrogen bond pattern

[24] R. Y. Tsien, "Fluorescent Proteins and Applications Conference," March 6, Palo Alto, CA, 1995.

[25] M. Chattoraj, B. A. King, G. U. Bublitz, and S. G. Boxer, *Proc. Natl. Acad. Sci. U.S.A.* **93,** 8362 (1996).

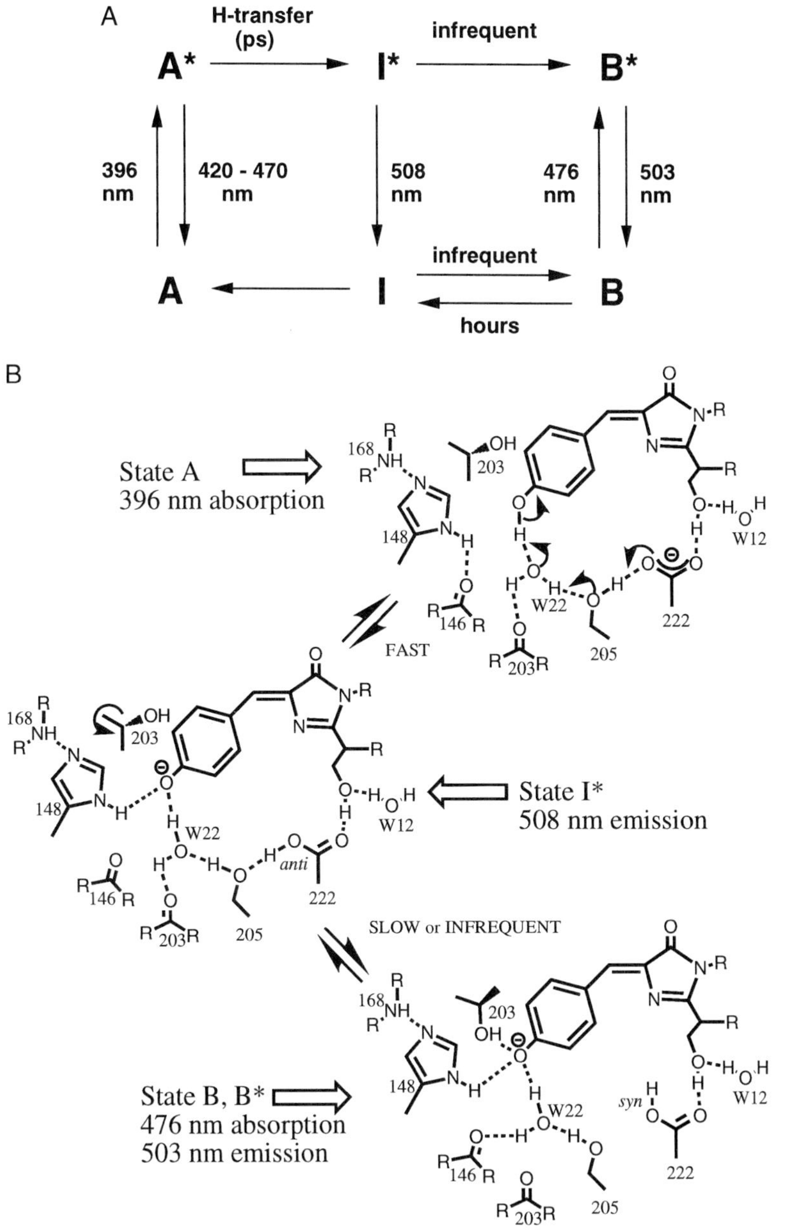

FIG. 6. (a) Diagram showing proposed proton transfer and interconversion of two different sets of ground and excited states of the chromophore of wild-type GFP, as proposed on the basis of picosecond fluorescence upconversion spectroscopy,[25] to explain the dual absorption maxima of wild-type GFP. (b) Atomic model for proton transfer and interconversion of A and B states based on a comparison of the three-dimensional structures of wild-type GFP and the S65T mutant.[71] Rapid proton transfer to Glu222 is proposed to precede emission at 508 nm, but occasionally the proton relay breaks and the chromophore becomes a relatively stable anion.

is disrupted and Glu222 is uncharged, making it possible for the chromophore to be permanently deprotonated. Thus, the 395-nm absorption band is absent in this mutant (and all other mutations at Ser65). We proposed[21] that occasionally in wild type, a rearrangement occurs that allows the proton on Glu222 to shift to the more stable *syn* configuration and that the energy barrier for return of the proton to the *anti* position (for return to the chromophore) is large. This satisfactorily accounts for the slow photoisomerization seen on irradiation of wild type in the near UV, but the mechanism will need to be tested, perhaps by mutation at Ser205.

Apparently, only serine at residue 65 can stabilize either the neutral or the charged form of Glu222. Note that as described earlier in this review, mutation of Glu222 to Gly eliminates the proton "sink" and the chromophore is normally protonated, showing only the 395-nm absorption. Palm *et al.*[22] proposed a similar proton relay mechanism based on their studies of mutant forms of GFP. It is interesting that in the monomeric wild-type structure, Thr203 is seen in a mixture of two conformations, one of which makes a hydrogen bond with the phenol oxygen and the other of which does not.[21] The phenol oxygen makes only one other hydrogen bond, to an internal water molecule. Because the anionic form would be stabilized by additional hydrogen bonds, it is reasonable to conclude that the chromophore is in a delicate equilibrium between protonated and unprotonated forms, and indeed, the ratio of occupancies of the two conformations of Thr203 was calculated to be about 6 : 1, favoring the nonhydrogen bonding conformation. This is about what one would expect if the two absorption maxima (395 and 475 nm) really do reflect an equilibrium between neutral and anionic forms of the chromophore. This conclusion is supported indirectly by the demonstration (again using ultrafast fluorescence upconversion) that in Y66H BFP, fluorescence emission is not coupled to proton motion,[23] thus the imidazole ring is a fairly good model for the absorption of the protonated phenol in wild type as the 388-nm absorption suggests.

On/Off Switching Behavior and Photoactivation

The advent of solid-state cameras that are capable of counting single-photon events has led to the imaging of single molecules trapped in gels. In applying these techniques to the study of single GFP molecules (specifically the mutant T203Y), the fascinating discovery was made that single molecules blink on and off[26,27] on time scales of the order of 1 sec, despite continuous irradiation at 488 nm. Presumably, this results from changes in

[26] R. M. Dickson, A. B. Cubitt, R. Y. Tsien, and W. E. Moerner, *Nature* **388,** 356 (1997).
[27] D. W. Pierce, N. Hom-Booher, and R. D. Vale, *Nature* **388,** 338 (1997).

the resonant frequency of the molecule due to local environmental effects. Furthermore, after emitting a certain number of photons, the molecules switch off into a dark state of a long lifetime (>5 min).[26] This dark state must be different than the dark state achieved during blinking as it is reversible on irradiation at 405 nm, suggesting that in this dark state the chromophore is neutral. This behavior does not seem to be related to the photodynamic behavior described in the previous section because illumination is at the lower energy absorption band of wild-type protein. That is, we have light being absorbed by the anionic form of the chromophore that somehow favors a state of permanent protonation, thus there must be several distinct structural states of the chromophore and its environment.

Finally, in addition to these fascinating reversible photodynamic behaviors, it has been discovered that on irradiation in an oxygen-free atmosphere, GFP very rapidly turns red[28,29] absorbing in the green at about 525 nm and emitting at about 600 nm. The structural basis for this photoactivation process is entirely unknown but may involve the formation of a free radical. Elowitz *et al.*[28] used this effect to measure the rate of diffusion of protein in the cytoplasm of a single living bacterium by irradiating GFP at one end of the cell.

The role of oxygen in chromophore formation and in these processes clearly invites further study, as the absence of photoactivation in the presence of oxygen clearly implies that oxygen has access to the chromophore. If so, why does oxygen not quench fluorescence under "normal" conditions?

Some Applications for Biotechnology

In addition to the obvious and ever growing use of GFP as a visible tag for both protein localization and gene expression,[11,30–32] use of GFP as a biosensor is potentially an equally significant application. As an example, two groups have taken advantage of the previously observed resonance energy transfer between blue and green GFP variants.[33,34] In both cases, Ca^{2+}-dependent conformational changes of calmodulin were harnessed to alter the physical separation of two physically linked GFPs (Y66H BFP

[28] M. B. Elowitz, M. G. Surette, P.-E. Wolf, J. Stock, and S. Liebler, *Curr. Biol.* **7,** 809 (1997).

[29] K. E. Sawin and P. Nurse, *Curr. Biol.* **7,** 606 (1997).

[30] H.-H. Gerdes and C. Kaether, *FEBS Lett.* **389,** 44 (1996).

[31] R. Rizzuto, M. Brini, F. De Giorgi, R. Rossi, R. Heim, R. Y. Tsien, and T. Pozzan, *Curr. Biol.* **6,** 183 (1996).

[32] S.-H. Park and R. T. Raines, *Protein Sci.* **6,** 2344 (1997).

[33] R. Heim and R. Y. Tsien, *Curr. Biol.* **6,** 178 (1996).

[34] R. D. Mitra, C. M. Silva, and D. C. Youvan, *Gene* **173,** 13 (1996).

and S65T GFP). Romoser *et al.*[35] engineered a construction in which BFP and GFP were linked by the calmodulin-binding domain of smooth muscle myosin light chain kinase. In the presence of $(Ca^{2+})_4$calmodulin, the resonant energy transfer between the two GFPs was reduced, changing the ratio of 440 to 505 nm fluorescence. Similarly, Miyakawi *et al.*[36] synthesized a gene that included both calmodulin and a calmodulin-binding peptide M13 (also from myosin light-chain kinase) along with BFP and GFP. In this case, collapse of the molecule in the presence of Ca^{2+} enhanced the resonance energy transfer between the flanking GFPs. These molecules have been called "cameleons" as they change color and have a long tongue (M13) that retracts and extends into and out of the calmodulin "mouth." Both groups demonstrated that the constructions are effective calcium indicators in living cells, in the latter case, by genetic modification of the calmodulin calcium ligands, over a Ca^{2+} concentration range of 10^{-8} to 10^{-2} *M*. Presumably, constructs based on other proteins that undergo large conformational changes in response to binding a ligand could be put to use to detect many other types of small molecules.

Similar constructions may also be useful for detecting other cellular activities. For example, in the original publications demonstrating resonance energy transfer between linked BFP and GFP, the linker was sensitive to protease action,[33,34] which greatly reduced energy transfer on cleavage. One can easily imagine designing linker sequences that would be sensitive to a particular protease or constructing nonpeptide linkers (esters, phosphodiesters, polyglycosides, etc.) between GFPs that would be sensitive to any other type of hydrolytic cleavage that one would wish to detect.

Unlike wild type, many GFP variants can be titrated with pH and/or salt, with large effects on absorption and emission spectra. This was first noticed by Wachter *et al.* with BFP,[23] who suggested that this property may be used to measure the pH, or changes in pH, of subcellular organelles by appropriately targeting the indicator. Although the absorption maximum of BFP shifts by only a few nanometers on titration (as does the emission maximum), there are large changes in the quantum efficiency of fluorescence. From this result it is not clear (and it seems unlikely) that the chromophore itself was being titrated. However, it has been found more recently[37] (R. Y. Tsien, personal communication) that the chromophores of both S65T GFP and T203Y YFP do directly titrate, with very large changes in both absorption and emission (Fig. 7). For YFP, the absorption

[35] V. A. Romoser, P. M. Hinkle, and A. Persechini, *J. Biol. Chem.* **272,** 13270 (1997).

[36] A. Miyawaki, J. Llopis, R. Heim, J. M. McCaffery, J. A. Adams, M. Ikura, and R. Y. Tsien, *Nature* **388,** 882 (1997).

[37] P. J. Kraulis, *J. Appl. Cryst.* **24,** 946 (1991).

[38] M. A. Elsliger, R. M. Wachter, G. T. Hanson, K. Kallio, and S. J. Remington, *Biochemistry* **38,** 5296 (1999).

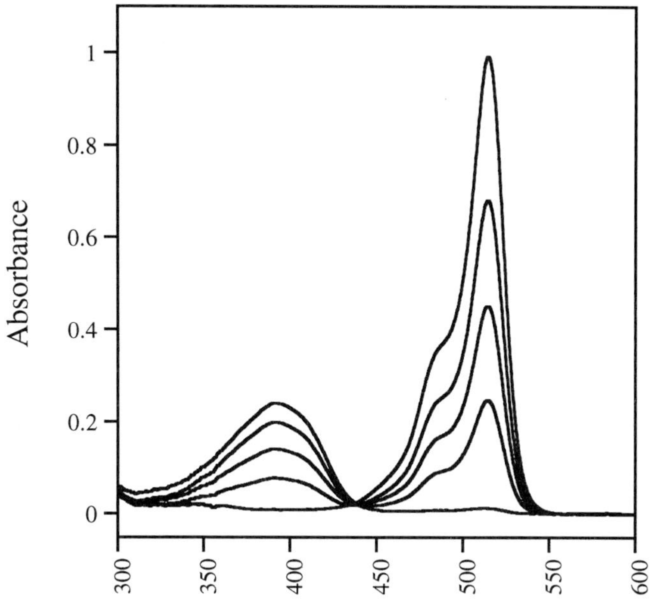

FIG. 7. Titration of the T203Y YFP chromophore between pH 5.0 and 8.6 (75 m*M* acetate or phosphate buffer, 140 m*M* NaCl) in the intact protein. The normalized absorption spectrum as a function of pH is plotted. At pH 5, the absorption maximum is approximately 395 nm, whereas at pH 8.6, the maximum is about 510 nm. Note the clean isosbestic point implying that only two species are present.

maximum is about 395 nm at pH 5.0 and shifts to about 510 nm at pH 8.0. We[38] have conducted high-resolution structural studies of S65T GFP at pH 4.5 and 8.0 and have found evidence that the chromophore phenol oxygen is directly titrated. This was indicated by localized perturbations in the vicinity of this oxygen atom (changes in the hydrogen-bonding pattern). Unfortunately, for reasons yet unknown, with these mutants fluorescence emission is abolished at low pH so it is not clear how useful these properties will be. Nevertheless, at this point there seems to be nearly unlimited possibilities for the use of this amazing molecule.

Note Added in Proof

Information in this review may have changed since its submission date of January 1998.

Acknowledgment

The work in our laboratory that is summarized here has been supported by National Science Foundation Grant MCB-9418479, which is gratefully acknowledged. Due to space limitations, the work of a number of others on structure/function relationships in GFP could not be included in this brief review.

[16] Large-Scale Purification of Recombinant Green Fluorescent Protein from *Escherichia coli*

By DANIEL G. GONZÁLEZ and WILLIAM W. WARD

General Introduction

The green fluorescent proteins (GFPs) are brilliantly fluorescent chromoproteins derived biologically from various species of bioluminescent cnidaria (classes Anthozoa and Hydrozoa).[1] The most studied of these have been the GFPs derived from the jellyfish *Aequorea victoria* and from the sea pansy *Renilla reniformis.* With the cloning of the gene for GFP from *A. victoria* the possibility has risen of producing the protein heterologously in large quantities.[2] The first reports of the expression of recombinant GFP were made in 1994, 2 years after its cloning.[3,4]

Before this time, GFP was isolated from the native source, and the need for substantial quantities of GFP for physical characterization led to the sacrifice of large numbers of animals. The purification of this protein to near homogeneity was reported as early as 1974 as a by-product of aequorin isolation and was first reported to be crystallized at this time (as part of their purification method).[5] The availability of the clone has made it possible to express and purify large amounts of GFP for the first time in *Escherichia coli.*[6] Here a mixed-mode (hydrophobic interaction and size exclusion) effect, using high salt conditions, was used to purify the protein chromatographically. Using this method the authors reported the purification of 21.8 mg of recombinant GFP.

Green fluorescent protein has been shown to form without the need for any cofactors or specific posttranslational processing from the native jellyfish in a cell-free translation system.[7] It has been demonstrated, however, that molecular oxygen is necessary for chromophore formation.[8] Its

[1] W. W. Ward, *Photochem. Photobiol. Rev.* **4,** 1 (1979).
[2] D. C. Prasher, V. K. Eckenrode, W. W. Ward, F. G. Pendergast, and M. J. Cormier, *Gene* **111,** 229 (1992).
[3] M. Chalfie, Y. Tu, G. Euskirchen, W. W. Ward, and D. C. Prasher, *Science* **263,** 802 (1994).
[4] S. Inouye and F. I. Tsuji, *FEBS Lett.* **341,** 277 (1994).
[5] H. Morise, O. Shimomura, F. H. Johnson, and J. Winant, *Biochemistry* **13,** 2656 (1974).
[6] J. R. Deschamps, C. E. Miller, and K. B. Ward, *Protein Express. Purif.* **6,** 555 (1995).
[7] V. A. Kolb, E. V. Makeyev, W. W. Ward, and A. S. Spirin, *Biotechnol. Lett.* **18,** 1447 (1996).
[8] D. F. Davis, W. W. Ward, and M. W. Cutler, *in* "Bioluminescence and Chemiluminescence: Fundamentals and Applied Aspects" (A. K. Campbell, L. J. Kricka, and P. E. Stanley, eds.), p. 596. Wiley, New York, 1994.

METHODS IN ENZYMOLOGY, VOL. 305
0076-6879/00 $30.00

fluorescence properties allow for easy visualization and quantitation of GFP in a nondestructive manner. As a result, GFP has been used increasingly as a fusion partner for use in monitoring expression, as, for example, by Poppenborg *et al.*,[9] who also describe the polyhistidine tagging of a GFP fusion protein as means for its purification. This indicates that the fluorescence from GFP may be used to assay for a valuable recombinant protein nondestructively. Affinity tagging of proteins for purification is not a novel idea. A polyhistidine tag was used to purify GFP for structure determination by X-ray crystallography.[10] Interestingly, the authors of this paper managed to crystallize the protein as a monomer by using a precipitant containing 50 m*M* $MgCl_2$. This is in contrast to the procedure of Yang *et al.*[11] who did not use $MgCl_2$ in their crystallization solvent and resolved the protein structure as a dimer.

The stability of GFP under a variety of conditions has been known for a long time.[12] For example, recombinant GFP is always purified at room temperature. Such stability, along with its visual properties, makes GFP one of the easiest proteins in nature to purify from a recombinant source. As a consequence, GFP can be used as a model protein to develop methods for protein purification. It can also be used to demonstrate protein purification concepts in an instructional setting.[13] There is now even an instructional kit on the market that allows children to learn how to purify protein using GFP (Bio-Rad, Hercules CA). The methods that follow are an amalgam of various preparations performed in our laboratory. Presented here are typical conditions necessary to purify sizable amounts of the protein to homogeneity from a recombinant source.

Assaying Green Fluorescent Protein

There are generally two methods used to assay for GFP in a protein purification attempt. Both methods are nondestructive and can be automated. The first relies on the extinction maxima (λ_{max}) in the visible and near ultraviolet absorption range of the GFP variant being purified. This

[9] L. Poppenborg, K. Friehs, and E. F. Flaschel, *J. Biotechnol.* **58,** 79 (1977).

[10] M. Ormö, A. B. Cubitt, K. Kallio, L. A. Gross, R. Y. Tsien, and S. J. Remington, *Science* **273,** 1392 (1996).

[11] F. Yang, L. G. Moss, and G. N. Phillips, *Nature Biotechnol.* (*London*) **14,** 1246 (1996).

[12] W. W. Ward, *in* "Bioluminescence and Chemiluminescence: Basic Chemistry and Analytical Applications" (M. A. DeLuca and W. D. McElroy, eds.), p. 235. Academic Press, New York, 1981.

[13] M. W. Cutler, D. F. Davis, and W. W. Ward, *in* "Bioluminescence and Chemiluminescence: Fundamentals and Applied Aspects" (A. K. Campbell, L. J. Kricka, and P. E. Stanley, eds.), p. 383. Wiley, New York, 1994.

value is compared to absorption maximum of the protein at $\lambda = 280$ nm. The ratio of $A_{\lambda max}/A_{280}$ is used as a determinant of purity. This method can be problematic for several reasons. Initially in the purification, light scattering and non-GFP-absorbing components of the preparation can generate falsely elevated values for λ_{max} values, especially if the concentration of GFP is very low, as it might be early in a purification attempt. Additionally, those GFPs exhibiting two absorbance peaks, such as the wild type *Aequorea* GFP ($\lambda_1 = 395$ and $\lambda_2 = 475$), do not conform to Beer's law ($A_\lambda = \varepsilon_\lambda 1c$). That is, as the concentration of these GFPs increases, the absorbance at λ_1 increases and the absorbance at λ_2 decreases, in a disproportional manner due to the dimerization of GFP at a higher concentration.[14]

A better measure of GFP concentration is fluorescence emission of the protein when excited at its λ_{max}. This is measured most conveniently using filter fluorometers. A standard can be generated from previously purified (ClonTech Laboratories) GFP by diluting it to the low microgram or high picogram level, depending on the sensitivity of the instrument. We usually use a standard at a concentration of 5 μg/ml in azide-containing Tris buffer at pH 8.0, which is kept sealed in a fluorometer cuvette and stored at room temperature in the dark. It is used to calibrate a fluorometer, such that the sample gives a reading that is between 50 and 75% of the maximum output of the fluorometer after it has been adjusted by manipulating the slit widths and masking the sample chamber. This value is recorded as the reading corresponding to the standard. At low concentrations, the aforementioned dimerization effect of GFP is not noticed in an absorption spectrum. Beer's law can now be applied to generate a reliable concentration and be used to ascertain purity as shown in Table I. Factors such as a meniscus or light-scattering particles in the sample can interfere with an accurate fluorescence measurement.

As the purification progresses and the GFP concentration increases, increasingly higher and higher dilutions of the preparation must be made to stay within the range of the standard. The extent of such dilutions can be estimated by observing the volume of the preparation at the step being assayed versus the volume previously assayed and of course by the reading on a fluorometer. It is always wise to have a diluted sample return a value as close to the standard reading as possible and to rerecord the value of the standard immediately before reading the sample to take into account

[14] M. W. Cutler, Ph.D. Thesis, Rutgers University, New Brunswick, NJ, 1995.

[15] D. G. González and W. W. Ward, *in* "Green Fluorescent Protein: Properties, Applications, and Protocols" (M. Chalfie and S. Kain, eds.), p. 289. Wiley-Liss, New York, 1998.

TABLE I
GFP FORMS AND ABSORBANCE CHARACTERISTICS WITH RATIOS USEFUL IN ESTABLISHING PURITY[a]

Protein	Chromophore λ_{max}	ε	λ_1/λ_2	Numerical ratio
Native *Aequorea* GFP	395	27,600	395/280	1.25
	470	12,000		
Recombinant wild-type GFP	397	27,000	397/280	1.25
	475	12,000		
S65T	489	56,000	489/280	2.25
Stemmer mutant	397	27,000	397/280	1.25
cycle 3 mut: (F100S, M154T, V164A)	475	12,000		
P4: (Y66H)	382	25,000	382/280	1.23

[a] Taken with permission from ref. 15.

any instrument drift. A measure of the amount of GFP in the preparation can then be generated using the following simple formula:

$$\left(\frac{(\text{sample reading} \times \text{dilution factor})}{\text{standard reading}}\right) \times 5\ \mu\text{g/ml} \times \text{volume of extract}$$

We have written two application notes for the TD-700 filter fluorometer made by Turner Designs (Turner Designs, Sunnyvale, CA), which describe this assay procedure. The first describes a method to estimate protein concentration in bacteria before extraction.[16] The second describes using the TD-700 to assay for GFP in cell extracts.[17] Both application notes are available from Turner Designs and are accessible on the World Wide Web at http://www.turnerdesigns.com/tech_library.htm#bio, along with information on their fluorometers. Both methods can be adapted readily for use in any filter fluorometer, such as the older Turner 110 series, which we still use in our laboratory.

Expression of GFP in *E. coli*

The number of constructs expressing GFP and the number of expression schemes for its expression are far too numerous to review here; this chapter

[16] A suggested method for the assaying of green-fluorescent protein in whole bacteria. Turner Designs application note.

[17] A suggested method for the quantitation of green-fluorescent protein. Turner Designs application note.

deals with the basic system used in our laboratory to purify GFP in large quantity. The initial construct available to us was the pET 3a-derived TU#58 originating from Chalfie.[3] This construct was originally expressed in the *E. coli* strain BL21(λDE3)Lys S.[18] The Lys S in this T7 expression system is a plasmid that encodes T7 lysozyme. This enzyme binds to and inhibits T7 RNA polymerase (which drives the system), which can be present in low concentrations before induction. It also has the effect of degrading the cell wall of the bacteria. The Lys S plasmid is used to prevent leaky expression and can be advantageous if the gene product we are interested in is extremely toxic to cells. Its presence is, however, stressful for the cells, as not only do they need to propagate the plasmid, but must also tolerate its cell wall degrading gene product.[19] When used directly to express GFP (as in Deschamps *et al.*,[6] who produced 3.6 mg/liter of *E. coli* culture), this system was not ideal for expression of GFP. Diane Davis in our laboratory removed the Lys S plasmid from the system producing BL21(λDE3)/TU#58, with which we began optimization for the mass expression of GFP for physical studies in our laboratory.

Initially, without the Lys S plasmid, the dynamics of the system were very difficult to work with. In low levels GFP was toxic to these cells and we began to observe cells that carried the plasmid but did not express GFP being propagated in liquid culture. Eventually we noticed that our frozen stocks were mostly this species. We began to mass screen for GFP expression visually on solid agar under induction (14 g Bacto-Agar/37 μg ampicillin/10 g lactose/liter LB-Miller). For many generations we were only able to visualize the extent of expression reported in Chalfie *et al.*,[3] independent of whether the plates were incubated at 37 or 28°. The colonies being produced were small and "greened" slowly, often maturing days later when stored at 4°, and colonies would appear spontaneously that were not expressing GFP, but expressing ampicillin resistance. These colonies contained a plasmid that when linearized was the same size as TU#58. After 2 months, spectacularly green colonies began to appear randomly. After a stable line of cells was established that overexpressed GFP in this manner, the strain was cured of all plasmid by liquid culturing the cells at 42° in LB-Miller without selection. These cells were retransformed with TU#58 plasmid DNA, prepared months earlier, and were found to possess the same high-yield phenotype.

These colonies on solid culture appeared to grow normally and were, for the most part, brilliantly green and matured within a day of growth. In liquid culture, we were able to produce 170 mg of GFP/liter of *E. coli*

[18] F. W. Studier and M. A. Moffat, *J. Mol. Biol.* **189,** 113 (1986).
[19] F. W. Studier, *J. Mol. Biol.* **219,** 37 (1991).

culture. Diane Davis has reported that she has obtained the highest level of expression at 30°, in the unoptimized *E. coli*.[20] We obtained this high-level expression at 37° and have yet to experiment at exactly 30°, but do grow most of our cells at 28°. A summary of our procedure to obtain high level expression is as follows: We inoculate a wide-mouth 2.8-liter erlenmeyer baffled flask containing 1 liter of LB-Miller/37 μg/ml ampicillin and 50 μl of Antifoam A (Sigma Chemical Company, St. Louis, MO) with a 50-ml overnight starter culture. This is grown at 37° to an $OD_{660} = 0.8$, at which point isopropyl-β-D-thiogalactopyranoside (IPTG) (CAS 367-93-1) is added (500 m*M* stock in deionized water, stored at −20°) to a final concentration of 0.5 m*M*. We then induce at 37° for 12 hr, collect the cells by centrifugation, and store the cells at 4° for a few days before extracting and purifying the GFP. We typically use this procedure at 28°, with an average yield of 100 mg/liter of culture, sufficient for our needs. We have produced hundreds of liters of culture at 28° and purified grams of GFP with this expression scheme.

Extraction of GFP from *E. coli*

The *E. coli* cells are compacted into a paste by centrifugation (10,000*g*, for 15 min, Sorvall GSA rotor or a CEPA benchtop Model LE continuous flow centrifuge). After expressing large quantities of GFP by the procedure just described, one should be able to observe the intense fluorescence being expressed by exciting the cell paste with a hand-held long-wave UV light. The overexpressed recombinant GFP from these cells is liberated by repeated cycles of freezing and thawing.[21] Freeze slowly (60 min at −20°, in a lab freezer) and thaw slowly (60 min at room temperature) the cell paste three times. This is followed by two to four cold (4°) buffer washes (20× cell paste volume) with a pH 8.0 buffer composed of 25 m*M* Tris and 1 m*M* 2-mercaptoethanol (note: the number of washes depends on the extent of GFP still present in the cells; if it appears that another wash will liberate more protein, by all means perform another wash); the first of these washes also contains $MgCl_2$ (100 m*M*). We have found that the Mg^{2+} ions in the first wash aid endogenous *E. coli* DNases to initially degrade any DNA that might leak out of these cells, permitting compaction of the cell paste between subsequent washings.

As the supernatants are washed from the cell paste (10,000*g*, 15 min) phenylmethylsulfonyl fluoride (CAS 329-98-6) is added to a final concentration of 1 m*M* (from a 1 *M* stock solution in ethanol) to prevent proteolytic

[20] D. Davis, personal communication.

[21] B. H. Johnson and M. H. Hecht, *Bio/Technology* (*London*) **12,** 1357 (1994).

cleavage of the protease-susceptible C-terminal "tail" of GFP. Each washing is collected separately and pooled at the end of the procedure. It is usually necessary to freeze the cell paste between buffer washes to aid in releasing all the GFP. Throughout this and all subsequent steps, *n*-butanol (CAS 71-36-3) is used (a drop is usually added to a foamy solution) to dissipate foam that interferes with accurate volume determinations. Although the freeze-thaw process is slower than other methods, such as sonication or lysozyme treatment, the freeze-thaw extracts are remarkably clean (low viscosity, low DNA content, and high GFP content—up to 10% of total soluble protein). More than 90% of all the expressed GFP can be released from cells by this method.

Precipitation of DNA from the *E. coli* Extract

Protamine sulfate (salmine) (CAS 53597-25-4) is a natural DNA-binding protein isolated from salmon sperm. It has been found to be a very efficient precipitation agent for most DNA and RNA complexes in a crude extract. It is best to make a stock solution of 50 mg/ml protamine sulfate in 10 m*M* Tris, pH 8.0, 0.02% NaN_3 in quantities of about 100 ml. At this concentration, protamine sulfate will form a two-phase system that, when agitated, appears cloudy. The protamine sulfate becomes completely miscible when the stock solution is warmed to ~60°; usually running hot water from a nearby sink suffices to redissolve the protamine sulfate. A concentration of 1 mg protamine sulfate per 100 OD_{260} (A_{260} × dilution factor × volume) units will suffice to precipitate most of the DNA in the extract, but because each extraction is different, we recommend that the optimal concentration of protamine sulfate be ascertained for each extraction. This is necessary to remove the maximum amount of DNA and to prevent the GFP from precipitating (it is impossible to resuspend later).

Optimization of the protamine sulfate concentration is best performed in six to eight numbered 1.5-ml microfuge tubes each containing 1 ml of extract. The protamine sulfate solution is added to each tube to produce a range from 1 up to 10 mg/ml final concentration (20 to 160 μl), not accounting for any volumetric displacement. One tube is left untreated and is used as the control. The solution will immediately become cloudy, as the precipitated DNA will give a milky appearance to the assays. The microfuge tubes are agitated vigorously and centrifuged at 10,000 rpm for 10 min at room temperature. The resulting pellets should be very compact.

Depending on the concentration of total protein and DNA, the resulting supernatants should be diluted roughly 10- to 20-fold for absorbance readings and 200- to 500-fold for fluorescence measurements. Compare the absorbance at 260 nm (nucleic acids), at 280 nm (proteins), at the chromo-

TABLE II
EFFECT OF PROTAMINE SULFATE CONCENTRATION ON DNA PRECIPITATION IN 1-ml OF A 45-ml EXTRACT OF wtGFP ($\lambda_{max} = A_{395}$)

Protamine sulfate (50-mg/ml stock)					Relative fluorescence
Volume (μl)	mg	A_{260}[a]	A_{280}[a]	A_{395}[a]	Turner 110[b]
0	0	0.84	1.29	0.19	70
20	1	2.30	2.30	1.20	67
40	2	1.40	1.70	0.39	77
60	3	0.74	1.60	0.20	70
80	4	0.56	0.76	0.12	41
100	5	0.43	0.62	0.07	23
120	6	0.38	0.55	0.45	18
140	7	0.34	0.48	0.03	10
160	8	0.34	0.29	0.19	9

[a] 1/10th dilution of extract.
[b] 1/100th dilution of extract.

phore maximum (A_{395} for wtGFP), and the fluorescence emission maxima. What one is looking for is a sharp drop in the A_{260} and a lesser drop at the A_{280} readings (note: initially these readings will rise sharply due to light scattering caused by the precipitating DNA), indicating that DNA is precipitating and not proteins. Readings at the absorbance maximum of the chromophore should increase, as a result of light scattering, and then should decrease to the level seen before protamine sulfate treatment. Fluorescence, A_{280}, and λ_{max} values should not fall below their original levels until the proteins (along with GFP) in the extract begin to precipitate. Should the GFP be inadvertently precipitated, a hand-held long wave UV lamp can be used to observe the compacted protein at the bottom of the microfuge tubes.

A typical example of an experiment monitoring the effect of a protamine sulfate concentration on an extract of *E. coli* containing GFP is shown in Table II. *Escherichia coli* cells expressing wtGFP were grown in 1 liter LB-Miller,[22] under IPTG induction; the cells were collected and subjected to slow freeze-thaw extraction as described earlier to produce a volume of 45 ml of extract. The optimal concentration of protamine sulfate to be used for precipitating the nucleic acids was determined as described earlier (Table II). It was ascertained that a 60-μl volume of the 50-mg/ml solution of

[22] J. H. Miller, *in* "Experiments in Molecular Genetics," p. 433. Cold Spring Harbor Laboratory, Cold Spring Harbor, NY, 1972.

protamine sulfate (3 mg total protamine sulfate) was sufficient to precipitate most of the DNA without precipitating any GFP in 1 ml of extract. Based on our original volume, a quantity of 2.7 ml of protamine sulfate solution was added incrementally to the extract while stirring vigorously. Any GFP-containing supernatants from the protamine sulfate assay were recombined with the preparation. This was allowed to stir for no more than 30 min and was then centrifuged (10,000 rpm, 15 min, room temperature, Sorvall SS-34 rotor) and the supernatant retained for further processing.

Ammonium Sulfate Precipitation

Solid $(NH_4)_2SO_4$ (CAS 7783-20-2) is added to the *E. coli* extract, to saturation. A volume of 500 m*M* Tris, 0.02% NaN_3, pH 8.0, equivalent to 10% of the volume of the extract is added to the extract at this point to stabilize its pH. The extract is stirred at a moderate speed for 1 hr and centrifuged at the highest centrifugal force possible. In our case, extracts are usually dispensed into 500-ml polycarbonate centrifuge bottles (Nalge Nunc International, Rochester, NY) and centrifuged at 14,000 rpm, 4° for 1 hr in a Sorvall GSA rotor. The rotor is allowed to come to a full stop without braking, as the precipitated protein pellets are not very compact and will resuspend or break up into the supernatant. The supernatants are decanted carefully and discarded.

The precipitated protein pellets are pooled by resuspending them in the smallest volume possible of 25 m*M* Tris, 0.02% NaN_3, pH 8.0 (this volume depends on the amount of pellet material present and varies from preparation to preparation). Precipitated wtGFP in the pellets appears yellow and will return to a green color as it goes into solution. The same buffer is then added carefully to the resuspended pellets until the resuspension becomes translucent to visual observation. Again the volume of added buffer varies depending on the amount of GFP precipitated; the resuspended material is then centrifuged as before. Precipitates that result are any remaining high molecular weight contaminants and debris. All the GFP should be in the supernatant. The extract is now "clarified" and sufficiently concentrated for column chromatography.

Preparative Gel Permeation

The reversible dimerization behavior that *Aequorea* GFP exhibits[14] can be exploited by using two gel permeation steps with other processing in between. The first column we run in our laboratory is a 10 × 120-cm (~8-liter bed volume), Bio-Gel P-100 medium gel (Bio-Rad). Diluted GFP extract (0.2 mg/ml or less) is applied in a running buffer containing 10 m*M* Tris–EDTA, 1 *M* ammonium sulfate, and 0.02% NaN_3, pH 8.0. GFP chromatographed in this fashion should behave as a monomer with a partial hydrophobic interaction in high salt causing the GFP to elute at an apparent molecular weight of 21,000.[15]

A word of caution about using P-100 Bio-Gel in this manner: the flow rates change during a run and are consistently lower and lower with each subsequent run. This is due to the large volume and weight of the column. As this matrix is very compressible, it tends to become compacted at the bottom of the column during chromatography and must be repacked between chromatographic runs. A more convenient but more expensive alternative is to replace the P-100 Bio-Gel with another matrix such as CL Sepharose 6B (Pharmacia Biotech), which is more rigid than P-100 Bio-Gel and may be used in place of P-100 Bio-Gel without the need to repour the bed between runs.

Concentration of Eluted GFP

Green fluorescent protein that is eluted from the first gel permeation column is usually loaded at room temperature onto an octyl Sepharose (Pharmacia Biotech) hydrophobic interaction column (usually 2.5 × 12 cm, a 50–60 ml packed bed volume) that has been preequilibrated with 1.5 *M* $(NH_4)_2SO_4$, 25 m*M* Tris, 0.02% NaN_3, pH 8.0, running buffer. The GFP is pulse eluted with 25 m*M* Tris, 0.02% NaN_3, pH 8.0, buffer [lacking $(NH_4)_2SO_4$]. Elution of GFP requires 100–500 ml of this second buffer solution and can take less than 2 hr to complete. The column is capable of binding more than 1 g of total protein. Another method of concentration that we have used at this step is ultraconcentration on a pressure-driven stirred cell ultraconcentrator (Millipore, Bedford, MA), according to the manufacturer's instructions.

A third method, and by far the easiest, is to concentrate the GFP by placing it into 12- to 14-kDa MWCO dialysis bags (Spectrum, Houston TX). The bags are placed in glass trays and either G-25 or carbowax (polyethylene glycol 6000) is sprinkled liberally onto the bags to absorb and remove a great quantity of the water content and salt in the preparation. After about an hour the dialysis bags will have shriveled up quite a bit and the G-25 or PEG-6000 will have become a paste; the GFP inside is now highly concentrated. This is the easiest method as the next step in the purification is exhaustive dialysis against 25 m*M* Tris, 0.02% NaN_3, pH 8.0. After concentrating the material in the 12–14 kDa MWCO dialysis bag, the G-25 or PEG-6000 paste can simply be rinsed off with deionized water and the sample is ready for dialysis. During dialysis, other contaminants precipitate, so we usually transfer the dialysate into 500-ml polycarbonate centrifuge bottles and centrifuge at 14,000 rpm, 4° for 30 min in a Sorvall GSA rotor to remove the precipitate.

Anion Exchange

An anion-exchange step is used between the two gel permeation columns. DEAE Sepharose Fast Flow (Pharmacia, Piscataway NJ) is our preferred matrix. We run a 2.5 × 17-cm column at room temperature after equilibration with 5 m*M* Tris–HCl, 0.02% NaN_3, pH 8.0, running buffer. The sample is loaded by gravity at a flow rate of 2–4 ml/min and eluted with a 2.0-liter gradient of 0–0.5 *M* NaCl in the same Tris buffer. Minor isoforms of GFP that differ by even one charged amino acid can be resolved quantitatively on DEAE Fast Flow. This column is also capable of processing up to 1 g of GFP at a time. The eluted GFP should now be concentrated further by one of the methods described previously.

Gel Permeation

A highly concentrated GFP sample (20–100 mg/ml), with or without $(NH_4)_2SO_4$, in 10 m*M* Tris–EDTA buffer, pH 8.0, can be chromatographed on a second gel permeation column. At such high protein concentrations, GFP dimerizes and elutes at an apparent molecular weight of 44,000.[15] Contaminants that eluted with GFP on the first column and are present after anion exchange are resolved on the second gel permeation column. We usually chromatograph the sample at room temperature on a 3 × 95-cm CL Sepharose 6B or a 3 × 120-cm (~0.75-liter bed volume) Bio-Gel P-100 medium gel (~1-liter bed volume), preequilibrated with 5 m*M* Tris–HCl, 0.02% NaN_3, pH 8.0, running buffer.

Final Processing

A final anion-exchange column is used as a finishing step and to somewhat concentrate the GFP. Again a 2.5 × 17-cm DEAE Sepharose Fast Flow column at room temperature is equilibrated with 5 m*M* Tris–HCl, 0.02% NaN_3, pH 8.0, buffer. The sample is loaded in the same buffer (by gravity at a flow rate of 2–4 ml/min) and eluted with a 2.0-liter gradient of salt (0–0.5 M NaCl) in the same Tris buffer. Fractions eluted from this chromatography step (except for the tails of the elution peak) should be >95% pure. We usually pool the purer fractions, dialyze away any residual salt (10 m*M* Tris, 0.02% NaN_3, pH 8.0, dialysis buffer), and again concentrate via stirred cell ultraconcentration to a convenient stock concentration.

Conclusion

Those fortunate enough to purify GFP for physical studies will appreciate the advantages the protein possesses. Even in a crude extract, wtGFP can be recognized easily by its bright green fluorescence, which is so bright that it can be discerned in ordinary room light. GFP can be assayed on the basis of absorbance and fluorescence nondestructively. It is expressed heterologously in a variety of systems and can be produced in huge quantities. The tremendous stability of the protein makes it possible to purify by means of a wide array of purification methodologies.[23] We have pointed out in this method that by processing the protein carefully during the extraction stage and utilizing the property of the protein to dimerize reversibly, one can easily purify large quantities of GFP to homogeneity.

Acknowledgments

The authors acknowledge Diane F. Davis for reading the manuscript in progress and contributing critical comments and suggestions and Scott Lang for his assistance with the development of the high-level expression system in liquid culture.

[23] W. W. Ward, *in* "Green Fluorescent Protein: Properties, Applications, and Protocols" (M. Chalfie and S. Kain, eds.), p. 45. Wiley-Liss, New York, 1998.

[17] Recombinant Obelin: Cloning and Expression of cDNA, Purification, and Characterization as a Calcium Indicator

By BORIS A. ILLARIONOV, LUDMILA A. FRANK, VICTORIA A. ILLARIONOVA, VLADIMIR S. BONDAR, EUGENE S. VYSOTSKI, and JOHN R. BLINKS

Introduction

The term "photoprotein" was introduced by Shimomura and Johnson[1] as a convenient general designation for certain self-contained bioluminescent systems that do not fit the classical pattern in which an enzyme (the luciferase) catalyzes the oxidation of a separate diffusible organic substrate molecule (the luciferin) with the creation of an excited state and the emis-

[1] O. Shimomura and F. H. Johnson, *in* "Bioluminescence in Progress" (F. H. Johnson and Y. Haneda, eds.), p. 495. Princeton Univ. Press, Princeton, 1966.

0076-6879/00 $30.00

sion of light.[2] Although other kinds of photoproteins have been described, the great majority of photoproteins now known to exist are stimulated to luminesce by calcium, and the term "calcium-activated photoproteins" was applied to them by Hastings and Morin.[3] Later, the term "calcium-regulated photoproteins" was suggested, first, because calcium regulates the function of these proteins, but is not essential for it, and second, because these proteins are similar in many ways to calcium-regulated effector proteins such as calmodulin and troponin C.[4]

The first Ca^{2+}-regulated photoprotein to be discovered was aequorin, isolated in 1961 by Shimomura *et al.*[5] from the jellyfish *Aequorea.* Similar but distinct calcium-regulated photoproteins have since been identified in a variety of luminescent marine organisms, mostly coelenterates.[6] The fact that certain marine hydroids of the genus *Obelia* contain such photoproteins was first reported by Morin and Hastings.[7] In luminescent species of *Obelia,* the photoprotein is contained in specialized cells called photocytes. However, the location of the photocytes is not the same in all species of this genus. The photocytes of *O. geniculata* are scattered in the gastroderm of the stolons, uprights, and pedicels, but do not occur in the hydranths[8]; in contrast, the photocytes of *O. longissima* are organized in clusters and localized in the gastroderm of the distal part of the hydranth.[9] The photocytes are identified most easily by the intense green fluorescence that they exhibit when illuminated with ultraviolet light.[8] As in *Aequorea,* this fluorescence is due to a "green fluorescent protein," which occurs in photocytes in association with obelin and is responsible for the fact that the bioluminescence of *Obelia* is green, whereas the light emitted by obelin is blue.[7,10] Luminescence occurs intracellularly in response to chemical, electrical, or gentle mechanical stimulation[9,11] of the colonies.

Like other Ca^{2+}-regulated photoproteins, obelin from *O. longissima*[10] consists of a single polypeptide chain of relatively small size (approximately 20 kDa) to which a low molecular weight chromophore (presumably coelen-

[2] M. J. Cormier, J. E. Wampler, and K. Hori, *Fortschr. Chem. Org. Naturst.* **30,** 1 (1973).
[3] J. W. Hastings and J. G. Morin, *Biochem. Biophys. Res. Commun.* **37,** 493 (1969).
[4] J. R. Blinks, *in* "The Sarcoplasmic Reticulum in Muscle Physiology" (M. L. Entman and W. B. Van Winkle, eds.), Vol. II, p. 73. CRC Press, Boca Raton, FL, 1986.
[5] O. Shimomura, F. H. Johnson, and Y. Saiga, *J. Cell Comp. Physiol.* **59,** 223 (1962).
[6] J. G. Morin, *in* "Coelenterate Biology: Reviews and New Perspectives" (L. Muscatine and H. M. Lenhoff, eds.), p. 397. Academic Press, New York, 1974.
[7] J. G. Morin and J. W. Hastings, *J. Cell. Physiol.* **77,** 305 (1971).
[8] J. G. Morin and G. T. Reynolds, *Biol. Bull.* **147,** 397 (1974).
[9] V. N. Letunov and E. S. Vysotski, *Zh. Obshchei Biologii* **49,** 381 (1988) (in Russian).
[10] V. S. Bondar, K. P. Trofimov, and E. S. Vysotski, *Biochemistry-Russia* **57,** 1481 (1992) (in Russian).
[11] J. G. Morin and I. M. Cooke, *J. Exp. Biol.* **54,** 689 (1971).

terazine) is tightly bound. By analogy with the more thoroughly studied photoprotein aequorin, it seems probable that the chromophore is oxidized during the light-yielding reaction, with the elimination of a mole of carbon dioxide and the generation of an excited state.[12,13] The oxygen required for this reaction is derived from within the photoprotein, where it is bound in some as yet undetermined way; the rate of the reaction is thus independent of the availability of molecular oxygen. Because the energy for light emission is derived from the oxidative degradation of the chromophore, each molecule of photoprotein can react only once. Unlike coelenterazine, the oxidized chromophore (coelenteramide) dissociates readily from apoaequorin when the latter is stripped of calcium ions. The apoprotein can be recharged by incubating it with synthetic coelenterazine under calcium-free conditions in the presence of molecular oxygen and a sulfhydryl-reducing agent.[14]

The main use of Ca^{2+}-regulated photoproteins has been for the detection of ionized calcium in biological systems.[15,16] Photoproteins have now been used successfully inside a great many different types of living cells both to estimate the intracellular [Ca^{2+}] under steady-state conditions and to study the role of calcium transients in the regulation of cellular function.[17,18] Both aequorin[17] and obelin[18–20] have been used in this way. However, only aequorin has been used widely, perhaps partly because it was the first photoprotein to be discovered, but more significantly because of its more general availability. Nonetheless, aequorin has a number of shortcomings that have limited its utility. Among the most important of these are that it responds too slowly to follow the most rapid intracellular calcium transients without distortion and that physiological concentrations of magnesium antagonize the effects of calcium and slow the kinetics even further. Other Ca^{2+}-regulated photoproteins do not necessarily share these defects in equal

[12] O. Shimomura and F. H. Johnson, *Biochemistry* **11,** 1602 (1972).

[13] M. J. Cormier, K. Hori, Y. D. Karkhanis, J. M. Anderson, J. E. Wampler, J. G. Morin, and J. W. Hastings, *J. Cell Physiol.* **81,** 291 (1973).

[14] O. Shimomura and F. H. Johnson, *Nature (London)* **256,** 236 (1975).

[15] J. R. Blinks, W. G. Wier, P. Hess, and F. G. Prendergast, *Prog. Biophys. Mol. Biol.* **40,** 1 (1982).

[16] J. R. Blinks, *Methods Enzymol.* **172,** 164 (1989).

[17] J. R. Blinks, *Techniq. Cell. Physiol.* **P126,** 1 (1982).

[18] M. B. Hallett and A. K. Campbell, *in* "Clinical and Biochemical Analysis" (L. J. Kricka and T. J. N. Carter, eds.), Vol. 12, p. 89. Dekker, New York, 1982.

[19] V. S. Bondar, E. S. Vysotski, O. M. Rozhmanova, and S. G. Voronina, *Biochemistry-Russia* **56,** 806 (1991) (in Russian).

[20] V. S. Bondar, E. S. Vysotski, I. A. Gamaley, and A. B. Kaulin, *Cytology-Russia* **33,** 50 (1991) (in Russian).

measure. For example, obelin is considerably faster,[21,22] and obelin[22] and halistaurin[23] [from the jellyfish *Halistaura* (*Mitrocoma*) *cellularia*] are relatively insensitive to physiological concentrations of Mg^{2+}. Although aequorin's range of calcium sensitivity suits it well for the majority of applications, there are circumstances in which it would be useful to have photoproteins with higher or lower calcium sensitivity. Phialidin [from the jellyfish *Phialidium* (*Clytia*) *gregarium*] is reportedly less sensitive.[24] However, only molecular cloning of cDNAs for apophotoproteins can be expected to provide such scarce photoproteins in useful quantities.

This chapter describes the approaches that we have used to clone the cDNA for an apophotoprotein from *O. longissima,* to express the cDNA in *Escherichia coli* cells, to purify the product, and to charge it with synthetic coelenterazine. We also describe some of the properties of the recombinant obelin and compare them with corresponding properties of recombinant aequorin. The general approach described here of using the potential for bioluminescence of the protein of interest for expression screening of the cDNA library could certainly be employed in efforts to clone the cDNAs for other Ca^{2+}-regulated photoproteins and might well prove useful in cloning cDNAs for proteins involved in other bioluminescent systems as well.

Molecular Cloning of Apoobelin cDNA

Construction of the cDNA Library

We collected *Obelia longissima* in Chupa Bay of the White Sea, at the Marine Biological Station of the Institute for Zoology, Russian Academy of Sciences. The collections were made in October and November, when the colonies reach maximum length. Terminal uprights (2–3 cm long), each of which normally contained several dozen hydranths, were cut from the stolon and immediately put into liquid nitrogen where they can be stored for at least 3 months without RNA degradation.

All of the procedures for RNA isolation were carried out in disposable plasticware sterilized with γ-irradiation or in glass/ceramicware that had been acid-washed, rinsed with sterile doubly distilled water, and sterilized by baking at 180° for 3 hr. After removal from liquid nitrogen, the tissues

[21] D. G. Stephenson and P. J. Sutherland, *Biochim. Biophys. Acta* **678,** 65 (1981).

[22] V. A. Illarionova, B. A. Illarionov, V. S. Bondar, E. S. Vysotski, and J. R. Blinks, *in* "Bioluminescence and Chemiluminescence: Molecular Reporting with Photons" (J. W. Hastings, L. J. Kricka, and P. E. Stanley, eds.), p. 427. Wiley, Chichester, 1997.

[23] J. R. Blinks and D. D. Caplow, *Physiologist* **34,** 110, (1991).

[24] O. Shimomura and A. Shimomura, *Biochem. J.* **228,** 745 (1985).

were immediately ground to a fine suspension with a ceramic mortar and pestle containing RNA extraction solution (7 *M* urea, 50 m*M* Tris–HCl, 10 m*M* EDTA, 1 g/liter CsCl, 0.5% 2-mercaptoethanol, 0.5% *N*-lauroylsarcosine, pH 8.0; approximately 7 ml solution per g of tissue). Debris was sedimented by centrifugation for 30 min at 38,000 *g* (20°). The supernatant was layered on top of a 5.7 *M* CsCl cushion in polyallomeric tubes (silanized and autoclaved) for the Beckman SW-55 ultracentrifuge rotor. RNA was spun down to the bottom of the tubes and collected as described by Glisin *et al.*[25] Normally 0.4–0.6 mg of RNA was obtained per gram of *O. longissima* tissue. The poly(A)$^+$ RNA fraction was isolated with an oligo(dT)-cellulose column[26]; typically 8–10 μg of poly(A)$^+$ RNA was obtained from 1 mg of total RNA.

Although commercially available kits for double-stranded cDNA synthesis and cloning might now be more convenient, we made our cDNA library from mRNA–cDNA hybrids after blunting their ends with RNase A and adding oligonucleotide [poly(dC)] tails.[27–29] This approach has the advantages that the problem of second strand synthesis priming is avoided and that the sequences cloned are normally those at the 5′ terminus of the mRNA.

The mRNA–cDNA hybrids were inserted into the plasmid cloning vector pUC18,[30] which confers ampicillin resistance on bacterial cells containing it. The vector also contains the lac promoter-operator, and in consequence the transcription of cDNA inserts can be induced by isopropylthiogalactoside (IPTG). The inserts were annealed to the *Pst*I-cut and poly (dG)-tailed pUC18 vector with methods similar to those described by Prasher *et al.*[31] for double-stranded aequorin cDNA, except that the temperature conditions for annealing were modified: the annealing mixture was incubated at 72° for 5 min and then the temperature was decreased to 12° at a rate of 1–2° per minute. Nucleic acids were precipitated from the annealing mixture with ethanol, dissolved in doubly distilled water at 0°, and immediately electroporated into electrocompetent *E. coli* JM103 cells with a GenePulser (Bio-Rad) set to the standard pulse characteristics recommended by the manufacturer.

The electroshocked cells were grown out on twenty 15-cm plates of LB agar (10 g/liter peptone, 5 g/liter yeast extract, 5 g/liter NaCl, 16 g/liter

[25] V. Glisin, R. Crkvenjakov, and C. Byus, *Biochemistry* **13,** 2633 (1974).
[26] H. Aviv and P. Leder, *Proc. Natl. Acad. Sci. U.S.A.* **69,** 1408 (1972).
[27] K. O. Wood and J. C. Lee, *Nucleic Acids Res.* **3,** 1961 (1976).
[28] S. Zain, J. Sambrook, R. J. Roberts, W. Keller, M. Fried, and A. R. Dunn, *Cell* **16,** 851 (1979).
[29] J. Sambrook, E. F. Fritsch, and T. Maniatis, "Molecular Cloning," p. 5.81. Cold Spring Harbor Laboratory Press, Cold Spring Harbor, NY, 1989.
[30] J. Norrander, T. Kempe, and J. Messing, *Gene* **26,** 101 (1983).
[31] D. Prasher, R. O. McCann, and M. J. Cormier, *Methods Enzymol.* **133,** 288 (1986).

Bacto-agar, pH 7.4) containing 50 μg/ml ampicillin. In the aggregate, approximately 400,000 independent recombinant clones were obtained on these 20 primary plates. The growth on each of the primary plates was scraped separately into liquid LB medium (10 g/liter peptone, 5 g/liter yeast extract, 5 g/liter NaCl, pH 7.4) containing 15% glycerol. The resulting cell suspensions were dispensed into 100-μl aliquots and stored at − 70°. A set consisting of one aliquot from each of the primary plates constituted one copy of the amplified cDNA library.

Isolation of a Full-length cDNA for Apoobelin by Bioluminescence Expression Cloning

In the absence of information about the amino acid sequence of obelin, we attempted (successfully, as it turned out) to use the approach of expression cloning to screen the cDNA library for clones capable of producing apoobelin. The following strategy was used in the initial effort[32]; it was later found that the procedure could be simplified considerably (see below).

The titer of viable *E. coli* cells in one of the aliquots constituting the amplified library was determined as described by Lech and Brent.[33] With this information, several standard 90-mm petri dishes containing LB agar with ampicillin (50 μg/ml) were plated from the aliquot to a density of approximately 5000 colonies per plate. After they had grown to a size of 0.5–1.0 mm, the colonies from each of these (secondary) plates were lifted on sterile nitrocellulose membranes as described by Sambrook *et al.*[34] Each membrane was cut into six to eight sectors. The cells from each sector were washed into liquid LB medium and mixed thoroughly. Enough of the resulting cell suspension was inoculated into 2 ml of liquid LB with ampicillin to give an initial OD_{590} of 0.05–0.1; the remainder of the cells from each sector were stored as a stock culture at 4° for not more than 2 days. (For longer term storage, stock cultures were frozen at −70° after the addition of 15% sterile glycerol.) The inoculated culture was shaken at 37° for 3 hr, IPTG was added to a final concentration of 0.5 m*M*, and shaking was continued for 1 hr. The cells were collected by centrifugation, washed once with sterile 0.9% saline, and resuspended in 0.2 ml of a solution containing 10 m*M* EDTA, 5 m*M* dithiothreitol (DTT) (or 5 m*M* 2-mercaptoethanol), and 20 m*M* Tris–HCl pH 7.0. The cell suspension was transferred into a

[32] B. A. Illarionov, S. V. Markova, V. S. Bondar, E. S. Vysotski, and J. I. Gitelson, *Proc. Russ. Acad. Sci.* **326,** 911 (1992) (in Russian).

[33] K. Lech and R. Brent, *in* "Current Protocols in Molecular Biology," p. 1.3.1. Wiley, New York, 1991.

[34] J. Sambrook, E. F. Fritsch, and T. Maniatis, "Molecular Cloning," p. 1.96. Cold Spring Harbor Laboratory Press, Cold Spring Harbor, NY, 1989.

1.5-ml polypropylene microcentrifuge tube, which was then placed in an ice bath. The microtip of a standard ultrasonic homogenizer was dipped into the suspension and the cells were lysed with ultrasound (5 sec ×5). Cell debris was removed by centrifugation (10,000*g*, 15 min, 4°).

To charge the apoobelin, we added 3 μl of 40 μM coelenterazine in methanol to the lysate and the mixture was incubated in the dark at 4° for at least 4 hr (preferably overnight). The coelenterazine concentration was determined from the extinction coefficient in methanol (8900 cm^{-1} M^{-1} at 434 nm[35]).

For the measurement of obelin luminescence, 10–20 μl of the mixture was added to 0.5 ml of an assay buffer containing 10 m*M* EDTA, 100 m*M* Tris–HCl, pH 8.8, and placed in a monitoring cell (cuvette) of a BLM 8801 luminometer ("Nauka," Krasnoyarsk). To elicit the luminescence, 0.2 ml of a solution containing 100 m*M* $CaCl_2$ in 100 m*M* Tris–HCl, pH 8.8, was injected into the cell. The luminometer was calibrated with the light standard of Hastings and Weber[36] and the threshold for detection was found to be about 5×10^6 photons/sec. (As it turned out, this was just barely sufficient for the job at hand.) Nearly 300 sectors containing approximately 200,000 colonies were tested in this way before the first light signal was registered. Note that these 200,000 colonies were from secondary plates (*after* amplification). They represent a very small fraction of the original library, which contained approximately 400,000 recombinant clones (*before* amplification).

Cells of the stock culture from the sector containing the light-yielding clone were plated on LB agar with ampicillin to a density of 500–1000 colonies per plate, and the procedures of apoobelin charging and light measurement were repeated until a new "luminescent" sector was found. This procedure was repeated until an individual light-yielding clone was isolated. We isolated two such clones (designated pOL1 and pOL101). cDNAs from the recombinant plasmids of these clones were sequenced by the method of Maxam and Gilbert[37] and found to be identical. Plasmid pOL101 was used for further investigation.

The use of bioluminescence expression cloning to screen a cDNA library has four advantages: no advance knowledge about the amino acid sequence of the protein is required, the measurement of luminescence is simple and exceedingly sensitive, only clones producing active photoproteins are identified, and no radioisotopes are required. We have since simplified the assay procedure further, having switched to using a more sensitive photometer.[16]

[35] A. K. Campbell, A. K. Patel, Z. S. Razavi, and F. McCapra, *Biochem. J.* **252,** 143 (1988).
[36] J. W. Hastings and G. Weber, *J. Opt. Soc. Am.* **53,** 1410 (1963).
[37] A. M. Maxam and W. Gilbert, *Methods Enzymol.* **65,** 499 (1980).

We now find that apoprotein concentrations sufficient for detection are usually produced without IPTG, and that apoobelin synthesized inside *E. coli* cells can be charged with coelenterazine and then discharged with Ca^{2+} without disrupting the cells. For the simplified assay, which may be carried out on just part of a single colony, the bacteria are suspended in a small quantity (e.g., 50 μl) of a charging solution (2 μM coelenterazine, 20 mM DTT, 1 M NaCl, 10 mM EDTA, 33 mM Tris base, pH 7.5) directly in the measuring cuvette. After overnight incubation at 4°, the cuvette is placed in the photometer and a small volume (e.g., 20 μl) of a concentrated calcium solution (0.5 M $CaCl_2$, 0.5 M PIPES buffer, pH 7.5) is injected. (The discharge is somewhat faster if the calcium solution contains 1% Triton X-100, but this is not necessary.) We have used this approach successfully to screen cDNA libraries from other marine coelenterates (unpublished). It requires considerably less time and coelenterazine than the original method and is facilitated greatly if the photometer can be provided with a large number of cuvettes for the overnight incubation. The luminometer that we have used for this purpose[16] accommodates inexpensive glass shell vials that can be used by the hundred and has a practical detection threshold of about 5×10^4 photons/s emitted isotropically in a 1.0-ml reaction cuvette.

Nucleotide Sequence Analysis of cDNA for Apoobelin

As has been reported elsewhere,[38] nucleotide sequence analysis revealed two long open reading frames (ORF) in the cDNA of clones pOL1 and pOL101. One of them codes for apoobelin (apoObl ORF) with a calculated isoelectric point (pI) of 4.62, and the other for a protein of 16kDa, with a calculated pI of 10.56. No structures resembling a bacterial ribosome-binding site (RBS) are apparent in the cloned cDNA upstream from the start codon for the apoobelin gene, so the reason for its expression in *E. coli* remains unclear. (The same was reported in the case of the apoaequorin gene.[39]) Two possible explanations may be considered. First, the apoobelin gene in pOL101 might reside in the same ORF as the gene for lacZ′, which is located upstream, in the pUC18 vector. The translation of a chimeric protein might start from the RBS for the lacZ′ gene and continue on to include apoobelin. This possibility could not be tested by analysis of the nucleotide sequence of the cloned cDNA and surrounding regions because the number of nucleotides in the oligo(dC/dG) tail between the apoobelin start codon and the vector molecule could not be determined precisely. (The readability of such repetitive sequences is poor.)

The second possibility relates to the extreme sensitivity of the biolumi-

[38] B. A. Illarionov, V. S. Bondar, V. A. Illarionova, and E. S. Vysotski, *Gene* **153,** 273 (1995).

[39] D. Prasher, R. O. McCann, and M. J. Cormier, *Biochem. Biophys. Res. Commun.* **126,** 1259 (1985).

nescence assay. All of the plasmid constructions that we have made so far with the apoobelin gene, with or without a RBS for that gene, have directed the synthesis of enough apoobelin in *E. coli* cells so that light signals could be detected without difficulty. This was true even though the apoobelin sometimes represented no more than 0.002% of the soluble protein. Although protein synthesis is normally directed by the interaction of ribosomes with ribosome-binding sites, it seems possible that *E. coli* ribosomes may also be able to initiate protein synthesis randomly, if not on any mRNA, then at least on the mRNA for apoobelin.

Whatever the reason, the expression of detectable amounts of the apoprotein has proved to be very useful during the construction of new plasmids harboring the apoobelin gene. Instead of requiring laborious restriction analysis of numerous plasmid DNA samples, identification of the colony containing the construct of interest can normally be accomplished by testing for luminescence after only 30 min of incubation in charging solution.

Engineering of an Overexpression Plasmid for Apoobelin cDNA

To rearrange the apoobelin cDNA, we recloned it in the pTZ18U vector[40] using the *Bam*HI and *Sph*I restriction sites. The resulting plasmid was designated pOL103. Methods described by Kunkel *et al.*[41] were used to prepare single-stranded uracil-containing DNA from this plasmid and to accomplish the oligonucleotide-directed mutagenesis of apoobelin cDNA. In this way the oligo(dC/dG) tail, as well as the nontranslated region upstream from the translation initiation codon, was removed from apoobelin cDNA as shown in Fig. 1A. Simultaneously, a synthetic RBS was generated upstream from the apoobelin (apoObl) ORF. We used the same method with oligonucleotides S1 (5′-GTGGCTTCTAGCTTGGC-3′) and S2 (5′-GAAGTCGCTAATTGTTT-3′) to generate two consecutive stop codons (at positions 223 and 346, respectively) in the structure of ORFII (Fig. 1B) in such a way that the amino acid sequence encoded by the apoObl ORF was not changed.[38] The oligo(dC/dG) tail located downstream from the apoobelin cDNA does not seem to interfere with transcription or translation. The plasmid resulting from these manipulations was designated pOL110 and was used for expression of the apoObl ORF after the transformation of the K12 C600 strain of *E. coli.* That strain also harbors a plasmid (pGP1-2[42]) bearing the gene for T7 RNA polymerase, which is under the control of the transcription promoter λ P_L, and the gene for the temperature-regulated CI857 repressor.

We also made an IPTG-inducible overproducing strain by using the

[40] D. A. Mead, E. Szszesna-Skorupa, and B. Kempner, *Protein Eng.* **1,** 67 (1986).

[41] T. A. Kunkel, K. Bebenek, and J. McClary, *Methods Enzymol.* **204,** 125 (1991).

[42] S. Tabor, *in* "Current Protocols in Molecular Biology," p. 16.2.1. Wiley, New York, 1991.

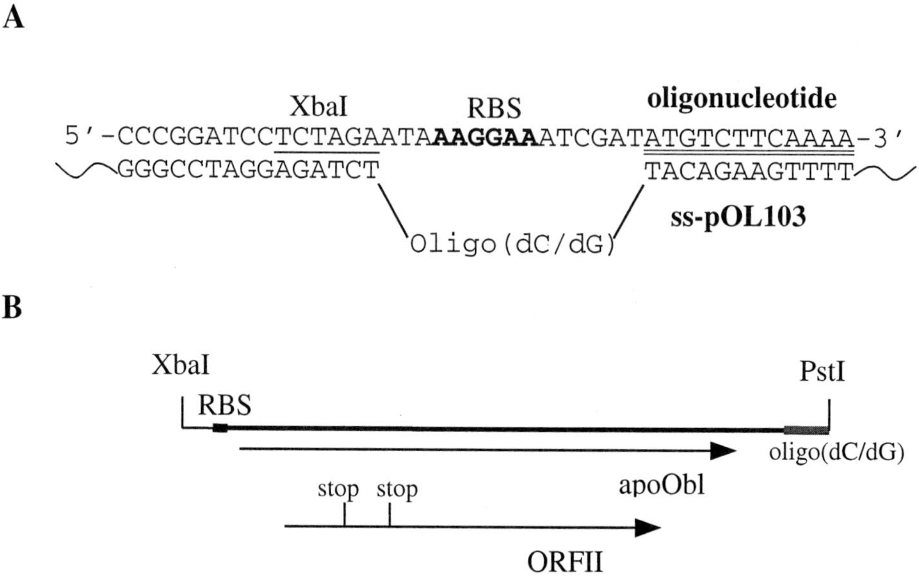

FIG. 1. Apoobelin cDNA rearranged for overexpression in *E. coli.* (A) Generation of a synthetic ribosome-binding site (RBS) upstream from the translation initiation codon for the apoobelin gene. (B) Structure of the rearranged cDNA. RBS, ribosome binding site (in bold); apoObl, open reading frame coding for apoobelin; ORFII, open reading frame coding for the 16-kDa protein; ss-pOL103, single-stranded deoxyuracil-enriched DNA of pOL103; oligonucleotide, synthetic oligonucleotide with arms complementary to the polylinker of pTZ18R and the first four codons of apoObl (an RBS is designed in the middle); stop, stop codon generated in ORFII. Nucleotides forming the *Xba*I restriction site are underlined; the first four codons of apoObl are double underlined.

*Xba*I and *Pst*I sites to reclone the rearranged DNA into another vector, pT7-7,[42] downstream from the phage T7 promoter. The resulting plasmid (termed pDO2) was used to transform the *E. coli* BL21 (DE3)[43] strain, in which the gene for T7 RNA polymerase resides on the chromosome, under control of the IPTG-inducible lacUV5 promoter of transcription.

Isolation and Purification of Recombinant Apoobelin

Growth and Harvesting of E. coli Cells Producing Apoobelin

The following procedures were used to grow and harvest *E. coli* BL21 and C600 strains in an LKB "Ultroferm 1601" fermenter containing 3 liters of medium.

[43] F. W. Studier and B. A. Moffat, *J. Mol. Biol.* **189,** 113 (1986).

IPTG-Induced Strain. The transformed BL21 strain of *E. coli* is grown up in LB or Terrific Broth (Sigma) at 37°. All media for growing this strain contain ampicillin, 200 mg per liter. A single colony from an LB agar plate is inoculated into 100 ml of liquid medium and grown overnight with vigorous agitation on an orbital shaker-water bath. Cells are then harvested by centrifugation, resuspended in fresh medium, and poured into the fermenter. Cells are grown to an OD_{590} of 1.0, and apoobelin synthesis is then induced by the addition of IPTG (100 mg per liter). Growth is continued for 3 hr, and the *E. coli* cells are then harvested by centrifugation (11,000 *g*, 20 min) and washed with 0.9% NaCl.

Temperature-Induced Strain. Escherichia coli C600 cells are grown at 30° in the presence of ampicillin (50 mg per liter) and kanamycin (50 mg per liter). The inoculum for the fermenter is prepared as described just above. When OD_{590} reaches 1.0–1.5, apoobelin synthesis is induced by a temperature shift up to 42° for 30 min. Growth is then continued for 3 hr at 37°. Harvesting and washing of the cells are carried out as for the other strain. The yield from such fermentations is generally between 12 and 15 g of packed wet cells. Until disrupted, the cells are stored in the centrifuge tubes at −20°.

Disruption of Cells and Solubilization of Inclusion Bodies

In most cases the overexpression of foreign genes in *E. coli* results in the accumulation of the recombinant protein within the host cell in the form of insoluble aggregates referred to as inclusion bodies.[44] The protein in the inclusion bodies occurs as a solid, in a relatively pure and concentrated state (the recombinant protein representing more than 60% of the protein in the inclusion bodies[45]). The inclusion bodies may be isolated from other cell components by cell disruption and centrifugation, which are simple operations. To disrupt the bonds holding the aggregate together, a strong chaotrope such as concentrated urea or guanidine hydrochloride is commonly applied. This is followed by removal of the chaotrope to allow the protein to refold into its active conformation. The refolding stage is the most complicated and unpredictable; the process is poorly understood and methods for renaturation are protein-specific.[46,47] Fortunately, we had no difficulties with protein refolding: after solubilization of the inclusion bodies

[44] R. C. Hockney, *Trends Biotechnol.* **12,** 456 (1994).

[45] J. F. Kane and D. L. Hartley, *in* "Purification and Analysis of Recombinant Proteins" (R. Seetharam and S. K. Sharma, eds.), p. 121. Dekker, New York, 1991.

[46] B. Fischer, I. Samner, and P. Goodenough, *Biotechnol. Bioengin.* **41,** 3 (1992).

[47] R. Rudolf and H. Lilie, *FASEB J.* **10,** 49 (1996).

and removal of the denaturant, the apoobelin could be charged readily with coelenterazine to produce an active photoprotein.

Apoobelin is extracted as follows: the cell paste is resuspended in 20 mM Tris–HCl, pH 7.0 (1 g cells to 5 ml solution). The cell suspension is then transferred to a glass beaker (for better heat transfer), which is then placed in an ice bath. The tip of an ultrasonic homogenizer is dipped into the suspension and the cells are lysed with ultrasound (20 sec, ×6). Between bursts of ultrasound, the cell suspension is mixed gently, without taking the beaker from the ice, in order to cool it. The mixture is then centrifuged (39,000*g*, 30 min), and the supernatant (which contains 25–30% of the luminescent activity) is usually discarded. (Luminescent activity is tested after charging with coelenterazine as described on p. 229.) The pellet containing the inclusion bodies is washed sequentially with the following solutions: (a) 150 mM NaCl, 20 mM Tris–HCl pH 7.0 (×2); (b) 1% Triton X-100, 20 mM Tris–HCl, pH 7.0 (×2); and (c) 5 mM $CaCl_2$, 20 mM Tris–HCl, pH 7.0 (×1) to remove contaminating substances. All the washing procedures are performed with centrifugation (39,000*g*, 30 min). According to measurements of luminescence activity, the total loss of apoobelin during the washing steps is no more than 6%. The final pellet is resuspended in 20 mM Tris–HCl, pH 7.0, containing 6 M urea and 5 mM $CaCl_2$ (3 ml of the solution per gram of cell paste), kept at 4° overnight with stirring, and then centrifuged (39,000*g*, 30 min). The supernatant contains 65–70% of the total luminescence activity originally present in the cells. The pellet contains relatively little luminescence activity, as was shown by repeated extraction.

We also tested other methods for disrupting *E. coli* cells. The cells can be disrupted by high-pressure dispersion with a French press[48] or with lysozyme treatment followed by osmotic lysis. The apoobelin can be extracted into guanidine hydrochloride or into urea. However, the procedure just described proved to be the simplest and most reproducible and yields apoobelin with good yield and purity.

Purification of Recombinant Apoobelin

The purification scheme that we currently employ for obtaining apoobelin of high purity includes the following steps: (1) ion-exchange chromatography on DEAE-Sepharose Fast Flow (Pharmacia, Sweden) in 6 M urea with salt gradient elution and (2) ion-exchange chromatography on a Mono P HR 5/20 column (Pharmacia, Sweden) with salt gradient elution, in the absence of urea. All the solutions used for apoobelin column chromatography contain 5 mM DTT and 5 mM $CaCl_2$; we find that the addition of

[48] V. S. Bondar, A. G. Sergeev, B. A. Illarionov, J. Vervoort, and W. R. Hagen, *Biochim. Biophys. Acta* **1231,** 29 (1995).

these substances improves protein separation and significantly increases the recovery of apoobelin and its stability. It has been reported that the stability of other Ca^{2+}-binding proteins (parvalbumin, troponin C, calmodulin) is increased when their Ca^{2+}-binding sites are occupied.[49,50] The solutions are bubbled with argon for 20–30 min to prevent DTT oxidation. All chromatography procedures are carried out at room temperature.

DEAE-Sepharose Fast Flow is used in preference to other anion exchangers such as DEAE-cellulose and QAE or DEAE-Sephadex because it has a high capacity for proteins, and the resin bed does not change volume with increasing salt concentration so that it is possible to separate proteins at high flow rates. DEAE-Sepharose chromatography is carried out on a 1.6 × 10-cm column with a FPLC system (Pharmacia, Sweden); the 6 *M* urea extract of the inclusion bodies from 5 g of *E. coli* cell paste can be loaded onto such a column at one time. The column is initially equilibrated with 6 *M* urea, 5 m*M* DTT, 5m*M* $CaCl_2$, and 20 m*M* Tris–HCl, pH 7.0 (solution A), and washed with this solution again after the sample has been loaded to remove unbound contaminants. The proteins are eluted with a linear salt gradient [0–0.5 *M* sodium acetate (NaOAc) in the same solution; 130 ml at 4 ml/min] (Fig. 2A). The apoobelin itself elutes between 0.25 and 0.43 *M* NaOAc. Fractions pooled for further processing contain 80–85% of the luminescent activity of the initial load applied to the column. To reduce the salt concentration in a sample eluted from DEAE-Sepharose Fast Flow, the sample is dialyzed overnight at 4° against solution A (not less than 100 volumes).

A Mono P column can be used instead of DEAE-Sepharose for the initial stage of ion-exchange chromatography in 6 *M* urea.[22] As is shown in Fig. 2B, the separation of proteins on this column is actually somewhat better than on DEAE-Sepharose Fast Flow under the same conditions. However, the crude 6 *M* urea extract of inclusion bodies tends to plug the Mono P column, making it necessary to carry out the full procedure for column regeneration after each run. For this reason we prefer to use DEAE-Sepharose Fast Flow for the initial stage.

The second stage of ion-exchange chromatography is on a Mono P HR 5/20 column; the solutions are the same as in the first chromatography run except that urea is omitted. The proteins are eluted with a linear salt gradient (20 ml, 0–0.5 *M* NaOAc, at a flow rate of 0.5 ml per min). The elution profile in Fig. 3 shows that apoobelin is eluted as a sharp peak at 0.25 *M* NaOAc. The protein peak determined by absorption at 280 nm

[49] E. A. Permyakov, "Parvalbumin and Related Calcium-Binding Proteins." Nauka, Moscow, 1985 (in Russian).

[50] S. Inouye, S. Zenno, Y. Sakaki, and F. I. Tsuji, *Protein Express. Purif.* **2,** 122 (1991).

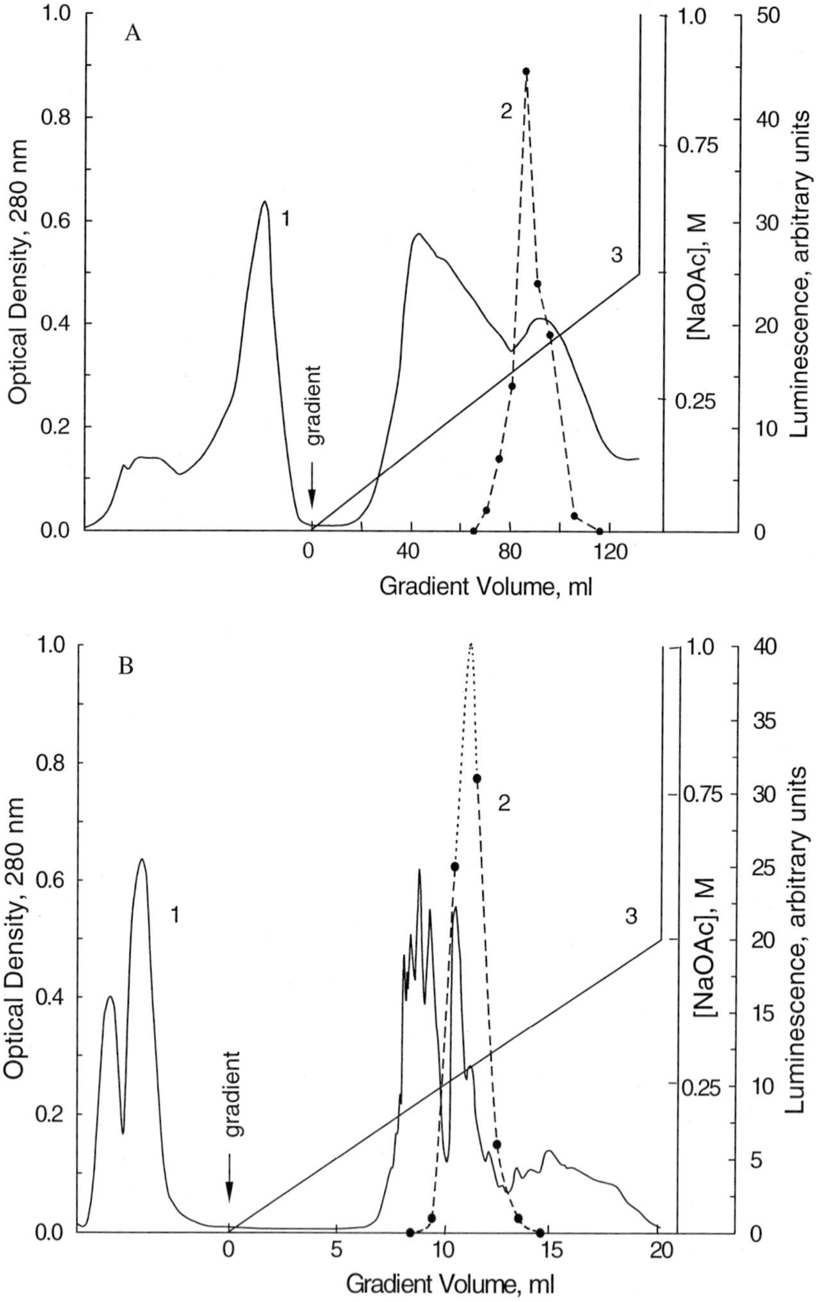
A
1
2
3
gradient
Optical Density, 280 nm
[NaOAc], M
Luminescence, arbitrary units
Gradient Volume, ml
B
1
2
3
gradient
Optical Density, 280 nm
[NaOAc], M
Luminescence, arbitrary units
Gradient Volume, ml

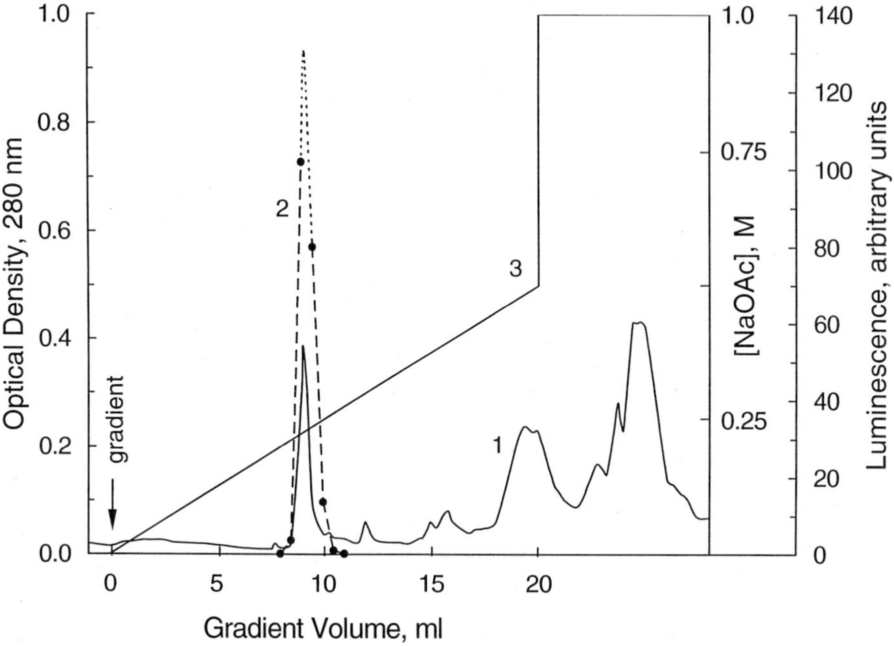

FIG. 3. Second-stage ion-exchange chromatography of apoobelin in the absence of urea. Pharmacia Mono P HR 5/20 FPLC column. Curve 1, OD at 280 nm; 2, luminescent activity after charging with coelenterazine; 3, sodium acetate concentration gradient. Arrow indicates beginning of gradient. Temperature 20°.

corresponded exactly to the luminescent peak. The fractions saved contain 85–90% of the luminescent activity initially loaded onto the column. It should be noted that the product of this purification stage is free of urea; it can be stored as it comes from the column for at least several weeks at 4° without loss of activity.

The purity of the apoobelin after these chromatography procedures was checked by SDS–PAGE, which revealed only a single protein band in the

---

FIG. 2. First-stage ion-exchange chromatography of apoobelin in 6 *M* urea. (A) DEAE exchanger (Pharmacia DEAE Sepharose Fast Flow in 1.6 × 10-cm column). (B) Mixed amine exchanger (Pharmacia Mono P HR 5/20 FPLC column). Parts A and B are presented as alternatives (see text). For both parts: curve 1, OD at 280 nm; 2, luminescent activity after charging with coelenterazine; 3, sodium acetate concentration gradient. Arrow indicates beginning of gradient. Temperature 20°.

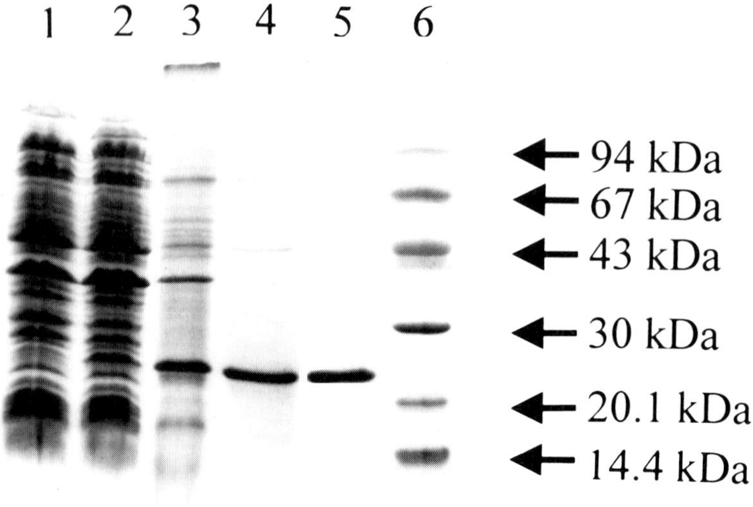

FIG. 4. SDS gel electrophoresis of apoobelin samples. Method of U. K. Laemmli, *Nature (London)* **227,** 680 (1970). Acrylamide concentration gradient is from 10 (top) to 22% (bottom). Gel stained with Serva Blue G. Lanes 1 and 2: cell lysates before and after IPTG induction, respectively. (Lysates were obtained from whole bacterial cells by extraction for 3 min with 2% SDS at 100°.) Lane 3: 6 *M* urea extract of inclusion bodies. Lane 4: pooled activity peak from first-stage DEAE-Sepharose chromatography (as in Fig 2A). Lane 5: pooled activity peak from second-stage Mono P chromatography (as in Fig. 3). Lane 6: Protein molecular weight calibration mixture (Pharmacia): phosphorylase *b* (94.0 kDa), bovine serum albumin (67.0 kDa), ovalbumin (43.0 kDa), carbonic anydrase (30.0 kDa), soybean trypsin inhibitor (20.1 kDa), α-lactalbumin (14.4 kDa).

sample taken after the second chromatography run (Fig. 4). After charging, the specific activity of the obelin was 6.5×10^{15} photons/mg of protein (protein concentration determined by the Lowry method[51]). The scheme described here is applicable to the purification of the recombinant apoobelin from both overproducing *E. coli* strains and probably for other apophotoproteins too.

The yield of crude apoobelin from the BL21 and C600 cells is, respectively, around 17–19 and 2–5 mg/g of wet cell paste, calculated according to the specific bioluminescent activity. The overall recovery of the apoprotein in the extraction and purification is 40–48%. Thus, the BL21 overproducing strain yields up to 100 mg of highly purified apoobelin from a single fermentation or approximately 8 mg of high-purity apoobelin/g of wet cell paste.

[51] O. H. Lowry, N. J. Rosebrough, A. L. Farr, and R. J. Randall, *J. Biol. Chem.* **193,** 265 (1951).

Characterization of Recombinant Obelin as a Biological Calcium Indicator: Comparison with Recombinant Aequorin

Our comparison is focused on results from two types of determinations: (1) full Ca^{2+} concentration–effect curves and (2) kinetic studies carried out with saturating [Ca^{2+}] in a rapid mixing stopped-flow machine. In both cases, measurements were made in the absence and in the presence of 1, 3, and 10 m*M* Mg^{2+}. Mg^{2+} evidently binds considerably more slowly to the photoprotein than does Ca^{2+} (a common circumstance among calcium-binding proteins). The full effect of Mg^{2+} on the response of the photoprotein to Ca^{2+} is apparent only if the protein has had time to come to equilibrium with the concentration of Mg^{2+} under study.[52] In the experiments reported here, the photoprotein was always equilibrated with the final concentration of Mg^{2+} before mixing with Ca^{2+}, either in the stopped-flow experiments or in the determination of Ca^{2+} concentration–effect curves. All solutions contained 5 m*M* PIPES buffer, pH 7.0, and 150 m*M* KCl, except that [KCl] was reduced enough to maintain constant ionic strength as [Ca^{2+}] and (or) [Mg^{2+}] were increased. Measurements were made at 20°.

Preparation of Photoproteins

Recombinant apoobelin produced and purified as described above is charged with coelenterazine as follows: A 600-μl sample containing approximately 0.5 mg of the apoprotein is taken from the protein peak of the final Mono P chromatography step (Fig. 3). As it comes from the column this sample is in a solution containing 5 m*M* $CaCl_2$, 5 m*M* DTT, 20 m*M* Tris–HCl pH 7.0, and (approximately) 0.25 *M* NaOAc. To this is added 25 μl 0.5 *M* EDTA, pH 7.7 (final concentration approximately 20 m*M*), enough fresh DTT (solid) to bring the total DTT concentration to 10 m*M*, and 15 μl methanol containing 0.125 mg coelenterazine/ml (final concentration 6.9 μ*M*). The mixture is incubated overnight at 4°; the charged photoprotein is then separated from the other components of the solution by gel filtration on a 1 × 25-cm column of Bio-Gel P6 (Bio-Rad) equilibrated and eluted with 150 m*M* KCl, 5 m*M* PIPES, pH 7.0, that has first been freed of contaminating Ca^{2+} by passage (twice) through 1 × 25-cm columns of Chelex-100 (Bio-Rad) prepared as described previously.[53]

The recombinant aequorin was a gift from Dr. Osamu Shimomura, who prepared it by charging recombinant apoaequorin with synthetic coelen-

[52] J. R. Blinks and E. D. W. Moore, *in* "Optical Methods in Cell Physiology" (P. de Weer and B. M. Salzberg, eds.), p. 229. Wiley, New York, 1986.

[53] J. R. Blinks, P. H. Mattingly, B. R. Jewell, M. van Leeuwen, G. C. Harrer, and D. G. Allen, *Methods Enzymol.* **57,** 292 (1978).

terazine.[54] The apoaequorin was produced by Dr. Satoshi Inouye, using the expression plasmid piP-HE,[55] which is derived from the cDNA clone designated AQ440.[56] This expression system differs from the one used for the production of apoobelin in that it causes the apoaequorin to be secreted by the bacterial (*E. coli*) cell rather than retained in the form of inclusion bodies.[50,55] The amino acid sequence of the recombinant apoaequorin is the same as the consensus sequence of native aequorin except for a slight modification at the N terminus, which results from the addition and subsequent cleavage of a signal peptide that causes the export of the protein. Thus, the N terminus, which is Val-Lys-Leu... in native aequorin, becomes Ala-Asn-Ser-Lys-Leu... in this particular recombinant version.[50,55] Like the recombinant obelin, the recombinant aequorin was freed of EDTA and other small molecules by gel filtration before it was used either for the determination of Ca^{2+} concentration–effect curves or for stopped-flow studies.

Expression of Results

In this article we frequently express light intensities in terms of a ratio we have termed the fractional rate of discharge (L/L_{int}). This requires explanation, as it represents a departure from previous practice. In the past, measurements of light intensity have commonly been expressed in terms of the ratio L/L_{max}, in which L is the light intensity measured from a sample of photoprotein under a particular set of conditions and L_{max} is the peak light intensity recorded from an identical sample of the photoprotein when it is mixed with a saturating [Ca^{2+}] under the same conditions.[57,58] Expressing light intensities in terms of the ratio L/L_{max} has the effect of normalizing the measurements for photoprotein concentration and for the optical efficiency of the measuring system, but it tends to obscure certain potentially significant characteristics of the light signal (such as the peak-to-integral ratio), which can differ considerably from one photoprotein to the next. Furthermore, the absolute light intensity at the peak of the flash may be influenced substantially by factors other than [Ca^{2+}] (by temperature and [Mg^{2+}], for example), and it is often useful to be able to display the

[54] O. Shimomura, S. Inouye, B. Musicki, and Y. Kishi, *Biochem. J.* **270,** 309 (1990).

[55] S. Inouye, S. Aoyama, T. Miyata, F. I. Tsuji, and Y. Sakaki, *J. Biochem.* **105,** 473 (1988).

[56] S. Inouye, M. Noguchi, Y. Sakaki, Y. Takagi, T. Miyata, S. Iwanaga, T. Miyata, and F. I. Tsuji, *Proc. Natl. Acad. Sci. U.S.A.* **82,** 3154 (1985).

[57] D. G. Allen, J. R. Blinks, and F. G. Prendergast, *Science* **195,** 996 (1977).

[58] D. G. Allen and J. R. Blinks, *in* "Detection and Measurement of Free Ca^{2+} in Cells" (C. C. Ashley and A. K. Campbell, eds.), p. 159. Elsevier, Amsterdam, 1979.

influence of such factors in a manner that does not obscure changes in L_{max} itself. A very serious drawback of attempting to express light intensities in terms of L/L_{max} is that sometimes (as in studies of photoprotein mutants with reduced Ca^{2+} sensitivity) L_{max} cannot be measured because it is not feasible to reach a saturating $[Ca^{2+}]$.

The ratio L/L_{int} avoids these problems. Here, L has the same significance as before, and L_{int} is the total amount of light emitted from the standard aliquot of the same photoprotein as it is discharged (the time integral of the flash). Like L/L_{max}, this ratio has the effect of normalizing light signals for the amount of active photoprotein present in the sample and for the optical efficiency of the recording system, but it does not obscure changes in L_{max} or in the peak-to-integral ratio. In switching from L/L_{max} to L/L_{int}, we are arguing, in part, that the total amount of light a sample is capable of emitting is a better indication of the amount of photoprotein present than is the peak light intensity in a saturating $[Ca^{2+}]$. The effect of Mg^{2+} illustrates that this is clearly the case: as $[Mg^{2+}]$ is increased from 0 to 10 m*M*, L_{max} is reduced substantially (see Figs. 6 and 8), but L_{int} remains constant within experimental error (results not shown). In plotting the Ca^{2+} concentration–effect curves shown here, we have normalized all light measurements to L_{int} determined in 5 m*M* Ca^{2+}. In effect, this involves the assumption that the quantum efficiency of the light-generating reaction is independent of $[Ca^{2+}]$. Although this appears to be true in the upper part of the Ca^{2+} concentration–effect curve, we do not know that it is so at low $[Ca^{2+}]$ because of the very long integration times required to measure L_{int}.

Because L is a measurement of light intensity, and L_{int} has the dimension of (light intensity) × (time), the ratio L/L_{int} has the units sec^{-1}. That is to say that L/L_{int} is a measure of rate, which is appropriate to the term "fractional rate of discharge." When a Ca^{2+} concentration–effect curve is plotted with effects expressed as L/L_{int}, the curve plateaus at the peak-to-integral (P/I) ratio.

Ca^{2+} concentration–effect curves expressed in the old way (i.e., with measurements normalized to L_{max}) can be distinctly misleading when changes in the P/I ratio occur. The normalization not only obscures the changes in L_{max}, it creates spurious changes in the lower part of the curve. An example of this sort of problem is shown in one of our previous publications,[22] in which results obtained in the presence of Mg^{2+} had to be normalized to L_{max} measured in the absence of Mg^{2+} to avoid giving the erroneous impression that Mg^{2+} increased the Ca^{2+}-independent luminescence. Problems of this sort do not arise if measurements of light intensity are expressed in terms of L/L_{int}. Each curve stands by itself, and results obtained at one time can be compared with results obtained at any other.

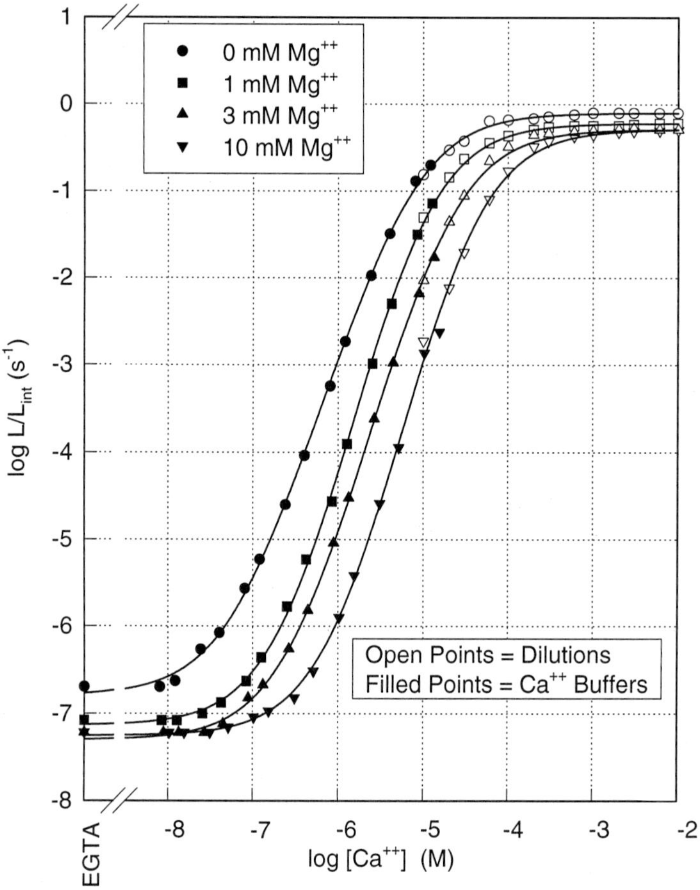

FIG. 5. Ca^{2+} concentration–effect curves for recombinant aequorin (AQ440). Photoprotein samples were freed of chelators by gel filtration and preequilibrated with the same concentration of Mg^{2+} added to the test solution. Ten-microliter aliquots of the photoprotein were injected forcefully into cuvettes containing 1 ml of test solution. The [Ca^{2+}] of the test solution was established either with Ca-EGTA buffers (lower concentrations) or by simple dilution of $CaCl_2$ (higher concentrations). See text for further details. Ca^{2+} buffers contained 2 m*M* EGTA (total) and were prepared by mixing CaEGTA and EGTA stock solutions in 14 volume ratios between 0.02 and 30. The CaEGTA stock solution was checked for stoichiometric balance and corrected as described by M. Klabusay and J. R. Blinks (*Cell Calcium* **20,** 227 (1996)). Ca^{2+}-independent luminescence was determined in the EGTA stock solution, with checks for significant Ca^{2+} contamination by further addition of 1 m*M* CDTA, and for Ag^{+} contamination by the addition of 1 m*M* NaCN [D. G. Allen, J. R. Blinks, and F. G. Prendergast, *Science* **195,** 996 (1977)]. The apparent equilibrium dissociation constant for CaEGTA was calculated and corrected for [Mg^{2+}] with the computer program MaxChelator [D. M. Bers, C. W. Patton, and R. Nuccitelli, *Methods Cell Biol.* **40,** 3 (1994)]. Light measurements were converted to units of L/L_{int} by first calculating L/L_{max} and then multiplying by the peak/integral ratio determined from stopped-flow measurements carried out under the same conditions and with the same specimen of aequorin (as in Fig. 7). Temperature 20°.

Ca^{2+} Concentration-Effect Relations

Measurements were made as described previously[16] by injecting 10-μl aliquots of the photoprotein solution into 1.0 ml of a test solution in a cuvette in a temperature-controlled photometer. Both calcium-EGTA buffer solutions (lower range of [Ca^{2+}]) and simple dilutions of $CaCl_2$ (upper range) were used as test solutions. Details are given in the legend to Fig. 5.

Figures 5 and 6 show Ca^{2+} concentration–effect curves (C-E curves) for recombinant aequorin and obelin determined in the presence of 0, 1, 3, and 10 m*M* Mg^{2+} and plotted on double-logarithmic scales. The curves for recombinant aequorin are very similar to many that have been published for native aequorin, except that here we have expressed all light intensities

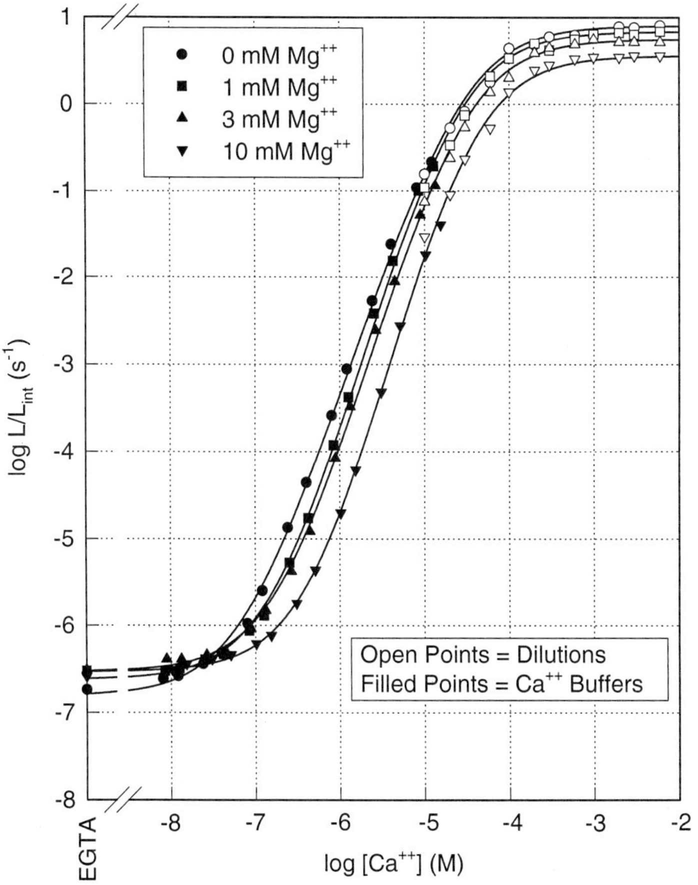

FIG. 6. Ca^{2+} concentration–effect curves for recombinant obelin. Same as for Fig. 5.

in terms of L/L_{int}. As is shown more clearly in Figs. 7 and 8, Mg^{2+} decreases L_{max} for both photoproteins, although the effect is more pronounced in the case of obelin. In the absence of Mg^{2+}, the calcium concentration–effect curves for the two photoproteins are remarkably similar. Both reveal a low level of Ca^{2+}-independent luminescence (CIL) at [Ca^{2+}] below approximately 10^{-8} *M*. Both reach saturation at [Ca^{2+}] above 10^{-4} *M*. Between these extremes, the curves both span vertical ranges of 6.5 to 7.5 log units and have maximum slopes of about 2.5. In the absence of Mg^{2+} the curve for aequorin lies somewhat to the left of that for obelin, but the difference is not great.

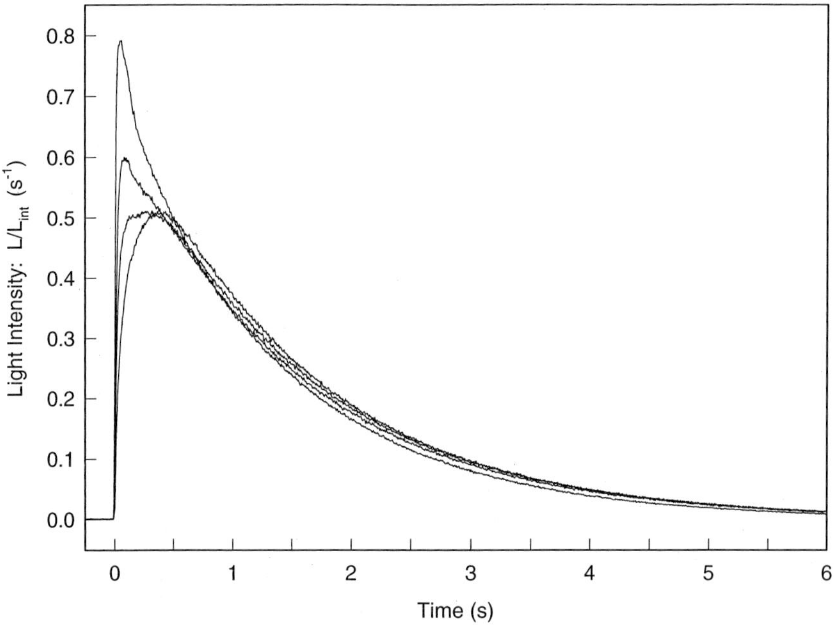

FIG. 7. Low-speed stopped-flow record for recombinant aequorin (AQ440). The apparatus used was a home-built adaptation of the stopped-flow machine described by Q. H. Gibson and L. Milnes [*Biochem. J.* **91,** 161 (1964)]. It is driven by compressed air, and with the air pressure used in these experiments, it had a dead time of 1.5 msec. The signal from an EMI 9635B photomultiplier was recorded through an operational amplifier wired as a current-to-voltage converter with a time constant of 0.1 msec; signals were captured with a Nicolet Model 4094C digital oscilloscope operating in the pretriggered mode, with digital sampling at 1-msec intervals (16K points per sweep). To reduce photomultiplier shot noise, 10 successive signals were averaged. Signals were determined from top (fastest) to bottom (slowest) in 0, 1, 3, and 10 m*M* Mg^{2+}. For display, each signal was normalized to its own time integral; the integration was performed digitally over a period of 15 sec. Temperature 20°. [Ca^{2+}] after mixing = 5 m*M*. Display begins before flow starts.

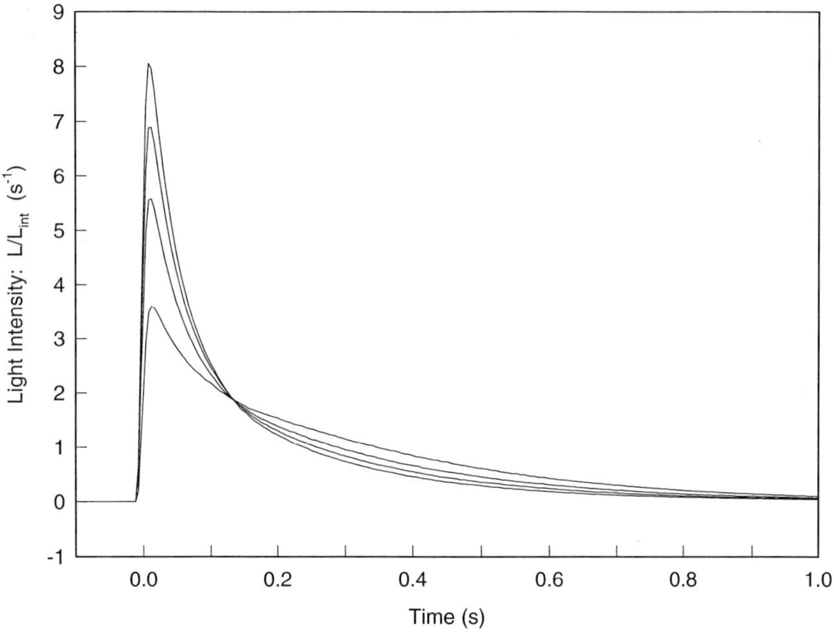

FIG. 8. Low-speed stopped-flow records for recombinant obelin. Same as in Fig. 7 except that the display is on a somewhat faster time scale. Sampling interval 1 msec; temperature 20°.

An important difference between the two photoproteins lies in the effect of Mg^{2+} on the Ca^{2+}-independent luminescence. If one takes a solution of aequorin in EGTA and adds 1, 3, and then 10 m*M* Mg^{2+} to it, the light intensity goes progressively down. If one does the same with obelin, the CIL does not change (the small apparent rise of CIL in Fig. 6 is not observed consistently and is probably due to experimental error). Even in concentrations as low as 1 m*M*, Mg^{2+} clearly shifts the Ca^{2+} concentration–effect curve for aequorin to the right; 10 m*M* Mg^{2+} shifts the curve by more than a log unit. The effect of Mg^{2+} on the C-E curves for obelin is more complicated. Primarily because Mg^{2+} produces a large decrease in the P/I ratio, there is also a Mg^{2+}-induced shift of the C-E curve for obelin. However, this is not a rightward shift, as is clearly the case for aequorin; it is primarily a downward shift, reflecting the Mg^{2+}-induced decrease of L_{max}.

Comparison of Kinetics: Rapid Mixing Stopped-Flow Measurements

More striking differences between the effects of Mg^{2+} on the two photoproteins are apparent in the stopped-flow records shown in Figs. 7–10, which show the results of rapidly mixing equal volumes of a calcium-free

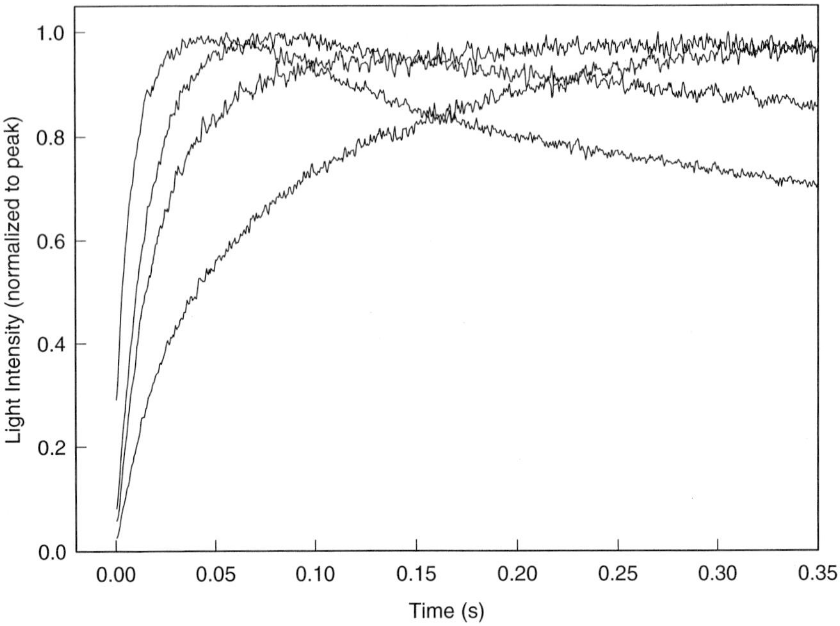

FIG. 9. Higher speed stopped-flow records for recombinant aequorin (AQ440) to display the rate of rise of luminescence. Same as in Fig. 7 except that the display begins at the time flow was stopped and each tracing has been normalized to its own peak to facilitate the comparison of rise times. Sampling interval 100 μsec; temperature 20°.

photoprotein solution and a solution containing 10 m*M* $CaCl_2$. As can be seen from Figs. 5 and 6, the final [Ca^{2+}] in the reaction mixture (5 m*M*) was saturating for both photoproteins in all the [Mg^{2+}] studied. The low sweep-speed records of Fig. 7 show that Mg^{2+} decreases the amplitude of the flash obtained by mixing aequorin with a saturating concentration of Ca^{2+}. At the same time the flash is prolonged, with the result that the total amount of light emitted in a flash is essentially unchanged. This effect of Mg^{2+} cannot be overcome by mixing with a higher concentration of Ca^{2+}.

The same sort of effect is observed in the case of obelin, but the Mg^{2+}-induced change in the amplitude of the flash is nearly twice as great as for aequorin (Fig. 8). (Note that the sweep speeds are different in Figs. 7 and 8; the flash of obelin is considerably briefer than that of aequorin.) More important than the overall duration of the flash is the speed of its rising phase, for this is what limits the ability of a photoprotein to follow rapid changes in [Ca^{2+}]. Figures 9 and 10 show stopped-flow tracings made with higher sweep speeds; to facilitate the comparison of the rising phases, each curve has been normalized to its own maximum. Rate constants calculated

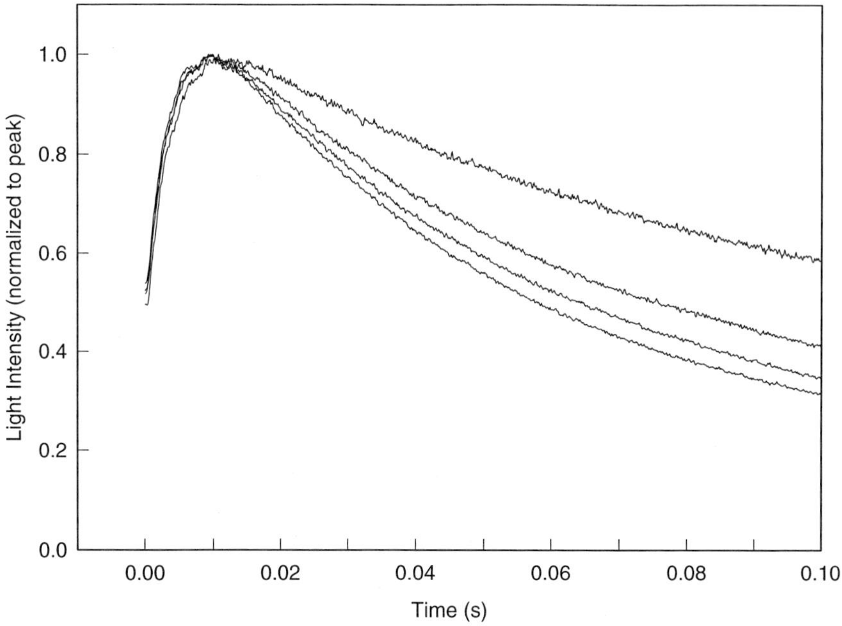

FIG. 10. Higher speed stopped-flow records for recombinant obelin. Same as in Fig. 9 except that the display is on a faster time scale. Sampling interval 20 μsec; temperature 20°.

for the rising phase of luminescence by computer-fitting the experimental data of each curve to a simple exponential are shown in Table I.

Mg^{2+} has at least four different effects on aequorin: it decreases the CIL, it decreases Ca^{2+} sensitivity, it decreases the rate of rise of luminescence, and it alters the time course of the flash (reducing its amplitude while increasing

TABLE I
RATE CONSTANTS FOR RISE OF LUMINESCENCE FOR AEQUORIN AND OBELIN[a]

	Rate constant (sec^{-1})	
$[Mg^{2+}]$ (m*M*)	Aequorin	Obelin
0	115	405
1	63	372
3	42	366
10	13	331

[a] Rate constants were determined from the rising phases of the stopped-flow records of Figs. 9 and 10 by computer-fitting experimental curves with a single exponential. Temperature 20°.

its duration). In the cases of all of these effects but the last, the effect of Mg^{2+} on obelin is much less than it is on aequorin, and one might reasonably attribute the difference between the photoproteins to a lower affinity for Mg^{2+} of the Ca^{2+}-binding sites of obelin. However, this simple general conclusion is not consistent with the fact that Mg^{2+} has a greater effect on the amplitude and time course of the flash in the case of obelin. This discrepancy suggests that Mg^{2+} must do more than simply compete with Ca^{2+} for binding sites on the photoprotein. It may be necessary to conclude that Mg^{2+} interacts with photoproteins at more than one site or category of sites.

Relative Merits of Recombinant Aequorin and Recombinant Obelin as Biological Calcium Indicators

Obelin has two apparent advantages over aequorin: it is faster and it is considerably less sensitive to Mg^{2+}. Because $[Mg^{2+}]$ has such a large effect on the speed of response of aequorin, the advantage of obelin in this regard increases with $[Mg^{2+}]$ and is in the range of 5 to 10-fold in the presence of physiological Mg^{2+} concentrations (1–3 m*M*).[59] Although the Ca^{2+}-sensitivity of aequorin is somewhat greater than that of obelin in the absence of Mg^{2+}, the difference is the reverse in 10 m*M* Mg^{2+}, and the two photoproteins are about equally sensitive in the presence of physiological concentrations of Mg^{2+}. In this comparison it is primarily the C-E curve for aequorin that shifts with changes in $[Mg^{2+}]$. The importance of the obelin curve's greater stability is mainly that the quantitative interpretation of light signals from living cells depends less on information about intracellular $[Mg^{2+}]$ levels.

Other differences between recombinant aequorin and recombinant obelin may well come to light as more experience is gained in their use within living cells. For example, on the basis of experience with a rather limited range of experimental conditions, we have the impression that recombinant obelin is considerably more stable than aequorin during prolonged storage at 4°. However, aequorin appears to tolerate freezing and lyophilization better. Carefully controlled comparisons remain to be done. The fact that obelin is not yet available commercially may be expected to inhibit its use for a while in "conventional" applications that depend on introducing complete photoproteins into living cells. However, there is no obvious reason why the cDNA for obelin should not be used now in applications involving genetic engineering.

Many different organisms are known or believed to contain calcium-regulated photoproteins. Although apparently similar in their general mech-

[59] L. A. Blatter, *Pflüg. Arch.* **416,** 238 (1990).

anism of luminescence, these various photoproteins may differ in important ways, as is demonstrated amply by the results shown in this article. Important differences of other kinds (some less obvious than those studied here) will presumably come to light as more photoproteins are studied in detail. Such studies may reveal properties that are useful in themselves. However, even if they do not, they may be expected to further our understanding of how photoproteins work and to provide clues to structure–activity relations that may be exploited in the design of new photoproteins that may ultimately be created in the laboratory by means of site-directed mutagenesis.

Acknowledgments

The authors are indebted to Michelle Woodbury for expert technical assistance and to Dr. Osamu Shimomura for the gift of recombinant aequorin. This work was supported in part by Grant 96-04-48489 from the Fundamental Research Foundation of the Russian Academy of Sciences, by Grants HL 12186 and TW 00412 from the National Institutes of Health (USA), and by Grant NA76RG0119, Project R/B-32 from the National Oceanic and Atmospheric Administration to Washington Sea Grant Program, University of Washington. The views expressed herein are those of the authors and do not necessarily reflect the views of NOAA or any of its subagencies.

[18] *Gonyaulax* Luciferase: Gene Structure, Protein Expression, and Purification from Recombinant Sources

By LIMING LI

Introduction

Gonyaulax polyedra is a photosynthetic marine dinoflagellate that emits light ($\lambda = 474$ nm) upon mechanical stimulation and is largely responsible for the so-called "phosphorescence of the sea." Three components are involved in the bioluminescent process: *Gonyaulax* luciferase (LCF, 137 kDa), the substrate (luciferin, an open-chain tetrapyrrole), and a luciferin-binding protein (LBP, 75 kDa), all of which are localized in unique organelles called scintillons (light-emitting unit).[1–5] At a pH of 7.5 or above,

[1] M. T. Nicolas, G. Nicolas, C. H. Johnson, J. M. Bassot, and J. W. Hastings, *J. Cell Biol.* **105,** 723 (1987).
[2] M. T. Nicolas, D. Morse, J. M. Bassot, and J. W. Hastings, *Protoplasma* **160,** 159 (1991).
[3] M. Desjardins and D. Morse, *Biochem. Cell Biol.* **71,** 176 (1993).
[4] C. H. Johnson, S. Inoue, A. Flint, and J. W. Hastings, *J. Cell Biol.* **100,** 1435 (1985).
[5] J. W. Hastings and J. C. Dunlap, *Methods Enzymol.* **133,** 307 (1986).

0076-6879/00 $30.00

luciferase is inactive, and luciferin is sequestered by LBP to prevent its interaction with luciferase.[6,7] *In vivo,* the light emission is believed to be triggered by a vacuolar action potential, causing a transient local pH decrease in scintillons, thereby activating luciferase and releasing the luciferin from LBP.[1]

The genes coding for LBP and LCF in *G. polyedra* were cloned and sequenced, revealing some unique features of these genes.[8–10] Expression of LCF in *Escherichia coli* has produced a functionally active protein, which becomes a valuable source for biochemical studies of this enzyme.[10] This article discusses some interesting structural features of *G. polyedra* luciferase and then presents a detailed protocol for its heterologous expression.

Gene Structure

The mRNA coding for *G. polyedra* luciferase is about 4100 nucleotides in length, including a coding region of 3723 nucleotides, a 5′-untranslated region (UTR) of 71 nucleotides, and a 3′-UTR of 242 nucleotides (Fig. 1A). The complete polypeptide of *G. polyedra* luciferase (136,994 Da) contains four regions: an N-terminal region of 110 amino acids with 50% identity with the N-terminal region of LBP and three contiguous highly homologous regions, each of which contains a functional catalytic domain.[10] Exon recombination[11] has been considered as a probable mechanism for the existence of homologous N-terminal regions in LCF and LBP, and unequal crossing-over might be responsible for the three repeats. The sequence identities among the three catalytic domains are 75% for domains 1 and 2, 75% for domains 1 and 3, and 80% for domains 2 and 3; the corresponding nucleotide identities are 74, 74, and 78%, respectively. The boundary regions among the repeats have nucleotide identities of ~50%, but their central regions are highly conserved, with nucleotide identities of ~94%. Interestingly, the nucleotide substitutions at the wobble positions in the central region of each repeat are also strongly constrained, which cannot be explained by the conservation of protein structure, as substitutions at wobble positions do not cause changes in the protein amino acid composition. The possibility that the suppression of nucleotide substitutions

[6] M. Fogel and J. W. Hastings, *Arch. Biochem. Biophys.* **142,** 310 (1971).

[7] D. Morse, A. M. Pappenheimer, and J. W. Hastings, *J. Biol. Chem.* **264,** 11822 (1989).

[8] D. Morse, P. M. Milos, E. Roux, and J. W. Hastings, *Proc. Natl. Acad. Sci. U.S.A.* **86,** 172 (1989).

[9] Y. M. Bae and J. W. Hastings, *Biochim. Biophys. Acta* **1219,** 449 (1994).

[10] L. Li, R. Hong, and J. W. Hastings, *Proc. Natl. Acad. Sci. U.S.A.* **94,** 8954 (1997).

[11] W. H. Li and D. Graur, "Fundamentals of Molecular Evolution." Sinauer, Sunderland, MA, 1991.

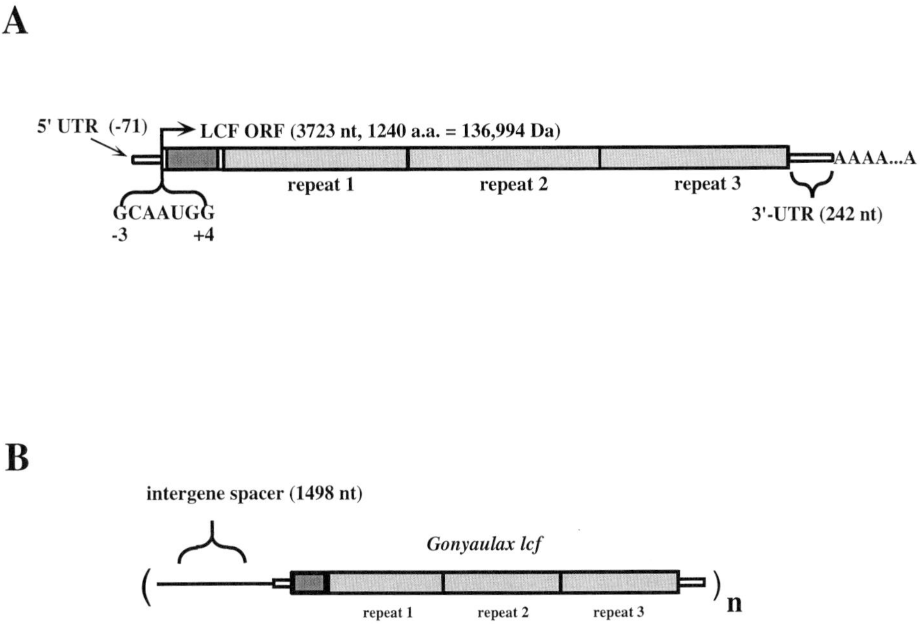

FIG. 1. *G. polyedra* luciferase gene structure and genomic organization. (A) Schematic diagram of the *lcf* cDNA structure. Thin unfilled boxes represent the untranslated regions, and filled boxes represent the four modular regions of the luciferase, with the three repeated regions indicated. The arrow indicates the start site and orientation of the translation. (B) Schematic diagram of the genomic organization of *lcf*. The untranslated regions and the four modular regions are indicated as in A. The line represents the intergene spacer region.

in those positions was caused by codon bias is considered unlikely, as a careful examination of the conserved amino acid residues and their corresponding codons in both boundary and central regions revealed no such a bias.[10] It is possible that the suppression of the nucleotide substitutions in the central regions was caused by a mechanism similar to gene conversion or that the mRNA structure of the central regions is maintained for some unknown function.

Sequence analysis of the corresponding genomic DNA of *G. polyedra lcf*, obtained through the inverse polymerase chain reaction (IPCR) technique, revealed another interesting structural feature: *G. polyedra lcf* genes are organized in a multiple tandem repeats (Fig. 1B). The copy number of the *lcf* gene was estimated to be about 100.[12] Identical sequences were obtained from the overlapping regions of several IPCR clones from differ-

[12] D. H. Lee, M. Mittag, and J. W. Hastings, unpublished data.

ent restriction enzyme digests, indicating that most (if not all) of the *lcf* genes have identical sequences, including the spacer regions. Genomic Southern analysis also suggests that there are many copies of *lcf* genes physically organized in tandem repeats,[13] although the actual number is not certain. Like the *lbp* gene, the *lcf* gene does not contain introns and there are no TATA-like or other known promoter elements in its 5′ flanking region.[13,14] However, there is a 13 nucleotide sequence found in both of the 5′ flanking regions of *G. polyedra lcf* and peridinin chlorophyll-binding protein (*pcp*) genes.[15] This sequence might be a potential dinoflagellate promoter element.

Recombinant Luciferase Expression and Activity Assay

The glutathione *S*-transferase (GST) expression vectors for *E. coli*[16] were chosen for *G. polyedra* luciferase expression based on the following considerations: GST is a small protein that keeps most of its expression partners in a soluble and active form; GST-fusion proteins can be purified through one single affinity purification step and elution can be done under nondenaturing conditions; the strong, isopropyl β-D-thiogalactoside (IPTG)-inducible *ptac* promoter usually gives a high or moderate expression yield of GST-fusion proteins; and the pGEX vectors can be transformed into any commercially available *E. coli* strains. The GST expression system has been used successfully in many applications.[17–19]

Expression Plasmid Construction and Transformation.

Several *lcf* cDNA fragments (Fig. 2), generated from convenient restriction enzyme digestions, were cloned in frame with either pGEX:2T (for pLLb) or pGEX:3X (for all others). In most cases, the cDNA fragments were blunt-ended and cloned into the pGEX vectors via the *Sma*I site. The resulting plasmid constructs were transformed into competent *E. coli cells,* JM 109, according to a heat-shock protocol.[20] To identify cells

[13] L. Li and J. W. Hastings, *Plant Mol. Biol.,* **36,** 275 (1998).

[14] D. H. Lee, M. Mittag, S. Sczekan, D. Morse, and J. W. Hastings, *J. Biol. Chem.* **268,** 8842 (1993).

[15] Q. H. Le, P. Markovic, J. W. Hastings, R. V. M. Jovine, and D. Morse, *Mol. Gen. Genet.* **255** (1997).

[16] D. B. Smith and K. S. Johnson, *Gene* **67,** 31 (1988).

[17] E. Fikrig, S. W. Barthold, F. S. Kantor, and R. A Flavell, *Science* **250,** 553 (1990).

[18] T. Chittenden, D. M. Livingston, and W. G. Kaelin, *Cell* **65,** 1973 (1991).

[19] U. Knippschild, D. M. Milne, L. E. Campbell, A. J. DeMaggio, E. Chreistenson, M. F. Hoekstra, and D. W. Meek, *Oncogene* **15,** 1727 (1997).

[20] J. Sambrook, E. F. Fritsch, and T. Maniatis, "Molecular Cloning: A Laboratory Manual," p. 1.80. Cold Spring Harbor Laboratory Press, Cold Spring Harbor, NY, 1989.

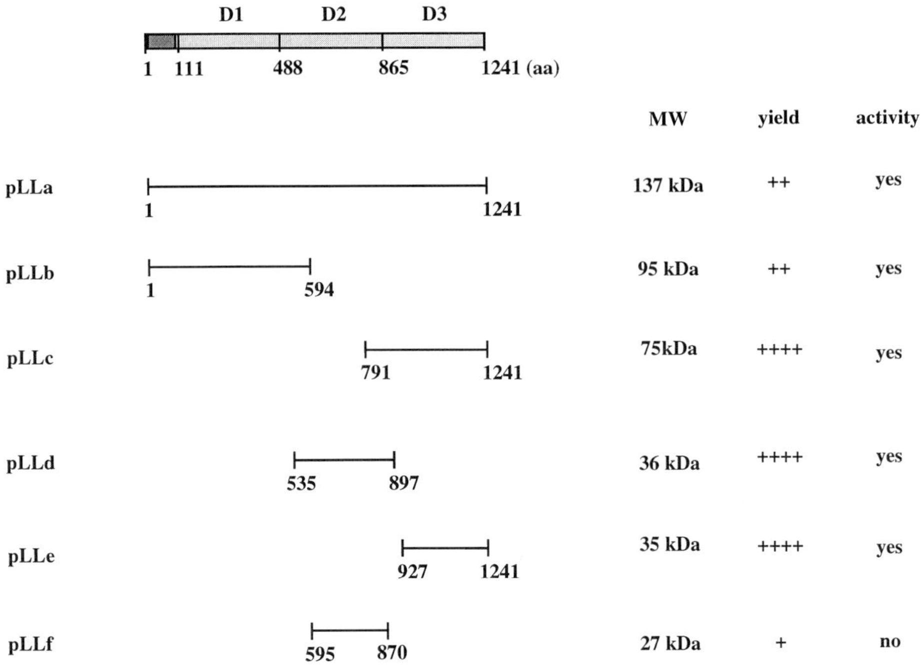

FIG. 2. Schematic diagram showing the expression constructs containing full-length (pLLa) or partial (pLLb, pLLc, pLLd, pLLe, and pLLf) sequences of LCF. The amino acid numbers and sizes of each peptide are indicated. The N-terminal fusion partner, GST (26 kDa), is not shown. The symbols +, ++, and ++++ indicate the expression yield of the corresponding fusion proteins. ++++, about 4 mg; ++, about 80 μg; and +, less than 40 μg purified fusion proteins can be obtained from a 1-liter culture, respectively.

containing plasmids with the *lcf*-cDNA insert, about 20–30 transformants were picked from each transformation. A quick and easy protocol[21] was used to screen their plasmid sizes using pGEX vector as control.

Small-Scale Expression of GST-LCF and Luciferase Activity Assay

Cells containing plasmids whose sizes are larger than the vector control, which are thus likely to contain an *lcf* insert, were picked to test whether they have luciferase activity after induction of GST-LCF expression. Recombinant *E. coli* cell growth and GST-LCF induction were basically performed according to the manufacturer's protocol[22] with minor modifica-

[21] J. Sambrook, E. F. Fritsch, and T. Maniatis, "Molecular Cloning: A Laboratory Manual," p. 1.32. Cold Spring Harbor Laboratory Press, Cold Spring Harbor, NY, 1989.

[22] GST Gene Fusion System, Pharmacia P-L Biochemicals Inc.

tions. Briefly, a single colony is inoculated into 3 ml of 2YT medium (16 g/liter of tryptone, 10 g/ liter of yeast extract, 5 g/liter of NaCl, pH 7.0) supplemented with 2% glucose (w/v) and 100 μg/ml ampicillin (2YT-G/amp) and grown overnight with vigorous shaking at room temperature. The overnight culture is then diluted into 10× volume of fresh 2YT-G/Amp and grown at room temperature for 3 hr before IPTG is added to a final concentration of 0.1 m*M*. After an additional 3-hr growth, the cells are harvested by centrifugation for 5 min at 5000 rpm. The supernatant is decanted and the cell pellet is resuspended in 300 μl of ice-cold 1× phosphate-buffered saline (PBS) in a 1.5-ml ependorf tube. After the addition of 5 μl of 200 m*M* phenylmethylsulfonyl fluoride (PMSF), cells are disrupted by sonicating three times, 10 sec each, with cooling on ice between sonications. After a 10-min centrifugation, the supernatant is carefully removed into a new tube and assayed for luciferase activity as described previously[9] by rapidly mixing *E. coli* extract and dinoflagellate luciferin with 1.5 ml of 0.1 *M* sodium citrate, pH 6.3, in a vial placed in a light-tight chamber of a photometer.[23] Typically, 5–10 μl of crude extract is enough for one luciferase assay.

After the luciferase assays, 20 μl of glutathione-Sepharose 4B (1:1 slurry in PBS: 140 m*M* NaCl, 2.7 m*M* KC1, 10 m*M* Na_2HPO_4, and 1.8 m*M* KH_2PO_4, pH 7.3) is added to the rest of the extract to purify the GST-LCF. The mixture is incubated for 30 min at 4° with as gentle shaking, followed by a brief centrifugation and three washes of the beads with 1 ml of PBS equilibrated at room temperature. After the addition of 20 μl of 2× SDS sample buffer and a 2-min incubation at 95°, the purified GST-LCF is analyzed by SDS–PAGE to check the relative yield and the molecular size of the fusion protein. Those constructs producing GST-fusion proteins with the correct molecular weights are subjected to nucleotide sequencing to verify their identities.

Large-Scale Preparation and Purification of GST-LCF

The just-described protocol can be scaled up for larger preparations. The manufacturer's protocol (Pharmacia) is also easy to follow and works well if the cells are grown at room temperature. Larger amounts of cell suspension can be disrupted using 3× 1-min sonication using larger sonication tips (the author uses a microtip for a volume smaller than 5 ml). One m*M* PMSF is enough to keep the fusion proteins from being proteolyzed during the cell disruption and purification. For GST-LCF purification, the

[23] G. Mitchell and J. W. Hastings, *Anal. Biochem.* **39,** 243 (1971).

cell lysates are usually incubated with the glutathione Sepharose 4B for 1–3 hr at 4° instead of room temperature for 30 min as the manufacturer suggests. This results in a better binding of the GST-LCF to the beads. In the hands of the author, a longer incubation time (up to overnight) does not cause a loss of luciferase activity or an increase of nonspecific binding. However, the washing should be conducted at room temperature with 1× PBS to reduce the binding of nonspecific proteins. The author also usually uses much larger volumes of washing solution for the washing step than suggested by the manufacturer.

The elution of GST-LCF is done according to the manufacturer's protocol, 10 min in 10 m*M* reduced glutathione at room temperature. One elution step is usually sufficient for the recovery of the fusion protein. As shown in Fig. 3, the amount of fusion protein (pLLc) recovered from the second elution is only about 20% of the first elution, and the third round of elution did not recover a significant amount of the fusion protein.

Sequence analysis indicates that the *G. polyedra* luciferase has many thrombin cleavage sites but does not contain a factor Xa cleavage site. Following expression and purification, most of the *lcf* cDNA fragments

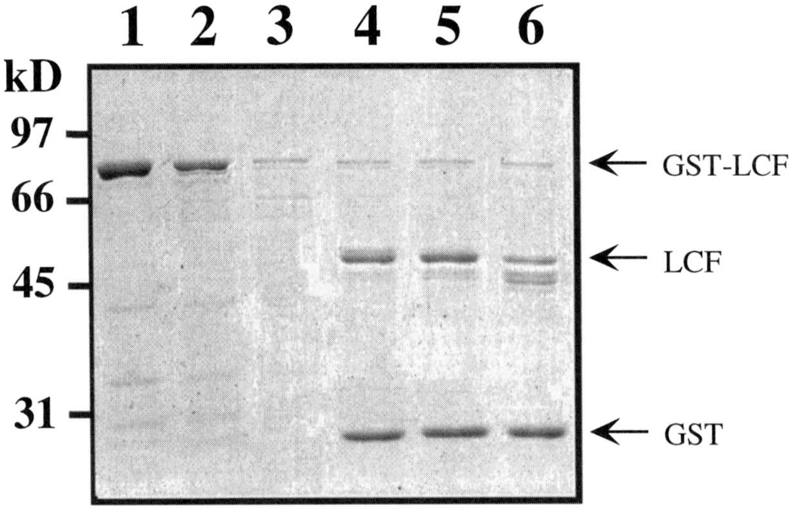

FIG. 3. Elution and factor Xa cleavage of purified GST-LCF. The GST-LCF fusion protein (75.8 kDa) was purified from *E. coli* cells containing pLLc. Lanes 1–3: first, second, and third elution of GST-LCF. Elution was done in 10 m*M* of reduced glutathione at room temperature, 10 min each. Lanes 4–6: 1-hr, 2-hr, and overnight incubation of the eluted GST-LCF with factor Xa. The proteolysis was carried out at room temperature with factor Xa to a substrate ratio of 1% as recommended (Pharmacia).

were therefore cloned into pGEX:3X as factor Xa can be used to generate recombinant luciferase without the GST domain. Figure 3 shows one representative experiment of factor Xa digestion of GST-LCF. The purified pLLc fusion protein was digested with factor Xa at room temperature for different times as indicated. Although the digestion mixture was still enzymatically active after overnight incubation with factor Xa using the manufacturer's suggested concentration, the activity was largely lost. Several breakdown products of luciferase also appeared (Fig. 3, lane 6), even though there are no factor Xa cleavage sites within the *G. polyedra* luciferase as judged by sequence analysis.

Optimization of GST-LCF Expression and Purification

In addition to the promoter strength and the copy number of an expression plasmid, there are many factors that affect the expression of recombinant proteins in *E. coli.* In examining the effects of temperature, inducer concentration, and induction time on GST-LCF expression, temperature has a dramatic effect on the solubility and activity of GST-LCF fusion proteins. By growing parallel cultures at 37°, 30°, and room temperature (22–23°), the author has observed that cells grown under room temperature usually produce GST-LCF with the highest activity. Although a higher yield of GST-LCF can be obtained when cells are grown at 37°, the luciferase activity obtained is 50% less than that of cells grown at room temperature. Cells containing the full-length GST-LCF construct (pLLa) need to grow at room temperature in order to produce active fusion protein, probably due to its larger size. Because *G. polyedra* cells grow in the ocean, a lower temperature may be optimal (they are usually kept at 19° in laboratories).

Escherichia coli cells containing expression constructs of pLLa and pLLc were harvested after different induction times, and whole cell lysates were prepared, followed by a SDS–PAGE analysis (Fig. 4a). As expected, the apparent MW of pLLa and pLLc were 161 and 75 kDa, respectively. The intensity of the corresponding bands (indicated by arrows) increased with the induction time. When purified GST-LCF fusion proteins from the same cell lysates were analyzed by SDS–PAGE, they remained soluble and intact after 3 hr of induction (Fig. 4B), even when a longer induction time (5 hr) and a larger inducer concentration (0.5 m*M*) were used.[24]

The GST-LCF constructs shown in Fig. 2 were tested for their expression in *E. coli* and their luciferase activities. The full-length luciferase-GST

[24] R. Hong, L. Li, and J. W. Hastings, unpublished results.

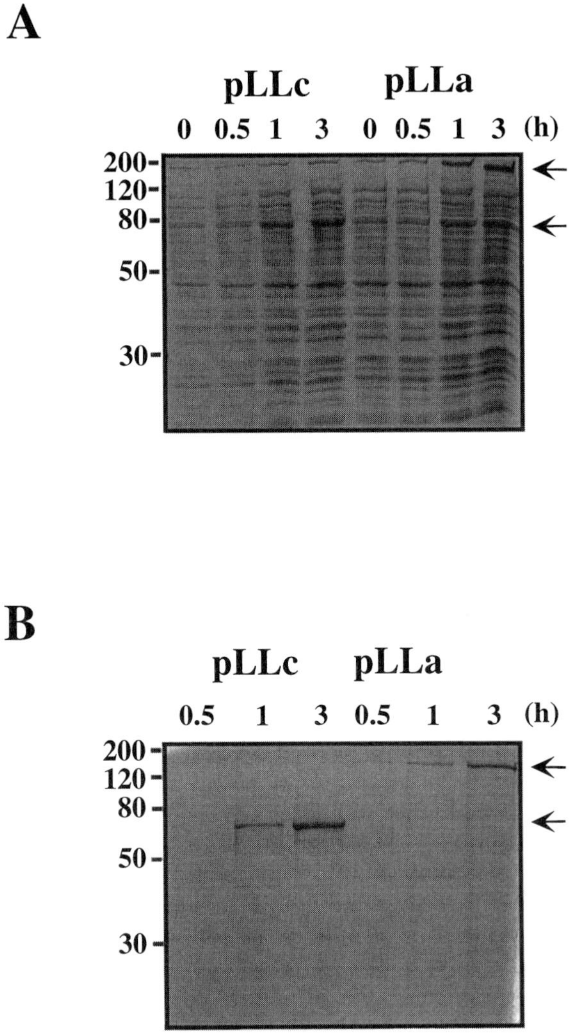

FIG. 4. The effect of induction time on the expression of GST-LCF. SDS–PAGE analysis of whole cell lysates (A) and purified GST-LCF (B) prepared from *E. coli* cells containing pLLa and pLLc at different induction time as indicated.

fusion protein, which has a molecular mass of 161 kDa, is soluble, catalytically active, and without noticeable degradation products (Fig. 4B). All other constructs also produce soluble fusion proteins of the expected sizes (Fig. 2). The constructs pLLa, pLLb, pLLc, pLLd, and pLLe are capable of producing GST-LCF fusion proteins with luciferase activity; however, pLLf produces a soluble fusion protein with the correct size but lacks activity. These studies have provided us with information on which part(s) of the protein is necessary for the light emission reaction. An important conclusion was drawn from these studies: the *G. polyedra* luciferase contains three functional catalytic domains.[10]

Acknowledgment

The author thanks Professor J. W. Hastings for his critical review and helpful comments on the manuscript.

[19] Dinoflagellate Luciferin-Binding Protein

By DAVID MORSE and MARIA MITTAG

Introduction

Light is produced from the dinoflagellate *Gonyaulax polyedra* when the substrate luciferin (LH_2), a high-energy tetrapyrrole (Fig. 1), is oxidized in a luciferase (LCF)-catalyzed reaction [Eq. (1)].

$$LH_2 + O_2 \xrightarrow{\text{LCF}} L = O + H_2O + h\nu\,(490\text{ nm}) \qquad (1)$$

In this enzyme-catalyzed oxidation, oxygen is added at the C-13^2 position of the LH_2.[1] In contrast, in samples exposed to oxygen without LCF, a blue compound is formed (Fig. 1) in a reaction that neither produces light nor results in oxygen addition to the LH_2 [Eq. (2)].

$$LH_2 + \tfrac{1}{2}O_2 \rightarrow L + H_2O \qquad (2)$$

LH_2 is quite sensitive to the presence of oxygen (the half-life of purified LH_2 is measured in hours), which may explain why most of the LH_2 in *Gonyaulax* is found bound to a luciferin-binding protein (LBP). One LBP dimer binds a single LH_2 with an association constant of about 5×10^7 M^{-1} at basic pH[2] [Eq. (3)].

$$LBP + LH_2 \rightarrow LBP\text{-}LH_2 \qquad (3)$$

LH_2 bound to LBP has a half-life measured in days.

[1] H. Nakamura, Y. Kishi, O. Shimomura, D. Morse, and J. W. Hastings, *J. Am. Chem. Soc.* **111,** 7607 (1990).

[2] D. Morse, A. M. Pappenheimer Jr., and J. W. Hastings, *J. Biol. Chem.* **264,** 11822 (1989).

0076-6879/00 $30.00

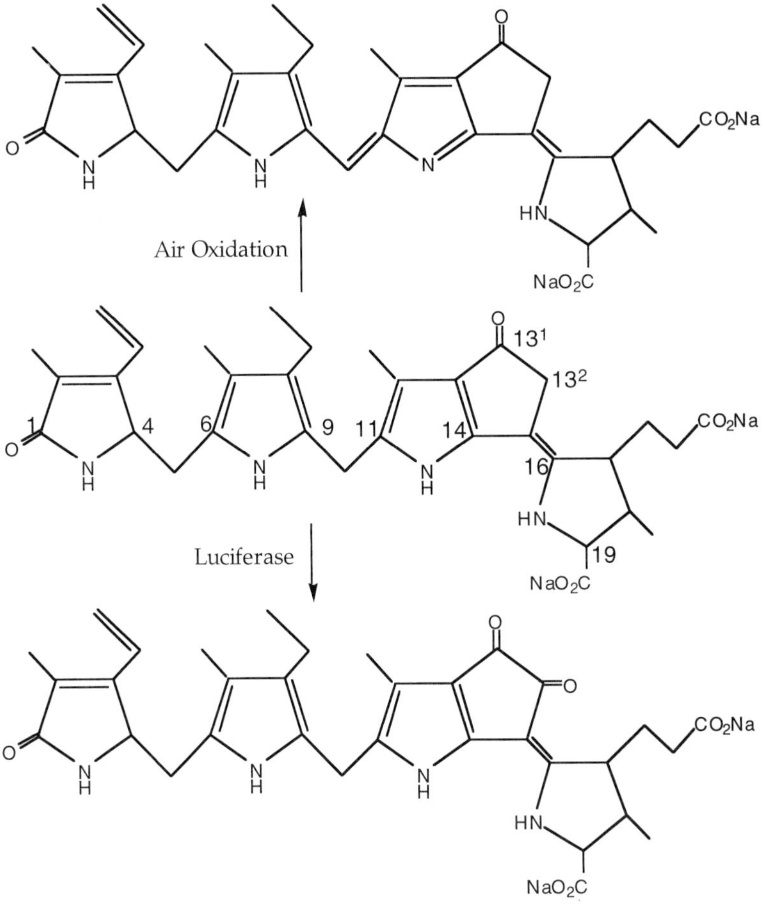

FIG. 1. Structures of dinoflagellate luciferin and two different oxidation products. When luciferin (center compound) oxidation is catalyzed by luciferase the oxidation produces a photon of blue-green light (490 nm) and a product that contains oxygen at the 13^2 position (lower compound). Luciferin is also oxidized in a nonluminescent process during the preparative procedures to yield a blue compound (upper compound).

To ensure that LH_2 oxidation is coupled with light production, *Gonyaulax* has exploited both structural and biochemical constraints. In terms of structure, LCF and LBP-LH_2 are compartmentalized into specialized organelles termed scintillons.[3,4] LBP is found primarily in the scintillons,[2,4]

[3] M. T. Nicolas, G. Nicolas, C. H. Johnson, J.-M. Bassot, and J. W. Hastings, *J. Cell Biol.* **105,** 723 (1987).

[4] M. T. Nicolas, D. Morse, J. M. Bassot, and J. W. Hastings, *Protoplasma* **160,** 159 (1991).

and purified scintillons contain only LCF and LBP as their protein components.[5] Scintillons are thus densely packed spheres where LCF and LBP are held in close proximity and LCF catalyzed oxidation will be favored kinetically. LH_2 can also be demonstrated in scintillons *in vivo* by virtue of its endogenous fluorescence.[5]

With respect to their biochemical properties, both LCF activity and LH_2 binding by LBP are regulated by pH.[2,6,7] LCF is inactive at pH 7.5 and maximally active at pH 6.3, whereas LH_2 binding by LBP is maximal at pH 7.5 and unmeasurable at pH 6.3. A decrease in pH thus simultaneously decreases binding of LH_2 to LBP and increases binding to LCF. The pH dependence of LH_2 binding to LBP suggests that four protons must bind to release the LH_2[2] [Eq. (4)].

$$\text{LBP-LH}_2 + 4\text{H}^+ \rightarrow \text{H}_4\text{LBP} + \text{LH}_2 \qquad (4)$$

These two strategies complement each other in regulation of light production because scintillons are located at the cell periphery in close contact with the vacuolar membrane.[3] As in many other plants, the *Gonyaulax* vacuole is acidic and can thus serve as a ready source of protons. Bioluminescence in dinoflagellates is believed to result from protons entering the cytoplasm from the vacuole in the vicinity of the scintillons.[3] The usual trigger for bioluminescence emission is mechanical stimulation, which results in a brief (0.1 sec) flash of high-intensity light (10^9 quanta/sec/cell, corresponding roughly to oxidation of the LH_2 on all LBP in the cell). In the dinoflagellate *Noctiluca,* mechanical agitation of the cells causes current flow across the vacuolar membrance equivalent to a cytoplasmic directed proton flow.[8,9] In addition to these bioluminescent "flashes," a low-intensity bioluminescent "glow" (10^4 q/sec/cell) can also be detected. The glow is observed over a period of several hours at the end of the night phase[10] and may be related to the daily degradation of scintillons.[11] As glow cannot be seen with the naked eye, the significance of this mode of light production is unclear.

At a totally different level of regulation, light production by the cell is restricted to the night phase of a diurnal cycle by the action of a circadian (daily) clock,[12] which regulates the amount of light produced after mechanical stimulation by controlling the number of scintillons within the cell.[11]

[5] C. H. Johnson, S. Inoue, A. Flint, and J. W. Hastings, *J. Cell Biol.* **100,** 1435 (1985).
[6] N. Krieger and J. Hastings, *Science* **161,** 586 (1968).
[7] M. Fogel and J. Hastings, *Proc. Natl. Acad. Sci. U.S.A.* **69,** 690 (1972).
[8] R. Eckert and T. Sibaoka, *J. Gen. Physiol.* **52,** 258 (1968).
[9] T. Nawata and T. Sibaoka, *J. Comp. Physiol.* **134,** 137 (1979).
[10] R. Krasnow *et al. J. Comp. Physiol.* **138,** 19 (1980).
[11] L. Fritz, D. Morse, J. W. Hastings, *J. Cell Sci.* **95,** 321 (1990).
[12] J. W. Hastings and B. M. Sweeney, *Biol. Bull.* **115,** 440 (1958).

Because the only proteins found within purified scintillons are LBP and luciferase,[13] the number of scintillons is proportional to the amount of LBP[14] and luciferase.[15,16] For LBP at least, circadian increases in protein levels are directly due to a transient increase in the rate of *lbp* mRNA translation.[14] It is not known how LBP and LCF assemble into scintillons, nor what factors tether the scintillon to the vacuolar membrane.

The *lbp* gene is expressed as a 2.3-kb mRNA. The longest open reading frame (ORF), which contains amino acid sequences corresponding to those determined by protein sequencing, comprises 2004 nucleotides and encodes a protein of 668 amino acids (ca. 75 kDa). The untranslated regions (UTRs) of the mRNA are 111 nucleotides (5′ UTR) and 158 nucleotides (3′ UTR), respectively.

The primary structure of the *lbp* cDNA revealed three interesting features. First, two major *lbp* cDNA isoforms (*lbp-α* and *lbp-β*) were found, which share 86% sequence identity at both nucleotide and protein levels. The presence of two protein isoforms *in vivo* has been confirmed by two-dimensional gel electrophoresis.[17] Second, a small upstream ORF (uORF) is present in the 5′ UTR of *lbp-α* mRNA. This uORF may play a key role in the regulation of *lbp* translation (see later). It is not known if the 5′ UTR of *lbp-β* mRNA contains this uORF, as this isoform has not been completely sequenced, and genomic clones are not available. Finally, *lbp* mRNA does not contain the typical eukaryotic polyadenylation signal (AAUAAAA). Both *lbp-α* and *lbp-β* share two short repeat sequences in their 3′ UTR, 'TGTGTGTTG' AND 'GCATRGCAT'. The former has now been identified as part of a regulatory sequence (GCTTTGTGTGTGTTGTGTGCAG) involved in the circadian control of gene expression.[18] The latter could represent a poly(A) signal from *Gonyaulax*, although it is not found in other genes from this organism.

Methods

Cell Culture

The growth of *G. polyedra* and *Pyrocystis lunula* cultures has been described in an earlier volume of this series.[19]

[13] M. Desjardins and D. Morse, *Biochem. Cell Biol.* **71,** 176 (1993).
[14] D. Morse, P. M. Milos, E. Roux, and J. W. Hastings, *Proc. Natl. Acad. Sci. U.S.A.* **86,** 172 (1989).
[15] J. -C. Dunlap and J. W. Hastings, *J. Biol. Chem.* **256,** 10509 (1981).
[16] C. H. Johnson, J. F. Roeber, J. W. Hastings, *Science* **223,** 1428 (1984).
[17] S. Machabee, L. Wall, D. Morse, *Plant Mol. Biol.* **25,** 23 (1994).
[18] M. Mittag, D.-H. Lee, J. W. Hastings, *Proc. Natl. Acad. Sci. U.S.A.* **91,** 5257 (1994).
[19] J. W. Hastings and J. C. Dunlap, *Methods Enzymol.* **133,** 307 (1986).

Bioluminescence Assay

Assay Buffer

0.2 *M* phosphate (pH 6.3), 0.25 m*M* EDTA, and 0.1 mg/ml bovine serum albumin (BSA)

Because only one LH_2 is bound per LBP dimer[2], the amount of LBP in a crude or partially purified extract of *Gonyaulax* can be estimated from the amount of LH_2 released from the protein when the pH of the medium is decreased rapidly in the presence of LCF. At pH 6.3, all the LH_2 is released from LBP and becomes a substrate for the LCF-catalyzed bioluminescence reaction. In a typical assay (Fig. 2), light intensity rises rapidly to a maximum and then decays exponentially with time. Light intensity (measured in quanta/sec) is a direct measure of the reaction rate at any time. The total amount of light produced [quanta, (Q)] is equivalent to the area under the curve and can be calculated as the product of the initial light intensity (I_o) and the rate constant for the exponential decay. In

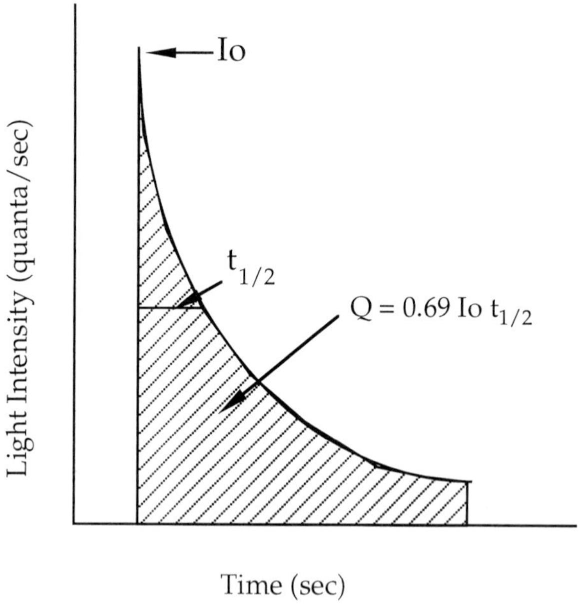

FIG. 2. LBP measurements by luminescence assays. In a typical assay, luciferase in low pH buffer is added to LBP charged with luciferin in the luminometer. The initial light intensity (I_o) is recorded and the decay in light intensity is followed for at least one half-time ($t_{1/2}$). The area under the curve, calculated by assuming an exponential decay, is a measure of the total quanta produced in the reaction.

practice, it is more convenient to measure the half-time ($t_{1/2}$), which is the time required for the light intensity to drop to one-half the I_o [Eq. (5)], and to calculate the total amount of light as shown in Eq. (5).

$$Q = I_o \cdot t_{1/2}/\ln 2 \tag{5}$$

The number of LH_2 molecules oxidized in the reaction can be calculated from the total quanta produced using a quantum yield of 0.22.[20] Measurements of total quanta of course require standardization of the light meter using standards of known light intensity,[21] although for routine assays where only relative values are needed (such as following LBP through a purification profile), such standardization is not necessary.

For assays in aqueous solution, the substrate concentration is always below the Michaelis constant of the luciferase (K_m) and thus increasing enzyme concentrations cause a decrease in the half-time and a concomitant increase in the I_o (such that the product of the two, the total quanta, is constant for a given substrate concentration[2]). This is convenient in that LCF concentrations do not have to be standardized to compare total quanta measurements taken on different days.

The most important factor in ensuring that LH_2 measurements reflect the amount of LBP accurately is the proportion of LBP charged with substrate. This is achieved by incubating LBP with an excess of LH_2 at pH 8 and separating the unbound LH_2 by gel filtration prior to assay.[22]

For assays of soluble LBP, typically 50 μl of a solution containing LBP charged with LH_2 is placed in a scintillation vial inside the luminometer. One milliliter of assay buffer is injected through the cap into the sample, and the light intensity produced is recorded on a chart recorder. Both the initial light intensity and the half-time for the decay are required to estimate the amount of LBP. To ensure that LBP is fully charged with LH_2, the LBP sample in a pH 8 buffer is incubated for 10 min with excess LH_2 on ice. A small (1 ml) gel filtration column (P10 or G50) is then used to separate the LH_2 bound to LBP (which elutes in the void volume) from the unbound LH_2. The amount of LH_2 needed to fully saturate the LBP present in the sample is usually determined empirically by incubating the LBP with several different volumes of LH_2. The saturation point is reached if the addition of further LH_2 does not increase the amount of bound LH_2.

For assays of LBP in scintillons, typically 1–10 μl of a solution containing scintillons is placed in a scintillation vial inside the luminometer. One

[20] J. C. Dunlap, J. W. Hastings, O. Shimomura, **123,** 273 (1981).

[21] J. W. Hastings and G. Weber, *J. Opt. Soc. Am.* **53,** 1410 (1963).

[22] F. M. Sulzman, N. R. Krieger, D. V. Gooch, and J. W. Hastings, *J. Comp. Physiol. A* **128,** 251 (1978).

milliliter of assay buffer is injected through the cap into the sample, and the light intensity produced is recorded on a chart recorder. The $t_{1/2}$ for scintillons is typically 0.1–0.4 sec. Because the $t_{1/2}$ of scintillons is constant at all stages of the purification,[13] I_o is the most convenient measure of scintillons.

Purification of Luciferase for Bioluminescence Assays

LCF is obtained most conveniently from samples during purification of LBP (see LBP purification later). LCF elutes from a DEAE column at lower salt concentrations than LBP (Fig. 3), and LCF-containing fractions can be pooled and frozen during each purification of LBP to maintain a supply of luciferase.

Purification of Luciferin for Bioluminescence Assays

Luciferin is purified from *P. lunula* as described.[19,23] Typically, luciferin after the DEAE column is pooled, saturated with argon, and frozen at −70° in 1-ml aliquots.

Purification of LBP

Buffers

Extraction buffer (EB): 100 m*M* Tris, pH 8, 10 m*M* EDTA, and 50 m*M* dithiothreitol (DTT)
EBS: extraction buffer containing 0.25 *M* sucrose
Running buffer (RB): 10 m*M* Tris, pH 8, 1 m*M* EDTA, and 15 m*M* 2-mercaptoethanol
RBN: running buffer containing 50 m*M* NaC1

The following procedure has been optimized for an LBP purification from 10 liters of *G. polyedra* (strain 70) cells grown to roughly 10^4 cells/ml (10^8 cells total). To begin, cells in the middle of the night phase are first exposed to light for 1 hr (150 $\mu E/m^2/sec$) in order to inhibit the flash response of the alga to mechanical stimulation and to ensure that LBP will contain bound LH_2. The cells are harvested by vacuum filtration on Whatman 541 paper, scraped from the paper, and resuspended in 35 ml EB. Cells can be broken in a French press (single pass at 6000 psi), a bead beater (4 × 15 sec), or a nitrogen bomb (15 min at 2000 psi). The extract is usually examined with a microscope to ensure that >95% of the cells are broken.

The crude extract is clarified by centrifugation, first for 10 min at 3000*g*

[23] J. C. Dunlap and J. W. Hastings, *Biochemistry* **20,** 983 (1981).

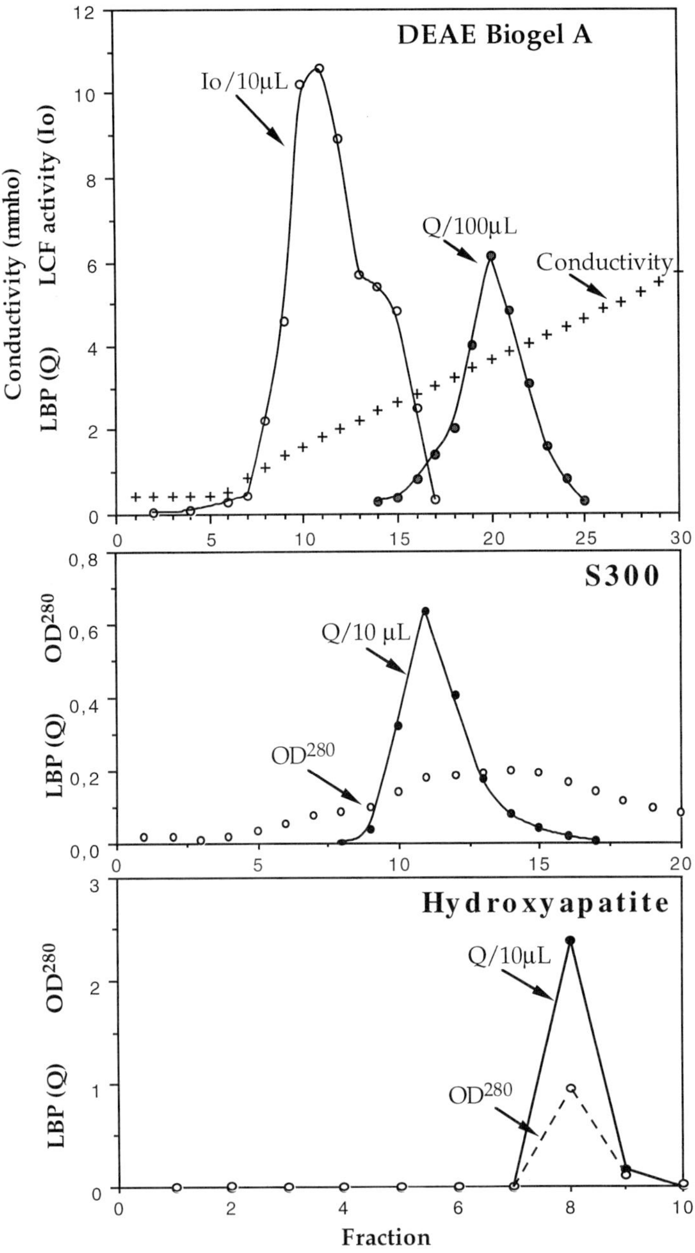

FIG. 3. A representative purification of LBP. LBP activity is measured in arbitrary units (Q) proportional to the number of luciferin molecules bound to the protein. Luciferase activity is measured as the initial light intensity resulting from a reaction with a standard amount of luciferin. The ammonium sulfate precipitation and desalting are not shown.

(5000 rpm in an SS34 rotor) and then after decanting the supernatant to a fresh tube for 10 min at 12,000*g* (10,000 rpm). The supernatant is decanted again and brought to 33% saturation with ammonium sulfate by adding 0.5 volume 100% ammonium sulfate in EB. The mixture is stirred gently at 4° for 15 min and then centrifuged for 10 min at 12,000*g* (10,000 rpm). The colored supernatant is decanted and brought to 52.5% saturation by adding 0.4 volumes 100% ammonium sulfate in EB. After gentle stirring at 4° for 15 min, the mixture is centrifuged for 10 min at 12,000*g* (10,000 rpm). The yellowish pellet is resuspended in 8 ml EB. At this point, 1 ml purified LH_2 can be added to the sample for a 10 min incubation on ice, although this step is usually not required to follow LBP through the procedure if the cells have been photoinhibited. The resuspended proteins are desalted on a 2.5 × 25-cm P10 column equilibrated and run with RB, and all fractions with color are pooled. It is important here to check the conductance of the eluant and dilute it, if necessary, to a conductance of ≤0.5 mmho. This low conductance ensures that luciferase as well as LBP can be recovered from the ion-exchange column.

The sample is then loaded onto a 2.5 × 25-cm DEAE Biogel A column, equilibrated and run with RB, and washed with RB until the eluant contains no protein. LCF and LBP are eluted at a flow rate of 1–2 ml/min with a 0–250 m*M* NaCl gradient (100 ml each) in RB (Fig. 3). LCF elutes first, and fractions containing >30% of the activity found in the peak fraction are combined, aliquoted, and frozen at −70° for use in LBP assays after adding glycerol to a final concentration of 15%. A separate pool of LBP is prepared using fractions that contain >30% maximal of the maximal peak activity. LBP in the pool is concentrated by adding 1 g solid ammonium sulfate for every 2-ml solution (70% saturation). The solution must be mixed well (gentle stirring for 15 min at 4°), after which the solution is centrifuged for 10 min at 20,000*g* (13,000 rpm).

The colored pellet is resuspended in 4 ml RBN and loaded onto a 2.5 × 100-cm S300 column equilibrated and run in RBN (Fig. 3). This column is conveniently run overnight at a flow rate of 0.5 ml/min, although it can be run during the day at 2.5 ml/min. As before, LBP activity is measured and fractions with >30% of the maximal peak activity are pooled.

The enriched LBP sample is now loaded directly onto a 2.5 × 2-cm hydroxyapatite column equilibrated and washed with RBN. LBP is eluted with a gradient from 0 to 500 m*M* potassium phosphate (pH 8.5) in RBN. A faint yellow, blue fluorescent band, which contains all the LBP, can be seen migrating down the column during the elution (Fig. 3). At this point, the sample should be essentially homogeneous when examined by SDS—PAGE (Fig. 4) and should contain 2–3 mg LBP (Table I).

If the components for the luminescent assay are not available, LBP

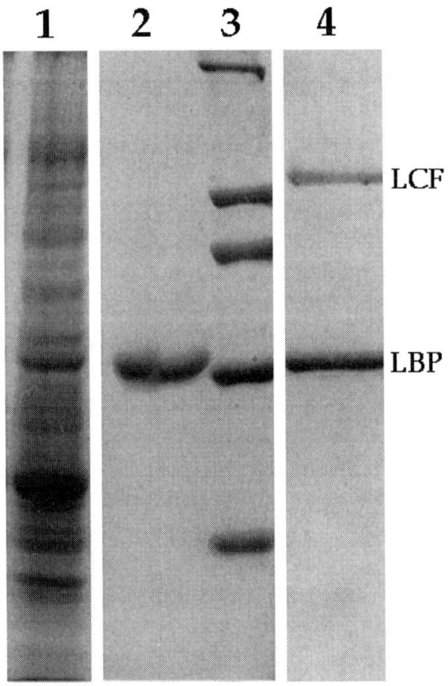

FIG. 4. LBP purity by SDS–PAGE on a 12% acrylamide gel. LBP prepared by column chromatography is homogeneous when purified to the stage of the hydroxyapatite column (lane 2). LBP purified from scintillons (lane 4) contains LCF as the only other protein contaminant. Molecular weight markers are shown in lane 3 (from top, 205,000, 116,000, 97,000, 68,000 and 45,000) and a crude extract of *Gonyaulax* is shown in lane 1.

TABLE I
PURIFICATION OF LBP FROM 10^8 CELLS[a]

	Total LBP ($10^{15}Q$)	Total LCF (10^{14} Q/sec)	Protein (total mg)
P10	0.7	92	80
DEAE	0.9	0.9	18
S300	1.2	0.2	4.2
HAp	1.1	n.d.	3.1

[a] In this representative experiment the amount of LBP is measured as total quanta produced in a luciferase-catalyzed reaction with LBP charged with luciferin. Luciferase is measured as the initial light intensity (Q/sec) produced with a standard amount of luciferin. Protein is measured with the Bio-Rad dye-binding assay.

purification can also be monitored using endogenous LH_2 fluorescence as a guide. Long wavelength UV light, from a standard hand-held UV lamp, will produce a blue fluorescence in samples that contain LH_2.

Purification of Scintillons

As for the purification of soluble LBP, scintillons are prepared from cells harvested in the middle of the night phase by filtration on Whatman 541 paper after a 1-hr exposure to bright light (150 $\mu E/m^2/sec$). The cells are weighed, scraped from the filter with a spatula, and resuspended in 1.5 ml ice-cold EBS/g wet weight. The cells are broken in the bead beater (10–15 sec) and should be >90% broken when examined under the microscope. The homogenate is pooled with two washes of the beads and diluted with EBS to give a final volume of 15 ml/g original wet weight. The pool is then diluted with 0.3 volumes 90% Percoll (Pharmacia) containing 0.25 *M* sucrose, mixed gently but thoroughly, and centrifuged for 20 min at 30,000*g* (15,000 rpm in an SS34 rotor). Fractions are collected from the bottom of the tube using either a peristaltic pump or a Pasteur pipette, which can be inserted inside a hollow glass tube placed into the centrifuge tube, and the scintillon activity of the fractions is determined (Fig. 5). Fractions containing >30% of the maximal peak activity are combined, diluted to 27 ml with EBS, and vortexed vigorously. To the solution is added 8 ml of 90% Percoll containing 0.25 *M* sucrose, and after gentle mixing, the solution is centrifuged for 20 min at 30,000*g*. Fractions are collected again from the bottom and the scintillon activity is determined. All fractions containing >30% maximal activity are pooled and the Percoll is removed by dilution with 10 volumes of EBS and centrifugation for 20 min at 30,000*g*. The supernatant is removed carefully and discarded while the loose pellet is resuspended in a total volume of 35 ml EBS and recentrifuged for 20 min at 30,000*g*. The final pellet is redissolved in 10 μl EBS/g original wet weight and should contain only luciferase and LBP when examined by SDS–PAGE (Fig. 4). Typically about 5% of the cellular LBP can be recovered as scintillons by this procedure (Table II).

Cloning of Genomic lbp by Inverse Polymerase Chain Reaction (IPCR)

Buffers

CTAB extraction buffer: 100 m*M* Tris (pH 8.0), 2% CTAB (w/v), 20 m*M* EDTA, 1.4 *M* NaC1, and 1% polyvinylpyrrolidine (M_r 40,000)

CTAB precipitation buffer: 50 m*M* Tris (pH 8.0), 1% CTAB, and 10 m*M* EDTA

TEN buffer: 10 m*M* Tris (pH 8.0), 1 m*M* EDTA, and 1 *M* NaC1

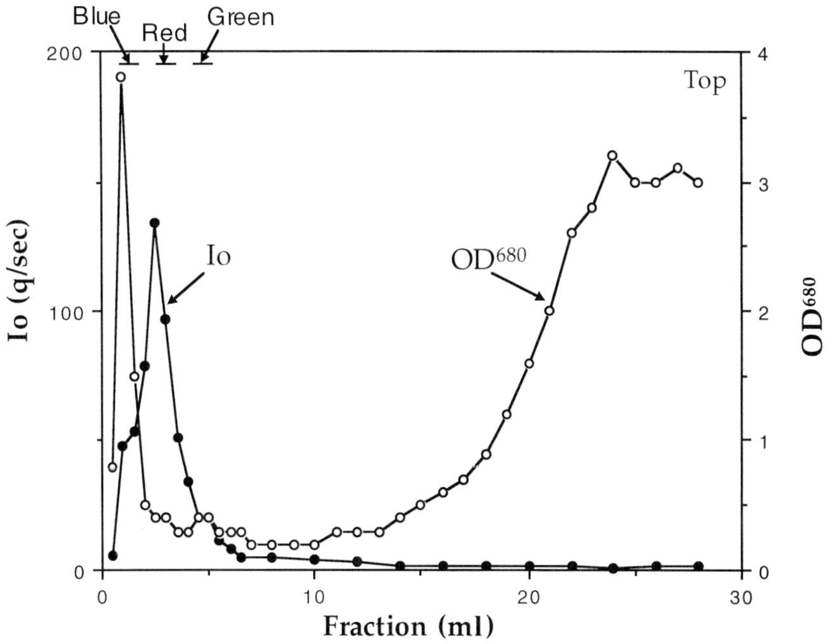

FIG. 5. A representative purification of scintillons. A crude extract mixed and centrifuged with Percoll (25% final concentration) separates scintillon activity (measured as I_o) from the bulk of the chloroplasts ($OD_{680\ nm}$). In subsequent repititions of the centrifugation, the Percoll concentration can be increased to move the scintillon peak closer to the middle of the gradient. The density markers [1.08 g/ml (blue), 1.065 g/ml (red), and 1.053 g/ml (green)] are obtained from Pharmacia.

TABLE II
PURIFICATION OF SCINTILLONS FROM 10^8 CELLS[a]

	Total Q ($10^{14}Q$)	I_o (10^{15} Q/sec)
Crude lysate	8.9	14
First Percoll	1.8	3.6
Second Percoll	1.4	2.8
Third Percoll	0.7	1.2

[a] In this representative experiment, the amount of LBP is typically measured as the initial light intensity (Q/sec) produced by injection of pH 6.3 buffer, but reported as total quanta by multiplying by the decay rate constant for the light-producing reaction catalyzed by scintillons ($\approx 2\ sec^{-1}$).

Ligation buffer: 50 mM Tris–HCl (pH 7.6), 10 mM $MgCl_2$
10 mM DTT, and 50 μg/ml BSA

The large amount of DNA in a *Gonyaulax* cell (~30 times that of the human nucleus) makes screening a genomic library difficult. However, cloning of *Gonyaulax* genes can be achieved readily by IPCR. The reported genomic *lbp* sequences obtained so far correspond to the *lbp-α* isoform and do not contain introns.

As a prerequisite for IPCR, genomic *Gonyaulax* DNA should be purified using a cetyltrimethylammonium bromide (CTAB) extraction.[24] In a typical procedure, frozen cells are ground with liquid nitrogen in a precooled mortar until a fine powder is obtained. This material is then transferred carefully to a 50-ml centrifuge tube and 1 volume of hot (65°) CTAB extraction buffer is added, mixed gently with 1 volume of chloroform/isoamyl alcohol (24:1), and centrifuged at 11,000g for 30 sec. The upper phase is transferred to a new tube and, after adding 0.1 volume of a solution containing 10% CTAB and 0.7% NaCl, another extraction with chloroform/isoamyl alcohol is performed. To separate nucleic acids from polysaccharides, 1 volume of CTAB precipitation buffer is added and the solution is mixed gently and centrifuged (1 min at 11,000g). The supernatant is discarded, and the pellet is dissolved in TEN buffer and reprecipitated with 2 volumes of 95% ethanol. The nucleic acids are recovered with a hooked glass rod, vacuum dried, and resuspended in TE buffer. To assure the purity of the DNA, the preparation is treated with 100 μg/mL RNase A for 1 hr at 37° and is extracted successively with equal volumes of phenol, phenol/chloroform, and chloroform, and the nucleic acids are finally precipitated in the presence of sodium acetate and ethanol.[25] The pellet is resuspended in TE and stored at −20°.

For IPCR cloning, genomic DNA is first digested with different restriction enzymes. Some care should be taken in choosing the enzymes as several enzymes digest dinoflagellate genomic DNA poorly or not at all.[24] This could be due to a paucity of the requisite sites or to the presence of modified nucleotides such as 5-hydroxymethyluracil, which replaces thymine residues in many dinoflagellates, including *Gonyaulax*.[26] Enzymes such as *Eae*I, Bsty, and *Nco*I cut *Gonyaulax* genomic DNA well.[24] Standard Southern blot analyses[25] with radiolabeled *lbp* as a probe should be used to check if the restriction fragment sizes obtained are in a suitable size range for

[24] D. H. Lee, M. Mittag, S. Sczekan, D. Morse, and J. W. Hastings, *J. Biol. Chem.* **268,** 8842 (1993).

[25] J. Sambrook, E. F. Fritsch, and T. Maniatis, "Molecular Cloning: A Laboratory Manual." Cold Spring Harbor Laboratory Press, Cold Spring Harbor, NY, 1989.

[26] P. Rizzo, *J. Protozool.* **38,** 246 (1991).

amplification by PCR (targets greater than 3 kb should be avoided with standard *Taq* polymerase). If necessary, DNA fragments containing *lbp* sequences (identified by Southern hybridization) can be excised from agarose gels[25] and used for circularization.

To prepare the circularized sample for IPCR, 1 μg of genomic DNA is first restricted for 8–16 hr at 37°. The sample is then extracted once with phenol/chloroform and precipitated with ethanol. The DNA is redissolved in 400 μl ligation buffer and circularized overnight at 19° by adding 1 unit of T4 DNA ligase (Bethesda Research Labs, Gaithersburg, MD). The DNA is extracted again with phenol/chloroform and precipitated with ethanol, and the DNA pellet is redissolved in 10 μl TE buffer. One microliter is used for each PCR reaction, together with gene-specific primers directed outward from the known sequence (toward each other in the circularized sample).

Thc standard DNA amplification protocol involvcs 30–40 cyclcs of dcnaturation for 1 min at 94°, annealing for 30 sec at 58°, and extension for 3 min at 72°. Final extensions are carried out at 72° for 5 min, followed by cooling to 4°. Each reaction, typically in a volume of 20 μl, contains 10 m*M* Tris—HCl (pH 8.4), 1.9 m*M* $MgCl_2$, 25 m*M* KCl, 2 μg BSA, 200 μM each of dNTP, 4 pmol of the appropriate primers, and variable amounts of template DNA. The reaction is initiated by the addition of *Taq* DNA polymerase at a final concentration of 0.025 U/ul. The PCR product in reaction buffer is treated for 15 min at 37° with 10 units T4 DNA polymerase, and the blunt end fragment is cloned into a *Sma*I-linearized vector and sequenced.

Copy Number Estimation of lbp Gene Sequences

The total number of genomic *lbp* sequences can be estimated using either Southern blot or slot blot type analyses. For Southern blots, 20 μg of genomic DNA is restricted (e.g., with *Eae*I or *BstyI) and electrophoresed in a 1% agarose gel along with different amounts of standard* (10^5 to 10^8 molecules of a plasmid containing an *lbp* insert). The gel is transferred to a nylon membrane (Pall Company, East Hills, NY) and probed with a ^{32}P-labeled RNA transcript complementary to the region of the *lbp* DNA used as a standard. The membrane is then exposed to X-ray film, and the hybridization signals of the genomic DNA are compared with those of the *lbp* standard using densitometry. For slot blots, 10 μg of genomic DNA along with different amounts of the standard DNA are immobilized on a nylon membrane utilizing a slot blot apparatus (BRL, Gaithersburg, MD), probed, exposed, and compared in the same way as described earlier. The total number of copies of the *lbp* sequence present in the *G. polyedra* genome (which contains about 200 pg DNA per cell) has been estimated to be roughly 1000 copies per cell.[24]

Southern blots analyses suggest that *lbp* genes are arranged in tandem repeats[17] as are other dinoflagellate genes such as *pcp*.[27] However, the sequence of a *lbp* spacer region has not yet been reported, and it is possible that this postulated spacer region is too large for the typical *Taq* DNA polymerase amplification using outwardly directed primers at the 5′ and 3′ ends of the coding sequence.

In Vivo Labeling of LBP

To incorporate [^{35}S]methionine efficiently into LBP, care must be taken to perform the labeling roughly 2 hr after the onset of darkness. Synthesis rates are controlled by the circadian clock, and little or no incorporation of radiolabel into LBP can be detected at other times.[14,28]

In a typical experiment, 50 ml of cell culture is concentrated to 2 ml by filtration on a 25-μm nylon (Nitex) filter. To reduce incorporation of the methionine into bacterial proteins (the cultures are unialgal, not axenic), the sample is incubated for 5 min after the addition of 2 μl 100 m*M* chloramphenicol in ethanol.[29] Strong labeling can be obtained in a 20-min labeling period using 200 μCi [^{35}S]methionine. The cell suspension is washed five times with culture medium by centrifugation in a microfuge. The final cell pellet is dissolved in 2 volumes SDS sample buffer and is prepared for electrophoresis by heating for 15 min at 95°.

LBP Expression and Regulation

Regulation of *lbp* mRNA translation is complex and involves both 5′ and 3′ UTR. A small 87 nucleotide ORF, situated in the *lbp* 5′ UTR, causes differential initiation of *lbp* mRNA, resulting in two size classes of LBP with different N termini (LBP-A: ~75 kDa and LBP-B: ~72 kDa).[30] Because the first initiation codon of *lbp* mRNA (AUG1 in Fig. 6) does not contain a perfect context for translational initiation,[31] some of the ribosomes scan over it and instead begin translation at the second initiation codon (AUG2). AUG2, with a purine in the −3 (adenine) and +4 position (guanine) with respect to the A of the AUG as the +1 position, has a perfect context for the initiation of translation. Translation initiation at AUG2 results in the

[27] Q. H. Le, P. Markovic, R. Jovine, J. Hastings, and D. Morse, *Mol. Gen. Genet.* **25,** 595 (1997).
[28] P. Markovic, T. Roenneberg, and D. Morse, *J. Biol. Rhythms* **11,** 57 (1996).
[29] W. Olesiak, A. Ungar, C. H. Johnson, J. W. Hastings, *J. Biol. Rhythms* **2,** 121 (1987).
[30] M. Mittag, C. Eckerskorn, K. Strupat, and J. Hastings, *FEBS Lett.* **411,** 245 (1997).
[31] M. Kozak, *J. Cell Biol.* **108,** 229 (1989).

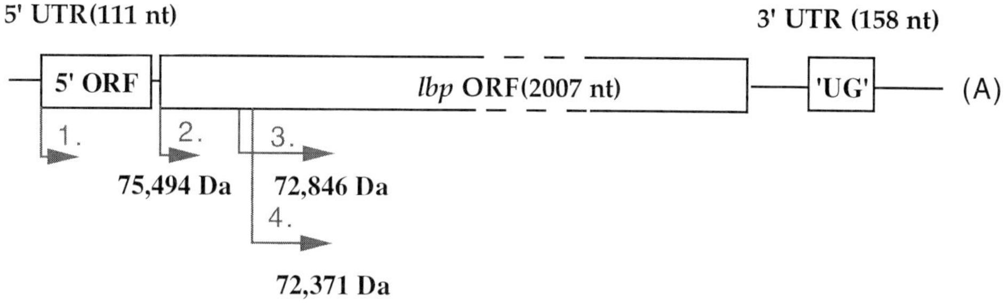

FIG. 6. The *lbp-α* mRNA contains four sites of translational initiation. Each potential translational initiation site is numbered, starting with the 5′ most AUG, and the molecular mass of the protein product (LBP-A or LBP-B) is indicated below. The stop codon of the small 5′ ORF is only two nucleotides upstream from AUG2. The UG-rich binding site for the *trans*-acting CCTR, situated in the 3′ UTR, is marked with a "UG."

formation of full-length LBP-A. In contrast, any ribosomes that initiate translation at AUG1 (and thus translate the uORF) reach a stop codon two nucleotides upstream from AUG2. This distance is insufficient to allow the ribosomes to reinitiate translation at AUG2, and these ribosomes continue to scan the mRNA and eventually reinitiate later, at either AUG3 or AUG4. These later initiation events produce the shorter length LBP-B. Western blots of *Gonyaulax* proteins from a crude extracts using a polyclonal antibody directed against LBP show that both LBP-A and LBP-B are present.[32] Interestingly, when different times of the circadian cycle are examined, it appears that LBP-A is strongly circadian controlled with a peak during the night phase, whereas LBP-B seems to be either constant during the day–night cycle or less abundant during the night.

Regulation at the translational level is often mediated by protein factors interacting with untranslated regions of the mRNA, and these interactions can be analyzed in mobility shift assays. With *lbp* mRNA, no factors have yet been reported that bind specifically to the *lbp* 5′ UTR. However, a factor designated circadian-controlled translational regulator (CCTR) has been identified that binds to a 22 base-long region of the *lbp* 3′ UTR.[18,33] This region contains seven U(U)G repeats and represents a novel RNA-binding motif. The binding activity of the CCTR changes during the day–night cycle with maximum binding from the end of the night until the end of the day. The CCTR thus behaves as a inhibitor of translation.

The translational control of *lbp* mRNA seems to involve a complex interaction of both 5′ and 3′ UTR. The observation that mRNAs can be

[32] D. Morse, L. Fritz, A. M. Pappenheimer Jr., and J. W. Hastings, *J. Biochem. Chemilum.* **3,** 79 (1989).

[33] M. Mittag, H. Waltenberger, *J. Biol. Chem.* **378,** 1167 (1997).

circularized *in vivo*, with the poly(A) binding protein binding simultaneously to the initiation factor eIF4G and the poly(A) tail of the mRNA,[34] may thus be relevant to the control of *lbp* translation. Interactions between the two ends of an mRNA may explain how regulatory elements that bind the different ends act synergistically to influence the initiation of translation.

Mobility Shift Assays

Buffers

Binding buffer: 40 m*M* NaCl, 10 m*M* Tris (pH 7.4), 0.1 m*M* EDTA (pH 8.0), 2 m*M* dithiothreitol, and 5% (v/v) glycerol

Elution buffer: 195 m*M* ammonium acetate, 0.1% SDS, and 1 m*M* EDTA (pH 8.0)

Electrophoresis buffer: 45 m*M* Tris/borate (pH 8.3) and 1.2 m*M* EDTA (pH 8.3)

Loading dye: 98% glycerol, 1% bromophenol blue, and 1% xylene cyanol

To prepare a crude extract of RNA-binding proteins, *Gonyaulax* cultures harvested by vacuum filtration on Whatman 541 filters and stored in liquid nitrogen are thawed on ice and resuspended in a small volume of binding buffer so that a viscous cell suspension is formed. Cells are broken by a 55-sec treatment with zirconium beads in a bead-beater. Cell debris is removed by centrifugation at 13,000*g* for 12 min, and only the supernatant is used for the mobility shift assay.

To synthesize an RNA transcript for use as a probe for binding factors, the appropriate cDNA sequence should be cloned in a pTZ plasmid (or any other plasmid that contains the T_7 promoter). The start site of the RNA probe is determined by the T_7 promoter, and the end is determined by cutting the plasmid with a suitable restriction enzyme. The transcript is prepared following the protocol of the supplier (Promega) with some modifications. The listed components are added at room temperature: 4 μl transcription buffer (5$\times$ concentrate, Promega), 2 μl 100 m*M* DTT, 0.5 μl RNAsin (20 U, Promega), 1 μl 100 μM UTP, 4 μl nucleotide mix (ATP, GTP, and CTP: each 2.5 m*M*), 0.7 μl linearized template (1–2 μg), 6.8 μl α ^{32}P-UTP (3000 Ci/mmol), and 1 μl T_7 RNA Polymerase (20 U, Promega). The reaction is incubated at 37° for 60 min. The *in vitro*-synthesized radiolabeled RNA is precipitated by the addition of 10 μl 7.5 *M* ammonium acetate and 70 μl ethanol, vacuum dried, and dissolved in DEPC-treated water. It is then purified by electrophoresis on a urea/6% polyacrylamide gel and eluted from a piece of the gel with elution buffer. After isolation the RNA

[34] S. J. Tarun, A. Sachs, *EMBO J.* **15,** 7168 (1996).

is again ethanol precipitated, resolved in DEPC-treated water, and heated at 70° for 10 min. The RNA is then slowly cooled down to room temperature to ensure uniform secondary structure of the transcript.

RNA–protein complexes are formed by incubating 28 μg protein from a *Gonyaulax* crude extract in binding buffer containing 28 μg of poly (G) as a nonspecific competitor RNA in a total volume of 16 μl.[18] After 20 min of incubation at room temperature (23°) the ^{32}P-labeled RNA is added (1–1.5 $\times$ 10^4 cpm) and the reaction mixture is incubated for another 20 min at 23°. To analyze the RNA–protein complexes formed, 1 μl loading dye is added and the samples are electrophoresed on a nondenaturing 4% polyacrylamide gel containing 10% glycerol. Typically, the gel is preelectrophoresed for 1–2 hr before the samples are loaded, and the gel is run for another 3 hr or until the bromophenol blue band has just run off the gel. The gel is transferred to Whatman paper, covered with Saran wrap, dried, and autoradiographed.

Immunocytochemistry and in Situ Hybridizations

Buffers

TBSB: 10 m*M* Tris–HCl (pH 7.5), 150 m*M* NaCl, and 3% BSA

Hybridization buffer: 2× SSC, 50% formamide, 10 m*M* DTT, 0.1% SDS, and 1× Denhardt's solution

20× SSC: 0.4 *M* NaCitrate (pH 7) and 3 *M* NaCl

Gonyaulax cells, concentrated by centrifugation, are fixed in 0.3 *M* phosphate buffer (pH 7.4) containing 3% glutaraldehyde at room temperature for 15 min. The fixed cells are washed twice in water and allowed to solidify together with 50 μl 2% agar, which is centrifuged to produce a small pellet. The agar block, cut into small pieces, is then carried through the dehydration process (5 min in each of 12, 25, 50, 75, 95, and 100%; the 100% ethanol is repeated two more times). For both immunocytochemistry and *in situ* hybridizations, at either the light or the electron microscopic level, the dehydrated samples are incubated for 24 hr in 100% LR White resin and then polymerized in a 50° oven overnight. If less resolution is required (i.e., for light microscopy only), the samples can be embedded in paraffin.

The blocks are sectioned to 0.1–0.2 μm thickness and allowed to dry on glass slides, which are coated with 0.5% gelatin or poly-L-lysine (drying takes several hours to overnight). The sections are hydrated in distilled water for 10 min and are then blocked by a 1-hr treatment with TBSB at room temperature. The slides are then placed into a humid environment with a drop (10–20 μl) of the primary antibody diluted in TBSB on top of the section. A humid environment is made easily by placing damp paper towels in a petri plate. After placing the slides gently on top, the petri plate is sealed with parafilm and incubated overnight at 4°.

The correct conditions for incubation with the primary antibody depend on the titer of the antibody and should be determined empirically. A useful rule of thumb is to start with 100 times the ELISA titer (or 10 times the titer used for Western blots) in a overnight incubation at 4°. The primary antibody is removed from the slide by three rinses in TBSB, and a drop (10–20 μl) of a suitable commercial secondary antibody, diluted according to the manufacturer's recommendations in TBSB, is incubated with the section for 1–2 hr at room temperature. The visualization method chosen will depend on the secondary antibody chosen.

For *in situ* hybridizations, digoxigenin-11-UTP (DIG; Boehringer Mannheim) is used as a substrate for T7 RNA polymerase to prepare DIG-labeled probes. These probes are convenient in that they can be stored indefinitely, and an immunological method using a commercially available antidigoxigenin can be used for detection. Sections on slides are first washed with 2× SSC for 10 min at room temperature and then prehybridized for 1 hr at 45° with hybridization buffer. DIG-labeled probes corresponding to sense and antisense strands of a cDNA are prepared according to the manufacturer's instructions, diluted in hybridization buffer, and hybridized overnight at 45° on separate slides. After hybridization, the slides are washed repeatedly with 2× SSC and finally blocked for 1 hr with TBSB. From this point on the samples are treated as for immunocytochemistry, using the commercial antidigoxigenin as the primary antibody.

Acknowledgments

This paper is dedicated to the late A. M. Pappenheimer, Jr., who was involved in the LBP story from his seventies on. Without his critical contributions and stimulating discussions, we would not have had so much fun in performing these studies. Our work is supported by the Deutsche Forschungsgemeinschaft (MM; habilitation fellowship Mi 373/2-1 and Schwerpunkt Grant Mi 373/3-1) and by the National Science and Engineering Research Council of Canada (DM).

Section IV

Bacterial Autoinduction System and Its Applications

[20] Assay of Autoinducer Activity with Luminescent *Escherichia coli* Sensor Strains Harboring a Modified *Vibrio fischeri lux* Regulon

By JERRY H. DEVINE and GERALD S. SHADEL

Introduction

Cell–cell signaling is essential for normal development and function of both prokaryotic and eukaryotic organisms. The signals involved in intercellular communication are diverse, and a specific cell or organism often synthesizes a number of chemically similar signaling molecules that elicit and modulate specific cellular responses (e.g., changes in gene expression). This complexity makes analysis of these types of signaling pathways both interesting and difficult. An experimentally tractable model for understanding intercellular communication is provided by bacteria that exhibit a cell density-dependent gene induction mechanism termed quorum sensing.[1–3] Quorum sensing was first documented in certain bioluminescent marine bacteria and was originally termed "autoinduction."[4,5] These early studies revealed that each species of bacteria produced a small signaling molecule, termed autoinducer, that accumulates in culture growth media and can stimulate light production when added back to an uninduced culture. Purification and structural characterization of the primary luminescence autoinducer produced by *Vibrio fischeri* revealed that it is an acyl-substituted homoserine lactone, *N*-(3-oxohexanoyl) homoserine lactone.[6] The *lux* gene regulon from *V. fischeri* was subsequently isolated and shown to confer regulated bioluminescence when transformed into *Escherichia coli.*[7,8] Through exploitation of this recombinant system, synthesis of the autoinducer was shown to require the *luxI* gene product, and its effects on *lux* gene transcription were shown to be mediated primarily through its binding and activation of the transcriptional regulatory protein LuxR. *luxR*

[1] W. C. Fuqua, S. C. Winans, and E. P. Greenberg, *J. Bacteriol.* **176,** 269 (1994).
[2] W. C. Fuqua, S. C. Winans, and E. P. Greenberg, *Annu. Rev. Microbiol.* **50,** 727 (1996).
[3] D. M. Sitnikov, J. B. Schineller, and T. O. Baldwin, *Mol. Microbiol.* **17,** 801 (1995).
[4] A. Eberhard, *J. Bacteriol.* **109,** 1101 (1972).
[5] K. H. Nealson, T. Platt, and J. W. Hastings, *J. Bacteriol.* **104,** 313 (1970).
[6] A. Eberhard, A. Burlingame, C. Eberhard, G. L. Kenyon, K. H. Nealson, and N. J. Oppenheimer, *Biochemistry* **20,** 2444 (1981).
[7] J. Engebrecht, K. Nealson, and M. Silverman, *Cell* **32,** 773 (1983).
[8] J. Engebrecht and M. Silverman, *Proc. Natl. Acad Sci. U.S.A* **81,** 4154 (1984).

0076-6879/00 $30.00

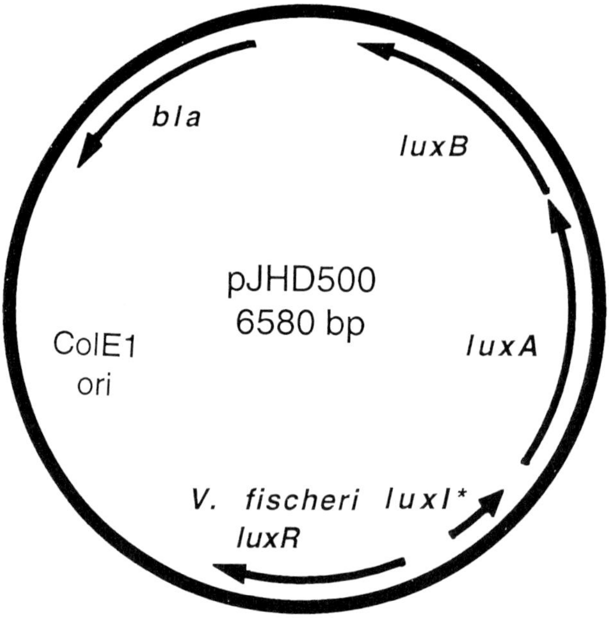

FIG. 1. Diagram of the reporter plasmid pJHD500. The relative locations of important genes in the plasmid are indicated by arrows. The *luxR–luxI* regulatory region from *V. fischeri* is inserted into the plasmid so that the major autoinducer-dependent promoter (*luxI* promoter) is fused transcriptionally to *luxAB* genes from *V. harveyi,* which encode bacterial luciferase. The *luxI* gene is truncated (denoted *luxI**) and produces no detectable autoinducer. The plasmid is a derivative of pBR322 with a ColE1 origin of replication and an ampicillin resistance gene (*bla*).

and *luxI* genes are located in two different operons that are transcribed in opposite directions from a common *lux* regulatory region (Fig. 1).[9–11] Because LuxI is responsible for autoinducer synthesis and transcription of its gene, as well as the *luxR* gene, is induced by LuxR in the presence of the autoinducer, a dual positive feedback circuit is initiated whenever the autoinducer reaches a threshold concentration.[12] A preinduction lag period, during which the autoinducer accumulates as a result of basal level expression of *lux,* and a large and rapid induction of bioluminescence, due to

[9] T. O. Baldwin, J. H. Devine, R. C. Heckel, J.-W. Lin, and G. S. Shadel, *J. Biolumin. Chemilumin.* **4,** 326 (1989).

[10] J. H. Devine, C. Countryman, and T. O. Baldwin, *Biochemistry* **27,** 837 (1988).

[11] J. Engebrecht and M. Silverman, *Nucleic Acids Res.* **15,** 10455 (1987).

[12] G. S. Shadel and T. O. Baldwin, *J. Bacteriol.* **173,** 568 (1991).

positive transcriptional feedback, are the hallmark characteristics of a quorum sensing response.[1–3] In recent years, LuxR and LuxI homologs have been found in several genera of bacteria; the quorum sensing signal transduction mechanism is used by a variety of prokaryotes as a means to achieve cell density-dependent gene expression.[1–3,13,14]

Analysis of various quorum sensing bacteria has revealed the existence of a relatively diverse family of autoinducers that differ in the type of acyl group attached to the homoserine lactone ring.[3] In addition, it is now clear that more than one autoinducer is produced by *V. harveyi, V. fischeri, P. aeruginosa,* and probably others, making simple models for quorum sensing based on one LuxR/LuxI-like protein pair largely inadequate.[15–18] Another complicating factor is that genes that exhibit no homology to *luxI* may also participate in the synthesis of compounds with autoinducer activity.[19] Finally, the detection of autoinducers in biofilms suggests that such molecules may also have signaling roles in the external environment.[20] Given the general use of *N*-(acyl) homoserine lactones as signaling molecules in nature and the complexity added to the study of quorum sensing by the presence of multiple autoinducers in each system, the development of simple assays for autoinducer activity has become increasingly important. This article describes a general strategy for assaying autoinducer activity using recombinant plasmids harboring an engineered *V. fisheri lux* regulon as a sensor in *E. coli.*

Experimental Procedures

Design of Autoinducer Sensor Strains

As already mentioned, *E. coli* harboring plasmids that contain the cloned *V. fisheri lux* regulon are bioluminescent and exhibit a quorum sensing

[13] G. P. C. Salmond, B. W. Bycroft, G. S. A. B. Stewart, and P. Williams, *Mol. Microbiol.* **16,** 615 (1995).

[14] R. A. Wirth, A. Muscholi, and G. Wanner, *Trends Microbiol.* **4,** 96 (1996).

[15] B. L. Bassler, M. Wright, and M. R. Silverman, *Mol. Microbiol.* **13,** 273 (1994).

[16] A. Kuo, N. V. Blough, and P. V. Dunlap, *J. Bacteriol.* **176,** 7558 (1994).

[17] J. P. Pearson, L. Passadsor, B. H. Iglewski, and E. P. Greenberg, *Proc. Natl. Acad. Sci. U.S.A* **92,** 1490 (1995).

[18] M. K. Winson, M. Camara, A. Latifi, M. Foglino, S. R. Chhabra, M. Daykin, M. Bally, V. Chapon, G. P. C. Salmond, B. W. Bycroft, A. Lazdunski, G. S. A. B. Stewart, and P. Williams. *Proc. Natl. Acad. Sci. U.S.A* **92,** 9427 (1995).

[19] I. Gilson, A. Kwo, and P. V. Dunlap, *J. Bacteriol.* **177,** 6946 (1995).

[20] R. J. McLean, M. Whiteley, D. J. Stickler, and W. C. Fuqua, *FEMS Microbiol. Lett.* **154,** 259 (1997).

response. This observation has led a number of investigators to study *lux* gene regulation in *E. coli.*[1–3] Numerous plasmid vectors have been engineered to facilitate such studies and a subset of these have used the sensitivity of a bioluminescence reporter to monitor *lux* gene transcription.[12,21–25] It is these plasmids that are the basis for the autoinducer sensor system described here. Important features of the sensor plasmid include the replacement of the *luxCDABEG* genes of *V. fischeri* with the *V. harveyi luxAB* genes and truncation or removal of the *luxI* gene (Fig. 1). The *V. harveyi luxAB* genes encode a stable luciferase enzyme that serves as a transcriptional reporter for the autoinducer response.[26] Under normal circumstances, this enzyme utilizes an aldehyde substrate that is made available through the actions of the *luxCDE* gene products; removal of these genes in our system allows the controlled assay of luciferase by the addition of an aldehyde substrate (usually *n*-decanal) to the system. Inactivation of the *luxI* gene by truncation or deletion is necessary to make the system responsive primarily to an autoinducer that is supplied from an external source (e.g., synthetic autoinducer or conditioned culture medium from an autoinducer-producing bacterial species). The plasmid pJHD500[22,25] is a pBR322-based plasmid that has these features and is depicted in Fig. 1. Reporter plasmids with similar *lux* gene configurations have been described by others.[27,28]

A plasmid-borne sensor configuration like pJHD500 has the advantage that different strains of *E. coli* can be tested for an ability to respond to various potential autoinducers. The choice of *E. coli* strain to be used as the sensor background is not critical at the outset. We have observed typical autoinducer responses in many common laboratory strains and thus several strains should be tested to achieve the best response. However, certain mutant genetic backgrounds (e.g., *crp, cya, htpR*) have been reported to affect luminescence in *E. coli* dramatically and should be avoided.[21,23,29]

[21] Y. Y. Adar, M. Simaan, and S. Ulitzur, *J. Bacteriol.* **174,** 7138 (1992).

[22] J. H. Devine, G. S. Shadel, and T. O. Baldwin, *Proc. Natl. Acad. Sci. U.S.A* **86,** 5688 (1989).

[23] P. V. Dunlap and E. P. Greenberg, *J. Bacteriol.* **164,** 45 (1985).

[24] G. S. Shadel and T. O. Baldwin, *J. Biol. Chem.* **267,** 7690 (1992).

[25] G. S. Shadel, R. Young, and T. O. Baldwin, *J. Bacteriol.* **172,** 3980 (1990).

[26] T. O. Baldwin, T. Berends, T. A. Bunch, T. F. Holzman, S. K. Rausch, L. Shamansky, M. L. Treat, and M. M. Zeigler, *Biochemistry* **23,** 3663 (1984).

[27] D. L. Milton, A. Hardman, M. Camara, S. R. Chhabra, B. W. Bycroft, G. S. A. B. Stewart, and P. Williams, *J. Bacteriol.* **179,** 3004 (1997).

[28] S. Swift, N. J. Bainton, B. W. Bycroft, P. F. Chan, S. R. Chhabra, P. J. Hill, C. E. D. Rees, G. P. C. Salmond, J. P. Throup, M. K. Winson, P. Williams, and G. S. A. B. Stewart, *Mol. Microbiol.* **10,** 511 (1993).

[29] G. S. Shadel and T. O. Baldwin, *J. Biol. Chem.* **267,** 7696 (1992).

Assay for Autoinducer Activity in Liquid Culture

This section describes two different types of assays for autoinducer activity. The general strategy is to monitor luciferase activity *in vivo* during growth of an autoinducer-sensor strain in the presence and absence of added autoinducer. The first assay is based on the ability of different autoinducers to stimulate light production in the sensor strain. This requires the molecule to behave essentially like a natural *V. fischeri* autoinducer and bind to LuxR to elicit a positive response. The ability of the *V. fischeri* sensor system to respond to autoinducer-like molecules other than *N*-(3-oxohexanoyl) homoserine lactone is shown in Fig. 2, where a positive response is obtained by the addition of a synthetic autoinducer-like mole-

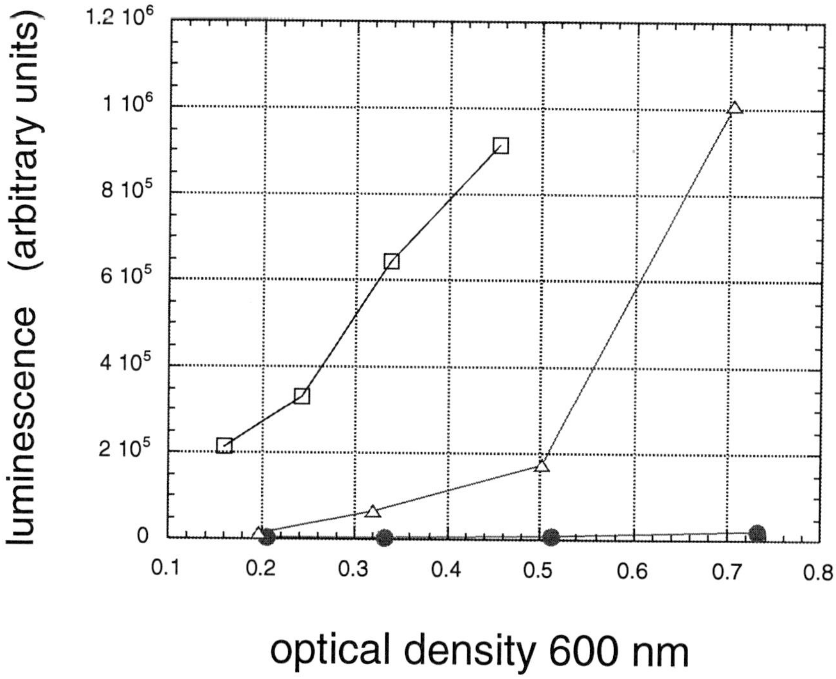

FIG. 2. An *E. coli* sensor strain containing pJHD500 responds to *N*-(3-oxodecanoyl) homoserine lactone. Culture luminescence (arbitrary units) is plotted against culture cell density (optical density at 600 nm). The presence of 100 μM *N*-(3-oxodecanoyl) homoserine lactone in culture medium (△) produces a strong luciferase signal from *E. coli* strain TB1 bearing pJHD500. The response to 10 μM natural *V. fischeri* autoinducer (□) or to the ethanol carrier alone (●) is also shown. All cultures were grown at 28° in LB medium containing 100 μg/ml ampicillin.

cule, *N*-(3-oxodecanoyl) homoserine lactone. As expected, the magnitude of the response to this compound was substantially less than that obtained with the natural *V. fischeri* autoinducer. Nonetheless, the ability to generate a significant response with molecules that are similar to, yet distinct from, natural *V. fischeri* autoinducers demonstrates the utility of the recombinant *V. fischeri lux* sensor system in an assay for diverse autoinducer activities. Analysis of various autoinducer analogs in *V. fischeri* revealed that certain autoinducer-like molecules can bind LuxR but not induce luminescence.[30] However, several of these molecules inhibited the action of the natural *V. fischeri* autoinducer. The ability of test autoinducer samples to antagonize the effects of a natural *V. fischeri* autoinducer in the *E. coli* sensor system is the basis for the second autoinducer assay.

Preparation of Autoinducer Test Samples. The assays described in this article should be useful for detecting autoinducers in a number of different sample types, including those that contain synthetic or naturally produced autoinducers. One common application of this type of sensor system is to assay for the production of autoinducers by other bacterial species. Because most autoinducers diffuse out of the cell readily, it is possible to use medium that has been conditioned by growth of the organism of interest. After growth to the near stationary phase, the culture should be centrifuged (5000–7000*g*, 5 min) to remove the majority of bacterial cells. The supernatant from this centrifugation is then sterilized by passing it through a 0.2-μm pore-size filter. The resulting sterile, conditioned medium can then be added directly to a sensor culture to assay for autoinducer activity. Once a sample is identified as potentially containing an autoinducer, the active component should be extracted and purified. Methods for the purification and structural elucidation of autoinducers and autoinducer analogs can be found elsewhere in this volume.[31]

lux Gene Stimulation Assay for Autoinducer Activity. The following procedure is an assay for autoinducer activity based on an ability to activate the *V. fischeri lux* system in *E. coli.* A seed culture of the *E. coli* sensor strain is prepared by inoculating 5 ml of LB ampicillin medium (yeast extract, 10 g/liter, tryptone, 5 g/liter, NaCl, 5 g/liter, ampicillin, 100 mg/liter) with a single colony picked from a fresh plate. The culture is then grown for 8–10 hr at 30° with shaking. Bacterial cells from a 1-ml sample of the culture are pelleted in a microcentrifuge (7000–10,000*g*, 1 min),

[30] A. L. Schaefer, B. L. Hanzelka, A. Eberhard, and E. P. Greenberg. *J. Bacteriol.* **178,** 2897 (1996).

[31] A. L. Schaefer, B. L. Hanzelka, M. R. Parsek, and E. P. Greenberg, *Methods Enzymol.* **305** [21] 2000 (this volume).

washed by resuspending the cells in 1 ml of fresh LB ampicillin medium, and centrifugation repeated. The washed pellet is then resuspended in 1 ml LB ampicillin medium. To begin an experiment, the washed seed culture is diluted (to an optical density at 600 nm, OD_{600} ~0.05) into LB ampicillin (preheated to 30°) in culture flasks and growth is continued at 30° with shaking (we normally grow the cultures at 200 rpm in an orbital-shaking water bath). The volume of LB ampicillin medium in the test cultures should be enough to allow withdrawal of several 1-ml samples during the course of the experiment (we typically use 20- to 50-ml cultures). For each putative autoinducer-containing sample to be tested, we suggest growing at least four cultures: one culture without added autoinducer to serve as a negative control, one culture containing *V. fischeri* autoinducer (10–2000 n*M*) to serve as a positive control, and at least two cultures containing different amounts of the putative autoinducer-containing sample. Similar results are obtained whether autoinducer is added to the cultures from the beginning of the experiment or added later after the cultures reach early log-phase growth. At various time points (taken at culture cell densities between OD_{600} 0.2 and 1.5), 1-ml samples are removed from the culture and used to determine the OD_{600} of the culture and to perform a luciferase assay *in vivo*. Commercially available instrumentation for the detection of light emission in the luciferase assay is reviewed elsewhere in this volume.[32] We typically use a calibrated photomultiplier-photometer that is outfitted to house a scintillation vial. Using a syringe and needle, 1 ml of a sonicated mixture of *n*-decanal (Sigma) in LB medium is injected through a rubber septum into a scintillation vial containing 1 ml of the test culture. After injection, peak luminescence is recorded. An autoinducer–response curve is then constructed by plotting the results from the entire growth experiment as luminescence versus OD_{600} of the culture. An example of a typical plot is given in Fig. 2. Samples resulting in activation of luminescence above the minus autoinducer control culture are interpreted to potentially contain an autoinducer activity. However, if the autoinducer test sample is complex in nature (e.g., conditioned growth medium), it is necessary to determine if the stimulatory action is due to a bona fide autoinducer and not some other component in the sample. This can be done by purifying the active component from the sample (see "Preparation of Autoinducer Test Samples") and testing it again in the activation assay.

lux Gene Inhibition Assay for Autoinducer Activity. The following procedure is an assay for autoinducer activity based on inhibition of the

[32] P. E. Stanley, *Methods Enzymol.* **305** [6] 2000 (this volume).

normal response of the *V. fischeri lux* sensor strain to *N*-(3-oxohexanoyl) homoserine lactone. Inoculation and growth conditions are precisely the same as described earlier in the stimulation assay. However, here it is necessary to first perform a titration of the sensor system with the natural *V. fischeri* autoinducer to determine minimal concentrations required to elicit a significant response. This is done to establish a set of conditions where the possibility for inhibition of the response by the addition of a test autoinducer sample is greatest (i.e., natural autoinducer is below saturating concentrations). Once these conditions are established, it is possible to test the putative autoinducer-containing sample for inhibitory effects. A set of cultures that contain the natural autoinducer at the predetermined concentration is prepared. A titration of the test autoinducer sample is then accomplished by adding different amounts of the sample to the flasks containing the natural *V. fischeri* autoinducer. Two controls, one sensor culture without added autoinducer and one containing the natural *V. fischeri* autoinducer only, are analyzed simultaneously. Growth curves and luciferase assays are performed as described in the stimulation assay. Those exhibiting a response significantly lower than the positive control culture (natural autoinducer only) are interpreted to potentially contain an autoinducer that has inhibitory activity. The influences of numerous pleiotropic regulatory systems (e.g., catabolite repression, heat shock, anaerobiosis) on the *lux* regulon, even in this artificial cloned setting in *E. coli,* necessitate that great care be taken in the interpretation of this inhibition assay. For this reason, purification of the active component from the sample is necessary to confirm that the observed inhibitory action is due to an autoinducer-like molecule.[31]

Special Considerations

We have described two assays for autoinducer activity that involve the growth of an *E. coli* sensor strain in liquid medium and measurement of a bacterial luciferase reporter that is linked to an autoinducer-responsive *V. fischeri lux* gene configuration. In *E. coli,* the *V. fischeri* LuxR protein is able to respond to autoinducer-like molecules that differ from either of its known natural autoinducers. For example, luminescence is stimulated when a nonnatural autoinducer, *N*-(3-oxodecanoyl) homoserine lactone, is used (Fig. 2), a compound that behaves as an apparent inhibitor of *V. fischeri* luminescence. Regardless of why the *V. fischeri lux* system is more promiscuous in the context of *E. coli,* it is advantageous for the assays described here because it allows a larger variety of autoinducers to be screened using a one sensor system. A similar autoinducer-sensor system has been described that utilizes a set of natural *V. harveyi* strains that respond differentially to each of the two known *V. harveyi* autoinducers.[33] This system was proved effective for assaying the production of *V. harveyi*-like autoinducer activity by a variety of bacterial species. As predicted by the authors, this system did not respond to *V. fischeri* autoinducers. This suggests that, like in natural *V. fischeri* strains, mechanisms exist in *V. harveyi* to prevent the utilization of certain types of autoinducers. Other autoinducer-sensor systems based on an *A. tumefaciens tral::lacZ* fusion strain or violacein pigment production by *C. violaceum* have also been reported.[18,20] Because the *V. fischeri lux* system in *E. coli* appears to be capable of bypassing some of the mechanisms that prevent recognition of a foreign autoinducer, the recombinant sensor system may be better suited for assaying a wider range of autoinducer activities than these other systems. However, it should be noted that the *V. fischeri lux*-based sensor system described here will almost certainly not be capable of detecting the full spectrum of autoinducers that exist. It may be possible to generate a minimal set of recombinant sensor strains using the regulatory genes from other quorum sensing bacteria to extend the assay capability to encompass virtually any autoinducer molecule.

It should be possible to modify the recombinant *V. fischeri lux* system to screen on solid medium for genes encoding LuxR and Luxl homologs in an expression library of interest (i.e., from other bacterial species). In fact, this method has been used successfully to isolate Luxl homologs from *E. carotovora, E. agglomerans,* and *V. anguillarum.*[27,28] A strategy similar to that used to isolate these Luxl homologs has not been utilized to isolate LuxR homologs based on LuxR activity. Such a strategy may be possible but would require the replacement of the *V. fischeri luxR*-dependent promoter on the reporter with the promoter of an autoinducer responsive gene from the organism of interest. An expression library could then be transformed into this strain and screened for the presence of a LuxR homolog that can stimulate luminescence in the presence of autoinducer. Altogether, the assays described here will facilitate the identification of new autoinducers and genes required for the synthesis of and response to this family of signaling molecules. The continued study of quorum sensing bacteria and the molecular interactions involved in achieving the observed specificity of these multicomponent systems will increase our understanding of complex cell–cell signaling mechanisms required for normal function and development of other organisms.

[33] B. L. Bassler, E. P. Greenberg, and A. M. Stevens, *J. Bacteriol.* **179,** 4043 (1997).

[21] Detection, Purification, and Structural Elucidation of the Acylhomoserine Lactone Inducer of *Vibrio fischeri* Luminescence and Other Related Molecules

By Amy L. Schaefer, Brian L. Hanzelka, Matthew R. Parsek, and E. Peter Greenberg

Introduction

Many gram-negative bacteria are known to use acylhomoserine lactones (acyl-HSLs) as intercellular signals, called autoinducers, in density-dependent gene expression.[1–4] Because cells are permeable to these signal molecules, autoinducers at high cell densities can achieve critical concentration and activate the expression of specific genes.

The first acylhomoserine lactone autoinducer, 3-oxohexanoylhomoserine lactone (VAI for *Vibrio* autoinducer, Fig. 1), was identified in the marine, luminescent bacterium *Vibrio fischeri* in 1982.[5] The *V. fischeri* membrane is permeable to the autoinducer so that the intracellular and environmental concentrations of this signal molecule are equal.[6] When present at sufficient concentrations the autoinducer interacts with a regulatory protein, LuxR, which can then activate the genes necessary for luminescence.[7,8] Similar signals and regulatory proteins (called R proteins) have since been found in a number of gram-negative bacteria, including *Pseudomonas aeruginosa*,[9–11] *Agrobacterium tumefaciens*,[12,13] *Rhizobium legu-*

[1] W. C. Fuqua, S. C. Winans, and E. P. Greenberg, *J. Bacteriol.* **176,** 269 (1994).
[2] W. C. Fuqua, S. C. Winans, and E. P. Greenberg, *Annu. Rev. Microbiol.* **50,** 727 (1996).
[3] G. P. C. Salmond, B. W. Bycroft, G. S. A. B. Stewart, and P. Williams, *Mol. Microbiol.* **16,** 615 (1995).
[4] D. M. Sitnikov, J. B. Schineller, and T. O. Baldwin, *Mol. Microbiol.* **17,** 801 (1995).
[5] A. Eberhard, A. L. Burlingame, C. Eberhard, G. L. Kenyon, K. H. Nealson, and N. J. Oppenheimer, *Biochemistry* **20,** 2444 (1981).
[6] H. B. Kaplan and E. P. Greenberg, *J. Bacteriol.* **163,** 1210 (1985).
[7] J. Engebrecht, K. H. Nealson, and M. Silverman, *Cell* **32,** 773 (1983).
[8] J. Engebrecht and M. Silverman, *Proc. Natl. Acad. Sci. U.S.A.* **81,** 4154 (1984).
[9] L. Passador, J. M. Cook, M. J. Gambello, L. Rust, and B. H. Iglewski, *Science* **260,** 1127 (1993).
[10] J. P. Pearson, K. M. Gray, L. Passador, K. D. Tucker, A. Eberhard, B. H. Iglewski, and E. P. Greenberg, *Proc. Natl. Acad. Sci. U.S.A.* **91,** 197 (1994).
[11] J. P. Pearson, L. Passador, B. H. Iglewski, and E. P. Greenberg, *Proc. Natl. Acad. Sci. U.S.A.* **92,** 1490 (1995).
[12] K. R. Piper, S. B. von Bodman, and S. K. Farrand, *Nature (London)* **362,** 448 (1993).
[13] L. Zhang, P. J. Murphy, A. Kerr, and M. E. Tate, *Nature (London)* **362,** 446 (1993).

METHODS IN ENZYMOLOGY, VOL. 305
0076-6879/00 $30.00

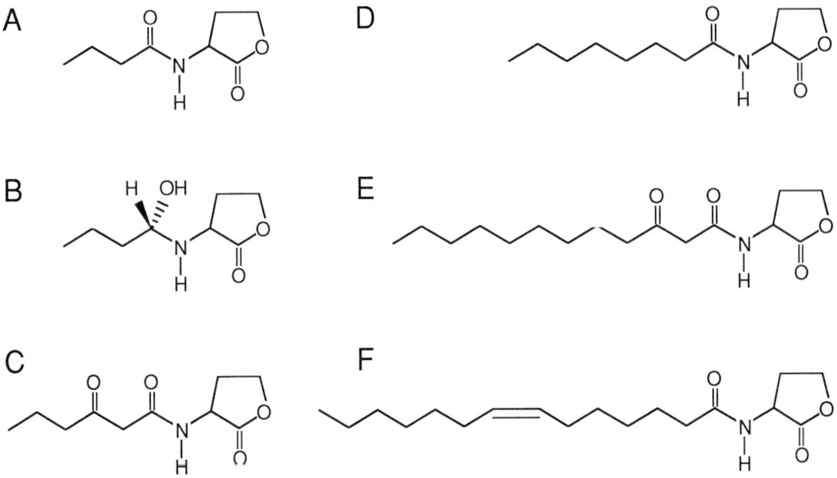

FIG. 1. Examples of known autoinducer signal molecules. (A) Butanoyl-HSL (PAI-2), (B) 3-hydroxybutanoyl-HSL (HAI), (C) 3-oxohexanoyl-HSL (VAI), (D) octanoyl-HSL (OHL), (E) 3-oxododecanoyl-HSL (PAI-1), and (F) 7,8-*cis*-tetradecenoyl-HSL.

minosarum,[14,15] and *Rhodobacter sphaeroides.*[16] Although acyl-HSL molecules from various bacteria are related in structure, these signal molecules can differ in the nature of the acyl side chain moiety (Fig. 1). Depending on the particular autoinducer, the acyl group varies from 4 to 14 carbons in length, possesses a hydroxyl, a carbonyl, or no substitution on the third carbon, and is either fully saturated or contains a single carbon–carbon double bond (Fig. 1).

This article discusses methods for detecting, purifying, and determining the structures of acyl-HSL autoinducers. We concentrate on methods employed in our laboratory and discuss other methods briefly.

Assay Methods

Principles

We rely on a series of bioassays to detect acyl-HSLs. The principle for each bioassay is the same: expression of a reporter gene dependent on an

[14] K. M. Gray, J. P. Pearson, J. A. Downie, B. E. A. Boboye, and E. P. Greenberg, *J. Bacteriol.* **178,** 372 (1996).

[15] J. Schripsema, K. E. E. de Rudder, R. B. van Vliet, P. P. Lankhorst, E. de Vroom, J. W. Kijne, and A. A. N. van Brussel, *J. Bacteriol.* **178,** 366 (1996).

[16] A. Puskas, E. P. Greenberg, S. Kaplan, and A. L. Schaefer, *J. Bacteriol.* **179,** 7530 (1997).

autoinducer-R protein interaction is used to detect exogenously added autoinducers (the exception is the HAI assay in which an R homolog is not involved). If a sample contains an acyl-HSL capable of interacting with the specific R protein, the reporter gene will be expressed to varying degrees depending on the concentration and structure of the added acyl-HSL. The different reporter strains, their genotypes, and the acyl-HSLs they are known to detect are listed in Table I. Because no one single bioassay can be used to detect the entire range of known autoinducer structures, we employ five bioassays in our general screening for the production of acyl-HSLs. A report shows that some bacteria can produce acyl-HSLs with a hydroxyl group on the third acyl carbon.[17] It is not clear how sensitive any of the five assays we employ are to hydroxylated acyl-HSLs, but these signals can be detected using the *Agrobacterium* AAI reporter (Table I).[12,17]

Sample Preparation

Samples are prepared in glass test tubes. To prepare samples that are dissolved in acidified ethyl acetate (see extraction protocol later), the appropriate volume is pipetted into the bottom of the test tube, and the solvent is removed by evaporation under a gentle stream of N_2 gas. For initial screenings of culture supernatant fluids for acyl-HSLs, 250 ml is extracted and concentrated (as described later). The extract is resuspended in 2.5 ml of acidified ethyl acetate (100× concentrate). Duplicate samples representing a range of extract concentrations are screened in each bioassay.

To prepare samples that are dissolved in methanol–water (e.g., HPLC fractions), evaporation of the methanol is usually not required unless the final methanol concentration in the bioassay is greater than 5%.

Standard Curves, Positive and Negative Controls

It is useful to include both negative and positive controls for each set of bioassays. We include a sample with no extract added (the reporter strain alone) as the negative control. The preferred positive control is an autoinducer signal, either purified from culture fluids of a known autoinducer-producing organism or synthesized chemically (VAI can be purchased from Sigma Chemical Co. as *N*-β-ketocaproyl-L-homoserine lactone). Although the linear range of response varies depending on the bioassay and the autoinducer structure, we usually generate standard curves

[17] P. D. Shaw, G. Ping, S. L. Daly, C. Cha, J. E. Cronan Jr., K. L. Rinehart, and S. K. Farrand *Proc. Natl. Acad. Sci. U.S.A.* **94,** 6036 (1997).

TABLE I

AUTOINDUCER BIOASSAY STRAINS AND ACYLHOMOSERINE LACTONE MOLECULES DETECTED BY EACH STRAIN

Assay	Reporter strain	Relevant characteristics	Acylhomoserine lactone(s) detected	Ref.
HAI	*V. harveyi* D1	Unknown mutation resulting in reduced autoinducer production	3-Hydroxybutanoyl-HSL, 3-hydroxyvaleryl-HSL	19, 26
PAI-2	*E. coli* XL-1Blue (pECP61.5)	*rhlA::lacZ* translational fusion and *ptac-rhlR* in pSW205; Ap^R	Butanoyl-HSL, hexanoyl-HSL[a]	10, 27
VAI	*E. coli* VJS533 (pHV200I$^-$)	*V. fischeri* ES114 *lux* regulon with inactivated *luxI* in pBR322; Ap^R	3-Oxohexanoyl-HSL, hexanoyl-HSL[a], 3-oxooctanoyl-HSL, octanoyl-HSL[a]	10, 28, 29
OHL	*R. solanacearum* AW1-AI8 (p395B)	Inactivated *solI,* p395B contains *aidA::lacZ* fusion; Nx^R, Sp^R, Tc^R	Unsubstituted acyl-HSLs with acyl chains of 8 carbons or longer	20
PAI-1	*E. coli* MG-4 (pKDT17)	*lasB::lacZ* translational fusion and *plac-lasR;* Ap^R	3-Hydroxy, 3-oxo, and unsubstituted acyl-HSLs with side chain lengths of 8–14 carbons	10, 30
AAI	*A. tumefaciens* NT1 (pJM749, pSVB33-23)	*traI::lacZ* and *traR* on separate plasmids; pTi cured; Cb^R, Km^R	3-Oxooctanoyl-HSL, octanoyl-HSL, and other acyl-HSLs[a]	12
CAI	*C. violaceum* CV0blu	*cviI*::Tn5*xylE* (inactivated *cviI,* an autoinducer synthase required for violacein production); Hg^R Km^R, Cm^R	Hexanoyl-HSL, butanoyl-HSL, 3-oxohexanoyl-HSL,[a] octanoyl-HSL[a]; acyl-HSLs with longer side chains can be detected by screening for inhibition of hexancyl-HSL-mediated violacein production	21, 22

[a] Indicates weak activation of the reporter.

Resistance to ampicillin (Ap^R), nalidixic acid (Nx^R), tetracycline (Tc^R), kanamycin (Km^R), carbenicillin (Cb^R), spectinomycin (Sp^R), and mercury (Hg^R).

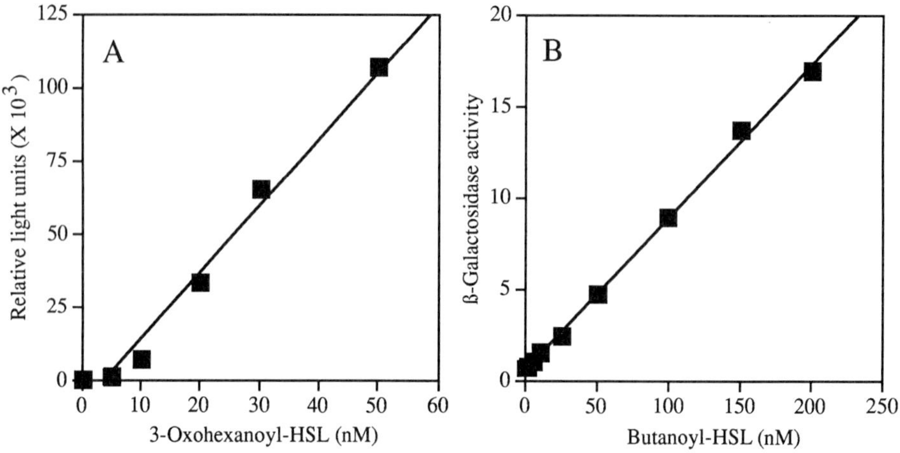

FIG. 2. Examples of standard curves for two acyl-HSL bioassays. (A) The relationship between 3-oxohexanoyl-HSL concentration and luminescence reporter activity in the VAI bioassay. (B) The relationship between butanoyl-HSL concentration and β-galactosidase reporter activity in the PAI-2 bioassay.

using 5 to 200 n*M* concentrations of a particular acyl-HSL. Examples of two standard curves are shown in Fig. 2.

Reagents for VAI Bioassay[10]

Escherichia coli VJS533 (pHV200I$^-$)

Bioassay medium[10]: 0.05% tryptone, 0.03% glycerol, and 100 m*M* NaCl. Dispense in 47-ml volumes; autoclave, allow to cool; add 2.5 ml of sterile 1 *M* $MgSO_4$ and 0.5 ml of 1 *M* potassium phosphate (pH 7.0)

Procedure for VAI Bioassay

Inoculate 50 ml of Luria broth[18] containing ampicillin (100 μg/ml) with 5 ml from an overnight culture of *E. coli* VJS533 (pHV2001$^-$). Incubate the culture at 30° with shaking. When the OD_{600} reaches 2.5, dilute the cells to an OD of 1.0 in bioassay medium and store 1-ml stocks in freezer vials at −70° until needed.

Remove and thaw (at room temperature) the stocks. Dilute the reporter cells 1 : 100 in fresh bioassay medium containing ampicillin (100 μg/ml) and incubate with shaking at 30° for 2 hr. Then add 1 ml of cell suspension per

[18] J. Sambrook, E. F. Fritsch, and T. Maniatis, "Molecular Cloning: A Laboratory Manual." Cold Spring Harbor Laboratory Press, Cold Spring Harbor, NY, 1989.

sample tube (prepared as described earlier), vortex well, and incubate at room temperature for 3 hr. After the incubation, measure luminescence (we generally use a 96-well microplate luminometer, Anthos Lucy1, Anthos, Austria).

Reagents for HAI Bioassay[19]

V. harveyi D1

Seawater complete (SWC) broth[6]: 750 ml reconstituted seawater, 250 ml distilled water, 0.5% tryptone, 0.3% yeast extract, 0.3% glycerol, pH to 7.0 with NaOH

Procedure for HAI Bioassay

Grow *V. harveyi* D1 in 5 ml of SWC broth overnight at room temperature with shaking. Use the overnight culture to inculate 25 ml of SWC broth (starting OD_{600} of 0.1). Incubate at room temperature with shaking for 1 hr (the OD_{600} should be between 0.2 and 0.3). Add 1 ml of the culture to each sample tube (prepared as described earlier), incubate for 2 hr at room temperature with shaking, and then measure light production.

Reagents for PAI-1/PAI-2 Bioassays[10]

E. coli MG4 (pKDT17) or *E. coli* XL1-Blue (pECP61.5)

A medium[18]: 60 m*M* K_2HPO_4, 33 m*M* KH_2PO_4, 7.5 m*M* $(NH_4)_2SO_4$, 1.7 m*M* sodium citrate $\cdot$ $2H_2O$, 0.4% glucose, 0.05% yeast extract, and 1 m*M* $MgSO_4$

Procedure for PAI-1/PAI-2 Bioassays

The PAI-1 and PAI-2 assay protocols are identical, differing only in the reporter strain used. *E. coli* MG4 (pKDT17) and *E. coli* XL1-Blue (pECP61.5) are reporter strains for the PAI-1 and PAI-2 assays, respectively. Incubate a culture overnight at 30° with shaking in A medium containing ampicillin (100 μg/ml). For the PAI-2 bioassy, 1 m*M* IPTG is included in the culture medium to induce RhlR expression (see Table I). Subculture *E. coli* MG4 (pKDT17) in A medium or *E. coli* XL1-Blue (pECP61.5) in A medium plus IPTG (initial OD_{600} of 0.1), add 1 ml of subculture to each sample tube, and incubate with shaking at 30° for 5–6 hr. Measure β-galactosidase activity in each sample.

[19] J. Cao and E. A. Meighen, *J. Bacteriol.* **175,** 3856 (1993).

Reagents for OHL Bioassay[20]

Ralstonia solanacearum AW1-AI8 (p395B)
BG medium[20]: 1% BactoPeptone, 0.5% glucose, 0.1% yeast extract, and 0.1% casamino acids

Procedure for OHL Bioassay

The protocol for this assay is exactly the same as PAI-1/PAI-2 bioassays except that BG medium containing tetracycline (10 μg/ml) and spectinomycin (100 μg/ml) is used and the reporter strain is *R. solanacearum* AW1-AI8 (p395B).

Other Assays

A number of other assays detailed in the literature are also useful, including the *A. tumefaciens*[12] (AAI) and *Chromobacterium violaceum*[21,22] (CAI) bioassays. The *A. tumefaciens* bioassay is similar to the PAI-1/PAI-2/OHL β-galactosidase bioassays described earlier. In *C. violaceum,* production of the pigment violacein is controlled by an acyl-HSL. The CAI bioassay uses the stimulation of pigment production in a nonautoinducer-producing mutant to detect acyl-HSLs of four to eight carbons in length.[21,22] Autoinducers with longer acyl side chains can also be detected by including hexanoyl-HSL (an acyl-HSL required for violacein synthesis) in the bioassay reaction mixture and screening for the inhibition of pigment production by the longer chained acyl-HSL.[21] We have found that the entire range of known autoinducer structures can be detected by at least one of the five bioassays described earlier, although we have not examined the sensitivities of our reporters to 3-hydroxy-acyl-HSLs in a systematic fashion.

Purification Procedures

The extraction and subsequent purification of the autoinducer signal should be performed relatively quickly (usually 2–3 days from extraction to structural analyses). Methanol–water fractions may be stored overnight at 4°.

[20] A. B. Flavier, L. M. Ganova-Raeva, M. A. Schell, and T. P. Denny, *J. Bacteriol.* **179,** 7089 (1997).

[21] A. M. Taylor, K. H. McClean, S. R. Chhabra, M. Camara, M. K. Winson, M. Daykin, B. W. Bycroft, G. S. A. B. Stewart, and P. Williams, *in* "Abstracts of the 96th General Meeting of the American Society for Microbiology 1996." American Society for Microbiology, Washington, DC, 1996.

[22] S. Swift, A. V. Karlyshev, L. Fish, E. L. Durant, M. K. Winson, S. R. Chhabra, P. Williams, S. MacIntyre, and G. S. A. B. Stewart, *J. Bacteriol.* **179,** 5271 (1997).

Growth of Organisms

Acyl-HSLs are usually extracted from late exponential phase cultures that have been grown in media and at temperatures appropriate for that organism. It is important to monitor the pH of the culture fluid, as the homoserine lactone ring will hydrolyze under basic conditions. For example, the half-life of 3-oxohexanoyl-HSL (VAI) is approximately 1 day at pH 7.0 and 2–3 hr at pH 8.0.[23] To screen for acyl-HSL production, a 250-ml culture volume is sufficient, but to purify and identify an autoinducer we generally start with 3 to 5 liters of culture fluid.

Step 1: Extraction of Acylhomoserine Lactones from Culture Fluid. Separate cells and culture fluid by centrifugation (10,000*g* for 10 min at 4°). Extract the acyl-HSLs from cell-free fluid at least twice with an equal volume of acidified ethyl acetate (0.1 ml glacial acetic acid per liter of ethyl acetate). Because acyl-HSLs partition rather evenly between acidified ethyl acetate and water, multiple extractions are essential. Dry the combined extracts with anhydrous $MgSO_4$. Remove the $MgSO_4$ by filtration through a Whatman No. 1 qualitative filter and transfer the clarified extract to a round-bottom flask. Remove the ethyl acetate by rotary evaporation at 40° (an oily residue should remain). Dissolve the residue in 3–5 ml of acidified ethyl acetate, transfer to a glass sample vial with a Teflon-sealed cap, evaporate the organic solvent using a stream of N_2 gas, and store at −20°.

Step 2: Prepurification Step (Optional). It is sometimes desirable to use an additional purification step, especially when the culture medium is rich or when the organism is known to produce large amounts of extracellular products. We have used C_{18}-BondElute cartridges (Varian Sample Preparation Products, San Diego, CA) prior to high-pressure liquid chromatography (HPLC) analysis.[14] Suspend the sample in methanol and apply to a BondElute column that has been washed with 5% methanol. Elute the autoinducer by adding 2–3 column volumes each of 5, 25, 50, 75, and 100% methanol in water over the column. Test each wash for acyl-HSL activity (see earlier discussion), pool all appropriate fractions, and remove the methanol–water by evaporation.

Step 3: C_{18} Reverse-Phase HPLC, Gradient Profile. Dissolve the sample in 0.2 ml of either 20% methanol in water (for those samples with activity in the PAI-2 and HAI bioassays) or 100% methanol (for those samples with activity in the VAI, OHL, and PAI-1 bioassays). The acyl-HSL, in a 0.2-ml volume, can be separated by HPLC using a C_{18} reverse-phase column equilibrated with 20% methanol in water. A linear, 20–100% (v/v) methanol in water gradient (a flow rate of 0.5 ml/min and a run duration of 140 min)

[23] A. Eberhard, personal communication.

can be used to separate a variety of autoinducer molecules (Table II). For maximum resolution, we collect 70 1-ml fractions. After testing each fraction for activity with the appropriate bioassay, pool the bioactive fractions (usually 1 or 2 fractions) and evaporate to dryness.

Step 4: C_{18} Reverse-Phase HPLC, Isocratic Profile. The methanol–water ratio for the isocratic profile is selected based on the elution profile of the active fractions in step 3. For example, in Step 3, 7,8-*cis*-tetradecenoyl-HSL (Fig. 1) is eluted in approximately 87% methanol (Table II), thus an isocratic profile of 70% methanol is used (approximately 10–15% less methanol than the concentration in which the molecule was eluted in the HPLC gradient profile). Suspend the sample in 0.2 ml of methanol–water at the appropriate ratio and separate on a C_{18} reverse-phase column (0.5 ml/min, collecting 30 1-ml fractions). Assay the fractions for acyl-HSLs, pool the active fractions (1 or 2 fractions), and evaporate to dryness. In our experience, samples prepared as described earlier are of sufficient purity for structural analyses.

Structural Elucidation

Mass Spectrometry (MS)

To determine the molecular mass of acylhomoserine lactones we have used both chemical ionization (CI) and high-resolution fast atom bombardment (FAB) mass spectrometry.[10,14,16] Electrical ionization (EI) MS has not been useful in the analysis of autoinducers with a carbonyl at the acyl

TABLE II
HPLC ELUTION PROFILES OF SOME ACYLHOMOSERINE LACTONES

Acylhomoserine lactone	% methanol in which the acyl-HSL is eluted[a]
Butanoyl-HSL	28–30
3-Oxohexanoyl-HSL	28–30
Hexanoyl-HSL	44–46
3-Oxooctanoyl-HSL	50–52
Octanoyl-HSL	63–65
3-Oxodecanoyl-HSL	64–66
3-Oxododecanoyl-HSL	74–75
3-Hydroxy-7,8-*cis*-tetradecenoyl-HSL	78–80
3-Oxotetradecanoyl-HSL	83–84
7,8-*cis*-Tetradecenoyl-HSL	87–88

[a] According to the HPLC gradient elution profile described in step 3 of the text.

3-C position, although we have used EI-MS in the analysis of butanoyl-HSL.[11] Because known acyl-HSL molecules do not carry a charge, techniques such as electrospray MS are of limited value. Acyl-HSLs contain a single nitrogen (bonded to two carbons and a hydrogen, Fig. 1) and thus have odd-numbered molecular masses.

In CI-MS, the sample in the gas (methane) phase becomes protonated at its most basic site so the molecular ion appears one mass unit higher than the molecular weight of the unprotonated autoinducer. This type of analysis gives not only the parental $M + H^+$ ion of the acyl-HSL (Fig. 3), but also a number of characteristic fragment ions with *m/z* intensities at 102 (homoserine lactone ring) and 143 (homoserine lactone ring plus one carbonyl group). If the autoinducer possesses a carbonyl at the 3-C position, often there is a *m/z* peak at 185 (homoserine lactone ring plus two carbonyl groups).

Although the nominal mass can be obtained by CI-MS, high-resolution FAB is used as a means to determine the exact mass of an acyl-HSL. This method allows precise elemental analysis of an acyl-HSL.

Nuclear Magnetic Resonance (NMR) Spectroscopy

^{1}H-NMR profiles of all acyl-HSLs (in $CDCl_3$), such as that of 7,8-*cis*-tetradecenoyl-HSL (Fig. 4), share several common proton peaks, including

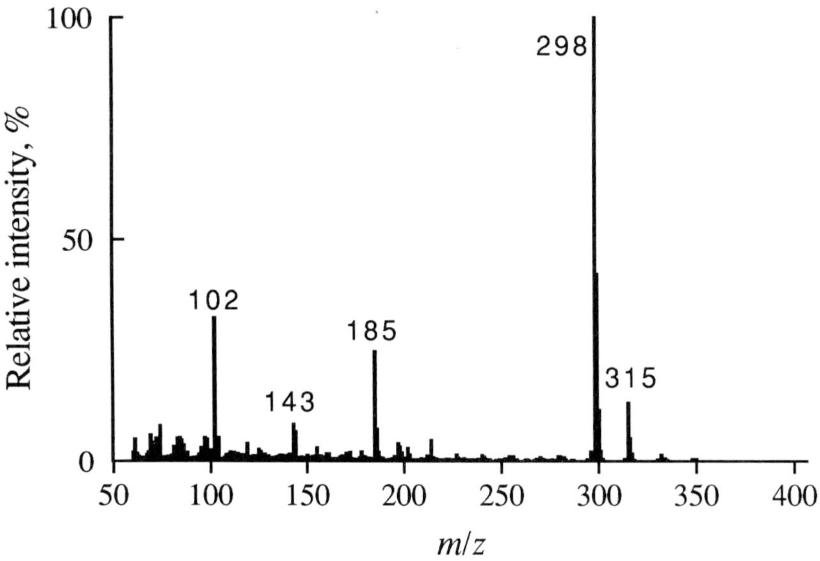

FIG. 3. Chemical ionization mass spectrum of 3-oxododecanoyl-HSL (PAI-1). Adapted from J. P. Pearson *et al., Proc. Natl. Acad. Sci. U.S.A.* **91,** 197 (1994).

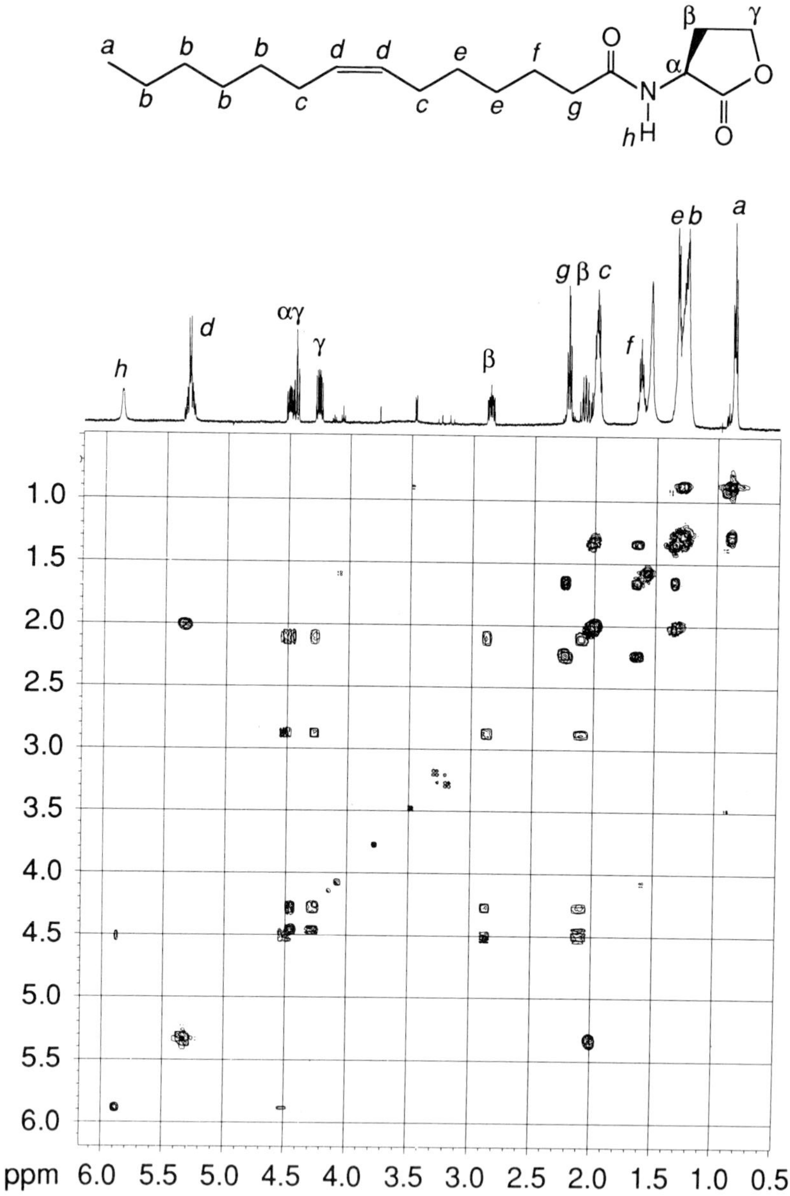

FIG. 4. Structure, proton NMR, and COSY spectrum of 7,8-*cis*-tetradecenoyl-HSL. Protons are indicated by italicized letters corresponding to the peaks in the ^{1}H-NMR ($CDCl_3$, 500 MHz) spectrum: δ_H 0.88 (3H, t, *a*), 1.27 (8H, m, *b*), 1.34 (4H, m, *e*), 1.54 (H_2O), 1.65 (2H, m, *f*), 2.01 (4H, m, *c*), 2.11 (1H, m, β), 2.24 (2H, m, *g*), 2.87 (1H, m, β), 4.27 (1H, m, γ), 4.46 (1H, m, γ), 4.52 (1H, m, α), 5.33 (2H, m, *d*), 5.89 (1H, broad d, *h*). Reprinted from A. Puskas *et al., J. Bacteriol.* **179,** 7530 (1997).

the terminal methyl triplet at ~0.86 ppm; methylene singlet(s) at ~1.2 to 1.3 ppm; the HSL ring β protons (multiplet) at ~2.2 and ~2.7 ppm; the HSL ring γ protons (multiplet) at ~4.27 and ~4.45 ppm; the HSL ring α proton (multiplet) at ~4.55 ppm; the CH_2CO protons (multiplet) at ~2.3; and the amide proton (broad doublet) is usually the most downfield signal, its position can vary from 5 to 8 ppm. NMR profiles of autoinducers possessing a carbonyl on the 3-C show a triplet peak at ~2.5 ppm from the CH_2CO protons and a singlet peak at ~3.4 ppm from the $COCH_2CO$ protons. A double triplet at ~5.4 ppm may indicate the presence of a double bond in the acyl side chain, and a broad singlet in the 3.5-ppm range could be evidence of a hydroxyl group. The presence of a hydroxyl can be demonstrated by performing NMR in D_2O (rather than $CDCl_3$) because the 3.5-ppm signal should disappear as the hydrogen is exchanged with deuterium.

Two-Dimensional Correlated Spectroscopy (COSY)

To date, two autoinducers have been found to contain an unsaturation in their acyl side chain. Acyl-HSLs from *R. leguminosarum*[14,15] and *R. sphaeroides*[16] (Fig. 1) contain a double bond between carbons 7 and 8 of the acyl group. Using conventional NMR techniques, one can detect the presence and configuration of such a double bond (see earlier discussion), but it is usually difficult to assign its exact position in the side chain. COSY is a useful technique because it allows one to trace the carbon skeleton of the spin system through ^1H–^1H couplings of the attached protons. The proton NMR spectrum appears along the diagonal as contours representing peak intensities (Fig. 4); the nondiagonal contours (found symmetrically on both sides of the diagonal) are the cross peaks. Horizontal lines drawn from one diagonal contour and vertical lines drawn from a separate diagonal contour that connect at a cross peak are said to be correlated.[24] This means that the protons from these two contours are coupled, indicating that their attached carbons are adjacent to one another.

In Fig. 4 the COSY spectrum of 7,8-*cis*-tetradecenoyl-HSL (protons are designated by italicized letters) shows the cross peak connectivity of the methyl signal at 0.88 ppm (*a*) coupling to the signal at 1.27 ppm (*b*), which couples to the 2.01 ppm (*c*) signal. This indicates the presence of four methylene groups (8H) between the terminal methyl group (*a*) and the $CH_2CH{=}CHCH_2$ group (*c*) at 2.01 ppm, which means the double bond occurs between the acyl carbons 7 and 8.

[24] R. M. Silverstein, "Spectrometric Identification of Organic Compounds," 5th Ed. Wiley, New York, 1991.

Other Methods of Acylhomoserine Lactone Characterization

Thin-layer chromatography (TLC) has been used to characterize acyl-HSLs from bacterial culture fluid extracts.[17] Here the bacterial culture extract is separated by TLC and the plate is then overlaid with an agar mixture containing a reporter, e.g., the *A. tumefaciens* reporter strain. Each acyl-HSL migrates with a characteristic R_f value. By comparing the R_f value of an unknown signal with those of known autoinducers, it is possible to tentatively identify an unknown acyl-HSL. This type of system has some limitations[17]; for instance, no single reporter strain can detect the whole range of known autoinducers and each reporter exhibits a wide range of sensitivities for different acyl-HSLs.

Organic Synthesis of Autoinducer

To confirm the assignment of an autoinducer structure, the molecule can be synthesized. If the synthetic autoinducer has biological activity and MS and NMR spectra that are identical to those of the natural compound, one can be confident of the structure. Organic syntheses of acyl-HSLs have been described previously.[5,10,25] Also, see Chemical Synthesis of Bacterial Autoinducers and Analogs by Eberhard and Schineller, this volume.

Autoinducer Stability

3-Oxohexanoyl-HSL (VAI) is a temperature-stable compound and can even be boiled without loss of activity. VAI is unstable in alkaline solutions, with the approximate half-time for decomposition being equal to (in days) $1/(1 \times 10^7 * [OH^-])$.[23] Decomposition is due to lactone ring hydrolysis and is irreversible in base due to carboxylate formation. The autoinducer is quite stable under acidic conditions though (unless treatment is so vigorous as to hydrolyze the amide bond).[23] The preferred storage of acyl-HSLs is dry at −20°. Most can also be stored in acidified ethyl acetate (unless the autoinducer possesses a hydroxyl group in its side chain; then it should be

[25] M. Dekhane, K. T. Douglas, and P. Gilbert, *Tetrahedron Lett.* **37,** 1883 (1996).
[26] J.-G. Cao, Z.-Y. Wei, and E. A. Meighen, *Biochem. J.* **312,** 439 (1995).
[27] J. P. Pearson, E. C. Pesci, and B. H. Iglewski, *J. Bacteriol.* **179,** 5756 (1997).
[28] A. Eberhard, C. A. Widrig, P. McBath, and J. B. Schineller, *Arch. Microbiol.* **146,** 35 (1986).
[29] A. L. Schaefer, B. L. Hanzelka, A. Eberhard, and E. P. Greenberg, *J. Bacteriol.* **178,** 2897 (1996).
[30] L. Passador, K. D. Tucker, K. R. Guertin, M. P. Journet, A. S. Kende, and B. H. Iglewski, *J. Bacteriol.* **178,** 5995 (1996).

stored dry). Those acyl-HSLs containing a double bond in the acyl side chain should be stored in the dark so that isomerization of the *cis* double bond does not occur.

Acknowledgments

We are grateful to Anatol Eberhard for sharing pH and temperature stability data of VAI and for continuing contributions to our efforts in the area of acyl-HSL purification and structural analysis. This work was supported by a grant from the Office of Naval Research (N00014-5-0190). A.L.S. and B.L.H. are predoctoral fellows supported by U.S. Public Health Service Training Grant 732 GM8365 from the National Institute of General Medical Sciences. M.R.P. is a National Research Science Award postdoctoral fellow supported by Grant GM18740-01A1 from the National Institutes of Health.

[22] Chemical Synthesis of Bacterial Autoinducers and Analogs

By ANATOL EBERHARD and JEFFREY B. SCHINELLER

Autoinducers consisting of homoserine lactone (HL) coupled to a fatty acid have been isolated from *Vibrio fischeri, V. harveyi, Erwinia carotovora, Enterobacter agglomerans, Pseudomonas aeruginosa, Agrobacterium tumefaciens,* and other gram-negative bacteria.[1–6] Since the first synthesis of the autoinducer of the marine bioluminescent bacterium *V. fischeri,*[3] the chemical synthesis of autoinducers has provided readily obtainable quantities of the signal molecules to further characterize the signaling mechanisms. The first literature report of the *in vitro* enzymatic synthesis of an autoinducer is that of *A. tumefaciens: N*-(3-oxooctanoyl)-L-homoserine lactone was synthesized using purified hexahistidinyl autoinducer synthase (H_6-TraI) with the substrates 3-oxooctanoyl-acyl carrier protein and *S*-adenosyl-

[1] W. C. Fuqua, S. C. Winans, and E. P. Greenberg, *J. Bacteriol.* **176,** 269 (1994).
[2] S. Swift, M. K. Wilson, P. F. Chan, N. J. Bainton, M. Birdsal, P. J. Reeves, C. E. D. Rees, S. R. Chhabra, P. J. Hill, J. P. Throup, B. W. Bycroft, G. P. C. Salmond, P. Williams, and G. S. A. B. Stewart, *Mol. Microbiol.* **10,** 511 (1993).
[3] A. Eberhard, A. L. Burlingame, C. Eberhard, G. L. Kenyon, K. H. Nealson, and N. J. Oppenheimer, *Biochemistry* **20,** 2444 (1980).
[4] J.-G. Cao and E. A. Meighen, *J. Biol. Chem.* **264,** 21670 (1989).
[5] S. R. Chhabra, P. Stead, N. J. Bainton, G. P. C. Salmond, G. S. A. B. Stewart, P. Williams, and B. W. Bycroft, *J. Antibiot.* **46,** 441 (1993).
[6] J. P. Pearson, K. M. Gray, L. Passador, K. D. Tucker, A. Eberhard, B. H. Iglewski, and E. P. Greenberg, *Proc. Natl. Acad. Sci. U.S.A.* **91,** 197 (1994).

0076-6879/00 $30.00

methionine.[7] At present, however, chemical synthesis of autoinducers is far more practical than enzymatic synthesis.

Autoinducers and Autoinducer Analogs

A large number of autoinducers and their analogs have been synthesized chemically by us.[8] Most of the classes of compounds reported on here have been made by several different routes with many minor modifications. Even closely related compounds were often made in slightly or even totally different ways. In order not to fill these pages with highly detailed instructions for the synthesis of each and every compound that we have made, we have attempted to generalize the description of the procedures for related types of compounds. The reader should thus be aware that a given compound may not, in fact, have been made by us using the procedures given here. We see no reason, however, that our descriptions should not work in general.

Essentially none of the synthetic schemes reported on here has been optimized for yield of product. This was mainly due to the fact that, for our purposes, the yield was largely irrelevant: Only such tiny amounts of a compound are needed to test its activity in a given system that we concentrated on purifying the products by preparative high-performance liquid chromatography (HPLC) and stopped collecting when we had obtained enough for analysis by nuclear magnetic resonance (NMR) and infrared (IR), plus a little more for our specific purposes. We then simply discarded the remainder of the impure material and thus rarely ever determined the yield that could have been obtained by the method.

The natural autoinducers that have been identified to date are 3-oxoacyl-HLs, 3-hydroxyacyl-HLs, and simple acyl-HLs. Interestingly, no α, β-unsaturated acyl-HLs have been identified so far. All of these, except the 3-oxo compounds, can be synthesized readily by directly coupling the acyl moiety to HL · HBr. We give examples of such syntheses later. The problem with 3-oxo compounds is that 3-oxo carboxylic acids and carboxylates decarboxylate readily. This problem can be circumvented by protecting the ketone group through formation of the ethylene glycol ketal, coupling with HL · HBr, and finally deprotecting. Alternatively, the 3-oxo group can be made in an entirely different way by using the Meldrum's acid method.[9] We have found the former method to be more useful for large-scale syntheses

[7] M. I. Moré, L. D. Finger, J. L. Stryker, C. Fuqua, A. Eberhard, and S. C. Winans, *Science* **272,** 1655 (1996).

[8] A. Eberhard, C. A. Widrig, P. McBath, and J. B. Schineller, *Arch. Microbiol.* **146,** 35 (1986).

[9] Y. Oikawa, K. Sugano, and O. Yonemitsu, *J. Org. Chem.* **43,** 2087 (1978).

whereas the latter seems better for small-scale preparations. Examples of both approaches are given.

Storage of Analogs

Solid samples of the analogs can be stored safely under refrigeration with desiccation. We routinely store dilute solutions (≤ 5 mM) of the compounds in ethyl acetate or acetonitrile. Methanol should be avoided due to the risk of ester exchange with opening of the lactone ring. We prefer ethyl acetate to acetonitrile because it is the easier solvent to remove under a stream of dry nitrogen gas. The solutions are kept in screw-cap culture tubes with PTFE-lined caps, with further sealing using Parafilm, either in the refrigerator or in the freezer. Great care should be taken to warm the samples to room temperature before opening in order to prevent condensation of water into the tubes. Water can lead to hydrolytic opening of the lactone ring. Addition of a trace of glacial acetic acid (10 μl per liter) to the ethyl acetate seems to reduce the chance of ring opening. Our analogs have been stable for decades under these conditions.

Synthesis of 3-Oxoacyl-HLs by the Meldrum's Acid Method

We present a general scheme (Fig. 1) that we have used for the synthesis of various 3-oxoacyl-HLs. Finely ground and dried 4-dimethylaminopyridine (DMAP; 0.01 mol) and Meldrum's acid (2,2-dimethyl-1,3-dioxane-4,6-dione; 0.005 mol) are dissolved under nitrogen in 10 ml of dry tetrahydrofuran (THF). The solution is cooled in an ice bath, and 0.005 mol of the desired

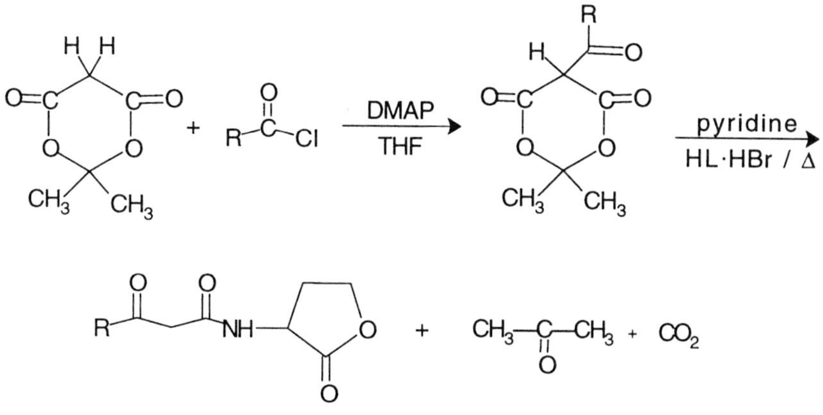

FIG. 1. Synthesis of 3-oxoacyl-HLs by the Meldrum's acid method.

acid chloride is added dropwise with stirring. The ice bath is removed and the mixture is allowed to stir at room temperature overnight. Methylene chloride (CH_2Cl_2; 30 ml) or α,α,α-trifluorotoluene is added and the mixture is extracted repeatedly with 50 ml of 1 *M* HCl until the extract is acidic and then repeatedly with water until the extract is near neutrality. The organic layer is dried with $MgSO_4$ and evaporated to dryness using a rotary evaporator, giving the crude acylated Meldrum's acid intermediate. For compounds with long alkyl chains, the crude product can sometimes be recrystallized from THF–hexane mixtures. An ^{1}H-NMR spectrum of the crude product should show the methyl groups derived from Meldrum's acid at about $\delta = 1.7$ as a singlet and the methylene group that was α to the carbonyl group of the acid chloride at about 3.0 as an unresolved doublet of doublets in addition to the remainder of the resonances derived from the alkyl portion of the acid chloride. The C–H group between the three carbonyl groups is usually completely enolized and thus is not recognizable.

The acylated Meldrum's acid intermediate is dissolved in 15 ml of dry pyridine and is then allowed to react under reflux for 1 hr with 0.005 mol of dried and finely ground DL-α-amino-γ-butyrolactone hydrobromide [3-amino-2-oxotetrahydrofuran hydrobromide or homoserine lactone hydrobromide (HL · HBr)]. After rotary evaporation to remove the pyridine, the residue is extracted several times with ethyl acetate, the evaporation of which gives the crude desired product. For compounds with long alkyl chains, the crude product can sometimes be recrystallized from water–acetonitrile mixtures, otherwise the product is purified by preparative HPLC on an octadecylsilane (ODS) column using water–acetonitrile as the solvent.

The synthesis can also be performed without the isolation of the acyl Meldrum's acid intermediate, but in our hands the products were purer and the yields were higher if the intermediate was isolated as described earlier. Meldrum's acid (0.02 mol) is allowed to dissolve in 25 ml of methylene chloride, and 3.2 ml (0.04 mol) of pyridine is added. To the cooled mixture under nitrogen, 0.022 mol of the acid chloride is added. The mixture is stirred for 1 hr at ice temperature and for 1 hr at room temperature and is then subjected to rotary evaporation. Pyridine (150 ml) and 0.02 mol of HL · HBr are added and the mixture is allowed to reflux for 1.5 hr. The pyridine is removed by rotary evaporation and the residue is extracted with ethyl acetate. The soluble fraction is purified by preparative HPLC.

The final product can be purified by passage through a silica gel column with elution by a suitable solvent such as acetone or ethyl acetate/hexane before resorting to preparative HPLC. We have successfully experimented in some cases with the use of solid-phase extraction cartridges (BondElut) to effect a preliminary purification of the acyl Meldrum's intermediate or

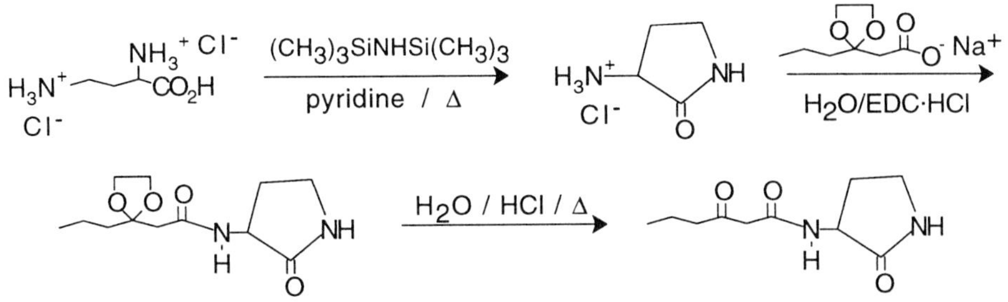

FIG. 2. Synthesis of the lactam analog of the *V. fischeri* autoinducer, 3-oxo-*N*-(2-oxo-3-pyrrolidinyl)hexanamide, using EDC coupling.

of the final product, but care must be taken, particularly with relatively more polar compounds, that the compound is actually being retained. We have also found that for low molecular weight or more polar compounds, the aqueous extracts during the workup of the intermediate must be minimized or replaced by rotary evaporation (using pyridine in place of DMAP and methylene chloride in place of THF). However, for high molecular weight nonpolar compounds, it is often advisable to extract the final products with chloroform in place of ethyl acetate.

Synthesis of 3-Oxoacyl-HLs or of Simple Acyl-HLs by Coupling with 1-Ethyl-3-(3-dimethylaminopropyl)carbodiimide Hydrochloride (EDC)

An example involving coupling with EDC[10] is the synthesis of 3-oxo-*N*-(2-oxo-3-pyrrolidinyl) hexanamide (Fig. 2): We give directions first for the cyclization of the commercially available 2,4-diaminobutanoic acid to the lactam analog of HL, then for an EDC coupling reaction with a ketal-protected 3-oxocarboxylate, and finally for the deprotection of the ketal. The coupling reaction can also be accomplished using "Woodward's reagent" (*N*-ethyl-5-phenylisoxazolium-3'-sulfonate).[3] Presumably the desired product could also be obtained by reaction of the lactam with the Meldrum's acid intermediate (see earlier discussion).

Cyclization to the Lactam

Using dried apparatus, to 20 ml of dry acetonitrile is added 0.70 g (0.0037 mol) of 2,4-diaminobutanoic acid dihydrochloride (Aldrich Chemical Co.),

[10] J. C. Sheehan and J. J. Hlavka, *J. Org. Chem.* **21,** 439 (1956).

0.30 ml (0.0037 mol) of pyridine, and 7.7 ml (0.0037 mol) of hexamethylene disilazane. The mixture is heated under reflux for 1 day after which it is poured into 50 ml of methanol and subjected to rotary evaporation. The residue is extracted with chloroform and the soluble portion is evaporated again, treated with 10 ml of 0.5 *M* HCl, and evaporated to dryness, giving the 2-aminobutyrolactam hydrochloride (3-amino-2-oxopyrrolidine hydrochloride).

Synthesis of Ethylene Glycol Ketal of Sodium 3-oxohexanoate

Commercially available ethyl butyrylacetate (ethyl 3-oxohexanoate; 20.4 ml, 0.13 mol) and ethylene glycol (0.89 mol, 22 ml) are refluxed with 0.23 g of *p*-toluenesulfonic acid in 500 ml of benzene for 28 hr. Water evolution is monitored using a Dean-Stark trap. After 2.4 ml (0.13 mol) of water has collected in the trap, the benzene solution is washed three times with 50 ml of saturated $NaHCO_3$ solution, once with saturated NaCl solution, and dried over $MgSO_4$. The benzene is removed *in vacuo* with a rotary evaporator to yield a pale yellow oil. This procedure is repeated four more times. The combined pale yellow oil is vacuum distilled and the fraction between 113 and 120° (1 mm Hg) is collected, giving 100 g (0.5 mol) of the ketal. The ketal is transferred to a 500-ml round-bottom flask (rbf) containing 100 ml of 5 *M* NaOH and refluxed for 1 hr. The solution is cooled and neutralized before washing three times with diethyl ether. The water is removed at 30° *in vacuo* with a rotary evaporator, equipped with an oil pump, to give the sodium salt as a white solid. Other protected salts of 3-oxocarboxylates can be made similarly, but starting with the corresponding ethyl 3-oxo ester, made by treating monoethyl malonic acid and an acid chloride with 2 mol of butyl lithium,[11] then proceeding as just described.

EDC Coupling

The 2-aminobutyrolactam hydrochloride (0.38 g, 0.0028 mol), or any other amine hydrochloride of pK_a about 7 or higher, is dissolved in 10 ml of water and the pH is adjusted to 7. The ethylene glycol ketal of sodium 3-oxohexanote (0.55 g; 0.0028 mol), or any other lithium, potassium, or sodium carboxylate, and 0.64 g (0.0034 mol) of EDC are added. We have found that EDC methiodide gives more satisfactory results than EDC hydrochloride. After reaction overnight, an excess of ethanol is added, giving a precipitate that is removed by centrifugation. After rotary evapora-

[11] H. Kaplan, A. Eberhard, C. Widrig, and E. P. Greenberg, *J. Labelled Compds. Radiopharm.* **22,** 387 (1985).

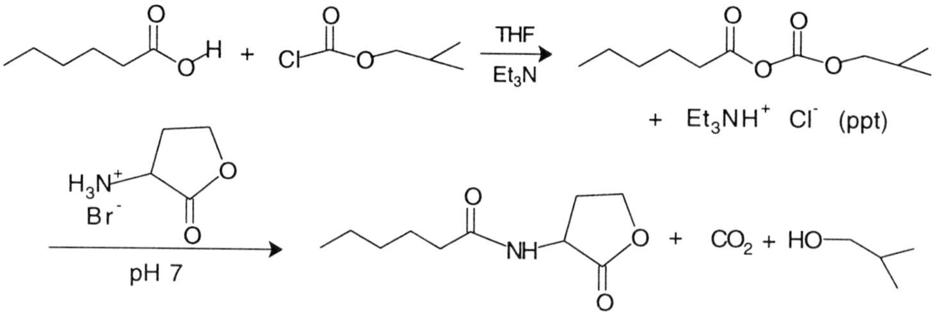

FIG. 3. Synthesis of hexanoyl-HL, a simple acyl-HL, by the chloroformate method.

tion, the residue is treated with ethyl acetate, clarified by centrifugation, and the solvent is evaporated again, giving the impure coupled protected compound, which is purified by preparative HPLC using an ODS column and eluting with 1 : 1 water : methanol or water : acetonitrile.

Deprotecting

The coupled protected compound is deprotected by refluxing for 20 min in 0.1 *M* HCl followed by rotary evaporation and preparative HPLC with the same system as before.

Synthesis of Simple Acyl-HLs by the Chloroformate Method (Fig. 3)[12]

Simple acyl-HLs can be prepared from the corresponding carboxylic acid and HL · HBr as follows: Tetrahydrofuran and triethylamine should be dried with 4-Å molecular sieves and the glassware should be flame-dried before use. A 50-ml rbf with a magnetic stir bar is fitted with a septum and flushed gently with dry nitrogen gas. Using a syringe, 1.03 ml (1.0 g; 0.0086 mol) of hexanoic acid is added with stirring, followed by 15 ml of THF, 1.35 ml of isobutylchloroformate (1.05 g; 0.0086 mol), and 1.38 ml (0.73 g; 0.0086 mol) of triethylamine. A white precipitate forms. The mixture is allowed to react with continued stirring under nitrogen for 1 hr. The slurry is then centrifuged to remove the precipitate. A solution of 1.56 g (0.0086 mol) of HL · HBr in 20 ml of water is carefully brought to pH 7 with 6 *M* NaOH (the pH should not be allowed to exceed 7). The supernatant from the centrifugation is added dropwise with stirring to this HL

[12] J. R. Vaughan, Jr., and R. L. Osato, *J. Am. Chem. Soc.* **74,** 676 (1952).

solution over about 20 min while the pH is kept at or slightly below 7 with the dropwise addition of 2 *M* NaOH. The reaction is allowed to proceed for 1 hr at room temperature, with the continued addition of 2 *M* NaOH as needed to keep the pH at 7. Small (5 ml) portions of THF are added if insoluble material begins to form. The resulting solution is passed through a 1 × 15-cm Dowex-50-Na^+ cation-exchange column, eluted with water into a 50-ml rbf, and evaporated with a rotary evaporator. The product can be purified by preparative HPLC with a 1 × 25-cm reverse-phase ODS column using 60% water : 40% acetonitrile (v : v) or purified more crudely by adsorption to a MegaBondElut ODS cartridge, washing with water, and then eluting with acetonitrile.

Such simple acyl-HLs as well as 3-hydroxyacyl-HLs, such as the *V. harveyi* autoinducer, can also be made by coupling of the sodium carboxylate with HL · HBr using EDC (see earlier discussion). The 3-hydroxyacyl groups are readily accessible through the Reformatsky reaction.[13] We have not yet tried preparative oxidation of 3-hydroxyacyl-HLs as an easy route to give 3-oxoacyl-HLs, although we do have evidence that 3-hydroxyhexanoyl-HL slowly oxidizes spontaneously to give 3-oxohexanoyl-HL.

Coupling with Dicyclohexylcarbodiimide (DCC)

Occasionally, one may wish to make an autoinducer analog of which either the analog of the HL ring or the acyl group is not soluble in water, thus making use of the EDC method impossible. In such cases, coupling with DCC in pyridine solution can be successful.[14] We give as an example the synthesis of 3-oxo-*N*-(2-oxo-3-oxazolidinyl) hexanamide (Fig. 4A): To 20 ml of pyridine is added 0.5 g (0.0028 mol) of the ethylene glycol ketal of lithium 3-oxohexanoate, 0.028 g (0.0014 mol) of 3-amino-2-oxazolidinone sulfate (Aldrich Chemical Co.), and 0.71 g (0.0034 mol) of DCC. After reaction overnight, a precipitate is removed by centrifugation, 40 ml of water is added to the supernatant, and another precipitate is removed by centrifugation. The supernatant is extracted several times with 20 ml of ether followed by removal of the ether using a rotary evaporator. The residue is treated successively with ethanol and ethyl acetate, with removal of any precipitate by centrifugation, followed by rotary evaporation. The crude protected compound thus obtained is deprotected as described ear-

[13] R. L. Shriner, *in* "Organic Reactions" (R. Adams, W. E. Bachmann, L. F. Fieser, J. R. Johnson, and H. R. Snyder, eds.), Vol. I, pp. 1–37.

[14] J. C. Sheehan and G. P. Hess, *J. Am. Chem. Soc.* **77,** 1067 (1955).

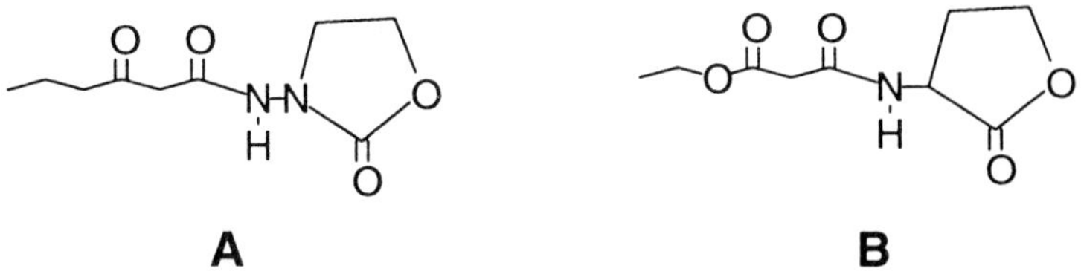

FIG. 4. Structures of 3-oxo-*N*-(2-oxo-3-oxazolidinyl)hexanamide (A) and of the 4-oxa analog of the *V. fischeri* autoinducer, 2-ethoxycarbonyl-*N*-(tetrahydro-2-oxo-3-furanyl)ethanamide (B).

lier. The product is purified by preparative HPLC using 1 : 1 water : methanol.

Large-Scale Synthesis of 3-Oxohexanoyl-HL

The ethylene glycol ketal of sodium 3-oxohexanoate (4.3 g, 0.022 mol; see earlier) is coupled in 40 ml of water with HL · HBr (3.64 g, 0.20 mol) using EDC (3.83 g, 0.20 mol) as described earlier. The resulting ketal of 3-oxohexanoyl-HL is deprotected as described earlier giving the crude 3-oxohexanoyl-HL as a thick oil.

Column Chromatography and Assay

The oil is dissolved in 5 ml of ethyl acetate, loaded onto a 2.5 × 40-cm silica gel column (grade 62, 60–200 mesh, 150 Å), and eluted with 250 ml of 1 : 1 ethyl acetate : hexane followed by 500 ml of ethyl acetate, collecting 10-ml fractions. The fractions are assayed using *Escherichia coli* HB101(pJHD500/pVFR901), which bears the bioluminescent genes from *V. fischeri* on a plasmid, but is deficient in autoinducer biosynthesis,[15] with the naturally autoinducer-deficient *V. fischeri* strain B-61,[16] or any other similar assay organism that has been designed to contain the gene for the desired receptor protein (i.e., *luxR*) along with the promoter region and a suitable reporter gene or genes (i.e., *luxA* and *luxB* or *lacZ*), but that is missing, or is defective in, the autoinducer synthase gene (i.e., *luxI*).[15] Aliquots (25 μl) are removed from each fraction, and the ethyl acetate is evaporated with a stream of N_2 gas. A 3-ml aliquot of a 100-fold diluted overnight culture of the assay organism is added. At an OD_{600} of 0.5, the light output of each fraction is read with a luminometer. The active fractions are pooled, and the ethyl acetate is removed by rotary evaporation. The

[15] J. H. Devine and G. S. Shadel, *Methods Enzymol.* **305** [20] 2000 (this volume).
[16] K. H. Nealson, *Arch. Microbiol.* **112,** 73 (1977).

resulting clear oil (3.88 g, 91%) is stored at 4°. The oil crystallizes slowly to a greasy white solid overnight. The solid is stored at −20 or −80°.

Other Analogs

The analog with an oxygen in place of a methylene group in position 4 of the acyl group of the *V. fischeri* autoinducer (see Fig. 4B) has been found to be almost as active as the natural autoinducer. This compound can be made readily by EDC coupling, as described previously, of the commercially available potassium ethyl malonate with HL · HBr. For higher homologs, the corresponding half esterified malonate must be synthesized from malonic acid and the alcohol.

For analogs with ring systems other than the HL ring, we have generally restricted ourselves to using commercially available amines having the desired ring (but see earlier discussion for synthesis of the lactam analog). Because we have concentrated on making analogs for testing with the luciferase system of *V. fischeri,* we have always combined such ring systems with the 3-oxohexanoyl acyl group. We have on hand a large quantity of the ethylene glycol ketal of sodium 3-oxohexanoate, and thus we have coupled this to the amines using the EDC reaction, as described earlier, followed by deprotection. In some cases, where solubility considerations did not allow reaction in water, the ethylene glycol ketal of lithium 3-oxohexanoate was used with the desired amine hydrobromide in pyridine solution with coupling by DCC (see earlier discussion). Amines with low pK_a values tended not to give the desired coupled products, particularly when using the EDC method.

Analogs derived from dicarboxylic acids and thus having two HL rings were made by methods similar to those used for mono-HL compounds. The larger non-3-oxo analogs proved to be quite insoluble in most solvents, thus resulting in very low yields. A mixture of dimethyl sulfoxide and chloroform was a useful solvent for these compounds in some cases. The D (or "R") enantiomers of the naturally occurring L-(or "S") 3-oxohexanoyl-HL can be made by first cyclizing the commercially available D-homoserine to D-HL by refluxing in 2 *M* HCl for 2 hr. After rotary evaporation, the solid is redissolved in water and acetone is added until the D-HL precipitates.

Spectral Characteristics

Infrared spectra (as KBr pellets) of HLs show the lactone peak at about 1778 cm^{-1} and the amide peaks at about 1650 and 1530 cm^{-1}. The 3-oxo group shows a peak at 1720 cm^{-1} (Fig. 5).

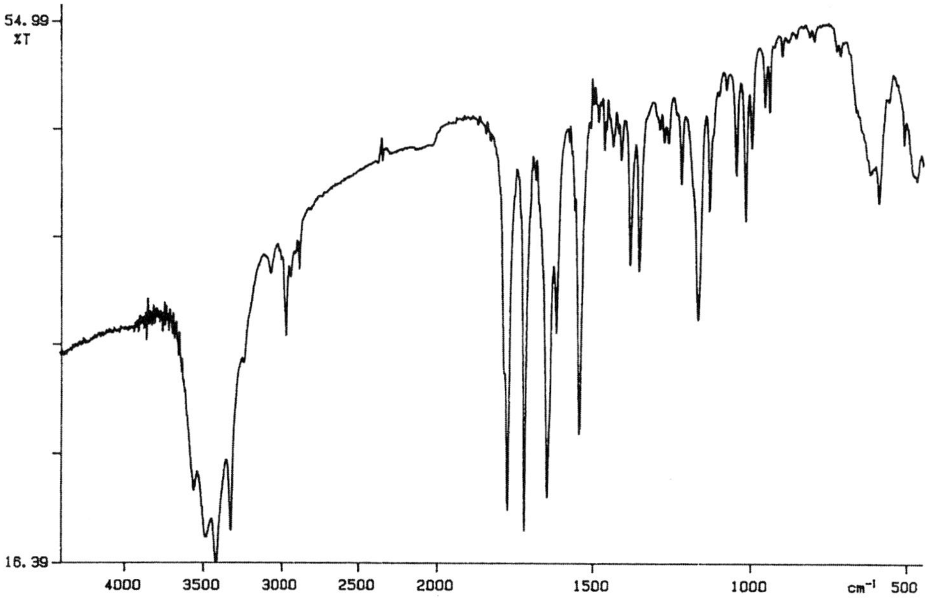

FIG. 5. Infrared spectrum of the *V. fischeri* autoinducer, 3-oxohexanoyl-HL. KBr die.

^{1}H-NMR spectra (in $CDCl_3$, Fig. 6) show homoserine lactone β-hydrogens as multiplets at $\delta = 2.2$ and 2.8 (1 H each), and the α and γ hydrogens also as multiplets at 4.3, 4.45, and 4.6 (1 H each). Acyl chains show the usual signals at δ = 0.9 (3 H, triplet) for the terminal methyl group, 1.3 (multiplet) for methylene groups more than three carbons away from a carbonyl group, 1.65 (2 H, quintet) for the methylene two carbons from a

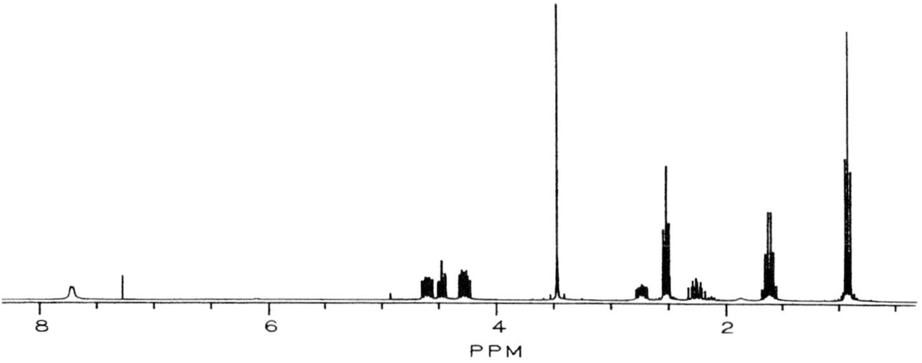

FIG. 6. ^{1}H-NMR spectrum of the *V. fischeri* autoinducer, 3-oxohexanoyl-HL, in $CDCl_3$.

carbonyl group, and 2.6 (2 H, triplet) for the methylene next to the ketone group in 3-oxo compounds or at 2.3 for simple acyl-HLs. For 3-oxo compounds, there is also a singlet at $\delta = 3.5$, which disappears due to exchange if the spectrum is taken in D_2O and a broad N-H signal at $\delta \approx 6.3$ (1 H), which also disappears in D_2O. If the sample has not been dried adequately, a broad singlet at about $\delta = 1.5$ will be present. All peaks are referenced to internal TMS ($CDCl_3$) or, for spectra taken with D_2O, the HOD peak was set to $\delta = 4.7$.

^{13}C-NMR spectra (Fig. 7) show a peak at 178 ppm for the carbonyl group of the HL ring, peaks at 48, 26, and 66 ppm for the α, β, and γ carbons, respectively, and a peak at 168 ppm for the carbonyl of the amide group. For 3-oxo compounds, the additional carbonyl is at 208 ppm, whereas the 2-methylene group (between the two carbonyls) has a peak at 48 ppm and the 4-methylene group at 44 ppm. For the remainder for the acyl chain the methyl group has a peak at 12 ppm, and any remaining methylene groups have their peaks between 15 and 40 ppm. All peaks are referenced to chloroform resonances at 77 ppm.

The 3-oxoacyl-HLs and higher molecular weight analogs may not give mass spectra using a GC-MS apparatus. For these, a direct-inlet probe, FAB-MS or other mass spectral methods can be used. Lower molecular weight simple analogs do give satisfactory CG-MS results. The parent ions are usually evident. In 3-oxo compounds, there is usually a small peak corresponding to the entire acyl group (m/e = 113 in Fig. 8) and a large one for cleavage between carbons number 2 and 3. There are also large peaks corresponding to HL and to its protonated form (m/e = 100 and 101 in Fig. 8) and smaller peaks from the loss of ethylene and of carbon monoxide (m/e = 186 in Fig. 8).

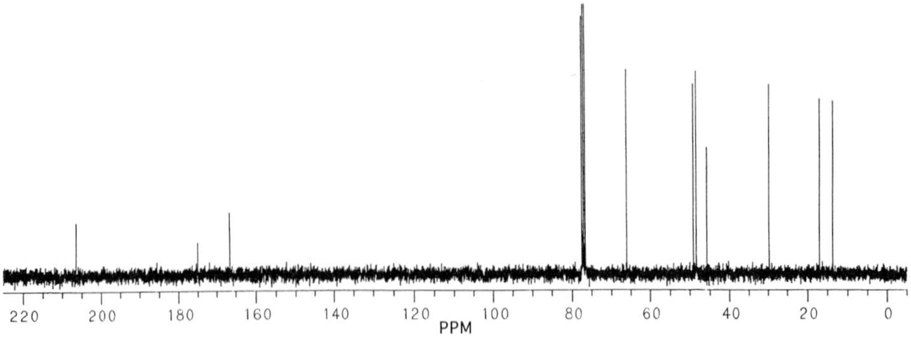

FIG. 7. ^{13}C-NMR spectrum of the *V. fischeri* autoinducer, 3-oxohexanoyl-HL, in $CDCl_3$.

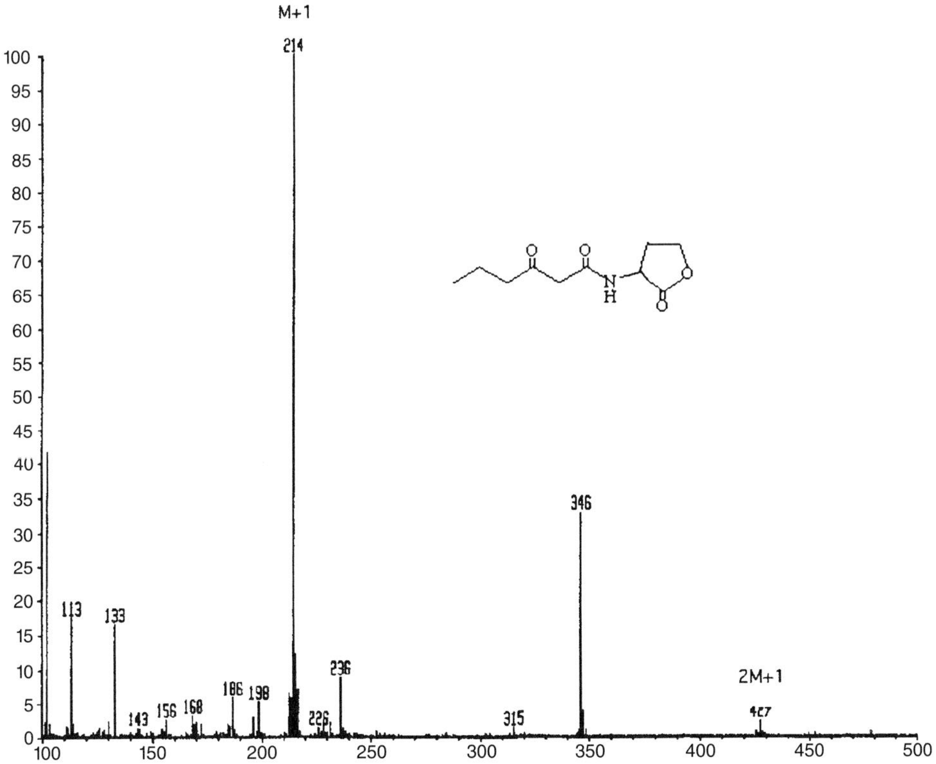

FIG. 8. FAB-MS of 3-oxohexanoyl-HL using a thioglycerol matrix. The peak at 214 represents the $(M + H)^+$ parent ion peak.

Assay of Analogs

Analogs can be tested using any number of assay organisms, such as those described earlier. We have been using *V. fischeri* strain B-61[16] to test analogs for their ability to act as agonists or antagonists of 3-oxohexanoyl-HL. B-61 makes only little, if any, of its own 3-oxohexanoyl-HL, but it responds to the added autoinducer with vigorous light production. This strain is easy to maintain and to work with. Because it grows in sea water or other high salt media, the risk of contamination is low and it requires no antibiotics. It can be grown in sea water or artificial sea water with 5 g/liter of tryptone or peptone, 3 g/liter of yeast extract, and 3 g/liter of glycerol with good aeration at room temperature. It can be maintained for months on slants of the same medium with 14 g/liter of agar, at 15°, or

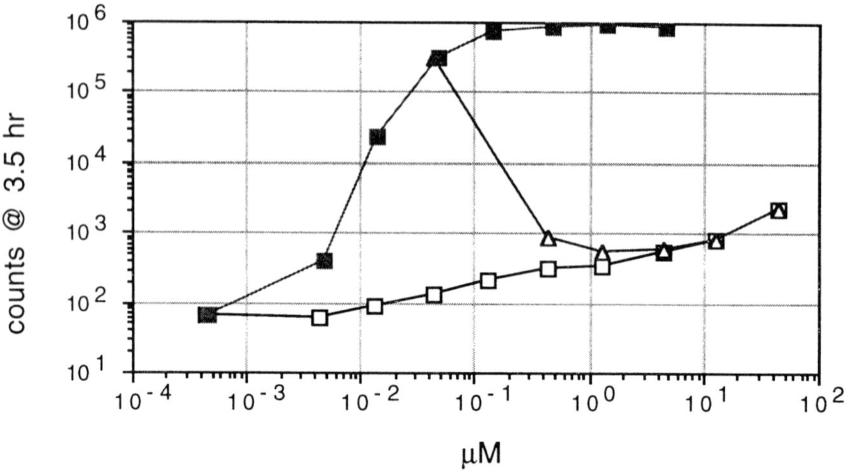

FIG. 9. Assay of octanoyl-HL with and without 3-oxohexanoyl-HL using *V. fischeri* B-61 in "NG" medium. ■, 3-oxohexanoyl-HL; □ octanoyl-HL; and △, 0.047 μM 3-oxohexanoyl-HL plus the amounts of octanoyl-HL indicated on the *X* axis. From J. B. Schineller, 1985, unpublished.

frozen in the same medium in liquid nitrogen and stored at −70°. A sample agonist/antagonist assay using B-61 follows (Fig. 9).

For the assay, an overnight culture of B-61 is diluted into fresh medium. When the OD_{660} has reached 0.08 to 0.1, 0.5 ml of it is put into 50 ml of "no growth" (NG) medium at pH 7 (0.3 *M* NaCl, 0.05 *M* $MgSO_4 \cdot 7H_2O$, and 0.04 *M* potassium phosphate buffer). The NG medium with the B-61 is dispensed (0.5 or 1.0 ml) at timed intervals into glass tubes containing the evaporated ethyl acetate solutions of the compounds to be tested. One set of tubes contains just various concentrations of the 3-oxohexanoyl-HL as a control; another set contains various concentrations of the compound to be tested for its agonist activity; and a third contains a fixed amount of 3-oxohexanoyl-HL with increasing amounts of the compound to be tested for its antagonist activity. Light production in the tubes is measured with a luminometer at timed intervals (see Fig. 9).

As Fig. 9 shows, light production is generally not proportional to autoinducer concentration. At first glance, the curve showing the combined effect of 3-oxohexanoyl-HL and octanoyl-HL seems to indicate that 0.44 μM octanoyl-HL has decreased light production and, by inference, autoinducer activity by over two orders of magnitude, from 2.5×10^5 to 8×10^2 counts, or to about 0.3% of the original activity. In fact, however, close inspection of the figure indicates that 0.44 μM octanoyl-HL decreases light production by 0.047 μM 3-oxohexanoyl-HL to a level about equivalent to that produced by 0.006 μM 3-oxohexanoyl-HL or to about 13% of the original activity. Also note that the inhibition by octanoyl-HL at higher concentrations is masked by its own activity as shown by the merging of the curves with and without 3-oxohexanoyl-HL. This is found to be generally true for any compound having both agonist and antagonist activity.

Acknowledgment

We thank Carolyn Eberhard for a critical reading of the manuscript.

[23] Overexpression of Foreign Proteins Using the *Vibrio fischeri lux* Control System

By MICHAEL D. THOMAS and ANITA VAN TILBURG

Introduction

Prokaryotic protein expression systems, particularly those based on *Escherichia coli*, compare favorably with yeast, insect, mammalian, and plant systems[1–3] for the expression of heterologous proteins. As many posttranslationally modified eukaryotic proteins are fully functional when not glycosylated, using prokaryotic systems for the expression of such proteins is becoming more common.[4] Advantages of using bacterial systems include (i) short generation time, (ii) achievement of extremely high biomass densities, (iii) ease of protein extraction, (iv) the ability to eliminate specific contaminating proteins by genetic means, and (v) lower cost of growth medium. In addition, use of microorganisms in biological research has greater public acceptance than does use of higher organisms for experimental purposes.

Protein Expression in *E. coli*

Culture Conditions

Within *E. coli* systems, one needs to consider the scale of protein production, promoter and vector characteristics, host genotype, and culture condi-

[1] G. Hanning and S. C. Makrides, *Trends Biotechnol.* **16,** 54 (1998).
[2] M. J. Weickert, D. H. Doherty, E. A. Best, and P. O. Olins, *Curr. Opin. Biotechnol.* **7,** 494 (1996).
[3] M. H. Marino, *BioPharm.* **2,** 18 (1989).
[4] S. C. Makrides, *Microbiol. Rev.* **60,** 512 (1996).

0076-6879/00 $30.00

tions for optimal expression. This article considers only research-scale protein expression. Bacteria are grown in shaker flask culture, gene expression is induced, and the cells are harvested. Commercial production normally utilizes fed-batch culture techinques in which the culture conditions are monitored continuously and nutrients and/or amendments are supplied to the culture on demand or according to an established program during incubation. Although fed-batch culture techniques can easily result in an enormous increase in protein production over shaker flask culture, research needs seldom justify the expense of a bioreactor or the time required to optimize fermentation culture conditions. Common dry weight yields for *E. coli* cells grown by various methods have been reported as 1–2 g/liter shaker flask culture, 10 g/liter bioreactor (batch culture), and more than 100 g/liter fed-batch culture.[5] In some cases, even higher yields can be obtained for batch culture (16.4 g/liter)[6] and fed-batch culture (174 g/liter).[7] The dry weight yield for fed-batch cultures can approach the theoretical maximum for *E. coli* grown in liquid culture, i.e., 200 g/liter.[7]

Shaker flask culture has a number of advantages over the other methods, including simplicity of experimental design, minimal capital outlay and recurring expenses, and few requirements for culture monitoring. Using the shaker flask culture conditions described in this article, we routinely achieve more than 6 g (dry weight) of cells per liter for most vector/strain combinations incorporating the *luxI* promoter [P*luxI,* also referred to as P_R (right promoter); Fig. 1]. If the recombinant protein comprises 50% of the total cellular protein, then the yield of recombinant protein will be greater than 1 g per liter culture medium. This level of expression is adequate for many research needs.

Promoter Characteristics

Vectors incorporating the *luxI* promoter of *Vibrio fischeri* appear to be among the best options available for the expression of foreign proteins in *E. coli.* The bioluminescence genes of *V. fischeri* are under the control of a small, freely diffusible molecule, *N*-(3-oxohexanoyl)-L-homoserine lactone,[8] which is called autoinducer (AI) because its presence stimulates its own production.[9] This regulation of bioluminescence involves cell-to-cell

[5] W. A. Knorre, W. D. Deckwer, D. Korz, H. D. Pohl, D. Riesenberg, A. Ross, E. Sanders, and V. Schulz, *Ann. N.Y. Acad. Sci.* **646,** 300 (1991).

[6] L. Strandberg, L. Andersson, and S. O. Enfors, *FEMS Microbiol. Rev.* **14,** 53 (1994).

[7] H. Markl, C. Zenneck, A. C. Dubach, and J. C. Ogbonna, *Appl. Microbiol. Biotechnol.* **39,** 48 (1993).

[8] A. Eberhard, A. L. Burlingame, C. Eberhard, G. L. Kenyon, K. H. Nealson, and N. J. Oppenheimer, *Biochemistry* **20,** 2444 (1981).

[9] K. H. Nelson, T. Platt, and J. W. Hastings, *J. Bacteriol.* **104,** 313 (1970).

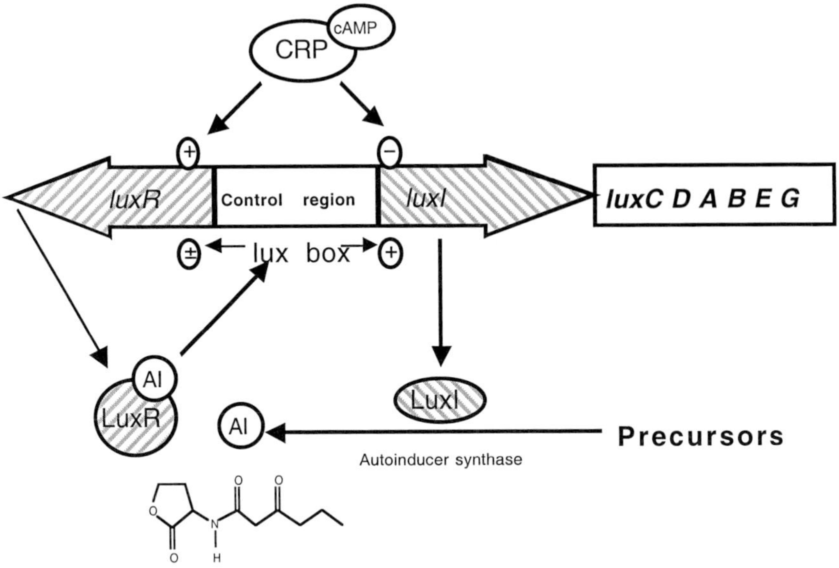

FIG. 1. Diagramatic representation of the *V. fischeri lux* control system. See text for details.

communication between bacteria, and *V. fischeri* provides a model system for quorum sensing regulation in a large number of pathogenic bacteria. Figure 1 presents the major features of autoinduction regulation. Briefly, the signaling molecule in *V. fischeri, N*-(3-oxohexanoyl)-L-homoserine lactone, is a positive regulator of its own synthesis. AI is produced at basal levels and accumulates both intra- and extracellularly as the bacterial population density increases. As the AI concentration increases, the chemical equilibrium shifts, favoring formation of a complex with the transcription factor LuxR. The LuxR:AI complex acts on a 20-bp palindromic sequence (the lux box) within the control region and induces expression of *luxI,* the gene encoding the biosynthetic enzyme for AI, as well as genes downstream of *luxI.* The concentration of AI rises even higher, leading to a further induction of genes controlled by LuxR:AI.

In *V. fischeri* there is evidence for regulatory mechanisms that prevent runaway induction. One negative regulatory system, also mediated by a homoserine lactone derivative, represses autoinduction even in the presence of LuxR:AI.[10,11] The induction mediated by LuxR:AI stimulates transcription from both the left and the right operons at low levels of LuxR and

[10] A. Eberhard, C. A. Widrig, P. McBath, and J. B. Schineller, *Arch. Microbiol.* **146,** 35 (1986).
[11] A. Kuo, S. M. Callahan, and P. V. Dunlap, *J. Bacteriol.* **178,** 971 (1996).

AI; at high levels, repression of the left operon is observed. This repression is dependent on a specific sequence in *luxD* that is similar to the lux box.[12] However, when portions of the *luxR–luxI* divergent operons are expressed in *E. coli,* some of these control mechanisms may not be present.

As measured by light output, induction of P*luxI* results in an exponential increase in expression on the order of 10^7-fold over basal levels in only a few hours. This level of expression allows the protein to accumulate to very high levels after induction. However, as the basal level is extremely low, the deleterious effects of toxic gene products on the bacterial cells are minimized prior to induction. For promoter induction, the AI is typically used at 1/1000 the concentration that isopropyl-β-D-thiogalactopyranoside (IPTG) is used, resulting in a very cost-efficient expression system. Although these characteristics of the *lux* system make it eminently suitable for high-level expression in *E. coli,* this regulatory system has not yet been fully exploited for the expression of recombinant proteins. It is not even listed as being used for this purpose in a recent comprehensive review.[4]

Vector Characteristics

Plasmid pJHD500 is an expression vector incorporating the *luxI* promoter; its construction is described by Devine *et al.*[13] A map of plasmid pJHD500 showing several unique restriction endonuclease sites is shown in Fig. 2. In pJHD500, *luxI* is truncated so that AI is not produced by the cells. This allows cultures of bacterial cells to be grown to high densities before induction of the gene of interest occurs by the addition of synthetic AI. This preinduction growth period is important for maximum protein yield, particularly when the foreign gene encodes a protein that is detrimental to the host cell. In pJHD500, the foreign genes expressed are *V. harveyi luxA* and *luxB,* encoding the α and β subunits of bacterial luciferase. Maximal expression of luciferase appears to be increased by the formation of stem loops in the mRNA corresponding to bp 991–1007 of pJHD500 (beginning of *luxI*) and bp 3462–3481 (after *luxB*) that may enhance the stability of the transcripts. The stem loop downstream of *luxB* has characteristics in common with *E. coli* transcription terminators.[14] Maximal P*luxI*-mediated expression appears to require some of the nucleotides of *luxI.* We have had greater success using transcriptional fusions with *luxI,* i.e., insertion of the foreign gene after the truncated *luxI* and downstream of the *luxA* Shine-Dalgarno sequence, than with complete replacement of *luxI.* Densitometry of Coomassie-stained SDS–polyacrylamide gels of clarified

[12] G. S. Shadel and T. O. Baldwin, *J. Bacteriol.* **173,** 568 (1991).
[13] J. H. Devine, G. S. Shadel, and T. O. Baldwin, *Proc. Natl. Acad. Sci. U.S.A* **86,** 5688 (1989).
[14] A. Das, *Annu. Rev. Biochem.* **62,** 893 (1993).

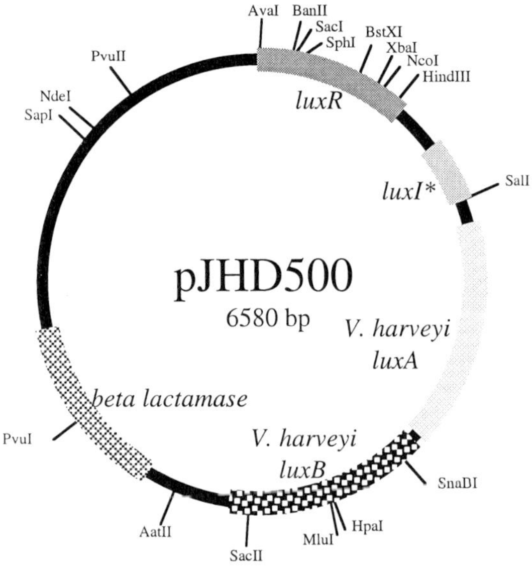

FIG. 2. Plasmid map of pJHD500. Selected unique restriction endonuclease sites are shown. The region containing *luxR* and *luxI** is from *V. fischeri; luxA* and *luxB* are from *V. harveyi;* and the remainder of the plasmid is derived from pBR322. The region containing *luxA* and *luxB* is replaced with the foreign gene(s) to be overexpressed.

lysates from cells containing pJHD500 and grown under inducing conditions shows that the expressed protein can constitute more than 50% of the total soluble proteins in the *E. coli* host cell.

It is possible to construct the expression vector utilizing translational fusion. Translational fusions can be used to increase the stability of a foreign protein or to attach tags, e.g., maltose-binding protein, polyhistidine, or Flag epitope, to facilitate purification of the recombinant protein. However, even if the tag is ultimately cleaved from the protein, translational fusions can alter protein folding and result in proteins that are inactive or active but in a nonnative conformation. The strength of the *luxI* promoter obviates many of the advantages of translational fusions as the proteins are usually produced rapidly enough to minimize degradation and at concentrations sufficiently high to facilitate purification. We have made a number of base change and deletion mutations that reduce basal expression and/or increase induced expression. These mutations can be incorporated on vectors to minimize preinduction protein expression and/or maximize protein expression after induction.

The transcriptional efficiency of many promoters is affected not only by specific *trans*-acting factors, but also by the topology of the promoter region.[15,16] The *lux* control region and the left and right coding regions of both operons contain significant intrinsic bends. Analyzing the expression of various plasmids containing *lux* DNA in topoisomerase mutant strains of *E. coli* demonstrated that both *V. fischeri* promoters respond to variations in DNA topology, which may be mediated by changes in the location of intrinsic bends, changes in cellular superhelicity, or a combination of the two.[17]

Host Genotype and Culture Conditions

Experiments conducted in our laboratory indicate that expression of the *lux* promoters is influenced by intrinsic bends within *luxAB* that are more than 2 kb downstream from the *luxI* promoter. We constructed pSPD43, a derivative of pJHD500, in which the only difference was the replacement of *V. harveyi luxAB* with *luxAB* from *Xenorhabdus luminescens.* The *luxAB* from *X. luminescens* also contains significant intrinsic bends; however, they are found at different locations than those in *V. harveyi* genes. The topological differences associated with the exchange of the *lux* genes had a very dramatic effect on expression. In some *E. coli* strains, expression levels of *V. harveyi* luciferase were similar to those of *X. luminescens* luciferase. However, in other strains, e.g., W3110, there was a greater than 10^5-fold difference in induced luciferase expression from pJHD500 and pSPD43. As the topology of the foreign gene can have a marked influence on its expression, it may be desirable to transform the expression vector into strains of *E. coli* that differ in their average cellular superhelicity and determine empirically in which strain the plasmid is expressed maximally.

The optimal strain can be identified by growing the transformants in small (250-ml) culture flasks. The *luxI* promoter is induced by adding *V. fischeri* autoinducer several hours after the flasks are inoculated. An aliquot of the culture is centrifuged 12 to 18 hr later, and the cells are resuspended in a lysis buffer and sonicated. The crude cell lysate is clarified by centrifugation, and an aliquot is prepared with SDS–PAGE sample buffer for electrophoresis on an appropriate gel. The clarified lysate can also be tested for enzymatic activity corresponding to the protein of interest. Stock cultures of the verified strains should be made and maintained at −80°.

Bacterial culture conditions can have a profound effect on the consistency of expression levels. Flasks rather than test tubes should be used for

[15] J. Pérez-Martín and V. de Lorenzo, *Annu. Rev. Microbiol.* **51,** 593 (1997).

[16] J. Pérez-Martín and M. Espinosa, *Science* **260,** 805 (1993).

[17] A. B. van Tilburg, M. D. Thomas, and T. O. Baldwin, in preparation (2000).

broth cultures. The flasks should contain a small volume of broth and be shaken sufficiently rapidly to ensure aerobic growth conditions. A speed of 300 rpm is appropriate for 50 ml of broth in a 250-ml flask, whereas 350–400 rpm appears to be better for 1–1.5 liters of broth in a 2.8-liter Fernbach flask when shaken in a Psycrotherm shaker (New Brunswick Scientific, Edison, NJ).

Our preferred medium for the production of foreign proteins is Terrific Broth, originally developed for increased plasmid production through support of enhanced cell growth.[18] This medium contains (per liter) 12 g Bacto-tryptone, 24 g Bacto-yeast extract, 4 ml glycerol, 2.3 g KH_2PO_4, and 12.5 g K_2HPO_4. Other digests of casein appear to work equally well, e.g., peptone and enzymatic (low NaCl) casein hydrolysate; the source of yeast extract does not seem to be important. To minimize caramelization, the phosphate buffer is autoclaved separately from the other medium ingredients in one-tenth of the final volume. The final phosphate buffer is 890 m*M* with a pH of approximately 8.

The *V. fischeri* autoinducer is very labile at pH > 7[19]; however, the presence of bacteria appears to protect the autoinducer such that a single addition of AI to these cultures fully induces expression of the heterologous protein. There is little experimental difference between expression levels achieved by dense cultures induced by 100 n*M* (final concentration) and 1 μM AI.[20] In our laboratory we routinely use AI at a final concentration of 400 n*M*.

Experimental Protocols

Vector Construction

A number of factors influence the design and construction of an expression vector. Important attributes include stability of the plasmid and the ease of inserting a foreign gene into a context where it can be expressed at a high level. Plasmid pJHD500 has unique restriction endonuclease sites for *Sal*I and *Aat*II; these allow convenient insertion of a gene downstream of the truncated *luxI*. However, expression of foreign genes inserted at this location has not been optimal in all cases. The reasons for suboptimal expression are not clear but may be due to the three-dimensional structure of the resultant plasmid. We have found that precisely inserting a gene in

[18] K. D. Tartof and C. A. Hobbs, *Bethesda Res. Lab.* **9,** 12 (1987).

[19] A. Eberhard, *J. Bacteriol.* **109,** 1101 (1972).

[20] M. D. Thomas, J. B. Schineller, and T. O. Baldwin, *in* "Bioluminescence and Chemiluminescence" (J. W. Hastings, L. J. Kricka, and P. E. Stanley, eds.), p. 109. Wiley, New York, 1996.

pJHD500 from the Shine-Dalgarno sequence preceding *luxA* to the stop codon of *luxB* gives more consistently reliable expression.

This cloning strategy utilizes oligonucleotide primers with class 2S restriction endonuclease recognition sequences incorporated at their 5′ ends. The gene of interest and pJHD500 are amplified by polymerase chain reaction (PCR) with primers that will be left with complementary single-stranded overhangs after digestion with the appropriate restriction endonuclease. This procedure is similar to a number of earlier PCR mutagenesis/cloning techniques and is a modification of the method of Stemmer and Morris.[21] This procedure does not require that specific indigenous restriction endonuclease recognition sequences be located within the vector and gene sequences, and thus allows precise insertion of the foreign gene at any location. However, the procedure does require PCR amplification of a vector fragment of 4.5 kb. Such amplification may be problematical in some laboratories as it requires the isolation of extremely pure DNA, long PCR extension times, and use of either highly processive thermal-stable DNA polymerases, e.g., *pfu* or Vent, and/or reaction amendments for enhanced amplification.

A number of restriction endonucleases form a four-base overhang outside their recognition site, e.g., *Alw*26I, *Bbs*I, *Bbv*I, *Bsa*I, *Bsm*BI, *Bsm*FI, *Bsp*MI, *Eco*NI, *Fok*I, and *Sfa*NI. Of these, only *Bsp*MI does not cut pJHD500. Therefore, it is the enzyme of choice for digesting pJHD500 to form the empty vector: the recognition sequences of *Bsp*MI are incorporated on the 5′ ends of the primers used to amplify the vector into which the amplified gene of interest will be inserted. We chose the Shine-Dalgarno sequence upstream of *luxA* as one four-base overhang and the stop codon and the next nucleotide of *luxB* as the other four-base overhang. *Bsp*MI or any other class 2S restriction endonuclease that leaves a 5′ four-base overhang can be used to cleave the gene insert as long as the four-base overhangs of the vector and the insert are designed to be compatible.

If a class 2S enzyme that will not cleave the coding region of the gene of interest cannot be found, *Eam*1104I can be used to digest both pJHD500 and the insert. *Eam*1104I is a class 2S enzyme that only recognizes and digests nonmethylated nucleotide sequences. Stratagene (La Jolla, CA) markets a kit for cloning with *Eam*1104I that includes 5-methyldeoxycytosine, allowing PCR amplification of fragments containing the *Eam*1104I recognition sequence within the coding region. *Eam*1104I will not cleave a PCR product incorporating 5-methyldeoxycytosine except within the nonmethylated primer sequences.

[21] W. P. Stemmer and S. K. Morris, *Bio Techniques* **13,** 214 (1992).

```
A.
Start of V. harveyi luxA in pJHD500
                                  MetLysPhe
'5-CGTAATACCAACAAATAA GGAAATGTTATGAAATTT
'3-GCATTATGGTTGTTTATTCCTT ATGTcgtccaTAA
35nt

B.
Start of V. fischeri luxA
                                  MetLys
   '5-AAAacctgcAAAA GGAATAGAGTATGAAG
'3-TTATTTAGTGGTTTTTCCTT ATCTCATACTTC
29nt

C.
End of V. harveyi luxB in pJHD500
 ValLysTyrHisSer
  '5-CAAacctgcTTCG TAACGTTTAACTGATGC
'3-CAGTTTATGAGAAGCATTG CAAATTGACTACG
30nt

D.
End of V. fischeri luxB
    GluLysAsnLeuPro
'5-GAAAAGAATCTACCC TAATAAAATTAAGGGCA
'3-CTTTTCTTAGATGGGATTg TTTTcgtccaCGT
30nt
```

FIG. 3. Primer design to replace *V. harveyi luxAB* in pJHD500 with *luxAB* from *V. fischeri.* Primer sequences are underlined, with single-stranded overhangs double underlined. Primers A and C amplify pJHD500 without *luxAB,* and primers B and D amplify *luxAB* from *V. fischeri.* Upon digestion with *Bsp*MI, primers A and B form compatible single-stranded overhangs as do primers C and D.

Primers designed to clone a gene into pJHD500 are shown in Fig. 3. The primability and stability of these primers were assessed using Amplify, a freeware program for the analysis of PCR primers.[22] PCR products resulting from the use of these primers are digested with *Bsp*MI, generating the compatible single-stranded ends. The vector and gene insert are then purified and ligated. The ligation mixture is transformed into an appropriate host strain, and transformants are analyzed for the desired plasmid restriction maps or for production of the protein of interest. Once the clone has been verified, it can be transformed into a number of *E. coli* strains that

[22] W. R. Engels, *Trends Biochem. Sci.* **18,** 448 (1993).

differ in their ability to express genes under control of the *V. fischeri luxI* promoter.

Detailed Protocol

Prepare template DNA from strains containing the P*luxI* expression vector and the gene encoding the protein of interest. Supercoiled template DNA is amplified efficiently by PCR; however, if the template molecules are carried through the procedure they may result in false-positive colonies during screening. It may be desirable to linearize the template with a restriction endonuclease that cleaves the DNA outside of the region to be amplified. The DNA should be of high purity and free of excess salts. If a resin/chaotropic salts kit, e.g., Qiaprep (Qiagen, Valencia, CA), is used for purification, then the host cell line should contain an *endA* mutation to minimize DNAse carryover and potential DNA degradation during subsequent enzymatic steps. Alternatively, an SDS/alkaline lysis procedure utilizing phenol/$CHCl_3$ extractions can be used. In either case, the wash steps or 70% ethanol rinses in the final stages of purification should be repeated to minimize salt carryover.

To ensure cloning efficiency, it should be verified that the cloning primers produce unique PCR products under the selected amplification conditions. The currently available algorithms for predicting annealing temperatures can be imprecise for long primers that have mismatches at the 5′ end. Therefore, it is recommended that the optimal parameters for amplification of the desired fragments be determined empirically.

The gene encoding the protein of interest is usually short enough to allow direct testing using both cloning primers. Parameters for the longer P*luxI* expression vector are best tested by amplifying short fragments using an additional primer pair. This is accomplished by pairing the cloning primers with primers that match the template perfectly and have been shown to amplify short fragments at elevated annealing temperatures. These preliminary tests do not require the amplification of the whole 4.5-kb vector fragment. Concentrations of the primers, template DNA, and Mg^{2+} in the reaction influence the efficiency of the PCR and the required number of amplification cycles. The primer and template concentrations suggested here work well with the buffers supplied with *Pfu, Taq,* and Vent polymerases. There is usually a wide latitude in effective primer and template concentrations; however, template concentrations should be minimized to reduce the potential for template plasmid carryover.

Although *Taq* polymerase can be used for the PCR, its use increases the possibility of introducing undesirable mutations in the cloned gene and decreases the efficiency of amplifying a full-length 4.5-kb fragment.

Consequently, use of a highly accurate and processive thermal stable DNA polymerase/reaction buffer combination is recommended. There are various products of this type available from a number of suppliers.

Amplification of the P*luxI* vector fragment usually requires a much longer extension time than that required to amplify the gene fragment. Therefore, these reactions are usually performed separately.

1. Set PCR parameters

Initial hold	95°	Denaturation	3 min
20 cycles	95°	Denaturation	60 sec
	40–60°	Annealing	30 sec
	60–75°	Extension	Varies with fragment length

   Annealing temperature is determined empirically; extension temperature is dependent on the polymerase used. Duration of extension is 1 min/500 bases.

2. Set up the following reaction:

Volume	Component	Final concentration
×μl	1 ng (20 fmol) pJHD500 or DNA containing the foreign gene	1 n*M*
0.4 μl	40 m*M* dNTP mix (10 m*M* each)	200 μ*M*
× μl	20 ng each primer (30 nucleotides)	100 n*M*
2 μl	10× PCR buffer	1×
× μl	H_2O to bring volume to 19.6 μl	

   Mix these components by flicking the microfuge tube and add 0.4 μl thermal stable polymerase (1 unit at 2.5 units/μl). Mix by gently pipetting up and down a few times.

   Note: Initial mispriming may lead to the amplification of undesired products. This may be avoided by keeping the reaction mixture on ice until the polymerase is added and then transferring it immediately to a thermal cycler that has already reached 95°. Alternately, the reaction mixture can be placed in the thermal cycler at 95° before the polymerase is added; any of a number of "hot start" products may be used or one can use a thermal-activated polymerase (Perkin-Elmer, Norwalk, CT).

3. Electrophorese 5 μl of the amplified sample on a 1% agarose gel to verify the efficiency of the reaction.
4. Purify the remainder of the sample with Qiaex II (Qiagen) or similar product or by ethanol precipitation.
5. Digest the sample with the class 2S restriction endonuclease specific for the recognition sites incorporated on the primers.
6. Purify the digested sample as in step 4. Resuspend the DNA in a

small volume of TE or H_2O. A concentration of at least 0.5 μM (2 μg/μl for the vector) is optimal for the subsequent ligation step.

7. Ligate the P*luxI* vector fragment with the gene insert fragment in a small volume (10–20 μl) with 1 to 10 Weiss units of T4 DNA ligase, 1× fresh T4 DNA ligase buffer, and a final concentration of DNA 5′ ends of 0.1–1 μM. Ligate at 16° for 4 to 16 hr unless the single-stranded ends of both the P*luxI* vector and the gene insert fragments were generated by the same restriction endonuclease. If the same class 2S enzyme was used for both fragments and the internal sequences are not cleaved by it, then add 2 to 5 units of the restriction endonuclease to the ligation reaction. Ligate for 30–60 min at 37° to prevent ligation of the digested primer fragments from interfering with the construction of the desired vector.
8. Transform or electroporate 0.5 μl of the ligation reaction into highly competent *E. coli* cells. Unless one can routinely prepare competent cells with a transformation efficiency of at least 10^8 transformants/μg DNA, use of commercial competent cells is recommended. Plate the cells out onto solid media containing an appropriate antibiotic.
9. Colonies appearing on selective media should be analyzed immediately by single-colony streaking them to fresh media. A plasmid miniprep procedure should then be performed and the construction of the plasmid verified by restriction mapping. Alternately, cells may be incubated in broth culture, induced with *V. fischeri* autoinducer, and screened by PAGE for production of the protein of interest. Colonies arising from wild-type pJHD500 transformants may be identified by transferring the colonies to solid media containing autoinducer at 100 nM, incubating the plates at 30°, and adding 5–10 μl decanal to the underside of the petri plate lid. These colonies will produce light and are easily observed in a darkroom.
10. After the plasmid construct has been verified, the bacterial culture containing the plasmid should be streaked to single colonies several times to ensure its homogeneity, and a glycerol stock should be established and maintained at −80°. Most of the antibiotics used for plasmid selection are bacteriostatic, not bactericidal. Therefore, contamination of the desired transformant by escapes or spontaneously resistant bacteria is common. This contamination can be minimized by starting cultures from the −80° stock on freshly made plates, streaking to single colonies, and minimizing incubation times at each step.

Culture Conditions for Maximizing Protein Expression

This protocol is for the production of protein expressed from genes under regulatory control of the *V. fischeri* autoinducer system in plasmids derived from pJHD500. It is designed for 9 liters of broth (1.5 liters in each of six Fernbach flasks). Terrific Broth is used to support enhanced cell densities,[17] and the expression period is minimized to reduce potential protein degradation. Sterilization times are minimized to avoid caramelization of the broth. Therefore, media should be prepared immediately before use as this pasteurization may allow microbial growth on extended storage. As cells can enter the stationary phase before they have exhausted the medium nutrients, the protocol calls for washing the cells by centrifugation and increasing their number in stages. At higher population densities, aeration is very important: covering the Fernbachs with foil or cotton plugs or shaking them at slower speeds results in reduced growth rates. Alternation between aerobic and anaerobic growth is likely to reduce expression of the heterologous protein and will increase the number of indigenous proteins present in the cell, complicating purification of the protein of interest. An OD_{600} of 15–20/ml is usually reached using this protocol. At an OD_{600} of 18, the 9 liters of culture usually yields about 200 g cell paste containing approximately 30 g of total protein or 15 g of the foreign protein, assuming that it is 50% of the total.

1. Day 1: Start a fresh plate culture of the desired strain from a −80° stock on medium containing the selective antibiotic at the appropriate concentration, e.g., 50 μg/ml carbenicillin. This seed culture should be used within 48 hr.
2. Day 2: Make 3.3 liters of 3× Terrific Broth as follows: 50 g glycerol, 114 g Bacto-tryptone (or peptone or enzymatic casein hydrolysate), 228 g Bacto (or other) yeast extract, and 3 liters H_2O. Dilute 150 ml of 3× Terrific Broth with 300 ml H_2O; dispense 45 ml into each of ten 250-ml culture flasks. Dispense 0.5 liters of 3× broth into each of six Fernbach flasks. Add 850 ml of H_2O to each Fernbach. Make 1 liter of buffer salt in a 2-liter flask: 23.1 g KH_2PO_4 125 g K_2HPO_4, and 950 ml H_2O. Cover the Fernbachs and the buffer salt flask with foil and autoclave them and the 250-ml culture flasks for 20 min on the liquid cycle. Do not autoclave longer. Remove the medium promptly to avoid caramelization. The medium should be used within 24 hr as it may only be pasteurized.
3. Late on Day 2: Start four overnight cultures of the desired strain by adding the appropriate antibiotic and 5 ml buffer salts to each of four 250-ml culture flasks. Inoculate each flask with a single colony from the seed culture made in step 1 and grow at 37° at 350 rpm.
4. Early on Day 3: Take the OD_{600} of the overnight cultures. Centrifuge the culture in 50-ml conical centrifuge tubes or Oak Ridge tubes for 5 min at 5000 rpm.

5. While the cultures are spinning, add enough of the selection antibiotic for 9.3 liters to the remaining buffer salt solution and dispense 5 ml into each of the six remaining 250-ml culture flasks and 150 ml into each Fernbach flask.
6. Resuspend the cell pellets in a small volume of Terrific Broth and divide equally among the six 250-ml culture flasks. Grow at 37° at 350 rpm for 8 hr. Take the OD_{600} of the cultures.
7. Centrifuge these cultures as in step 4.
8. Resuspend the cells in each centrifuge tube in a few milliliters of fresh broth and pour all of the cells from one centrifuge tube into each Fernbach.
9. Grow for 5 hr at 350 rpm. The incubation temperature may need to be altered if the protein of interest is not stable or does not fold properly at 37°. For example, for optimal expression of luciferase encoded by *V. harveyi luxAB*, cultures are grown at 25–30°. Do not cover the flasks tightly during incubation. Optimally, the flasks should remain uncovered. However, phage infection is a potential problem with uncovered flasks, and some bacterial strains produce aromatic secondary products that may be offensive to some laboratory workers.
10. After 5 hr, determine the OD_{600}. If OD_{600} is not ≥ 4, incubate 1–2 hr more and measure the OD_{600} again. When an $OD_{600} \geq 4$ is reached, add 15 μl of 40 m*M* *V. fischeri* autoinducer to each 1.5 liters of culture. This is a final concentration of 400 n*M* *N*-(3-oxohexanoyl)-DL-homoserine lactone.
11. Continue incubating the cultures at 350 rpm for 12–24 hr. It is usually advantageous to determine the period required for maximal accumulation of the protein of interest by doing a trial run during which samples are withdrawn at various times. Take the OD_{600}, centrifuge the cells for 15 min at 5000 rpm in 1-liter centrifuge bottles (one of which has had its empty weight recorded), and weigh the final combined cell pellet.
12. Resuspend the cells in an appropriate lysis buffer at the rate of 1.25 ml buffer/g cells. These cells are somewhat fragile: to avoid premature cell lysis, the buffer can be supplemented with 100 m*M* NaCl.
13. Using this procedure, it seems to work best to lyse the cells immediately rather than freeze the pelleted cells. Cell disruption by French press is extremely efficient with cells produced by this method.
14. Centrifuge the crude cell lysate for 45 min at 10,000 or 14,000 rpm (Beckman JA10 or JA14 rotor, respectively). Proceed with the purification procedure designed for the protein of interest. Ammo-

nium sulfate precipitation of the cleared lysate is typically a first step in protein purification and has the advantage of stabilizing most proteins. For comparison of protein production between individual runs, cleared lysate samples may be taken before proceeding with purification. An aliquot of 0.5 μl is a good starting point as the amount of sample to load on a polyacrylamide gel that will subsequently be stained with Coomassie blue.

Concluding Remarks

An advantage of P*luxI* expression vectors is the speed with which the proteins accumulate. Very high levels of protein are produced 12–18 hr after induction, even at reduced incubation temperatures. Although proteinases are more a problem in eukaryotic expression systems than prokaryotic ones, the speed of P*luxI* regulated protein accumulation increases the potential for expressing and recovering heterologous proteins having limited proteolytic stability. Basal expression levels are frequently low enough to allow cloning of genes whose products are detrimental to the host cell.[23] In such cases the speed of expression may allow recovery of proteins that are toxic to the host cell.

However, the expression levels are sufficiently high that with some proteins misfolding and formation of inclusion bodies may occur, thus reducing the recovery of the biologically active protein. Expression levels can be moderated by the choice of the bacterial host cell. Alternatively, expression vectors can be transformed into cells that produce elevated levels of protein chaperones.

[23] G. S. Shadel, R. Young, and T. O. Baldwin, *J. Bacteriol.* **172,** 3980 (1990).

Section V

Luminescence-Based Assays *in Vitro*

[24] Application of Bioluminescence and Chemiluminescence in Biomedical Sciences

By LARRY J. KRICKA

Introduction

Chemiluminescent (CL), bioluminescent (BL), and electrochemiluminescent methods have become established in both routine analysis and for research applications in the biomedical sciences. This article briefly reviews the current and emerging applications of these technologies and the progress that has been made since the publication of the previous volume on this topic in 1986. The reader is referred to a number of books, reviews,[1-13] the Proceedings of the International Symposia on Bioluminescence and Chemiluminescence,[14-18] and a regular literature

[1] A. R. Bowie, M. G. Sanders, and P. J. Worsfold, *J. Biolumin. Chemilumin.* **11,** 61 (1996).
[2] L. J. Kricka, *Anal. Biochem.* **175,** 14 (1988).
[3] M. Cunningham, A. Simmonds, and I. Durrant, *Curr. Innovations Mol. Biol.* **2,** 49 (1995).
[4] A. Tsuji, M. Maeda, and H. Arakawa, *Seibutsu Butsuri Kagaku* **39,** 301 (1995).
[5] L. J. Kricka and G. H. G. Thorpe, *in* "Complementary Immunoassays—II" (W. P. Collins, ed.), p. 169. Wiley, Chichester, 1988.
[6] H. A. H. Rongen, R. M. W. Hoetelmans, A. Bult, and W. P. Van Bennekom, *J. Pharm. Biomed. Anal.* **12,** 433 (1994).
[7] L. J. Kricka, *Clin. Biochem.* **26,** 325 (1993).
[8] L. J. Kricka, *in* "Immunoassay" (E. P. Diamandis and T. K. Christopoulos, eds.), p. 337. Academic Press, San Diego, 1996.
[9] L. J. Kricka *Clin. Chem.* **40,** 347 (1994).
[10] I. Bronstein and L. J. Kricka, *J. Clin. Lab. Anal.* **3,** 316 (1989).
[11] K. Van Dyke and R. Van Dyke (eds.), "Luminescence Immunoassay and Molecular Applications." CRC Press, Boca Raton, FL, 1990.
[12] L. J. Kricka, G. H. G. Thorpe, and R. A. W. Stott, *in* "Principles and Practice of Immunoassay" (C. P. Price and D. J. Newman, eds.), p. 417. Macmillan, London, 1991.
[13] D. Champiat and J.-P. Larpent, "Bio-ChimiLuminescence." Masson, Paris, 1993.
[14] J. W. Hastings, L. J. Kricka, and P. E. Stanley (eds.), "Bioluminescence and Chemiluminescence: Molecular Reporting with Photons." Wiley, Chichester, 1997.
[15] A. K. Campbell, L. J. Kricka, and P. E. Stanley (eds.), "Bioluminescence and Chemiluminescence: Fundamental and Applied Aspects." Wiley, Chichester, 1994.
[16] A. A. Szalay, P. E. Stanley, and L. J. Kricka (eds.), "Bioluminescence and Chemiluminescence: Status Report." Wiley, Chichester, 1993.
[17] P. E. Stanley and L. J. Kricka (eds.), "Bioluminescence and Chemiluminescence: Current Status." Wiley, Chichester, 1991.
[18] M. Pazzagli, E. Cadenas, L. J. Kricka, A. Roda, and P. E. Stanley (eds.), "Bioluminescence and Chemiluminescence: Studies and Applications in Biology and Medicine." Wiley, Chichester, 1989.

0076-6879/00 $30.00

survey[19–38] for further information on the fundamental and applied aspects of both CL and BL during this period.

The benefits of luminescent technologies (chemiluminescence, bioluminescence, and electrochemiluminescence) include ultra sensitivity, rapid assay procedures, and a diversity of analytical applications. These technologies are now used routinely by clinical laboratories for immunoassay and DNA probe assays. Concurrently, the scope of applications for luminescent technologies in medical research has continued to expand, and the most active growth has been in reporter gene assays, nitric oxide studies, imaging of luminescent reactions, and cellular luminescence. A factor in the increasing popularity of bioluminescent techniques has been the rapid strides in the molecular biology of bioluminescent organisms and the cloning of the genes for luciferases and photoproteins. The main areas of application for luminescence are listed in Table I, and selected applications in the biomedical sciences are reviewed in subsequent sections of this article.

Laboratory Medicine

Immunoassay represents the most important routine application of CL in laboratory medicine (e.g., clinical chemistry, toxicology, virology, and endocrinology). CL acridinium ester labels[39] and CL detection techniques for horseradish peroxidase and alkaline phosphatase labels have been com-

[19] L. J. Kricka and P. E. Stanley, *J. Biolumin. Chemilumin.* **2,** 45 (1988).
[20] L. J. Kricka and P. E. Stanley, *J. Biolumin. Chemilumin.* **3,** 159 (1989).
[21] L. J. Kricka and P. E. Stanley, *J. Biolumin. Chemilumin.* **6,** 45 (1991).
[22] L. J. Kricka and P. E. Stanley, *J. Biolumin. Chemilumin.* **6,** 203 (1991).
[23] L. J. Kricka and P. E. Stanley, *J. Biolumin. Chemilumin.* **7,** 47 (1992).
[24] O. Nozaki, L. J. Kricka, and P. E. Stanley, *J. Biolumin. Chemilumin.* **7,** 263 (1992).
[25] L. J. Kricka, O. Nozaki, and P. E. Stanley, *J. Biolumin. Chemilumin.* **8,** 169 (1993).
[26] L. J. Kricka and P. E. Stanley, *J. Biolumin. Chemilumin.* **9,** 87 (1994).
[27] L. J. Kricka and P. E. Stanley, *J. Biolumin. Chemilumin.* **9,** 379 (1994).
[28] L. J. Kricka and P. E. Stanley, *J. Biolumin. Chemilumin.* **10,** 133 (1995).
[29] L. J. Kricka and P. E. Stanley, *J. Biolumin. Chemilumin.* **10,** 361 (1995).
[30] L. J. Kricka and P. E. Stanley, *J. Biolumin. Chemilumin.* **10,** 301 (1995).
[31] L. J. Kricka and P. E. Stanley, *J. Biolumin. Chemilumin.* **11,** 39 (1996).
[32] L. J. Kricka and P. E. Stanley, *J. Biolumin. Chemilumin.* **11,** 107 (1996).
[33] L. J. Kricka and P. E. Stanley, *J. Biolumin. Chemilumin.* **11,** 289 (1996).
[34] L. J. Kricka and P. E. Stanley, *J. Biolumin. Chemilumin.* **11,** 271 (1996).
[35] P. E. Stanley, *J. Biolumin. Chemilumin.* **11,** 175 (1996).
[36] L. J. Kricka and P. E. Stanley, *J. Biolumin. Chemilumin.* **12,** 113 (1997).
[37] L. J. Kricka, *Anal. Chem.* **65,** 460R (1993).
[38] L. J. Kricka, *Anal. Chem.* **67,** R499 (1995).
[39] M. J. Pringle, *Adv. Clin. Chem.* **30,** 89 (1993).

TABLE I
RANGE OF APPLICATIONS OF CHEMILUMINESCENT, BIOLUMINESCENT, AND ELECTROCHEMILUMINESCENT METHODS IN BIOMEDICAL SCIENCES

Antioxidant assays
Assays of enzymes and substrates via coupled reactions
Cellular chemiluminescence (phagocytosis)
Cytotoxicity testing
Drug discovery
Drug susceptibility testing
Genetic disease screening
High-throughput drug screening
Human aging
Immunoassay
Inflammation
Oxidative stress
Membrane lipid dynamics
Mutagenesis
Nitric oxide measurements
Nucleic acid analysis (probe assays, PCR, sequencing)
Protein blotting
Rapid microbiology (cell enumeration, drug suceptibility)
Reporter genes (luciferase, β-galactosidase, glucuronidase)
Single cell studies
Tumor chemosensitivity

mercialized, and for some analytes more than 20% of clinical laboratories in the United States use these types of immunoassay.[40] The impact in immunoassay of the new dioxetane-based substrates for enzyme labels, most notably alkaline phosphatase, has been considerable. Stabilized adamantyl 1,2-dioxetane substrates for alkaline phosphatase (e.g., AMPPD, CSPD) represent one of the most significant advances in CL technologies in the past decade. These substrates provide simple, one reagent assays for alkaline phosphatase labels with a limit of detection of 1 zeptomole (10^{-21} mole) of enzyme. Intelligent design has produced a family of substrates (5-chloro adamanty- and 5-choloradamanty-5-chloro phenyl derivatives) and enhancers (charged polymers such as poly[vinylbenzyl(benzyldimethylam-

[40] L. J. Kricka, *in* "Bioluminescence and Chemiluminescence: Fundamental and Applied Aspects." (A. K. Campbell, L. J. Kricka, and P. E. Stanley, eds.), p. 171. Wiley, Chichester, 1994.

TABLE II
CHEMILUMINESCENT AND ELECTROCHEMILUMINESCENT IMMUNOASSAY AND DNA PROBE ANALYZERS

Alkaline phosphatase label: Chemiluminescent detection (dioxetane substrate)
Access (Sanofi)
Immulite and Immulite 2000 (Diagnostics Products Corp.)
Lumipulse 1200 (Fujirebio)
Acridinium ester or sulfonamide label: Chemiluminescent detection
ACS : 180, ACS : 180 Plus, and ACS : Centaur (Chiron Diagnostics Corp.)
Advantage (Nichols Institute Diagnostics)
PACE 2 (GenProbe)
Horseradish peroxidase label: Enhanced chemiluminescent detection (luminol-based reagent)
Vitros ECi (Johnson & Johnson Clinical Diagnostics)
Glucose oxidase label: Chemiluminescent detection (isoluminol-microperoxidase reagent)
Luminomaster (Sankyo)
Ruthenium trisbipyridyl label: Electrochemiluminescent detection
Elecsys 1010 and 2010 (Roche Diagnostics)

monium chloride)]) with different signal characteristics (light intensity, kinetics of light emission).[41–46]

The commercialization of immunoassay reagent test kits and immunoassay analyzers has been extensive (Table II), and several companies offer automatic immunoassay analyzers and comprehensive menus of the commonly performed clinical tests (e.g., thyroid function, fertility, cancer markers, therapeutic drug monitoring, myocardial damage, hepatitis, specific proteins, steroids). In addition, a large number of the emerging clinical tests have been developed in a CL or BL format [e.g., cholecystokinin,[47] cytokines (IL-4, IL-5, IL-6, IL-10),[48] endothelin-1 (0.5 ng/liter),[49] epidermal

[41] I. Bronstein and L. J. Kricka, *in* "Nonradioactive Labeling and Detection of Biomolecules" (C. Kessler, ed.), p. 168. Springer-Verlag, Berlin, 1992.

[42] I. Bronstein, B. Edwards, and A. Sparks, PCT Int. Patent Appl. 94,26,726 (1994).

[43] I. Bronstein, B. Edwards, and J. C. Voyta, PCT Int. Patent Appl. 94,21,821 (1994).

[44] I. Bronstein, B. Edwards, and R.-R. Juo, U.S. Patent 5,326,882 (1994).

[45] I. Bronstein, B. Edwards, and J. C. Voyta, PCT Int. Patent Appl. 96,25,667 (1996).

[46] I. Bronstein, R.-R. Juo, J. C. Voyta, and B. Edwards, *in* "Bioluminescence and Chemiluminescence: Current Status" (P. E. Stanley and L. J. Kricka, eds.), p. 73. Chichester, Wiley, 1991.

[47] K. Ito, R. Kodama, M. Maeda, and A. Tsuji, *Anal. Lett.* **28,** 797 (1995).

[48] T. Satoh, D. J. Tollerud, L. Guevarra, Y. Rakue, T. Nakadate, and J. Kagawa, *J. Arerugi* **44,** 661 (1995).

[49] R. Iwata, T. Hayashi, Y. Nakao, M. Yamaki, T. Yoshimasa, H. Ito, Y. Saito, M. Mukoyam, and K. Nakao, *Clin. Chem.* **42,** 1155 (1996).

growth factor (2 ng/ml),[50] granulocyte colony–stimulating factor (1 ng/ml),[51] TP53,[52] interferon-γ (0.2 pg/ml),[53] pituitary adenylate cyclase polypeptide,[54] and vascular endothelial growth factor (1 pg/ml)].[55]

In immunoassay, the combination of a nonseparation (homogeneous) assay format and ultra sensitivity has been elusive, and an important recent development has been the luminescent oxygen channeling immunoassay (LOCI). This assay exploits the *in situ* production of a CL compound due to singlet oxygen transfer between two different antibody-coated microbeads (250 nm diameter) brought into contact as a result of specific binding with the test antigen. One of the antibody beads is loaded with bromosquaraine I (a sensitizer dye for production of singlet oxygen) and the other is loaded with a mixture of thioxene (a precursor of a CL dioxetane) and a europium chelate (a fluorescent acceptor for sensitized CL). Rapid and very sensitive immunoassays ($<$25 min) have been developed for a range of clinically important analytes, including thyrotropin (TSH) (1.25 U/liter), hepatitis B surface antigen (5 ng/liter), and digoxin, and the LOCI technique is applicable in nucleic acid detection.[56–58]

Progress in the molecular biology of BL organisms, and the availability of the genes for many BL proteins, has stimulated interest in BL immunoassays. Commercial immunoassays have been developed[59] using the recombinant photoprotein aequorin (detection limit for TSH is 1 amol in 0.2 ml of serum),[60] and experimental immunoassays described are based on lucifer-

[50] N. Liu, F. Wang, Z. Jiang, L. Zhang, Z. Xu, J. Chen, H. Wang, A. Chen, and M. Li, *Mianyixue Zazhi* **11,** 259 (1995).

[51] A. Ishiguro, K. Inoue, T. Nakahata, H. Nishihira, S. Kojima, K. Ueda, Y. Suzuki, and T. Shimbo, *J. Pediatr.* **128,** 208 (1996).

[52] H. H. De Witte, J. A. Foekens, J. Lennerstrand, M. Smid, M. P. Look, J. G. M. Klijn, T. J. Benraad, and E. M. J. J. Berns, *Int. J. Cancer* **69,** 125 (1996).

[53] S. Alkan, C. Akdis, and H. Towbin, *J. Immunoassay* **15,** 217 (1994).

[54] K. Ito, R. Kodama, M. Maeda, and A. Tsuji, *Kidorui* **26,** 408 (1995).

[55] M. Hanatani, Y. Tanaka, S. Kondo, I. Ohmori, and H. Suzuki, *Biosci. Biotechnol. Biochem.* **59,** 1958 (1995).

[56] E. F. Ullman, H. Kirakossian, A. C. Switchenko, J. Ishkanian, M. Ericson, C. A. Wartchow, M. Pirio, J. Pease, B. R. Irvin, S. Singh, R. Singh, R. Patel, A. Dafforn, C. Davalian, C. Skold, N. Kurn, and D. B. Wagner, *Clin. Chem.* **42,** 1518 (1996).

[57] E. F. Ullman, H. Kirakossian, S. Singh, B. R. Irvin, J. D. Irvine, and D. B. Wagner, *in* "Bioluminescence and Chemiluminescence: Fundamental and Applied Aspects" (A. K. Campbell, L. J. Kricka, and P. E. Stanley, eds.), p. 16. Wiley, Chichester, 1994.

[58] E. F. Ullman, H. Kirakossian, S. Singh, Z. P. Wu, B. R. Irvin, J. S. Pease, A. C. Switchenko, J. D. Irvine, A. Dafforn, and C. N. Skold, *Proc. Natl. Acad. Sci. U.S.A.* **91,** 5426 (1994).

[59] R. E. Geiger, D. Gabrijelcic, and W. Miska, *in* "Immunoassay" (E. P. Diamandis and T. K. Christopoulos, eds.) p. 355. Academic Press, San Diego, 1996.

[60] D. F. Smith and N. L. Stults, *Proc. SPIE-Int. Soc. Opt. Eng.* **2680,** 156 (1996).

TABLE III
FUSION CONJUGATES OF BIOLUMINESCENT PROTEINS

Aequorin: IgG heavy chain
Aequorin: 5-HT1A receptor
Aequorin: Nuclear translocation signal from rat glucocorticoid receptor
Aequorin: Protein A
Luciola lateralis luciferase: Biotin acceptor peptide
Marine bacterial luciferase (β-subunit): Protein A
Obelin: Protein A
Photinus pyralis luciferase: Protein A
P. pyralis luciferase: RNA-binding protein
Pyrophorus plagiophthalamus luciferase: Protein A
P. plagiophthalamus luciferase: Streptavidin
Vargula hilgendorfii luciferase: Protein A
V. hilgendorfii luciferase: Protein G

ases, e.g., recombinant firefly luciferase,[61] *Vargula* luciferase,[62] or luciferins (e.g., *Cypridina* luciferin derivatives) as labels.[63] The availability of fusion conjugates has overcome the problem of label inactivation encountered using conventional chemical coupling reagents, and there are now a range of fusion conjugates of luciferases and photoproteins with protein A and other binding agents (Table III).[64–70] In addition, a DNA fragment coding for firefly luciferase has been attached to streptavidin and used to detect an antigen. Expression of the bound DNA label in a cell-free expression system produced active firefly luciferase, which was measured using the conventional ATP-luciferin reaction (detection limit 50,000 molecules of antigen).[71]

[61] A. Doi and H. Takahashi, Jpn. Patent 07,244,049 (1995).

[62] Y. Maeda, H. Ueda, H. Hara, T. Kazami, J. Kawano, G. Suzuki, and T. Nagamune, *BioTechniques* **20,** 116 (1996).

[63] H. Sawada, Y. Totani, M. Mitani, H. Ichikawa, and T. Matsumoto, Jpn. Patent 05,286,976 (1993).

[64] J. Casadei, M. J. Powell, and J. H. Kenten, *Proc. Natl. Acad. Sci. U.S.A.* **87,** 2047 (1990).

[65] E. Kobatake, T. Iwai, Y. Ikariyama, and M. Aizawa, *Anal. Biochem.* **208,** 300 (1993).

[66] C. Lindbladh, K. Mosbach, and L. Bulow, *J. Immunol. Methods* **137,** 199 (1991).

[67] C. Oker-Blom, A. M. Suomalainen, K. Akerman, Z. Oi, C. Lindqvist, A. Kuusisto, and M. Karp, *BioTechniques* **14,** 800 (1993).

[68] M. Karp, C. Lindqvist, R. Nissinen, S. Wahlbeck, K. Akerman, and C. Oker-Blom, *BioTechniques* **20,** 452 (1996).

[69] Y. Maeda, H. Ueda, J. Kazami, G. Kawano, E. Suzuki, and T. Nagamune, *Anal. Biochem.* **249,** 147 (1997).

[70] S. V. Matveev, J. C. Lewis, and S. Daunert, *Anal. Biochem.* **270,** 69 (1999).

[71] N. H. Chiu and T. K. Christopoulos, *Anal. Chem.* **68,** 2304 (1996).

Bioluminescent reactions also provide a sensitive method for measuring a range of enzyme labels. Work with an acetate kinase label, detected using a coupled firefly luciferase–luciferin reaction (detection limit 60 zmol; 1.2 amol/ml), illustrates the potential of this type of label and end point.[72]

Electrochemiluminescent (or electro-generated) immunoassays based on a ruthenium trisbipyridyl label are alternatives to immunoassays using the more familiar CL labels, such as the acridinium esters.[73] Light emission is generated at an electrode as a result of a redox reaction involving tripropylamine. Two different electrochemiluminescent automatic immunoassay analyzers are available commercially (Table II) and this technique is now being used in a number of routine clinical laboratories.

Chemiluminescent-based assays have had a significant impact on all forms of nucleic acid diagnostic techniques in combination with digoxigenin: antidigoxigenin and biotin:streptavidin labeling schemes and the polymerase chain reaction (PCR) and are slowly replacing isotopic assays in microbiology and molecular diagnostics. The nonseparation chemiluminescent hybridization protection procedure for detecting specific DNA sequences is now used extensively to test for a range infectious agents (e.g., *Chlamydia*).[74] A CL end point is also utilized in the branched DNA assay for hepatitis C virus RNA (Quantiplex, Chiron Diagnostics). In this assay, target probes hybridize both to target nucleic acid and to the immobilized capture probe.[75] The bound target probe then hybridizes to branched DNA reporter probes (up to 15 branches). Finally, each branch of the reporter probe reacts with up to three alkaline phosphatase-labeled probes to introduce a high degree of amplification into the assay. Bound enzyme labels are detected using a 1,2-dioxetane substrate (e.g., AMPPD).

A CL dioxetane substrate has also been selected for the detection of alkaline phosphatase label in the hybrid capture assay.[76] This assay procedure uses immobilized antibodies to capture RNA probe : DNA target or DNA probe:RNA target hybrids onto the inside surface of a microwell. Captured hybrids are then reacted with an alkaline phosphatase-labeled detection antibody, and the bound enzyme label is measured using an

[72] M. Maeda, H. Ikeda, A. Tsuji, S. Murakami, S. Ito, and S. Kamada, *Anal. Lett.* **28,** 383 (1995).

[73] N. R. Hoyle, B. Eckert, and S. Kraiss, *Clin. Chem.* **42,** 1576 (1996).

[74] L. J. Arnold Jr., P. W. Hammond, W. A. Weise, and N. C. Nelson, *Clin. Chem.* **35,** 1588 (1989).

[75] M. Urdea, T. Horn, T. Fultz, M. Anderson, J. Running, S. Hamren, D. Ahle, and C.-A. Chang, *Nucleic Acids Res. Symp. Ser.* **24,** 197 (1991).

[76] J. M. Pawlotsky, A. Bastie, I. Lonjon, J. Remire, F. Darthuy, C. J. Soussy, and D. Dhumeaux, *J. Virol. Methods* **65,** 245 (1997).

adamantyl 1,2-dioxetane substrate (Hybrid Capture assay, Digene Corporation).

A BL tumor chemosensitivity assay is being used on an experimental basis to determine the most efficacious drug or drug combinations for the treatment of cancers.[77] Tumor cells obtained from a biopsy specimen are grown in culture and are exposed to candidate anticancer drugs (e.g., 5-fluorouracil, deoxorubicin, vincristine). Cell survival, and hence the effectiveness of the drug or drug combination, is assessed by measuring ATP using the BL firefly luciferase–luciferin assay.

Biomedical Research

General

An extensive range of CL and BL assays and protocols has been published for the analysis of clinically important analytes in body fluids and tissues, but none of these have made the transition into routine clinical laboratory practice. BL calcium assays[78] using either aequorin or obelin[79] offer excellent sensitivity and specificity for ionized calcium. Other BL research assays exploit coupled detection schemes based on the assay of ATP using the firefly luciferase reaction (e.g., adenylate kinase for cell enumeration)[80] or the assay of NADH using the marine bacterial luciferase–oxidoreductase reaction [e.g., serum lactate dehydrogenase (LDH).][81] Examples of CL assays include urinary iodide (80 ng)[82] and *N*-acetyl-β-D-glucosaminidase,[83] cholesterol (1 mg/ml),[84] vitamin C in seminal plasma,[85] and blood spot screening tests for phenyl ketonuria (PKU), galactosemia,

[77] P. E. Andreotti, I. A. Cree, C. M. Kurbacher, D. M. Hartmann, D. Linder, G. Harel, I. Gleiberman, P. A. Caruso, S. H. Ricks, and H. W. Bruckner, *in* "Bioluminescence and Chemiluminescence: Fundamental and Applied Aspects" (A. K. Campbell, L. J. Kricka, and P. E. Stanley, eds.), p. 403. Wiley, Chichester, 1994.

[78] M. Brini, R. Marsault, C. Bastianutto, J. Alvarez, T. Pozzan, and R. Rizzuto, *J. Biol. Chem.* **270,** 9896 (1995).

[79] G. I. Gitel'zon, V. A. Tugai, and A. N. Zakharchenko, *Ukr. Biokhim. Zh.* **62,** 69 (1990).

[80] M. J. Murphy, D. J. Squirrell, M. F. Saunders, and B. Blasco, *in* "Bioluminescence and Chemiluminescence: Molecular Reporting with Photons" (J. W. Hastings, L. J. Kricka, and P. E. Stanley, eds.), p. 319. Wiley, Chichester, 1997.

[81] I. Rusanova, V. Savov, and A. Ismailov, *Fiz. Fak.* **85,** 111 (1993).

[82] J. L. Burguera, M. R. Brunetto, Y. Contreras, M. Burguera, M. Gallignani, and P. Carrero, *Talanta* **43,** 839 (1996).

[83] K. Sasamoto, R. Zenko, K. Ueno, and Y. Ohkura, *Chem. Pharm. Bull.* **39,** 1317 (1991).

[84] H. Sasamoto, M. Maeda, and A. Tsuji, *Anal. Chim. Acta* **310,** 347 (1995).

[85] J. J. Thiele, H. J. Freisleben, J. Fuchs, and F. R. Ochsendorf, *Hum. Reprod.* **10,** 110 (1995).

and maple syrup urine disease.[86] The new understanding of the importance of nitric oxide as a messenger molecule has stimulated interest in assays for this substance. A CL reaction between nitric oxide and ozone forms the basis of a simple and sensitive assay procedure, which has emerged as an important research application for CL.[87] Luminescent reactions have provided new and relatively simple assays for substances with antioxidant properties (e.g., vitamin C, vitamin E, proteins).[88] For example, light emission from the enhanced CL reaction between horseradish peroxidase–luminol and 4-iodophenol is abolished in the presence of an antioxidant. Light emission resumes after consumption of the antioxidant, the duration of the delay is proportional to the concentration of the antioxidant, and the kinetic profile of the light emission is characteristic of the particular antioxidant.

In biomedical research there is continued interest in the coupling of CL with biomolecular separation techniques[89]: for example, CL detection of PCR products in an assay for hepatitis C virus DNA[90] separated by capillary electrophoresis; CL postcolumn detection (peroxyoxalate reaction) of phosphatidylcholine and phosphatidylethanolamine hydroperoxides in red blood cells separated by HPLC[91]; and CL assays for choline containing phospholipids in serum (1.3 pmol) separated by flow injection.[92]

Immunoassay and Blotting

Most new luminescent reagent development has been directed at immunoassay, blotting, and DNA probe applications (Western blotting procedures for proteins, Southern blotting for DNA, and Northern blotting for RNA).[93,94] CL detection reagents have replaced isotopic methods in many institutions, and reagent kits are available commercially (e.g., from Amer-

[86] V. K. Mahant and R. A. Gabardy, *in* "Bioluminescence and Chemiluminescence: Fundamental and Applied Aspects" (A. K. Campbell, L. J. Kricka, and P. E. Stanley, eds.), p. 257. Wiley, Chichester, 1994.

[87] K. Kikuchi, T. Nagano, H. Hayakawa, Y. Hirata, and M. Hirobe, *Anal. Chem.* **65,** 1794 (1993).

[88] T. P. Whitehead, G. H. G. Thorpe, and S. R. J. Maxwell *Anal. Chim. Acta* **266,** 265 (1992).

[89] A. R. Bowie, M. G. Sanders, and P. Worsfold, *J. Biolumin. Chemilumin.* **11,** 61 (1996).

[90] T. A. Felmlee, P. S. Mitchell, K. J. Ulfelder, D. H. Persing, and J. P. Landers, *J. Capillary Electrophor.* **2,** 125 (1995).

[91] T. Miyazawa, T. Suzuki, K. Fujimoto, and M. Kinoshita, *Mech. Aging Dev.* **86,** 145 (1996).

[92] M. Wada, K. Nakashima, N. Kuroda, S. Akiyama, and K. Imai, *J. Chromatogr. B Biomed. Appl.* **678,** 129 (1996).

[93] P. Trayhurn, J. S. Duncan, A. Nestor, M. E. A. Thomas, N. C. Eastmond, and D. V. Rayner, *in* "Bioluminescence and Chemiluminescence: Fundamental and Applied Aspects" (A. K. Campbell, L. J. Kricka, and P. E. Stanley, eds.), p. 211. Wiley, Chichester, 1994.

[94] J. S. Duncan, S. Lyke, and P. Trayhurn, *Biochem. Soc. Trans.* **22,** 161S (1994).

sham, DuPont, Pierce Chemical, Tropix). Significant improvements have been demonstrated in detection over colorimetric blotting methods.[95] Two detection methods predominate: enhanced chemiluminescent detection of horseradish peroxidase labels and dioxetane substrates for alkaline phosphatase labels.[41,96] New CL labels include trioxane derivatives,[97] isothiocyanatoisoluminol,[98] phenanthridinium, quinolinium, and isoquinolinium compounds,[99] pyridazino-[4,5-g]quinoxaline derivatives,[100] and a range of acridinium esters and related molecules.[100–103] For alkaline phosphatase labels, new CL substrates include phenacyl phosphate,[104] 3-indolyl phosphate,[105] and 2-(*N*-phenylphthalimidyl) benzothiazolyl-5-phosphate.[106] New options for the detection of horseradish peroxidase (HRP) labels include aryl *N*-alkylacridan carboxylates and thiocarboxylate substrates.[107] Also, further examples of enhancers for the peroxidase label-catalyzed oxidation of luminol [e.g., 4-biphenylboronic acid,[108,109] 3′-chloro-4′- hydroxyacetanilide,[110] 4,4′-thiazolyl phenol,[111] and 4-(3-thienyl)phenol,[112] and signal stabilizers, such as β-cyclodextrins[113]] have been tested.

Chemiluminescent techniques have found application in nucleic acid-based assays for viruses and other infectious disease agents testing in combi-

[95] N. T. Constantine, J. Bansal, X. Zhang, K. C. Hyams, and C. Hayes, *J. Virol. Methods* **47,** 153 (1994).

[96] R. Budowle, F. S. Baechtel, C. T. Comey, A. M. Guisi, and L. Klevan, *Electrophoresis* **16,** 1559 (1995).

[97] S. Yamada, H. Sasaki, J. Yokoyama, M. Mitani, and H. Ichikawa, Jpn. Patent 06,256,217 (1995).

[98] Q. Wang, H. Zhuang, F. Zhang, and L. Zhang, *Fuzhou Daxue Xuebao Ziran Kexueban* **23,** 75 (1995).

[99] R. R. Renotte, G. N. Sarlet, R. Lejeune, and P. A. Ghislain, PCT Int. Patent Appl. 95,19,976, (1995).

[100] J. Ishida, H. Arakawa, M. Takada, and M. Yamaguchi, *Analyst* **120,** 1083 (1995).

[101] N. Sato, H. Mochizuki, and T. Kanamori, Eur. Patent. Appl. 609,885 (1994).

[102] F. McCapra and I. Beheshti, U.S. Patent 5,284,951 (1994).

[103] I. Beheshti and H. Koelling, U.S. Patent 5,290,936 (1994).

[104] H. Sasamoto, M. Maeda, and A. Tsuji, *Anal. Chim. Acta* **306,** 161 (1995).

[105] T. Ikegami, H. Arakawa, M. Maeda, and A. Tsuji, *Anal. Sci.* **10,** 831 (1994).

[106] K. Sasamoto, G. Deng, T. Ushijima, Y. Ohkura, and K. Ueno, *Analyst* **120,** 1709 (1995).

[107] H. Akhavan-Tafti, R. Desilva, and Z. Arghavani, U.S. Patent 5,523,212 (1996).

[108] L. J. Kricka, U.S. Patent 5,512,451 (1996).

[109] O. Nozaki, X. Ji, and L. J. Kricka, *in* "Bioluminescence and Chemiluminescence: Fundamental and Applied Aspects" (A. K. Campbell, L. J. Kricka, and P. E. Stanley, eds.), p. 52. Wiley, Chichester, 1994.

[110] T. R. Kissel, S. A. Fingar, and A. E. Friedman, Eur. Patent Appl. 704,539 (1996).

[111] H. Ito, H. Suzuki, Y. Miki, T. Hayashi, M. Iwata, and M. Yamaki, Jpn. Patent 08,038,196 (1996).

[112] Y. Mitoma, K. Hara, and S. Kumakura, Jpn. Patent 07,311,197 (1995).

[113] M. Sugyama, Jpn. Patent 07,313,195 (1995).

nation with nucleic acid amplification techniques, such as the PCR [e.g., *Bacteroides fragilis,*[114] CMV,[115] hepatitis B,[116] herpes simplex,[117] Lassa fever virus (reverse transcriptase–polymerase chain reaction method),[118] papillomavirus in cervical scrapes,[119] and *Trichomonas vaginalis* (detect <12 organisms)[120]]. Nucleic acid assays based on BL are at an earlier stage of development, and examples include assays using a pyruvate kinase label in conjunction with PCR (7 amol of target DNA from HIV-1)[121] and a hybridization assay for PSA mRNA using streptavidin–aequorin and a biotinylated detection probe (detected PSA mRNA from a single cell in the presence of 10^6 cells).[122]

Reporter Gene-Based Assays

One of the most successful research applications for BL and CL is in conjunction with reporter genes developed as alternatives to the traditional chloramphenicol acetyl transferase gene (CAT) (Table IV).[123–126] Detection techniques are now available for the expression products of genes for placental alkaline phosphatase, β-galactosidase (*GAL*), and glucuronidase (*GUS*) based on either CL dioxetane derivatives or BL firefly luciferin derivatives. In addition, expression products from bacterial luciferase (*lux*) and firefly luciferase (*luc*) genes can be detected with rapid BL assays. A

[114] R. Jotwani, N. Kato, H. Kato, K. Watanabe, and K. Ueno, *Curr. Microbiol.* **31,** 215 (1995).
[115] M. Musiani, A. Roda, M. Zerbini, P. Pasini, G. Gentilomi, G. Gallinella, and S. Venturoli, *Am. J. Pathol.* **148,** 1105 (1996).
[116] M. L. Choong and S. H. Ton, *Proc. Malays. Biochem. Soc. Conf. 19th,* 258 (1994).
[117] M. Vesanen, H. Piiparinen, A. Kallio, and A. Vaheri, *J. Virol. Methods* **59,** 1 (1996).
[118] A. H. Demby, J. Chamberlain, D. W. G. Brown, and C. S. Clegg, *J. Clin. Microbiol.* **32,** 2898 (1994).
[119] O. Valdes, M. Marrero, M. Alvarez, M. Mune, G. Diaz, and E. Prieto, *Biotechnol. Appl.* **10,** 61 (1993).
[120] J. U'Ren and J. Virosco *in* "Bioluminescence and Chemiluminescence: Fundamental and Applied Aspects" (A. K. Campbell, L. J. Kricka, and P. E. Stanley, eds.), p. 60. Wiley, Chichester, 1994.
[121] N. Zammatteo, P. Moris, I. Alexandre, D. Vaira, J. Piette, and J. Remacle, *J. Virol. Methods* **55,** 185 (1995).
[122] B. Galvan and T. K. Christopoulos, *Anal. Chem.* **68,** 3545 (1996).
[123] I. Bronstein, J. Fortin, P. E. Stanley, G. S. A. B. Stewart, and L. J. Kricka, *Anal. Biochem.* **219,** 169 (1994).
[124] P. J. Hill, G. S. A. B. Stewart, and P. E. Stanley, *J. Biolumin. Chemilumin.* **8,** 267 (1993).
[125] B. Sherf, S. Navarro, R. Hannah, and K. Wood, *in* "Bioluminescence and Chemiluminescence: Molecular Reporting with Photons" (J. W. Hastings, L. J. Kricka, and P. E. Stanley, eds.), p. 228. Wiley, Chichester, 1997.
[126] I. Bronstein, J. Fortin, J. C. Voyta, C. E. M. Olesen, and L. J. Kricka, *in* "Bioluminescence and Chemiluminescence: Fundamental and Applied Aspects" (A. K. Campbell, L. J. Kricka, and P. E. Stanley, eds.), p. 20. Wiley, Chichester, 1994.

TABLE IV
REPORTER GENES ASSAYED USING CL OR BL METHODS

Reporter gene	Detection
Aequorin	BL
Firefly luciferase	BL
Marine bacterial luciferase	BL
Obelin	BL
Renilla luciferase	BL
Alkaline phosphatase	BL, CL
β-Galactosidase	BL, CL
Glucuronidase	CL
Dual detection	
Firefly luciferase + *Renilla* luciferase	BL/BL
Firefly luciferase + β-galactosidase	BL/CL
Firefly luciferase + glucuronidase	BL/CL
Firefly luciferase + placental alkaline phosphatase (PLAP)	BL/CL
PLAP + β-galactosidase	CL/CL

recent trend has been the development of dual reporter gene assays (e.g., firefly luciferase *luc* gene and β-galactosidase *GAL* gene).[126] Also, the gene encoding green fluorescent protein from the BL organism *Aequoria victoria*[127] is gaining in popularity, although the product of this gene is measured by direct fluorescence emission as opposed to a BL reaction.

Cellular CL and Luminol- and Lucigenin-Enhanced CL

Cellular CL from polymorphonuclear neutrophils, phagocytes, whole blood, and other biological fluids enhanced with luminol or lucigenin remains a widely used tool for cell-based studies, including investigations of toxicity associated with chemotherapy,[128] and the respiratory burst.[129,130] Pholasin, the photoprotein from the mollusk *Pholas dactylus,* is also useful in cell studies.[131] It reacts with superoxide, hydroxy, or ferryl radicals and has been used to investigate activation of the NADPH system and degranulation in polymorphonuclear cells.

[127] M. Chalfie, Y. Tu, G. Euskirchen, W. W. Ward, and D. C. Prasher, *Science* **263,** 802 (1994).
[128] M. Lejeune, E. Sariban, B. Cantinieaux, A. Ferster, C. Devalck, and P. Fondu, *Pediatr. Res.* **39,** 835 (1996).
[129] H. Lundqvist and C. Dahlgren, *Free Radic. Biol. Med.* **20,** 785 (1996).
[130] A. S. Al-Tuwaijri and A. D. Al-Dohayan, *Microbios* **83,** 167 (1995).
[131] P. A. Roberts, J. Knight, and A. K. Campbell, *Anal. Biochem.* **160,** 139 (1987).

Imaging

Imaging of light emission using charge-coupled device cameras and related devices has become popular, although the instrumentation is still relatively expensive. Blots, microwells, entire 96-well microplates, and even single cells[132] can be imaged and temporal and spatial information on reactions accumulated.[133]

Therapeutics

There are no direct uses of CL or BL compounds as therapeutic agents. However, chemical light from a peroxyoxalate reaction has been tested as an alternative to laser light to activate hematoporphyrins used as photodynamic therapeutic agents in the treatment of adenocarcinomas.[134] Also, CL and BL methods are competing with other technologies for a role in the current development of high-throughput drug screening systems (via assays for reporter genes, kinases, and proteases).[135]

Conclusions

Progress in chemi-, bio-, and electrochemiluminescent methods since 1986 has been extensive, and they are now established in routine clinical analysis and as tools in biomedical research. Currently, CL acridinium ester labels, the luminol-based assay for peroxidase labels, and dioxetane-based assays for alkaline phosphatase labels are the most widely used in immunoassay and nucleic acid assays. In biomedical research, a wider range of methods are in use, most notably cellular luminescence assays using luminol or lucigenin enhancement, nitric oxide assays, and a range of CL and BL reporter gene assays. Further advances in the molecular biology of bioluminescent organisms are anticipated to lead to further analytical techniques based on recombinant bioluminescent proteins.

[132] M. R. H. White, *J. Biolumin. Chemilumin.* **11,** 53 (1996).

[133] A. Roda, P. Pasini, M. Musiani, C. Robert, M. Baraldini, and G. Carrea, *in* "Bioluminescence and Chemiluminescence: Molecular Reporting with Photons" (J. W. Hastings, L. J. Kricka, and P. E. Stanley, eds.), p. 307. Wiley, Chichester, 1997.

[134] M. J. Phillip and P. P. Maximuke, *Oncology* **46,** 266 (1989).

[135] C. Lehal, S. Daniel-Issakani, M. Brasseur, and B. Strulovici. *Anal. Biochem.* **244,** 340 (1997).

[25] Use of Firefly Luciferase in ATP-Related Assays of Biomass, Enzymes, and Metabolites

By ARNE LUNDIN

General Considerations

Introduction

The ATP assay based on firefly luciferase has become an important analytical tool in science and, in recent years, also in industry. The assay is used frequently in a nonoptimal or even erroneous way. The main aim of this article is to describe the analytical considerations that should be taken when setting up or developing assays based on firefly luciferase. Various application areas are illustrated with references mainly to the author's own work. The reader is encouraged to contact the author via e-mail (biothema@swipnet.se) or to visit www.biothema.com for specific information.

Firefly Luciferase Reaction

Firefly luciferase catalyzes a reaction among ATP, D-luciferin, and O_2 to form AMP, inorganic pyrophosphate (PP_i), oxyluciferin, CO_2, and light. The quantum yield is 0.88,[1] i.e., almost one photon per ATP. The reaction proceeds in several steps:

$$\text{E} + \text{ATP} + \text{D-luciferin} \rightarrow \text{E(luciferyl-adenylate)} + \text{PP}_\text{i} \quad (1)$$
$$\text{E(luciferyl-adenylate)} + \text{O}_2 \rightarrow \text{E(oxyluciferin*; AMP)} + \text{CO}_2 \quad (2)$$
$$\text{E(oxyluciferin*; AMP)} \rightarrow \text{E(oxyluciferin; AMP)} + \text{photon} \quad (3)$$
$$\text{E(oxyluciferin; AMP)} \rightarrow \text{E} + \text{oxyluciferin} + \text{AMP} \quad (4)$$

In the absence of enhancers, reaction 4 is very slow and the enzyme–product complex, i.e., E(oxyluciferin; AMP), may be purified by gel chromatography.[2] In the presence of low concentrations of PP_i,[3] SH compounds

[1] H. H. Seliger and W. D. McElroy, *Arch. Biochem. Biophys.* **88,** 136 (1960).
[2] B. J. Gates and M. DeLuca, *Arch. Biochem. Biophys.* **169,** 616 (1975).
[3] A. Lundin, *in* "Luminescent Assays: Perspectives in Endocrinology and Clinical Chemistry" (M. Serio and M. Pazzagli, eds.), p. 29. Raven Press, New York, 1982.

METHODS IN ENZYMOLOGY, VOL. 305
0076-6879/00 $30.00

such as DTT[4] and coenzyme A[5,6] or some nucleotides,[7] reaction 4 may be enhanced strongly. If this is not done the enzyme is product inhibited and a flash of light is all that can be obtained. The oxidative decarboxylation of luciferyl-adenylate (reaction 2) involves a conformational change of the enzyme. In the presence of enhancers like PP_i, reaction 2 is rate limiting. Even then firefly luciferase has a low turnover number.

Optimized Reaction Conditions

Analytical methods should be optimized. The inherently high-sensitivity of the luminometric ATP assay has resulted in a poor interest in optimization. If a higher sensitivity is needed, it is obviously easier to increase the luciferase level than to optimize all the parameters of the assay. However, optimization in, e.g., clinical chemistry, is generally not driven by an interest to increase the sensitivity. The most important argument for optimization is the need to avoid a situation in which small changes in reaction conditions result in large changes in signal. This is the same type of argument used in various quality systems in industry, i.e., a small change in the production process should not give a large change in product performance.

Luciferase activity depends on a number of parameters, including D-luciferin, magnesium ions, buffer (type, concentration, and pH), activators (e.g., PP_i), inhibitors (e.g., luciferin analogs), and compounds added to make lyophilization possible or to stabilize the reconstituted reagent. Low-quality preparations of D-luciferin frequently contain inhibitors that may affect the light emission even in levels below 0.1%. With high-quality preparations of D-luciferin,[8] the optimum concentration is around 0.4 m*M*. The concentration of magnesium ions should be in excess of frequently added chelators such as EDTA, as the substrate in the luciferase reaction is Mg-ATP rather than ATP. The buffer should be chosen to give a constant pH with all samples to be assayed. A 50 or 100 m*M* tris(hydroxymethyl) aminomethane buffer adjusted to pH 7.75 with acetic acid is often a good

[4] A. Lundin, *in* "Bioluminescence and Chemiluminescence" (A. Szalay, L. J. Kricka, and P. E. Stanley, eds.), p. 291. Wiley, Chichester, 1993.

[5] R. L. Airth, W. C. Rhodes, and W. D. McElroy, *Biochim. Biophys. Acta* **27,** 519 (1958).

[6] K. W. Wood, *in* "Bioluminescence and Chemiluminescence: Current Status" (P. E. Stanley and L. J. Kricka, eds.), p. 11. Wiley, Chichester, 1991.

[7] S. R. Ford, K. H. Chenault, L. S. Bunton, G. J. Hampton, J. McCarthy, M. S. Hall, S. J. Pangburn, L. M. Buck, and F. R. Leach, *J. Biolumin. Chemilumin.* **11,** 149 (1996).

[8] A. Lundin, *in* "A Practical Guide to Industrial Uses of ATP-Luminescence in Rapid Microbiology" (P. E. Stanley, R. Smither, and W. J. Simpson, eds.), p. 55. Cara Technology Ltd., Lingfield, Surrey, 1997. More detailed information can be obtained from the author or at www.biothema.com.

choice. The acetate counterion is less inhibitory to luciferase than most other ions. With the addition of 0.5–2 m*M* EDTA this buffer gives highly stable ATP solutions and is not destroyed by microbial growth. When using activators like PP_i, it is important to use an optimum concentration, taking into account the possibility that the activator might be degraded in the reagent or by the sample. The optimum concentration of PP_i is around 1 μM, whereas higher concentrations are inhibitory. Luciferase contains SH groups, and reagent manufacturers often add SH compounds such as DTT to protect the enzyme; however, DTT may change the kinetics. Furthermore, oxidized DTT will inactivate luciferase rapidly. An alternative protective agent is bovine serum albumin (BSA), which also protects luciferase from other types of inactivation. Thus 0.1% BSA is included frequently in ATP reagents. Higher concentrations are inhibitory and should not be used.

Different Types of ATP Reagents

The firefly luciferase reaction can be used to measure ATP in several ways. Before the 1970s it was generally accepted that the reaction resulted in a flash of light that could be estimated by various techniques. However, with the reagents available in those days, only the peak height gave reliable results.[9] Introduction of the ATP-monitoring concept[3,10,11] changed the picture. This concept is based on the idea that during the measurement the luciferase activity should be constant and degrade only a negligible fraction of the ATP concentration (<1%/min). Under these conditions the ratio between the intensity of the light and the ATP concentration is essentially the same throughout the measurement. If ATP is constant, as in a biomass assay, the light is constant. This obviously simplifies light measurements. If ATP is formed or degraded as in coupled assays of enzymes or metabolites, the light goes up or down in parallel. Consequently, ATP can be monitored in the same way (but at much lower levels) as NAD(P)H can be monitored at 340 nm in the spectrophotometer. The ATP monitoring concept proved very useful, and today most manufacturers of ATP reagents advocate constant or stable light as an advantage of their reagents.

When sensitivity is an important issue, the ATP monitoring concept has a serious disadvantage. Because only negligible amounts of ATP are degraded per minute, the light intensity must remain low. We have looked

[9] A. Lundin and A. Thore, *Anal. Biochem.* **66,** 47 (1975).

[10] A. Lundin, A. Rickardsson, and A. Thore, *Anal. Biochem.* **75,** 611 (1976).

[11] A. Lundin, *in* "Bioluminescence and Chemiluminescence: Basic Chemistry and Analytical Applications" (M. DeLuca and W. McElroy, eds.), p. 187. Academic Press, New York, 1981.

into this problem in an effort to develop more sensitive reagents mainly for the detection of low numbers of microbes. Three types of reagent kinetics were identified, each with fairly distinct application areas, as summarized in Table I:

1. Stable light-emitting reagents. These reagents should have a decay rate <1%/min corresponding to a rate constant of <0.01 min^{-1} and can be used for ATP monitoring, e.g., in assays of enzymes and metabolites. The detection limit is typically 1000 amol, which is sufficient for the measurement of cytotoxicity and cell proliferation but not for low numbers of bacterial cells. The stable light makes it possible to mix reagent and sample outside the luminometer and to measure for any convenient time.
2. Slow decay reagents. These reagents typically have decay rates around 10%/min corresponding to a rate constant of 0.1 min^{-1}. These reagents can be formulated to degrade their own ATP blank during preparation. Detection limits are frequently in the 10–100 amol range corresponding to 5–50 bacterial cells. Reagent and sample may be mixed outside the luminometer, but the measurement must be done as soon as possible, as the light decays. For accurate assays, injection

TABLE I
PROPERTIES AND APPLICATIONS OF DIFFERENT TYPES OF ATP REAGENTS[a]

	Stable light reagent	Slow decay reagent	Flash reagent
Typical rate constants[a]	0.005 min^{-1}	0.10 min^{-1}	2.35 min^{-1}
$t_{1/2}$ of decay[a,b]	139 min	6.9 min	0.3 min
$t_{9/10}$ of decay[a,b]	461 min	23 min	1 min
Reagent blank per ml[a]	>1000 amol	10–100 amol	<1 amol
Detection limit[a]	>500 bacterial cells	5–50 bacterial cells	<0.5 bacterial cells
Sterility testing	No	No	Yes
Cytotoxicity/cell proliferation	Yes	Yes	Not required
Hygiene monitoring	Yes	Yes	Not required
ATP monitoring	Yes	No	No
Reagent dispenser required	No	No	Yes
Light measurement	Intensity	Intensity	Total integrated light
Affected by luciferase activity	Yes	Yes	No

[a] Kinetic properties and detection limits refer to ATP reagent SL, ATP reagent HS, and ATP reagent SS (BioThema, Dalarö, Sweden).

[b] $t_{1/2}$ mean that 50% of the ATP and the light intensity remains. $t_{9/10}$ means that 90% of the ATP is consumed and that 10% of the light intensity remains.

of the reagent in the measuring position of the luminometer is recommended.

3. Flash reagents. These reagents typically have decay rates >100%/min corresponding to a rate constant of >1 min^{-1}. With such a decay rate the reagent blank is no longer a problem. A detection limit around 1 amol corresponding to half a bacterial cell has been achieved. With injection of the reagent in the measuring position and at least two measurements of the light intensity on the decay of the luminescence, the total integrated light can be estimated as shown later. This is a completely new way of performing ATP assays with several interesting analytical characteristics, as is described later.

Kinetics of Firefly Luciferase Reaction

The time course of the decay of the light emission from flash reagents has not been studied previously in detail. It turns out that the kinetics can be utilized to obtain completely new measurement principles. An understanding of these principles is therefore important. In the following discussion, two assumptions apply:

1. Luciferase is neither inactivated by the formation of enzyme–product complex nor by sample components, e.g., extractants.
2. There are no ATP-degrading enzymes other than firefly luciferase in the reaction mixture.

Assuming that firefly luciferase follows the Michaelis–Menten equation, the rate of ATP degradation can be expressed by Eq. (5):

$$v/V_{max} = S/(S + K_m) \tag{5}$$

In most ATP assays the ATP concentration (S) is very low compared to the Michaelis–Menten constant (K_m), and Eq. (5) can be simplified to

$$v/S = V_{max}/K_m \quad \text{or} \quad v = k\,S \tag{6}$$

where $k = V_{max}/K_m$ is the first order rate constant for the degradation of ATP. The unit of k is min^{-1}; k is a measure of the fraction of ATP degraded per minute, which is independent of the ATP concentration at ATP $\ll K_m$. Because $v = -d[ATP]/dt$ and $S = [ATP]$, Eq. (6) may be written

$$-d[ATP]/dt = k[ATP] \quad \text{or} \quad -d[ATP]/[ATP] = k\,dt \tag{7}$$

The integrated form of Eq. (7) is

$$[ATP_t] = [ATP_0]\,e^{-kt} \tag{8}$$

Assuming that the factor between light intensity (I) and ATP concentration is the same throughout the measurement (cf. above), Eq. (8) may be replaced by

$$I_t = I_0\, e^{-kt} \tag{9}$$

The exponential decay of the light emission will have the same k value regardless of where on the curve it is measured, as is shown in Fig. 1. Therefore k can be calculated from light emission values obtained at any two times ($t1$ and $t2$) by Eq. (10):

$$k = (\ln I_{t1} - \ln I_{t2})/(t2 - t1) \tag{10}$$

Using the k value from Eq. (10) and measuring the light emission I_{t1} at time $t1$, an extrapolated peak light intensity, I_0, can be calculated from Eq. (9) in the form:

$$I_0 = I_{t1}/e^{-kt1} \tag{11}$$

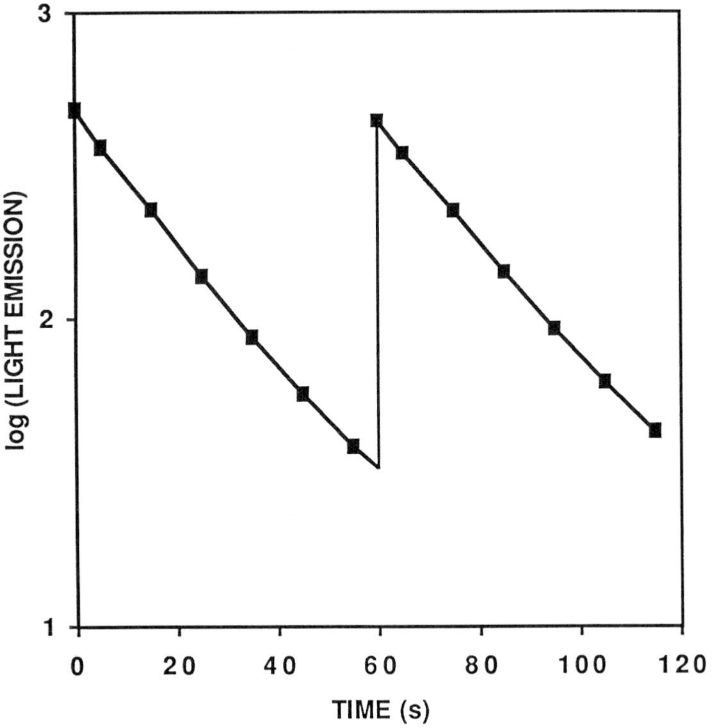

FIG. 1. Kinetics of decay of light emission after two subsequent additions of ATP (10^{-14} mol). The logarithmic scale is used to show exponential decay.

The extrapolated peak light intensity would theoretically be slightly higher than the actual peak height. The latter is affected by the rate-limiting steps in the luciferase reaction. Even when measured with stopped flow, the luciferase reaction shows a 25-msec lag before any light appears and a further 0.3-sec time to reach the peak emission.[12] Adding the firefly reagent to an ATP solution with a normal dispenser will result in an even longer time to reach the peak emission. The change of the environment from reagent to reaction mixture may also affect luciferase activity. Dispensing often causes an extra flash on top of the extrapolated peak emission. Therefore the actual peak emission may be higher as well as lower than the extrapolated peak emission. The actual peak emission is also more difficult to estimate accurately.

The total light emitted (I_{tot}) from peak to extinction can be obtained from the integrated form of Eq. (9):

$$I_{tot} = I_0/k \qquad (12)$$

Equation (12) written in the form $I_0 = kI_{tot}$ is another way of saying $-d[ATP_0]/dt = k[ATP_0]$, which follows from Eq. (7). The extrapolated peak light intensity I_0 and $-d[ATP_0]/dt$ are measures of reaction rates. Total integrated light I_{tot} is a measure of the total amount of light emitted from the reaction mixture as the ATP concentration goes from $[ATP_0]$ to zero. Multiplying $[ATP_0]$ by the reaction volume and the quantum yield will give the total number of photons emitted. Consequently, a measurement of I_{tot} can be used to calibrate the luminometer in photons per second rather than counts per second or relative light units (rlu).

Equation (12) emphasizes the fact that if the quantum yield is constant, all inhibitors and activators will have the same effect on extrapolated peak light intensity (I_0) and decay rate (k). Measuring I_t at various t makes it possible to calculate k, I_0, and I_{tot}. At a certain luciferase activity the k value should be the same for all ATP concentrations well below K_m, whereas I_0 and I_{tot} should be proportional to the ATP concentration. k and I_0 should be affected by inhibitors and activators of the luciferase reaction, whereas I_{tot} should only be affected by changes in the quantum yield. The addition of an ATP standard will give a new set of I_0 and I_{tot} values. These values may be used as internal standard values after subtraction of the remaining light emission from the unknown sample ATP. The remaining light intensity from the sample may be obtained by extrapolating to the time of ATP standard addition. The remaining total light emission may be obtained from the extrapolated total light emission minus the measured light emission until the time of ATP standard addition. I_0 and I_{tot} are extrapolated values

[12] M. DeLuca and W. D. McElroy, *Biochemistry* **13,** 921 (1974).

based on the decay part of the light emission curve. Thus the measurement requires neither rapid mixing to collect the peak nor an instrument with an electronic background absolutely stable over long times.

Figure 2 shows an experiment with different amounts of added ATP using ATP reagent SS (BioThema, Dalarö, Sweden) and a 1251 luminometer (Bio-Orbit Oy, Turku, Finland). Every assay was calibrated by secondary addition of an internal ATP standard, as discussed in the preceding paragraph. Measured ATP values were calculated from both I_0 and I_{tot} values as described in the preceding paragraph, with identical results. The major advantage of using I_{tot} rather than I_0 becomes apparent when internal ATP standards are not used. I_0 is affected by changes in luciferase activity, but I_{tot} is not. (I_{tot} is affected only by changes in quantum yield.)

The previous discussion applies only to assays of ATP. In assays of firefly luciferase (reporter gene technology), the ATP level is saturating

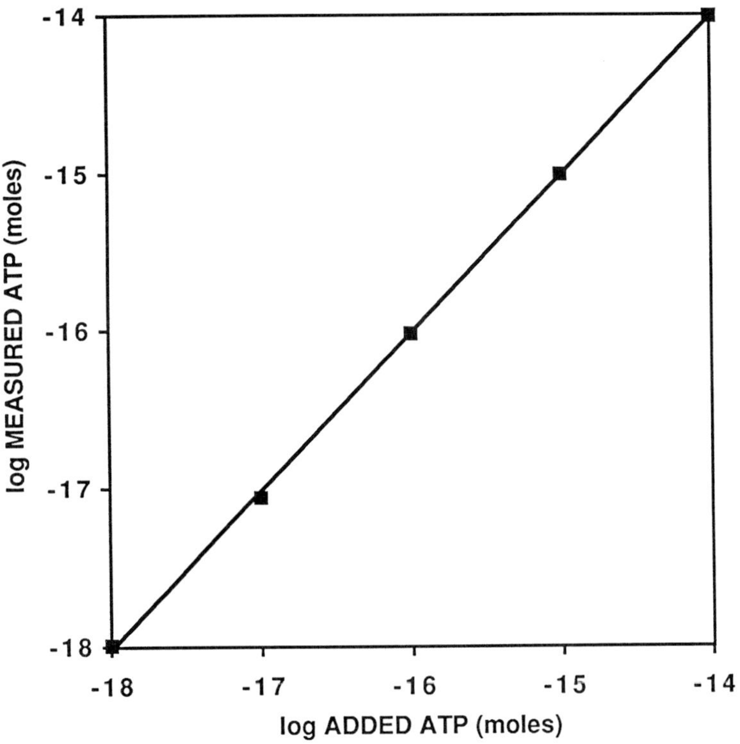

FIG. 2. Measured ATP as a function of added ATP based on the new method for utilizing the kinetics of light emission. Each assay was calibrated with a secondary addition of an internal ATP standard (10^{-14} mol). Amounts of ATP are plotted on logarithmic scales.

and the decay of the light is normally an effect of luciferase being inactivated. While developing optimized assays of firefly luciferase[4] it was found that reaction conditions leading to maximum light emission always resulted in some decay of the light during the reaction (approximately 4%/min under optimized conditions). After the light had gone down, a second addition of luciferase had the same initial activity as the luciferase added at the start of the reaction. Although the optimized assay included enhancers for the release of the enzyme from the enzyme–product complex, it cannot be ruled out that accumulation of the enzyme–product complex is the reason for the loss of activity.

Calibration with ATP Standard

The result of an assay of a chemical substance should be expressed in moles or grams. The luminometer provides only relative measures (cps, rlu, or mV). These measures can only be related to moles or grams by calibrating the measurement with a standard, i.e., a known amount of the analyte. The use of an ATP standard compensates for the following variations:

1. Luminometer sensitivity. Luminometers from the same manufacturer and even of the same model may vary. Furthermore, the luminometer may change its sensitivity for a variety of reasons. One example is chemicals splashing on reflectors changing the light collection efficiency.
2. Reagent sensitivity. Despite the efforts of manufacturers to standardize reagent preparation, certain variations must be accepted. Furthermore, the ATP reagent that is absolutely stable after reconstitution still remains to be developed. Even if luciferase can be more or less stabilized, the D-luciferin is sensitive to light and oxygen.
3. Inhibition from sample components. Biological samples often contain substances that are highly inhibitory to the luciferase reaction. Furthermore, the emitted light may be partially quenched by colored substances or particles. Swabs used in surface hygiene monitoring assays left in the cuvette will have a quenching effect. Inhibition from extractants used for releasing intracellular ATP is a special problem that will be discussed later.

Calibration with ATP standards is facilitated greatly by the fact that, as far as we know, there is an absolute linearity between light emission and ATP concentrations $\ll K_m$, i.e., up to 1 μM. The luminometer gives a signal that should be related linearly to light emission up to a certain limit. These two assumptions are checked easily by running an ATP standard curve covering several orders of magnitude. A double logarithmic plot is prepared after subtracting blanks. The slope of the plot should be 1 within experimental error. A deviation from linearity is often seen (particularly with photon-counting luminometers) at the upper end of the standard curve. If the linearity between ATP concentration and luminometer signal has been confirmed, all future experiments can be performed using only one accurately determined ATP standard concentration.

The calibration may be performed in three different ways depending on the application. If variations from all three sources described earlier are to be compensated, it is, in general, necessary to use an internal ATP standard. Such a standard can often be added to the reaction mixture after measurement of the signal from the sample. The increase of the light emission due to the addition of the ATP standard is used to calculate a ratio between rlu and ATP concentration. The ratio is used to convert rlu values to ATP concentration in the sample. The volume of the ATP standard must be small in order not to dilute the reaction mixture significantly, thereby changing the reaction conditions. The light emission is proportional to the concentrations of ATP, luciferase, and, if nonoptimized conditions are used, luciferin. Depending on the light collection properties of the luminometer, the light signal may also be proportional to the reaction volume. If the luciferin concentration is optimized, the volume before addition of the ATP standard is set to 1, and the volume fraction of the ATP standard is v, the total effect will be a factor $(1 + v)^{-2}(1 + v) \approx 1 - v$. However, with a luciferin level well below K_m and no light collected from the increased volume, the total effect will be a factor $(1 + v)^{-3}(1 + 0) \approx 1 - 3v$. The effect of an ATP standard volume being 10% of the total reaction volume may decrease the anticipated light signal by 10–30%. The anticipated increased light signal from the ATP standard is measured on top of the signal before the addition of standard. The dilution will obviously affect the signal from both sample and standard ATP. If the sample ATP is comparable to or higher than the standard ATP, the error in estimating the ratio between rlu and ATP concentration may be considerable. The dilution effects may be compensated mathematically. However, an internal ATP standard is only advantageous if sample components result in variable inhibition or activation. These effects cannot, in general, be compensated mathematically. The conclusions are as follows:

1. The level of D-luciferin should be optimized.
2. The total reaction volume should not exceed the limits of the light collection properties of the luminometer.
3. An internal ATP standard should be added in a volume that is negligible compared to the total reaction volume. Accurate volumes

below 10 μl cannot be obtained with conventional dispensers. Consequently, we have to accept that with most luminometers the ATP standard should be added as 10 μl in a total reaction volume of 1 ml.

4. The ATP standard should be as high as allowed by the linear range of the assay (limit usually set by the linear range of the luminometer and always <1 μM final ATP concentration in reaction mixture).

If these requirements are not fulfilled, it is necessary to run two assays for every sample. The first assay would be a sample diluted in buffer. The second assay would be a sample diluted in ATP standard dissolved in the same buffer. This technique is often referred to as spiking. The technique is obviously more laborious, as each sample leads to two measurements. However, the use of an automatic microplate luminometer for adding the buffer and the ATP standard partially compensates for this. Furthermore, the reagent consumption is lower in a microplate luminometer compared with that in a cuvette luminometer (2×0.2 ml as compared to a 1-ml reaction volume).

If no interference from sample components is expected, it is only necessary to compensate for variations from luminometer sensitivity and from reagent activity. This is best done by running an ATP standard diluted in the same medium as the actual samples. Such a standard should be included in each series of assays. This is obviously the most convenient procedure but is limited to situations where there is no analytical interference from samples or where such interference is identical in all samples. If this is not the case, one of the other two methods must be used. Alternatively, one can dilute the sample to avoid analytical interference and use a more sensitive reagent.

Biomass Assays

Introduction

The central role of ATP in intracellular metabolism makes it necessary for the cell to maintain a fairly constant intracellular concentration of ATP, which is similar in all cell types. The amount of ATP per cell is therefore proportional to the intracellular volume per cell. However, the correlation between ATP and colony-forming units (cfu) is often rather poor. The major explanation is the clumping of cells to each other or to debris in the sample. This effect has been described even in simple model systems such as a culture of *Escherichia coli* growing overnight in nutrient broth. Consequently, one colony-forming unit may consist of several cells. Another reason for poor correlation is failure to remove extracellular ATP by an appropriate pretreatment of the sample. In the following discussion, a number of methodological aspects will be described, followed by a few application examples.

Collection and Pretreatment of Samples

When designing sample collection procedures, one must be aware of the rapid turnover of the intracellular ATP pool. If the cells are subjected to unfavorable conditions, their ATP concentration may fall dramatically within a second. However, the intracellular metabolism tries to compensate for any such changes with the same speed. With the exception of physiological studies aimed at measuring precisely the intracellular ATP concentration, it is generally adequate simply to try to avoid changes to nonphysiological conditions. The concentration of cells by centrifugation results in a pellet in which conditions are far from physiological. A better alternative is to suck up the sample in a syringe and collect the cells on a disposable filter. If washing is necessary, it should be done with a sterile and ATP-free medium that will not affect the intracellular ATP concentration. Finally, the intracellular ATP is extracted from the filter by a flow of extractant.

The removal of nonmicrobial ATP may sometimes be accomplished using an ATP-degrading enzyme such as apyrase. If the sample contains somatic cells, these may sometimes be lysed with a mild detergent like Triton X-100. Pretreatment with apyrase and Triton X-100 can be performed in one step and has been shown to be effective in bacteriuria testing.[13–19] However, it does not work with all somatic cells, particularly if the cells are not present as individual cells. It is necessary to confirm that this pretreatment works with the type of cells and samples that are to be assayed.

In surface hygiene monitoring, samples are collected by swabs. It is known from bacteriology that swabbing is an extremely poor way of collecting bacterial cells. It particularly does not work with biofilms.

[13] A. Thore, S. Ånséhn, A. Lundin, and S. Bergman, *J. Clin. Microbiol.* **1,** 1 (1975).

[14] A. Thore, A. Lundin, and S. Ånséhn, *J. Clin. Microbiol.* **17,** 218 (1983).

[15] A. Lundin, H. Hallander, A. Kallner, U. Karnell Lundin, and E. Österberg, *in* "Rapid Methods and Automation in Microbiology and Immunology" (K.-O. Habermehl, ed.), p. 455. Springer-Verlag, Berlin, 1985.

[16] H. Hallander, A. Kallner, A. Lundin, and E. Österberg, *Acta Path. Microbiol. Immunol. Scand. B* **94,** 39 (1986).

[17] B. Gästrin, R. Gustafsson, and A. Lundin, *Scand. J. Infect. Dis.* **21,** 409 (1989).

[18] A. Lundin, H. Hallander, A. Kallner, U. Karnell Lundin, and E. Österberg, *J. Biolumin. Chemilumin.* **4,** 381 (1989).

[19] E. Österberg, H. Hallander, A. Kallner, A. Lundin, and H. Åberg, *Eur. J. Clin. Microbiol.* **10,** 70 (1991).

Extraction of Intracellular ATP

The extraction of intracellular ATP has two objectives: (1) the release of ATP from within the cell to the outside medium and (2) the immediate inactivation of all ATP-converting enzymes.

The first objective is fairly straightforward, although the potential existence of several more or less strongly bound ATP pools may be difficult to evaluate. The second objective follows from the rapid turnover of the ATP pool. If the ATP-converting enzymes are not inactivated simultaneously with the opening of the cell membranes, ATP will be degraded rapidly by the high concentrations of ATP-degrading enzyme systems in the cells. Furthermore, the long-term stability of ATP in the extract can only be achieved if ATP-converting enzymes are inactivated. These enzymes should preferably be inactivated irreversibly rather than simply inhibited, as the activity may otherwise come back during the assay. This is particularly important if metabolites other than ATP are to be assayed, e.g., ADP and AMP. The most reliable extractant is one that opens up the cell walls and inactivates the enzymes immediately. Such agents include strong acids with a chaotropic anion, e.g., trichloroacetic acid. It is, however, possible to inhibit ATP-converting enzymes rapidly by including EDTA in the extractant solution and using an extractant that inactivates the enzymes more slowly, e.g., quaternary ammonium compounds.

Extraction of ATP is the most overlooked problem in ATP technology and the literature is crowded with serious mistakes. The problem is also difficult to asses, as we cannot perform an ATP assay easily without extracting the cells. As soon as the extractant solution is added, the intracellular enzymes will start to degrade the ATP. Consequently, the best extractant can only be found by comparing several different types of extractants each used in several concentrations. Over the years this has been done for more than 20 different cell types and several types of mixed samples in the author's laboratory. Parts of these data have been published.[20–23] The general conclusions are:

1. An optimum extractant concentration giving a maximum ATP yield can be found for each combination of cell type and extractant. The

[20] A. Lundin and A. Thore, *Appl. Microbiol.* **30,** 713 (1975).

[21] A. Lundin, *in* "Analytical Applications of Bioluminescence and Chemiluminescence" (L. Kricka, P. Stanley, G. Thorpe, and T. Whitehead, eds.), p. 491. Academic Press, New York, 1984.

[22] A. Lundin M. Hasenson, J. Persson, and Å. Pousette, *Methods Enzymol.* **133,** 27 (1986).

[23] S. E. Hoffner, C. A. Jimenez-Misas, and A. Lundin, *in* "Bioluminescence and Chemiluminescence" (A. K. Campbell, L. J. Kricka, and P. E. Stanley, eds.), p. 399. Wiley, Chichester, 1994.

optimum concentration is dependent on the medium in which the cells are suspended. Extractants based on acids are obviously counteracted by a high buffering strength. Extractants based on protein-inactivating detergents are neutralized by protein. Consequently, minor changes in the sample composition may, if the extractant composition is not optimal, change the ATP yield considerably.

2. The most reliable extractant is trichloroacetic acid (TCA). One should compare 10, 5, and 2.5% TCA and use the optimum concentration as the reference extraction method. Two exceptions to this rule have been found. With mycobacteria, hot dodecyl trimethyl ammonium bromide (DTAB) with EDTA gives a higher ATP yield than TCA. With some algae, perchloric acid gives a higher yield than TCA.
3. Several other extractants, e.g., DTAB with EDTA, can be used for the extraction of ATP with most types of cells. With each new type of cells or type of sample the optimum extractant concentration must be determined and compared with the optimum TCA concentration.

Neutralization of Extractants

A reliable extractant should inactivate all ATP-converting enzymes. Therefore it is an inherent property of a good extractant to interfere with the luciferase reaction. Extracts are normally diluted to some extent in the assay. The interference can then be either inhibition or a time-dependent inactivation of luciferase. A moderate degree of inhibition, as with TCA, can be tolerated if assays are calibrated by the use of an internal ATP standard. A time-dependent inactivation, as with some detergents, is more difficult to compensate, although performing two assays with and without spiking with a known amount of ATP can be used. An internal ATP standard cannot be used easily, as this would be added when the activity has already gone down and would give too low a value. This was a serious problem for a long time, as quaternary ammonium compounds such as DTAB are among the most reliable extractants and are less hazardous to work with than TCA.

In 1978, addition of a protein was suggested as a means of protecting luciferase from inactivation by the effect of quaternary ammonium compounds.[24] However, the levels of BSA that had to be used were strongly inhibitory. Simpson and Hammond[25] were more successful using neutral detergents to neutralize the effect of quaternary ammonium compounds.

[24] S. Ånséhn, A. Lundin, L. Nilsson, and A. Thore, *in* "Proceedings: International Symposium on Analytical Applications of Bioluminescence and Chemiluminescence" (E. Schram and P. E. Stanley, eds.), p. 438. State Printing & Publishing, Inc., Westlake Village, CA, 1979.

[25] W. J. Simpson and R. M. Hammond, U.S. Patent 5,004,684 (1991).

An even better alternative was the use of cyclodextrins.[23,26,27] Cyclodextrins (α-, β-, and γ-) consist of six to eight glucose entities in a doughnut-shaped molecule. The interior of the ring is hydrophobic and was known to form strong complexes with the hydrophobic tails of detergents. It was found[26] that the inactivating effect of cationic, anionic, and amphoteric detergents could be obviated by adding an appropriate amount of cyclodextrin. When the appropriate cyclodextrin (α-, β-, or γ-) was chosen, depending on the size of the detergent, the reaction was like a stoichiometric titration and could be performed in a single cuvette by adding aliquots of detergent to a known amount of cyclodextrin dissolved in a stable light-emitting ATP reagent containing ATP. The cyclodextrin should be added to the buffer in which the extract is diluted. Because the reaction between cyclodextrin and detergent is rapid, it is also possible to have the cyclodextrin in the ATP reagent, provided that the extract and ATP reagent are mixed rapidly. Cyclodextrins also form complexes with D-luciferin and this should be considered when optimizing the system. However, the binding affinity of cyclodextrins for detergents is stronger than with D-luciferin, and the cyclodextrin should only be present in a small excess over the detergent.

Sterility Testing

With a detection limit of 1 amol using flash-type reagents, the possibility for sterility testing is obvious, as most bacterial species contain approximately 2 amol per cell.[13] However, the single cell to be detected must be present in a volume that is lower than the total reaction volume, i.e., at least <1 ml. If sterility means less than 1 cell per 1 ml, the bacterial cells must be concentrated. This can be achieved on a disposable microfilter unit. With such a unit it is also possible to remove free ATP by washing with ATP-free medium. Finally, the bacterial ATP is released from the bacteria on the filter by running extraction solution through the filter. The minimum volume should be used. It may be difficult to elute all ATP by running the extractant in one direction only. Consequently, it may be necessary to run the solution back and forth through the filter. The ATP concentration is determined in an aliquot of the extract.

Bacteriuria Assays

Bacteriuria testing is perhaps the most frequent bacteriological assay (>200 million assays per year). Urine contains free ATP, somatic cell ATP,

[26] A. Lundin, J. Anson, and P. Kau, *in* "Bioluminescence and Chemiluminescence" (A. K. Campbell, L. J. Kricka, and P. E. Stanley, eds.), p. 399. Wiley, Chichester, 1994.
[27] A. Lundin, J. Anson, and P. Kau, U.S. Patent 5,558,986 (1996).

and bacterial ATP. Several groups have shown that the combined treatment with apyrase and Triton X-100 removes nonbacterial ATP and that the remaining bacterial ATP can be successfully used to pick out positive samples.[13–19] ATP cutoff limits for a positive assay generally correspond to higher cell numbers than would be expected from the cutoff limit in colony-forming units per milliliter. Possible reasons include (1) the removal of nonbacterial ATP was incomplete, or (2) the average colony-forming unit consists of more than one bacterial cell. Regardless of the explanation, only a small number of patients with urinary tract infections produce samples with bacterial levels close to the cutoff limits.

In a study performed at four health care centers, 781 samples were assayed with the following techniques: bacterial ATP, dipslide, nitrite, granulocyte esterase, microscopic identification of bacteria, and white blood cells.[16] A conventional culture was used as the reference method. Bacterial ATP was found to give the highest diagnostic efficiency (percentage of correctly classified samples as compared to the reference method).

A rapid bacteriuria test is required when the doctor needs the result before deciding on diagnosis, e.g., at health care centers or in the doctor's office. Unfortunately, most studies on bacteriuria have been performed at hospitals. At the hospital, most of the samples have been sent in from outside the hospital just to confirm the diagnosis already made. At the hospital laboratory, an identification and often sensitivity testing are requested. The hospital laboratory is also where various companies have tried to introduce kits for bacterial ATP.

Antibiotic Assays and Sensitivity Testing

Antibiotics prevent sensitive bacterial strains from growing or may even lyse the cells. Consequently, the concentration of an antibiotic may be determined by adding the sample to a standardized culture of a sensitive strain, incubating for 1–2 hr, and finally measuring bacterial ATP[28] or, in some situations, extracellular ATP.[28] Similarly, susceptibility testing can be done by culturing the unknown strain in the presence of various antibiotics followed by measuring bacterial ATP.[28] Although not as rapid as direct chemical methods, ATP technology is a bioassay suitable for high-throughput screening.

Cytotoxicity and Cell Proliferation Assays

Cytotoxicity and cell proliferation assays are done by adding the test substance to a cell culture and comparing the growth measured as ATP to

[28] L. Nilsson, *in* "Rapid Methods and Automation in Microbiology and Immunology" (K.-O. Habermehl, ed.), p. 448. Springer-Verlag, Berlin, 1985.

a control without the substance. The rapid technology can be used in deciding the best therapeutic agents against cancer cells obtained from patients.[29–30] It will also become a valuable tool in high-throughput screening in the development of new drugs.

Hygiene Monitoring

Hazard Analysis Critical Control Point (HACCP) is being introduced in the food industry all over the world. In many countries, e.g., those of the European Union, HACCP is implemented by law. The objective is to identify critical control points in which hazards may affect food safety. Such critical points should be monitored using methods that are rapid enough to allow corrective actions to be taken. The hygiene of food contact surfaces after cleaning and disinfection, before starting a new production batch, is such a critical control point. Previously the hygiene has been estimated by visual inspection (absence of food residues) and by bacteriological methods. The former method is rapid but the sensitivity is not adequate in most situations. The second method requires incubation for 3–7 days before results are available and is completely unsuitable for HACCP. The ATP method has therefore become the major alternative, providing both instant results and an adequate sensitivity. Total ATP is measured, i.e., no effort is made to distinguish between ATP from food residues and from microbial cells. In most situations, microbial ATP is only a minor fraction of the total ATP measured. Food residues act as a growth medium for microbes and counteract disinfection. Removal of food residues is therefore required to avoid reinfection, e.g., from the air. However, critical limits for acceptable ATP levels have to be established and can obviously not be calculated from acceptable levels of cfu.

Most kits for hygiene monitoring do not include ATP standards. Critical ATP levels expressed in rlu cannot be recalculated easily to different kits or different luminometers. The experiences obtained in one food factory can therefore not be shared with other factories. Consequently, each factory has to establish its own critical ATP levels in each sampling location, a factor that seriously limits the spreading of ATP technology. Regulatory authorities will most likely have to intervene to resolve this issue.

Hygiene monitoring can be done using swabs to collect samples from surfaces or by liquid sampling from rinsing water in cleaning-in-place plants. Swabbing is an inherently inaccurate technology in terms of the efficiency

[29] A. S. Rhedin, U. Tidefelt, K. Jönsson, A. Lundin, and C. Paul, *Leukemia Res.* **17,** 271 (1993).

[30] P. E. Andreotti, D. Linder, D. M. Hartmann, I. A. Cree, M. Pazzagli, and H. W. Bruckner, *J. Biolumin. Chemilumin.* **9,** 373 (1994).

with which the amount of material on the surface is taken up by the swab. Furthermore, there is a considerable variation of food residues actually being present on adjacent surfaces. The sampling technique presently used to swab a 10 × 10-cm area obviously needs refinement. From bacteriology swabbing is known to collect only a few percent of the bacteria present on a surface. It is impossible to collect a good sample from biofilms using swabs that have been wetted with saline as is recommended with many kits. Adding the extractant directly on the surface, e.g., with a dropper bottle, will ideally release ATP even from the biofilm. The extracted ATP may then be collected almost quantitatively with a swab. Rotating the swab in the ATP reagent will eventually release most of the ATP from the swab. Disinfectant on the surface may interfere with the luciferase reaction but can be counteracted by additives to the reagents. Therefore each series of assays can be calibrated easily by running an external ATP standard and results can be expressed as picograms or femtomoles of ATP.

Because total ATP rather than bacterial ATP is measured, one should not expect a correlation with cfu. However, for statistical reasons, if the hygiene is improved as measured with ATP, it will also be improved as measured with culturing techniques. The most relevant correlation one can do is not to measure ATP and cfu at the same time, but to assay ATP directly after cleaning and disinfection and to do a culture at an adjacent sampling site immediately before starting the next production batch. This comparison demonstrates the major advantage of the ATP test: this test provides an early warning for future bacterial growth on poorly cleaned surfaces. A high ATP level, even if the surface is sterile, indicates food residues that may result in considerable growth before production is started the next morning.

Assays of Enzymes and Metabolites

Introduction

Any enzyme or metabolite that can be coupled to a reaction forming or degrading ATP can be assayed using firefly luciferase. Complex reactions such as oxidative phosphorylation, photophosphorylation, lysis of cells, or platelet aggregation may be measured in the same way. The ATP-converting reaction can proceed in a separate reaction mixture with subsequent mixing of an aliquot with an ATP reagent. Whenever possible it is less laborious and more accurate to allow the ATP-converting reaction to proceed in the presence of a stable light-emitting ATP reagent, monitoring the light emission continuously. The calibration of the assay can be done by adding an internal ATP standard or by making a standard curve for the analyte. The following discussion illustrates some examples of assays of metabolites (end point and kinetic), assays of enzymes, and monitoring of more complex reactions.

ATP/ADP/AMP

This end point assay has been described in detail in a previous volume of this series.[22] The assay is performed in four steps in a single cuvette containing the sample:

1. Addition of a stable light ATP reagent (with the addition of potassium ions) to the sample, with measurement of the light emission corresponding to the ATP level.
2. Addition of pyruvate kinase plus phosphoenolpyruvate (PEP) to convert ADP to ATP (ADP + PEP → ATP + pyruvate), measuring the light emission corresponding to the level of ATP + ADP.
3. Addition of adenylate kinase plus CTP to convert AMP to ADP (AMP + CTP → ADP + CDP), followed by the conversion of ADP to ATP by the pyruvate kinase reaction. The measurement of the light corresponds to ATP + ADP + AMP.
4. Addition of an internal ATP standard.

A few procedural notes should be observed:

1. Without the additon of CTP, low concentrations of ATP as well as AMP will result in a long time to reach the end point of the adenylate kinase reaction (AMP + ATP → ADP + ADP). CTP is used to reach the end point more rapidly and in the same time for all samples. Advantage is taken of the poor specificity of adenylate kinase at the ATP-binding site and the high specificity of firefly luciferase for ATP.
2. The factor between rlu and moles of ATP obtained in step 4 is valid for step 3. If the additions of pyruvate kinase and adenylate kinase affect the luciferase activity, it is necessary to obtain correction factors for this by running a blank with an internal ATP standard added after steps 1, 2, and 3.
3. Commercial preparations of the auxiliary enzymes and metabolites contain significant levels of adenine nucleotides (particularly CTP). A blank is therefore required in each series of assays.
4. Detailed instructions and software for running the ATP/ADP/AMP assay can be obtained from the author.

ATP/Phosphocreatine

In muscle tissue the ATP level is maintained at an essentially constant level by the creatine kinase reaction (ADP + phosphocreatine → ATP +

creatine). Muscular activity will degrade the phosphocreatine pool. Two types of muscle fibers exist, with different ways of regenerating ATP and creatine phosphate (glycolysis and oxidative phosphorylation). Lyophilization of muscle samples allows single muscle cells to be isolated under the microscope. Isolated cells can be divided into two parts and weighed on a special balance. A method to determine the amounts of ATP and phosphocreatine in one of the parts has been developed.[31] This end point assay is similar to the assay of ATP/ADP/AMP and is performed automatically in a single cuvette as follows:

1. Add the extract containing ATP and phosphocreatine to a stable light reagent and measure light intensity.
2. Add ADP substrate also containing AMP and diadenosine pentaphosphate (to inhibit adenylate kinase) and measure the increase in light intensity from contaminating ATP.
3. Add creatine kinase and measure the increase in light intensity.
4. Add an internal ATP standard.

Glycerol

Lipolysis results in the formation of glycerol and free fatty acids. In fat tissues the glycerol (in contrast to the free fatty acids) is not metabolized. The rate of glycerol formation, e.g., in suspensions of fat cells, is therefore a good measure of lipolysis and is influenced by a large number of hormones. An automated assay of glycerol for this type of study has been developed using stable light ATP reagents.[32,33] The assay is kinetic and is based on the first-order degradation of ATP in the glycerol kinase reaction: ATP + glycerol → ADP + glycerol phosphate. The assay is performed as follows:

1. A stable light reagent is added to the glycerol samples and loaded in the cuvettes in an automatic luminometer.
2. The luminometer adds ATP and glycerol kinase. During the degradation of ATP the light intensity is measured twice (12 and 72 sec after the addition of glycerol kinase).
3. The rate of the reaction is calculated from two light measurements using the k value (percentage of ATP consumed per minute) obtained by Eq. (10) as a measure of reaction rate.

Each day a glycerol standard curve is measured. Instead of plotting

[31] R. Wibom, K. Söderlund, A. Lundin, and E. Hultman, *J. Biolumin. Chemilumin.* **6,** 123 (1991).

[32] A. Lundin, P. Arner, and J. Héllmer, *Anal. Biochem.* **177,** 125 (1989).

[33] J. Hellmér, P. Arner, and A. Lundin, *Anal. Biochem.* **177,** 132 (1989).

reaction rate (v) versus glycerol concentration (S), advantage is taken of the fact that the reaction follows the Michaelis–Menten equation. The standard curve is done in a way that V_{max} and K_m can be determined. Subsequently, v/V_{max} is plotted versus $S/(S + K_m)$. This kind of plot is linear going from the point (0, 0) to (1, 1) and can also be plotted on a log–log graph. In the glycerol assay a linear plot was confirmed in the range of 0.07–100 μM (K_m for the glycerol kinase used was 40 μM). All the calculations are performed automatically by a Microsoft Excel macro that can be obtained from the author together with a protocol for reagent preparation, setting up the luminometer, and performing the assay. The plot of v/V_{max} versus $S/(S + K_m)$ can be used for any kinetic assay of a metabolite.

Urea

ATP-hydrolyzing urease catalyzes the following reaction: urea + ATP → $2NH_3$ + CO_2 + ADP + P_i. The assay of urea[34] can be performed like the glycerol assay described earlier. The addition of 1,2-propanediol was found to reduce nonurea-dependent ATPase activity (v_{blank}) and to increase urea-dependent ATPase activity (V). The assay was optimized using statistical design methodology and multivariate data analysis.

Creatine Kinase Isoenzymes

Creatine kinase (CK) catalyzes the reaction ADP + phosphocreatine → ATP + creatine. The enzyme is made up of two subunits (M and B) and exists in the forms MM, MB, and BB. The MM isoenzyme is the predominant form in muscles and the BB isoenzyme is found in the brain. The heart muscle contains mainly MM but also a low percentage of MB isoenzyme. In serum, B subunit activity (CK-B) is a very specific indicator of acute myocardial infarction (AMI) and can be measured by adding an M-subunit inhibiting antibody. Kits for total CK and CK-B using a stable light-emitting ATP reagent were the first clinical kits based on luminometry.[35] The CK-B kit can be used to measure the activity in healthy individuals. An increase in the serum CK-B level in a patient is an extremely specific indicator of AMI and can be measured within a very short time after the onset of the infarction. The corresponding spectrophotometric assay of CK-B could only be performed at the peak serum value appearing 12–24 hr after the onset of the AMI. Furthermore, among patients that do not have

[34] B. Näslund, L. Ståhle, A. Lundin, B. Anderstam, P. Arner, and J. Bergström, *Clin. Chem.* **44,** 1964 (1998).

[35] A. Lundin, B. Jäderlund, and T. Lövgren, *Clin. Chem.* **28,** 609 (1982).

an AMI by conventional criteria, those patients that are likely to undergo an AMI within 1 year can be picked out with a high probability using luminometric CK assays (unpublished results). The CK kits were mostly used for research purposes, despite excellent diagnostic results, because a single luminometric assay did not fit in the routines of clinical laboratories in those days. The assay may be more interesting today when clinical laboratories often have luminometers. The optimized analytical procedure is as follows:

1. Add serum to CK reagent containing a stable light-emitting reagent and *N*-acetylcysteine (a CK activator). In the assay of CK-B the CK reagent also contains M-subunit inhibiting antibodies.
2. Add ADP substrate, including AMP and diadenosine pentaphosphate for inhibition of adenylate kinase, and measure residual adenylate kinase activity by measuring the rate of increase of the light signal.
3. Add phosphocreatine and measure CK activity (including residual adenylate kinase activity) by measuring the rate of increase of light signal.
4. Add an internal ATP standard.

An assay of total CK in dried blood spots for screening for Duchenne muscular dystrophy (DMD) has also been developed.[36] DMD is one of the most frequent hereditary diseases affecting 1 in 4000–5000 boys. The test is used in a voluntary CK screening program in Germany to prevent secondary cases in affected families. The assay is performed on blood spots collected on filter paper and sent to the laboratory by mail. The optimized analytical procedure is as follows:

1. Punch out a disk near the center of the blood spot and add a stable light-emitting reagent containing *N*-acetylcysteine (CK activator) and AMP. The degradation of ATP from the blood cells is due to adenylate kinase from the blood (ATP + AMP → ADP + ADP).
2. Add ADP substrate containing diadenosine pentaphosphate for inhibition of adenylate kinase and measure remaining adenylate kinase activity.
3. Add phosphocreatine and measure CK activity.
4. Add an internal ATP standard.

PP_i and DNA Sequencing

ATP sulfurylase catalyzes the reaction PP_i + adenosine 5′-phosphosulfate → ATP + SO_4^{2-}. In the presence of adenosine 5′-phosphosulfate and ATP sulfurylase and using a stable light reagent not containing PP_i, it is possible to do an end point assay of PP_i or monitor PP_i synthesis continu-

[36] G. Scheuerbrandt, A. Lundin, T. Lövgren, and W. Mortier, *Muscle Nerve* **9,** 11 (1986).

ously.[37] Pyrophosphatase activity can be measured as the amount of PP_i remaining after a fixed incubation time.

The PP_i assay can be used to measure RNA and DNA polymerase activity, as these reactions result in the formation of PP_i.[38] By stepwise addition of the four nucleotides (dCTP, dTTP, dGTP, and dATP) in the presence of (exo-) Klenow DNA polymerase, it is possible to perform real-time DNA sequencing.[39] Whenever the right nucleotide is added to the reaction mixture, an amount of PP_i corresponding to the amount of DNA template present is formed and results in ATP formation. The procedure may be automated adding the four nucleotides in different order to four parallel samples. The DNA template may also be bound to a solid support such as magnetic particles, thereby allowing removal of unincorporated nucleotides between different cycles.[40]

Oxidative Phosphorylation and Photophosphorylation

Monitoring of photophosphorylation and ATPase activity in *Rhodospirillum rubrum* chromatophores using a stable light reagent has been described previously in this series.[41] A special luminometer allowing illumination (continuous or flash) during luminescence measurement was used.[42] The flash illumination caused a single turnover of the cyclic electron transport and the corresponding amount of ATP could be measured.[43]

Oxidative phosphorylation in mitochondria using a stable light reagent has been described.[44] The stable light reagent was reconstituted in a solution containing 180 mM sucrose, 35 mM potassium phosphate buffer, and 1 mM EDTA, adjusting pH to 7.5 with NaOH. Commercially available ADP is normally contaminated with 1% ATP and has to be specially purified for this assay. Needle biopses, e.g., from human muscle, weighing as little as 40 mg, were used to prepare mitochondrial suspensions. Before the assay the mitochondrial suspension was diluted 500× in the reconstituted reagent. The analytical procedure is as follows:

1. Prepare cuvettes containing 950 μl stable light reagent, 5 μl purified

[37] P. Nyrén and A. Lundin, *Anal. Biochem.* **151,** 504 (1985).

[38] P. Nyrén, *Anal. Biochem.* **167,** 235 (1987).

[39] M. Ronaghi, S. Karamouhamed, B. Pettersson, M. Uhlén, and P. Nyrén, *Anal. Biochem.* **242,** 84 (1996).

[40] P. Nyrén, B. Pettersson, and M. Uhlén, *Anal. Biochem.* **208,** 171 (1993).

[41] A. Lundin and M. Baltscheffsky, *Methods Enzymol.* **57,** 50 (1978).

[42] A. Lundin, M. Baltscheffsky, and B. Höijer, *in* "Proceedings: International Symposium on Analytical Applications of Bioluminescence and Chemiluminescence" (E. Schram and P. E. Stanley, eds.), p. 339. State Printing & Publishing, Inc., Westlake Village, CA, 1979.

[43] A. Lundin, A. Thore, and M. Baltscheffsky, *FEBS Lett.* **79,** 73 (1977).

[44] R. Wibom, A. Lundin, and E. Hultman, *Scand. J. Clin. Invest.* **50,** 143 (1990).

ADP, 35 μl substrate solution, and finally 10 μl diluted mitochondrial suspension. Different substrates and combinations of substrates can used in different cuvettes, e.g., when screening for metabolic defects.

2. Cuvettes are read repeatedly in an automatic luminometer for approximately 20 min.
3. Add an internal ATP standard and calculate the rate of ATP formation.

Summary and Conclusions

The kinetics of ATP reagents not affected by product inhibition or other forms of inactivation of luciferase during the measurement time has been clarified. Under these conditions the decay rate of the light emission expressed as percentage per minute is a measure of luciferase activity and can be given as the rate constant k (min^{-1}), directly reflecting the degradation of ATP in the luciferase reaction. Three types of reagents with different analytical characteristics and different application possibilities have been identified.

Stable light-emitting reagents are suitable for measurements of ATP down to 1000 amol. This is the only type of reagent suitable for monitoring ATP-converting reactions, i.e., assays of enzymes or metabolites, assays of oxidative phosphorylation, photophosphorylation, and so on.

A higher luciferase activity resulting in a slow decay of the light emission by approximately 10% per minute ($k = 0.1\ min^{-1}$) gives a reagent suitable for measurements down to 10–100 amol. The slow decay of light emission allows use of manual luminometers without reagent dispensers.

A further increase of the luciferase activity resulting in a decay rate of approximately 235% per min ($k = 2.35\ min^{-1}$) and only 10% of the light emission remaining after 1 min is suitable for measurements down to 1 amol corresponding to half a bacterial cell. With this type of flash reagent the total light emission can be calculated from two measurements of the light intensity on the decay part of the light emission curve. This new measure is not affected by moderate variations in luciferase activity, but only by changes in quantum yield and self-absorption of the light in the sample. Flash-type reagents require the use of reagent dispensers. The stringent requirements for ATP-free cuvettes, pipette tips, and contamination-free laboratory techniques make it unlikely that flash reagents would be useful in nonlaboratory surroundings. A potential application for this type of reagent is sterility testing.

In general, it is concluded that one should select the ideal ATP reagent carefully for each application. Obviously the reagents used in a particular application do not have to match the decay rates given earlier exactly.

However, various applications of the ATP technology and the properties of manual and automatic luminometers fall quite nicely into categories corresponding to the properties of the three reagents described. The rapidly growing interest in ATP technology has already resulted in the development of a greater variety of luminometers, from hand-held instruments to high-throughput systems. The continuation of efforts in both reagent and instrument development will undoubtedly result in many new applications.

[26] Chemiluminescent Methods for Detecting and Quantitating Enzyme Activity

By LARRY J. KRICKA, JOHN C. VOYTA, and IRENA BRONSTEIN

Introduction

This article surveys the range of chemiluminescence (CL) methods for the detection and measurement of enzyme activities, with an emphasis on methods developed since 1986.[1] The principal advantage of CL enzyme assays is superior sensitivity to conventional spectrophotometric assay methods. Detection and quantitation of enzyme labels in immunoassays, in protein and nucleic acid-blotting assays (e.g., Western, Northern, Southern blotting), and in reporter gene assays have been the main applications or intended applications for most of the newly developed CL substrates.[2,3] CL enzyme assays have not replaced spectrophotometric assays for serum or plasma enzymes in clinical laboratories. Most automatic clinical chemistry analyzers are not adapted easily for light emission measurements, and ultrasensitivity is generally not required for the commonly measured enzymes, e.g., serum aspartate aminotransferase and urine amylase.

Assay Strategies

Two general types of CL enzyme activity assays have been developed. In the first, an enzyme catalyzes the CL decomposition of a substrate or releases a CL molecule from a specially designed enzyme substrate to produce light emission. The second type of assay uses a CL compound, either directly or indirectly, as an indicator of a reaction between an enzyme

[1] M. A. DeLuca and W. D. McElroy, *Methods Enzymol* **133** (1986).

[2] O. Nozaki, L. J. Kricka, and P. E. Stanley, *J. Biolumin. Chemilumin.* **7,** 263 (1992).

[3] I. Bronstein, J. Fortin, P. E. Stanley, G. S. A. B. Stewart, and L. J. Kricka, *Anal. Biochem.* **219,** 169 (1994).

METHODS IN ENZYMOLOGY, VOL. 305 0076-6879/00 $30.00

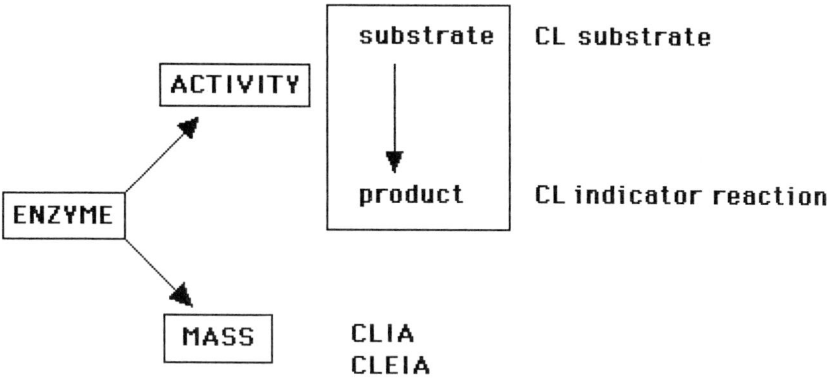

FIG. 1. Chemiluminescent assay strategies for enzymes.

and an enzyme substrate (e.g., to detect peroxide production). In addition, CL immunoassays (enzyme mass as opposed to an enzyme activity assay) are available for some clinically important enzymes (e.g., creatine kinase MB isoenzyme) (Fig. 1). The mass assays are based on either CL acridinium ester labels or enzyme labels (alkaline phosphatase, horseradish peroxidase, glucose oxidase) detected with a CL assay.[4–6] Immunoassays for enzymes that involve CL will not be considered in this article, and the reader is referred to other reviews for further information.[3,7,8] Table I[9–102] surveys

[4] U. Piran, D. W. Kohn, L. S. Uretsky, D. Bernier, E. H. Barlow, C. A. Niswander, and M. Stastny, *Clin. Chem.* **33,** 1517 (1987).

[5] P. G. Johnston, J. C. Drake, S. M. Steinberg, and C. J. Allegra, *Biochem. Pharmacol.* **45,** 2483 (1993).

[6] I. Ramasamy, *Clin. Biochem.* **28,** 519 (1995).

[7] L. J. Kricka, *Clin. Chem.* **40,** 347 (1994).

[8] M. J. Pringle, *Adv. Clin. Chem.* **30,** 89 (1993).

[9] D. C. Williams, G. F. Huff, and W. R. Seitz, *Anal. Chem.* **48,** 1003 (1976).

[10] M. Maeda, H. Arakawa, and A. Tsuji, *J. Biolumin. Chemilumin.* **4,** 140 (1989).

[11] H. Akhavan-Tafti, R. DeSilva, Z. Arghavani, R. A. Eickholt, R. S. Handley, B. A. Schoenfelner, and A. P. Schaap, *in* "Bioluminescence and Chemiluminescence: Molecular Reporting with Photons" (J. W. Hastings, L. J. Kricka, and P. E. Stanley, eds.), p. 501. Wiley, Chichester, 1997.

[12] K. Sekiya, Y. Saito, T. Ikegami, M. Yamamoto, Y. Sato, M. Maeda, and A. Tsuji, *in* "Bioluminescence and Chemiluminescence: Current Status" (P. E. Stanley and L. J. Kricka, eds.), p. 123. Wiley, Chichester, 1991.

[13] K. Nakashima, N. Kuroda, S. Kawaguchi, M. Wada, and S. Akiyama, *J. Biolumin. Chemilumin.* **10,** 185 (1995).

[14] A. Baret, V. Fert, and J. Aumaille, *Anal. Biochem.* **187,** 20 (1990).

[15] L. Bruun and G. Houen, *Anal. Biochem.* **233,** 130 (1996).

[16] A. Hinkkanen, F. E. Maly, and K. Decker, *Hoppe Seyler's Z. Physiol. Chem.* **364,** 407 (1983).

[17] I. Minkenberg and E. Ferber, *J. Immunol. Methods.* **71,** 61 (1984).

TABLE I
CL ASSAYS FOR ENZYMES

EC number	Enzyme	Reagents	Detection limit	Reference
1.	Oxidoreductases			
1.1.1.28	Lactate dehydrogenase	Methylene blue/luminol	—	9
1.1.1.49	Glucose-6-phosphate dehydrogenase	$NADP^+$/1-MPMS/isoluminol/MP	3 zmol	10
1.1.3.4	Glucose oxidase	Acridan-9-carboxylate	2–7 amol	11
		Luminol/MP	6.2 amol	12
1.1.3.6	Cholesterol oxidase	Peroxylate/fluorophore	5 mU/ml	13
1.1.3.17	Choline oxidase	Acridan-9-carboxylate	<4.8 fmol	11
		Peroxylate/fluorophore	0.5 mU/ml	13
1.1.3.22	Xanthine oxidase	Luminol/Fe-EDTA	3 amol	14
		Peroxylate/fluorophore	0.5 mU/ml	13
1.4.3.6	Diamine oxidase	Luminol enhancer	<0.1 U	15
1.5.3.5	6-Hydroxy-L-nicotine oxidase	Luminol	50 amol	16
1.5.3.6	6-Hydroxy-D-nicotine oxidase	Luminol	1 fmol	16
1.6.-.-	NAD(P)H oxidase	Lucigenin	—	17
1.7.3.3	Uricase	Peroxylate/fluorophore	1 mU/ml	13
1.11.1.6	Catalase	Luminol	—	18
1.11.1.7	Horseradish peroxidase	Luminol/enhancer	25 amol	19–37
		5-Hydroxyluminol	—	38
		Pyrogallol	4.18 pmol	39
		Purpurogallin	4.18 pmol	39
		Purpurogallincarboxylate	—	40
		Isobutene	—	41
		Homogentisic acid-γ-lactone	1 fmol	42,43
		Cypridina luciferin analog	137 pmol	44
		Acridan-9-carboxylate	—	45
	Arthromyces peroxidase	Luminol	<0.02 μg/ml	46
	Coprinus cinereus peroxidase	Luminol	—	46
	Lactoperoxidase	Luminol/enhancer	—	47

	Myeloperoxidase	Luminol	—	48,49
		Tyrosine	—	50
1.13.11.12	Lipoxygenase	Luminol	<3 mU	51–53
1.15.1.1	Superoxide dismutase	*Cypridina* luciferin analog	—	54
		Luminol	—	55–57
		Lucigenin/adrenalin	—	59
		Lucigenin	0.17 fmol	57
2.	Transferases			
2.3.1.6	Choline acetyltransferase	Luminol/MP	—	60
2.7.1.-	Protein kinase A	CLEIA	20 ng/ml	61
2.7.1.-	src kinase	CLEIA	—	61
2.7.1.37	Protein kinase C	CLEIA	—	61
2.7.1.123	Calmodulin-dependent protein kinase II (CAM kinase II)	CLEIA	—	61
3.	Hydrolases			
3.1.1.1	Esterase	Propenyl ester	2 pmol	41
		Dioxetane ester	—	62
3.1.1.3	Lipase	Propenyl ester	—	41
3.1.1.7	Cholinesterase	2-Naphthyl acetate/luminol	8 U/ml	63, 64
3.1.1.7	Acetyl cholinesterase		—	65
3.1.3.-	Calcineurin	Dioxetane	300 pmol/liter	66
3.1.3.-	Phosphomonoesterase	Luminol phosphate	2.5 U/liter	67
3.1.3.1	Alkaline phosphatase	Dioxetane phosphate	1 zmol	68–73
		Cortisol phosphate/lucigenin	87.5 zmol	74
		Glycerol phosphate/lucigenin	0.25 amol	75
		Glucose 1-phosphate/GOD/isoluminol/MP	2 amol	10
		Galactose 1-phosphate/1-MPMS/isoluminol/MP	2.5 pg	10
		Phenacyl phosphate	34 zmol	76
		BCIP/isoluminol/MP	0.7 amol	77
		$FADP^+$/luminol	0.4 amol	78
		$NADP^+$/lucigenin	0.125 amol	75

(*continued*)

TABLE I (*continued*)

EC number	Enzyme	Reagents	Detection limit	Reference
		$NADP^+$/1-MPS/isoluminol/MP	—	10
		$NADP^+$/ADH/1-MPMS/SOD/luminol	—	79
		Acridan phosphate	3.5 zmol	80,81
		Indolyl derivatives	—	82
		Luminolbenzothiazolyl phosphate	—	83
		Luminol phosphate	100 U/tube	84
		4-Iodophenyl phosphate/luminol	100 amol	85
		2-Naphthyl phosphate/luminol	—	85
		PNPP/luminol	1 pmol	85
3.1.3.16	Phosphoprotein phosphatases	Dioxetane phosphate	—	66
3.1.4.3	Phospholipase C	Luminol	—	86
3.1.4.4	Phospholipase D	Luminol	—	86
		Luminol/4-iodophenol/CHO/HRP	0.167 mIU	87
		Luminol	0.417 mIU	87
3.1.4.10	Phosphatidyl inositol-specific phospholipase C	Dioxetane inositol phosphate	10 pg	88
3.1.6.1	Aryl sulfatase	2-Naphthyl sulfate/luminol	—	85
3.2.1.1	Amylase	Amylose-peroxidase/luminol	—	89
3.2.1.21	β-Glucosidase	Dioxetane glucoside	—	90
		BCIP/isoluminol/MP	0.5 fmol	77
		6-Bromo-2-naphthyl glucopyranoside/luminol	—	85
3.2.1.23	β-Galactosidase	Dioxetane galactoside	59 zmol	91–93
		X-Gal	3 amol	94
		XGal/isoluminol/MP	0.2 amol	77
		$NADP^+$/1-MPMS/isoluminol/MP	—	10
		Resorufin	<1000 molecules	95
		Cypridina luciferin analog	—	96
		Luminol galactoside	50 μU/tube	84
		6-Bromo-2-naphthyl galactopyranoside/luminol	—	85

		2-Naphthyl-galactopyranoside/luminol	—	85
		o-Nitrophenyl galactoside/1-MPMS/isoluminol/MP	33 zmol	10
		Lactose/GOD/isoluminol/MP	5 fmol	10
		Lactose/GOD/TCPO/ANS	0.2 fmol	10
3.2.1.26	Invertase	Sucrose/lucigenin	0.74 fmol	97
3.2.1.31	β-Glucuronidase	Dioxetane glucuronide	<0.1 pg	91
3.2.1.30	Acetyl glucosaminidase	Luminolglucosaminide	<0.3 IU/liter	98
3.4.-.-	Protease	Isoluminol peptide	2 μg/liter	99–101
3.4.21.1	α-Chymotrypsin	Isoluminol peptide	0.27 μg/liter	99
		Isoluminol peptide	0.1 ng	101
3.4.21.4	Trypsin	Isoluminol peptide	40 μg/liter	99
			0.01 ng	101
3.4.21.5	Thrombin	Isoluminol peptide	0.5 ng	101
4.	Lyases			
4.1.3.18	Acetolactate synthase	Mn^{2+}/pyruvate	—	102
5.	Isomerases			
6.	Ligases			

[a] ADH, alcohol dehydrogenase; ANS, 8-amino-2-naphthalenesulfonic acid; BCIP, 5-bromo-4-chloro-3-indolyl phosphate; CHO, choline oxidase; CLEIA, chemiluminescent enzyme immunoassay; EDTA, ethylenediaminetetraacetic acid; FADP, flavin adenine dinucleotide phosphate; GOD, glucose oxidase; HRP, horseradish peroxidase; MP, microperoxidase; 1-MPMS, 1-methoxyphenazinium methyl sulfate; $NADP^+$, nicotinamide adenine dinucleotide phosphate; PNPP, 4-nitrophenyl phosphate; SOD, superoxide dismutase; TCPO, bis(2,4,6-trichlorophenyl)oxalate; X-Gal, 5-bromo-4-chloro-3-indolyl galactopyranoside.

[18] S. Mueller, H. D. Riedel, and W. Stremmel, *Anal. Biochem.* **245,** 55 (1997).

[19] G. H. G. Thorpe and L. J. Kricka, *Methods Enzymol.* **133,** 331 (1986).

[20] O. Nozaki, X. Ji, and L. J. Kricka, *in* "Bioluminescence and Chemiluminescence: Fundamentals and Applied Aspects" (A. K. Campbell, L. J. Kricka, and P. E. Stanley, eds.), p. 48. Wiley, Chichester, 1994.

[21] O. Nozaki, X. Ji, and L. J. Kricka, *in* "Bioluminescence and Chemiluminescence: Fundamentals and Applied Aspects" (A. K. Campbell, L. J. Kricka, and P. E. Stanley, eds.), p. 52. Wiley, Chichester, 1994.

[22] X. Ji and L. J. Kricka, *in* "Bioluminescence and Chemiluminescence: Molecular Reporting with Photons" (J. W. Hastings, L. J. Kricka, and P. E. Stanley, eds.), p. 473. Wiley, Chichester, 1997.

[23] X. Ji and L. J. Kricka, *in* "Bioluminescence and Chemiluminescence: Molecular Reporting with Photons" (J. W. Hastings, L. J. Kricka, and P. E. Stanley, eds.), p. 477. Wiley, Chichester, 1997.

[24] T. R. Kissel, S. A. Fingar, and A. E. Friedman, Eur. Patent Appl. 704,539 (1996).

[25] L. J. Kricka, UK Patent Appl. 2,276,721 (1994).

[26] A. E. Friedman and T. R. Kissel, Eur. Patent Appl. 603,953 (1994).

[27] L. J. Kricka, US Patent 5,512,451 (1996).

[28] R. Iwata, H. Ito, T. Hayashi, Y. Sekine, N. Koyama, and M. Yamaki, *Anal. Biochem.* **231,** 170 (1995).

[29] H. Ito, H. Suzuki, Y. Miki, T. Hayashi, M. Iwata, and M. Yamaki, Japanese Patent 07,327,694 (1995).

[30] M. Iwata, T. Hayashi, and M. Yamaki, Japanese Patent 08,005,560 (1996).

[31] H. Ito, H. Suzuki, Y. Miki, T. Hayashi, M. Iwata, and M. Yamaki, Japanese Patent 08,038,196 (1996).

[32] M. Iwata, T. Hayashi, and M. Yamaki, Japanese Patent 08,051,997 (1996).

[33] M. Sugyama, Japanese Patent 07,099,996 (1995).

[34] Y. Sekine and K. Obara, Japanese Patent 07,143,899 (1995).

[35] Z. Zhang, S. Zhang, and X. Zhang, *Chem. Res Chin. Univ.* **9,** 314 (1993).

[36] H. Ito, H. Suzuki, Y. Miki, T. Hayashi, M. Iwata, and M. Yamaki, Japanese Patent 08,038,195 (1996).

[37] T. Schlederer and G. Himmler, *in* "Bioluminescence and Chemiluminescence: Molecular Reporting with Photons" (J. W. Hastings, L. J. Kricka, and P. E. Stanley, eds.), p. 533. Wiley, Chichester, 1997.

[38] L. J. Kricka, X. Ji, G. H. Thorpe, B. Edwards, J. Voyta, and I. Bronstein, *J. Immunoassay* **17,** 67 (1996).

[39] O. Nozaki, X. Ji, and L. J. Kricka, *J. Biolumin. Chemilumin.* **10,** 151 (1995).

[40] T. Segawa, T. Ooizumi, T. T. Matsubara, M. Kamidate, and H. Watanabe, *Anal. Sci.* **11,** 209 (1995).

[41] L. H. Catalani, N. G. Malta, and A. Campa, *in* "Bioluminescence and Chemiluminescence: Molecular Reporting with Photons" (J. W. Hastings, L. J. Kricka, and P. E. Stanley, eds.), p. 493. Wiley, Chichester, 1997.

[42] A. Campa, A. C. Andrade, and L. H. Catalani, *Photochem. Photobiol.* **63,** 742 (1996).

[43] T. Segawa, T. Ooizumi, T. Yoshimura, M. Tohma, M. Kamidate, and H. Watanabe, *Anal. Sci.* **11,** 581 (1995).

[44] M. Mitani, Y. Yokoyama, S. Ichikawa, H. Sawada, T. Matsumoto, K. Fujimori, and M. Kosigi, *J. Biolumin. Chemilumin.* **9,** 355 (1994).

[45] H. Akhavan-Tafti, Z. Arghavani, and R. Desilva, PCT Intl. Appl. WO 9,607,912 (1996).

[46] K. Akimoto, Y. Shinmen, M. Sumida, S. Asami, T. Amachi, H. Yoshizumi, Y. Saeki, S. Shimizu, and H. Yamada, *Anal. Biochem.* **189,** 182 (1990).

[47] S. Girotti, E. Ferri, S. Ghini, A. Roda, J. Navarro, and E Ortega, *in* "Bioluminescence and Chemiluminescence: Fundamentals and Applied Aspects" (A. K. Campbell, L. J. Kricka, and P. E. Stanley, eds.), p. 44. Wiley, Chichester, 1994.
[48] T. C. Carmine, G. Bruchelt, M. Zipfel, and D. Niethammer, *in* "Bioluminescence and Chemiluminescence: Fundamentals and Applied Aspects" (A. K. Campbell, L. J. Kricka, and P. E. Stanley, eds.), p. 223. Wiley, Chichester, 1994.
[49] R. Cooray, C. G. Petersson, and O. Holmberg, *Veterin. Immunol. Immunopathol.* **38,** 261 (1993).
[50] A. Nishida, M. Kondou, and M. Nakano, *Ensho* **15,** 161 (1995).
[51] E. M. Lilius, P. Turunen, and S. Laakso, *in* "Analytical Applications of Bioluminescence and Chemiluminescence" (L. J. Kricka, and P. E. Stanley, G. H. G. Thorpe, and T. P. Whitehead, eds.), p. 393. Wiley, Chichester, 1984.
[52] S. Laakso, E. M. Lilius and, P. Turunen, *Methods Enzymol.* **105,** 126 (1984).
[53] Y. Kondo, Y. Kawai, T. Miyaza, H. Matsui, and J. Mizutani, *Biosci. Biotechnol.* **58,** 421 (1994).
[54] Y. Tokuda, T. Uozumi, and T. Kawasaki, *Neurochem. Int.* **23,** 107 (1993).
[55] I. N. Popov, G. Lewin, and R. von Baehr, *Biomed. Biochim. Acta* **46,** 775 (1987).
[56] P. H. Trung, C. Marquetty, C. Pasquier, and J. Hakim, *Anal. Biochem.* **142,** 467 (1984).
[57] R. Bensinger and C. Johnson, *Anal. Biochem.* **116,** 142 (1981).
[58] Z. P. Cheremisina, T. B. Suslova, and L. G. Korkina, *Klin. Lab. Diagn.* **1,** 22 (1994).
[59] J. K. Laihia, C. T. Jansen, and M. Ahotupa, *Free Radic. Biol. Med.* **14,** 457 (1993).
[60] E. J. Menzel, L. R. Walzer, G. Meissl, and H. Millesi, *in* "Bioluminescence and Chemiluminescence: New Perspectives" (J. Scholmerich, R. Andreesen, A. Kaap, M. Ernst, and W. G. Woods, eds.), p. 579. Wiley, Chichester, 1987.
[61] C. Lehel, S. Daniel-Issakani, M. Brasseur, and B. Strulovici, *Anal. Biochem.* **244,** 340 (1997).
[62] A. P. Schaap, R. S. Handley, and B. P. Giri, *Tetrahedron Lett.* **28,** 935 (1987).
[63] A. Navas Diaz, F. Garcia Sanchez, J. A. Gonzalez Garcia, and V. Bracho Del Rio, *J. Biolumin. Chemilumin.* **10,** 285 (1995).
[64] R. Lippmann *in* "Bioluminescence and Chemiluminescence: Basic Chemistry and Analytical Applications" (M. DeLuca and W. D. McElroy, eds.), p. 633. Academic Press, New York, 1981.
[65] S. Birman, *Biochem. J.* **225,** 825 (1985).
[66] D. J. O'Kane, B. J. Hallaway, and F. Rusnak, *in* "Bioluminescence and Chemiluminescence: Fundamentals and Applied Aspects" (A. K. Campbell, L. J. Kricka, and P. E. Stanley, eds.), p. 24. Wiley, Chichester, 1994.
[67] J. Sugama, K. Katayama, K. Kuroiwa, and T. Nagasawa, Japanese Patent 07,053,525 (1993).
[68] B. Edwards, A. Sparks, J. C. Voyta, and I. Bronstein, *in* "Bioluminescence and Chemiluminescence: Fundamentals and Applied Aspects" (A. K. Campbell, L. J. Kricka, and P. E. Stanley, eds.), p. 56. Wiley, Chichester, 1994.
[69] I. Bronstein, R. R. Juo, J. C. Voyta, and B. Edwards, *in* "Bioluminescence and Chemiluminescence: Current Status" (P. E. Stanley and L. J. Kricka, eds.), p. 73. Wiley, Chichester, 1991.
[70] I. Bronstein, J. C. Voyta, Y. Vant Erve, and L. J. Kricka, *Clin. Chem.* **37,** 1526 (1991).
[71] I. Bronstein, J. C. Voyta, G. H. G. Thorpe, L. J. Kricka, and G. Armstrong, *Clin. Chem.* **35,** 1441 (1989).
[72] I. Bronstein and L. J. Kricka, *in* "Nonradioactive Labeling and Detection of Biomolecules" (C. Kessler, ed.), p. 168. Springer-Verag, Berlin, 1992.
[73] H. Akhavan-Tafti, *Diss. Abstr. Int. B* **56,** 2626 (1995).

[74] M. Maeda, A. Okumura, and A. Tsuji, *in* "Bioluminescence and Chemiluminescence: Fundamentals and Applied Aspects" (A. K. Campbell, L. J. Kricka, and P. E. Stanley, eds.), p. 289. Wiley, Chichester, 1994.

[75] M. Kitamura, M. Maeda, and A. Tsuji, *J. Biolumin. Chemilumin.* **10,** 1 (1995).

[76] H. Sasamoto, M. Maeda, and A. Tsuji, *Anal. Chim. Acta* **306,** 161 (1995).

[77] H. Arakawa, M. Maeda, and A. Tsuji, *Anal. Biochem.* **199,** 238 (1991).

[78] M. Fisher, S. Harbon, and B. R. Rabin, *Anal. Biochem.* **227,** 73 (1995).

[79] T. Mori, *Rinsho Kagaku* **24,** 132 (1995).

[80] H. Akhavan-Tafti, Z. Arghavani, R. DeSilva, R. A. Eickholt, R. S. Handley, B. A. Schoenfelner, S. Siripurapu, K. Sugioka, and A. P Schaap, *in* "Bioluminescence and Chemiluminescence: Molecular Reporting with Photons" (J. W. Hastings, L. J. Kricka, and P. E. Stanley, eds.), p. 311. Wiley, Chichester, 1997.

[81] H. Akhavan-Tafti, Z. Arghavani, and R. DeSilva, PCT Intl. Appl. WO 9,607,911 (1996).

[82] T. Ikegami, H. Arakawa, M. Maeda, and A. Tsuji, *Anal. Sci.* **10,** 831 (1994).

[83] K. Sasomoto, G. Deng, T. Ushijima, Y. Ohkura, and K. Ueno, *Analyst* **120,** 1709 (1995).

[84] M. Nakazono, H. Nohta, K. Sasamoto, and Y. Ohkura, *Anal. Sci.* **8,** 779 (1992).

[85] L. J. Kricka, D. Schmerfeld-Pruss, and B. Edwards, *J. Biolumin. Chemilumin.* **1,** 231 (1991).

[86] M. Lucas, V. Sanchez-Margalet, C. Pedrera, and M. Luz Bellido, *Anal. Biochem.* **231,** 277 (1995).

[87] P. Rauch, E. N. Ferri, S. Girotti, H. Rauchova, G. Carrea, R. Bovara, F. Fini, and A. Roda, *Anal. Biochem.* **245,** 133 (1997).

[88] M. Ryan, J. C. Huang, O. H. Griffith, J. F. W. Keana, and J. J. Volwerk, *Anal. Biochem.* **214,** 548 (1993).

[89] L. J. Kricka, J. M. Marcinkowski, P. Wilding, and S. Lekhakula, *Talanta* **37,** 971 (1990).

[90] J. C. Voyta, J. Y. Lee, C. E. M. Olesen, R.-R. Juo, B. Edwards, and I. Bronstein, *in* "Bioluminescence and Chemiluminescence: Molecular Reporting with Photons" (J. W. Hastings, L. J. Kricka, and P. E. Stanley, eds.), p. 529. Wiley, Chichester, 1997.

[91] I. Bronstein, J. Fortin, J. C. Voyta, C. E. M. Olesen, and L. J. Kricka, *in* "Bioluminescence and Chemiluminescence: Fundamentals and Applied Aspects" (A. K. Campbell, L. J. Kricka, and P. E. Stanley, eds.), p. 20. Wiley, Chichester, 1994.

[92] C. S. Martin, C. E. M. Olesen, B. Liu, J. C. Voyta, J. L. Shumway, R. R. Juo, and I. Bronstein, *in* "Bioluminescence and Chemiluminescence: Molecular Reporting with Photons" (J. W. Hastings, L. J. Kricka, and P. E. Stanley, eds.), p. 525. Wiley, Chichester, 1997.

[93] E. Beale, E. Deeb, R. Handley, H. Akhavan-Tafti, and A. Schaap, *BioTechniques* **12,** 320 (1992).

[94] H. Arakawa, M. Maeda, and A. Tsuji, *in* "Bioluminescence and Chemiluminescence: Molecular Reporting with Photons" (J. W. Hastings, L. J. Kricka, and P. E. Stanley, eds.), p. 489. Wiley, Chichester, 1997.

[95] T. Schlederer and P. G. Fritz, PCT Int. Appl. WO 9,427,154 (1994).

[96] M. Mitani, S. Sakai, Y. Koinuma, Y. Toya, and M. Kosugi, *Anal. Sci.* **10,** 813 (1994).

[97] S. Shimizu, H. Arakawa, M. Maeda, and A. Tsuji, *Bunseki Kagaku* **37,** 123 (1988).

[98] K. Sasamoto and Y. Ohkura, *Chem. Pharm. Bull* (*Tokyo*) **38,** 1323 (1990).

[99] R. Edwards, A. Townshend, and B. Stoddart, *Analyst* **120,** 117 (1995).

[100] B. R. Branchini, J. D. Hermes, F. G. Salituro, N. J. Post, and G. Claeson, *Anal. Biochem.* **111,** 87 (1981).

[101] B. R. Branchini, J. D. Hermes, F. G. Salituro, and N. J. Post, *Biochem. Biophys. Res. Commun.* **97,** 334 (1980).

[102] J. Durner, V. Gailus, and P. Boger, *FEBS Lett.* **354,** 71 (1994).

the scope of CL assays for the six classes of enzymes, and subsequent sections discuss the different assay strategies developed to assay enzyme activity.

Chemiluminescence Substrates

Substantial progress has been made in the design of CL substrates for enzymes, particularly hydrolases. A CL molecule is modified to include a protecting group that is recognized by an enzyme. Enzymatic removal of the group releases a CL molecule or an intermediate that decomposes with the emission of light. Emission intensity, total light emission, or the rate of light emission is measured and related to enzyme activity.

Luminol and Analogs

The cyclic diacyl hydrazide, luminol (Fig. 2, I), in common with its isomer isoluminol and analogs such as naphthalene-1,2-dicarboxylic acid hydrazides, is a cosubstrate for horseradish peroxidase.[19] It undergoes peroxidase-catalyzed oxidation and the intensity of the glow-type light emission (425 nm) is proportional to peroxidase activity. Incorporation of enhancers into the reaction buffer improves the analytical characteristics of the reaction (i.e., increased signal and decreased background light emission). Kinetic studies indicate that preferential oxidation of enhancers by peroxidase intermediates (compounds I and II) and the rapid formation of enhancer radicals underlie the signal enhancement effect.[103] The original enhancers were firefly luciferin, substituted phenols, naphthols, and 6-hydroxybenzothiazoles.[19] Subsequently, other molecules have been discovered to have enhancer properties, e.g., aromatic amines,[104] substituted boronic acids,[20–23,25,39,105] acetanilides,[24,26] indophenols, and indocresols (Fig. 2).[106] A new luminol-based detection reagent with an as yet undisclosed formulation has been commercialized (SuperSignal and SuperSignal ULTRA, Pierce, Rockford, IL). Comparative studies with these reagents versus the ECL system (Amersham, Arlington Heights, IL) reveal that the light emission with horseradish peroxidase is brighter (2-fold) in the first 20 min, but eventually decays to the same intensity >100 min after initiation. In a Western blot assay for β-actin, better performance (more intense band

[103] W. Sun, X. Ji, L. J. Kricka, and H. B. Dunford, *Can. J. Chem.* **72,** 2159 (1994).

[104] L. J. Kricka, A. M. O'Toole, G. H. G. Thorpe, and T. P. Whitehead, U.S. Patent 4,729,950 (1988).

[105] L. J. Kricka, and X. Ji, *J. Biolumin. Chemilumin.* **10,** 49 (1995).

[106] H. Ito, H. Suzuki, Y. Miki, T. Hayashi, M. Iwata, and M. Yamaki, Japanese Patent 08,038,195 (1996).

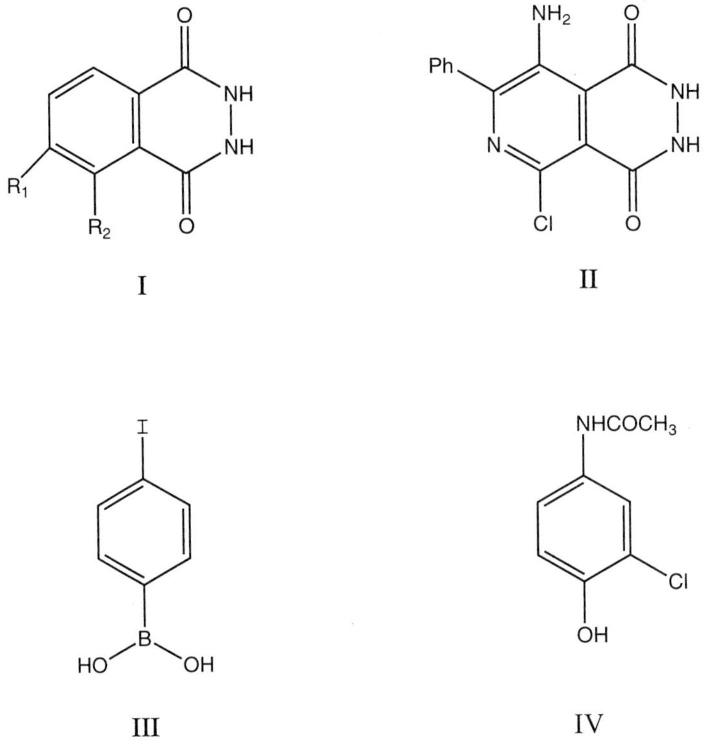

FIG. 2. Substrates [I, luminol, R_1 = H, R_2 = NH_2; isoluminol R_1 = NH_2, R_2 = H; II, 8-amino-5-chloro-7-phenylpyrido[3,4-*d*]pyridazine-1,4(2*H*,3*H*)dione] and enhancers (III, 4-iodophenylboronic acid; IV, 3′-chloro-4′-hydroxyacetanilide) for the chemiluminescent assay of horseradish peroxidase.

on X-ray film) was observed at 5 μg of this protein.[107] Alternatives to luminol have also been explored, most notably the pyridopyridazine analogs.[108,109] These provide more intense light emission as compared to luminol, but background light emission is also increased. Testing of derivatives, such as 8-amino-5-chloro-7-phenylpyrido[3,4-d]pyridazine-1,4(2*H*, 3*H*)dione (Fig. 2, II), in enhanced CL assays (phenol, boronate enhancers) for horseradish peroxidase revealed increases in light emission of up to 10-fold.[108,109]

The analytical utility of luminol as a substrate has been extended by

[107] D. L. Mattson and T. G. Bellehumeur, *Anal. Biochem.* **240,** 306 (1996).

[108] H. Masuya, K. Kondo, Y. Aramaki, and Y. Ichimori, Eur. Patent Appl. 491,477 (1992).

[109] M. Ti, H. Yoshida, Y. Aramaki, H. Masuya, T. Hada, M. Terada, M. Hatanaka, and Y. Ichimori, *Biochem. Biophys. Res. Commun.* **193,** 540 (1993).

synthesizing luminol galactoside and phosphate substrates for β-galactosidase and alkaline phosphatase, respectively. Enzyme action releases luminol, which is detected with hexacyanoferrate(III) in the presence of base or by a mixture of peroxide and horseradish peroxidase.[67,84] Similarly, protease assays (trypsin, α-chymotrypsin, thrombin) were developed using either immobilized isoluminol derivatives or insoluble isoluminol tripeptide derivatives.[98–100] Soluble CL isoluminol-containing products were detected, for example, in a flow assay using the CL cobalt(II)-catalyzed isoluminol reaction.[99]

Dioxetanes

In recent years, one of the major developments in CL enzyme assays has been the design of families of enzyme substrates based on adamantyl-stabilized aryl-1,2-dioxetanes.[68–72,110–112] The design of this class of substrate incorporates an aryl group modified with a specific enzyme cleavable group, a four-membered 1,2-dioxetane ring as a source of chemical energy for a CL reaction, and an adamantyl or substituted adamantyl group to stabilize the 1,2-dioxetane ring. Substrates, such as AMPPD [3-(4-methoxyspiro[1,2-dioxetane-3,2-tricyclo[3.3.1.1^{3,7}]decan]4-yl) phenyl phosphate] (Tropix, Bedford, MA) (Fig. 3, I), undergo an alkaline phosphatase-catalyzed dephosphorylation to form a phenoxide intermediate that decomposes with the emission of a prolonged glow of light. Subsequently, families of dioxetane substrates have been synthesized for a range of enzymes (e.g., phosphatases, phospholipases, proteases, hydrolases) (Fig. 3, II, IV). Structural modification has produced new dioxetanes with improved properties, e.g., CSPD and CDP-*Star* (Fig. 3, II, III).[68] Also, enhancement reagents designed to increase light intensity are available and these include solutions of polymers [e.g., polyvinylbenzyl(benzyldimethylammonium) chloride],[111] fluorescent additives (e.g., fluorescein), and detergent : fluorophore mixtures [e.g., fluorescein derivatives + cetyltrimethylammonium bromide (CTAB)].[24] With these systems, as little as 1 zmol (10^{-21} mol) of alkaline phosphatase can be detected.

Acridan Carboxylates

This class of molecule undergoes a CL oxidation via a dioxetane or dioxetanone intermediate.[113] *N*-Methylacridan derivatives can now be pur-

[110] H. P. Josel, R. Herrmann, C. Klein, and D. Heindl, German Patent 4,210,759 (1993).
[111] J. C. Voyta, B. Edwards, and I. Bronstein, US Patent 5,145,772 (1992).
[112] A. P. Schaap, H. Akhavan, and L. J. Romano, *Clin. Chem.* **35,** 1863 (1989).
[113] K.-D. Gunderman and F. McCapra, "Chemiluminescence in Organic Chemistry." Springer-Verlag, Berlin, 1987.

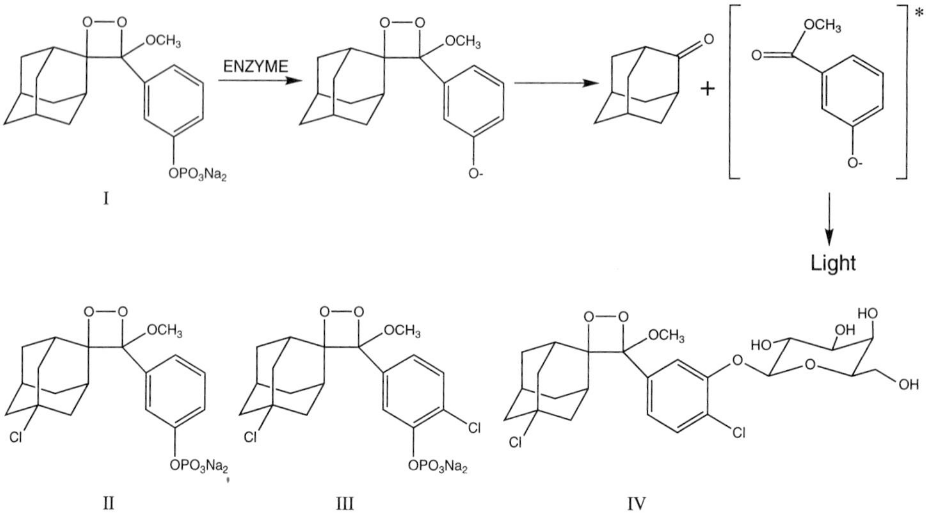

FIG. 3. Enzyme-catalyzed decomposition of 3-(4-methoxyspiro[1,2-dioxetane-3,2′-tricyclo[3.3.1.1^{3,7}]decan]4-yl) phenyl phosphate (AMPPD) (I) and related adamantyl 1,2-dioxetane substrates for alakaline phosphatase (CSPD, II; CDP-*Star,* III) and β-galactosidase (Galacton-*Star,* IV).

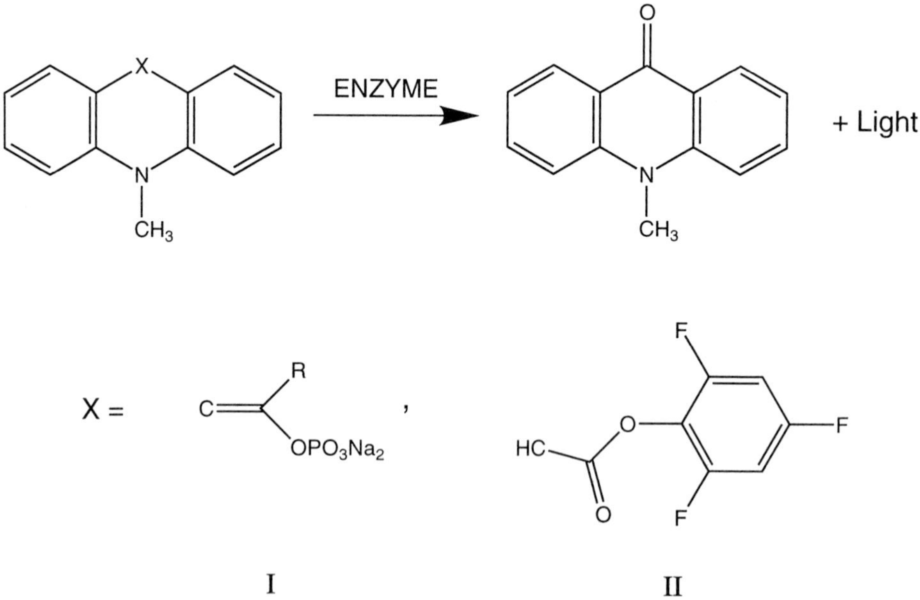

FIG. 4. Enzyme-catalyzed decomposition of an acridan-based substrate.

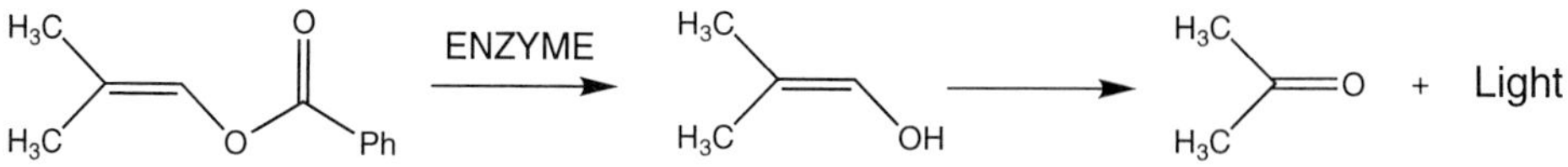

FIG. 5. Enzyme assay using a 2-methyl-1-propenyl benzoate substrate.

chased for oxidase (e.g., horseradish peroxidase) and hydrolase assays (e.g., alkaline phosphatase) (Lumigen, Southfield, MI).[114] Unsubstituted and 3-methoxy and 1,6-dimethoxy-substituted acridan-9-carboxylate 2,3,6-trifluorophenoxy esters are oxidized by peroxidase in the presence of peroxide and a phenolic enhancer (e.g., 4-iodophenol) to produce a glow of light. Comparative studies have shown improved sensitivity for horseradish peroxidase detection (limit of detection, 0.1 amol). The alkaline phosphatase substrate shown in Fig. 4 also produces a rapid onset glow of light following dephosphorylation and as little as 305 zmol of the enzyme can be detected.[80]

Propenyl Esters

2-Methyl-1-propenyl benzoate, acetate, and laurate derivatives have been synthesized as substrates for esterase and lipase enzymes. For example, 2-methyl-1-propenyl benzoate is hydrolyzed by esterase to propenol. This is in turn oxidized by a mixture of HRP and hydrogen peroxide to form triplet excited state acetone, which decays to the electronic ground state, and energy is released as a broad emission centered at 450 nm (Fig. 5). The reaction is initiated by adding the aldehyde to a mixture of the sample and other reagents. The CL signal is increased up to fivefold by adding a fluorophore, e.g., sodium dibromoanthracene sulfonate.[41,115]

Indoles

β-Galactosidase has been measured in a reaction in which the CL substrate and the CL indicator molecule are identical. 5-Bromo-4-chloro-3-indolyl galactopyranoside (X-Gal) (Fig. 6A, I R = β-D-galactopyranoside) is converted by the enzyme to an indoxyl that oxidizes to form an indigo dye and hydrogen peroxide. In the presence of horseradish peroxidase the

[114] H. Akhavan-Tafti, R. DeSilva, Z. Arghavani, R. A. Eickholt, R. S. Handley, and A. P. Schaap, *in* "Bioluminescence and Chemiluminescence: Fundamentals and Applied Aspects" (A. K. Campbell, L. J. Kricka, and P. E. Stanley, eds.), p. 199. Wiley, Chichester, 1994.

[115] B. Yavo, A. Campa, and L. H. Catalani, *Anal. Biochem.* **234,** 215 (1996).

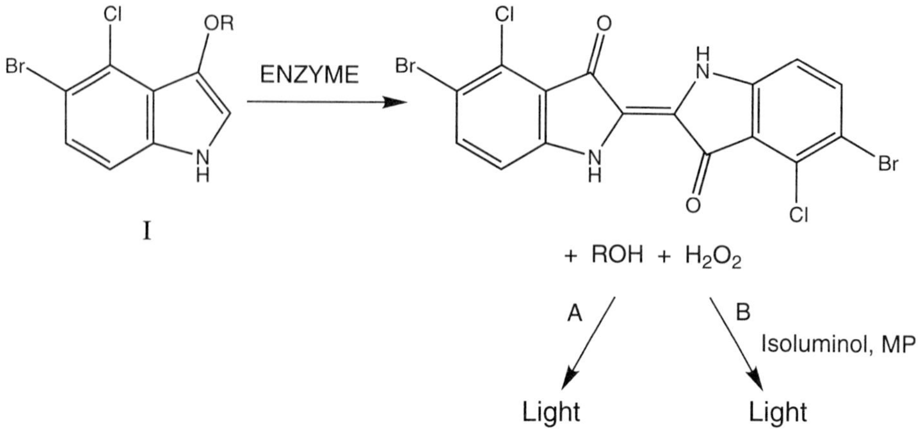

FIG. 6. Chemiluminescent assay schemes (A and B) using 5-bromo-4-chloro-3-indolyl derivatives as enzyme substrates (MP, microperoxidase).

peroxide reacts with residual X-Gal to produce an intense CL emission (detection limit, 3 amol).[94]

Imidopyrazines

2-Methyl-6-(4-methoxyphenyl)-3,7-dihydroimidazo[1,2-a]pyrazin-3-one (Fig. 7, I), an analog of *Cypridina* luciferin, reacts with peroxide in the presence of horseradish peroxidase (type VI) to produce a light emission centered at 462 nm. This reaction is the basis of an assay for the enzyme in the range of 100 pmol/L to 100 nmol/liter.[44,54] This derivative has also been used to assay superoxide dismutase.[54] Attachment of a β-D-galactopyranoside group at the 3 position produces a substrate effective in a CL assay for β-galactosidase.[96]

Chemiluminescent Indicator Reactions

Much interest has focused on devising assays in which the reaction of an oxidoreductase or hydrolase is coupled to a CL indicator reaction for peroxide or reducing substances.[116] In addition, immunoassays have been adapted to measure products of enzyme action.

[116] V. K. Mahant, US Patent 5,624,813 (1997).

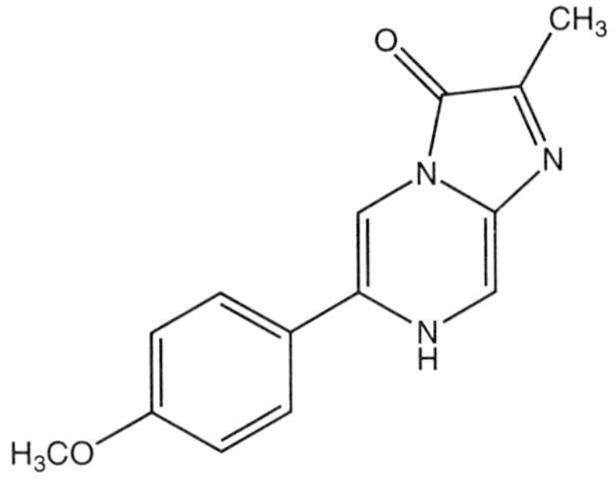

I

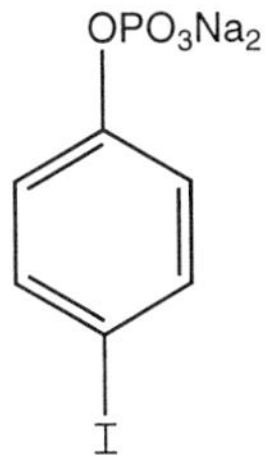

II

FIG. 7. Peroxidase substrate, 2-methyl-6-(4-methoxyphenyl)-3,7-dihydroimidazo[1,2-a]pyrazin-3-one (I), and the proenhancer, 4-iodophenyl phosphate (II).

Coupled Enzyme Reactions

Detection of peroxide formation using a coupled peroxidase-luminol or peroxyoxalate indicator reaction is a common strategy for the detection of oxidase enzymes, e.g., glucose oxidase, diamine oxidase.[15] In the case of xanthine oxidase, peroxide production is quantitated using an iron EDTA complex and luminol.[14] Hydrolases are assayed using a variant of this general strategy as illustrated by the phospholipase D assay shown in Fig. 8A.[87] Consumption of peroxide can also be monitored by a CL assay, which provides the basis of a flow assay for catalase in tissue homogenates and cell suspensions.[18]

Ultrasensitive CL assays for NADH extend the CL enzyme assay protocols to dehydrogenases. NADH will reduce oxygen to superoxide anion and hydrogen peroxide in the presence of an electron mediator (e.g.,

Scheme A

phospholipase D

lecithin + H_2O -----------------------> phosphatidic acid + choline

choline oxidase

choline + O_2 ------------------------> betaine + H_2O_2

horseradish peroxidase

H_2O_2 + luminol + 4-iodophenol -----------------------------> LIGHT

Scheme B

glucose-6-phosphate dehydrogenase

glucose 6-phosphate + NAD ---------> 6-phospho-D-gluconate + NADH

NADH + H^+ + 1-$MPMS^+$ ----------------> NAD^+ + 1-$MPMSH_2$

1-$MPMSH_2$ + O_2 ------> 1-$MPMS^+$ + H_2O_2

microperoxidase

H_2O_2 + isoluminol -----------------------------> LIGHT

FIG. 8. Chemiluminescent enzyme assay schemes for phospholipase (A), glucose-6-phosphate dehydrogenase (B), and alkaline phosphatase (C and D). FAD, flavin adenine dinucleotide; FADP, 3′-phosphorylated form of FAD, from London Biotechnology Ltd.; NAD^+, nicotinamide adenine dinucleotide; NADH, reduced form of NAD^+; 1-MPMS, 1-methoxyphenazinium methyl sulfate; 1-$MPMSH_2$, reduced form of 1-MPMS; P_i, inorganic phosphate.

Scheme C

alkaline phosphatase

glycerol-3-phosphate -------------------> glycerol + H_3PO_4

glycerol dehydrogenase

glycerol + NAD^+ -------------------------> dihydroxyacetone + NADH

dihydroxyacetone + lucigenin -------------> LIGHT

Scheme D

alkaline phosphatase

$FADP^+$ ---------------------------> FAD^+ + Pi

FAD^+ + apo-D-amino acid oxidase ----> D-amino acid oxidase

D-amino acid oxidase

proline + O_2 -------------------> H_2O_2 + 5-oxo-5-amino pentanoic acid

horseradish peroxidase

H_2O_2 + luminol + 4-hydroxycinnamic acid -------------------> LIGHT

FIG. 8. (*continued*)

methylene blue, 1-methoxyphenazinium methyl sulfate).[117] Peroxide is then detected using isoluminol and microperoxidase (Fig. 8B). Glucose-6-phosphate dehydrogenase has been assayed using this procedure down to 1 amol (1-methoxyphenazinium methyl sulfate mediator). Likewise, hy-

[117] A. Tsuji, M. Maeda, and H. Arakawa, *in* "Luminescence Immunoassay and Molecular Applications" (K. Van Dyke and R. Van Dyke, eds.), p. 157. CRC Press, Boca Raton, 1990.

drolases such as β-galactosidase can be assayed by linking it to a NADH-producing system. This is achieved by coupling the β-galactose produced in the β-galactosidase-catalyzed cleavage of *o*-nitrophenyl galactopyranoside to a β-D-galactose dehydrogenase reaction.[117] A more complex coupled system has been devised for the assay of alkaline phosphatase. The enzyme releases NAD^+ from an $NADP^+$ substrate.[118] The NAD^+ is converted to NADH in a coupled alcohol dehydrogenase reaction. NADH is then detected using the 1-methoxyphenazinium methylsulfate-mediated peroxide-producing reaction, linked in turn to a CL isoluminol-microperoxidase detection reaction. Less sensitive variants of this assay for alkaline phosphatase include the use of glucose 1-phosphate (glucose oxidase-coupled reaction) and galactose 1-phosphate substrates (galactose dehydrogenase-coupled reaction).

Hydrolases can also be assayed using 5-bromo-4-chloro 3-indolyl derivatives.[77,94] For example, 5-bromo-4-chloro 3-indolyl phosphate (BCIP) reacts with alkaline phosphatase (2-hr incubation) to produce an intermediate indoxyl that is oxidized in air to a mixture of an indigo dye and hydrogen peroxide. The peroxide produced in this reaction is measured using a mixture of isoluminol-microperoxidase (Fig. 6B; R = phosphate). This assay has a detection limit of 0.1 amol. A similar assay for β-galactosidase using X-Gal detected 0.1 amol of enzyme.[77]

The well-known CL molecule lucigenin reacts with reducing compounds under basic conditions to produce a light emission at 500 nm. This has been exploited as an indicator reaction in a series of assays that produce products that are reductants. For example, invertase (β-D-fructofuranosidase) can be assayed using lucigenin to detect the reducing sugar glucose, released from a sucrose substrate under the action of the enzyme. The assay involves an overnight incubation and detects 125 μU (equivalent to 0.75 fmol) of invertase.[117] Similarly, assays for alkaline phosphatase have been devised based on the detection of ascorbate, cortisol, and phenacyl alcohol, released by the dephosphorylation of ascorbic acid 2-phosphate, cortisol phosphate, and phenacyl phosphate,[76] respectively. In the case of the ascorbic acid phosphate substrate, the assay detected 0.1 amol of alkaline phosphatase.[119] The same general strategy can also be used with a glycerol 3-phosphate or an $NADP^+$ substrate. Phosphatase action supplies glycerol or NAD^+ to a coupled glycerol dehydrogenase reaction to form the reductant, dihydroxy acetone, which is detected in a CL lucigenin reaction (Fig. 8C). An indoxyl-lucigenin reagent for the measurement of phosphatase activity is now available commercially (KPL, Gaithersburg, MD).

[118] M. Kawamoto, H. Arakawa, M. Maeda, A. Tsuji, *Bunseki Kagagu* **40,** 537 (1991).

[119] M. Maeda, A. Tsuji, K. H. Yang, and S. Kamada, *in* "Bioluminescence and Chemiluminescence: Current Status" (P. E. Stanley, and L. J. Kricka, eds.), p. 119. Wiley, Chichester, 1991.

The enzyme-catalyzed release of a component of a CL reaction can be used to detect certain hydrolases. For example, an amylose substrate to which the enzyme horseradish peroxidase is immobilized is the basis of a CL assay for amylase. The enzyme cleaves the substrate and releases soluble peroxidase-labeled fragments into solution. After centrifugation to remove residual insoluble substrate, peroxidase activity is measured in the supernatant using the 4-iodophenol-enhanced luminol reaction.[89]

Enzyme Generated Activators, Enhancers, and Inhibitors

Another analytical strategy for CL enzyme assay involves enzyme-generated activators, enhancers, or inhibitors of an enzyme-based detection system. The prosthetogenesis assay detects alkaline phosphatase via the generation of FAD from FADP (an FAD analog phosphorylated at the 3′ position of the adenylate), and the subsequent activation of an apoenzyme, apo-D-amino acid oxidase.[78] Peroxide production in the oxidase reaction is detected with a 4-hydroxycinnamic acid-enhanced luminol reaction. The sensitivity of this assay stems from the combination of activation of amplifying molecules (inactive apo enzyme converted to an active holo enzyme) and a CL detection reaction for the product of the enzyme reaction (Fig. 8D). Although this assay is rapid and very sensitive (0.4 amol, 5-min assay), the dynamic range is only 1.5 orders of magnitude of enzyme concentration, which restricts its potential applications.

The feasibility of an assay for alkaline phosphatase based on the generation of an enhancer of a CL reaction from an inactive proenhancer has been demonstrated using 4-iodophenol phosphate (Fig. 7, II). This molecule is converted to 4-iodophenol by enzyme-catalyzed dephosphorylation. 4-Iodophenol is a potent enhancer of the horseradish peroxidase-catalyzed oxidation of luminol, and the enhancement in light emission is directly related to alkaline phosphatase activity (detection limit, 100 amol).[85] Analogous assays were developed for aryl sulfatase, β-D-glucosidase, β-D-galactosidase, and cholinesterase using 2-naphthyl sulfate, 6-bromo-2-naphthyl glucopyranoside, 6-bromo-2-naphthyl galactopyranoside, and 2-naphthyl acetate, respectively.[63,85] A similar strategy was adopted in an assay for hydrolytic enzymes using a protected phenolic enhancer. Enhancer released as a result of enzyme action was detected using an *N*-alkyl-acridan-9-carboxylic acid derivative.[81]

An alternative strategy based on the generation of an inhibitor of a CL reaction was less sensitive and had a limited dynamic range. In one example, alkaline phosphatase generated 4-nitrophenol, an inhibitor of an enhanced CL reaction, from a 4-nitrophenol phosphate substrate, and the inhibition in light emission was quantitated (detection limit 1 pmol).[85]

The luminescence-enhanced reagent (LERS) system detects horseradish peroxidase by first converting dihydrorhodamine G (in the presence of hydrogen peroxide) to rhodamine G. This fluorophore then participates in a sensitized CL reaction with bis(2,4,6-trichlorophenyl)oxalate (TCPO) and hydrogen peroxide. In this assay, light is emitted as a flash ($t_{1/2}$ 1.5 sec), which has the advantage of minimizing cross-talk in microplate-based assays. Comparative studies indicate that this CL assay is 10 times more sensitive than the colorimetric assay (3,3′,5,5′-tetramethylbenzidine substrate) for peroxidase.[37]

Chemiluminescent Enzyme Immunoassay

Immunological assay of the product of the enzyme substrate reaction provides an alternative method of detecting enzyme activity. This analytical protocol has been used to assay protein kinase A and C, calmodulin-dependent protein kinase II, src tyrosine kinase, and receptor interacting protein kinase. The enzyme substrate is a biotinylated peptide captured onto a streptavidin-coated microwell. The substrate is phosphorylated by the kinase and the phosphorylated peptide is then detected by sequential reaction with a specific antiphosphoserine monoclonal antibody (this recognizes the phosphorylated peptide), an antimouse IgG antibody alkaline phosphatase conjugate, and finally a CL dioxetane substrate. The CL assay was shown to be superior to colorimetric or radioactive detection systems.[61] Reaction products in an enzyme reaction can also be measured by means of the biotin : avidin reaction. This is exemplified by an assay for reverse transcriptase. The biotinylated reaction product is spotted onto a nylon membrane, and incorporation of the biotin-21-deoxyuridinetriphosphate (biotin-21-dUTP) is detected with an alkaline phosphatase streptavidin conjugate and a CL dioxetane substrate.[120]

Conclusions

Chemiluminescence assays are available for all of the major classes of enzymes except isomerases and ligases. Generally, CL enzyme assays are more sensitive than colorimetric, fluorimetric, or radiometric assays, and many of the assays have subattomole detection limits. The success of CL assays for enzyme labels in immunoassay and nucleic acid assays has been the spur for the continued exploration of different CL reactions with the goal of developing enzyme assays with improved sensitivity (zeptomole amounts of enzyme), convenience, and versatility.

[120] R. F. Cook, S. J. Cook, and C. J. Issel, *BioTechniques* **13,** 380 (1992).

[27] Chemiluminescence Assay of Serum Alkaline Phosphatase and Phosphoprotein Phosphatases

By BRENDA J. HALLAWAY and DENNIS J. O'KANE

Introduction

Several neutral and alkaline phosphatase activities can be demonstrated in human tissues, cells, and body fluids. These phosphatase activities can be subdivided into those nonspecific enzyme activities that dephosphorylate small organic molecules (acid, neutral, and alkaline phosphatases) and those that dephosphorylate specific phosphorylated amino acid residues in proteins (phosphoprotein phosphatases) (PPases). Nonspecific phosphatases, such as alkaline phosphatase (ALP), are present in many tissues, including gastrointestinal mucosa, liver, spleen, vascular endothelium, renal tubules, thyroid, placenta, myeloid, and osteoblasts.[1,2] PPases are involved in the regulation of cellular signaling pathways and have the opposite effect(s) of protein kinase coregulators.[3]

Alkaline Phosphatase

Introduction

Assay of ALP is important in formulating clinical diagnosis in hepatobiliary disease, bone disorders, and malignant tumors.[4,5] The highest levels of ALP are observed in Paget's disease. Clinical assay of total serum alkaline phosphatase (sALP) is usually performed colorimetrically using *p*-nitrophenyl phosphate and alkaline conditions. Although more sensitive bioluminescence and chemiluminescence assays for ALP have been described,[6–10] their

[1] J. P. Bretaudiere and T. Spillman, *in* "Methods of Enzymatic Analysis" (H. U. Bergmeyer, ed.), p. 75. Verlag Chemie, Weinheim, 1984.

[2] D. W. Moss, *Clin. Chem.* **38,** 2486 (1992).

[3] E. Hafen, *Science* **280,** 1212 (1998).

[4] V. O. Van Hoof, A. T. Van Oosterom, L. G. Lepoutre, and M. E. De Broe, *Clin. Chem.* **38,** 2546 (1992).

[5] J. T. Deng, M. F. Hoylaerts, V. O. Van Hoof, and M. E. De Broe, *Clin. Chem.* **38,** 2532 (1992).

[6] S. Girotti, E. Ferri, S. Ghini, R. Budini, D. Patrono, L. Incorvara, and A. Roda, *Anal. Lett.* **27,** 323 (1994).

[7] I. Bronstein and L. J. Kricka, *J. Clin. Lab. Anal.* **3,** 316 (1989).

0076-6879/00 $30.00

clinical implementations have been problematic. The most popular assays utilize substituted dioxetane phosphate substrates. However, several problems exist with these assays that must be overcome before it would be feasible to implement chemiluminescence (CL) sALP assay. First, CL ALP assays are very sensitive, which necessitates dilutions of some samples in order to measure levels of sALP in all patient samples. In serial dilutions, the serum matrix is also diluted out by buffer and causes the loss of linearity of the signal response with concentrations of sALP. A sample matrix is needed that will maintain the serum matrix constant and allow dilution of samples. Second, the time to achieve a stable glow emission with dioxetane phosphates is longer than with colorimetric procedures, which increases the turnaround time for assay results. Conditions could be optimized to make the CL emission more stable and rapid by manipulating the effect of pH and temperature on sALP. Third, polymeric enhancers utilized to increase signal response in one commercial substrate cause the precipitation of the serum matrix and increase the time to a stable glow emission. These problems associated with measuring sALP are circumvented by incorporating a fluorescein-labeled serum as a combined sample matrix diluent and fluorescent enhance to replace the polymeric enhancers used in ALP assays.[11]

Materials and Methods

Reagents for sALP Method

Fluorescein isothiocyanate (FITC) on celite obtained from Sigma Chemicals (St. Louis, MO)
$NaHCO_3$ (10 m*M*) buffer for dialysis (pH 8.0)
Carbonate buffer (pH 10.0) for substrate preparation containing Na_2CO_3 (0.1 *M*), $MgCl_2$ (1 m*M*), and NaN_3 (0.05%)
CSPD® (a 1,2-dioxetane phosphate substrate, 25 m*M*) as received from Tropix (Bedford, MA)

---

[8] Y. Vant Erve, J. C. Voyta, B. Edwards, L. J. Kricka, and I. Bronstein, *in* "Bioluminesence and Chemiluminescence" (A. A. Sazalay, L. J. Kricka, and P. E. Stanley, eds.), p. 306. Wiley, New York, 1993.

[9] S. Girotti, E. Ferrie, S. Ghini, A. Roda, J. Navarro, and E. Ortega, *in* "Bioluminescence and Chemiluminescence" (A. K. Campbell, L. J. Kricka, and P E Stanley, eds.), p. 44. Wiley, New York, 1994.

[10] M. Maeda, M. Kitamura, and A. Tsuji, *in* "Bioluminescence and Chemiluminescence" (A. A. Szalay, L. J. Kricka, and P. E. Stanley, eds.), p. 356. Wiley, New York, 1993.

[11] B. J. Hallaway, M. F. Burritt, and D. J. O'Kane, *in* "Bioluminescence and Chemiluminescence" (J. W. Hastings, L. J. Kricka, and P. E. Stanley, eds.), p. 521. Wiley, Chichester, 1997.

Spectra/Por membrane MWCO 50,000 (Spectrum, Houston, TX) for dialysis

Matrix Preparation

A matrix for serum samples was prepared by heat inactivating a unit of serum for several hours at 55° to denature endogenous alkaline phosphatase. Several aliquots of the heat-inactivated serum were stored at −20° and used as sample diluent for linearity studies.

FITC-Labeled Serum Preparation

FITC on celite is added to the remaining heat-inactivated serum at a concentration of 10 g/liter and mixed together on a rotator for 4 hr at room temperature. Centrifuge the FITC-labeled serum for 10 min to sediment the celite. Dialyze the supernatant against bicarbonate buffer overnight (for three changes) to remove free FITC. Heat inactivate the FITC-labeled serum for a second time for 1 hr at 55° in the presence of 2 m*M* EDTA to remove ALP-bound zinc and dialyze again in bicarbonate buffer for 2 days with at least six changes to remove EDTA. Store the dialyzed FITC-labeled serum in aliquots at −20°.

ALP Substrate Preparation

The ALP substrate is prepared in carbonate buffer (pH 10.0). Add CSPD (2 μl/ml) and FITC-labeled serum (50 μl/ml) to the carbonate buffer. Prepare the substrate immediately prior to use in the assay.

sALP Assay Method

1. Add 100 μl buffered substrate to a 12 × 75-mm polystyrene test tube and preincubate at 37° for 2 min.
2. Add 2.5 μl serum or serum dilution to the buffered substrate and incubate for an additional 10 min.
3. Detect sALP activity by CL emission for 1 sec using a Berthold AutoLumat 953 luminometer maintained at 37°.

Discussion

Chemiluminescence emission from dioxetane phosphates results in a glow emission that can persist for several minutes to several hours. The immediate goal for determination of sALP is to define conditions of assay where constant CL emission is attained rapidly and maintained for a long time. This requires optimizing conditions of buffer pH, temperature, and substrate.

Effect of Temperature

Assay temperature has a profound effect on the stability of ALP CL emission. Constant CL emission was attained for 30 min at 23°; 20 min at 30°; and 8 min at 37°. CL emission peaked immediately at 42° but declined rapidly. The optimum assay temperature for sALP assay was determined to be 37° and was utilized in subsequent procedures.

Effect of pH

pH affects the sALP assay in two ways: (i) the time to constant emission is pH dependent and (ii) the CL emission intensity is pH dependent. CL emission is greater at pH < 10.0 (9.0, 9.5, and 9.7) but is constant for only 5 min. This was in contrast to results obtained at pH 10.0 where CL emisson remains constant for up to 25 min although at a lower intensity compared to pH values <10.0. At pH 10.5 the CL emission is lower still and only stable for 15 min. Because sensitivity is not a significant problem with the sALP assay, the assay is performed at pH 10.0 because of the lengthy duration of constant CL emission.

Interferences

Hemoglobin, bilirubin, and triglycerides interfere with sALP colorimetric determination. Bilirubin and triglycerides do not interfere with sALP determination by CL. Hemoglobin at >5 mg/ml concentration decreases the CL emission from sALP. Consequently, noticeably hemolyzed serum samples are diluted in heat-inactivated serum to minimize the inhibitory effect of hemoglobin and then reassayed.

Linearity, Precision, and Recovery

When assayed at pH 10 and 37°, serial dilutions (1 : 2 to 1 : 128) in serum maintain constant CL emission for the same length of time and produce a linear dose–response. The coefficient of variation (CV) for intraassay precision, determined by running 10 replicates of 8 different serum samples within the same assay, ranged from 3.7 to 7.3%. The CVs for interassay precision CV, determined on duplicate measurements of 5 different sera run over 5 consecutive days, ranged from 2.9 to 6.5%.

A recovery study was run by mixing three different sets of two sera samples together in concentrations of 2 : 1, 1 : 1, and 1 : 2, then tested in the assay. The observed values of CL emission were compared with the ex-

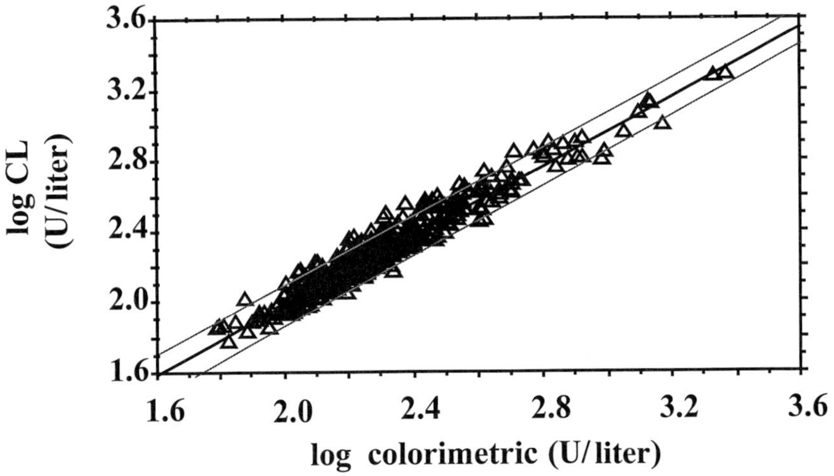

FIG. 1. Relationship between sALP measured by CL procedure and sALP measured colorimetrically. Reproduced with permission from B. J. Hallaway, M. F. Burritt, and D. J. O'Kane, *in* "Bioluminescence and Chemiluminescence: Molecular Reporting with Photons" (J. W. Hastings, L. J. Kricka, and P. E. Stanley, eds.), p. 521. Wiley, Chichester. © John Wiley & Sons Limited, 1996.

pected values and calculated for percentage of recovery. Recoveries ranged from 97.1 to 105%.

Method Comparison

A comparison of CL sALP assay results with those obtained with routine clinical colorimetric assays is shown in Fig. 1. A conversion factor was devised to relate relative light units (RLU) to units of enzyme activity: 342 RLU = 1 IU/liter under the conditions of assay. The calculated regression equation is log IU/liter (CL) = 0.978 × log IU/liter (colorimetric) + 0.096 with $r^2 = 0.931$ ($n = 498$). Twenty-five values were expected to fall outside the 95% confidence intervals; 27 values were observed.

Phosphoprotein Phosphatases

Introduction

PPase activities are implicated in a wide variety of cellular regulatory phenomena, including oncogenesis[12] and immunosuppression.[13,14] A tyro-

[12] J. S. Kovach, *J. Natl. Cancer Inst.* **84,** 515 (1992).

[13] A. Haddy, S. K. Swanson, T. L. Born, and F. Rusnak, *FEBS Lett.* **314,** 37 (1992).

[14] D. A. Fruman, C. B. Klee, B. E. Bierer, and S. J. Burakoff, *Proc. Natl. Acad. Sci. U.S.A.* **89,** 3686 (1992).

TABLE I
CHARACTERISTICS OF PHOSPHOPROTEIN PHOSPHATASES[a]

Phosphoprotein phosphatase	Source	(Inhibitor)	Activator P-Ser	Activator P-Thr	Activator P-Tyr
PP-1	Erythrocyte	Cation^{++} (microcystin)	+	+	−
PP-2A	Kidney	Thiols (okadaic acid)	+	+	− (+)[b]
PP-2B (CN)	Brain/T cell	Calmodulin (cyclosporin)	+	+	−
PP-2C	Liver/muscle		+	+	−
PTP-1B	Lymphocytes	(*p*-Bromotetramisole)	−	−	+
PTP	*Yersinia*				

[a] Reproduced with permission from D. J. O'Kane, B. J. Hallaway, and F. Rusnak, *in* "Bioluminescence and Chemiluminescence: Fundamentals and Applied Aspects" (A. K. Campbell, L. J. Kricka, and P. E. Stanley, eds.), p. 24. Wiley, Chichester. © John Wiley & Sons Limited, 1994.
[b] Following incubation with ATP.

sine phosphatase domain has been identified in PTEN, a proposed tumor suppressor homologous to tensin.[15] Properties of PPases are presented in Table I. Phosphotyrosine phosphatase (PTP-1B) is tyrosine specific and is the proposed target of levamisole chemotherapeutic agents.[12] PP-1, PP-2A, PP-2B, and PP-2C are serine/threonine-specific protein phosphatases. PP-2B (calcineurin) is a target of immunosuppressive drugs, such as cyclosporin A and FK506, which form complexes with cellular protein receptors referred to as immunophillins. Immunosuppression is accomplished by drug–receptor binding and inhibition. While serum levels of immunosuppressive drugs can be measured, these do not always correlate well with actual immunosuppression. A functional, enzymatic assay of the inhibition of T-cell calcineurin (CN) by immunosuppressive drugs would be advantageous. The amount of CN in T cells is low, necessitating the development of sensitive assays.

The major difficulty in measuring PPase activities by CL is the interference by alkaline phosphatases that hydrolyze the same substrates. Consequently, selective inhibitors of ALP must be incorporated in the assay to permit the measurement of PPases. Alkaline phosphatase activities can be

[15] J. Li. C. Yen, D. Liaw, K. Podsypanina, S. Bose, S. I. Wang, J. Puc, C. Miliaresis, L. Rodgers, R. Mccombie, S. H. Bigner, B. C. Giovanella, M. Ittmann, B. Tycko, H. Hibshoosh, M. H. Wigler, and R. Parsons, *Science* **275,** 1943 (1997).

TABLE II
INHIBITION OF PHOSPHOPROTEIN PHOSPHATASES[a]

Compound	*E. coli* ALP	Calf intestine ALP	Milk ALP	Calcineurin (PP-IIB)
L-Cysteine	97	98	99	5
2-Mercaptoethanol	10	17	18	15
meso-DTA	96	90	99	0
Dithiothreitol	97	97	98	30
Levamisole	0	0	79	7
o-Phenanthroline	30	45	70	40
Sodium fluoride	0	0	0	61
Sodium *o*-vanadate	50	87	79	50
Manganese chloride	20	40	10	20
Zinc acetate	0	0	0	48

[a] Reproduced with permission from D. J. O'Kane, B. J. Hallaway, and F. Rusnak, *in* "Bioluminescence and Chemiluminescence: Fundamentals and Applied Aspects" (A. K. Campbell, L. J. Kricka, and P. E. Stanley, eds.), p. 24. Wiley, Chichester. © John Wiley & Sons Limited, 1994.

inhibited efficiently by thiol compounds[16] (Table II). Cysteine and *meso*-dithioadipamide (DTA) inhibit ALP activity without inhibiting calcineurin activity. Several PPases were evaluated for reactivity toward phosphorylated substrates capable of producing CL,[17–21] including dioxetane-phosphates, firefly luciferyl-*o*-phosphate (LOP), and phosphoascorbic acid. Assay conditions are optimized for the phosphatases that were able to hydrolyze each substrate.

Materials and Methods

Reagents for Dioxetane Phosphate Methods for Assaying Calcineurin

MOPS buffer (3-[*N*-morpholino]propanesulfonic acid), 25 m*M*, pH 7.0, containing $MgCl_2$ (1 m*M*), and $CaCl_2$ (0.1 m*M*)

[16] H. Van Belle, *Biochim. Biophys. Acta* **289,** 158 (1972).

[17] H. Arakawa, M. Maeda, and A. Tsuji, *Anal. Biochem.* **199,** 238 (1991).

[18] M. Maeda, A Tsuji, K. H. Yang, and S. Kamada, *in* "Bioluminescence and Chemiluminescence: Current Perspectives (P. E. Stanley and L. J. Kricka, eds.), p. 119. Wiley, New York, 1991.

[19] W. Miska and R. Geiger, *J. Biolum. Chemilum.* **4,** 119 (1989).

[20] M. Isobe, Y. Sugiyama, T. Ito, I. I. Ohtani, Y. Toya, Y. Nishigohri, and A. Takai, *Biosci. Biotechnol. Biochem.* **59,** 2235 (1995).

[21] Y. Sugiyama, K. Fujimoto, I. I. Ohtani, A. Takai, and M. Isobe, *Biosci. Biotechnol. Biochem.* **60,** 1260 (1996).

Calcineurin (PP-2B) was prepared chromatographically as described previously[13] as a 1.41 μM stock in MOPS buffer
Calmodulin (100 μM stock)
L-Cysteine (10 mM stock)
Lumi-Phos Plus (pH adjusted to 7.5) (Lumigen Inc., Southfield, MI)
Bio-Rad CL substrate dilution buffer (Bio-Rad, Hercules, CA)

Dioxetane Phosphate Method[22]

1. Prepare a 1:10 dilution of stock CN in MOPS buffer.
2. Preincubate 10 μl CN, 10 μl calmodulin, and 20 μl cysteine with 164 μl Lumi-Phos Plus (pH 7.5) for 30 min in a 12 $\times$ 75-mm polstyrene test tube.
3. Adjust the pH of the assay to 9.5 by the addition of 10 μl Bio-Rad CL substrate buffer. CL emission is measured in a Turner TD-20e luminometer. Maximum activity is obtained immediately and decreases twofold over 2 hr.

Discussion

Calcineurin (PP-2B) can be assayed selectively from other phosphoprotein phosphatase activities by using dioxetane phosphate substrates. The Ser/Thr-specific, cation-independent PP-1A and PP-2A, as well as PTP-1B, failed to hydrolyze dioxetane phosphates. Calcineurin from several sources is active with dioxetane phosphate substrates. Bovine brain and recombinant calcineurin have comparable enzyme activities. Appreciable activity is obtained from the recombinant catalytic α subunit of calcineurin alone. Only the recombinant regulatory β subunit of calcineurin had no activity with dioxetane phosphate substrates. Initial activity at pH 9.5 is proportional to CN concentration. The assay background in the presence of cysteine is less than 2.2 total light units, resulting in a minimum detectable concentration of less than 180 fmol of CN/0.1 ml of CN solution. This is more sensitive than isotopic CN assays.

Membrance Blotting Method for Detecting Calcineurin[22]

EQUIPMENT

MilliBlot Systems slot blot (Millipore, Bedford, MA)
Immun-Lite blotting membrane (Bio-Rad, Richmond, CA)

[22] D. J. O'Kane, B. J. Hallaway, and F. Rusnak *in* "Bioluminescence and Chemiluminescence" (A. K. Campbell, L. J. Kricka, and P. E. Stanley, eds.), p. 24. Wiley, Chichester, 1994.

METHOD

1. Prewet nylon membrane and filter paper in MOPS buffer containing 1 mM cysteine. Assemble membrane in slot blot apparatus.

2. Prepare serial dilutions of CN ranging from 1000 to 30 fmol in MOPS buffer. Add 100 μl CN dilution to each corresponding slot and include buffer blanks. Apply vacuum for 1 min to aspirate CN into the nylon membrane.

3. Remove nylon membrane from slot blot apparatus and place in a flat-bottom container. Apply 1 ml Lumi-Phos Plus (pH 7.5) containing 1 mM cysteine that has been filtered through a 0.2-μm filter to the nylon membrane.

4. Transfer the membrane and CL substrate liquid to a plastic bag and seal. Expose the membrane to X-ray film for 5–24 hr.

Discussion

Calcineurin activity assayed on nylon blots with Lumi-Phos Plus in the absence of thiol compounds is obscured by the activities of contaminating ALP in the solutions. The addition of 1 mM cysteine abolishes ALP background CL. The lowest concentration of CN applied (30 fmol) was detected easily in the 17-hr exposure of the X-ray film. Dithiothreitol (DTT), another ALP inhibitor, inhibits CN activity as well as the activity of ALP contaminants.

Phosphotyrosine Protein Phosphatase Activities

In a screen of phosphorylated CL substrates utilized by various PPases, it was observed that phosphotyrosine protein phosphatases selectively hydrolyzed firefly luciferyl-*o*-phosphate.

Reagents for the LOP Method

Imidazole/BSA buffer (pH 7.0): Prepare imidazole (25 mM) and autoclave. Add 100 μg/ml BSA to the prepared buffer and heat treat at 55° for 1 hr to deactivate contaminating alkaline phosphatase

d-Luciferyl-*o*-phosphate (BioAss, Diesen, Germany): Working LOP is diluted 1:16 in imidazole/BSA buffer

HEPES buffer, 0.1 M (pH 7.0), containing $MgCl_2$ (0.1 M) and glycerol (25%)

Firefly luciferase-ATP (Sigma Chemical, St. Louis, MO): A stock solution of firefly luciferase-ATP (900 μM) is prepared in HEPES buffer and stored in aliquots at −20°

PTP-1B (Upstate Biotechnology, Lake Placid, NY)

TABLE III
PHOSPHOPROTEIN PHOSPHATASE HYDROLYSIS OF LOP[a]

Phosphoprotein phosphatase	TLU[b]/μg protein
PTPP-1 (GST fusion protein)	5776
PTP (*Yersinia enterocolitica*)	343
Protein phosphatase-1	24
Protein phosphatase-2A	62
Protein phosphatase-2B (calcineurin)	0.2

[a] Reproduced with permission from D. J. O'Kane, B. J. Hallaway, and F. Rusnak, *in* "Bioluminescence and Chemiluminescence: Fundamentals and Applied Aspects" (A. K. Campbell, L. J. Kricka, and P. E. Stanley, eds.), p. 24. Wiley, Chichester. © John Wiley & Sons Limited, 1994.

[b] Total luminescence unit, Turner luminometer.

Yersinia enterocolitica PTP (Calbiochem, La Jolla, CA)
Dithithreitol (50 m*M* stock)

LOP method

1. Prepare a 1:50 dilution of the PTPs in imidazole/BSA buffer.
2. Add 25 μl LOP, 10 μl DTT, 10 μl FF luciferase-ATP (180 μ*M* final concentration), and 4 μl imidazole/BSA buffer together in a 12 × 75-mm polystyrene test tube.
3. Add 1 μl PTP to start the reaction. CL emission is read in a Turner luminometer at 5 min. The activity decreases slowly over time.

Results

LOP is hydrolyzed by the PTP-1B catalytic subunit and can be detected in a coupled assay using firefly luciferase. Recombinant human PTP-1B fusion protein is the most active with LOP. The PTP catalytic domain from *Y. enterocolitica* is only 6% as active as PTP-1B (Table III). Little hydrolysis activity is found with PP-1 and PP-2A. These activities can potentially be inhibited selectively with microcystin-LR or 100 μ*M* okadaic acid. PP-2A activity may be due to ATP in the assay that has been shown to affect PP-2A substrate specificity.[23] PP-2B (calcineurin) has negligible activity with this substrate.

The hydrolysis of LOP by PTP is dose dependent with respect to the enzyme when the initial velocity is measured. Minimum detectable levels are <0.02 pmol PTP-1B and <0.2 pmol *Yersinia* PTP. Attempts to blot

[23] G. D. Amick, S. A. G. Reddy, and Z. Damuni, *Biochem. J.* **287,** 1019 (1992).

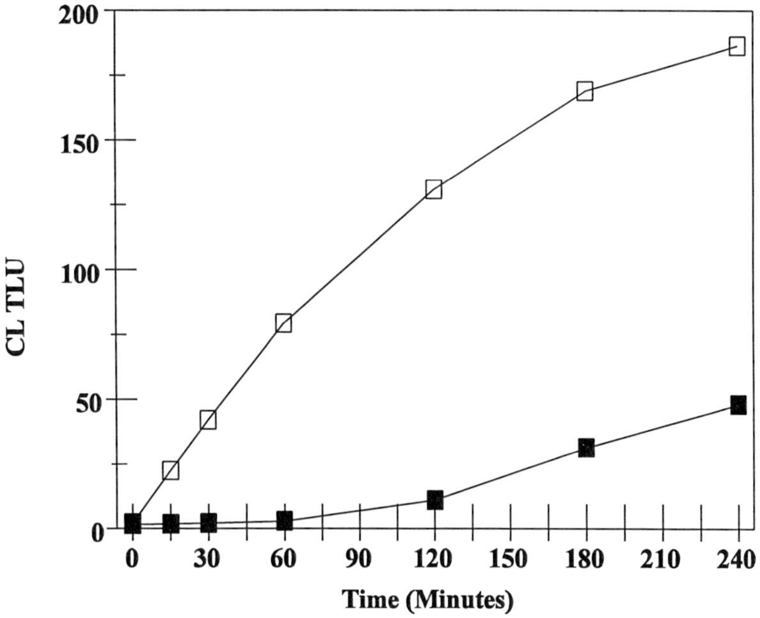

FIG. 2. Inhibition of sALP activity by cysteine. CL emission from sALP was measured in the presence (■) or absence (□) of 1 m*M* cysteine. Reproduced with permission from B. J. Hallaway, M. F. Burritt, and D. J. O'Kane, *in* "Bioluminescence and Chemiluminescence: Molecular Reporting with Photons" (J. W. Hastings, L. J. Kricka, and P. E. Stanley, eds.), p. 521. Wiley, Chichester. © John Wiley & Sons Limited, 1996.

PTP and detect them on nylon or nitrocellulose membranes have so far been unsuccessful.

Conclusions

The CL assay for sALP is relatively fast, easy to perform, and compares well with the colorimetric method. The coefficients of variation for the CL assay are low with good recoveries of activities. The addition of FITC-labeled serum improves the assay linearity by minimizing the dilution effects on the serum matrix and serves as a fluorescence enhancer as well.

Alkaline phosphatase activity can be inhibited efficiently by thiol compounds to permit the selective detection of PPase activities. Cysteine is most effective in calcineurin assays and produces the fewest problems, such as enzyme inhibition and background CL (Fig. 2). DTT is the most effective in luciferyl-*o*-phosphate assays. Sensitive nonisotopic CL assay development for assessing the efficacy of immunosuppression at the cellular level appears to be promising.

[28] Chemiluminescence Screening Assays for Erythrocytes and Leukocytes in Urine

By VALERIE J. BUSH, BRENDA J. HALLAWAY, THOMAS A. EBERT, DAVID M. WILSON, and DENNIS J. O'KANE

Introduction

The use of chemiluminescence as a detection system within laboratory medicine, especially applied toward immunoassays, provides the advantages of high specificity and sensitivity, rapid analysis, low cost, and high stability.[1] Only recently, however, has chemiluminescence become practical as a screening system in urinalysis.[2] Current, clinical methods used to detect erythrocytes and leukocytes in urine include manual microscopic analysis, dipsticks, and automated imaging.[3–5] These procedures are time-consuming, expensive, prone to error, and cannot exclude normal samples efficiently from analysis. Applying chemiluminescence screening systems for erythrocytes and leukocytes in urine has the potential to reduce the number of normal urine samples requiring additional diagnostic testing and may facilitate reprioritizing laboratory resources through the use of automation.

The routine application of automation to urinalysis, however, is compromised by cell sedimentation in the sample vessels over time, giving rise to heterogeneous sampling artifacts, as well as by variation in urine composition, patient to patient. This impacts the establishment of stringent guidelines for the validation of method performance and clinical utility. Additional difficulties for the validation of chemiluminescence method performance in urines include stability of the urine matrix, quantification and uniformity of controls and standards, effective cell lysis, and the effects of hyperosmolality on cellular components in the assays.[6–8]

[1] H. A. H. Rongen, R. M. W. Hoetelmans, A. Bult, and W. P. Van Bennekom, *J. Pharm. Biom. Anal.* **12,** 433 (1994).

[2] B. J. Hallaway, M. E. Copeman, B. S. Stevens, T.S. Larson, D. M. Wilson, and D. J. O'Kane, *in* "Bioluminescence and Chemiluminescence: Molecular Reporting with Photons" (J. W. Hastings, L. J. Kricka and P. E. Stanley, eds.), p. 517. Wiley, Chichester, 1997.

[3] D. J. Holland, K. J. Bliss, C. D. Allen, and G. L. Gilbert, *Pathology* **27,** 91 (1995).

[4] R. C. Bartlett, D. A. Zern, I. Ratkiewicz, and J. Z. Tetreault, *Arch. Pathol. Lab. Med.* **118,** 1096 (1994).

[5] M. McGinley, L. L. Wong, J. H. McBride, and D. O. Rodgerson, *J. Clin. Lab. Anal.* **6,** 359 (1992).

[6] S. Kubo, T. Matsumoto, M. Sakumoto, O. Mochida, Y. Abe, and J. Kumazawa, *Renal Failure* **20,** 75 (1998).

0076-6879/00 $30.00

Screening for Erythrocytes in Urine

Introduction

The presence of red blood cells (hematuria) and/or hemoglobin (hemoglobinuria) in urine can indicate significant health problems and patient morbidity. Hematuria results from any disease or trauma to the kidney and/or urinary tract and is also seen with the presence of calculi, neoplasms, and infections or with the use of certain nephrotoxic drugs. Hemoglobinuria is a less common clinical finding and results from either acquired hemolytic disease or genetic defects. However, certain conditions (pH, specific gravity, or osmolality) within the urine result in the lysis of erythrocytes leading to hemoglobinuria. Lysis of erythrocytes in the urine can occur in the bladder or in the specimen container after the urine is voided.

A sensitive, screening chemiluminescence (CL) assay has been developed for determining hemoglobinuria/hematuria. The pseudoperoxidase activity of hemoglobin reacts with a cyclic hydrazide substrate in the presence of hydrogen peroxide to form a superoxide radical that proceeds spontaneously to form a product with the emission of light (Fig. 1).

Materials and Methods

Assay Reagents Required

Cetyltrimethylammonium chloride (CTAC), 25% solution (Aldrich Chemicals, Milwaukee, WI)
Dimethyl sulfoxide (DMSO): minimum 99.5% (GC) grade
Sodium acetate: anhydrous, ACS reagent
Sodium chloride
Acetic acid: glacial, +99% purity
Nitric acid, 70%, ACS reagent
Hydrogen peroxide, ACS reagent, assay, 30.0%
Hemoglobin (Hb), human, purified, crystallized, dialyzed, and lyophilized (Sigma Chemicals, St. Louis, MO)

Assay Solutions Required

7-Dimethylaminonaphthalene-1,2-dicarbonic acid hydrazide (7-DNH) (Assay Designs, Inc., Ann Arbor, MI): Dissolve 5.0 mg 7-DNH in

[7] M. Qian, J. W. Eaton, and S. P. Wolff, *Biochem. J.* **326,** 159 (1997).

[8] T. Matsumoto, K. Takahashi, Y. Mizunoe, N. Ogata, M. Tanaka, J. Kumazawa, M. Nozaki, and P. van der Auwera, *Eur. Urol.* **20,** 232 (1991).

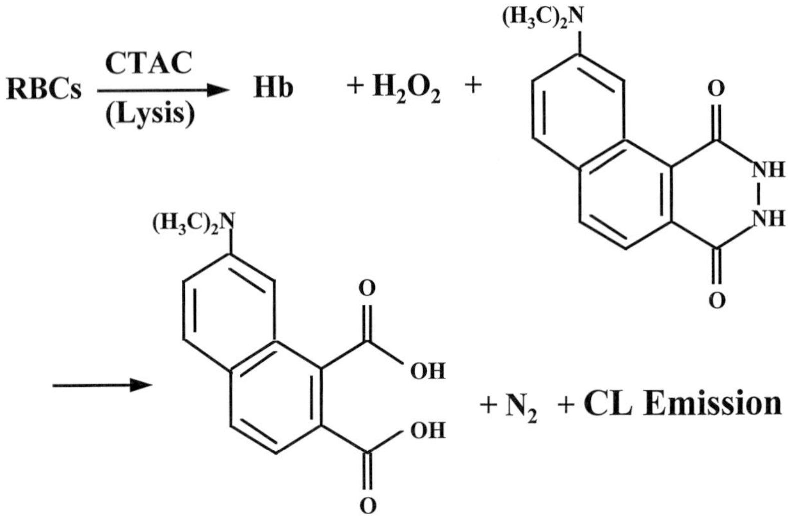

FIG. 1. Assay of erythrocytes in urine with 7-DNH. Erythrocytes are first lysed with detergent, liberating hemoglobin. Hemoglobin pseudoperoxidase activity catalyzes the oxidation of 7-DNH, resulting in CL emission.

5.0 ml of DMSO and vortex. Dilute to 200 μg/ml by adding 20 ml of DMSO and vortex. Perform a spectrographic analysis, comparing absorbance at 324 and 414 nm, to any previous 7-DNH stocks. If the new and previous stock have <10% coefficient of variation (CV), the new stock is acceptable for routine assay use. Aliquot 1.0 ml 7-DNH stocks and wrap containers or tubes with aluminum foil to block light. Store at −20°. The 7-DNH stock substrate can be frozen and thawed repeatedly without detrimental effects.

7-DNH substrate buffer: Dissolve 4.1 g of sodium acetate and 292.2 g NaCl in 1.0 liter of distilled H_2O to make a 50 m*M* sodium acetate and 5 *M* NaCl solution. Adjust to pH 5.5 with 1 to 2 drops of glacial acetic acid. The substrate buffer will be used to dilute the 7-DNH to 68 ng/ml, or approximately 1:3000. Diluting 3.4 μl of 7-DNH stock substrate into 10.0 ml of substrate buffer is sufficient for 34 tubes. Add 34 μl CTAC to the 7-DNH substrate solution to permit prelysis of erythrocytes.

1% nitric acid: In a 15-ml polypropylene tube, dilute 100 μl of nitric acid into 10.0 ml of distilled H_2O for a 1.0% solution. Store at room temperature for up to 6 months, then discard.

Chemiluminescence trigger solution: Dilute 166 ml of 30.0% H_2O_2 with 834 ml of distilled H_2O (5% H_2O_2). Adjust the pH with 1% nitric acid to 3.25 ± 0.02, using a pipettor if necessary. Transfer 5.0% H_2O_2 to the large trigger reservoir and place in the luminometer. One liter is sufficient for 3250 injections. Store at room temperature for 2 weeks, then discard.

Negative urine matrix. The requirements for selecting suitable urine samples for use as a negative urine matrix are that the samples are negative by microscopy for erythrocytes, leukocytes, casts, and crystals and are negative for free hemoglobin and leukocyte esterase by dipstick analyses. Also, the urine samples must have less than 15 mg/dl glucose, preferably 1–5 mg/dl; that the samples have less than 8 mg/dl of protein, preferably 1–5 mg/dl; that the samples have a normal pH in the range of 5.0 to 7.0; and that the samples have a normal osmolality in the range of 300–800 mOsm/kg. Combine selected urine samples to form at least a 2.0 liter pool and recheck the parameters. Prepare 10- to 12-ml aliquots of the pooled negative urine matrix and store at −20° for up to 6 months, then discard. Thaw a single aliquot for each assay run, preparing other assay reagents during this time. Four urine blanks are assayed at the beginning of each run.

Hb standard dilution buffer: 0.15 *M* sodium chloride (NaCl). Dissolve 87.6 mg into 10.0 ml of distilled H_2O. Vortex and adjust pH to 5.0. Store at room temperature for up to 3 months, then discard.

Hb standards: Dissolve 1.0 mg of Hb into 1.0 ml of 0.15 *M* sodium chloride, pH 5.0. Vortex and incubate for 15 min. Vortex again. Wrap tube or container with aluminum foil to block light and store Hb stock standard for up to 1 week, then discard. Prepare a new set of Hb standards for each assay run using the negative urine matrix. Five standards, at concentrations of 32, 160, and 800 ng/ml and 4 and 20 μg/ml, are assayed at the end of each run and are prepared from the stock Hb standard of 1.0 mg/ml as a fivefold, serial dilution, using the negative urine matrix.

Instrumentation. The assay is performed using a CLS-ID chemiluminescence detection system (Nichols Institute Diagnostics, San Juan Capistrano, CA), which includes a four probe Tecan diluter and a luminometer with injectors for assay solutions and triggers. All assay procedural steps are programmed using the CLS-ID system.

Assay Procedures. Urine samples are prelysed by the addition of Triton X-100 to 0.01% final concentration. Ten microliters of prelysed urine is added to 290 μl of the 7-DNH substrate solution in a 12 × 75-mm polystyrene tube. Three hundred microliters of 5% H_2O_2 is added by the luminometer and chemiluminescence emission is detected. CL decision levels must be determined with each luminometer. Under the conditions of the assay used in these studies, samples emitting ≤2000 relative light units (RLU) are reported as normal. Samples emitting >2000 RLU are considered positive for erythrocytes and are considered potentially abnormal. Positive samples are examined further by manual microscopic cell analysis or by hemocytometer counting.

Results and Discussion

Kinetics of CL Emission. The kinetics of CL was determined by monitoring the signal generated from the substrate over a 20-sec time period added to filtered urine in the presence and absence of Hb (background RLU). Background CL emission was low with the 7-DNH substrate in contrast to other cyclic hydrazide derivatives investigated. The low background was in part attributed to the low pH of the assay (pH 5.5). Maximum background emission occurred within the first second after the addition of H_2O_2 followed by a rapid decline to the level of instrument noise. The addition of hemoglobin to the filtered urine altered the course of CL such that the emission peaked at 2 sec with a more gradual decay. A high signal-to-noise ratio was obtained by integrating photons emitted in the first 5 sec.

Interferences. Uric acid, ascorbic acid, and bilirubin do not affect assay performance. Myoglobin (Mb) at >5.0 μg/ml and myeloperoxidase at >2.5 μg/ml provided positive interferences that are detected by the substrate. High levels of bacteria do not affect the CL emission.

Method Comparison. A patient study was performed on urine samples from 500 patients (246 females and 254 males) and compared to values obtained by automated imaging. The sensitivity and specificity of the assay were 95 and 95.5% (>3 RBC/high power microscopic field) using a CL emission cutoff of >2000 RLU. The normal reference range for urinary Hb is <150 ng/ml (calculated from the number of erythrocytes per high power field and assuming 1 RBC = 30 pg of Hb). Urines emitting CL >2000 RLU (or >150 ng/ml Hb) are reflexed for automated or manual microscopic analysis. The false negative rate was low (3/246 females). The CL screen, manual microscopic analysis, and dipstick were negative for these patients. The apparent false-positive rate was 12/246 females and 8/254 males.

Screening for Leukocytes in Urine

High levels of leukocytes in urine may be an indication of urinary tract infection (UTI), cystitis, nephritis, various autoimmune diseases, and physiological responses to drugs. Leukocytes, including monocytes and

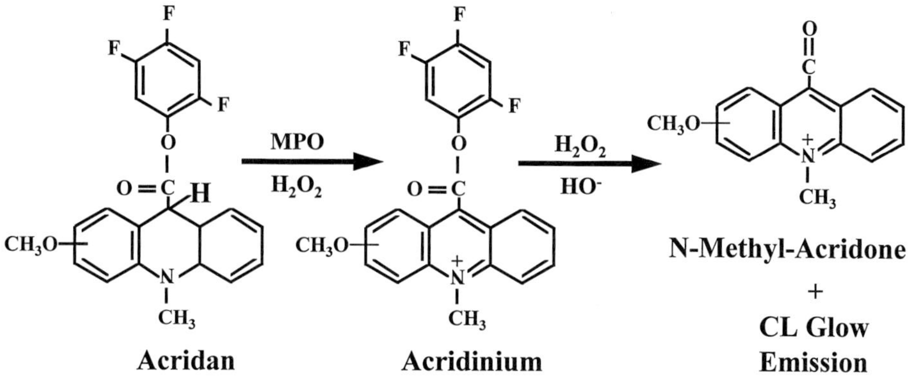

FIG. 2. Assay of leukocytes in urine with an acridan substrate. Leukocytes are first prelysed with detergent, liberating myeloperoxidase. Myeloperoxidase catalyzes the dehydrogenation of the acridan to form an acridinium species that decomposes spontaneously with CL emission.

granulocytes, express myeloperoxidase (MPO) activity. MPO has been assayed with luminol derivatives,[9,10] but luminol also reacts with other heme proteins found in urine, such as hemoglobin (Hb) and myoglobin (Mb).[11,12] A rapid, inexpensive CL screening procedure has been developed using the peroxidase triggering of acridinium ester[13] by MPO present in leukocytes[2] (Fig. 2).

Materials and Methods

Assay Reagent Solutions Required

Dimethyl sulfoxide: minimum 99.5% (GC) grade
Triton X-100 stock solution: 1% by volume
Methimazole stock solution (optional): dissolve 11.4 mg methimazole (Sigma Chemicals) in 10.0 ml of DMSO and store in 0.5-ml aliquots at 0.01 *M*. Dilute to 5.0 m*M* for working stock.
Acridan substrate (PS-1, solutions A and B, Lumigen Inc., Southfield, MI): prior to use add 0.01% Triton (v/v) to solution A. Use this

[9] G. Briheim, O. Stendahl, and C. Dahlgren, *Infect. Immun.* **45,** 1 (1984).
[10] P. Roschger, W. Graninger, and H. Klima, *Biochem. Biophys. Res. Commun.* **123,** 1047 (1984).
[11] T. Olsson, K. Bergstrom, and A. Thore, *Clin. Chim. Acta* **122,** 125 (1982).
[12] T. Olsson, K. Bergstrom, and A. Thore, *Clin. Chim. Acta* **138,** 31 (1984).
[13] H. Akhavan-Tafti, K. Sugioka, Z. Arghavani, R. DeSiva, R. S.Handley, Y. Sugioka, R. A. Eickholt, M. P. Perkins, and A. P. Schaap, *Clin. Chem.* **41,** 1368 (1995).

solution to dilute solution B 1:100 for immediate use. Optional add methimazole to 50 μM. This additive selectively inhibits the activities of most peroxidase activities, but is a poor inhibitor of MPO.

Negative urine matrix: prepared as described earlier for screening for erythrocytes.

Peroxidase controls: Soybean peroxidase (SBP) (Pierce Chemicals, Rockford, IL): dissolve 1.0 mg into 0.911 ml of negative urine matrix for a concentration of 1.1 mg/ml. Vortex, invert, and wrap tube with aluminum foil or place control in a light-sensitive container to block light. Incubate for 15 min at ambient temperature and vortex again. Store stock concentration of 1.1 mg/ml at 4° for 4 weeks. Prepare a new set of controls for each assay run, using the corresponding urine matrix. Three controls at concentrations of 22, 55, and 110 ng/ml are assayed at the end of each run and are prepared from a stock SBP of 1.1 mg/ml and the negative urine matrix. Soybean peroxidase is utilized instead of MPO because of its greater stability in urine.

Instrumentation. The assay is performed using a CLS-ID chemiluminescence detection system (Nichols Institute Diagnostics), which includes a four probe Tecan diluter and a luminometer with injectors for assay solutions and triggers. All assay procedural steps are programmed using the CLS-ID system.

Assay Procedure. Urine samples are prelysed with 0.01% Triton X-100 and diluted 1:8 (v/v) with distilled H_2O. Ten microliters of dilute urine is added to 50 μl of PS-1 substrate and incubated at ambient temperature for 10 min. MPO activity is detected by counting the CL emissions for 1 sec. CL decision levels must be determined for each luminometer and each assay condition. CL emission <1500 RLU under the conditions of assay used for these studies are considered normal (≤ 5 leukocytes per high power microscope field). Samples with CL emission >1500 RLU are considered abnormal and are considered as potentially positive for leukocytes. Potentially positive samples are examined further by hemacytometer counting, or by a manual microscopic analysis, to determine the number of leukocytes per milliliter.

Results and Discussion

Luminol derivatives react with a wide variety of heme proteins. This lack of specificity limits its application in urinalysis where Hb and Mb may be present and have peroxidase activities. The reactivity of PS-1 with heme proteins was assessed to determine if this substrate had a more restricted specificity. Several peroxidase activities are detected with the PS-1 substrate, including eosinophil peroxidase (EPO), MPO, lactoperoxidase (LPO), and thyroid peroxidase (TPO). The addition of 50 μM methimazole, an inhibitor of EPO,[14] reduces activities of EPO, TPO, and LPO by >90%, whereas MPO is inhibited by only 50%. The addition of methimazole to PS-1 assay allows MPO activity to be detected selectively. Alternatively, omitting methimazole permits the detection of eosinophils in urine but with an increased background CL.

Prelysing Cells. A major advantage of this assay is the use of automated pipettors and sample identification. However, while the primary sample tube is in queue for sampling and dilution, cells in the urine are settling due to gravity, causing decreased signal levels and potentially resulting in false-negative results. The samples must be prelysed to overcome this difficulty. Several lysis reagents were investigated, including saponin, mellitin, Tween 20, CHAPS, and Triton X-100. Triton X-100 proved to be the most useful in that it lysed leukocytes and erythrocytes, did not interfere in the leukocyte assay, and did not interfere in the erythrocyte assay, permitting one prelysed sample to be used in both assays.

Osmolality. Hyperosmolal urine samples (>600 mOsm) do not give as high a CL signal as lower osmolal samples with the same number of leukocytes present. Hyperosmolal urine samples inhibit MPO activity.[8] Consequently, the effect of urine osmolality on the acridan CL assay was investigated. Filtered urine samples with osmolalities ranging from 260 to 1053 mOsm are spiked with the same amount of MPO. The quenching effect of hyperosmolal urines is minimized at an eightfold dilution with water.[2]

Interferences. Uric acid, ascorbic acid, bilirubin, and trolox (water-soluble vitamin E analog) do not affect assay performance. Hemoglobin at >10 μg/ml and myoglobin at >2 μg/ml are detected by the PS-1 substrate. Samples populated grossly with bacteria caused slight inhibition of the MPO activity.

Method Comparison. A patient study was performed on samples from 99 males and 89 females to compare cell quantification by automated imaging with MPO CL emission with the PS-1 substrate.[2] Arbitrary set points to define abnormal urines were established from literature values as >10,000 WBC/ml for males and >16,000 WBC/ml for females. Assay sensitivity and specificity are 92 and 86% (≥5 WBC/high power microscopic field). The specificity of the assay is lower because of the apparent false-positive results (11/89 females and 12/99 males). Apparent false-negative rates were 3/99 (males) and 2/89 (females). The number of samples submitted

[14] A. Taurog and M. L. Dorris, *Arch. Biochem. Biophys.* **296,** 239 (1992).

for WBC quantification by manual microscopic analysis was reduced by 70%.

Conclusions

Erythrocyte and leukocyte prescreening assay sensitivities for both assays are equivalent to those obtained by either manual microscopic analysis or automated imaging. Only 30% of urine samples need to be analyzed by quantitative methods for the presence of erythrocytes and leukocytes. The proportion of false-negative samples is small. Rapid, inexpensive prescreening for erythrocytes and leukocytes allows for high throughput of samples and fast turnaround times. This also allows focusing laboratory resources on the analysis of abnormal samples.

[29] Immunoassay Protocol for Quantitation of Protein Kinase Activities

By JENNIFER MOSIER, CORINNE E. M. OLESEN, JOHN C. VOYTA, and IRENA BRONSTEIN

Introduction

Phosphorylation and dephosphorylation play a critical role in many cellular signal transduction processes, including control of cell cycle, response to growth factors, apoptosis, cellular transformation by oncogenes, and infection.[1–3] Protein kinases, which catalyze phosphoryl transfer from ATP to a protein or peptide substrate, are critical regulatory enzymes in many pathways and are the focus of biomedical and pharmaceutical research. These enzymes are important therapeutic targets for cancer, immune response disorders, infectious disease, and cardiovascular disease.[4]

Protein kinase assays typically measure phosphorylation of a peptide or protein substrate. Traditionally, kinase assays are performed with radioactively labeled ATP by the measurement of ^{32}P incorporation into the

[1] T. Hunter, *Cell* **80,** 225 (1995).

[2] J. Cleaveland, P. Kiener, D. Hammond, and B. Schacter, *Anal. Biochem.* **190,** 249 (1990).

[3] G. Rijksen, B. van Oirschot, and G. Staal, *Anal. Biochem.* **182,** 98 (1989).

[4] J. G. Foulkes, M. Chow, C. Gorka, A. R. Frackelton, Jr., and D. Baltimore, *J. Biol. Chem.* **260,** 8070 (1985).

0076-6879/00 $30.00

substrate. The substrate can be captured on phosphocellulose paper or in a microplate ELISA format. These assays generate radioactive waste, involve excessive washes, and require a peptide substrate concentration that is much higher than physiological concentrations. They are also conducted at ATP concentrations much lower than physiologically relevant, which leads to difficulties in identifying specific *in vivo* kinase inhibitors.[5]

Nonisotopic assays have been demonstrated with peptide substrates using immunological detection of the phosphorylated substrate. Peptide capture is possible in microtiter plate ELISA[2,6] or dot-blot[3] formats, followed by detection with an antiphosphopeptide antibody and secondary antibody enzyme conjugate and colorimetric or chemiluminescent enzyme substrates. Chemiluminescent assays with 1,2-dioxetanes have been performed for the quantitation of protein kinase A (PKA), protein kinase C (PKC), calmodulin-dependent protein kinase II (CAMKII), and receptor interacting protein kinase (RIP) activities. Chemiluminescent detection is more sensitive with reaction conditions close to physiological conditions than radioactive or colorimetric assays.[5]

The chemiluminescent signal from the alkaline phosphatase-activated 1,2-dioxetanes CSPD or CDP-*Star* exhibits glow light emission kinetics. Enzymatic dephosphorylation in the presence of excess substrate occurs at a constant rate proportional to enzyme concentration. Upon substrate dephosphorylation, the resulting anion decomposes with the emission of light. Maximum light emission is reached in 5–20 min. Highly sensitive detection with 1,2-dioxetane substrates has been demonstrated for the quantitation of a variety of analytes with both sandwich and competitive immunoassay formats.[7–12] Additional applications include the calculation

[5] C. Lehel, S. Daniel-Issakani, M. Brasseur, and B. Strulovici, *Anal. Biochem.* **244,** 340 (1997).

[6] I. Lazaro, M. Gonzalez, G. Roy, L. Villar, and P. Gonzalez-Porque, *Anal. Biochem.* **192,** 257 (1991).

[7] I. Bronstein, J. C. Voyta, G. H. G. Thorpe, L. J. Kricka, and G. Armstrong, *Clin. Chem.* **35,** 1441 (1989).

[8] G. H. G. Thorpe, I. Bronstein, L. J. Kricka, B. Edwards, and J. C. Voyta, *Clin. Chem.* **35,** 2319 (1989).

[9] I. Nishizono, S. Lida, N. Suzuki, H. Kawada, H. Murakami, Y. Ashihara, and M. Okada, *Clin. Chem.* **37,** 1639 (1991).

[10] F. Legris, J. Martel-Pelletier, J.-P. Pelletier, R. Colman, and A. Adam, *J. Immunol. Methods* **168,** 111 (1994).

[11] S. Fimbel, H. Déchaud, C. Grenot, L. Tabard, F. Claustrat, R. Bador, and M. Pugeat, *Steroids* **60,** 686 (1995).

[12] T. Jordan, L. Walus, A. Velickovic, T. Last, S. Doctrow, and H. Liu, *J. Pharm. Biomed. Anal.* **14,** 1653 (1996).

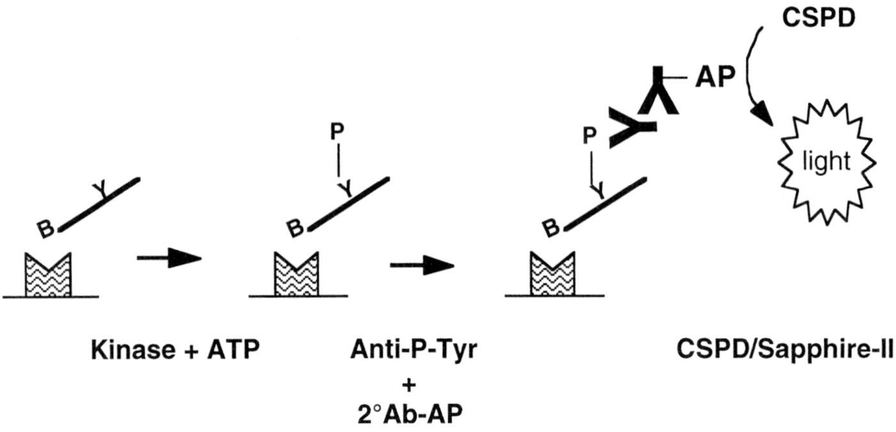

FIG. 1. Chemiluminescent immunoassay detection of protein kinase activity.

of antibody-binding constants,[13] receptor-binding assays,[14] quantitation of oligonucleotide : protein binding,[15] and immunodetection of nucleic acid capture/hybridization assays, including detection with an anti-DNA : RNA duplex antibody.[16]

We have performed the chemiluminescent kinase immunoassay with src protein tyrosine kinase. src Kinase is a member of a large family of nonreceptor tyrosine kinases, including multiple cellular forms and a viral transforming form.[17] src Kinases are found in many types of tissues, especially in the nervous system and in platelets, and have been shown to be involved in bone resorption.[18]

Principles of Method

A biotinylated peptide substrate is first captured in a NeutrAvidin-coated microplate well. Next, a kinase reaction is performed with captured peptide, protein kinase, and ATP. Immunodetection is then performed by incubating the phosphorylated peptide with an antiphosphopeptide primary

[13] L. Ge, A. Lupas, S. Peraldi-Roux, S. Spada, and A. Plückthun, *J. Biol. Chem.* **270,** 12446 (1995).

[14] M. Sanicola, C. Hession, D. Worley, P. Carmillo, C. Ehrenfels, L. Walus, S. Robinson, G. Jaworski, H. Wei, R. Tizard, A. Whitty, R. B. Pepinsky, and R. L. Cate, *Proc. Natl. Acad. Sci. U.S.A.* **94,** 6238 (1997).

[15] D. W. Drolet, L. Moon-McDermott, and T. S. Romig, *Nature Biotech.* **14,** 1021 (1996).

[16] W. R. Carpenter, T. E. Schutzbank, V. J. Tevere, K. R. Tocyloski, N. Dattagupta, and K. K. Yeung, *Clin. Chem.* **39,** 1934 (1993).

[17] H-C. Cheng, H. Nishio, O. Hatase, S. Ralph, and J. H. Wang, *J. Biol. Chem.* **267,** 9248 (1992).

[18] J. Cooper and B. Howell, *Cell* **73,** 1051 (1993).

antibody, followed by an alkaline phosphatase-conjugated secondary antibody. Finally, the phosphorylated peptide is quantitated by the addition of a 1,2-dioxetane alkaline phosphatase substrate (Fig. 1). Alternatively, peptide capture and the kinase reaction can be performed simultaneously. The peptide concentration should be optimized for the desired assay format, as this parameter may differ between the two (Fig. 2). Presumably, this is due to the efficiency of phosphorylation of a bound peptide compared to that of a peptide in solution.

Reagents

Microlite-1 multiwell strips: opaque, white microplate stripwells (Dynex Laboratories Inc., Chantilly, VA)
NeutrAvidin: biotin-binding protein (Pierce, Rockford, IL)
Biotinylated peptide II: biotin-EGPWLEEEEEAYGWMDF, a syn-

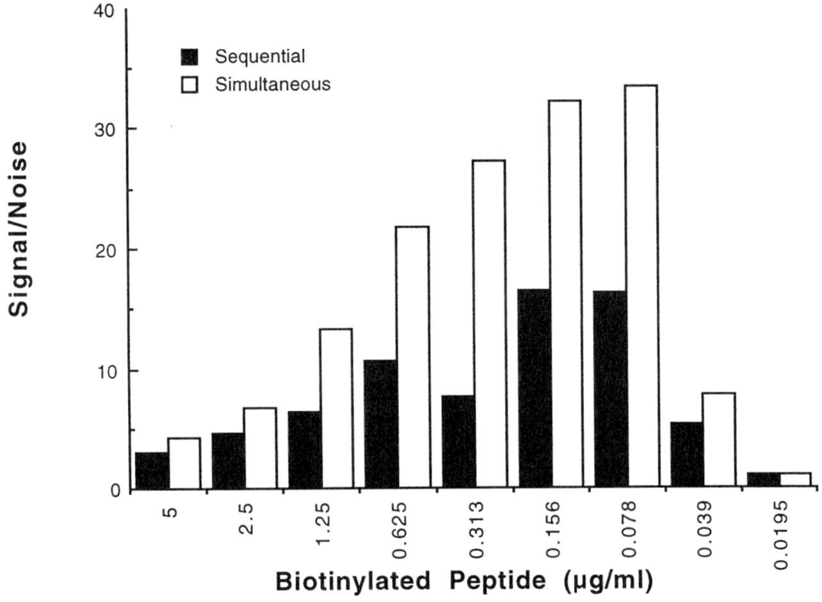

FIG. 2. Optimization of peptide concentration. Dilutions of biotinylated peptide were captured (100 μl/well) in NeutrAvidin-coated wells. Purified src kinase was diluted to a final concentration of 1 unit/well. In the sequential assay, 10 μl of diluted kinase, with 90 μl of kinase buffer, was added per well after peptide capture. In the simultaneous assay, 5 μl of 2× diluted kinase and 45 μl of 2× kinase buffer were added to 50 μl of 2× diluted peptide. Wells were then incubated with 4G10 antiphosphotyrosine, followed by goat antimouse IgG + IgM-alkaline phosphatase conjugate. Finally, CSPD/Sapphire-II was added and light emission was measured 45 min later.

thetic peptide derived from amino acids 1–17 of human gastrin (Pierce)

Protein tyrosine kinase: $p60^{c\text{-}src}$ kinase (Calbiochem, San Diego, CA)

Antiphosphotyrosine: monoclonal antibody, clone 4G10 (Upstate Biotechnology Inc., Lake Placid, NY)

Goat antimouse IgG + IgM-alkaline phosphatase conjugate: secondary antibody (Tropix, Inc., Bedford, MA)

CSPD/Sapphire-II: 0.4 m*M* CSPD/10% Sapphire-II Ready-to-use (Tropix)

I-Block blocking reagent: highly purified casein (Tropix)

Recipes

Wash buffer: 0.05% Tween-20, 1× phosphate-buffered saline (PBS)

Blocking buffer: 0.2% I-Block, 0.05% Tween-20, 1× PBS. Add I-Block to 1× PBS, microwave for 40 sec, then stir, or heat on a hot plate. *Do not boil.* The solution will remain slightly opaque, but particles should be dissolved. Cool to room temperature and add Tween-20.

Sample buffer: 20 m*M* HEPES, pH 7.4, 2 m*M* $MnCl_2$, 100 μg/ml bovine serum albumin (BSA), 6.3 m*M* β-mercaptoethanol 2× sample buffer: 40 m*M* HEPES, pH 7.4, 4 m*M* $MnCl_2$, 200 μg/ml BSA, 12.6 m*M* β-mercaptoethanol

Kinase buffer: Mix 90 μl of 20 m*M* HEPES, pH 7.4, 2 m*M* $MnCl_2$, 0.25 μM $Na_3\,VO_4$ with 10 μl 10× ATP mix. (10× ATP mix is diluted 1:10 into kinase buffer 5–10 min prior to use.) 10× ATP mix: 1 m*M* ATP, 67 m*M* $MgCl_2$, store aliquots at −20°

2× kinase buffer: Mix 40 μl of 40 m*M* HEPES, pH 7.4, 4 m*M* $MnCl_2$, 0.5 μM $Na_3\,VO_4$ with 10 μl 10× ATP mix. (10× ATP mix is diluted 1:5 into 2× kinase buffer 5–10 min prior to use.)

Assay buffer: 20 m*M* Tris–HCl, pH 9.8, 1 m*M* $MgCl_2$

Methods

All incubations are performed at room temperature on an orbital shaker unless otherwise indicated. Plate should be sealed with Parafilm, or an equivalent, during all incubations.

Plate Coating

Incubate multiwell strips overnight with 100 μl/well of 5 μg/ml NeutrAvidin, diluted in sterile PBS. Wash three times with wash buffer, fill completely with blocking buffer, and incubate for 1 hr. Wash wells three times with wash buffer, allow to dry overnight, and store desiccated at 4°.

Rehydrate coated strips by filling wells with wash buffer for 5 min and then wash twice with wash buffer before use.

Simultaneous Peptide Capture and Kinase Reaction

Dilute biotinylated peptide to twice the desired concentration (typically 0.5 μg/ml) in blocking buffer and add 50 μl of diluted peptide to each well. Dilute src kinase in 2× sample buffer and add 5 μl of diluted enzyme and 45 μl 2× kinase buffer to each well. Incubate for 2 hr and then wash three times with wash buffer.

Sequential Peptide Capture and Kinase Reaction

Dilute biotinylated peptide to desired concentration, typically 0.25 μg/ml in blocking buffer, and add 100 μl of diluted peptide to each well. Incubate for 1 hr and wash three times with wash buffer. Dilute src kinase in sample buffer and add 10 μl of diluted enzyme and 90 μl kinase buffer to each well. Incubate for 1 hr and wash three times with wash buffer.

Chemiluminescent Immunodetection

Dilute antiphosphopeptide to recommended concentration in blocking buffer, add 100 μl to each well, and incubate for 1 hr. Wash wells three times with wash buffer. Dilute goat antimouse IgG + IgM-alkaline phosphatase conjugate 1:20,000 in blocking buffer, add 100 μl to each well, and incubate for 1 hr. Wash plate three times with wash buffer and twice with assay buffer. Finally, add 100 μl of CSPD/Sapphire-II per well and measure light emission in a microplate luminometer after 20–30 min.

Troubleshooting

Several controls should be performed during optimization of an assay. These include omitting peptide to quantitate nonspecific phosphorylation; omitting kinase to quantitate the binding of primary antibody to nonphosphorylated peptide; omitting ATP to determine that activity is dependent on the transfer of the phosphate group from ATP; and omitting primary antibody to determine nonspecific binding of the secondary antibody conjugate. In this protocol, with this particular antiphosphopeptide antibody, casein as a blocking reagent was found not to interfere with the detection of the phosphorylated peptide substrate. However, with a different antibody, BSA might be considered as an alternative blocking reagent.

Summary

Quantitation of at least two orders of magnitude of kinase enzyme concentration is achieved with detection of less than 0.1 U/well of src kinase

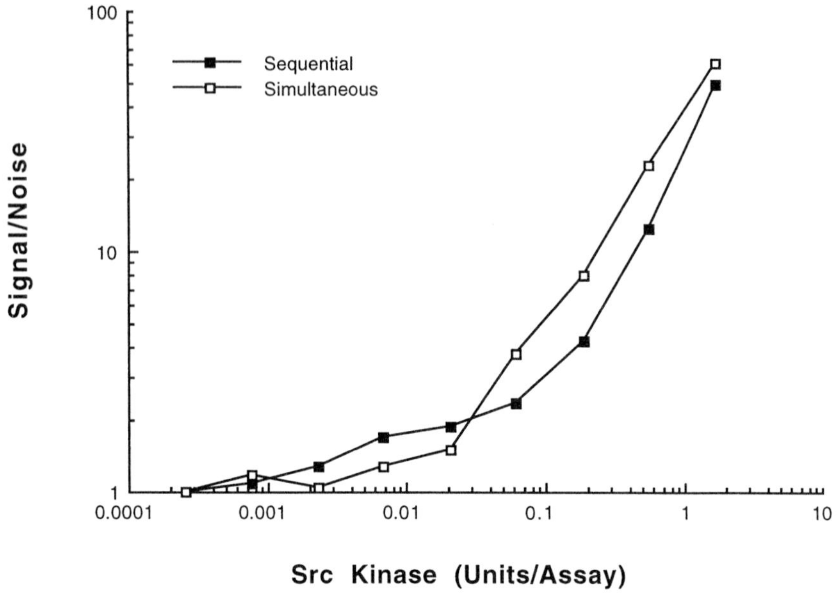

FIG. 3. Quantitation of src protein tyrosine kinase activity with simultaneous and sequential peptide capture/kinase reaction. The biotinylated peptide was diluted in blocking buffer to 0.5 μg/ml for the simultaneous assay and to 0.25 μg/ml for the sequential assay. In the sequential assay, 100 μl of peptide was added per well. Then 10 μl of diluted src protein kinase, with 90 μl of kinase buffer, was added per well. In the simultaneous assay, 50 μl of diluted peptide, 5 μl of 2× diluted src protein kinase, and 45 μl of 2× kinase buffer were added to each well. In both assays, wells were then incubated with 4G10 antiphosphotyrosine, followed by goat antimouse IgG + IgM-alkaline phosphatase conjugate. Finally, CSPD/Sapphire-II was added and light emission was measured 45 min later.

activity (Fig. 3). A comparison between a sequential protocol, in which biotinylated peptide substrate is captured prior to incubation with the kinase enzyme, and a simultaneous protocol, in which peptide capture and the kinase reaction proceed concurrently, demonstrates that the simpler simultaneous protocol provides similar detection sensitivity. These have also been demonstrated with 0.1 μM peptide substrate in a protein kinase A assay.[5] Quantitation of protein kinase activity with chemiluminescent detection has been demonstrated with several different protein kinases, including both tyrosine and serine/threonine kinases.[5] An immunoassay format provides high sensitivity and can be performed under conditions that most closely mimic physiological substrate and ATP concentrations with chemiluminescent detection. This assay format is also automated easily for use in high-throughput screening.

[30] Chemiluminescent Immunodetection Protocols with 1,2-Dioxetane Substrates

By CORINNE E. M. OLESEN, JENNIFER MOSIER, JOHN C. VOYTA, and IRENA BRONSTEIN

Introduction

Immunoblot analysis of proteins is one of the most extensively used research tools. Immunoblotting combines the resolution of gel electrophoresis with the specificity of an antigen : antibody interaction for the identification of specific proteins immobilized on a membrane. This technique has also become widely used for analysis of the phosphorylation state of a protein with the increasing availability of highly specific antibodies directed to phosphorylated or nonphosphorylated protein epitopes. The use of enzyme-based detection methods for these assays is due primarily to their advantages over radioisotopic labels. Chemiluminescent enzyme substrates offer distinct advantages over colorimetric substrates, including higher sensitivity, speed, blot reuse, and ease of imaging. Chemiluminescent immunodetection systems for alkaline phosphatase, horseradish peroxidase, and β-galactosidase enzyme labels are available commercially.

1,2-Dioxetane substrates, including AMPPD, CSPD, and CDP-*Star* substrates[1–3] for alkaline phosphatase and Galacton-*Star* substrate[4] for β-galactosidase, generate a prolonged luminescent signal lasting from hours to days on membranes. Enzymatic dephosphorylation or deglycosylation of the substrate produces a metastable anion intermediate, which decomposes with a concomitant emission of light. The luminescent signal is imaged on X-ray or photographic film, on a chemiluminescence phosphor imaging

[1] I. Bronstein, J. C. Voyta, O. J. Murphy, L. Bresnick, and L. J. Kricka, *BioTechniques* **12,** 748 (1992).

[2] I. Bronstein, C. E. M. Olesen, C. S. Martin, G. Schneider, B. Edwards, A. Sparks, and J. C. Voyta, *in* "Bioluminescence and Chemiluminescence: Fundamentals and Applied Aspects" (A. K. Campbell, L. J. Kricka, and P. E. Stanley, eds.), p. 269. Wiley, Chichester, 1994.

[3] C. E. M. Olesen and I. Bronstein, *J. NIH Res.* **8,** 58 (1996).

[4] C. S. Martin, C. E. M. Olesen, B. Liu, J. C. Voyta, J. L. Shumway, R.-R. Juo, and I. Bronstein, *In* "Bioluminescence and Chemiluminescence: Molecular Reporting with Photons" (J. W. Hastings, L. J. Kricka, and P. E. Stanley, eds.), p. 525. Wiley, Chichester, 1997.

0076-6879/00 $30.00

screen,[5] or with a camera imaging system.[6,7] The long-lived light signal generated with 1,2-dioxetanes enables multiple imaging exposures to be performed. The CDP-*Star* substrate provides the highest intensity light emission and is ideal for rapid film exposures and imaging with low light sensitive cameras. Chemiluminescent detection with 1,2-dioxetane substrates can be performed on polyvinylidene fluoride (PVDF), nitrocellulose, and nylon membranes. The use of 1,2-dioxetanes has been widely demonstrated for immunodetection with enzyme-labeled secondary antibodies and the detection of biotinylated proteins on membranes. Additional immunodetection applications include detection with a recombinant single-chain fragment of the variable domains (scFv) fused to alkaline phosphatase[8] and detection of cell surface proteins on human epithelial cells imprinted on nylon membrane.[9]

1,2-Dioxetanes have also been used in related protein-blotting applications for the identification of protein–protein interactions using a calmodulin–alkaline phosphatase conjugate to detect calmodulin-binding proteins[10] and to detect the expression of interleukin-1β converting enzyme (ICE) proteases during apoptosis using a biotinylated peptide inhibitor of ICE.[11]

Immunoblot Detection

Blot Preparation

Electrophoretic gel separation and electrotransfer of proteins to a blotting membrane are performed using standard protocols. A PVDF membrane is prewet prior to electrotransfer by immersion in 100% methanol. Deionized H_2O is added slowly while shaking until the methanol concentration is reduced to approximately 20% of the total volume. Then the membrane is equilibrated in transfer buffer for 15 min prior to electrophoretic transfer. A recommended buffer is 20 m*M* Tris, 60 m*M* glycine, and 20% methanol; however, transfer buffers and conditions should be optimized

[5] Q. Nguyen and D. M. Heffelfinger, *Anal. Biochem.* **226,** 59 (1995).
[6] C. S. Martin and I. Bronstein, *J. Biolumin. Chemilumin.* **9,** 145 (1994).
[7] G. Milosevich, *J. NIH Res.* **9,** 52 (1997).
[8] P. Lindner, K. Bauer, A. Krebber, L. Nieba, E. Kremmer, C. Krebber, A. Honegger, B. Klinger, R. Mocikat, and A. Plückthun, *BioTechniques* **22,** 140 (1997).
[9] R. E. Gibson-D'Ambrosio, T. Brady, and S. M. D'Ambrosio, *BioTechniques* **19,** 784 (1995).
[10] R. G. Walker, A. J. Hudspeth, and P. G. Gillespie, *Proc. Natl. Acad. Sci. U.S.A.* **90,** 2807 (1993).
[11] A. Srinivasan, L. M. Foster, M.-P. Testa, T. Örd, R. W. Keane, D. E. Bredesen, and C. Kayalar, *J. Neurosci.* **16,** 5654 (1996).

for the particular proteins being studied. Membranes must be handled carefully and never touched with ungloved hands.

Chemiluminescent Immunodetection

All volumes refer to processing of a single minigel blot (100 cm^2). All incubations and washes are performed at room temperature with agitation on an orbital shaker. Phosphate-buffered saline (PBS) or (Tris-buffered saline (TBS) buffers can be used with nitrocellulose or PVDF membranes. With nylon membranes, the background is slightly higher with TBS than PBS buffers. The indicated dilutions of secondary antibody are those recommended with reagents supplied by Tropix. For alternative suppliers, the optimal dilutions may be different and should be determined for the best results.

REAGENTS. All reagents listed, except for the nitrocellulose membrane and X-ray film, are available from Tropix, Inc. (Bedford, MA). Alternate suppliers are available for membranes, blocking reagents, and secondary antibodies; however, results can vary dramatically and should be tested carefully for optimal performance with chemiluminescent detection.

Nitrocellulose membrane: Optitran BA-S 85 (Schleicher & Schuell, Keene, NH) is recommended.

PVDF membrane: Tropifluor (Tropix) or Immobilon-P (Millipore Corporation, Bedford, MA) is recommended.

Nylon membrane: Tropilon-Plus positively charged nylon (Tropix) or Biodyne A neutral nylon (Pall Corporation, Glen Cove, NY) is recommended.

I-Block blocking reagent: highly purified casein (store dry at room temperature).

BSA blocking reagent: Fraction V bovine serum albumin (store at 4°).

PY20: antiphosphotyrosine monoclonal antibody (store at 4°) (optional).

Secondary antibody AP conjugate: goat antimouse IgG + IgM, goat antirabbit IgG, goat antihuman IgG, goat antirat IgG + IgM, rabbit antichicken IgY, or rabbit antisheep IgG AP conjugate (store at 4° or optimally at −20°).

Biotinylated secondary antibody: biotinylated goat antirabbit IgG or biotinylated goat antimouse IgG + IgM (store at 4° or optimally at −20°).

Avid*x*-AP: streptavidin alkaline phosphatase conjugate (store at 4° or optimally at −20°).

Secondary antibody β-galactosidase conjugate: goat antirabbit IgG or goat antimouse IgA + IgG + IgM β-galactosidase conjugate (store at 4°).

10× assay buffer: 200 m*M* Tris–HCl, pH 9.8, 10 m*M* $MgCl_2$. Dilute 1:10 with Milli-Q H_2O or equivalent (store at 4°).

CSPD Ready-to-Use substrate solution (0.25 m*M*) (store at 4°).

CDP-*Star* Ready-to-Use substrate solution (0.25 m*M*) (store at 4°).

Nitro-Block and Nitro-Block-II: 20× chemiluminescence enhancers. Nitro-Block provides signal enhancement on nitrocellulose and PVDF membranes (store at 4°).

Galacton-*Star* substrate concentrate (10 m*M*).

Galacto-*Star* reaction buffer: contains Sapphire-II chemiluminescence enhancer, which provides signal enhancement on nitrocellulose and PVDF membranes.

Development folders: 14 × 19-cm clear polypropylene sheet folders.

X-ray film: X-AR 5 (Eastman Kodak Company, Rochester, NY) or other autoradiography films designed for chemiluminescence imaging.

RECIPES. All solutions should be made with filtered, deionized water. 10× PBS or TBS should be kept sterile at room temperature. Blocking and wash buffers should be made fresh to prevent possible contamination from bacterial alkaline phosphatase. Blocking buffer may be stored at 4° for short periods with the addition of 0.02% sodium azide.

10× PBS[11a]: 0.58 *M* Na_2HPO_4, 0.17 *M* $NaH_2PO_4 \cdot H_2O$, 0.68 *M* NaCl.

10× TBS[11a]: 0.2 *M* Trizma base, 1.37 *M* NaCl; adjust pH to 7.6 with HCl.

Blocking buffer (all membranes except positively charged nylon): 0.2% I-Block, 1× PBS or TBS, 0.1% Tween 20. For positively charged nylon: 3% I-Block, 1× PBS, 0.1% Tween 20. Add I-Block to 1× PBS or TBS, microwave for 40 sec, and then stir or heat on hot plate. *Do not boil.* The solution will remain slightly opaque, but particles should be dissolved. Cool to room temperature and add Tween 20. For nitrocellulose and nylon, prepare at least 30 ml. For PVDF, prepare at least 150 ml.

Wash buffer: 1× PBS or 1× TBS, 0.1% Tween 20. For nitrocellulose and nylon, prepare at least 100 ml.

2-Mercaptoethanol stripping buffer: 62.5 m*M* Tris–HCl, pH 6.8, 2% SDS, 100 m*M*. 2-Mercaptoethanol.

Glycine stripping buffer: 0.2 *M* glycine, 0.1% SDS, 1.0% Tween 20, pH 2.2.

[11a] Alternative PBS and TBS recipes may also be used.

Protocal Ia: Immunodetection with AP-Conjugated Secondary Antibody

1. Following protein transfer, wash blot briefly with 1× PBS or 1× TBS and then incubate in blocking buffer for 30–60 min (10 ml).
2. Dilute primary antibody in blocking buffer. Incubate with primary antibody for 30–60 min (5–10 ml).
3. For nitrocellulose and nylon, wash for a minimum of 2 × 5 min in wash buffer. For PVDF, wash for a minimum of 2 × 5 min in blocking buffer (at least 20 ml per wash).
4. Dilute secondary antibody–alkaline phosphatase conjugate 1:5000 in blocking buffer. Incubate with diluted conjugate solution for 30–60 min (5 ml).
5. For nitrocellulose and nylon, wash 3 × 5 min in wash buffer. For PVDF, wash 3 × 5 min in blocking buffer (at least 20 ml per wash).
6. Wash 2 × 2 min with 1× assay buffer (20 ml per wash).
7. Drain blot by touching an edge on paper towel and place on Saran wrap on a flat surface (without letting blots dry).
8. Pipette a thin layer of substrate solution, approximately 3 ml, onto the blot. For nitrocellulose membrane, incubate for 5 min in CDP-*Star* or CSPD Ready-to-Use substrate solution containing 1:20 Nitro-Block-II (CSPD can be used with either Nitro-Block or Nitro-Block-II, and CDP-*Star* is used with Nitro-Block-II only). For PVDF and nylon membranes, incubate for 5 min in CDP-*Star* or CSPD (1:20 Nitro-Block or Nitro-Block-II can be included with CSPD for higher signal intensity on PVDF) Ready-to-Use substrate solution.
9. Drain blot by touching an edge on paper towel and place in development folder or Saran wrap. Smooth out any bubbles or wrinkles and image.

Protocol Ib: Immunodetection Using an Antiphosphospecific Primary Antibody

Immunodetection of phosphotyrosine-containing proteins has been widely used for the analysis of tyrosine kinase substrates and activities, and the use of 1,2-dioxetane alkaline phosphatase substrates for chemiluminescent detection has been demonstrated.[12–15] The antiphosphotyrosine mono-

[12] B. Hundle, T. McMahon, J. Dadgar, C.-H. Chen, D. Mochly-Rosen, and R. O. Messing, *J. Biol. Chem.* **272,** 15028 (1997).

[13] P. Lazarovici, M. Oshima, D. Shavit, M. Shibutani, H. Jiang, M. Monshipouri, D. Fink, V. Movsesyan, and G. Guroff, *J. Biol. Chem.* **272,** 11026 (1997).

[14] T. M. Miller, M. G. Tansey, E. M. Johnson, Jr., and D. J. Creedon, *J. Biol. Chem.* **272,** 9847 (1997).

[15] T. F. Zioncheck, L. Richardson, J. Liu, L. Chang, K. L. King, G. L. Bennett, P. Fügedi, S. M. Chamow, R. H. Schwall, and R. J. Stack, *J. Biol. Chem.* **270,** 16871 (1995).

clonal antibody, clone PY20,[16] is widely cited for this application. Several other antiphosphotyrosine antibodies have been used for immunoblot detection, and different antibodies may result in distinct staining patterns due to differential reactivity to specific phosphotyrosine-containing epitopes.[17] This protocol can also be used with other antiphosphospecific primary antibodies, including other antiphosphotyrosine, antiphosphoserine, antiphosphothreonine, and dual specificity antibodies.

RECIPES

Blocking buffer: 1.0% BSA, 1× TBS, 0.1% Tween 20. Dissolve BSA in 1× TBS and then add Tween 20. For best results, allow BSA to dissolve into TBS. Do not heat or stir BSA, as it may clump and thus dissolve with more difficulty.

Wash buffer: 1× TBS, 0.1% Tween 20.

PVDF wash buffer: 0.5% BSA, 1× TBS, 0.1% Tween 20.

PROTOCOL. Follow Protocol Ia, using the alternative recipes described earlier for detection with antiphosphospecific primary antibody (e.g., PY20). BSA is recommended as the blocking agent because milk proteins, including casein (I-Block), can contain phosphorylated residues, absorbing the antiphosphospecific antibody and thus causing a signal reduction on the blot. Using TBS instead of PBS is recommended. This detection protocol can be used with either PVDF or nitrocellulose membranes.

Protocol II: Immunodetection with Biotinylated Secondary Antibody or Detection of Biotinylated Protein

A tertiary detection scheme using a biotinylated secondary antibody, followed by streptavidin–alkaline phosphatase, can result in an increased signal strength compared to systems utilizing an enzyme-linked secondary antibody.[18] In addition, it enables simultaneous antigen and biotinylated molecular weight marker detection on the same membrane or detection with biotinylated primary antibodies. Alternatively, this protocol can be used for the detection of biotinylated proteins,[19–21] labeled chemically, or

[16] J. R. Glenney, Jr., L. Zokas, and M. P. Kamps, *J. Immunol. Methods* **109,** 277 (1988).

[17] B. M. Sefton, *in* "Current Protocols in Molecular Biology" (F. M. Ausubel, R. Brent, R. E. Kingston, D. M. Moore, J. G. Seidman, J. A. Smith, and K. Struhl eds.), p. 18.4.1. Wiley, New York, 1995.

[18] E. Harlow and D. Lane, *in* "Antibodies: A Laboratory Manual," p. 421. Cold Spring Harbor Laboratory, Cold Spring Harbor, NY, 1988.

[19] P. G. Gillespie and A. J. Hudspeth, *J. Cell Biol.* **112,** 625 (1991).

[20] P. G. Gillespie and A. J. Hudspeth, *Proc. Natl. Acad. Sci. U.S.A.* **88,** 2563 (1991).

[21] R. C. Armstrong, T. J. Aja, K. D. Hoang, S. Gaur, X. Bai, E. S. Alnemri, G. Litwack, D. S. Karanewsky, L. C. Fritz, and K. J. Tomaselli, *J. Neurosci.* **17,** 553 (1997).

in one application by cross-linking a biotinylated peptide to a cell surface protein.[22] Antigen labeling with biotin, followed by immunoprecipitation and chemiluminescent detection, offers a highly sensitive alternative to radioactive labeling.[23–25] In addition, a biotinylated protein or peptide[11] can be used as a probe for detection of proteins or protein–protein interactions on membranes.

PROTOCOL

1. Following protein transfer, wash blot briefly with 1× PBS or 1× TBS and then incubate in blocking buffer for 30–60 min (10 ml).
2. Dilute primary antibody in blocking buffer. Incubate with blot for 30–60 min (5–10 ml).
3. For nitrocellulose and nylon, wash for a minimum of 2 × 5 min in wash buffer. For PVDF, wash for a minimum of 2 × 5 min in blocking buffer (at least 20 ml per wash).
4. Dilute biotinylated secondary antibody 1:10,000 in blocking buffer and incubate with blot for 30 min (5–10 ml).
5. For nitrocellulose and nylon, wash for a minimum of 2 × 5 min in wash buffer. For PVDF, wash for a minimum of 2 × 5 min in blocking buffer (at least 20 ml per wash).
6. Dilute Avid*x*-AP conjugate 1:20,000 in blocking buffer and incubate with blot for 20 min (5–10 ml).
7. For nitrocellulose and nylon, wash 3 × 5 min in wash buffer. For PVDF, wash 3 × 5 min in blocking buffer (at least 20 ml per wash).
8. Wash 2 × 2 min with 1× assay buffer (20 ml per wash).
9. Drain blot by touching an edge on paper towel and place on Saran wrap on a flat surface (without letting blots dry).
10. Pipette a thin layer of substrate solution, approximately 3 ml, onto the blot. For the nitrocellulose membrane, incubate for 5 min in CDP-*Star* or CSPD Ready-to-Use substrate solution containing 1:20 Nitro-Block-II (CSPD can be used with either Nitro-Block or Nitro-Block-II, and CDP-*Star* is used with Nitro-Block-II only). For PVDF and nylon membranes, incubate for 5 min in CDP-*Star* or CSPD (1:20 Nitro-Block or Nitro-Block-II can be included with CSPD for higher signal intensity on PVDF) Ready-to-Use substrate solution.

[22] P. L. J. M. Zeeuwen, W. Hendricks, W. W. de Jong, and J. Schalkwijk, *J. Biol. Chem.* **272,** 20471 (1997).

[23] A. C. Chang, D. R. Salomon, S. Wadsworth, M.-J. P. Hong, C. F. Mojcik, S. Otto, E. M. Shevach, and J. E. Coligan, *J. Immunol.* **154,** 500 (1995).

[24] S. K. H. Gillespie and S. Wasserman, *Mol. Cell. Biol.* **14,** 3559 (1994).

[25] L. M. Lantz and K. L. Holmes, *BioTechniques* **18,** 56 (1995).

11. Drain blot by touching an edge on a paper towel and place in development folder or Saran wrap. Smooth out any bubbles or wrinkles and image.

Protocol IIIa: Immunodetection with β-Galactosidase-Conjugated Secondary Antibody

The Galacton-*Star* substrate enables the use of β-galactosidase as an enzyme label for membrane-blotting applications.[4] With Galacton-*Star* and CSPD substrates, dual detection of two protein antigens on a single membrane blot can be performed sequentially using β-galactosidase and alkaline phosphatase-conjugated antibodies (Protocol IIIb)[26] without the need to strip and reprobe the blot.

PROTOCOL

1. Following protein transfer, wash blot briefly with 1× PBS or 1× TBS and then incubate in blocking buffer for 30–60 min (10 ml).
2. Dilute primary antibody in blocking buffer. Incubate with primary antibody for 30–60 min (5–10 ml).
3. For nitrocellulose and nylon, wash for a minimum of 2 × 5 min in wash buffer. For PVDF, wash for a minimum of 2 × 5 min in blocking buffer (at least 20 ml per wash).
4. Dilute secondary antibody–β-galactosidase conjugate 1:5000 in blocking buffer. Incubate with diluted conjugate solution for 30–60 min (5 ml).
5. For all membranes, wash 3 × 5 min in wash buffer (at least 20 ml per wash).
6. Drain blot by touching an edge on a paper towel and place on Saran wrap on a flat surface (without letting blots dry).
7. Pipette a thin layer of Galacton-*Star*/Enhancer solution, approximately 3 ml, onto the blot. Prepare Galacton-*Star*/Enhancer solution by diluting Galacton-*Star* 1:100 in reaction buffer.
8. Drain blot by touching an edge on a paper towel and place in development folder or Saran wrap. Smooth out any bubbles or wrinkles and image.

Protocol IIIb: Dual Immunodetection with AP- and β-Galactosidase-Conjugated Secondary Antibodies

1. Following protein transfer, wash blot briefly with 1 × PBS and then incubate in blocking buffer for 30–60 min (10 ml).

[26] I. Bronstein, C. S. Martin, C. E. M. Olesen, and J. C. Voyta, *in* "Bioluminescence and Chemiluminescence: Molecular Reporting with Photons" (J. W. Hastings, L. J. Kricka, and P. E. Stanley eds.), p. 451. Wiley, Chichester, 1997.

2. Dilute both primary antibodies together in blocking buffer. Incubate with primary antibodies for 30–60 min (5–10 ml).

3. For nitrocellulose and nylon, wash for a minimum of 2 × 5 min in wash buffer. For PVDF, wash for a minimum of 2 × 5 min in blocking buffer (at least 20 ml per wash).

4. Dilute secondary antibody–β-galactosidase and secondary antibody–AP conjugates 1:5000 together in blocking buffer. Incubate with conjugate solution for 30–60 min (5 ml).

5. For all membranes, wash 3 × 5 min in wash buffer (at least 20 ml per wash).

6. Drain blot by touching an edge on a paper towel and place on Saran wrap on a flat surface (without letting blots dry).

7. Pipette a thin layer of Galacton-*Star*/Enhancer solution, approximately 3 ml, onto the blot. Prepar Galacton-*Star*/Enhancer solution by diluting Galacton-*Star* 1:100 in reaction buffer.

8. Drain blot by touching an edge on a paper towel and place in development folder or Saran wrap. Smooth out any bubbles or wrinkles. Expose to X-ray film to image β-galactosidase-labeled immune complex.

9. Remove blot from report cover and wash 2 × 2 min with 1× assay buffer (prepared by diluting 10× assay buffer 1:10 with Milli-Q H_2O or equivalent).

10. Pipette a thin layer of CSPD Ready-to-Use substrate solution, approximately 3 ml, onto the blot.

11. Drain blot by touching an edge on a paper towel, place in development folder, and expose to image both β-galactosidase and alkaline phosphatase-labeled immune complexes.

Protocol IV: Blot Stripping

1. Incubate blot for 30 min at 70° in 2-mercaptoethanol stripping buffer or 2 × 30 min in glycine stripping buffer (10 ml).

2. Wash 3 × 5 min with wash buffer (at least 20 ml per wash).

3. Follow the detection protocol, beginning with incubation in blocking buffer.

Note: To verify removal of the detection antibodies from the blot, repeat the detection protocol, omitting primary antibody incubation. The complete removal of detection antibodies is confirmed by the absence of an image on film. To reprobe, begin at step 2. Results may vary depending on membrane, primary antibody affinity, and use of Nitro-Block-II. Incubation time and choice of stripping buffer may need to be optimized. The 2-mercaptoethanol/SDS method has been shown to effectively strip antibody complexes from PVDF membrane, although with some loss of protein or reactivity with repeated stripping.[27]

Imaging

Blots may be imaged by placing the wrapped membrane in direct contact with standard X-ray film or on Polaroid instant photographic black and white film. Initial exposures of 5 min are recommended to assess the optimum exposure time. Exposure times with PVDF and nylon membranes are shorter than with nitrocellulose. If the film is overexposed, the exposure time may need to be shortened appreciably.

For imaging with phosphor storage screen instrumentation, the Bio-Rad Molecular Imaging Screen-CH is optimized for the imaging and quantitation of the visible light produced by chemiluminescent 1,2-dioxetane substrates. The screen chemistry is sensitive to 400–500 nm emission and is designed for use with the Bio-Rad GS-525 Molecular Imager laser scanner. The sensitivity is similar to that of X-ray film, and the linear dynamic range is greater than four orders of magnitude.[5]

Charge-coupled device camera systems for low-light level imaging can also be used for chemiluminescent blot imaging with sensitivities similar to X-ray film.[6,7] Some commercially available instrumentation systems include the ChemiImager (Alpha Innotech Corporation, San Leandro, CA) and the Lumi-Imager workstation (Boehringer Mannheim, Indianapolis, IN). Manufacturers can be contacted for specific information on chemiluminescence imaging with these systems.

Troubleshooting

These protocols have been optimized with alkaline phosphatase-conjugated antibodies, blocking agents, Nitro-Block-II, and assay buffer supplied in Tropix western blotting systems. With other enzyme conjugates or blocking reagents, results may vary. Because 1,2-dioxetane substrates provide extremely sensitive detection of alkaline phosphatase activity, it is recommended that only ultrapure water and other reagents free of alkaline phosphatase and bacterial contamination be used.

If expected sensitivity is not attained:

1. Lengthen the film exposure time as much as possible.
2. Increase the concentrations of primary antibody or alkaline phosphatase conjugate. This may, however, contribute to increased nonspecific binding and a subsequent higher background.

[27] J. Tesfaigzi, W. Smith-Harrison, and D. M. Carlson, *BioTechniques* **17,** 268 (1994).

3. Increase incubation times with primary antibody and alkaline phosphatase conjugate.
4. Optimize gel electrophoresis conditions, transfer buffer, and electrotransfer conditions.

If the nonspecific background is too high (e.g., the film is overexposed or the image is uneven or spotty):

1. Decrease the exposure time until the appropriate resolution is achieved.
2. Splotchy images may result from bacterial contamination of the membrane or from incomplete or uneven removal of methanol from the membrane prior to blocking. Make sure that all reagents and buffers are kept free of contamination and that the membrane is free of fingerprints.
3. Filter antibodies to remove aggregates that may result in small dark spots. Dilute antibody to 2× the final working dilution in wash buffer and filter through a 0.45-μm filter. Then dilute filtered antibody 1:1 with blocking buffer to get the final 1× working dilution.
4. Increase wash steps to 15 min. Increasing the concentration of Tween 20 in the wash buffer to 0.3%–0.5% may also help.
5. Blocking may be improved by overnight incubation in blocking buffer at 4°.
6. Increase the dilution of primary antibody or secondary antibody–alkaline phosphatase conjugate.
7. If TBS-based solutions have been used, try PBS. TBS results in a higher background on nylon membranes and may result in a slightly higher background on PVDF and nitrocellulose.
8. For best results, buffers should be prepared fresh daily.

Summary

Chemiluminescent 1,2-dioxetane enzyme substrates provide a highly sensitive and versatile detection method for immunoblots and other membrane-based detections. 1,2-Dioxetane substrates, coupled with either alkaline phosphatase or β-galactosidase enzyme labels, generate glow light emission kinetics, with a signal duration that is significantly longer than most enhanced luminol/horseradish peroxidase chemiluminescent detection systems. The long-lived, high-intensity light signal is ideal for imaging using a variety of formats, including X-ray film, photographic film, chemiluminescence phosphor imaging screens, and the rapidly expanding selection of camera imaging systems.

[31] Chemiluminescent Reporter Gene Assays with 1,2-Dioxetane Enzyme Substrates

By CORINNE E. M. OLESEN, CHRIS S. MARTIN, JENNIFER MOSIER, BETTY LIU, JOHN C. VOYTA, and IRENA BRONSTEIN

Introduction

Reporter gene assays are widely used for studying gene regulation and function in cell biology.[1,2] Reporter genes are used as indicators for the identification of sequences and factors that control gene expression at the transcriptional level. A reporter gene construct contains gene regulatory elements of interest, sequence requirements for transcription of a functional mRNA, and the coding sequence for a reporter protein. Introduction of a reporter gene construct into cells, followed by quantitation of the expressed protein or its activity, provides an indirect measure of gene expression. Reporter quantitation enables mapping of promoter and enhancer regions and the identification of factors, mechanisms, or compounds that can alter gene expression levels. Several genes have been adapted as reporters of transcriptional activity, including chloramphenicol acetyltransferase,[3] β-galactosidase,[4] β-glucuronidase,[5,6] placental alkaline phosphatase,[7] human growth hormone,[8] firefly luciferase,[9,10] green fluorescent protein,[11] and others. Analytical techniques utilizing radioisotopic, colorimetric, fluorescent, and immunoassay methods have all been used for the measurement of reporter gene products. 1,2-Dioxetane substrates for several reporter enzymes, including β-galactosidase, β-glucuronidase, and alkaline phosphatase, provide highly sensitive alternatives to either colorimetric or fluores-

[1] J. Alam and J. L. Cook, *Anal. Biochem.* **188,** 245 (1990).

[2] I. Bronstein, J. Fortin, P. E. Stanley, G. S. A. B. Stewart, and L. Kricka, *Anal. Biochem.* **219,** 169 (1994).

[3] C. M. Gorman, L. F. Moffat, and B. H. Howard, *Mol. Cell. Biol.* **2,** 1044 (1982).

[4] V. K. Jain and I. T. Magrath, *Anal. Biochem.* **199,** 119 (1991).

[5] R. A. Jefferson, T. A. Kavanagh, and M. W. Bevan, *EMBO J.* **6,** 3901 (1987).

[6] "GUS Protocols: Using the GUS Gene as a Reporter of Gene Expression" (S. R. Gallagher, ed.). Academic Press, San Diego, 1992.

[7] J. Berger, J. Hauber, R. Hauber, R. Geiger, and B. R. Cullen, *Gene* **66,** 1 (1988).

[8] R. F. Selden, K. B. Howie, M. E. Rowe, H. M. Goodman, and D. D. Moore, *Mol. Cell. Biol.* **6,** 3173 (1986).

[9] A. R. Brasier, J. E. Tate, and J. F. Habener, *BioTechniques* **7,** 1116 (1989).

[10] T. M. Williams, J. E. Burlein, S. Ogden, L. J. Kricka, and J. A. Kant, *Anal. Biochem.* **176,** 28 (1989).

[11] M. Chalfie, Y. Tu, G. Euskirchen, W. W. Ward, and D. C. Prasher, *Science* **263,** 802 (1994).

0076-6879/00 $30.00

cent substrates. In addition, 1,2-dioxetane substrates are used in combination with firefly luciferase detection for the measurement of two reporter enzymes in a single sample.

β-Galactosidase Reporter Gene Assay

Background

The bacterial β-galactosidase gene is widely employed as a reporter enzyme in many organisms and, more recently, for the identification of protein : protein interactions and in an assay for cell fusion. Chemiluminescent 1,2-dioxetane substrates for β-galactosidase, including Galacton®, Galacton-Plus®, and Galacton-*Star*®, provide highly sensitive enzyme detection and a wide dynamic range, enabling detection from 2 fg to 20 ng of β-galactosidase.[4,12–14] These substrates have been incorporated into the Galacto-Light™, Galacto-Light Plus™, and Galacto-*Star*™ reporter gene assays, respectively. Chemiluminescent reporter assays with 1,2-dioxetane substrates have been performed with extracts from mammalian cell culture,[15–19] tissues,[20–29] and microinjected frog embryos.[30] In addition, 1,2-

[12] I. Bronstein, J. Fortin, J. C. Voyta, C. E. M. Olesen, and L. J. Kricka, *in* "Bioluminescence and Chemiluminescence: Fundamentals and Applied Aspects" (A. K. Campbell, L. J. Kricka, and P. E. Stanley, eds.), p. 20. Wiley, Chichester, 1994.

[13] C. S. Martin, C. E. M. Olesen, B. Liu, J. C. Voyta, J. L. Shumway, R. -R. Juo, and I. Bronstein, *in* "Bioluminescence and Chemiluminescence: Molecular Reporting with Photons" (J. W. Hastings, L. J. Kricka, and P. E. Stanley, eds.), p. 525. Wiley, Chichester, 1997.

[14] I. Bronstein, C. S. Martin, J. J. Fortin, C. E. M. Olesen, and J. C. Voyta, *Clin. Chem.* **42,** 1542 (1996).

[15] M. Hu, C. B. Bigger and P. D. Gardner, *J. Biol. Chem.* **270,** 4497 (1995).

[16] J. P. McMillan and M. F. Singer, *Proc. Natl. Acad. Sci. U.S.A.* **90,** 11533 (1993).

[17] D. Ng, M. J. Su, R. Kim, and D. Bikle, *Front. Biosci.* **1,** 16 (1996).

[18] F. U. Reuss and J. M. Coffin, *Proc. Natl. Acad. Sci. U.S.A.* **92,** 9293 (1995).

[19] R. Schmidt-Ullrich, S. Mémet, A. Lilienbaum, J. Feuillard, M. Raphaël, and A. Israël, *Development* **122,** 2117 (1996).

[20] A. Bosch, P. B. McCray, Jr., S. M. W. Chang, T. R. Ulich, W. S. Simonet, D. J. Jolly, and B. L. Davidson, *J. Clin. Invest.* **98,** 2683 (1996).

[21] D. DiSepio, A. Jones, M. A. Longley, D. Bundman, J. A. Rothnagel, and D. R. Roop, *J. Biol. Chem.* **270,** 10792 (1995).

[22] L. J. Feldman, P. G. Steg, L. P. Zheng, D. Chen, M. Kearney, S. E. McGarr, J. J. Barry, J.-F. Dedieu, M. Perricaudet, and J. M. Isner, *J. Clin. Invest.* **95,** 2662 (1995).

[23] U. R. Hengge, P. S. Walker, and J. C. Vogel, *J. Clin. Invest.* **97,** 2911 (1996).

[24] P. D. Kessler, G. M. Podsakoff, X. Chen, S. A. McQuiston, P. C. Colosi, L. A. Matelis, G. J. Kurtzman, and B. J. Byrne, *Proc. Natl. Acad. Sci. U.S.A.* **93,** 14082 (1996).

[25] S. W. Lee, B. C. Trapnell, J. J. Rade, R. Virmani, and D. A. Dichek, *Circul. Res.* **73,** 797 (1993).

[26] K. L. O'Connor and L. A. Culp, *BioTechniques* **17,** 502 (1994).

[27] K. L. O'Connor and L. A. Culp, *Oncol. Rep.* **1,** 869 (1994).

dioxetanes have been used to measure β-galactosidase activity in yeast extracts,[31,32] including the two-hybrid system[33–39] to study protein : protein interactions and a one-hybrid system[40] to study DNA : protein interactions, protozoan parasites,[41] and in bacteria.[42] A novel mammalian two-hybrid system, based on β-galactosidase peptide complementation to report protein : protein interactions, has been performed with Galacton-Plus.[43,44] β-Galactosidase peptide complementation has also been utilized in a chemiluminescent assay for myoblast cell fusion in cell culture.[43,45] A cytotoxicity assay, measuring the release of β-galactosidase reporter enzyme from transfected cells, has been performed with Galacton.[46]

The cell lysate is incubated with Galacton (Galacton-Plus) for 15 min to 1 hr, during which enzyme cleavage forms an excited state intermediate. The sample is then placed in a luminometer and a light emission accelerator, which introduces a chemiluminescence enhancer at a high pH, is added, terminating the enzyme activity and accelerating breakdown of the excited

[28] H. Oswald, F. Heinemann, S. Nikol, B. Salmons, and W. H. Günzburg, *BioTechniques* **22,** 78 (1997).

[29] N. Shaper, A. Harduin-Lepers, and J. H. Shaper, *J. Biol. Chem.* **269,** 25165 (1994).

[30] C. Gove, M. Walmsley, S. Nijjar, D. Bertwistle, M. Guille, G. Partington, A. Bomford, and R. Patient, *EMBO J.* **16,** 355 (1997).

[31] J. E. Remacle, G. Albrecht, R. Brys, G. H. Braus, and D. Huylebroeck, *EMBO J.* **16,** 5722 (1997).

[32] V. R. Stoldt, A. Sonneborn, C. E. Leuker, and J. F. Ernst, *EMBO J.* **16,** 1982 (1997).

[33] Y. Bourne, M. H. Watson, M. J. Hickey, W. Holmes, W. Rocque, S. I. Reed, and J. A. Tainer, *Cell* **84,** 863 (1996).

[34] L. A. Carver and C. A. Bradfield, *J. Biol. Chem.* **272,** 11452 (1997).

[35] R. Groisman, H. Masutani, M.-P. Leibovitch, P. Robin, I. Soudant, D. Trouche, and A. Harel-Bellan, *J. Biol. Chem.* **271,** 5258 (1996).

[36] J. B. Hogenesch, W. K. Chan, V. H. Jackiw, R. C. Brown, Y.-Z. Gu, M. Pray-Grant, G. H. Perdew, and C. A. Bradfield, *J. Biol. Chem.* **272,** 8581 (1997).

[37] A. N. Hollenberg, T. Monden, J. P. Madura, K. Lee, and F. E. Wondisford, *J. Biol. Chem.* **271,** 28516 (1996).

[38] M.-J. Lee, M. Evans, and T. Hla, *J. Biol. Chem.* **271,** 11272 (1996).

[39] T. Ulmasov, R. M. Larkin, and T. J. Guilfoyle, *J. Biol. Chem.* **271,** 5085 (1996).

[40] S. S. Wolf, K. Roder, and M. Schweizer, *BioTechniques* **20,** 568 (1996).

[41] J. K. Beetham, K. S. Myung, J. J. McCoy, M. E. Wilson, and J. E. Donelson, *J. Biol. Chem.* **272,** 17360 (1997).

[42] E. D'Haese, H. J. Nelis, and W. Reybroeck, *Appl. Environ. Microbiol.* **63,** 4116 (1997).

[43] W. A. Mohler and H. M. Blau, *Proc. Natl. Acad. Sci. U.S.A.* **93,** 12423 (1996).

[44] F. Rossi, C. A. Charlton, and H. M. Blau, *Proc. Natl. Acad. Sci. U.S.A.* **94,** 8405 (1997).

[45] C. A. Charlton, W. A. Mohler, G. L. Radice, R. O. Hynes, and H. M. Blau, *J. Cell Biol.* **138,** 331 (1997).

[46] H. Schäfer, A. Schäfer, A. F. Kiderlen, K. N. Masihi, and R. Burger, *J. Immunol. Methods* **204,** 89 (1997).

state with the concomitant emission of light. Galacton has a half-life of light emission of approximately 4.5 min after addition of accelerator and should be used with luminometers with automatic injectors. Galacton-Plus light emission has a half-life of approximately 180 min after addition of Accelerator-II and is ideal for luminometers without automatic injectors or with scintillation counters. In conjunction with Accelerator-II, Galacton-Plus shows a higher signal-to-noise ratio than Galacton. Galacto-*Star* provides a simplified method for detecting β-galactosidase. The cell lysate is incubated with Galacton-*Star* and Sapphire-II enhancer (in a single reagent) until maximum light emission is reached (approximately 60–90 min). Light emission remains constant for nearly 1 hr. Light emission is measured in a luminometer or scintillation counter without the need of injectors.

Chemiluminescent reporter assays may be conducted in cells or tissues that have endogenous mammalian β-galactosidase activity. In this case, it is important to assay the level of endogenous enzyme background with nontransfected cell extracts. A significant reduction of endogenous activity can be achieved by heat inactivation.[29,47]

Reagents

The following reagents are available as components of either Galacto-Light™/Galacto-Light Plus™ or Galacto-*Star*™ reporter gene assay systems (Tropix, Inc., Bedford, MA).

- Lysis solution (mammalian cells): 100 m*M* potassium phosphate, pH 7.8, 0.2% Triton X-100 (store at 4°); dithiothreitol (DTT) can be added fresh prior to use at a final concentration of 0.5 m*M*.
- 5× Z buffer (yeast cells): 0.5 *M* sodium phosphate, pH 7.1, 50 m*M* KCl, 5 m*M* $MgSO_4$ (store at 4°); dilute 1:5 with Milli-Q H_2O or equivalent; DTT can be added fresh prior to use at a final concentration of 0.5 m*M*.
- Galacto-*Star* reaction buffer diluent: 100 m*M* sodium phosphate, pH 7.5, 1 m*M* $MgCl_2$, 5% Sapphire-II (store at 4°).
- Galacton-*Star* substrate: 50× concentrate (store at 4°).
- Galacto-Light/Galacto-Light Plus reaction buffer diluent: 100 m*M* sodium phosphate, pH 8.0, 1 m*M* $MgCl_2$ (store at 4°).
- Galacton/Galacton-Plus substrate: 100× concentrate (store at 4°).
- Light Emission Accelerator or Accelerator-II: luminescence accelerator reagent (store at 4°).

[47] D. C. Young, S. D. Kingsley, K. A. Ryan, and F. J. Dutko, *Anal. Biochem.* **215,** 24 (1993).

Other Reagents

β-Galactosidase (G-5635, Sigma Chemical Co., St. Louis, MO): positive control enzyme (optional).

PMSF (phenylmethylsulfonyl fluoride) (P-7626, Sigma) or AEBSF (4-(2-aminoethyl)-benzenesulfonyl fluoride) (A-8456, Sigma): protease inhibitor (optional).

Leupeptin (L-2884, Sigma): protease inhibitor (optional).

Phosphate-buffered saline (PBS): 75 m*M* sodium phosphate, pH 7.2, 68 m*M* NaCl (sterilize and store at room temperature) or use any standard PBS recipe.

Enzyme dilution buffer: 0.1 *M* sodium phosphate, pH 7.0, 0.1% bovine serum albumin (BSA) (Fraction V).

Protocols

Protocol I: Preparation of Mammalian Cell Extracts from Tissue Culture Cells

Prepare extracts at 4° (on ice), if possible.

1. Add fresh DTT to 0.5 m*M* to sufficient lysis solution (250 μl of lysis solution per 60 mm culture plate).

Note: DTT can be added to stabilize β-galactosidase activity. However, concentrations of reducing agents such as 2-mercaptoethanol and DTT higher than 0.5 m*M* will decrease the half-life of light emission of 1,2-dioxetanes. If the half-life of light emission is critical, reducing agents should be omitted from the lysis solution. If lysis buffer containing excess DTT has been used, the addition of hydrogen peroxide to the accelerator at a final concentration of 10 m*M* (add 1 μl of 30% H_2O_2 per 1 ml of accelerator) or to the Galacto-*Star* reaction buffer will prevent the rapid decay of signal half-life.

2. Rinse cells twice with PBS.
3. Add lysis solution to cover cells (250 μl per 60-mm culture plate).
4. Detach cells from plate using a cell scraper. Nonadherent cells should be pelleted and lysis solution added sufficient to cover cells. Resuspend cells by pipetting.
5. Transfer lysate to a microcentrifuge tube and centrifuge for 2 min to pellet any debris.
6. Transfer supernatant to a fresh microcentrifuge tube. Extracts may be used immediately or stored at −70°. Proceed to Protocol III or IV.

Protocol II: Preparation of Yeast Cell Extracts (Freeze/Thaw)

1. Dilute 5× Z buffer 1:5 in Milli-Q H_2O or equivalent. Add fresh DTT to 0.5 m*M*. (One sample typically requires 1.8–2 ml of 1× Z buffer.)

2. Transfer 1.5 ml of a freshly grown yeast culture (OD_{600} = 0.4–0.6) to a microcentrifuge tube. Centrifuge at 12,000*g* for 30 sec.

Note: If OD_{600} is less than 0.4, use more than 1.5 ml culture.

3. Remove supernatant and resuspend yeast pellet in 1.5 ml of 1× Z buffer.
4. Centrifuge at 12,000*g* for 30 sec.
5. Remove supernatant and resuspend pellet in 300 μl of 1× Z buffer.
6. Vortex cell suspension, and transfer 100 μl to a fresh microcentrifuge tube. Store the remainder at 4°.
7. Perform freeze/thaw cycle two times: Place tube in liquid nitrogen until cells are frozen. Thaw cells in a 37° water bath for 30–60 sec.
8. Centrifuge at 12,000*g* for 5 min at 4° in a microcentrifuge.
9. Transfer supernatant to a fresh tube and store on ice. Proceed to Protocol III or IV.

Protocol IIIa: Detection Protocol for Tube Luminometers (Galacto-Star)

Perform all assays in triplicate at room temperature.

1. Dilute sufficient Galacton-*Star* 1:50 with reaction buffer diluent to make reaction buffer (300 μl/tube). Galacton-*Star* should be diluted fresh each time.
2. Equilibrate reaction buffer to room temperature.
3. Aliquot 2–20 μl of cell extracts into luminometer tubes. Lysis solution or 1× Z buffer should be added to give the same total volume in each tube.

Note: The amount of cell extract required may vary depending on the expression level and instrumentation. Use 5 μl of extract for positive controls and 10–20 μl of extract for potentially low levels of enzyme.

4. Add 300 μl of reaction buffer to each tube, mix gently, and incubate for 60–90 min until maximum light emission is reached.

Note: Measurements can be performed in as short as 20–30 min if the time between reaction buffer addition and light signal measurement is identical for all samples.

5. Place tube in a luminometer and measure for 5 sec.

Protocol IIIb: Detection Protocol for Tube Luminometers (Galacto-Light/Galacto-Light Plus)

Perform all assays in triplicate at room temperature.

1. Dilute sufficient Galacton (Galacton-Plus) 1:100 with reaction buffer diluent to make reaction buffer (200 μl/tube).

2. Equilibrate reaction buffer and light emission accelerator to room temperature.
3. Aliquot 2–20 μl of cell extracts into luminometer tubes. Lysis solution or 1× Z buffer should be added to give the same total volume in each tube.

Note: The amount of cell extract required may vary depending on the amount of expression and instrumentation used. Use 5 μl of extract for positive controls and 10–20 μl of extract for potentially low levels of enzyme.

4. Add 200 μl of reaction buffer to each tube, mix gently, and incubate for 60 min. Incubations can be as short as 15 min, but the linear range of the assay may decrease.

Note: Light intensities are time dependent. Reaction buffer should be added to samples in the same time frame as they are measured. For example, if it takes 10 sec to measure a sample completely, then reaction buffer should be added to tubes every 10 sec.

5. Place tube in luminometer. Inject 300 μl of Light Emission Accelerator (or Accelerator-II). After a 2- to 5-sec delay following injection, measure for 5 sec. If manual injection is used, the accelerator should be added in the same consistent time frame as the reaction buffer is added. This is critical when using Galacton.

Note: Reaction components are scaled down if a smaller volume injector is used. For example, with a 100-μl injector, use 2–10 μl of lysate, 70–100 μl of reaction buffer, and 100 μl of accelerator.

Protocol IVa: Detection Protocol for Microplate Luminometers (Galacto-Star)

Perform all assays in triplicate at room temperature.

1. Dilute sufficient Galacton-*Star* 1 : 50 with reaction buffer diluent to make reaction buffer (100–200 μl/well).
2. Equilibrate reaction buffer to room temperature.
3. Aliquot 2–10 μl of cell extracts into microplate wells. Lysis solution or 1× Z buffer should be added to give the same total volume in each well.

Note: The amount of cell extract required may vary depending on the expression level and instrumentation. Use 2–5 μl of extract for positive controls and 10 μl of extract for potentially low levels of enzyme.

4. Add 100–200 μl of reaction buffer and mix gently. Incubate for 60–90 min until maximum light emission is reached.

Note: Measurements can be performed in as short as 20–30 min if the time between reaction buffer addition and light signal measurement is identical for all samples.

5. Place microplate in luminometer and measure each well for 0.1–1 sec.

Protocol IVb: Detection Protocol for Microplate Luminometers (Galacto-Light/Galacto-Light Plus)

Perform all assays in triplicate at room temperature.

1. Dilute sufficient Galacton (Galacton-Plus) 1 : 100 with reaction buffer diluent to make reaction buffer (70 μl/well).
2. Equilibrate reaction buffer and accelerator to room temperature.
3. Aliquot 2–20 μl of cell extracts into microplate wells. Lysis solution or 1× Z buffer should be added to give the same total volume in each well.

Note: The amount of cell extract required may vary depending on the amount of expression and instrumentation used. Use 5 μl of extract for positive controls and 10–20 μl of extract for potentially low levels of enzyme.

4. Add 70 μl of reaction buffer to each well, mix gently, and incubate for 60 min. Incubations can be as short as 15 min, but the linear range of the assay may decrease.

Note: Light intensities are time dependent. Reaction buffer should be added to samples in the same time frame as they are measured. For example, if it takes 10 sec to measure a sample completely, then reaction buffer should be added to tubes every 10 sec.

5. Place microplate in luminometer. Inject 100 μl of Light Emission Accelerator (or Accelerator-II). After a 1- to 2-sec delay following injection, measure each sample for 1–5 sec. If manual injection is used (with Galacton-Plus only), Accelerator-II should be added in the same consistent time frame as the reaction buffer is added, and the entire plate can be measured for 0.1–1 sec per well.

Protocol V: Direct Lysis Protocol for Microplate Cultures of Adherent Cells (Galacto-Star)

Perform all assays in triplicate at room temperature. Heat inactivation is not recommended.

1. Dilute sufficient Galacton-*Star* 1:50 with reaction buffer diluent to make reaction buffer (100 μl/well).

2. Equilibrate reaction buffer to room temperature.
3. Rinse cells once with PBS.
4. Add 10 μl of lysis solution to each well and incubate for 10 min.
5. Add 100 μl of reaction buffer to each well and incubate for 60–90 min to reach maximum light emission.

Note: Measurements can be performed in as short as 20–30 min if the incubation time in the reaction buffer is identical for all samples.

6. Place microplate in luminometer and measure each well for 0.1–1 sec.

Protocol VI: Preparation of Controls

Positive Control. Reconstitute lyophilized β-galactosidase to 1 mg/ml in 0.1 *M* sodium phosphate, pH 7.0, 0.1% BSA. Store at 4°. Generate a standard curve by serially diluting stock enzyme in lysis buffer or Z buffer containing 0.1% BSA. Two to 20 ng of enzyme should be used for the high end detection limit.

Negative Control. Assay a volume of mock transfected cell extract equivalent to that of experimental cell extract.

Protocol VII: Heat Inactivation of Endogenous β-Galactosidase

Some cell lines may exhibit relatively high levels of endogenous β-galactosidase activity. This may lead to background, which will decrease the overall sensitivity of the assay by lowering the signal-to-noise ratio. A procedure for heat inactivation of endogenous β-galactosidase activity has been described.[29,47] A modified version of this protocol has also been described[29] in which a cocktail of protease inhibitors is used in conjunction with heat inactivation for reducing endogenous β-galactosidase activity in tissue extracts.

INACTIVATION OF ENDOGENOUS β-GALACTOSIDASE ACTIVITY IN CELL EXTRACTS

1. Following extract preparation, heat extract at 48° for 50 min.
2. Proceed with detection protocol (Protocol III or IV).

INACTIVATION OF ENDOGENOUS β-GALACTOSIDASE ACTIVITY IN TISSUE EXTRACTS

1. To the lysis solution, add PMSF or AEBSF to a final concentration of 0.2 m*M* and leupeptin to 5 μg/ml.
2. Heat extract at 48° for 60 min.
3. Proceed with detection protocol (Protocol III or IV).

Note: AEBSF is a water-soluble serine protease inhibitor similar to PMSF.

β-Glucuronidase Reporter Gene Assay

Background

The bacterial β-glucuronidase (GUS) gene has become widely used as a reporter for the analysis of gene expression in plants and can also be used in mammalian cells.[6] The 1,2-dioxetane substrate, Glucuron®, provides highly sensitive chemiluminescent detection of β-glucuronidase activity.[12,48,49] The GUS-Light™ reporter gene assay incorporates Glucuron and Emerald™ luminescence enhancer and provides a wide dynamic range, enabling detection of 60 fg to 2 ng of enzyme.

The cell lysate is incubated with Glucuron substrate for 1 hr, during which time enzyme cleavage forms an excited state intermediate. The sample is then placed in a luminometer and accelerator is added, which raises the pH and introduces a chemiluminescence enhancer, terminating the β-glucuronidase activity and accelerating breakdown of the excited state with the concomitant emission of light. Glucuron has a half-life of light emission of approximately 4.5 min after the addition of accelerator and should be used with luminometers with automatic injectors. Manual additions may be performed, but signal intensities must be measured at the same interval following accelerator addition. Alternative lysis buffers should be evaluated to ensure optimum results.

Bacterial contamination of plant material will cause high background. Best results will be obtained with sterile preparations. Chlorophyll in concentrated samples may interfere with the chemiluminescent signal intensity. Therefore, if high levels of chlorophyll are present, several dilutions of extract should be assayed.

Reagents

The following reagents are available as components of the GUS-Light reporter gene assay system (Tropix).

Lysis solution (plant cells or tissue): 50 m*M* sodium phosphate, pH 7, 10 m*M* EDTA, 0.1% sodium lauryl sarcosine, 0.1% Triton X-100 (store at 4°); 2-mercaptoethanol should be added prior to use at a final concentration of 10 mM.

Lysis solution (mammalian cells): 100 m*M* potassium phosphate, pH 7.8, 0.2% Triton X-100 (store at 4°); DTT can be added prior to use at a final concentration of 0.5 m*M*.

[48] I. Bronstein, J. J. Fortin, J. C. Voyta, R.-R. Juo, B. Edwards, C. E. M. Olesen, N. Lijam, and L. J. Kricka, *BioTechniques* **17,** 172 (1994).

[49] G. Hansen and M.-D. Chilton, *Proc. Natl. Acad. Sci. U.S.A.* **93,** 14978 (1996).

Glucuron substrate: 100× concentrate (store at 4°).
Reaction buffer diluent: 0.1 *M* sodium phosphate, pH 7.0, 10 m*M* EDTA (store at 4°).
Light Emission Accelerator: luminescence accelerator reagent (store at 4°).

Other Reagents

β-Glucuronidase (G-7896, Sigma): positive control enzyme (optional).
Enzyme dilution buffer: 0.1 *M* sodium phosphate, pH 7.0, 0.1% BSA (Fraction V).

Protocols

Protocol I: Preparation of Cell Extracts

Prepare extracts at 4° (on ice), if possible.

1. Add 2-mercaptoethanol or DTT to the required amount of lysis solution (250 μl of lysis solution per 60-mm culture plate).
2. Prepare sample: rinse cells to remove culture media or prepare plant material.
3. Add lysis solution to cover cells (250 μl of lysis solution per 60-mm culture plate) or plant material (250 μl of lysis solution per 25 mg of plant material).
4. Detach cells from culture plate using a cell scraper. For plant material, homogenize (grind) cells or tissue in a microhomogenizer.
5. Centrifuge sample in a microfuge for 2 min to pellet any debris.
6. Transfer supernatant (cell extract) to a fresh microfuge tube.

Protocol II: Detection Protocol for Tube Luminometers

Perform all assays in triplicate at room temperature.

1. Dilute sufficient Glucuron 1 : 100 with reaction buffer diluent to make reaction buffer (180 μl/tube).
2. Equilibrate reaction buffer and Light Emission Accelerator to room temperature.
3. Aliquot 2–20 μl of cell extracts into luminometer tubes. Lysis solution should be added to give 20 μl final volume in each tube.

Note: The amount of cell extract required may vary depending on the expression level and instrumentation. Use 5 μl of extract for positive controls and 10–20 μl of extract for potentially low levels of enzyme.

4. Add 180 μl of reaction buffer to each tube and mix gently. Incubate

for 60 min. Incubation may be as short as 15 min (especially if high levels of expression are expected), but the linear range of the assay may decrease.

Note: Light intensities are time dependent. Reaction buffer should be added to samples in the same time frame as they are measured. For example, if it takes 10 sec to measure a sample completely, then reaction buffer should be added to tubes every 10 sec.

5. Place tubes in luminometer. Inject 300 μl of Light Emission Accelerator. After a 2- to 5-sec delay following injection, measure for 5 sec. If manual injection is used, the accelerator should be added in the same consistent time frame as the reaction buffer is added.

Note: Reaction components are scaled down if a smaller volume injector is used. For example, with a 100-μl injector, use 2–10 μl of lysate, 70–100 μl of reaction buffer, and 100 μl of accelerator.

Protocol III: Detection Protocol for Microplate Luminometers

Perform all assays in triplicate at room temperature.

1. Dilute Glucuron 1 : 100 with reaction buffer diluent to make reaction buffer (70 μl/well required).
2. Equilibrate reaction buffer and Light Emission Accelerator to room temperature.
3. Aliquot 2–20 μl of cell extracts into microplate wells. Lysis solution should be added to give 20 μl final volume in each well.

Note: The amount of cell extract required may vary depending on the expression level and instrumenttion. Use 5 μl of extract for positive controls and 10–20 μl of extract for potentially low levels of enzyme.

4. Add 70 μl of reaction buffer to each well, mix gently, and incubate for 60 min. Incubations can be as short as 15 min (especially if high levels of expression are expected), but the linear range of the assay may decrease.

Note: Light intensities are time dependent. Reaction buffer should be added to samples in the same time frame as they are measured. For example, if it takes 10 sec to measure a sample completely, then reaction buffer should be added to tubes every 10 sec.

5. Place microplate in luminometer. Inject 100 μl of Light Emission Accelerator. After a 1- to 2-sec delay following injection, measure each sample for 1–5 sec. If manual injection is used, the accelerator should be added in the same consistent time frame as the reaction buffer is added.

Protocol IV: Preparation of Controls

Positive Control. Reconstitute lyophilized β-glucuronidase to 1 mg/ml in 0.1 M sodium phosphate, pH 7.0, 0.1% BSA. Store at 4°. Generate a standard curve by serially diluting stock enzyme in lysis buffer containing 0.1% BSA. Two nanograms of enzyme should be used for the high end detection limit.

Negative Control. Assay a volume of mock-transfected extract equivalent to that of experimental extract.

Placental Alkaline Phosphatase Reporter Gene Assay

Background

Secreted placental alkaline phosphatase (SEAP) has been utilized increasingly as a eukaryotic reporter gene. A gene contruct bearing a mutation in the membrane localization domain of human placental alkaline phosphatase causes the normally membrane-bound protein to be secreted from the cell.[7] Thus, SEAP detection is performed with a sample of cell culture medium while the cell population remains intact. The chemiluminescent 1,2-dioxetane, CSPD, a substrate for alkaline phosphatase, can be used for highly sensitive detection of either SEAP or nonsecreted placental alkaline phosphatase. The Phospha-Light reporter gene assay system incorporates CSPD and Emerald luminescence enhancer for high sensitivity and wide dynamic range.[12,14,48] CSPD has been used for the detection of secreted placental alkaline phosphatase reporter enzyme in cell culture media[50–57] and for the quantitation of nonsecreted placental alkaline phosphatase in both cell and tissue extracts.[18,26,27,58–60]

[50] A. R. Brooks, D. Shiffman, C. S. Chan, E. E. Brooks, and P. G. Milner, *J. Biol. Chem.* **271,** 9090 (1996).

[51] D. Chen, M. P. A. Davies, P. S. Rudland, and R. Barraclough, *J. Biol. Chem.* **272,** 20283 (1997).

[52] C. Feehan, K. Darlak, J. Kahn, B. Walchek, A. F. Spatola, and T. K. Kishimoto, *J. Biol. Chem.* **271,** 7019 (1996).

[53] S.-U. Gorr, *J. Biol. Chem.* **271,** 3575 (1996).

[54] P. O. Ilyinskii and R. C. Desrosiers, *J. Virol.* **70,** 3118 (1996).

[55] R. E. Jones, D. Defeo-Jones, E. M. McAvoy, G. A. Vuocolo, R. J. Wegrzyn, K. M. Haskell, and A. Oliff, *Oncogene* **6,** 745 (1991).

[56] D. Shiffman, E. E. Brooks, A. R. Brooks, C. S. Chan, and P. G. Milner, *J. Biol. Chem.* **271,** 12199 (1996).

[57] R. E. Means, T. Greenough, and R. C. Desrosiers, *J. Virol.* **71,** 7895 (1997).

[58] V. Andrés, S. Fisher, P. Wearsch, and K. Walsh, *Mol. Cell. Biol.* **15,** 4272 (1995).

[59] K. Guo and K. Walsh, *J. Biol. Chem.* **272,** 791 (1997).

[60] E. Suzuki, K. Guo, M. Kolman, Y.-T. Yu, and K. Walsh, *Mol. Cell. Biol.* **15,** 3415 (1995).

Secreted placental alkaline phosphatase is measured from 48 to 72 hr after cell transfection.[61] Cell culture medium or cell lysate is incubated with reaction buffer (containing CSPD and Emerald) until maximum light emission is reached (approximately 20 min). The light signal output is then measured in a luminometer.

Chemiluminescent reporter assays for placental alkaline phosphatase may be conducted in cells that have endogenous alkaline phosphatase activity. Endogenous nonplacental enzyme activity is reduced significantly by using a combination of heat inactivation and differential inhibitors that do not inhibit the transfected placental isozyme significantly. It is important to determine the level of endogenous enzyme in media or extracts from nontransfected cells to establish assay background. Certain cell lines, such as HeLa and others derived from cervical cancers and some other tumor cell lines, may express placental alkaline phosphatase, which may produce high assay backgrounds.[62] Therefore, the use of placental alkaline phosphatase as a reporter system in these cell lines is generally not recommended.

Reagents

The following reagents are available as components of the Phospha-Light reporter gene assay system (Tropix).

- 5× dilution buffer: dilute to 1× with Milli-Q H_2O (store at 4°).
- Assay buffer: a mixture of differential alkaline phosphatase inhibitors (store at 4°).
- CSPD substrate: 20× concentrate (store at 4°).
- Reaction buffer diluent: contains Emerald luminescence enhancer (store at 4°).
- Human placental alkaline phosphatase: 3 μg/ml (0.75 U/ml) (store at −20°).

Other Reagents

- Phosphate-buffered saline: 75 m*M* sodium phosphate, pH 7.2, 68 m*M* NaCl (sterilize and store at room temperature) or any standard PBS.
- Human placental alkaline phosphatase (P-3895, Sigma): lyophilized positive control enzyme (optional).
- Enzyme reconstitution buffer: 1× dilution buffer, 0.1% BSA (Fraction V), 50% glycerol (optional).

[61] B. Cullen and M. Malim, *Methods Enzymol.* **216,** 362 (1992).
[62] F. J. Benham, J. Fogh, and H. Harris, *Int. J. Cancer* **27,** 637 (1981).

Protocols

Protocol Ia: Detection Protocol for Tube Luminometers (Secreted Placental Alkaline Phosphatase)

Perform all assays in triplicate at room temperature, unless otherwise indicated.

1. Dilute sufficient CSPD 1:20 with reaction buffer diluent to make reaction buffer (100 μl/tube). Dilute 5× dilution buffer to 1× with H_2O (300 μl/sample).
2. Equilibrate sufficient assay buffer (100 μl/tube) and reaction buffer to room temperature.
3. Aliquot 100 μl of culture medium into a microcentrifuge tube, add 300 μl of 1× dilution buffer (sufficient for triplicates), and heat at 65° for 30 min.
4. Cool to room temperature by placing on ice briefly.
5. Add 100 μl of heat-treated, diluted culture media to three luminometer tubes (triplicates).
6. Add 100 μl of assay buffer and incubate for 5 min.

Note: Adding assay buffer to warm culture media or heating culture media with assay buffer may result in decreased sensitivity.

7. Add 100 μl of reaction buffer and incubate for 20 min.

Note: Light intensities are time dependent. Reaction buffer should be added to samples in the same time frame as they are measured. For example, if it takes 10 sec to measure a sample completely, then reaction buffer should be added to tubes every 10 sec.

8. Place tubes in luminometer and measure for 5 sec.

Protocol Ib: Detection Protocol for Microplate Luminometers (Secreted Placental Alkaline Phosphatase)

Perform all assays in triplicate at room temperature, unless otherwise indicated.

1. Dilute sufficient CSPD 1:20 with reaction buffer diluent to make reaction buffer (50 μl/well). Dilute 5× dilution buffer to 1× with H_2O (150 μl/sample).
2. Equilibrate sufficient assay buffer (50 μl/well) and reaction buffer to room temperature.
3. Aliquot 50 μl of culture medium into a microfuge tube, add 150 μl of 1× dilution buffer (sufficient for triplicates), and heat at 65° for 30 min.

4. Cool to room temperature by placing on ice briefly.
5. Add 50 μl of heat-treated culture media to three microplate wells (triplicates).
6. Add 50 μl of assay buffer and incubate for 5 min.

Note: Adding assay buffer to warm culture media or heating culture media with assay buffer may result in decreased sensitivity.

7. Add 50 μl of reaction buffer and incubate for 20 min.

Note: Light intensities are time dependent. Reaction buffer should be added to samples in the same time frame as they are measured. For example, if it takes 10 sec to measure a sample completely, then reaction buffer should be added to tubes every 10 sec.

8. Place microplate in a luminometer and measure each well for 0.1–1 sec.

Protocol II: Extract Preparation (Nonsecreted Placental Alkaline Phosphatase)

Prepare extracts at 4° (on ice), if possible.

1. Dilute sufficient 5× dilution buffer to 1× with H_2O. Add Triton X-100 to a final concentration of 0.2% (v/v). (One 60-mm culture plate typically requires 250 μl of 1× dilution buffer.)
2. Rinse cells twice with PBS.
3. Add 1× dilution buffer/0.2% Triton X-100 to cover cells.
4. Detach cells from culture plate using a cell scraper. Prepare extract by repeated pipetting and transfer to a microcentrifuge tube. Cell extracts may be used immediately or stored at −70°.

Note: Nonadherent cells should be pelleted and sufficient 1× dilution buffer/0.2% Triton X-100 added to cover cells. Resuspend and lyse cells by repeated pipetting.

5. Aliquot 30 μl of cell extract into a microfuge tube and add 370 μl of 1× dilution buffer (for triplicates with a single tube luminometer) or use 15 μl cell extract with 185 μl of 1× dilution buffer (for triplicates with a microplate luminometer). Heat tube to 65° for 30 min.
6. Proceed with detection protocol for tube luminometer (Protocol Ia) or microplate luminometer (Protocol IIb), starting at step 4.

Protocol III: Direct Assay Protocol for Microplate Cultures of Adherent Cells (Nonsecreted Placental Alkaline Phosphatase)

Perform all assays in triplicate at room temperature. Heat inactivation should not be performed with this protocol.

1. Dilute sufficient CSPD 1:20 with reaction buffer diluent to make reaction buffer (50 μl/well).
2. Equilibrate sufficient assay buffer (50 μl/well) and reaction buffer to room temperature. Dilute 5× dilution buffer to 1× with H_2O (50 μl/well).
3. Rinse wells once with PBS.
4. Add 50 μl of 1× dilution buffer to each well and incubate for 5 min.
5. Add 50 μl of assay buffer and incubate for 5 min.
6. Add 50 μl of reaction buffer and incubate for 20 min.
7. Place microplate in luminometer and measure each well for 0.1–1 sec.

Protocol IV: Preparation of Controls

Positive Control. Generate a standard curve by serially diluting stock enzyme in 1× dilution buffer or mock-transfected cell culture media. A 10-μl aliquot of stock enzyme should be used for the high end detection limit. Alternatively, a stock enzyme can be prepared by reconstituting lyophilized human placental alkaline phosphatase to 1 mg/ml in 1× dilution buffer containing 0.1% BSA and 50% glycerol. Store at −20°.

Negative Control. Assay a volume of culture medium from mock-transfected cells equivalent to that of experimental cell culture media.

Dual Luciferase/β-Galactosidase Reporter Gene Assay

Background

Cotransfection of two reporter gene constructs, one experimental and one as a constitutively expressed control, is used frequently to normalize experimental reporter activity for transfection efficiency. Luciferase and β-galactosidase reporter genes are ideal for cotransfections due to the availability of covenient assays with similar sensitivities, dynamic ranges, and luminescent readouts. The Dual-Light luminescent reporter gene assay system combines these reactions for highly sensitive, sequential detection of luciferase and β-galactosidase in a single extract sample.[14,63–67] Extremely

[63] T. Bourcier, G. Sukhova, and P. Libby, *J. Biol. Chem.* **272,** 15817 (1997).

[64] I. Bronstein, C. S. Martin, C. E. M. Olesen, and J. C. Voyta, *in* "Bioluminescence and Chemiluminescence: Molecular Reporting with Photons" (J. W. Hastings, L. J. Kricka, and P. E. Stanley, eds.) p. 451. Wiley, Chichester, 1997.

[65] A. N. Hollenberg, V. S. Susulic, J. P. Madura, B. Zhang, D. E. Moller, P. Tontonoz, P. Sarraf, B. M. Spiegelman, and B. B. Lowell, *J. Biol. Chem.* **272,** 5283 (1997).

[66] C. S. Martin, P. A. Wight, A. Dobretsova, and I. Bronstein, *BioTechniques* **21,** 520 (1996).

[67] H. Moessler, M. Mericskay, Z. Li, S. Nagi, D. Paulin, and J. V. Small, *Development* **122,** 2415 (1996).

high sensitivity and a wide dynamic range are attained, with detection of 1 fg to 20 ng of purified luciferase and 10 fg to 20 ng of purified β-galactosidase.

The Dual-Light reporter assay incorporates the luminescent substrates luciferin and Galacton-Plus. Cell lysate is mixed with buffer A, the luciferase reaction buffer. The luciferase signal is measured immediately after the addition of buffer B, which contains luciferin and Galacton-Plus. This enhanced luciferase reaction produces a light signal that decays with a half-life of approximately 1 min. The light signal from the β-galactosidase reaction is initially negligible due to the low pH (7.8) and absence of a luminescence enhancer. After a 30- to 60-min incubation, the light signal from the accumulated product of the β-galactosidase/Galacton-Plus reaction is initiated by the addition of a light emission accelerator, which raises the pH and provides Sapphire-II enhancer to increase light intensity. Light emission from the β-galactosidase reaction exhibits glow kinetics with a half-life of up to 180 min. Residual light emission from the luciferase reaction is minimal due to the rapid kinetic decay and the quenching effect of the accelerator. Generally, only very high luciferase concentrations (1 ng or greater of enzyme) may interfere with the low-level detection of β-galactosidase. A longer delay after the addition of accelerator prior to the measurement of light intensity will result in decreased levels of residual luciferase signal when extremely high levels of luciferase are present. However, it is important to maintain consistent timing of addition of buffer B and measurement of the β-galactosidase signal after the addition of the accelerator. A dual assay for the quantitation of both luciferase and β-glucuronidase activities in a single extract can be performed by substituting the Glucuron substrate for Galacton-Plus (Protocol llb).

The Dual-Light system is suitable for use with luminometers with automatic injectors. If only a single injector is available, rinse the injector thoroughly between injection of buffer B and accelerator. Manual injection may be performed if samples are measured at the same interval after adding the accelerator. The volume of cell extract should be 2–10 μl. The lysis solution may be substituted with alternative lysis solutions and lysis protocols. However, high levels of reducing agents interfere with the 1,2-dioxetane substrate, causing rapid signal decay. Alternative lysis solutions should be evaluated to ensure optimum performance.

High levels of endogenous β-galactosidase activity in cells or tissues may interfere with the measurement of β-galactosidase reporter enzyme. Endogenous enzyme activity is minimized at the pH of the Dual-Light reaction[4]; however, it is important to assay the level of endogenous enzyme with nontransfected cell extracts. Heat inactivation to reduce endogenous β-galactosidase activity should not be performed due to the detrimental effect on luciferase activity. When high endogenous β-galactosidase activity necessitates heat inactivation, assays for luciferase and β-galactosidase should be performed individually.

Reagents

The following reagents are available as components of the Dual-Light reporter gene assay system (Tropix).

Lysis solution: 100 m*M* potassium phosphate pH 7.8, 0.2% Triton X-100 (Store at 4°).

Note: DTT should be added fresh prior to use at a final concentration of 0.5 m*M* to stabilize luciferase activity. However, higher concentrations of reducing agents such as 2-mercaptoethanol and DTT will decrease the half-life of light emission of Galacton-Plus. If the extended half-life of light emission from Galacton-Plus is critical, reducing agents should be omitted from the lysis solution. If lysis buffer containing excess DTT has been used, the addition of hydrogen peroxide to Accelerator-II at a final concentration of 10 m*M* (add 1 μl of 30% H_2O_2 per 1 ml of Accelerator-II) will prevent rapid decay of signal half-life.

Buffer A: Lyophilized luciferase reaction buffer, pH 7.8. Reconstitute in 5 ml of sterile deionized or Milli-Q H_2O. Store at −20° before reconstitution. After reconstitution, store at 4° for 1 week or aliquot and store at −20°.

Buffer B: Lyophilized luciferin. Reconstitute in 22 ml of sterile deionized or Milli-Q H_2O. Store at −20° before reconstitution. After reconstitution, store at 4° for 1 week or aliquot and store at −20°. Add Galacton-Plus (or Glucuron) diluted 1:100 immediately before use. Do not store buffer B and Galacton-Plus (Glucuron) mixtures for longer than 24 hr.

Galacton-Plus: 100× concentrate (store at 4°).

Glucuron: 100× concentrate (store at 4°), available separately.

Accelerator-II: luminescence accelerator containing Sapphire-II enhancer (store at 4°).

Other Reagents

β-Galactosidase (G-5635, Sigma): positive control enzyme (optional).

Luciferase (L-1759, Sigma; Analytical Luminescence Laboratory, Sparks, MD): positive control enzyme (optional).

β-Glucuronidase (G-7896, Sigma): positive control enzyme (optional).

Phosphate-buffered saline: 75 m*M* sodium phosphate, pH 7.2, 68 m*M*

NaCl (sterilize and store at room temperature), or any standard PBS recipe.

Enzyme dilution buffer: 0.1 *M* sodium phosphate, pH 7.0, 0.1% BSA (Fraction V).

Protocols

Protocol I: Preparation of Cell Extracts from Tissue Culture Cells

Prepare extracts at 4° (on ice), if possible.

1. Add fresh DTT to 0.5 m*M* to the required volume of lysis solution.
2. Rinse cells twice with PBS.
3. Add lysis solution to cover cells (250 μl of lysis buffer for a 60-mm culture plate).
4. Detach cells from culture plate using a cell scraper.
5. Transfer cell lysate to a microcentrifuge tube and centrifuge for 2 min to pellet any debris.
6. Transfer supernatant to a fresh microcentrifuge tube. Extracts may be used immediately or frozen at −70°.

Note: Because the volume of cell extract assayed for each enzyme is identical, the ratio of control reporter vector to experimental vector used in a transfection should be adjusted to ensure that individual enzyme signal intensities are within the detection range of the instrument used for the measurement.

Protocol IIa: Detection Protocol for Tube or Microplate Luminometers (Luciferase/β-Galactosidase)

Perform all assays in triplicate at room temperature.

1. Equilibrate buffers A and B to room temperature.
2. Dilute sufficient Galacton-Plus 1:100 with buffer B.
3. Aliquot 2–10 μl of cell extracts into luminometer tubes or microplate wells.

Note: If using less than 10 μl of extract, the lysis solution should be added to bring the total volume up to 10 μl to ensure that the concentration of the reducing agent(s) and other components is identical in each sample.

Note: The amount of cell extract required may vary depending on the level of expression and instrumentation used. It is important to adjust the concentration of extract to keep the signal intensity within the linear range.

4. Add 25 μl of buffer A to each sample.

5. Within 10 min, inject 100 μl of buffer B. After a 2-sec delay, measure the luciferase signal for 5 sec.

Note: Signal intensities are time dependent. The time between the addition of buffer B and measurement should be consistent from sample to sample. Instruments with automatic injection will eliminate this concern. Longer or shorter measurements and delay times may be utilized but the same timing should be used when measuring the β-galactosidase signal after the addition of Accelerator-II.

6. Incubate samples for 30–60 min.
7. Inject 100 μl of Accelerator-II. After a 2-sec delay, measure the β-galactosidase signal for 5 sec or follow the same delay and measurement time used for the addition of buffer B.

Note: If manual injection is used, then Accelerator-II should be added in the same consistent time frame as buffer B addition.

Protocol IIb: Detection Protocol for Tube or Microplate Luminometers (Luciferase/β-Glucuronidase)

Perform all assays in triplicate at room temperature.

1. Equilibrate buffers A and B to room temperature.
2. Dilute sufficient Glucuron 1:100 with buffer B.
3. Aliquot 2-10 μl of cell extracts into luminometer tubes or microplate wells.

Note: If using less than 10 μl of extract, lysis solution should be added to bring the total volume up to 10 μl to ensure that the concentration of the reducing agent(s) and other components is identical in each sample.

Note: The amount of cell extract required may vary depending on the level of expression and instrumentation used. It is important to adjust the concentration of extract to keep the signal intensity within the linear range.

4. Add 25 μl of buffer A to each sample.
5. Within 10 min, inject 100 μl of buffer B. After a 2-sec delay, measure the luciferase signal for 5 sec.

Note: Signal intensities are time dependent. The time between the addition of buffer B and measurement should be consistent from sample to sample. Instruments with automatic injection should be utilized with the Glucuron substrate for accurate results. Longer or shorter measurements and delay times may be utilized but the same timing should be used when measuring the β-glucuronidase signal after the addition of Accelerator-II.

6. Incubate for 30–60 min.
7. Inject 100 μl of Accelerator-II. After a 2-sec delay, measure the β-glucuronidase signal for 5 sec or follow the same delay and measurement time used for the addition of buffer B. If manual injection is used, then Accelerator-II should be added in the same consistent time frame as buffer B addition.

Protocol III: Preparation of Controls

Positive Control. Reconstitute lyophilized β-galactosidase (β-glucuronidase) to 1 mg/ml in 0.1 *M* sodium phosphate, pH 7.0, 0.1% BSA. Store at 4°. Generate a standard curve by serially diluting in lysis solution containing 0.1% BSA. Reconstitute lyophilized luciferase to 1 mg/ml in 0.1 *M* sodium phosphate, pH 7.0, 0.1% BSA. Small aliquots should be stored at −20° and used only once after thawing. Prepare serial dilutions as described earlier. Two to 20 ng of β-galactosidase (β-glucuronidase) and 1–10 ng of luciferase should be used for the high end detection limit.

Negative Control. Assay a volume of mock-transfected cell extract equivalent to that of the experimental cell extract.

Instrumentation

Luminometers

Chemiluminescent reporter gene assays are ideally performed with a luminometer for maximum sensitivity. For optimum results with the assays described, a standard curve should be performed with the purified enzyme to determine the linear assay range that can be achieved with the particular instrument. The use of opaque white microplates is recommended for microplate assays. Black microplates can also be used, although the signal will be much lower due to absorbance of the emitted light.

Scintillation Counters

A liquid scintillation counter may be used as a substitute for a luminometer; however, sensitivity may be lower.[68–70] When using a scintillation

[68] R. Fulton and B. Van Ness, *BioTechniques* **14,** 762 (1993).
[69] V. T. Nguyen, M. Morange, and O. Bensaude, *Anal. Biochem.* **171,** 404 (1988).
[70] G. Erkel, U. Becker, and T. Anke, *J. Antibiotics* **49,** 1189 (1996).

counter, it is necessary to turn off the coincident circuit to measure chemiluminescence directly (single photon-counting mode). The instrument manufacturer should be contacted to determine how this is done. If it is not possible to turn off the coincident circuit, a linear relationship can be established by taking the square root of the counts per minute measured minus the instrument background:

$$\text{Actual} = (\text{measured-background})^{1/2}$$

Summary

1,2-Dioxetane chemiluminescent substrates provide highly sensitive, quantitative detection with simple, rapid assay formats for the detection of reporter enzymes that are widely used in gene expression studies. Chemiluminescent detection methodologies typically provide up to 100–1000× higher sensitivities than can be achieved with the corresponding fluorescent or colorimetric enzyme substrates. The varieties of 1,2-dioxetane substrates available provides assay versatility, allowing optimization of assay formats with the available instrumentation, and are ideal for use in gene expression assays performed in both biomedical and pharmaceutical research. These assays are amenable to automation with a broad range of instrumentation for high throughput compound screening.

[32] Clinical Application of Southern Blot Hybridization with Chemiluminescence Detection

By KATHLEEN S. TENNER and DENNIS J. O'KANE

Introduction*

Southern blot hybridization[1] is a membrane-based assay technique used for the detection and quantification of DNA. Following electrophoresis,

* Abbreviations: CL, chemiluminescence; PCR, polymerase chain reaction; TcR, T-cell receptor; LMP, low melting point; $J\beta_2$ probe, DNA probe to the second joining region of the β-TcR gene; PWS, Prader–Willi syndrome; AS, Angelman syndrome; HRP, horseradish peroxidase; ALP, alkaline phosphatase; CCD, charge-coupled device; CSPD, (disodium 3-(4-methoxyspiro{1,2-dioxetane-3,2′-(5′-chloro)tricyclo[3.3.1.1(3,7)]decan}-4-yl)phenyl phosphate); SSC, standard saline citrate; SDS, sodium dodecyl sulfate; TBS, Tris-buffered saline; CDP-*Star,* disodium 2-chloro-5-(4-methoxyspiro{1,2-dioxetane-3,2′-(5′-chloro)tricyclo[3.3.1.1^{3,7}]decan}-4-yl-1-phenyl phosphate.

[1] E. M. Southern, *J. Mol. Biol.* **98,** 503 (1975).

0076-6879/00 $30.00

DNA fragments are transferred from the separating gel to a nitrocellulose or nylon membrane and hybridized with a labeled nucleotide probe. In addition to quantification of target DNA, information about the size of the target nucleic acid fragment is obtained, which permits confirming the identity of the DNA fragment. This may be of critical importance for clinical assays where artifact amplicon bands and confounding restriction fragments must be resolved from the DNA fragments of interest in order to render a diagnosis.[2,3]

Probe labeling with ^{32}P-nucleotides is used frequently in Southern blot hybridization because it is reliable and sensitive. However, the use of radioisotopes has several significant disadvantages. Radioactive probes require special handling and disposal procedures and decay rapidly, necessitating the investment of extra personnel time for frequent probe labeling and quality control procedures. In addition, DNA probes are obtained generally by the cumbersome procedure of plasmid amplification and insert isolation. Fortunately, several nonisotopic methodologies and commercial products are available for use in a wide variety of Southern blot hybridization applications that minimize the difficulties experienced with isotopic protocols.[4–7]

Chemiluminescence detection in Southern blot hybridization offers several advantages compared with isotopic detection and has been reviewed recently.[8] Utilizing hapten-labeled probes with secondary enzyme or avidin conjugates and chemiluminogenic substrates has proved to be as sensitive as their isotopic counterparts.[9–14] Probes have been labeled with digoxigenin, biotin, and fluorescein by random priming, end labeling, PCR incorporation

[2] T. E. Mifflin, *J. Clin. Ligand Assay* **19,** 27 (1996).
[3] K. F. Kelly, *Proc. Nutr. Soc.* **55,** 591 (1996).
[4] H. J. Holtke, G. Sagner, C. Kessler, and G. Schmitz, *Biotechniques* **12,** 104 (1992).
[5] L. J. Kricka, "Nonisotopic Probing, Blotting, and Sequencing." Academic Press, San Diego, 1995.
[6] E. S. Mansfield, J. M. Worley, S. E. McKenzie, S. Surrey, E. Rappaport, and P. Fortina, *Mol. Cell Probes* **9,** 145 (1995).
[7] P. G. Isaac, J. Stacey, and C. M. Clee, *Mol. Biotechnol.* **3,** 259 (1995).
[8] I. Bronstein, J. C. Voyta, O. J. Murphy, R. Tizzard, C. W. Ehrenfels, and R. L. Cate, *Methods Enzymol.* **217,** 398 (1993).
[9] Q. Nguyen, F. Witney, and A. Tumolo, *Biotechniques* **13,** 116 (1992).
[10] U. Reischl, R. Ruger, and C. Kessler, *Mol. Biotechnol.* **1,** 229 (1994).
[11] B. Rihn, C. Coulais, M. C. Bottin, and N. Martinet, *J. Biochem. Biophys. Methods* **30,** 91 (1995).
[12] B. Rihn, M. C. Bottin, C. Coulais, and N. Martinet, *J. Biochem. Biophys. Methods* **30,** 103 (1995).
[13] K. A. Hodges, C. M. Kosciol, W. N. Rezuke, E. C. Abernathy, W. T. Pastuszk, and G. J. Tsongalis, *Ann. Clin. Lab. Sci.* **26,** 114 (1996).
[14] G. Enger-Blum, M. Meier, J. Frank, and G. A. Muller, *Anal. Biochem.* **210,** 235 (1993).

and photo fixation.[5] CL-labeled probes are stable for weeks to months, which decreases time spent in preparation and quality controlling probes.

Clinically relevant examples of Southern blot hybridization with CL detection are collected in Table I. The major use of Southern blot hybridization is to verify or confirm results obtained with amplified nucleic acids. Although amplified nucleic acids may be detected by a variety of techniques, including homogeneous CL procedures,[15] a lingering question is whether the detected signal results from the amplification of the correct sequence or from an irrelevant, confounding sequence due to nonspecific amplification or amplification of a similar but clinically irrelevant sequence. Consequently, primer-based procedures are often verified by secondary Southern blot hybridization to ensure that the amplicons have the correct molecular size. This is used routinely in clinical microbiology, especially for the detection of viruses (Table I). Another major application of Southern blot hybridization is in the diagnosis of genetic disease and malignancy resulting from gene rearrangements or translocations. These applications often involve the detection of a change in a single-copy gene.[13,14,16,17] For example, the translocation of the *abl* protooncogene on chromosome 8 to the *bcr* gene on chromosome 22 is associated with chronic myelogenous and acute lymphocytic leukemia.[18] Similarly, the T-cell receptor gene may be rearranged in various T-cell lymphomas.[19] Because the chromosomal break points are not always the same in each patient, nucleic acid amplification may not always be appropriate as the rearranged sequences may not have primer annealing sequences and so may not amplify. Southern blot hybridization may be employed to detect these single gene changes in nonamplified samples from patients suspected of having the condition. Accordingly, any CL detection method must be sufficiently sensitive for use with unamplified genomic DNA. A second requirement for the use of Southern blot hybridization in testing for hematologic malignancy is following the number of clonal malignant cells during the course of treatment. Consequently, the CL detection methodology must be sufficiently sensitive not only to detect single copy genes in genomic DNA, but also to detect the reduction in genomic equivalents during and posttherapy. Furthermore, this must be accomplished utilizing small amounts of DNA isolated from a patient's blood sample, which is generally a few milliliters.

[15] N. C. Nelson and D. L. Kacian, *Clin. Chim. Acta* **194,** 73 (1990).

[16] J. A. Hopfenbeck, J. A. Holden, C. T. Wittwer, and C. R. Kjeldsberg, *Am. J. Clin. Pathol.* **97,** 639 (1992).

[17] G. Sachdeva, G. Kaur, and R. Bameazi, *Indian J. Exp. Biol.* **33,** 173 (1995).

[18] A. Lesieru, S. Naber, S. McKenzie, and H. J. Wolfe, *Diagn. Mol. Pathol.* **3,** 75 (1994).

[19] D. K. Ryan, H. D. Alexander, and T. C. M. Morris, *J. Clin. Pathol. Med. Pathol.* **50,** 77 (1997).

TABLE I

CLINICALLY RELEVANT APPLICATIONS OF CHEMILUMINESCENCE DETECTION IN SOUTHERN BLOT HYBRIDIZATION

Application	Detection method[a]	Substrate	Reference
PCR (RT-PCR) confirmation			
Borrelia burdorfi (Lyme disease)	ALP-probe	CSPD	*b*
	HRP-probe	Luminol/ECL	*c*
Human immunodeficiency virus	HRP-probe	Luminol	*d*
	ALP-probe	Dioxetan-phosphate	*e*
Human papilloma virus	HRP-probe	Luminol/ECL	*f*
	Dig-probe/anti-Dig-ALP	Lumi-Phos 530	*g*
Human parvovirus	Dig-probe/anti-Dig-ALP	Lumi-Phos 530	*h*
Hepatitis C virus	Dig-probe/anti-Dig-ALP	Lumigen PPD	*i*
	Dig-probe/anti-Dig-HRP	Luminol/ECL	*j*
Respiratory syncitial virus	BrdU-probe/mouse anti-BrdU/goat antimouse IgG-HRP	Luminol/ECL	*k*
Norwalk-like viruses	Dig-probe/anti-Dig-ALP	Lumi-Phos 530	*l*
Lassa virus	Fluor-probe/anti-Fluor-HRP	Luminol/ECL	*m*
Infectious organisms			
Babesia species	HRP-probe	Luminol/ECL	*n*
	HRP-probe	Luminol/ECL	*o*
Mycobacterium species	Dig-probe/anti-Dig-ALP	Dioxetan-phosphate	*p*
	Bt-probe/avidin-ALP	Dioxetan-phosphate	*q*
Clostridium difficile	Bt-probe/steptavidin-HRP	Luminol	*r*
Single copy genes			
Interleukin-2 receptor	Dig-probe/anti-Dig-ALP	AMPPD	*s*
Philadelphia chromosome (chronic myelogenous leukemia)	Dig-probe/anti-Dig-ALP	CSPD	*t*
T-cell receptor rearrangement	Dig-probe/anti-Dig-ALP	AMPPD	*s*
(T-cell lymphomas)	Dig-probe/anti-Dig/ALP	Lumi-Phos 530	*u*
	Dig-probe/anti-Dig/ALP	Lumi-Phos 530	*v*
	Bt-probe/streptavidin-HRP	Acridan (PS-1)	*w*

(*continued*)

TABLE I (*continued*)

Application	Detection method[a]	Substrate	Reference
Prader–Willi/Angelman syndromes	Dig-probe/anti-Dig-ALP	CDP-tar	*x*
	Dig-probe/anti-Dig-ALP	CDP-Star	*y*
Trinucleotide repeats			
Fragile X syndrome	Bt-probe/streptavidin-ALP	Dioxetan-phosphate	*z*
	Dig-probe/anti-Dig-ALP	CSPD	*aa*
	Dig-probe/anti-Dig-ALP	Lumi-Phos 530	*bb*
	ALP-probe	Dioxetan-phosphate	*cc*
Huntington's disease	Fluor-probe/antifluorescein-HRP	Luminol/ECL	*dd*
Spinocerebellar ataxia type I	Fluor-probe/antifluorescein-HRP	Luminol/ECL	*dd*
Protooncogenes			
ptc (papillary thyroid cancer)	Dig probe/anti-Dig-ALP	Lumi-Phos 530	*ee*
Immunoglobulins			
IgA	HRP-probe	Luminol/ECL	*ff*
HLA typing	ALP-probe	Lumi-Phos 480	*gg*
Peptide hormones			
Gastrin mRNA	Dig-probe/anti-Dig-ALP	AMPPD	*hh*

[a] Detection format is designated by probe labeling configuration followed where appropriate by the detector-conjugate. Abbreviations: ALP, alkaline phosphatase; HRP, horseradish peroxidase; Dig, digoxigenin; anti-Dig, digoxigenin antibody; BrdU, bromodeoxyuridine; anti-BrdU, BrdU antibody; fluor, fluorescein; antifluor, fluorescein antibody; Bt, biotin.

[b] M. Delaviuda, M. Fille, J. Ruiz, and J. Aslanzadeh, *J. Clin. Microbiol.* **34,** 3115 (1996).

[c] M. M. Picken, R. N. Picken, D. Han, Y. Cheng, and F. Strle, *Eur. J. Clin. Microbiol. Infect. Dis.* **15,** 489 (1996).

[d] G. Levee, D. Boulay, G. Pialoux, and F. Laure, *Int. Conf. AIDS* **7,** 350 (1991).

[e] B. Conway, K. E. Adler, L. J. Bechtel, J. C. Kaplan, and M. S. Hirsch, *J. Acquir. Immune Defic. Syndr.* **3,** 1059 (1990).

[f] M. Marrero, O. Valdes, M. Alvarez, G. Diaz, A. Otero, and G. Roges, *Int. Conf. AIDS* **8,** 19 (1992).

[g] F. H. Sarkar, W. A. Sakr, Y. W. Li, P. Sreepathi, and J. D. Crissman, *Prostate* **22,** 171 (1993).

[h] E. L. Durigon, D. D. Erdman, B. C. Anderson, B. P. Holloway, and L. J. Anderson, *Mol. Cell. Probes* **8,** 199 (1994).

[i] P. Canepari, C. F. Distefano, and M. M. Lleo, *Microbiologica* **17,** 9 (1994).

[j] P. Komminoth, V. Adams, A. A. Long, J. Roth, P. Saremaslani, R. Flury, M. Schmid, and P. U. Heitz, *Pathol. Res. Practice* **190,** 1017 (1994).

[k] A. Dakhama and R. G. Hegele, *Mod. Pathol.* **9,** 849 (1996).

[l] T. Ando, S. S. Monroe, J. R. Gentsch, Q. Jim, D. C. Lewis, and R. I. Glass, *J. Clin. Microbiol.* **33,** 64 (1995).

[m] A. H. Demby, J. Chamberlain, D. W. G. Brown, and C. S. Clegg, *J. Clin. Microbiol.* **32,** 2898 (1994).

[n] J. W. Thomford, P. A. Conrad, S. R. Telford III, D. Mathiesen, B. H. Bowman, A. Spielman, M. L. Eberhard, B. L. Herwaldt, R. E. Quick, and D. H. Persing, *J. Infect. Dis.* **169,** 1050 (1994).

[o] P. A. Conrad, J. W. Thomford, A. Marsh, S. R. Telford, J. F. Anderson, A. Spielman, E. A. Sabin, I. Yamane, and D. H. Persing, *J. Clin. Microbiol.* **30,** 1210 (1992).

[p] R. A. Ghossein, D. G. Ross, R. N. Salomon, and A. R. Rabson, *Diagn. Mol. Pathol.* **1,** 185 (1992).

[q] F. S. Nolte, B. Metchock, J. E. Mcgowan, A. Edwards, O. Okwumabua, C. Thurmond, P. S. Mitchell, B. Plikaytis, and T. Shinnick, *J. Clin. Microbiol.* **31,** 1777 (1993).

[r] P. H. Gumerlock, Y. J. Tang, F. J. Meyers, and J. Silva, Jr., *Rev Infect. Dis.* **13,** 1053 (1991).

[s] G. Sachdeva, G. Kaur, and R. Bamezai, *Indian J. Exp. Biol.* **33,** 173 (1995).

[t] A. Lesieur, S. Naber, S. Mckenzie, and H. J. Wolfe, *Diagn. Mol. Pathol.* **3,** 75 (1994).

[u] K. A. Hodges, C. M. Kosciol, W. N. Rezuke, E. C. Abernathy, W. T. Pastuszak, and G. J. Tsongalis, *Ann. Clin. Lab. Sci.* **26,** 114 (1996).

[v] J. A. Hopfenbeck, J. A. Holden, C. T. Wittwer, and C. R. Kjeldsberg, *Am. J. Clin. Pathol.* **97,** 638 (1992).

[w] K. S. Tenner, M. Karst, S. Thibodeau, and D. J. O'Kane, *in* "Bioluminescence and Chemiluminescence: Molecular Reporting with Photons" (J. W. Hastings, L. J. Kricka, and P. E. Stanley, eds.), p. 513. Wiley, Chichester.

[x] K. S. Tenner, S. N. Thibodeau, and D. J. O'Kane, *Clin. Chem.* **43,** S269 (1997).

[y] D. N. Radu and K. C. S. Chiang, *Clin. Chem.* **43,** S268 (1997).

[z] E. Nanba, Y. Kohno, A. Matsuda, M. Yano, C. Sato, K. Hashimoto, T. Koeda, K. Yoshino, M. Kimura, Y. Maeoka, T. Yamamoto, Y. Maegaki, I. Eda, and K. Takeshita, *Brain Dev.* **17,** 317 (1995).

[aa] A. A. El-Aleem, I. Bohm, S. Temtamy, M. El-Awady, M. Awadalla, J. Schmidtke, and M. Stuhrmann, *Hum. Genet.* **96,** 577 (1995).

[bb] O. T. Mueller, J. K. Hartsfield, Jr., M. J. A. Amar, L. A. Gallardo, and B. G. Kousseff, *Am. J. Med. Genet.* **60,** 302 (1995).

[cc] W. T. Brown, G. E. Houck, Jr., A. Jeziorowska, F. N. Levinson, X. Ding, C. Dobkin, N. Zhong, J. Henderson, S. S. Brooks, and E. C. Jenkins, *JAMA* **270,** 1569 (1993).

[dd] S. Castellvi-Bel, T. Matilla, M. I. Banchs, H. Kruyer, J. Corral, M. Mila, and X. Estivill, *J. Med. Genet.* **31,** 654 (1994).

[ee] R. K. Martin, K. T. Archer, and R. M. Tuttle, *Diagn. Mol. Pathol.* **3,** 233 (1994).

[ff] M. Shimomura, N. Yoshikawa, K. Iijima, H. Nakamura, M. Miyazaki, and H. Sakai, *Clin. Nephrol.* **43,** 211 (1995).

[gg] Medintz, L. Chiriboga, L. Mccurdy, and L. Kobilinsky, *J. Foren. Sci.* **39,** 1372 (1994).

[hh] G. Monges, P. Biagini, J. F. Cantaloube, C. Chicheportiche, V. Frances, D. Brandini, P. Parc, J. F. Seitz, M. Giovannini, R. Sauvan, and J. Hassoun, *J. Mol. Endocrinol.* **11,** 223 (1993).

Several probe/detector pairs and CL substrates have been reported in the detection of genetic diseases, including detecting single-copy genes and hematologic malignancies (Table I). Digoxigenin-labeled probes, in combination with CL substrates, have been used in the following diagnostic assays: (i) gene rearrangements in hematologic malignancy[20] and TcR gene rearrangements,[16,17,21] (ii) Philadelphia chromosome translocation in chronic myelogenous leukemia,[18] and (iii) fragile X syndrome.[22,23] Biotin probes have been used in the diagnosis of fragile X syndrome,[24] and fluorescein probes have been used in the diagnosis of Huntington's disease and spinocerebellar ataxia type I.[25]

A number of factors contribute to the ultimate decision to utilize a probe/detector pair and a CL substrate in a given diagnostic test, including probe concentration required, ease of probe preparation, labeled probe stability, nonspecific background, length of time required for the procedure, and intensity and stability of the chemiluminescence signal, as well as unique clinical requirements for each assay. Two examples of the use of CL detection in Southern blot hybridization are presented. The first application describes the assay system with the best signal-to-noise ratio when used in an existing TcR gene rearrangement assay where less than a genomic equivalent of a single-copy gene must be detected. The second application describes the use of PCR for simultaneous generation and labeling of probe for Southern blot hybridization for detecting the Prader–Willi/Angelman syndrome, a rare genetic disease.

T-Cell Receptor Gene Rearrangement Assay

Introduction

The T-cell receptor complex is encoded by four separate single-copy genes.[26] These genes have segments that encode for variable, joining, and constant regions of the TcR subunits. In addition, the β-TcR gene has a

[20] J. E. Coad, D. J. Olson, T. A. Lander, and R. C. McGlennen, *Mol. Diagn.* **2,** 67 (1997).

[21] J. Cossman and M. Uppenkamp, *Clin. Lab. Med.* **8,** 45 (1988).

[22] A. A. el-Aleem, I. Bohm, S. Temtamy, M. el-Awady, M. Awadaila, J. Schmidtke, and M. Stuhrmann, *Hum. Genet.* **95,** 577 (1995).

[23] O. T. Mueller, J. K. Hartsfield, Jr., M. J. A. Amar, L. A. Gallardo, and B. G. Kousseff, *Am. J. Med. Genet.* **60,** 302 (1995).

[24] E. Nanba, Y. Kohno, A. Matsuda, Yano, C. Sato, K. Hashimoto, T. Koeda, K. Yoshino, M. Kimura, Y. Maeoka *et al., Brain Dev.* **17,** 317 (1995).

[25] Caste, S. vo-Bel, T. Matilla, M. I. Banchs, H. Kruyer, J. Corral, M. Mila, and X. Estivill, *J. Med. Genet.* **31,** 654 (1994).

[26] J. Cossman, M. Uppenlamp, J. Sundeen, R. Coupland, and M. Raffeld, *Arch. Pathol. Lab. Med.* **112,** 117 (1988).

segment encoding for diversity. Rearrangements of the genes encoding the β and γ subunits are generally used for the Southern blot hybridization assessment of clonality, which distinguishes T-cell malignancies from most T-cell reactive processes. Every cell within a malignant clonal expansion will have the same T-cell receptor gene rearrangement detected by Southern blot hybridization; benign reactive processes may have a number of TcR gene rearrangements, none of which may predominate. It is important clinically to be able to distinguish these TcR gene rearrangement differences on Southern blots.

Assay Principle[27]

Patient sample DNA is digested with *Eco*RI and separated by agarose gel electrophoresis. TcR gene rearrangements are detected by Southern blot hybridization using a probe to the gene encoding the second joining region of the β-TcR gene with chemiluminescence detection.

Reagents

Low melting point agarose (FMC, Rockland, ME)
High prime labeling mix for digoxygenin and for biotin (Boehringer-Mannheim Corporation, Indianapolis, IN)
Puregene kit for genomic DNA (Gentra Systems & Corp., Minneapolis, MN)
Formamide, deionized (Oncor, Gaithersburg, MD)
Yeast tRNA (Sigma Chemicals, St. Louis, MO)
Lumi-Phos Plus (Lumigen, Southfield, MI)
PS3 acridan substrate (Lumigen, Southfield, MI)
Kodak blocker (Kodak Chemicals, Rochester, NY)

Solutions

Dextran sulfate stock solution, 50% (Oncor, Gaithersburg, MD)
Denhardt's solution (100×): dissolve 5 each g of Ficoll, polvinylpyrrolidone, and nuclease-free bovine serum albumin in a total volume of 250 ml H_2O. Filter sterilize and store in aliquots at $-20°$.
Standard saline citrate, 20× stock solution: 0.3 *M* sodium citrate containing 3 *M* NaCl, pH 7.0 (Sigma Chemicals)

[27] K. S. Tenner, M. Karst, S. Thibodeau, and D. J. O'Kane, *in* "Bioluminescence and Chemiluminescence: Molecular Reporting with Photons" (J. W. Hastings, L. J. Kricka, and P. E. Stanley, eds.), p. 513. Wiley, Chichester.

Sodium dodecyl sulfate (SDS) stock solution, 10% (Sigma Chemicals)
Tris-buffered saline (TBS), 10× stock solution (Boehringer-Mannheim Corporation)
DNA denaturation solution: 0.4 *M* NaOH, 0.6 *M* NaCl
Neutralizing solution: 0.5 *M* Tris–HCl, 1.5 *M* NaCl, pH 7.5

Probe Generation/Dot Blots. A 4.4 kb probe insert encoding the second joining region of the β-TcR gene (Jβ_2) is isolated from plasmid DNA by *Eco*RI digestion and is separated by electrophoresis on a 0.6% low melting point agarose gel. The probe is heat denatured and labeled by random priming with digoxigenin or biotin using High-Prime mix (Boehringer-Mannheim). Other labeling methods explored are direct labeling with horseradish peroxidase (Amersham, Arlington Heights, IL) and nonenzymatic labeling with psoralen by the Rad-Free system (Schleicher & Schuell, Keene NH) or with biotin by the Fast-Tag system (Vector Laboratories, Inc., Burlingame CA) according to manufacturers' directions.

Southern Blot Transfer. Human genomic DNA is extracted by salt precipitation using the Puregene kit according to manufacturer's directions. Sample DNA (2.5 μg) is digested overnight with *Eco*RI and is separated by electrophoresis on a 0.8% agarose gel at 55 V. After electrophoresis, the DNA gel is denatured (0.4 *M* NaOH, 0.6 *M* NaCl), neutralized (0.5 *M* Tris–HCl, 1.5 *M* NaCl, pH 7.5), and transferred overnight to a Magnagraph positively charged nylon membrane (MSI, Westboro, MA). The membrane is baked at 70° for 2 hr and is stored in the dark at room temperature until hybridization with the Jβ_2 probe is initiated.

Hybridization and Stringency Washes. The membrane-bound DNA is prehybridized for 2 hr at 45° in 20 ml of prewarmed hybridization buffer (50% formamide, 5× SSC, 1× Denhardt's solution, 0.6% sodium dodecyl sulfate, 10% dextran sulfate, 0.2 mg/ml yeast t-RNA). Fifty nanograms (2.5 ng/ml) of the denatured, labeled Jβ_2 probe is added to the prehybridization buffer and incubated overnight in a roller bottle at 45°. The membrane is washed twice in 2× SSC, 0.1% SDS, once in 0.2× SSC, 0.1% SDS, and once in 0.1× SSC, 0.1% SDS at 60°, 30 min each.

Chemiluminescent Signal Generation and Detection. After high stringency washes, the membrane is equilibrated for 5 min in 1× TBS and blocked for 1.5 hr in 50 ml of 1× TBS, 0.5% Kodak blocker, 0.1% SDS at 37°. The membrane is incubated for 0.5 hr with antidigoxigenin or streptavidin coupled with either HRP or alkaline phosphatase and washed (1× TBS, 0.5% SDS) four times, 10 min each. Signal is generated with Lumi-Phos Plus, CSPD (Tropix, Bedford MA), or PS-3 (Lumigen) substrates and is detected by a 5-min exposure with a charged couple device camera cooled to $-35°$.[28,29]

Discussion

Efficiency of Probe Labeling. Dot blot dilution series are used to determine the efficiency of biotin, digoxigenin, HRP, and nonenzymatic labeling of the $J\beta_2$ probe, which has been separated by electrophoresis in LMP agarose. Dilution series of labeled control DNA are compared to 10-fold dilution series of the labeled $J\beta_2$ probe. Digoxigenin and biotin High-Prime probe labeling procedures are consistently the most efficient. ALP conjugates are detected with the dioxetane-based substrates Lumi-Phos Plus or CSPD, whereas HRP conjugates are detected with PS-3, an acridan-based substrate. Lumi-Phos Plus and PS-3 produce the most intense signals of the substrates tested (Fig. 1).

Southern Blot Hybridization. The biotin-HRP system used with the PS-3 substrate produces a more intense signal than the other probe-conjugated substrate combinations when used for visualizing TcR gene rearrangements by Southern blot hybridization (Fig. 2). Even an overnight incubation with Lumi-Phos Plus (Fig. 2B) does not produce an intensity comparable to a 5-min incubation with PS-3 (Fig. 2A). The biotin HRP-PS-3 system enabled resolution and identification of bands, indicating a polyclonal benign disease process, as well as a clonal gene rearrangement from bands of a normal patient (Fig. 2A). This high resolution was striking because the amount of DNA used in this protocol is approximately five-fold less than what is typically reported used in TcR gene rearrangements detected by Southern blotting procedures (Table II). The sensitivity of chemiluminescence detection was assessed by analyzing decreasing amounts of normal control DNA by Southern blot. The biotin-HRP system was able to detect the germline restriction fragment in 0.3 μg of total genomic DNA.[27] Comparison of the optimized HRP-PS-3 chemilumigraph (5-min exposure) with a ^{32}P autoradiograph (overnight exposure) on identical Southern blots demonstrated that the PS-3 system is as sensitive as isotopic detection (Fig. 2D).

The use of Southern blot hybridization will be used less frequently in the future for the assessment of TcR gene rearrangements. The reason for this is that nucleic acid amplification protocols are being improved and made more reliable.[19] Consequently, the same clinical information may be obtained in a few days by amplification strategies as compared to up to 2

[28] C. S. Matrin and I. Bronstein, *J. Biolumin. Chemilumin.* **9,** 145 (1994).

[29] H. Akhavan-Tafti, P. A. Schaap, A. Arghavani, R. DeSilva, R. A. Eickholt, R. S. Handley, B. A. Schoenfelner, K. Sugioka, and Y. Sugioka, *J. Biolumin. Chemilumin.* **9,** 155 (1994).

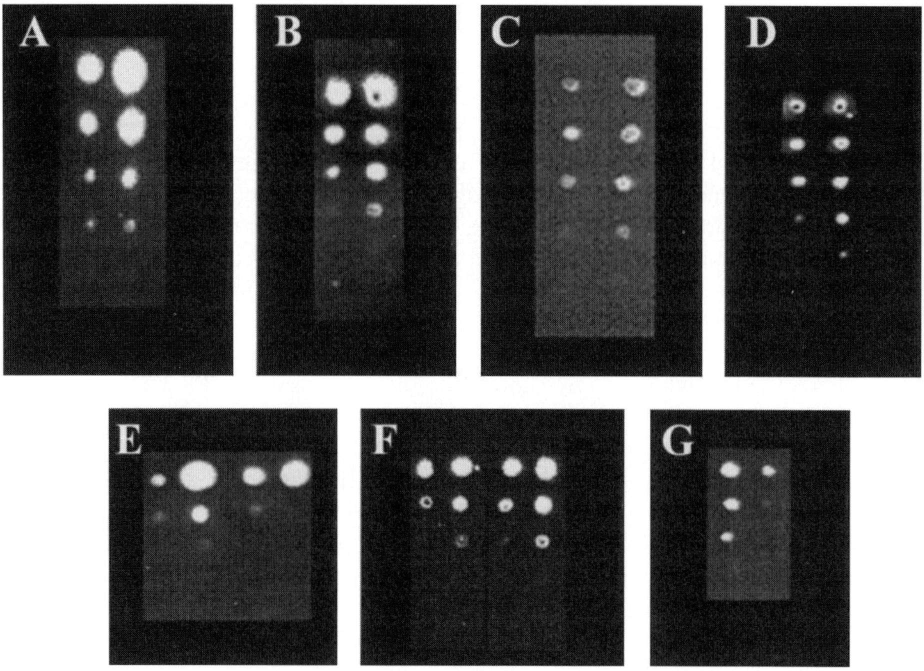

FIG. 1. Dot blots of dilution series of labeled control DNA and 10-fold dilution series of labeled $J\beta_2$ detected with chemiluminescent substrates for estimating the efficiency of labeling. Lane I, from the top: 1 ng, 50 pg, 10 pg, 2 pg, and 0.5 pg of labeled control DNA. Lane II, from the top: 1 : 10, 1 : 100, 1 : 1000, 1 : 10,000, and 1 : 100,000 dilution of labeled $J\beta_2$. $J\beta_2$ probe biotin labeled using High-Prime mix and detected with PS-3 (A) and Lumi-Phos Plus (B). Dot blot identical to A, but $J\beta_2$ probe biotin labeled by High-Prime mix (C) or Fast Tag (D); digoxigenin labeled with High-Prime mix and detected by PS-3 (E) or Lumi-Phos Plus (F); direct labeled with HRP and detected with PS-3 (G). Adapted with permission from K. S. Tenner, M. Karst, S. Thibodeau, and D. J. O'Kane, *in* "Bioluminescence and Chemiluminescence: Molecular Reporting with Photons" (J. W. Hastings, L. J. Kricka, and P. E. Stanley, eds.), p. 513. Wiley, Chichester. © John Wiley & Sons Limited.

weeks by Southern blot hybridization. However, Southern blot hybridization will still be utilized on a smaller number of patient samples that do not amplify because the rearrangement does not include one or both primer annealing sites.

Prader–Willi/Angelman Syndrome

Introduction

The Prader–Willi and Angelman syndromes are distinct, rare genetic disorders that may result from several different defects involving chromo-

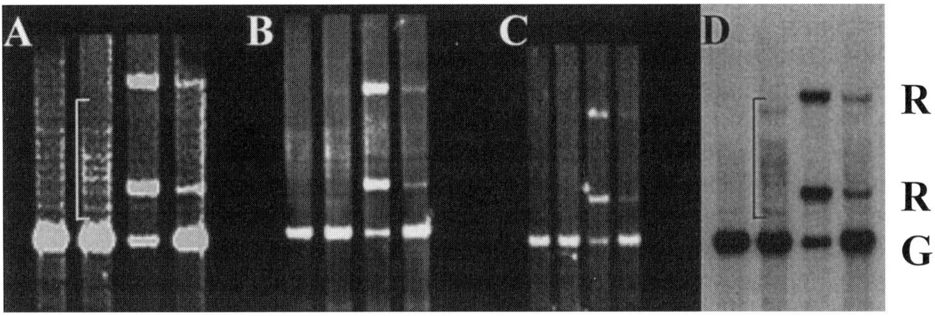

FIG. 2. Chemilumigrams of Southern blots of normal patient DNA (lane 1), polyclonal benign disease process (lane 2), 50% TcR clonal gene rearrangement (lane 3), and 10% TcR clonal gene rearrangement (lane 4) probed with biotin-labeled Jβ_2 probe and detected with PS-3/streptavidin-HRP (A) and Lumi-Phos Plus/streptavidin-ALP (B). Chemilumigram of Southern blot identical to A, but probed with digoxigenin-labeled Jβ_2 and detected with PS-3/anti-Dig-HRP (C). (D) Autoradiogram of blot identical to A, but probed with ^{32}P-labeled Jβ_2 probe. Vertical bars in A and D (lane 2) indicate the region of confounding polyclonal bands. G, germline; R, rearranged TcR band. Adapted with permission from K. S. Tenner, M. Karst, S. Thibodeau, and D. J. O'Kane, *in* "Bioluminescence and Chemiluminescence: Molecular Reporting with Photons" (J. W. Hastings, L. J. Kricka, and P. E. Stanley, eds.), p. 513. Wiley, Chichester, © John Wiley & Sons Limited.

some 15 abnormalities. These syndromes are manifested in early child development but are difficult to diagnose by clinical signs and symptoms alone. The Prader–Willi syndrome results in 70% of cases from loss of a gene or genes from the paternally derived chromosome at 15q11-13; 28% of the time from the maternal disomy of chromosome 15; and in <2% of

TABLE II
AMOUNT OF DNA USED IN SOUTHERN BLOT HYBRIDIZATION FOR TcR GENE REARRANGEMENTS[a]

Reference	μg/lane	Detection limit (μg)
14	10	NA
32	10–15	0.20
13	10	0.25
16	5	0.50
27	2.5	0.31

[a] Adapted with permission from K. S. Tenner, M. Karst, S. Thibodeau, and D. J. O'Kane, *in* "Bioluminescence and Chemiluminescence: Molecular Reporting with Photons" (J. W. Hastings, L. J. Kricka, and P. E. Stanley, eds.), p. 513. Wiley, Chichester. © John Wiley & Sons Limited.

cases from an abnormality in the imprinting process, which is detectable through methylation analysis. The Angelman syndrome results in 70% of case from maternally derived deletions of a gene or genes from 15q11-13; 3 to 5% from paternal disomy of chromosome 15; approximately 10% from an imprinting abnormality detectable though methylation analysis; and the remainder do not demonstrate paternal-specific inheritance or methylation imprinting abnormalities. The challenge is to develop a Southern blot hybridization assay that can detect the parent-of-origin methylation imprint differences.

Assay Principle[30,31]

Patient sample DNA is codigested with *Hpa*II, a methylation-sensitive restriction nuclease, plus *Hin*dIII, and the restriction fragments produced are separated by agarose gel electrophoresis. The methylation specific changes are detected by Southern blot hybridization using chemiluminescence detection and are visualized as the parental loss of a restriction fragment.

Materials and Methods

Reagents and Solutions

Puregene kit (Gentra Systems & Corp.)
Dig Easy-Hyb solution (Boehringer-Mannheim Corporation)
CDP-*Star* (Boehringer-Mannheim Corporation)

Stock Solutions

Standard saline citrate, 20× stock solution: 0.3 *M* sodium citrate containing 3 *M* NaCl, pH 7.0 (Sigma Chemicals)
Sodium dodecyl sulfate stock solution, 10% (Sigma Chemicals)
Maleic acid buffer: 0.1 *M* maleic acid containing 0.15 *M* NaCl, adjusted to pH 7.5 with solid NaOH

Probe Generation/Dot Blots. A plasmid-encoded 365-bp PW71b probe insert was isolated by *Eco*R1 digestion of plasmid DNA and separated by electrophoresis on a 0.6% LMP agarose gel. PCR was used to simultaneously amplify and digoxigenin label large quantities of PW71b.

[30] D. N. Radu and C. S. Chiang, *Clin. Chem.* **43,** S268 (1997).
[31] K. S. Tenner, S. N. Thibodeau, and D. J. O'Kane, *Clin. Chem.* **43,** S269 (1997).

Southern Blot Hybridization. Human genomic DNA is extracted by salt precipitation using the Puregene kit according to manufacturer's directions. Sample DNA (2.5 μg) is digested overnight with *Hpa*II/*Hin*dIII and is separated by electrophoresis on a 0.8% agarose gel at 55 V. After electrophoresis, the DNA gel is denatured (0.4 *M* NaOH, 0.6 *M* NaCl), neutralized (0.5 *M* Tris–HCl, 1.5 *M* NaCl, pH 7.5), and transferred overnight to a positively charged nylon membrane (Boehringer Mannheim). The membrane is baked at 70° for 2 hr fixed by UV cross-linking, and stored in the dark at room temperature until use.

Hybridization and Stringency Washes. The membrane-bound DNA is prehybridized for 0.5 hr at 45° in 20 ml of prewarmed Easy-Hyb buffer. Forty microliters of denatured digoxigenin-labeled PW71b PCR product is added to the prehybridization buffer (2 μl/ml) and incubated 3 hr in a roller bottle at 45°. The membrane is washed twice in 2× SSC, 0.1% SDS, and once in 0.5× SSC, 0.1% SDS at 60°, for 15 min each.

Chemiluminescent Signal Generation and Detection. After high stringency washes, the membrane is equilibrated for 5 min in maleic acid wash buffer and blocked for 1 hr in 50 ml of maleic acid buffer containing 1% casein at room temperature. The membrane is incubated for 0.5 hr with antidigoxigenin ALP and washed (maleic acid buffer containing 0.5% Tween 20) four times for 10 min each. Signal is generated with CDP-*Star* and detected by a 5-min exposure with a CCD camera cooled to −35°.

Results

Dot Blot. Dot blot dilution series are used to determine the efficiency of digoxigenin labeling of the PW71b probe by PCR amplification. Dilution series of labeled control DNA are compared to 10-fold dilution series of labeled PW71b probe. Dot blots are visualized using 10-fold dilutions of probe and a 5-min substrate incubation with CL substrates. CDP-*Star* and PS-3 substrates produce more intense signals than the other substrates tested. In a 17-hr incubation with substrate, the intensity of the CDP-*Star* signal continued overnight in comparison to that of PS-3, which decreases after a short period of time (Fig. 3). This characteristic of the antidigoxigenin-ALP/CDP-*Star* system enables visualization for longer periods of time and permits reexamining a blot at a nearly constant signal-to-noise ratio.

The digoxigenin-ALP/CDP-*Star* system enables distinguishing the absence of a band indicating PWS or AS, in contrast to the bands present of a normal patient (Fig. 4). All 12 patients with suspected PWS/AS were

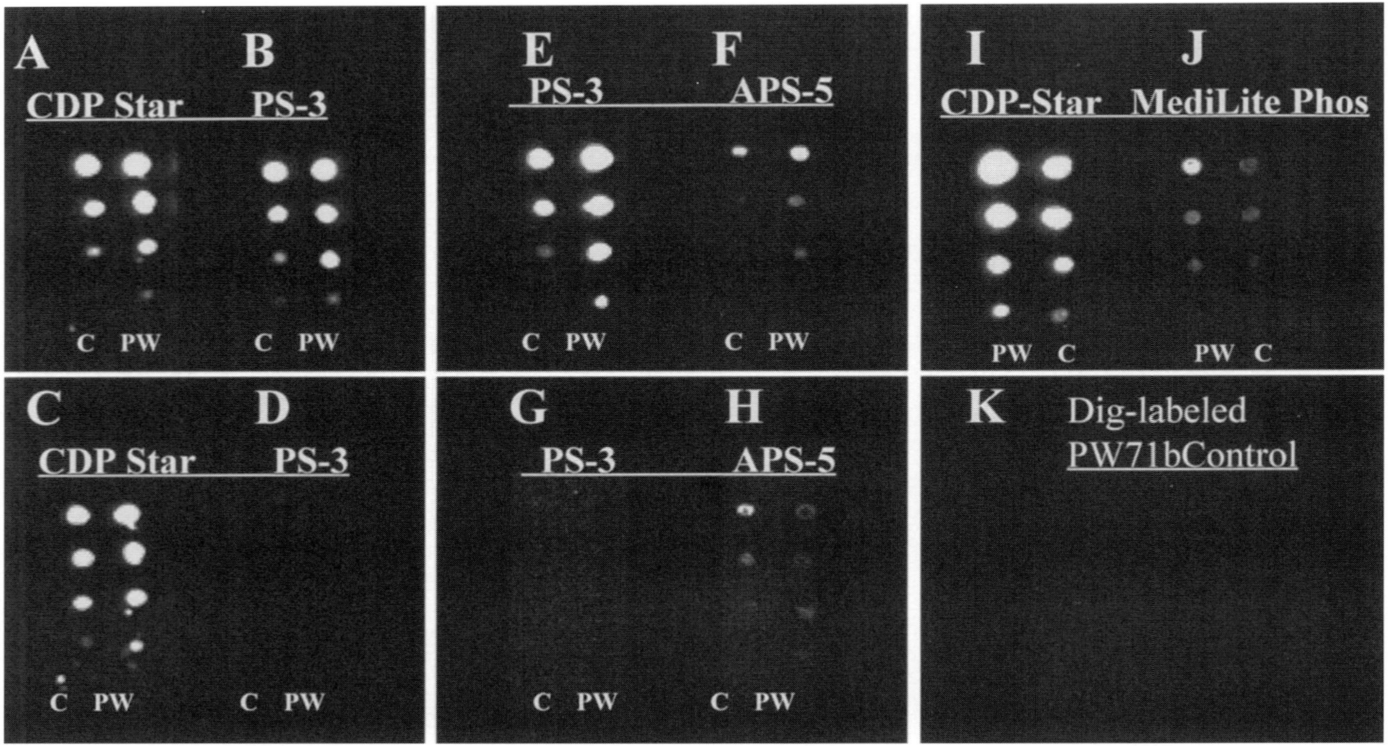

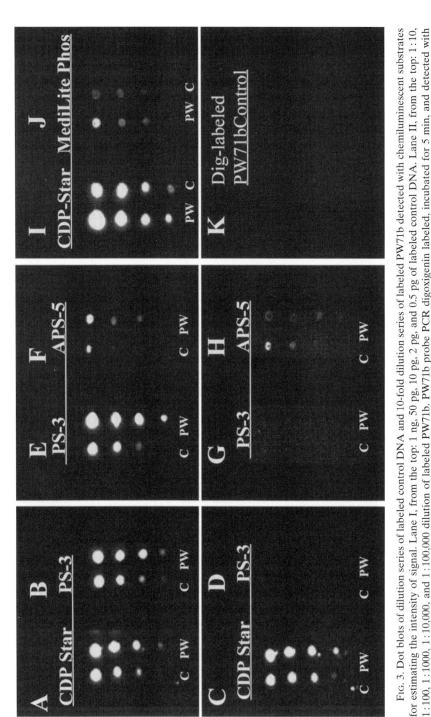

FIG. 3. Dot blots of dilution series of labeled control DNA and 10-fold dilution series of labeled PW71b detected with chemiluminescent substrates for estimating the intensity of signal. Lane I, from the top: 1 ng, 50 pg, 10 pg, 2 pg, and 0.5 pg of labeled control DNA. Lane II, from the top: 1:10, 1:100, 1:1000, 1:10,000, and 1:100,000 dilution of labeled PW71b. PW71b probe PCR digoxigenin labeled, incubated for 5 min, and detected with CDP-*Star* (A) and PS-3 (B) and incubated overnight and detected with CDP-*Star* (C) and PS-3 (D). Dot blots identical to A, but detected with PS-3 (E) and APS-5 (Lumigen) (F) following a 5-min incubation and an overnight incubation (G and H). Dot blots identical to A, but detected by CDP-*Star* (I) and MediLitePhos (J). Control probe alone (K).

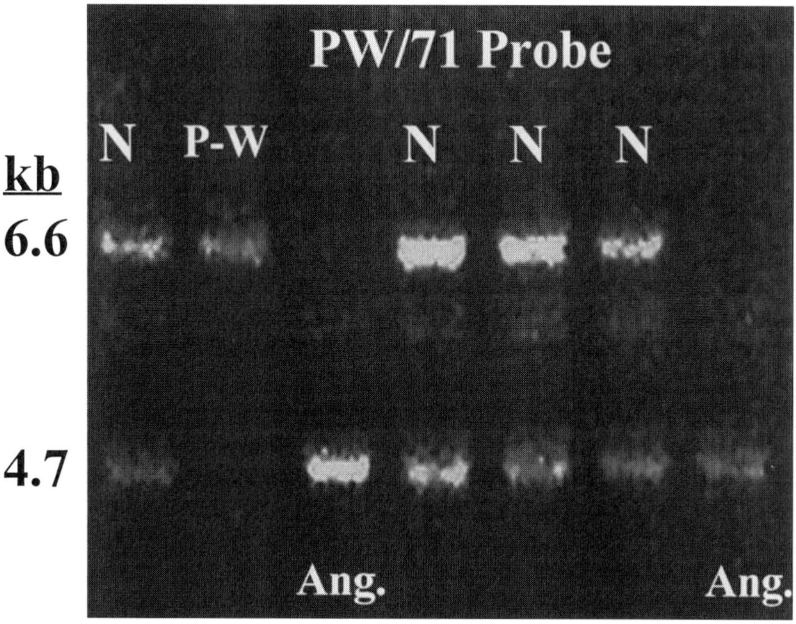

FIG. 4. Chemilumigram of Southern blots of DNA from normal patients (N), patient with Prader–Willi syndrome (PW, lane 2), and patient with Angelman syndrome (Ang., lane 3). The membrane was hybridized with the PW71b probe, which was labeled with digoxigenin during PCR, and detected with anti-Dig/ALP conjugate and CDP-*Star*.

TABLE III
COMPARISON OF TIME REQUIRED FOR SOUTHERN BLOT DETECTION METHODS[a]

	PS-3	CDP-*Star*	^{32}P
Probe labeling	5 min (batch)	PCR (batch)	2 hr (weekly)
Hybridization	17 hr	3 hr	17 hr
Stringency wash	2 hr	1 hr	2 hr
Signal generation	2.5 hr	2.5 hr	—
Exposure	5 min	5 min	1–3 days
Turnaround time	5 days	4 days	7–10 days

[a] Adapted with permission from K. S. Tenner, M. Karst, S. Thibodeau, and D. J. O'Kane, *in* "Bioluminescence and Chemiluminescence: Molecular Reporting with Photons" (J. W. Hastings, L. J. Kricka, and P. E. Stanley, eds.), p. 573. Wiley, Chichester. © John Wiley & Sons Limited.

diagnosed correctly. Two were diagnosed with AS, 1 was diagnosed with PWS, and 9 were normal.

Discussion

Resistance to the conversion of radioactive assay procedures to nonradioactive methods is understandable because the implementation of new procedures usually requires radical modifications and reoptimization of entire protocols. In both the T-cell gene rearrangement and PWS/AS detection procedures we developed, modifications are kept to a minimum and are advantageous.

In the TcR gene rearrangement detection procedure, CL detection is as sensitive as radioactive detection. In addition to the high sensitivity obtained with chemiluminescence detection, turnaround time and laboratory testing efficiency are improved. A comparison of times required for isotopic and chemiluminescence detection methods (Table III) illustrates these time savings. Batch labeling and a long shelf-life of nonisotopic probes result in decreased personnel time spent on the tedious task of probe labeling. High-Prime labeling also saves personnel time because hands-on time required for this method is approximately 5 min as compared to 2 hr each week required for labeling probes with ^{32}P.

The PWS/AS detection procedure decreases the time spent by personnel on probe labeling because of PCR high-volume batch labeling of probes. Hybridization, signal generation, and detection are completed within 1 day, reducing turnaround time further. Biohazards and disposal fees that coincide with isotope use are eliminated. Chemiluminescent detection methods are a convenient, cost effective, and safe alternative to isotopic detection in the Southern blot analysis of PWS/AS and the TcR gene rearrangements.

[33] Quantitative Polymerase Chain Reaction and Solid-Phase Capture Nucleic Acid Detection

By CHRIS S. MARTIN, JOHN C. VOYTA, and IRENA BRONSTEIN

Introduction

Quantitative polymerase chain reaction (PCR) is the process of determining the number of target DNA molecules by correlation to the quantity of amplified product. Quantitation of RNA is possible using reverse

METHODS IN ENZYMOLOGY, VOL. 305
0076-6879/00 $30.00

transcription PCR (RT-PCR) in which the RNA message is converted to the complementary DNA sequence by reverse transcription and then amplified. RT-PCR is commonly used to measure levels of gene expression. Traditional methods for the quantitation of PCR products include gel electrophoresis with fluorescent staining or autoradiography and liquid scintillation counting of excised gel fragments.

This article describes methods for solid-phase capture of PCR products and chemiluminescent detection, which can be performed in tubes or microplates. The PCR product is quantitated by measuring the amount of product bound to a solid support. As shown in Fig. 1, target DNA is amplified using one biotinylated and one unlabeled primer. Subsequently, the PCR product is captured on a streptavidin-coated solid support, and the nonbiotinylated strand is removed by denaturation. An internal nested fluorescein-labeled detection probe is hybridized to the bound biotinylated DNA strand and is detected with an antifluorescein alkaline phosphatase conjugate and CSPD or CDP-*Star* chemiluminescent substrate for alkaline phosphatase. Light emission is then measured with a luminometer. The high sensitivity exhibited by this chemiluminescent detection method enables the accurate quantitation of target DNA because the product can be measured within the exponential range of PCR amplification. The dynamic range of the assay, over three orders of magnitude of PCR product concentration, simplifies the process of determining the number of amplification cycles necessary for the accurate quantitation of target molecules.

The assays described utilize CSPD or CDP-*Star,* chemiluminescent 1,2-dioxetane substrates for alkaline phosphatase.[1,2] Upon enzymatic dephosphorylation, CSPD or CDP-*Star* decomposes, resulting in a prolonged, constant light emission with a maximum at 463 or 461 nm in Sapphire-II enhancer, respectively. Sapphire-II, a polymeric enhancer, improves the signal-to-noise ratio significantly. The light emission from alkaline phosphatase-activated 1,2-dioxetane substrates is in the form of a "glow." Maximum light emission is reached in 10 to 20 min. Constant light emission persists as long as excess substrate is available. Enzymatic dephosphorylation in the presence of excess substrate occurs at a constant rate proportional to enzyme concentration. Upon dephosphorylation of the substrate, the anion produced decomposes with a finite half-life ($t_{1/2}$).

To perform the assay, PCR amplification primers specific for the DNA

[1] I. Bronstein, R. R. Juo, J. C. Voyta, and B. Edwards, *in* "Bioluminescence and Chemiluminescence: Current Status" (P. Stanley and L. J. Kricka, eds), p. 73. Wiley, Chichester, 1991.

[2] B. Edwards, A. Sparks, J. C. Voyta, and I. Bronstein, *in* "Bioluminescence and Chemiluminescence: Fundamentals and Applied Aspects" (A. K. Campbell, L. J. Kricka, and P. E. Stanley, eds.), p. 56. Wiley, Chichester, 1994.

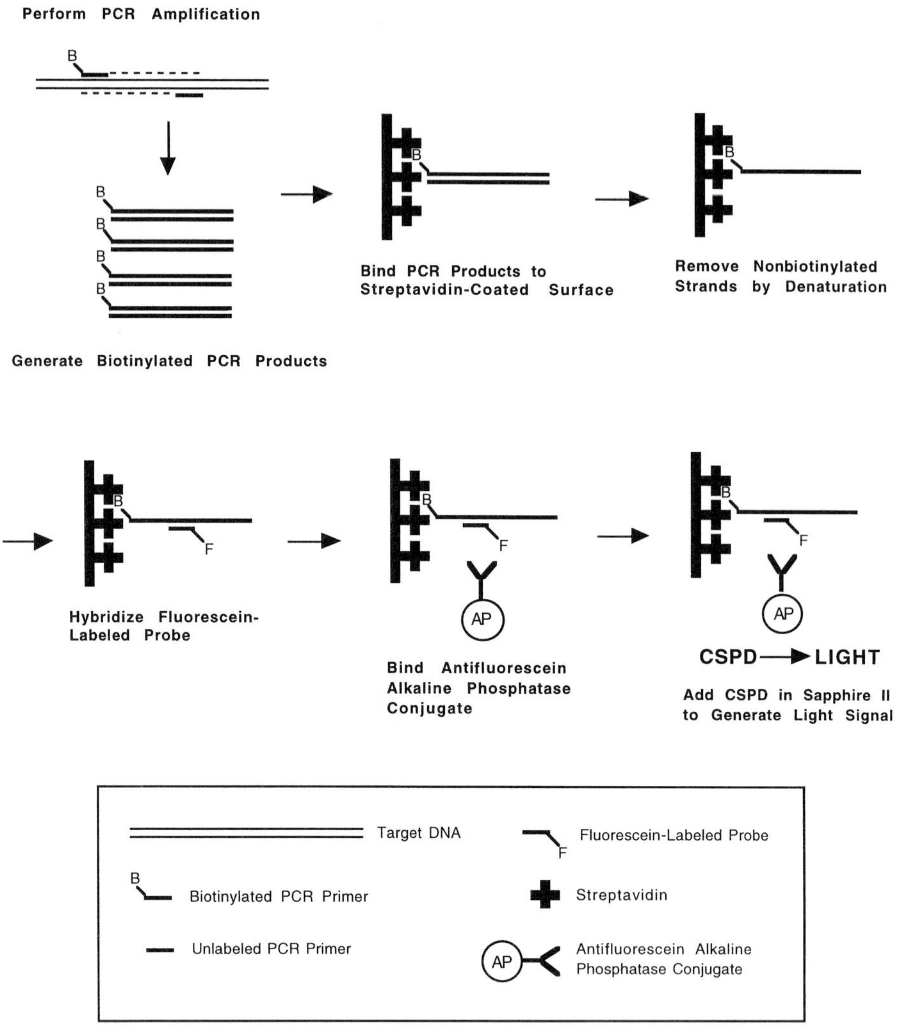

FIG. 1. The PCR-Light assay for the quantitation of PCR products.

target of interest are required. One primer is labeled with biotin at the 5′ end to facilitate solid-phase capture of the PCR product. Biotin-labeled oligonucleotides are available commercially from most suppliers, or biotin phosphoramidities may be used in synthesis. Standard conditions for PCR amplification may be used with biotinylated primers. For this assay protocol, the concentration of biotinylated primer used in each amplification should be limited to 10 pmol because the unincorporated biotinylated primer will compete for biotin-binding sites on the streptavidin support and reduce the capacity of the support to bind the biotinylated PCR product.

For detection of the bound PCR product, a 5′-fluorescein-labeled internal oligonucleotide probe is used in the assay. Probes 18–22 bases in length with a single fluorescein label are recommended. 3′-Labeling of an oligonucleotide with fluorescein is possible using terminal transferase and fluorescein-labeled nucleotides; however, a higher background in the assay can result if long nucleotide tails are added to probes. Alternatively, a single label can be placed on the 3′ end using terminal transferase and a fluorescein-labeled dideoxynucleotide. Optimization of the hybridization and wash temperatures may be necessary with some probes. However, the tetramethylammonium chloride (TMAC) buffer stringency wash eliminates some of the variation in melting temperature exhibited by different oligonucleotides.[3] A non-complementary fluorescein-labeled probe can be used as a negative control.

An initial study should be conducted to determine the number of amplification cycles necessary to be within the assay range for the number of target molecules. For example, if accurate quantitation of a viral copy number in infected cells is desired, cell extracts should be mixed with known quantities of viral target DNA and amplified with a range of amplification cycles, e.g., 18–35. Each amplification reaction is then assayed. For a given number of target molecules, the signal will eventually reach a plateau with an increasing number of cycles. It is important, for accurate quantitation, to assay for PCR products only during the exponential phase for amplification (i.e., linear range). Once these curves have been determined for a range of target molecule concentrations, the number of amplification cycles can be set. If samples containing fewer target molecules are quantitated, more cycles will be necessary and vice versa.[4–11] Alternatively, for highly accurate quantitation utilizing a kinetic approach,[12] a range of cycle numbers for each sample can be assayed.

[3] A. G. DiLella and S. L. C. Woo, *Methods Enzymol.* **152,** 447 (1987).

[4] P. Alard, O. Lantz, M. Sebagh, C. F. Calvo, D. Weill, G. Chavanel, A. Senik, and B. Charpentier, *BioTechniques* **15,** 730 (1993).

[5] J. A. Carcillo, R. A. Parise, and M. Romkes-Sparks, *PCR Methods Appl.* **3,** 292 (1994).

[6] D. Coen, *in* "Current Protocols in Molecular Biology" (F. M. Ausubel *et al.,* eds.), Chapter 15. Wiley, New York, 1989.

[7] F. Ferre, *PCR Methods Appl.* **2,** 1 (1992).

[8] W. C. Gause and J. Adamovicz, *PCR Methods Appl.* **3,** S123 (1994).

[9] U. Reischl, and B. Kochanowski, *Mol. Biotechnol.* **3,** 55 (1995).

[10] A. M. Wang, M. V. Doyle, and D. F. Mark, *Proc. Natl. Acad. Sci. USA* **86,** 9717 (1989).

[11] A. M. Wang, and D. F. Mark, *in* "PCR Protocols: A Guide to Methods and Applications" (M. A. Innis *et al.,* eds.), p. 70. Academic Press, San Diego, 1990.

[12] S. W. Umlauf, B. Beverly, O. Lantz, and R. H. Schwartz, *Mol. Cell. Biol.* **15,** 3197 (1995).

A variety of alternative solid support capture methods for the quantitation of PCR products have been published, including methods utilizing alkaline phosphatase-conjugated oligonucloetide detection probes,[13] oligonucleotide capture probes,[3,14] and incorporation of hapten labels into PCR products, which are subsequently captured.[14,15] All of these methods can be adapted for chemiluminescent detection with simple procedural modifications that should increase sensitivity and dynamic range for these assays.

Required Reagents

Streptavidin/avidin-coated microplates, magnetic beads or polystyrene beads (see protocol III for plate-coating procedure)
5′-Biotinylated PCR primer
Unlabeled PCR primer
Fluorescein-labeled oligonucleotide probe

The following reagents are available from Tropix, Inc. (Bedford, MA) as components of the PCR-Light kit.

Chemiluminescent Detection Reagents

Chemiluminescent substrate: CSPD and CDP-*Star* are supplied as 25 m*M*, 60× concentrates. This stock solution should be diluted in assay buffer prior to use. Store at 4° in the dark.
Blocking reagent (I-Block): Highly purified casein that has been screened for low alkaline phosphatase contamination and is an efficient blocking reagent. Store dry at room temperature.
Antifluorescein alkaline phosphatase conjugate (Fluor*x*-AP): Dilute 1:5000 prior to use. Store at 4°. Do not freeze.
99% diethanolamine (DEA): If material solidifies during storage, warm at 37–65° to melt. Dilute as described to make assay buffer. Store at room temperature.
Sapphire-II: Polymeric enhancer supplied as a 10× concentrate. Store at 4°.

[13] R. J. Cano, S. R. Rasmussen, G. S. Fraga, and J. C. Palomares, *J. Appl. Bacteriol.* **75,** 247 (1993).

[14] M. Shindo, A. M. Di Bisceglie, J. Silver, T. Limjoco, J. H. Hoofnagle, and S. M. Feinestone, *J. Virol. Methods.* **48,** 65 (1994).

[15] I. Psikal, B. Smid, R. Kubalikova, L. Valicek, L. Rodak, and E. Kosinova, *Vet. Microbiol.* **57,** 55 (1997).

Hybridization and Wash Buffers

Hybridization buffer: 0.9 *M* NaCl, 50 m*M* $NaPO_4$ (pH 7.7), 5 m*M* EDTA, 10% 50× Denhardt's solution (1% each of Ficoll type 400, polyvinylpyrrolidone, and BSA Fraction V). Store at 4°.

Wash buffer 1: 0.9 *M* NaCl, 50 m*M* $NaPO_4$ (pH 7.7), 5 m*M* EDTA. Store at room temperature.

Wash Buffer 2: 3 *M* TMAC, 50 m*M* Tris–HCl, pH 8.0, 2 m*M* EDTA, 0.025% Triton X-100. Store at room temperature.

Note: TMAC is highly toxic and solutions should be handled with appropriate safety measures.

Control PCR Product and Probe

Control PCR product: 2 pmol of 246-bp biotin-labeled PCR product, 50 fmol/μl, in 10 m*M* Tris–HCl, pH 7.5, 1 m*M* EDTA. Store at −20°.

Control probe: 50 pmol of 5′-fluorescein-labeled 18-mer oligonucleotide, 1 pmol/μl, in 10 m*M* Tris–HCl, pH 7.5, 1 m*M* EDTA. Store at −20°.

Recipes

Note: Prepare all solutions with sterile, deionized H_2O.

Binding buffer: 10 m*M* Tris–HCl, pH 8.0, 0.5 *M* NaCl, 1 m*M* EDTA

10× phosphate-buffered saline (PBS): 0.58 *M* Na_2HPO_4, 0.17 *M* $NaH_2PO_4 \cdot H_2O$, 0.68 *M* NaCl

Note: Upon dilution, a 1× PBS solution should have a pH of 7.3–7.4.

Blocking buffer: 0.2% I-Block reagent, 1× PBS, 0.1% Tween 20

Add I-Block reagent to 1× PBS and heat to 70° with stirring for 5 min. Do not boil. Add Tween 20. The solution will remain slightly opaque. Cool to room temperature before use.

Detection wash buffer: 1× PBS, 0.1% Tween 20

Assay buffer: 0.1 *M* DEA, pH 10, 1 m*M* $MgCl_2$. Dissolve DEA in 100 ml of deionized H_2O and adjust pH to 10.0 with HCl. Add $MgCl_2$.

Substrate solution: 0.4 m*M* CSPD or CDP-*Star*, 10% Sapphire-II in assay buffer. Prepare immediately before use. Do not store diluted solutions. Always prepare the Enhancer dilution first and then add substrate to prevent precipitation of the chemiluminescent substrate. Alternatively, ready-to-use (RTU) solutions of CSPD or CDP-*Star* in Sapphire-II are available from Tropix.

Protocol Ia: Assay with Streptavidin-Coated Microplates

The following procedure is performed in white streptavidin or avidin-coated microplates or strips. Microplates or strips may be coated with streptavidin or avidin. See protocol III for plate coating. Perform all steps at room temperature except where indicated otherwise. For these steps the plate can be placed in a 55° incubator. Prewarming of the hybridization and wash buffer is not necessary. Shaking is not required for any step. Each PCR reaction should be sampled in triplicate to maximize accuracy.

Capturing PCR Product

1. Add 100 μl 2× PBS to each well and incubate for 10 min to rehydrate the streptavidin.
2. Mix 10 μl of each PCR amplification with 90 μl of binding buffer.
3. Remove 2× PBS solution. Add 100 μl diluted PCR amplification/well and incubate for 60 min.
4. Remove diluted PCR product. Add 100 μl 0.25 *N* NaOH and incubate for 10 min.

Hybridization of Fluorescein-Labeled Probe

1. Remove NaOH and wash with 100 μl hybridization buffer.
2. Dilute fluorescein-labeled probe (0.5 pmol/100 μl hybridization buffer). Add 100 μl to each well and incubate at 55° for 15 min. Covering the plate will avoid evaporation.
3. Wash twice with wash buffer 1.
4. Wash once at room temperature with wash buffer 2.
5. Wash twice at 55° for 5 min with 100 μl of wash buffer 2.

Chemiluminescent Detection

1. Incubate with 100 μl blocking buffer for 10 min.
2. Dilute Fluor*x*-AP conjugate 1:5000 in blocking buffer.
3. Remove blocking buffer. Add 100 μl of diluted Fluor*x*-AP and incubate for 60 min.
4. Wash three times with detection wash buffer.
5. Wash twice with assay buffer.
6. Incubate with 100 μl substrate solution for at least 10 min and read on plate luminometer.

Protocol Ib. Assay with Streptavidin-Coated Polystyrene Beads

The following procedure is performed with single beads in luminometer tubes. Perform all steps with shaking (140–170 rpm). Liquids should be removed from the tubes by aspiration. Perform all steps at room temperature except where indicated. For these steps, the tubes can be placed in a

55° incubator with shaking. Prewarming of the hybridization and wash buffer is not necessary. Each PCR reaction should be sampled in triplicate to maximize accuracy. Because of the accumulation of materials on the inside surface of the tube during the assay, transferring the beads to clean tubes after the first assay buffer wash will eliminate the background signal from this source.

Capturing PCR Product

1. Add 500 μl 2× PBS to one bead in each tube and incubate for 10 min to rehydrate the streptavidin.
2. Mix 10 μl of each PCR amplification to 490 μl of binding buffer.
3. Remove 2× PBS solution. Add 500 μl of each diluted PCR amplification to each tube and incubate for 60 min.
4. Remove diluted PCR product. Add 500 μl 0.25 *N* NaOH and incubate for 10 min.

Hybridization of Fluorescein-Labeled Probe

1. Remove NaOH and wash with 500 μl hybridization buffer.
2. Dilute fluorescein-labeled probe (0.5 pmol/100 μl hybridization buffer). Add 500 μl to each tube. Incubate at 55° for 15 min.
3. Wash twice with 500 μl wash buffer 1.
4. Wash once at room temperature with 500 μl wash buffer 2.
5. Wash twice at 55° for 5 min with 500 μl wash buffer 2.
6. Wash one time with excess 1× PBS to remove crystallized salt from inside of tube.

Chemiluminescent Detection

1. Incubate in 500 μl of blocking buffer for 10 min.
2. Dilute Fluor*x*-AP conjugate 1:5000 in blocking buffer.
3. Remove blocking buffer. Add 500 μl of diluted Fluor*x*-AP and incubate for 60 min.
4. Wash three times with 500 μl detection wash buffer.
5. Wash twice with 500 μl assay buffer.

Note: After the first assay buffer wash, the bead can be placed in a clean tube to reduce the background signal.

6. Incubate with 500 μl substrate solution for 10 min and read on a luminometer.

Protocol Ic. Assay with Streptavidin-Coated Paramagnetic Particles

The following procedure was developed for use with Dynal M-280 streptavidin-coated magnetic beads (Oslo, Norway) in 1.5-ml microcentrifuge tubes. A magnetic separator, available from Dynal, is required. Perform all steps at room temperature except where indicated. For these steps, the tubes can be placed in a 55° incubator *with* shaking or rocking. Prewarming of the hybridization and wash buffer is not necessary. Each PCR reaction should be sampled in triplicate for greater accuracy.

Capturing PCR Product

1. Add 50 μg of magnetic beads to each tube and place in magnetic separator. Remove liquid from beads.
2. Wash beads twice in 40 μl binding buffer. Resuspend in 20 μl binding buffer.
3. Add 10 μl PCR amplification to 40 μl binding buffer.
4. Add 50 μl diluted PCR amplification to each tube and incubate for 30 min. Separate and remove liquid.
5. Add 30 μl of 0.1 *N* NaOH and incubate for 5 min. Wash with 50 μl of 0.1 *N* NaOH.

Hybridization of Fluorescein-Labeled Probe

1. Wash beads with 100 μl hybridization buffer.
2. Add fluorescein-labeled probe (0.5 pmol/100 μl hybridization buffer) in 50 μl to each microcentrifuge tube. Incubate at 55° for 15 min with rocking or shaking.
3. Wash twice with 100 μl wash buffer 1.
4. Wash once at room temperature with 100 μl wash buffer 2.
5. Wash twice for 5 min at 55° with 100 μl wash buffer 2 with rocking or shaking.

Chemiluminescent Detection

1. Incubate in 100 μl blocking buffer for 10 min with rocking or shaking.
2. Dilute Fluor*x*-AP conjugate 1:5000 in blocking buffer.
3. Remove blocking buffer. Add 100 μl of diluted Fluor*x*-AP and incubate for 60 min with rocking or shaking.
4. Wash three times with 100 μl detection wash buffer.
5. Wash twice with 100 μl assay buffer.
6. Resuspend beads in 10 μl assay buffer and transfer to luminometer tubes.

7. Incubate with 100 μl substrate solution for 10 minutes. Read on a luminometer.

Protocol II. Control Assay

To demonstrate assay linearity and dynamic range, an assay should be performed on a dilution series of a biotinylated PCR product. Ideally, this can be accomplished by amplifying and preparing a sufficient quantity of PCR product using a plasmid clone of the target and quantifying by UV absorbance. It is important to remove excess primers by gel filtration chromatography or by a spin column before measuring absorbance. Alternatively, a control biotinylated PCR product and fluorescein-labeled probe could be used (available from Tropix). We recommend diluting the control product from 300 to 0.2 fmol with successive 1:3 dilutions in the binding buffer and assaying each in triplicate. The diluted product is added to the solid support (100 μl/well for plate assays, 500 μl/tube for polystyrene bead assays, or 50 μl/tube for paramagnetic bead assays) and is assayed as described earlier for the appropriate streptavidin-coated solid support. As recommended for all probes, use 0.5 pmol/100 μl of hybridization buffer when diluting the control fluorescein-labeled probe. It is important to include a control that contains no PCR product to provide the background signal for the assay.

Protocol III. Coating Microplates with Avidin/Streptavidin

Prepare a solution of avidin (Sigma) at 100 μg/ml in 0.1 *M*, pH 9.6, Na_2CO_3 buffer. Add 100 μl to each well and incubate at 37° for 2 hr. Wash three times with 1× PBS, 1% Tween 20. Add 300 μl 1% I-Block in 0.1 *M*, pH 9.6, $NaCO_3$ buffer and incubate for 2 hr at 37°. The plates can be stored for several weeks at −20° with the I-Block solution. Before use, plates should be washed twice with 1× PBS, 1% Tween 20.

Summary

The combination of PCR amplification and chemiluminescent detection of PCR products provides a highly sensitive system for the quantitation of DNA and RNA. The broad dynamic range of the chemiluminescent detection assay simplifies the selection of cycling and concentration parameters critical to harnessing the quantitative aspects of PCR amplification. Detection of 200 amol of PCR product is attained using the described procedures.[16] The tube or microplate format of the assay avoids many of the limitations associated with other methods of PCR quantitation involving gel electrophoresis. This detection methodology can be applied to a variety of quantitative nucleic acid assays, including viral load and gene expression analysis.

[16] C. S. Martin, L. Butler, and I. Bronstein, *BioTechniques* **18,** 908 (1995).

Section VI

Luminescence Monitoring *in Vivo*

[34] Targeted Bioluminescent Indicators in Living Cells

By GRACIELA B. SALA-NEWBY, MICHAEL N. BADMINTON, W. HOWARD EVANS, CHRISTOPHER H. GEORGE, HELEN E. JONES, JONATHAN M. KENDALL, ANGELA R. RIBEIRO, and ANTHONY K. CAMPBELL

Introduction

Cell signaling plays a central role in determining events such as cell movement, secretion, division, transformation, development, defense, and death in all pro- and eukaryotic cells.[1-3] External stimuli or pathogens initiate an intracellular signal that than leads to covalent modification and translocation of proteins. A key issue is how these signaling pathways determine the threshold for an end response.[3] In order to unravel the mechanisms responsible for cell signaling it is necessary to measure and manipulate the initial signal, together with the components that follow, in defined compartments of live cells. The cloning and engineering of bioluminescent proteins enable this to be achieved for the first time. Bioluminescent indicators are superior to fluors for quantifying global cell responses temporally in large numbers of individual cells, whole organs, and intact organisms. Imaging is possible, provided great attention is paid to the capturing of the maximum number of photons by the detector. Modern fiber optics enable this to be achieved. Fluors are superior to bioluminescent indicators for the spaciotemporal imaging of individual optical sections of single cells using confocal microscopy.

Intracellular Ca^{2+} is a universal signal in all animal and plant cells and also appears to play a role as a signal in many bacteria.[1,4] The main source of the Ca^{2+} signal may be from outside the cell or from internal stores, but in most cases both internal Ca^{2+} from the endoplasmic reticulum (ER) and external Ca^{2+} are required to maintain a full cell response. There are two key questions: (1) How is a global Ca^{2+} signal initiated and then propagated

[1] A. K. Campbell, "Intracellular Calcium: Its Universal Role as Regulator." Wiley, Chichester, 1983.

[2] A. K. Campbell, *in* "Chemiluminescence: Principles and Applications in Biology and Medicine." Horwood/VCH, Chichester, 1988.

[3] A. K. Campbell, "Rubicon: the Fifth Dimension of Biology." Duckworth, London, 1994. Available from www.uwcm.ac.uk/uwcm/mb/ISCCG.

[4] V. Norris, S. Grant, P. Freestone, J. Canvin, T. Sheikh, I. Toth, M. Trinei, K. Modha, and R. I. Norman, *J. Bacteriol,* **178,** 3677 (1996).

0076-6879/00 $30.00

to produce waves or oscillations? (2) How does Ca^{2+} act to cause the end response and, in particular, how does it trigger processes in other subcellular compartments? We have focused on the role of Ca^{2+} loss from the ER as a central signal in controlling the intranet of the cell. The loss of ER Ca^{2+} initiates the opening of Ca^{2+} channels in the plasma membrane (SOC), degradation of ER proteins, and activation of calreticulin and chaperone gene expression. We have also investigated whether there is a Ca^{2+} barrier across the nuclear–cytosol membrane and whether there could be independent regulation of Ca^{2+} in the nucleus. To answer these questions, methods are required to measure and image Ca^{2+} in the cytosol, ER, nucleus, mitochondria, and at the inner *surface* of the plasma membrane. Targeted aequorin enables this to be achieved.[5–14]

Recombinant Ca^{2+}-activated photoprotein aequorin has given us the opportunity to develop techniques based on the genetic transformation of bacteria, yeast, plants, and animal cells. Particularly exciting has been the development of methods to measure $[Ca^{2+}]$ in discrete subcellular domains of living cells, including the mitochondria, nucleus, endoplasmic reticulum, and plasma membrane. This has been possible because the mechanisms that control the targeting of proteins to particular subcellular organelles have been uncovered. Targeting relies on the recognition of short sequences of amino acids either at the termini or within the protein sequence. The expression of aequorin inside live bacteria enables the role of Ca^{2+} as a potential signaling molecule in prokaryotes to be investigated properly for the first time.[15,16] Fluors such as fura2AM do not load well into prokaryotes.

ATP is the main energy currency of the cell. However, ATP itself and its metabolites ADP and AMP perform regulatory roles in cellular

[5] J. M. Kendall, R. L. Dormer, and A. K. Campbell, *Biochem. Biophys. Res. Commun.* **189,** 1008 (1992).

[6] J. M. Kendall, G. Sala-Newby, V. Ghalaut, R. L. Dormer, and A. K. Campbell, *Biochem. Biophys. Res. Commun.* **187,** 1091 (1992).

[7] R. Rizzuto, A. W. M. Simpson, M. Brini, and T. Pozzan, *Nature* **358,** 325 (1992).

[8] M. Brini, M. Murgi, L. Pasti, D. Picard, T. Pozzan, and R. Rizzuto. *EMBO J.* **12,** 4813 (1993).

[9] R. Rizzuto, M. Brini, M. Murgia, and T. Pozzan, *Science* **262,** 744 (1993).

[10] R. Rizzuto, P. Pinton, W. Carrington, F. S. Fay, K. E. Fogarty, L. M. Lifshitz, R. A. Tuft, and T. Pozzan, *Science* **280,** 1763 (1998).

[11] M. N. Badminton, J. M. Kendall, G. N. Sala-Newby, and A. K. Campbell, *Exp. Cell. Res.* **216,** 236 (1995).

[12] M. N. Badminton, A. K. Campbell, and C. M. Rembold, *J. Biol. Chem.* **271,** 31210 (1996).

[13] J. M. Kendall, M. N. Badminton, G. B. Sala-Newby, G. W. G. Wilkinson, and A. K. Campbell, *Cell Calcium* **19,** 133 (1996).

[14] P. E. M. Martin, C. H. George, C. Castro, J. M. Kendall, J. Capel, A. K. Campbell, A. Revilla, L. C. Barrio, and W. H. Evans, *J. Biol. Chem.* **273,** 1719 (1998).

[15] N. J. Watkins, M. R. Knight, A. J. Trewavas, and A. K. Campbell, *Biochem. J.* **306,** 865 (1995).

[16] H. E. Jones, I. B. Holland, and A. K. Campbell, *Cell Calcium* **25,** 265 (1999).

metabolism and can affect ion channels at the plasma membrane. Hence, measuring and locating changes in ATP concentration in living cells are important. DNA synthesis and chaperones require ATP. Thus it is important to measure ATP in subcellular compartments. Firefly luciferase catalyzes the oxidative decarboxylation of a luciferin in the presence of Mg-ATP. It has been used extensively to assess biomass. The enzyme is located in the peroxisomes of the insect light organ and contains the targeting signal at the C terminus comprising the last three amino acids (SKL). Its removal did not affect activity and the luciferase distributed uniformly throughout the cytosol.[17]

Principles of Methods

Two different approaches have been employed to target aequorin to subcellular organelles. These involve either the generation of a chimera with a protein whose localization has been demonstrated unequivocally or the addition of minimal targeting sequences to the protein. Proteins and peptides can be engineered onto both N and C termini of firefly luciferase without damaging biological activity. However, we have shown that the C-terminal proline of aequorin plays a critical role in the stability of the photoprotein. As a result, engineering peptides such as KDEL or proteins on the C terminus of aequorin result in an increased Ca^{2+}-independent light emission and decreased measured specific activity.[5,6,18,19]

Endoplasmic Reticulum

Two strategies have been used to target aequorin to the ER. (a) A calreticulin signal peptide at the N terminus and the KDEL retention sequence at the C terminus. This results in aequorin being targeted to the lumen of the ER. Without KDEL, 90% of the expressed protein appears in the extracellular fluid.[5] (b) HLA class 2–aequorin chimera, which targets to the membrane of the ER and does not require the KDEL retention sequence. Because of the high level of free Ca^{2+} in the ER, it is necessary to adapt aequorin as a luciferase[18] in order to follow free Ca^{2+} in the ER for several minutes. A low-affinity mutant with a 10-fold reduction in apparent affinity may allow aequorin to be used as a photoprotein at μM free Ca^{2+}.[6]

[17] G. B. Salsa-Newby and A. K. Campbell, *FEBS Lett.* **307,** 241 (1992).

[18] J. M. Kendall, M. N. Badminton, G. Sala-Newby, A. K. Campbell, and C. R. Rembold, *Biochem J.* **318,** 383 (1996).

[19] N. J. Watkins and A. K. Campbell, *Biochem. J.* **292,** 181 (1993).

Nucleus

Proteins destined for the cell nucleus are transported through the nuclear pore by a specific, ATP-dependent process that recognizes a nuclear localization signal (NLS) within the protein sequence. Nucleoplasmin is a soluble nuclear protein from *Xenopus laevis* that mediates the assembly of nucleosomes and is found throughout the nucleus and nucleolus. It is directed to the nucleus by a bipartite sequence consisting of two short basic sequences separated by 10 spacer residues.[11] Our strategy was to construct a nucleoplasmin–aequorin chimera.

Plasma Membrane

The minimal sequence requirements for the recruitment of proteins to the plasma membrane are still not completely known. As a consequence, targeting of proteins to the plasma membrane involves the construction of chimeric proteins with an integral plasma membrane protein or the addition of a protein acylation motif that direct protein–membrane interactions, causing the association of hydrophilic proteins with the cytoplasmic face of the plasma membrane. This particular feature may confer a real advantage as it avoids the complexities associated with trafficking through the secretory pathway. The fusion targeting strategy will be exemplified by the construction of a chimera between aequorin and the integral membrane protein connexin 43 (Cx43),[14] and we have engineered the N-terminal sequence of Lck (Lck-N): MGCVCSSNPE in which Gly2 is myristoylated and the cysteines become post-translationally *S*-palmitoylated to N termini of the firefly luciferase-aequorin chimera.

In order to monitor changes in free cytosolic Ca^{2+} inside the live bacteria,[15,16] the aequorin gene has been inserted into a multihost-range expression vector that is functional in any *Escherichia coli* strain, including those defective in various aspects of their ability to handle calcium. A β-lactamase chimera appears to target aequorin to the periplasmic space.

We have used one- or two-step PCR to engineer the targeted luminescent proteins (Fig. 1).

Materials and Reagents

Plasmids and Bacterial Strains

pLAC18 is used as the template for luciferase-aequorin.[13] pSVAEQ or pNSA3 carry the aequorin cDNA corresponding to different isoforms (sequence available on request), pSVB6 contains a mutated aequorin cDNA with lower Ca^{2+} affinity[6] and pA10,6.1, our own firefly luciferase, cDNA for bioluminescent/targeting partners

↓

PCR (1 or 2 stages)

↓

cDNA for targeted indicator (*in vitro* transcription/translation)

↓

ligation

↓

transformation of *E.coli*

↓

Selection of insert-containing clones (check for orientation)

↓

(*in vitro* transcription/translation of plasmid from clone)

↓

transfection of eukaryotic cells (check for orientation)

↓

Assessment of targeting efficiency (immunofluorescence, subcellular fractionation)

↓

Luminescence measurements/imaging

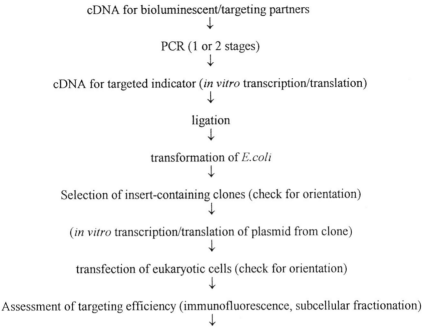

FIG. 1. Steps in the generation of targeted indicators.

cDNA is available on request. Several of our constructs are also available from Molecular Probes. High-efficiency competent cells used were (Promega), JM109, TOP10F[17] (Invitrogen), or DH5α (GIBCO-BRL).

Cloning Vectors

The fastest methods to clone PCR products take advantage of the fact that *Taq* polymerase and also BIO-X-ACT polymerase among the ones with proofreading capability add 1 to 3 dAs to the 3′ends of PCR products. Vectors have been developed that are supplied linearized with a single 3′T overhang to allow the ligation of PCR products. We recommend transfer vectors such as pCR2.1 (Invitrogen) and pGEM-T (Promega), which include at least one viral promoter (T7 or SP6), and a multiple cloning site within the coding sequence for the enzyme β-galactosidase, allowing the identification of inserts containing bacterial clones (ampicillin resistant) by their white color on indicator plates (the enzyme that generates a blue product becomes inactivated by the insertion). For eukaryotic expression, we recommend vectors such as pTargeT (Promega), which include all the features, mentioned previously plus the CMV IE promoter for constitutive expression, neomycin for selection in cell culture, and pCR 3.1 (Invitrogen) that lacks the blue/white selection. The presence of a viral promoter is useful for sequencing, for the characterization of the protein coded by the inserted cDNA using *in vitro* transcription–translation, and for the determination of the orientation of the inserts with respect to the eukaryotic promoter. Replication-deficient adenovirus vectors (RAd) have many advantages over recombinant plasmids. These vectors are highly efficient and flexible systems for gene transfer, both *in vivo* and *in vitro.*

When a vector with other features is desirable, e.g., presence of antigenic tags (myc or V5 epitope) for immunolocalization, the transfer vectors mentioned earlier should allow the excision of the insert using the restriction sites present or others that can be engineered. However, it should be remembered that C-terminal additions to aequorin will affect its activity. For bacterial expression, we have used the multihost plasmid pMMB66EH,[16] based on a ColE1 replicon and carrying the *tac* promoter, a functional hybrid between *lac* and *trp* promoters.[20]

Oligonucleotides

Oligonucleotide primers are prepared using a DNA synthesizer (Beckman Instruments or Applied Biosystems) and are purified as "trityl off" oligonucleotides. They are designed when possible to contain a Kozak sequence at the 5′ end of the cDNA (AXXATGA/G) to bring the first methionine into the best context to start translation.

LaseF: GGTAAAATGGAAGACGCCAAAAAC
LaseaeqR: TGATGTAAGCTTGACCTTTCCGCCCTTCTT
LaseaeqF: AAGAAGGGCGGAAAGGTCAAGCTTACATCA
AeqR: CCCATC**AGATCT**TTAGGGGACAGCTCC
LaseR: CTGCTT**GAGCTCGTCGA**CTTACTTTCCGCCCTTCTTG
AeqcallF:
GTGCCGCTGCTGCTCGGCCTGCTCGGCCTGGCCGC
*CGCC*GCCGTCAAGCTTACATCA
Aeqcal2F: CACCTAATACGACTCACTATAGGGAGA
ATGCTGCTCCCTGTGCCGCTGCTGCTC
AeqKDELR: **GTCGAC**TTACAGCTCATCCTTGGGGACAGCTC
CACCGTA

[20] J. P. Furste, W. Pansegrau, R. Frank, H. Blocker, P. Scholz, M. Bagdasarian, and E. Lanka, *Gene* **48,** 119 (1986).

NP1F: CAGCTAATACGACTCACTATAGGGAGAATGGCCTC TACAGTG
NP3F: *CCGGCTGCTAAGAAG*GTCAAGCTTACATCA
NP4R: TGATGTAAGCTTGAC*CTTCTTAGCAGCCGG*
C43F: CAGCGATTTAGGTGACACTATAGAGA**TCTAGA** *ATGGGTGACTGGAGT*
C43R: TGATGTAAGCTTGAC*AATCTCCAGGTCATC*
AeqC43F: *GATGACCTGGAGATT*GTCAAGCTTACATCA
Lck-LF: **AGATCT***ATGGGCTGTGTCTGCAGCTCAAACCCG* GAAGACGCCAAAAAC
BacaeqF: CGC**GTCGAC**ATGGTCAAGCTTACATCAGA CTTCGAC
Aeqio: CTCCTTGAGCTCGTCGACTTAGGGGACAGCTCCAC
BacaeqR: GCG**CTGCAG**TTAGGGGACAGCTCCACC
T7 primer (for sequencing): TAATACGACTCACTATAGGG
SP6 primer (for sequencing): TATTTAGGTGACACTATAG

Underlined bases are bases in or complementary to aequorin or luciferase sequence; bold represents restriction sites; and italics represent targeting sequences.

Buffers, Solutions, and Media

For aequorin measurements the reconstitution buffer for apoaequorin is 10 m*M* Tris–HCl, pH 7.5, 500 m*M* NaCl, 1 m*M* EDTA, 5 m*M* 2-mercaptoethanol, 0.1% gelatin, and 1–5 μM coelenterazine; for total aequorin bioluminescence measurement in cells or extract the buffer is 200 m*M* Tris–HCl, 0.5 m*M* EDTA, pH 7.4 and calcium-Nonidet buffer (to measure total remaining chemiluminescence in measurements for rate constants): 50 m*M* $CaCl_2$, 2% Triton X-100.

For luciferase (in extracts), a typical 2× assay buffer is 20 m*M* Tris–acetate buffer, pH 7.75, 0.3 m*M* dithiothreitol, 0.2 m*M* EDTA, 0.1% bovine serum albumin (BSA), 20 m*M* magnesium acetate, 1 m*M* ATP, 0.2 m*M* luciferin, and 8 μM sodium pyrophosphate.

Eukaryotic cells are cultured in DMEM supplemented with 10% fetal calf serum, 2 m*M* glutamine, 100 μg/ml streptomycin, 50 μg/ml carbenicillin, and 2.5 μg/ml amphotericin.

Lysis buffer (used to extract apoaequorin from tissue culture cells): 20 m*M* Tris–HCl, 0.5 m*M* EDTA, 5 m*M* 2-mercaptoethanol, pH 7.4 (for luciferase extracts replace Tris acetate at pH 7.75).

Modified Krebs-Ringer HEPES buffer (KRH): 120 m*M* NaCl, 4.8 m*M* KCl, 1.2 m*M* KH_2PO_4, 1.2 m*M* $MgSO_4$, 1.3 m*M* $CaCl_2$, 25 m*M* HEPES adjusted to pH 7.4 at 37° with 2 *N* NaOH. When stated, KRH is supplemented with 5.5 m*M* glucose.

KRH minus Ca^{2+}: 1 m*M* EGTA in place of 1.3 m*M* $CaCl_2$.

Phosphate-buffered saline (PBS): 10 m*M* phosphate, 138 m*M* NaCl, 2.7 m*M* KCl, pH 7.4.

TAE (Tris-Acetate-EDTA): used for running agarose gels diluted from a 50× concentration stock containing 242 g Tris base, 57.1 ml glacial acetic acid, 100 ml 0.5 *M* EDTA (pH 8.0) made up to 1 liter pH 8.3.

TE: 10 m*M* Tris–HCl, 1 m*M* EDTA at pH 8 is used for resuspending DNA and pH 7.5 for resuspending RNA.

Measurement of Light Emission

Chemiluminescence is measured routinely in a home-built chemiluminometer or a Berthold Biolumat LB9500T luminometer. Light emission from cells is imaged[21] using an intensified CCD camera (Photek). Two cameras are available: 50 or 60 Hz and 385 × 288 or 512 × 512 pixels. The sensitivity range can be adjusted over five orders of light intensity by gating the photocathode voltage. This is necessary when comparing low-light emission from intact cells with the total aequorin, when the light intensity may saturate the detector.

Reagents

Coelenterazine is purchased from either Molecular Probes (Eugene, OR) or Prolume Inc. (California) and D-luciferin-free acid and FuGENE are from Boehringer Mannheim. Tissue culture reagents and all other chemicals of AnalaR grade are from Sigma (UK). Restriction enzymes and T4 DNA ligase are purchased from Boehringer Mannheim, Promega, or GIBCO-BRL. *Taq* DNA polymerase and coupled transcription translation systems (TNT) are from Promega; BIO-X-ACT DNA polymerase is from Bioline (London, UK); dNTPs are obtained from Pharmacia; high-quality grade agarose is from Bio-Rad; Qiaquick gel extraction kit is from Qiagen; and Lipofectamine is from GIBCO-BRL.

Methods

PCR Protocols

Tables I and II detail the targeting strategies used. The section describes those oligonucleotides used for the different procedures carried out. The 5′ forward primer (F) is named first, bases complementary to the luminescent protein cDNA are underlined, whereas those complementary to targeting

[21] A. K. Campbell, A. J. Trewavas, and M. R. Knight, *Cell Calcium* **19,** 211 (1996).

TABLE I
TARGETING STRATEGIES

Subcellular compartment	Luminescent protein	Strategy	Method
Cytosol	Aequorin	Firefly luciferase fusion	2 × PCR F
	Firefly luciferase	Removal of C-terminal SKL	1 × PCR
ER	Aequorin	N-terminal calreticulin signal peptide plus C-terminal KDEL	2 × PCR E
		Invariant chain HLA class 2 N-terminal fragment	2 × PCR F
Nucleus	Aequorin	Nucleoplasmin fusion	2 × PCR F
Plasma membrane	Aequorin	Connexin 43 fusion	2 × PCR F
	Luciferase–aequorin	N-terminal dual acylation motif from LckN addition	1 × PCR

TABLE II
PCR PROTOCOLS

Oligonucleotides	Design to produce	Targeted to	From template
LaseF and LaseaeqR	1 × PCR fragment of luciferase = 1	Cytosol	pA10.6.1
LaseaeqF and AeqR	1 × PCR fragment of aequorin = 2		Aeq cDNA
LaseF and AeqR	Luciferase-Aequorin		Fragments 1 and 2
LaseF and LaseR	Luciferase	Cytosol	pA10.6
Aeqcal1F and AeqR	Calreticulin signal peptide Aeq = 3	Endoplasmic reticulum	Fragment 3
Aeqcal2F and AeqKDELR	Calreticulin signal peptide Aeq-KDEL		
NP1F and NP4R	1 × PCR fragment of nucleoplasmin = 4	Nucleus	Nucleoplasmin cDNA
NP3F and AeqR	1 × PCR fragment of Aequorin = 5		Aeq cDNA
NP1F and AeqR	Nucleoplasmin-aequorin		Fragments 4 and 5
C43F and C43R	1 × PCR fragment of connexin = 6	Plasma membrane	Cx43
AeqC43F and AeqR	1 × PCR fragment of aequorin = 7		Aeq cDNA
C43F and AeqR	Connexin-aequorin		Fragments 6 and 7
Lck-NLF and LaseR	Luciferase-Aequorin	Plasma membrane	pLAC18
BacaeqF and BacaeqR	Aequorin	Bacteria (pMMB66EH)	pSVAEQ

partners are in italics (the overlaps should not be less than 12 bases and are usually 15 to 20 bases), and restriction sites are shown in bold. Clamps are included if the PCR product is to be cut by restriction enzymes, but the use of TA cloning now makes them redundant. Coding sequences for peptides and amino acid codons are obtained by reverse translation using the most probable human codons.

One-Stage PCR Protocol (1 × PCR)

When the length of the oligonucleotide primers required is less than 60 bp, a single PCR reaction is required, i.e., addition or removal of small targeting sequences at either termini. The forward oligonucleotide primers should include at its 3′ end an overlapping sequence to the 5′ end of the luminescent cDNA for additions and the overlapping area should leave behind any coding sequence to be deleted. The reverse oligonucleotide primers (R) should contain the complementary sequence of amino acids to be added plus the bases overlapping to the 3′ end of the luminescent cDNA and should avoid any coding sequence to be deleted. Plasmid containing the cDNA for the luminescent protein is amplified as follows: 0.2–2 ng/μl of DNA and 0.5 pmol of each primer/μl, 200 μM of each dNTP, 2 mM $MgCl_2$ is usually optimum, and thermostable DNA polymerase and activity buffer is according to the manufacturers instructions. Each mixture is overlaid with 50 μl of light mineral oil. The cycling reactions are carried out in a Perkin-Elmer thermal cycler. An initial 2 min at 95° is followed by 25 cycles of 1 min at 94° (denaturation) and 1 min at 55° (for annealing of primers, this temperature should be as high as possible). However, if a lower annealing temperature is required at the beginning, the annealing temperature can be increased after a few cycles: 2 min at 72° plus a 5-sec extension on each cycle (polymerization), followed by one cycle of 10 min at 72° to allow *Taq* DNA polymerase to complete chains. The cycling conditions for BIO-X-ACT DNA polymerase are slightly different and the manufacturers conditions should be followed. The products are characterized by agarose gel electrophoresis.

Two-Stage PCR Protocol

The first-stage PCR generates fragments that need to be polished. The Klenow fragment of *E. coli* DNA polymerase (2 U/50 μl PCR) is added at the end of the polymerization cycles to fill in unfinished chains and to remove 3′dA additions (37°, 30 min). Oligonucleotide primers must be removed by ultrafiltration through Centricon 100 cartridges or by using resin-based systems such as Qiagen PCR preps following the manufacturers instructions. Avoid elution of DNA bands from agarose gels if possible as this may lead to unsuccessful PCR.

Protocol to Generate a Chimera: (2 × PCR F). Two fragments are generated in the first stage. They will have a complementary region added by the inside primers. They are mixed in equimolar amounts for second-stage PCR, and polymerase is added in the presence of external primers. A forward primer (F) hybridizes to the 5′ end of the targeting protein cDNA, and a reverse primer (R) hybridizes to the 3′ end of aequorin cDNA. The second stage of the PCR begins with a cycle to allow the overlapping sequences to anneal. This is then followed by a number of amplification cycles. DNA (0.2 μg in total) is denatured (1 min at 94°), allowed to reanneal by decreasing the temperature at a rate of 5.7°/min to 37° in 50 μl of amplification mixture containing the external primers, and extended for 1 min at 72° to generate DNA that is the sum of the two fragments. This fragment is prepared by 10 to 15 cycles of amplification under the conditions described for 1 × PCR.

Protocol to Add Upstream Sequences: (2 × PCR E). When the length of the sequence to be added made it impractical to use a one-stage protocol (i.e., the oligonucleotides required are longer than 60 bases), part of the sequence is added in a second stage using oligonucleotide primers with at least 12 bases of sequence homology. The cleaned product of first-stage PCR (0.2–2 ng/μl) is amplified for 25 cycles as described for 1 × PCR.

Examples of Engineering Protocols

Additional details of some of the engineering are presented. Table II summarizes all the protocols.

Engineering Nucleoplasmin–Aeq cDNA. The nucleoplasmin–aequorin construct is generated by 2 × PCR F using the oligonucleotide primers NP1, 5′ sense to nucleoplasmin with the T7 RNA polymerase promoter sequence, NP3, a 5′ sense primer to aequorin with a 15 base nucleoplasmin sequence overlap, NP4, a 3′ antisense oligonucleotide to nucleoplasmin with a 15 base aequorin sequence overlap, and AEQ10.

Engineering Cx43-Aeq cDNA. Overlapping fragments encoding aequorin and Cx43[14] are generated using 58° for annealing and *Taq* polymerase and oligonucleotides (Table II). The second PCR stage joined the fragments using primers C43F and AeqR.

Engineering LckN-lase-Aeq cDNA. Luciferase–aequorin (lase-aeq) is extended at the 5′ end with the sequence encoding the Lck-N 1 × PCR using oligonucleotide primers LCK2 and PET2.

Aequorin for Expression in Bacteria

Aequorin is cloned unidirectionally using the *Sal*1 and *Pst*1 sites in the multiple cloning site.

Characterization of PCR Products

Normally, PCR constructs should be sequenced to check for any mutations. However, our strategy for plasmid selection is based on the exquisite sensitivity of detection of luminescent proteins. The criterion we use to select suitable recombinants is based on their ability to express active proteins *in vitro* and in live cells. The *in vitro* characterization relies on the use of viral promoters (T3, T7, and SP6) to generate abundant mRNA that can be translated using rabbit reticulocyte lysate (RRL) or wheat germ extracts in a procedure that can now take place in a single tube using the coupled transcription–translation system (TNT). DNA downstream of a viral promoter is transcribed and translated following suppliers guidelines. Recombinant proteins generated in the TNT reaction are then reconstituted to luminesce. Ten microliters of TNT routinely generates 1–10 ng of aequorin. This produces several million chemiluminescent counts.

Activation of Recombinant Aequorin

Cell or TNT extracts are incubated in 2–5 volumes of reconstitution buffer in the presence of a final concentration of 1 μM coelenterazine overnight at 4° or for 2 h at room temperature. The chemiluminescent activity of the recombinant protein is monitored, in bioluminescence buffer, following injection of an equal volume of 50 mM $CaCl_2$ into the tube in a luminometer.

Assay of Luciferase Activity

Equal volumes of luciferase assay buffer are mixed with cell or TNT extracts and luciferase activity is measured.

Calculating Specific Activity

In vitro transcription–translation for the calculation of specific activities of recombinant luminescent proteins is performed in the presence of [^{35}S]methionine. Approximately 400 ng template DNA generated by PCR or plasmid containing it is incubated in a 20-μl lysate reaction according to the manufacturer's instructions. The chemiluminescent (CL) activity is measured in 10 μl of reaction mixture following reconstitution and is expressed as CL counts/10 sec. To determine the specific activity of the photoprotein, a 2-μl aliquot is analyzed by polyacrylamide gel electrophoresis in the presence of 10% SDS, and the ^{35}S labeled proteins are detected by fluorography. The band corresponding to the luminescent variant is excised and ^{35}S incorporation is determined by scintillation counting.

The specific activity (CL counts/10 sec pmol protein) is calculated using the specific activity (SA) of [^{35}S]methionine/corrected by the total methionine present. The pmol protein synthesized = dpm in eluted band/SA × (mol methionine/mol protein). The specific activity of the bioluminescent protein is chemiluminescent counts/10 sec 2-μl reaction mixture/pmol photoprotein/2-μl reaction mixture.

In Vitro Calibration of Ca^{2+}-Dependent Light Emission of Aequorin Variants

Any new apoaequorin variants intended for use in Ca^{2+} measurements should be calibrated. Apoaequorin can be extracted from expressing cells that are washed twice with PBS, scraped from the culture plates, and resuspended in lysis buffer. Cells are subjected to three freeze thaw cycles, centrifuged at 12,000 rpm for 5 min at room temperature, and the pellet discarded. The supernatants are spin dialyzed in a 10,000 NMWL Ultrafree-MC filter (Millipore, UK) at 4° with three exchanges of zero Ca^{2+} buffer containing 100 m*M* KCl, 1 m*M* $MgSO_4$, 10 m*M* EGTA, 10 m*M* MOPS, pH 7.2, at 37° (Molecular Probes). The resulting cell extracts are stored at −70°. The apoproteins are reconstituted on ice for a minimum of 3 hr by adding an equal volume of zero Ca^{2+} buffer containing 2 μM coelenterazine.

Chemiluminescent light production is measured in a luminometer that can integrate for variable periods. Approximately 10^6 CL counts (measured at saturating Ca^{2+}) are injected into 200 μl of prewarmed Ca^{2+} buffer (Molecular Probes) placed in a heated sample housing maintained at 37°. The CL counts are recorded at 1-sec intervals between 0 and 157 n*M* Ca^{2+}, at 0.1-sec intervals for [Ca^{2+}] between 419 and 942 n*M*, and at 0.02-sec intervals for [Ca^{2+}] between 5 and 39 μM. Total CL counts and machine background are measured at the beginning, half-way through, and at the end of each set of Ca^{2+} calibration measurements, and all measurements are made in triplicate. Photoproteins produce spurious kinetics at 37° in low protein because the photoprotein is irreversibly inactivated in a low quantum yield reaction. Thus it may be necessary to add a protein carrier, e.g., 0.1% gelatin.

When the free Ca^{2+} is constant, the aequorin decay in light emission should be first order, thus the apparent rate constants, k_{app} (sec^{-1}), are determined for each of the free Ca^{2+} concentrations from log counts against time. Alternatively, providing that mixing is rapid, it is possible to estimate the rate constant from the ratio of the initial peak in luminescence divided by the peak in saturating Ca^{2+}. For values between 0 and 942 n*M* free Ca^{2+}, the mean CL counts per second (CL cps) are determined, as the decay is relatively slow. The rate constant is calculated by dividing mean CL cps by total CL counts. This model is only valid if there is no significant photoprotein consumption over the period of measurement (i.e., <1% consumption) as this would affect the rate constant estimation.[2] When significant photoprotein consumption occurs, the rate constant may be calculated either from the log of counts versus time or by determining the cps and dividing by the total active photoprotein remaining (total CL counts minus consumed CL counts) or from the slope of the decay in photoprotein light emission. This method is therefore used to calculate the apparent rate constant for $[Ca^{2+}]$ greater than 1 μM.[2]

General Expression Methods

Ligation of cDNAs

The PCR product is gel purified (from a high-quality grade agarose gel; Bio-Rad) using the Qiaquick gel extraction kit (Qiagen). The gel-purified PCR products are ligated to the TA cloning vectors at a 3:1 molar ratio of insert to vector and using 50–100 ng of linearized vector at 4 to 14° overnight in the presence of 3 U of T4 DNA ligase in 10 μl.

Transformation of Highly Competent E. coli

Aequorin is cloned unidirectionally into pMMB66EH using the *Sal*I and *Pst*I sites in the multiple cloning site (see primers) and is under control of the *tac* promoter.[15,16] The expression of aequorin, which is inducible with IPTG, varies considerably between different bacterial strains, and a time course for induction and coelenterazine incubation should be performed in order to maximize expression, although a basic protocol is detailed. pMMB66EH was originally constructed by Furste and co-workers.[20] This vector system is based on a ColE1 replicon and, carrying the *tac* promoter, a functional hybrid between *lac* and *trp* promoters is modified, resulting in the multi host plasmid.

The ligation mix (2–5 μl) is used to transform competent *E. coli*. White bacterial colonies growing on ampicillin/β-Gal indicator plates or just ampicillin-resistant colonies, when selection is not available, are screened for the presence of insert by PCR (using *Taq* DNA polymerase from Promega), and the respective plasmids are isolated form bacterial cultures using the QIAprep miniprep kit (Qiagen).

Basic Protocol for Expression of Aequorin in Live Bacteria

1. Transform bacteria with a plasmid containing aequorin and plate on LB agar plates containing carbenicillin (50 μg/ml final concentration).

2. Take a single colony and grow in 10 ml LB broth plus antibiotic at 37° and shaking at 240 rpm overnight.

3. For experiments such as monitoring Ca^{2+} inside bacteria, take 1 ml of the overnight culture and add 24 ml of LB broth plus antibiotic.

4. Incubate with shaking at 37° until the OD_{600} reaches 0.3–0.6.

5. Add IPTG from a 100 m*M* stock to a final concentration of 1 m*M* and continue incubation at 30° for a further 2 hr.

6. Harvest the cells by centrifugation at 3000 rpm for 5 min. Resuspend the cells in 5 ml of buffer A (25 m*M* HEPES, 1 m*M* $MgCl_2$, and 125 m*M* NaCl), centrifuge as before, and repeat wash.

7. Resuspend cells in 1 ml buffer A containing coelenterazine (2 μM final concentration) and leave in the dark at room temperature for 1 hr.

8. Centrifuge at 3000 rpm to harvest cells and wash cell pellet twice with buffer A (as detailed in step 6) to remove excess coelenterazine.

9. In order to measure total light counts in a luminometer, resuspend the cell pellet in 1 ml buffer A. Add 100 ml of the resuspended pellet to 400 ml buffer A, and the reconstituted aequorin is estimated by discharging an equal volume of 50 m*M* $CaCl_2$/4% Triton X-100.

10. The correct sized chimeric DNA (~1.6 kb) is excised from a 1% (w/v) agarose gel, purified, and ligated into PCR3 (Invitrogen) using T4 ligase (overnight at 14°).

Sequencing confirmed that no mutations had been introduced by PCR.

Plasmid Preparation

This is carried out using Quiagen or Hybaid kits.

Transfection of Eukaryotic Cells

COS-7/HeLa cells, cultured routinely in supplemented DMEM, are transfected by several methods, including calcium phosphate, FuGENE, and lipofectamine.

Assessing the Intracellular Localization of Targeted Photoproteins

A number of techniques are used routinely to establish the localization of intracellular proteins, e.g., selective plasma membrane permeabilization, subcellular fractionation, and immunolocalization. Because these methods alone will not show unequivocally that a protein is located in a particular site, localization data can be strengthened by employing more than one of the techniques described earlier. The methods we use routinely to establish intracellular localization are described below.

Measurement of Subplasma Membrane Calcium Environments

The chimeric photoprotein is activated by adding coelenterazine (final concentration, 2 μM) to cells in Ca^{2+}-free culture medium (KRH) containing 1 mM EGTA at least 4 hr prior to experiments. The coverslips are then mounted in a perfusion chamber at 37° and brought into contact with a fiber optics bundle attached to a photon-counting camera.[21] Cells are perfused for at least 10 min at 37° in Krebs–Ringer–Heinsleit (KRH) solution containing no Ca^{2+} (KRH-Ca^{2+}) ([(in mM): NaCl, 120; KCl, 4.8; KH_2PO_4, 1.2; $MgSO_4$, 1.2; EGTA, 1; HEPES, 25; pH 7.4)] to remove excess coelenterazine prior to monitoring light emission. Cells are perfused with KRH-Ca^{2+} for several minutes, and plasma membrane-associated Cx43-Aeq is triggered by the addition of KRH in which EGTA is replaced by 1.3 mM Ca^{2+}. Chemiluminescent data are used to generate rate constants (k)[2,22], but following consumption of >90% aequorin, rate constants generated are inaccurate and are excluded from calculations. Rate constants are converted into calcium concentrations using an equation generated by the *in vitro* calibration of Cx43-Aeq [$-\log \text{Ca} = 0.426\ (-\log k) + 4.819)$].

Despite the low concentrations of free calcium occurring in the bulk cytoplasm ($[Ca^{2+}]_c$) (~100 nM), highly localized regions of elevated $[Ca^{2+}]_c$ have been reported to surround calcium-rich organelles such as the ER, Golgi, and mitochondria, thereby creating gradients of calcium concentration within the cytoplasm. Such gradients of $[Ca^{2+}]_c$ may be involved in cellular processes such as protein trafficking and oligomerization. These cytoplasmic "hot spots" of free Ca^{2+} raise intriguing questions as to the concentration of cytoplasmic Ca^{2+} directly beneath the plasma membrane ($[Ca^{2+}]_{pm}$) when, under physiological conditions, the concentration of Ca^{2+} in the extracellular environment is approximately 1 mM.

Targeting aequorin to the plasma membrane as a partner in a fusion protein has allowed Ca^{2+} microdomains under the plasma membrane to be measured more precisely, as it has been estimated that such targeting strategies position aequorin within 50 nm of the target membrane.

A fusion protein of luciferase and aequorin, designed to circumvent the problem caused by aequorin diffusion into the nucleus due to its small size (21 kDa), is used to image and measure changes in ATP and

[22] P. H. Cobbold and J. A. C. Lee, *in* "Cellular Calcium: a Practical Approach" (J. G. McCormack, and P. C. Cobbold, eds.), p 55. Oxford Univ. Press, Oxford, 1991.

cytosolic-free Ca^{2+} in HeLa cells attacked by complement using a Photek camera.[21,23]

Immunolocalization

COS7 or HeLa cells expressing the targeted luminescent proteins are seeded onto glass coverslips, and the cell density is assessed approximately 18 hr the cells after transfection and maintained in culture for a further 24 hr. The cells are fixed in 2% paraformaldehyde/0.1% gluteraldehyde in PBS [containing (in mM): NaCl, 120; KCl, 2.7; $Na_2HPO_4 \cdot 2H_2O$, 10, pH 7.4] for 10 min. After a thorough washing (3 × 5 min in PBS), the cell membranes are permeabilized by treatment for 30 min with 1% Triton X-100 in PBS containing 1% normal goat serum (NGS). The cells are then incubated overnight at 4° with a rabbit polyclonal antiserum to aequorin and are diluted appropriately in PBS containing 0.1% Triton and 1% NGS (for best results, an overnight incubation at 4° is recommended; this period may, however, be reduced to 1 hr at 37°). After further extensive washing in PBS, the cells are incubated with an FITC-labeled second antibody (e.g., FITC-conjugated goat antirabbit IgG) for 30 min at room temperature. Cells are then mounted (e.g., Citifluor mountant), and staining is observed on a fluorescence microscope. Ideally, the microscope should be equipped with an X63 oil immersion lens and epifluorescent illumination in order to observe details of any subcellular staining.

Selective Permeabilization of the Plasma Membrane

Cells expressing the targeted recombinant proteins are incubated in culture medium containing 2.5 μM coelenterazine at 37° for 2 hr (see later). The cells are then removed from their substrate (this can be achieved either enzymatically with trypsin or mechanically with a rubber policeman), centrifuged, and washed in PBS containing 1 mM EDTA. The plasma membranes are selectively permeabilized for 3 min. in PBS (+1 mM EDTA) containing 20 μM digitonin. The cells are then centrifuged rapidly (1 min in microcentrifuge) and divided into supernatant (cytosolic) and pellet (organelle) fractions. These fractions are then assayed for aequorin activity (see relevant section).

Luminescence Measurements from Cells

Imaging Ca^{2+} and ATP

Imaging of bioluminescent reactions is required to detect events such as Ca^{2+} signaling in individual cells and to investigate the complex spatial

[23] G. B. Sala Newby, K. T. Taylor, M. N. Badminton, C. R. Rembold, and A. K. Campbell, *Immunology* **934** 601 (1998).

and temporal patterns of these processes. This requires an imaging device that is true photon counting, sensitive, quantitative, and capable of amplifying the signal by several orders of magnitude, has a fast response time (ideally video rate, i.e., 50 or 60 cycles per second), and has good spatial resolution. For real-time bioluminescence, an intensified CCD camera with two or three microchannel plates has a superior signal to noise compared with a cooled CCD camera (288 $\times$ 385 or 512 $\times$ 512 pixels), which can be used for reporter gene imaging integrating for many minutes. A crucial factor is capturing as many photons as possible. The optimum way we have chosen is to exploit modern fiber optics, which have bundles 6 μm in diameter. This means that if two cells are at least 100 μm apart, then the photons will be collected into separate pixels. Thus cells are imaged for chemiluminescence via direct contact of the obverse side of the coverslip with the fiber optic array.[13] Data are stored as a photon count event file at video rates and analyzed using both supplied and custom software.[21]

Typically, photons are collected from at least two areas of 41 $\times$ 41 pixels and averaged or integrated every 10 sec for population studies. Data for individual cells can be extracted by selecting the appropriate pixels from an image sequence. In order to measure cytosolic Ca^{2+} concentration in cells using aequorin, the recombinant apoaequorin is reconstituted with 2 μM coelenterazine in culture medium at 34° for 2 hr prior to activation of the complement.[23] The unconsumed aequorin is determined at the end of each experiment by exposing cells to 5 mM $CaCl_2$ in water, which causes the cells to lyse slowly and avoids the use of detergents. When measuring ATP using firefly luciferase, 50 μM luciferin and 1 mg/ml BSA are included in the buffer. The light emitted by firefly luciferase in HeLa cells is normalized to that observed for the control cells at each time point. COS 7/HeLa cells are detached in cold PBS containing 1 mM EDTA, and 8×10^4 are seeded onto 22 $\times$ 22-mm glass coverslips (in approximately 100 μl).

For the measurement of cytosolic [Ca^{2+}] in cells infected with RAdLA, cytosolic recombinant apoaequorin is converted to the photoprotein by adding coelenterazine (final concentration 2 μM) at least 4 hr prior to experiments. The coverslips are then inverted over the reservoir of a plastic perfusion chamber. The perfusion chamber is placed on a heated microscope stage (maintained at 37°) in a purpose built dark box, and the coverslips are brought into contact with a fiber optics bundle, attached to the photon-counting camera. Light production is then monitored through a 55-mm macro lens attached to a Photek three microchannel plate-intensified CCD camera (Photek 216, Photek Ltd., Hastings, UK). The camera stores the photon image as an array of 385 $\times$ 288 pixels at video rates (50 Hz[21]).

Cells are perfused for at least 10 min in KRH at 37° to remove excess coelenterazine prior to monitoring light emission. The fractional discharge of aequorin is calculated from the total light emitted by the photoprotein at the end of each experiment by exposing the cells to water containing 5 mM $CaCl_2$ to discharge unconsumed aequorin. This value is used to convert light emission into free Ca^{2+}.

Aequorin as a Luciferase[18]

Cells expressing the ER targeted apoaequorin are not exposed to coelenterazine until the start of the experiment, when 2 μM coelenterazine is included in all perfusion medium. All cellular manipulations are delayed for at least 12 min after adding coelenterazine to allow the stabilization of light production. Light is measured from the whole area of the coverslip and is normalized to light production after adding coelenterazine.

Results and Conclusions

Results show that highly efficient targeting of bioluminescent indicators to the cytosol, ER, mitochondria, and nucleus can be achieved in animal cells and to chloroplasts[25] or tonoplast in plant cells. However, efficient targeting to the plasma membrane or gap junction is more difficult to achieve. It will therefore be necessary to have a way of either distinguishing between targeted or untargeted aequorin by kinetics or color or by selective destruction of one component. It is important to remember that many bioluminescent proteins have relatively short half-lives in mammalian cells (cytosolic apoaequorin ca. 20 min and firefly luciferase ca. 3 hr). Intriguingly, apoaequorin in the ER is very stable until Ca^{2+} is lost, when it degrades very rapidly. The half-life can be increased several fold by engineering chimeras. The stability of the full photoprotein aequorin appears entirely dependent of the concentration of Ca^{2+} to which it is exposed.

We have shown that the ER-free Ca^{2+} is at least several micromolar and that a relatively small decrease can still lead to large changes in cytosolic free Ca^{2+}. Free Ca^{2+} does not appear to be the direct signal for store-operated Ca^{2+} entry, although calreticulin can affect this. However, loss of ER-free Ca^{2+} does activate a proteolysis mechanism rapidly. The absolute level of free Ca^{2+} in the ER remains controversial. This controversy requires a ratiometric ER Ca^{2+} indicator. We have also shown that there is a barrier to Ca^{2+} between the cytosol and the nucleus. This is lost in semipermeabilized cells. The free Ca^{2+} at the inner surface of the plasma membrane appears to be several micromolar, resulting in the rapid consumption of aequorin. This must be taken into account when estimating free Ca^{2+}, particularly if the aequorin is not fully targeted. We have also shown how powerful the combination of an intensified CCD camera with fiber optics is. Single cell analysis in real time can be achieved by enabling the individuality of cell responses to be imaged and quantified (Fig. 2). ATP loss can be imaged in the cytosol following complement attack.[23]

We have also developed kinase indicators by engineering firefly luciferase,[17] but chemiluminescence resonance energy transfer (CRET) now appears to be the platform technology for other intracellular signals, such as cyclic nucleotides, inositol phosphates, kinases, proteases, and other covalent modifications, and protein–protein or DNA–protein interaction. CRET generates "rainbow proteins"[24] that change color when the ligand interacts with the active site in the engineered protein. GFP-linker aequorin chimeras have already been engineered that show spectra similar to GFP, not free aequorin. It also will enable multicolored indicators for Ca^{2+} and other signals to be used in different subcellular compartments of the same cell at the same time. Our imaging system has been developed to monitor the complete spectrum of the rainbow proteins even when the flash lasts only a few seconds, and dual wavelength imaging for two indicators for ratiometric measurement. This technology will also allow better calibration in cells and will correct for varying levels of expression or degradation of the indicator, which occurs, for example, with aequorin in the cytosol or ER when Ca^{2+} is released. The half-time for apoaequorin can be as short as 20 min. Dual wavelength imaging also allows the measurement of aequorin (465 nm) and firefly luciferase (565 nm) for the simultaneous estimation of Ca^{2+} and ATP. This will also enable dual reporters such as *Renilla* (480 nm) and firefly (565 nm) luciferases to be imaged simultaneously.

In conclusion, targeted bioluminescent indicators are set to revolutionize our understanding of how cell function is controlled in health and disease.

Acknowledgments

We thank the Medical Research Council, the Arthritis and Rheumatism Council (C0091), The British Heart Foundation, Subprograma Cencia e Technologia do 2° quadro Counitario de Apoia (Praxis XXI), and the Department of Medical Biochemistry, University of Wales College of Medicine for financial support.

[24] A. K. Campbell, British Patent Application 8916806.6. International patient application PCT/GB90/01131. U.S. patent 5,683,888. (1989 and 1997).

[25] C. H. Johnson, M. R. Knight, T. Kondo, P. Masson, J. Sedbrook, A. Haley, and A. Trewavas, *Science* **269,** 1863.

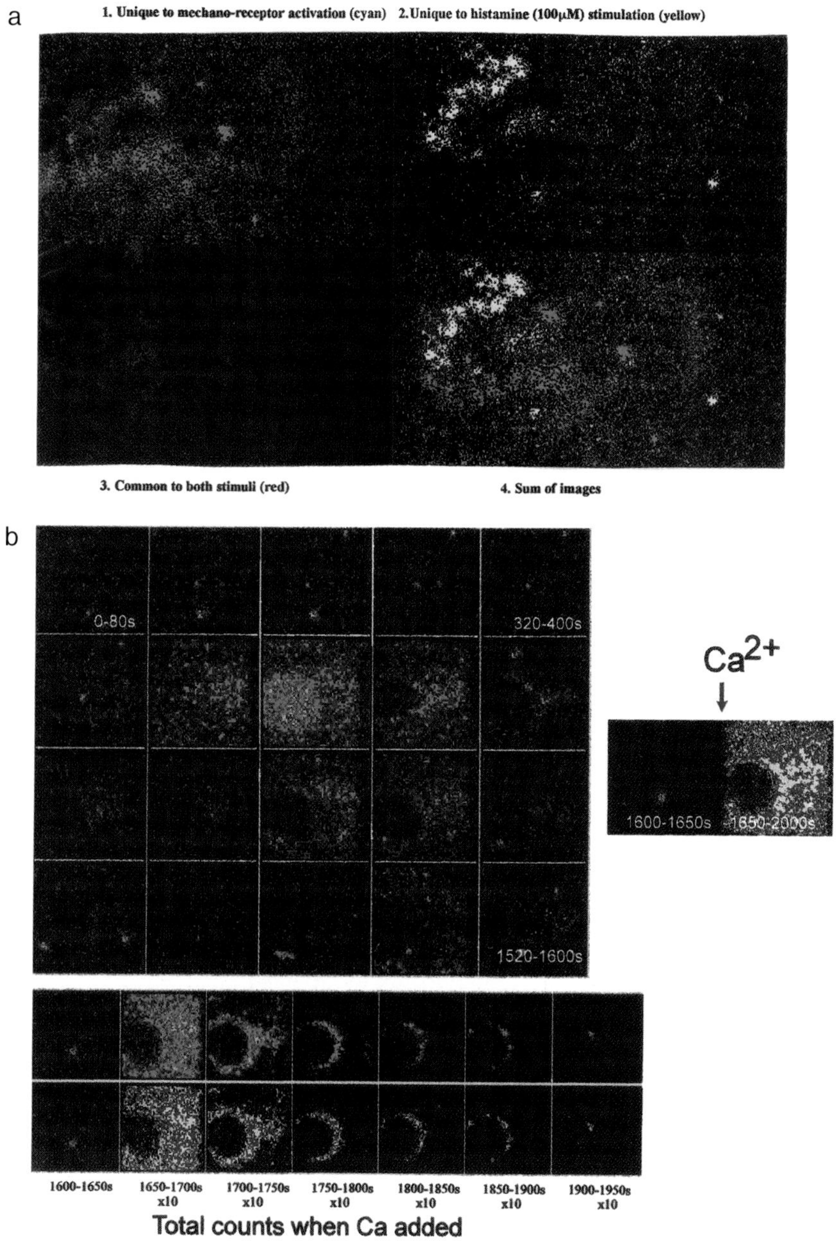

FIG. 2. Individuality of Ca^{2+} signals in HeLa cells. The software enables multiple image sequences to be imaged and analyzed. Using image processing it is possible to determine whether specific pixels (cells) have fired under each condition. (a) HeLa cell exposed first to mechano then histamine stimulus. (b) HeLa cells exposed to a dose response of ATP and then a variety of other Ca^{2+} agonists. Total aequorin is assessed at the end of the experiment by cell lysis. The photon counts are converted to color via a look-up table: black = 0 photons, and the scale is then log from blue, green, yellow, red, and white (1–255). A bubble stimulates mechanoreceptors and can be ignored by using odd area sets over cells not in the bubble.

[35] Green Fluorescent Protein as a Reporter of Transcriptional Activity in a Prokaryotic System

By DEBORAH A. SIEGELE, LISA CAMPBELL, and JAMES C. HU

Introduction

Fluorescent and chemiluminescent molecules are widely used as probes to localize biological molecules. The green fluorescent protein (GFP) from *Aequorea victoria* has become an important fluorescent reporter used in a wide variety of systems. Because it is a protein that can be expressed in living cells and because development of its fluorophore requires no additional cofactors, fusion proteins containing GFP have been applied to many questions in cell biology (e.g., see Stearns[1] or Gerdes and Kaether[2]). GFP has been used to localize specific proteins, DNA sequences, viruses, organelles, and cells.

The focus of this review is to describe how we have used GFP to address a common question in bacterial physiology: How is gene expression distributed among the individual cells in a culture? Intermediate levels of gene expression in cultures do not always mean that gene expression is uniform with respect to individual cells. In extreme cases, either (1) every cell in the culture makes the same percentage of the fully induced level as is seen for the culture as a whole or (2) the culture consists of a mixture of fully induced and completely uninduced cells. In the latter case, the intermediate expression levels observed in the culture reflect the proportion of cells that are fully induced, rather than intermediate expression in any individual cell.

Determining the actual distribution requires measuring gene expression in individual cells. For example, the distribution of β-galactosidase levels in individual cells has been examined by microscopic, single cell β-galactosidase assays[3] and by flow cytometry.[4] However, these assays, which involve fluorogenic substrates, require manipulations to limit the reaction products to single cells. When membrane-permeable substrates are used, the cell suspensions are dispersed into microdroplets in an oil emulsion to prevent diffusion of the fluorescent reaction products.[3] Membrane-impermeable

[1] T. Stearns, *Curr. Biol.* **5,** 262 (1995).
[2] H.-H. Gerdes and C. Kaether, *FEBS Lett.* **389,** 44 (1996).
[3] P. C. Maloney and B. Rotman, *J. Mol. Biol.* **73,** 77 (1973).
[4] F. Russo-Marie, M. Roederer, B. Sager, L. A. Herzenberg, and D. Kaiser, *Proc. Natl. Acad. Sci. U.S.A.* **90,** 8194 (1993).

0076-6879/00 $30.00

β-galactosidase substrates must be loaded into cells by an osmotic shock step.[4] In contrast, a GFP reporter allows a simpler and more convenient assay. GFP levels in a culture can be measured in a fluorometer,[5] whereas levels in individual cells can be followed by either fluorescence microscopy[6] or flow cytometry.[7,8] The following sections describe GFP reporter constructs and methods used to measure GFP fluorescence for populations in culture or for individual cells. In each section, the experimental details focus on our own experience. These protocols can undoubtedly be adapted to different instrumentation and software, and we will try to point out areas where protocols need to be optimized for specific instruments.

Green Fluorescent Protein Reporter Plasmids

The green fluorescent protein is a nonenzymatic reporter for gene expression that is detected by the fluorescence of a chromophore formed in the folded protein by cyclization and oxidation of residues 65–67 to a 4-(*p*-hydroxybenzylidene)-imidazoladin-5-one.[9] The mechanism of chromophore formation has two important consequences for the use of GFP as a reporter for gene expression in bacteria. First, the chemical steps in chromophore formation have been shown to occur *in vitro* with half-times of the order of hours.[5,10] In our experiments, the manipulation of samples before measurement seems to allow adequate time for development of the fluorescence signal, which seems to be stable for hours to days. However, "real-time" measurements of GFP fluorescence *in vivo* will reflect a substantial lag between the time a molecule is synthesized and when it is detected by fluorescence. Second, the oxygen dependence of chromophore formation should be kept in mind if GFP is to be used as a reporter under anaerobic conditions. GFP appears to fold normally in the absence of oxygen, and prefolded GFP can be allowed to develop the chromophore in the presence of air after samples have been taken. However, this obviously precludes the use of GFP as a reporter for sorting viable cells of any obligate anaerobe.

Wild-type GFP as originally cloned from *A. victoria* has been subjected to protein engineering by many laboratories, and a wide variety of GFP

[5] R. Heim, D. C. Prasher, and R. Y. Tsien, *Proc. Natl. Acad. Sci. U.S.A.* **91,** 12501 (1994).
[6] D. A. Siegele and J. C. Hu, *Proc. Natl. Acad. Sci. U.S.A.* **94,** 8168 (1997).
[7] R. H. Valdivia, A. E. Hromockyj, D. Monack, L. Ramakrishnan, and S. Falkow, *Gene* **173,** 47 (1996).
[8] B. P. Cormack, R. H. Valdivia, and S. Falkow, *Gene* **173,** 33 (1996).
[9] C. W. Cody, D. C. Prasher, W. M. Westler, F. G. Prendergast, and W. W. Ward, *Biochemistry* **32,** 1212 (1993).
[10] B. G. Reid and G. C. Flynn, *Biochemistry* **36,** 6786 (1997).

variants that are better suited for use as reporters for gene expression have been described. Many of these are available commercially. Although the purified, folded protein is stable in a variety of denaturants, including 1% sodium dodecyl sulfate (SDS) and 8 *M* urea,[11] GFP has a tendency to form inclusion bodies when expressed in *Escherichia coli.* Many of the variant GFP forms have "improved folding" properties; mutations in these engineered forms appear to change the partitioning of folding intermediates between nonfluorescent aggregates and soluble fluorescent protein without changing intrinsic folding rates *in vitro.*[10] This partitioning into nonfluorescent aggregates may account for the observation that cells expressing wild-type GFP show greater fluorescence when they are grown at 24° than at 37° or 42°.[12]

Green fluorescent protein variants have also been isolated that change the excitation and emission maxima of the protein. We used a GFP variant selected by Cormack *et al.*[8] by fluorescence-activated cell sorting. This variant, GFPmut2, is a triple mutant containing the changes S65A, V68L, and S72A. Like many of the extant variants, GFPmut2 does not form inclusion bodies. In addition, the maximum of the major excitation wavelength for GFPmut2 is shifted to around 488 nm. When excited at this wavelength, GFPmut2 gives 19-fold brighter emission than wild-type GFP.[8] These properties are well suited to the use of filter sets that are widely available for use with fluorescein isothiocyanate (FITC).

A plasmid expressing GFPmut2 was kindly provided by Dr. W. Margolin. In this plasmid and in pDS439, which expresses GFPmut2 from the inducible *araBAD* promoter, GFPmut2 is present on an *Xba*I–*Hin*dIII fragment (Fig. 1).[8] Transcription of the GFP gene is from the *Xba*I end to the *Hin*dIII end.

Colonies containing pDS439 are visibly green under fluorescent room lights when grown on LB ampicillin plates containing 0.2% arabinose and are bright green under illumination by a long wavelength (366 nm) hand-held UV lamp. In the absence of arabinose, the colonies are uniformly dark under UV illumination. Plasmid pDS439 is not maintained stably in *E. coli* under conditions where high levels of GFP are expressed; cells plated on LB arabinose plates in the absence of ampicillin form colonies with a variable number of dark and fluorescent sectors.

Fluorescence Measurements Using Cultures

With a fluorometer, measuring gene expression in *E. coli* cultures using GFP is even easier than the simple enzyme assays available for β-galactosi-

[11] S. H. Bokman and W. W. Ward, *Biochem. Biophys. Res. Commun.* **101,** 1372 (1981).
[12] K. R. Siemering, R. Golbik, R. Sever, and J. Haseloff, *Curr. Biol.* **6,** 1653 (1996).

A

```
XbaI                    NdeI
TCTAGATTTAAGAAGGAGATATACATATGAGTAAAGGAGAAGAACTTTTCACTGGAGTTGT
AGATCTAAATTCTTCCTCTATATGTATACTCATTTCCTCTTCTTGAAAAGTGACCTCAACA
                             >MetSerLysGlyGluGluLeuPheThrGlyValVa

CCCAATTCTTGTTGAATTAGATGGTGATGTTAATGGGCACAAATTCTCTGTCAGTGGAGAG
GGGTTAAGAACAACTTAATCTACCACTACAATTACCCGTGTTTAAGAGACAGTCACCTCTC
lProIleLeuValGluLeuAspGlyAspValAsnGlyHisLysPheSerValSerGlyGlu

GGTGAAGGTGATGCAACATACGGAAAACTTACCCTTAAATTTATTTGCACTACTGGAAAGC
CCACTTCCACTACGTTGTATGCCTTTTGAATGGGAATTTAAATAAACGTGATGACCTTTCG
GlyGluGlyAspAlaThrTyrGlyLysLeuThrLeuLysPheIleCysThrThrGlyLysL

TACCTGTTCCATGGCCAACACTTGTCACTACTTTCGCGTATGGTCTTCAATGCTTTGCGAG
ATGGACAAGGTACCGGTTGTGAACAGTGATGAAAGCGCATACCAGAAGTTACGAAACGCTC
euProValProTrpProThrLeuValThrThrPheAlaTyrGlyLeuGlnCysPheAlaAr

            NdeI
ATACCCAGATCATATGAAACAGCATGACTTTTTCAAGAGTGCCATGCCCGAAGGTTATGTA
TATGGGTCTAGTATACTTTGTCGTACTGAAAAAGTTCTCACGGTACGGGCTTCCAATACAT
gTyrProAspHisMetLysGlnHisAspPhePheLysSerAlaMetProGluGlyTyrVal

CAGGAAAGAACTATATTTTTCAAAGATGACGGGAACTACAAATCACGTGCTGAAGTCAAGT
GTCCTTTCTTGATATAAAAAGTTTCTACTGCCCTTGATGTTTAGTGCACGACTTCAGTTCA
GlnGluArgThrIlePhePheLysAspAspGlyAsnTyrLysSerArgAlaGluValLysP

TTGAAGGTGATACCCTCGTTAATAGAATTGAGTTAAAAGGTATTGATTTTAAAGAAGATGG
AACTTCCACTATGGGAGCAATTATCTTAACTCAATTTTCCATAACTAAAATTTCTTCTACC
heGluGlyAspThrLeuValAsnArgIleGluLeuLysGlyIleAspPheLysGluAspGl

AAACATTCTTGGACACAAAATGGAATACAACTATAACTCACACAATGTATACATCATGGCA
TTTGTAAGAACCTGTGTTTTACCTTATGTTGATATTGAGTGTGTTACATATGTAGTACCGT
yAsnIleLeuGlyHisLysMetGluTyrAsnTyrAsnSerHisAsnValTyrIleMetAla

GACAAACAAAAGAATGGAATCAAAGTTAACTTCAAAATTAGACACAACATTGAAGATGGAA
CTGTTTGTTTTCTTACCTTAGTTTCAATTGAAGTTTTAATCTGTGTTGTAACTTCTACCTT
AspLysGlnLysAsnGlyIleLysValAsnPheLysIleArgHisAsnIleGluAspGlyS

GCGTTCAACTAGCAGACCATTATCAACAAAATACTCCAATTGGCGATGGCCCTGTCCTTTT
CGCAAGTTGATCGTCTGGTAATAGTTGTTTTATGAGGTTAACCGCTACCGGGACAGGAAAA
erValGlnLeuAlaAspHisTyrGlnGlnAsnThrProIleGlyAspGlyProValLeuLe

ACCAGACAACCATTACCTGTCCACACAATCTGCCCTTTCCAAAGATCCCAACGAAAAGAGA
TGGTCTGTTGGTAATGGACAGGTGTGTTAGACGGGAAAGGTTTCTAGGGTTGCTTTTCTCT
uProAspAsnHisTyrLeuSerThrGlnSerAlaLeuSerLysAspProAsnGluLysArg

GATCACATGATCCTTCTTGAGTTTGTAACAGCTGCTGGGATTACACATGGCATGGATGAAC
CTAGTGTACTAGGAAGAACTCAAACATTGTCGACGACCCTAATGTGTACCGTACCTACTTG
AspHisMetIleLeuLeuGluPheValThrAlaAlaGlyIleThrHisGlyMetAspGluL

                                HindIII
TATACAAATAAXXCTGCAGGCATGCAAGCTT
ATATGTTTATTXXGACGTCCGTACGTTCGAA
euTyrLys•••
```

B

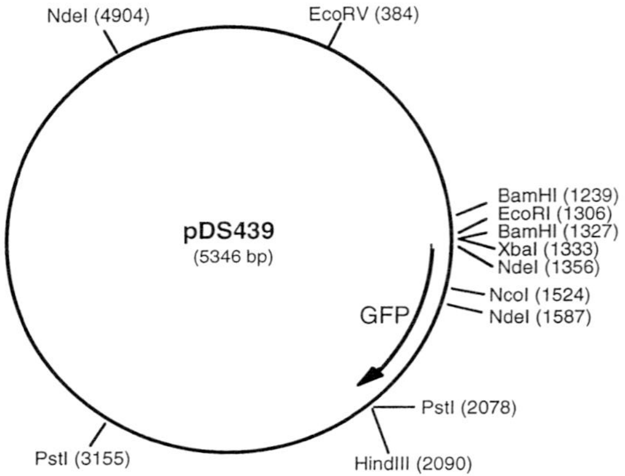

FIG. 1. (A) Sequence of the region encoding GFPmut2 in pDS439.[6] (B) Map of plasmid pDS439. pDS439 was constructed by cloning the *Xba*I to *Hin*dIII fragment shown in A into the same sites in pBAD18.[15]

dase, galactokinase, or chloramphenicol acetyl transferase. One simply puts a known amount of a cell suspension in the cuvette and measures the fluorescence intensity. The cells do not have to be lysed and can be grown in a variety of rich or defined media. However, rich media such as LB contain compounds that fluoresce strongly when excited at the wavelengths used to assay GFP. Cells grown in LB should be spun down and resuspended in a nonfluorescent buffer. For most of our experiments, cells were grown in M63 minimal medium[13] with 0.5% glycerol as the carbon source and were supplemented with amino acids at 40 μg/ml. Note that if the same samples are going to be assayed by flow cytometry, then all media and dilution buffers should be filtered through 0.2-μm filters to remove particulate matter.

For measuring GFP levels from many samples, it is useful to halt the synthesis of GFP to allow all of the samples to be read at the end of the experiment. We found that in the *E. coli* strain we used, MC1061,[14] the addition of chloramphenicol to 20 μg/ml allowed us to store samples at 4° with little or no change in fluorescence intensity for several days. Aliquots of *E. coli* cultures were pelleted and resuspended in M63 salts with 20 μg/ml chloramphenicol, and the OD_{600} of the suspension was adjusted to be $\leq$0.3 in a volume of $\geq$3 ml. The 3-ml samples were transferred to a 1 $\times$ 1-cm quartz cuvette, and fluorescence was measured in an SLM Aminco 8000 fluorometer. We excited culture suspensions at 420 nm and recorded the fluorescence signal at 520 nm.

Fluorescence was measured as arbitrary units (a.u.). Due to variations in lamp intensity, absolute units should not be compared between experiments performed on different days. For this reason, we always include one fully induced culture as a standard for each experiment.

Although the sensitivity for detecting GFP in cultures will vary from instrument to instrument and will be affected by lamp intensity and detector properties, we could easily monitor the kinetics of induction of GFP expression driven by the addition of the inducer arabinose to cells containing pDS439 (Fig. 2). Fully induced cultures gave a fluorescence signal that was about 50-fold over the background for uninduced cells. While this signal-to-noise ratio was adequate for our experiments, the basal level of expression is undoubtedly much lower. The pBAD vectors are known for the tight regulation of basal expression and are recommended for the expression of toxic genes.[15] The same vector expressing an alkaline phosphatase reporter

[13] J. H. Miller, "Experiments in Molecular Genetics." Cold Spring Harbor Laboratory Press, Cold Spring Harbor, NY, 1972.

[14] M. J. Casadaban and S. N. Cohen, *J. Mol. Biol.* **138,** 179 (1980).

[15] L.-M. Guzman, D. Belin, M. J. Carson, and J. Beckwith, *J. Bacteriol.* **177,** 4121 (1995).

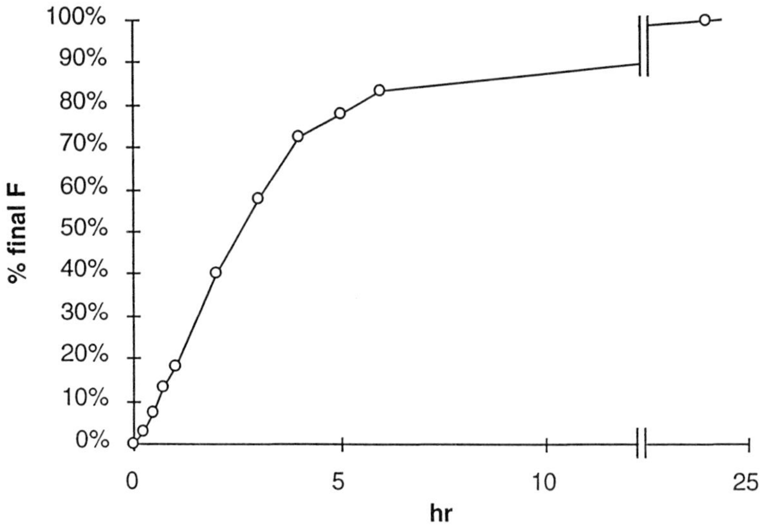

FIG. 2. Induction kinetics of GFP expression measured in cell suspensions. Cells containing pDS439 were grown at 37° in a shaking water bath to OD_{600} 0.1 to 0.2 at which time arabinose was added to 1.33 m*M*. At various times, samples were taken, and the fluorescence of cell suspensions was measured using a fluorometer as described in the text. Fluorescence is expressed in arbitrary units. Reproduced from D. A. Siegele and J. C. Hu, *Proc. Natl. Acad. Sci. U.S.A.* **94,** 8168 (1997) with permission of the publisher. Copyright 1997 National Academy of Sciences.

gave an induction ratio of more than 300-fold.[15] This difference clearly is due to our inability to detect very low levels of GFP in whole cells in the cuvette.

Fluorescence Measurements of Individual Cells by Microscopy and Image Quantitation

Microscopy is an obvious way to examine how expression of the GFP reporter is distributed among the cells in a population. By comparing a field seen with epifluorescent illumination to the same field visualized using a method that will show every cell, it is readily apparent whether all the cells or only a subset of cells are expressing GFP. However, obtaining quantitative information about GFP levels in individual cells requires additional image processing. This section describes how we quantitated the fluorescence of individual cells by microscopy.

Two- to 5-μl samples of cell suspensions were spotted onto glass slides. If necessary, cells can be concentrated severalfold by centrifugation to increase the number of cells visible in each microscope field. Coverslips were placed on the spots, and excess liquid was removed by placing a Kimwipe at the edge of the coverslip. Removing the excess liquid puts the cells in a thinner section and minimizes cell movement. *Escherichia coli* strain MC1061 is motile and when excess liquid was present some cells would swim. Because different exposures will be taken of the same field of cells, it is essential to minimize cell movement.

The slides were examined with an Olympus BX50 microscope using a UPlan Fl 100X/1.3 oil immersion objective. Cells were viewed using both Nomarski (DIC) optics and epifluorescent illumination. For detection of GFP fluorescence, a DM500 dichroic mirror/BP470-490 exciter filter/BA515 barrier filter set was used. Photomicrographs were taken with Kodak Ektachrome 100 film.

The fluorescence of individual cells was quantitated from photographs taken using epifluorescent illumination. Dramatic differences in the fluorescence intensity of cells grown under comparable conditions were observed as the lamp aged and was eventually replaced, which made it difficult to compare absolute values measured on different days. The inclusion of fluorescent beads in each sample is recommended to provide an internal standard. Suppliers of standardized fluorescent beads include Spherotech Inc. (Libertyville, IL) and Molecular Probes (Eugene, OR). Multiple exposures were taken of each field to assure that the fluorescent signal was within the linear range of the film and the scanner. Slides were scanned using a Nikon LS-1000 film scanner attached to an Apple Macintosh computer running Adobe Photoshop version 3.0.5 software. Slides were scanned in a color-positive mode using the factory default settings at 100% scale and were then converted to gray scale images. To reduce the size of the files to be quantitated, images were cropped in Photoshop. The cropped images, each of which typically showed 10 to 20 well-separated cells, were saved as Photoshop files and were then imported into MacBAS version 2.5 software (Fuji Photo Film Co., Ltd.) for quantitation. MacBAS processes images as 8-bit gray scale images with 256 gradations per pixel. The total intensity of an image is the sum of the digital intensity values of the pixels that make up the image.

Because cells at different points in the division cycle are different sizes, the relative expression of GFP is reflected in the concentration of the protein rather than in the total amount of GFP per cell. Thus, expression is reflected in the average intensity of a cell's image rather than its total intensity, which is a two-dimensional projection of the three-dimensional cell. This assumes that differences in image size reflect differences in cell size rather than the aspect from which the cells are viewed. Thus, we only quantitated cells that looked like rods. GFP expression in individual cells was expressed as fluorescence intensity (in arbitrary units) per unit area. The boundaries of each cell image were defined automatically using the magic wand tool provided in MacBAS. The total intensity and area within the boundary were reported by the software. To correct for background, a copy of the cell boundary was dragged to a nearby area with no cells. The intensity within the copy was subtracted from the intensity over the cell. The corrected intensity was then normalized to the size of the cell image to correct for differences in cell size. In this manner, the average fluorescence intensity of every well-separated cell in an image was quantitated. This process was repeated for different exposures of the same field and data were used only if the signal from a particular cell increased linearly as a function of expsoure time.

To compare fluorescence over a wide range of GFP expression, it is necessary to compare images taken at different exposure times. To allow these comparisons to be made, the final average intensity per cell was expressed as (arbitrary units − background)/(area × exposure time). This assumes that intensity is a linear function of exposure time and that GFP is not bleached significantly over the time of the exposure(s). This seemed to be the case for exposures up to 32 sec; however, this should be determined empirically as it may vary with lamp intensity.

To give a complete description of the distribution of GFP levels in the sample population, it is necessary to include both fluorescent and nonfluorescent cells in the analysis. Comparing the Nomarski and fluorescence images allowed us to identify any cells that were too dim to appear in the latter. Figure 3 shows a set of histograms plotting the numbers of cells with different levels of GFP as a function of time after addition of the inducer. As induction progresses, the number of cells with measurable fluorescence increases, and their average intensity also increases. The intensities of the brightest measurable cells were about 20-fold higher than the faintest cells. At each time point, the average intensities are distributed over a range of about two- to threefold. The width of the distribution reflects both the error in the measurements and real differences in the expression of GFP from cell to cell, presumably due to variations in the segregation of the multicopy plasmid carrying the GFP reporter and the stochastic nature of the induction process.

Sensitivity is a function of both lamp intensity and the sensitivity of the scanner. On the dimmest images, we could easily see fluorescent cells on the film by eye, but some of these cells gave no signal above background when the slide was scanned. Very long exposures led to nonlinearity due to bleaching of the GFP. This prevented us from estimating the basal expression of GFP in uninduced cells containing pDS439. It should be possible to increase sensitivity and eliminate the scanning step by using a digital CCD camera instead of film to record the cell images.

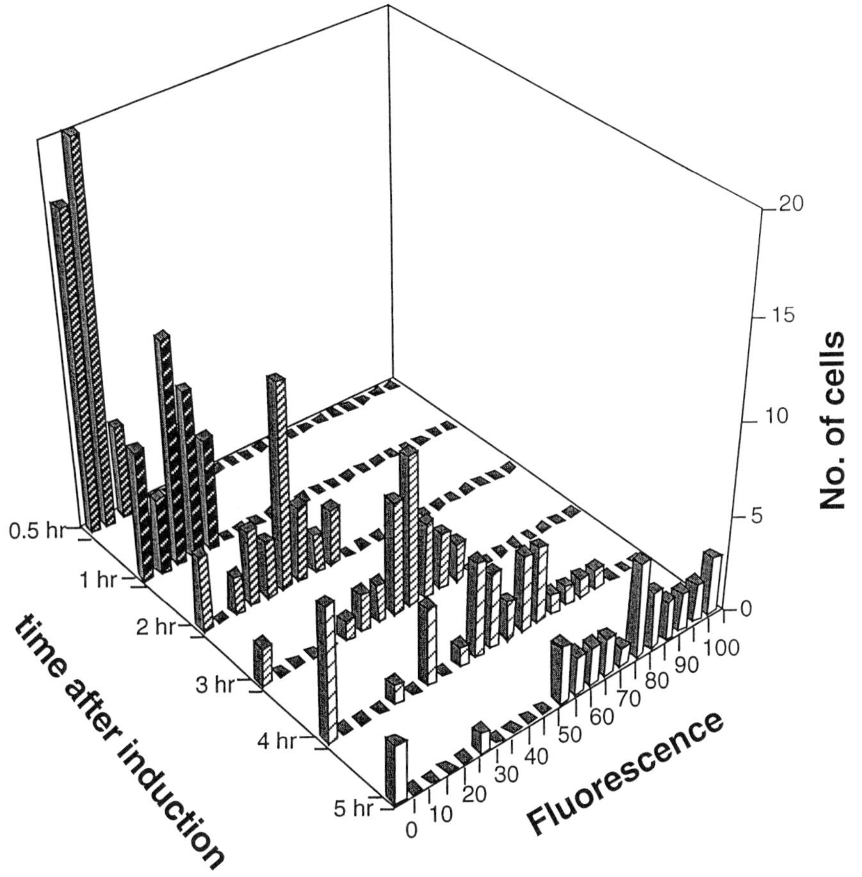

FIG. 3. Induction kinetics of GFP expression measured in individual cells. Cultures were grown as described in the legend to Fig. 2. The fluorescence intensities of 25 to 30 individual green fluorescent cells were quantitated by microscopy, and image analysis was done as described in the text. The histogram shows the distribution of fluorescence intensities for individual cells at each time point. Reproduced from D. A. Siegele and J. C. Hu, *Proc. Natl. Acad. Sci. U.S.A.* **94,** 8168 (1997) with permission of the publisher. Copyright 1997 National Academy of Sciences.

Fluorescence Measurements of Individual Cells by Flow Cytometry

The number of cells that can be examined by microscopy is limited by the need to examine each cell individually and in multiple exposures. Flow cytometry allows more rapid, quantitative analysis of cellular fluorescence. For example, a flow cytometer can analyze 50,000 cells in a few minutes compared to several days needed to analyze a few hundred cells by microscopy. The larger number of cells examined allows much higher confidence levels for statistical analysis of the data.

The most obvious disadvantage of flow cytometry compared to microscopic analysis is that flow cytometers are less widely available than fluorescence microscopes. In addition, because of their small size and proportionately lower signals, *E. coli* and other bacterial cells are near the sensitivity limits of many flow cytometers, particularly when measuring light scattering. Therefore, careful precautions in preparing particle-free reagents (e.g., sheath fluid, medium) and in optimizing the flow cytometer optics are necessary. A full discussion of the issues involved in the application of flow cytometry to bacteria is beyond the scope of this article; however, several reviews on this topic are available.[16–18] The Purdue University Cytometry Laboratories web site (http://www.cyto.purdue.edu/index.htm) includes information on applying flow cytometry to microorganisms, as well as links to many other flow cytometry sites.

For flow cytometric analysis of GFP fluorescence, cells were grown in filter-sterilized medium, treated with chloramphenicol as described earlier, and were diluted immediately before analysis to between 1×10^5 and 1×10^6 cells ml^{-1} in either filter-sterilized deionized water or M63 salts. The concentration of cells should be adjusted to obtain an event rate of <1000 events sec^{-1} at the sample flow rate used (typically 0.6 to 1 $\mu l\ sec^{-1}$). Standard fluorescent beads ($2 \pm 0.2\ \mu m$ diameter, multiple fluorophore purple/yellow low intensity beads from Spherotech, Inc., Libertyville, IL) were added to each diluted sample at a concentration so that about 1000 beads would be recorded during the run.

Cell suspensions were analyzed using a FACSCalibur (Becton Dickinson) instrument equipped with a 488-nm argon laser. Light-scattering and green fluorescence data were collected for at least 15,000 cells in each sample and stored as listmode files. Data were analyzed using WinMDI version 2.6, a Microsoft Windows-based public domain program, which is available at the Scripps Research Institute FACS Core Facility web site (http://facs.scripps.edu/software.html). Another public domain program for analyzing flow cytometry data is CYTOWIN, which is available from the Station Biologique de Roscoff web site (www.sb-roscoff.fr/Phyto/cyto.html).

Determining the distribution of GFP expression in a population requires counting total cells and then determining the fraction of the population that shows green fluorescence. Typical flow cytometers measure forward

[16] P. Fouchet, C. Jayat, Y. Hechard, M. H. Ratinaud, and G. Frelat, *Biol. Cell* **78,** 95 (1993).
[17] D. Lloyd (ed.), "Flow Cytometry in Microbiology." Springer-Verlag, London, 1993.
[18] H. M. Davey and D. B. Kell, *Microbiol. Rev.* **60,** 641 (1996).

angle light scatter (FSC), right angle (or side) light scatter (SSC), and fluorescence (FL). The user specifies one of these parameters as an event trigger. A particle signal must be greater than the threshold value set for this parameter to be recorded as an event. For our analysis of GFP expression, events were triggered by SSC; each event record contains data for FSC, SSC, and green fluorescence (530 ± 15 nm). Because of their small size, the FSC and SSC signals for *E. coli* are not always significantly larger than noise, i.e., a large fraction of the recorded events will be due to random instrument and laser noise or to contaminating particulates. Moreover, variations in cell size and the orientation of the cell as it passes through the laser beam make it difficult or impossible to preset gating parameters to count all of the cells without including a substantial fraction of events due to noise. However, after data have been recorded, gates can be set during analysis to eliminate most of the noise while still sampling a substantial fraction of the population.

This process is illustrated in Fig. 4, which shows flow cytometry data for cells containing pDS439 grown under conditions of full induction where every cell should be expressing GFP. This positive control culture was used to define a region in the FSC vs SSC plot where all of the events are cells.

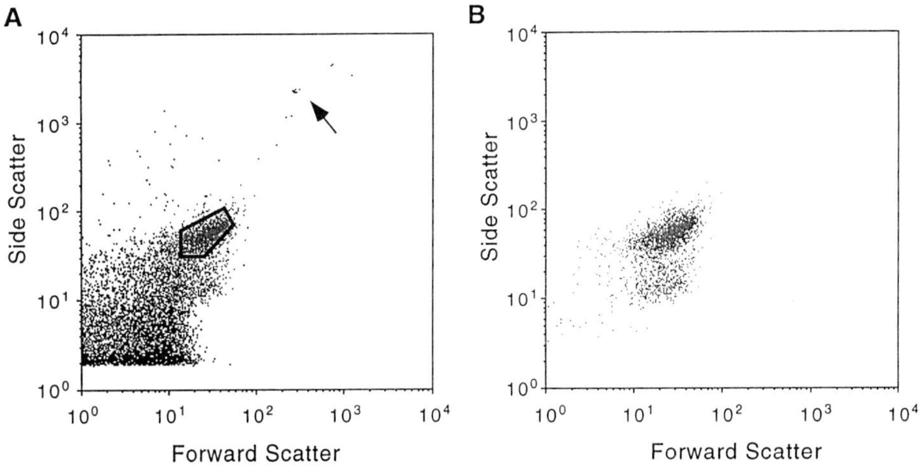

FIG. 4. Two-dimensional dot plots showing light-scattering data for a fully induced culture grown with 13.3 m*M* arabinose analyzed by flow cytometry. (A) SSC vs FSC dot plot before gating. Only the first 20,000 events collected are plotted. The dark line shows the gate (R1) that was selected to encompass the major peak in the plot. The arrow marks the position of the standard beads. (B) Distribution of scattering for events gated for a green fluorescence threshold (≥1 a.u.).

Figure 4A shows the distribution of scattering for 20,000 events recorded in the run, whereas Fig. 4B shows the scattering for only those events that gave a green fluorescence signal above a background level of 1 a.u. Because a fully induced culture was used, the points in Fig. 4B show the scattering distribution for cells, whereas Fig. 4A shows both cells and noise. Note that although the majority of the noise gives low intensity signals in the scattering channels, the distribution of events due to cells overlaps with the events due to noise. The dark line in Fig. 4A shows a gate (R1) that was defined to encompass the major peak in the FSC vs SSC plot. In the experiment shown, the gated area sampled about 60% of the total population of green fluorescent cells. Moreover, within this gated area, 95% of the recorded events also had strong green fluorescence signals, indicating that they were due to cells, and <5% of the events gated to be inside R1 are due to background noise or cells that have lost the plasmid. This result validated the identification of events in mixed samples as cells based on light-scattering signals only (see later).

This approach should obviously be used with caution when using GFP to report on the regulation of genes associated with changes in cell size or shape. For example, cells in stationary phase are known to be much smaller than cells growing in rich medium; this could bias measurements of the distribution of cells expressing GFP under the control of a stationary-phase inducible promoter.

Figure 5C shows how fluorescence intensity is distributed in the subset of the population within the gated area. The shape and position of the major peak in the histogram are identical to those observed if fluorescence

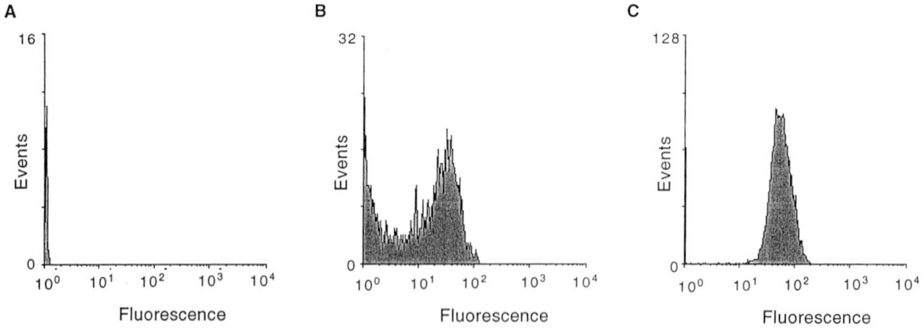

FIG. 5. Histograms showing the distribution of fluorescent intensities for cells from (A) an uninduced culture grown in the absence of arabinose, (B) a culture grown with a suboptimal concentration of arabinose (4.3 μM), and (C) a fully induced culture grown with a high concentration of arabinose (13.3 mM).

intensity is plotted for all events (not shown). The intensity values within the peak vary and are distributed over about a 10-fold range. The width of the distribution is much larger than intrinsic instrument error, as is apparent from the low coefficient of variation for the distribution of fluorescence intensities for the standard beads. Although this distribution is broader than that observed in the measurements made by microscopy (Fig. 3), the difference largely reflects that the intensity values from microscopy are normalized to cell size, whereas the fluorescence intensities from the flow cytometer are not.

This approach allowed us to examine the distribution of green fluorescent and nonfluorescent cells in a mixed population generated by autocatalytic induction of the *araBAD* promoter. The R1 gate shown in Fig. 4A was applied to data collected from cells containing pDS439 that were grown with different concentrations of arabinose. The results are shown graphically in Fig. 5 and are summarized in Table I. Figure 5 shows histograms for the fluorescence intensity of events within the R1 gate. In each case only the subset of cells that gave fluorescence ≥1 a.u. are shown; Table I shows the fraction of the gated cells that gave detectable fluorescence. In the absence of inducer (Fig. 5A), 99.8% of the cells gave a fluorescence signal below 1 a.u. and the few that were ≥1 a.u. are all close to background. Figure 5B shows the distribution of fluorescence intensities of cells from a culture grown with a suboptimal inducer concentration (4.3 μM arabinose). At this concentration of inducer, only 30% of the cells have a green fluorescence signal ≥1 a.u. The cells clearly fall into two populations with respect to fluorescence intensity. The majority of cells have intensities below 5 a.u., whereas a second population has intensities distributed around 40 a.u. With the high inducer concentration, 95% of the cells were brightly fluorescent (Fig. 5C).

An alternative to identifying cells by light scattering would be to use an event trigger based on a second fluorescent label that can be distinguished

TABLE I
GREEN FLUORESCENT CELLS AS A FUNCTION OF ARABINOSE CONCENTRATION

Ara concentration	% of cells in R1 that are fluorescent[a]	No. events within R1
0	0.2	8169
4.3 μM	33	7553
13.3 mM	95	8404

[a] Cells within the gated areas that were expressing GFP were defined as those that gave a green fluorescence signal above background (≥1 a.u.).

from GFP. For example, membrane-permeable dyes are available that stain nucleic acids. However, most of the available appropriate dyes cannot be visualized using a 488-nm argon laser; detection of the fluorescence from these dyes requires excitation by a UV light source.

Concluding Remarks

The green fluorescent protein provides an alternative to the use of enzymes as a reporter for gene expression. The presence of GFP is assayed directly by virtue of its intrinsic fluorescence rather than as a function of an enzymatic activity as is the case for most other commonly used reporters such as β-galactosidase, galactokinase, alkaline phosphatase, chloramphenicol acetyl transferase, or luciferase. Therefore, the signal amplification inherent in any enzymatic reporter is not available. Although it is technically possible to detect single molecules of purified GFP,[19,20] the sensitivity with which GFP can be detected remains one of the most serious limitations in using it routinely as a reporter for gene expression. However, the widespread use of GFP as a research tool should stimulate further work to overcome this problem through a combination of novel instrumental approaches and continued improvements in the brightness of GFP through protein engineering.

Despite this sensitivity problem, the ease with which moderate to high levels of gene expression can be followed in single cells makes GFP the reporter of choice in examining the distribution of gene expression within a population of cells. Protein engineering has already produced a wide variety of GFP variants with shifted excitation and emission properties.[8,21,22] The simultaneous use of two different variants has been combined with flow cytometry to examine the expression of different genes in mammalian cells.[23] Similar approaches should be possible in bacteria.

The sorting capability of flow cytometers provides an additional advantage to GFP reporter systems. Cell sorting allows the physical selection of cells expressing either higher or lower levels of GFP.[24] Moreover, unlike selections based on many other selectable or counterselectable genetic

[19] D. W. Pierce, N. Hom-Booher, and R. D. Vale, *Nature* **388,** 338 (1997).

[20] R. M. Dickson, A. B. Cubitt, R. Y. Tsien, and W. E. Moerner, *Nature* **388,** 355 (1997).

[21] A. Crameri, E. A. Whitehorn, E. Tate, and W. P. C. Stemmer, *Nature Biotechnol* **14,** 315 (1996).

[22] R. Heim and R. Y. Tsien, *Curr. Biol.* **6,** 178 (1996).

[23] M. T. Anderson, I. M. Tjioe, M. C. Lorincz, D. R. Parks, L. A. Herzenberg, G. P. Nolan, and L. A. Herzenberg, *Proc. Natl. Acad. Sci. U.S.A.* **93,** 8508 (1996).

[24] R. H. Valdivia and S. Falkow, *Mol. Microbiol.* **22,** 367 (1996).

markers, cell sorting allows the threshold for selection to be adjusted easily.

Acknowledgments

This work was supported by funds from NSF Grant MCB-9305403 to JH and a Faculty Research and Development Program Award from the College of Agriculture and Life Sciences at Texas A&M University. The authors thank D. Wade Lehmann for technical assistance with flow cytometry and Greg Reinhart for helpful discussions.

[36] Bacterial *lux* Genes as Reporters in Cyanobacteria

By F. FERNÁNDEZ-PIÑAS, F. LEGANÉS, and C. PETER WOLK

Introduction

Bacterial luciferase is a heterodimer ($\alpha\beta$) encoded by the genes *luxA* and *luxB*. The bioluminescent reaction catalyzed by bacterial luciferase involves an oxidation of $FMNH_2$ and a long-chain fatty aldehyde to produce the corresponding oxidized flavin (FMN) and long-chain fatty acid, and blue-green light (490 nm),[1,2] which can be quantitated easily. The fatty aldehyde is synthesized by a reductase, a transferase, and a synthetase that are encoded by *luxC, luxD,* and *luxE.* The *luxC, luxD,* and *luxE* genes flank the *luxAB* genes in different species of luminescent bacteria in the order *luxCDABE.*[3]

luxA and *luxB* have been transferred successfully by transformation, transduction, or conjugation into a number of bacteria and have rendered them luminous on addition of a fatty aldehyde (usually *n*-decanal).[3] In certain cases, the addition of *luxCDABE* as a whole, or *luxC, luxD,* and *luxE* expressed independently from *luxAB,* has eliminated the need to add exogenous aldehyde.[3,4] $FMNH_2$, another substrate of the reaction, is usually present in bacteria at concentrations that are nonlimiting for production of light by the reaction.[3,5] *Vibrio fischeri* or *Vibrio harveyi lux* genes have

[1] S. Ulitzur and J. W. Hastings, *Methods Enzymol.* **57,** 189 (1978).
[2] E. A. Meighen and P. V. Dunlap, *Adv. Microbiol. Physiol.* **34,** 1 (1993).
[3] E. A. Meighen, *Microbiol. Rev.* **55,** 123 (1991).
[4] F. Fernández-Piñas and C. P. Wolk, *Gene* **150,** 169 (1994).
[5] G. Schmetterer, C. P. Wolk, and J. Elhai, *J. Bacteriol.* **167,** 411 (1986).

0076-6879/00 $30.00

been used the most in cyanobacteria and other bacteria because they were the first to be cloned.[6,7]

This article indicates the main applications of *lux* genes in cyanobacteria, summarizes certain useful *lux* constructions, and describes procedures for *in vivo* assays of luminescence.

Applications of *lux* in Cyanobacteria

lux Genes as Reporters of Gene Expression during Cellular Differentiation

The use of transcriptional reporters greatly assists elucidation of the mechanisms that underlie cellular differentiation. Bacterial luciferase has been used widely as a reporter of gene expression[8] because light emission can be visualized and quantitated with great sensitivity and without metabolic disruption, allowing *in vivo* analysis of transcription in real time.

Because of the deep and fluorescent pigmentation of cyanobacteria, colorimetric or fluorescent dyes have been less generally usable in them than in heterotrophic bacteria. Bacterial luciferase, mainly from *V. fischeri,* has been used extensively to determine the timing and spatial localization of transcription from promoters of interest[9] in *Anabaena* sp. strain PCC 7120. The expression of P_{nifH} was visualized within heterocysts of PCC 7120 and in *Anabaena* sp. strain PCC 7118, which fails to form heterocysts, within certain cells that may have been abortive heterocysts. In PCC 7120, P_{rbcLS}::*luxAB* led to expression of LuxAB only in vegetative cells, whereas P_{glnA}::*luxAB* led to expression in all cells.

Tn*5* transposes in *Anabaena* sp. strain PCC 7120.[10] Derivatives of Tn*5* have been developed[11] that bear both *V. fischeri luxAB* as a transcriptional reporter and an *Escherichia coli* origin of replication to facilitate the recovery of DNA contiguous with the transposon. Such a construct, Tn*5*-1063, present in conjugally transferable suicide vector pRL1063a, has facilitated the identification of a variety of environmentally responsive genes in *Anabaena* sp. strain PCC 7120.[11,12] This and related transposons allowed the isolation of mutants that respond to nitrogen deprivation at different stages

[6] J. Engebrecht, K. H. Nealson, and M. Silverman, *Cell* **32,** 773 (1983).

[7] D. H. Cohn, R. C. Ogden, J. N. Abelson, T. O. Baldwin, K. H. Nealson, M. I. Simon, and A. J. Mileham, *Proc. Natl. Acad. Sci. U.S.A.* **80,** 120 (1983).

[8] G. S. A. B. Stewart and P. Williams, *J. Gen. Microbiol.* **138,** 1289 (1992).

[9] J. Elhai and C. P. Wolk, *EMBO J.* **9,** 3379 (1990).

[10] D. Borthakur and R. Haselkorn, *J. Bacteriol.* **171,** 5759 (1989).

[11] C. P. Wolk, Y. Cai, and J.-M. Panoff, *Proc. Natl. Acad. Sci. U.S.A.* **88,** 5355 (1991).

[12] Y. Cai and C. P. Wolk, *J. Bacteriol.* **179,** 258 (1997).

or times of development. Epistatic (dependency) relationships and transcriptional cascades could be identified once particular mutations were tested in other mutant backgrounds.[13,14] Subsequently, transposon derivatives were generated bearing, as reporter, *luxCDABE* from *V. fischeri, V. harveyi, Photobacterium leiognathi,* or *Xenorhabdus luminescens.*[4] Most colonies of *Anabaena* sp. strain PCC 7120 derived from the transposition of such transposons showed aldehyde-limited luminescence (discussed further later).

Anabaena sp. strain PCC 7120 has been widely used as a model organism for the study of heterocyst differentiation, but unlike certain other heterocyst-forming strains, shows no differentiation of hormogonia (motile, heterocyst-free filaments) or akinetes (a type of cell that resists certain environmental stresses).[15] The first direct demonstration of genetic activity in akinetes made use of *Nostoc ellipsosporum,* in which a *hetR::luxAB* fusion was actively expressed.[16]

Nostoc punctiforme strain ATCC 29133, a strain in which heterocysts, hormogonia, and akinetes differentiate,[17] establishes a symbiosis with the bryophyte *Anthoceros punctatus.* A hormogonium-regulating locus, *hrmVA,*[17,18] was identified upon mutagenesis of *N. punctiforme* with pRL1063a, and transcriptional fusions of *luxAB* from *V. fischeri* to *hrmV* and to *hrmA* were used to study the temporal pattern of transcription in response to the presence of an aqueous extract of *A. punctatus.* The results were consistent with the interpretation that *hrmVA* leads to the inhibition of hormogonium formation via metabolism of an unknown hormogonium-regulating metabolite. A new locus, *hglE,* was also identified that appears to be involved in the synthesis of heterocyst envelope glycolipids.[19]

Use of coliphage T7 RNA polymerase[20] permits amplified expression from a weakly expressed promoter.[21] The promoter of interest is fused to T7 gene 1 that encodes the T7 DNA-dependent RNA polymerase, which serves as an initial reporter. The T7 RNA polymerase that is produced acts on a specific T7 promoter (ϕ10) to which *luxAB* is fused transcriptionally.[21]

[13] Y. Cai and C. P. Wolk, *J. Bacteriol.* **179,** 267 (1997).
[14] C. P. Wolk, *Annu. Rev. Genet.* **30,** 59 (1996).
[15] R. Rippka, J. Deruelles, J. B. Waterbury, M. Herdman, and R. Y. Stanier, *J. Gen. Microbiol.* **111,** 1 (1979).
[16] F. Leganés, F. Fernández-Piñas, and C. P. Wolk, *Mol. Microbiol.* **2,** 679 (1994).
[17] M. F. Cohen, J. G. Wallis, E. L. Campbell, and J. C. Meeks, *Microbiology* **140,** 3233 (1994).
[18] M. F. Cohen and J. C. Meeks, *Mol. Plant. Microbe Interact.* **10,** 280 (1997).
[19] E. L. Campbell, M. F. Cohen, and J. C. Meeks, *Arch. Microbiol.* **167,** 251 (1997).
[20] S. Tabor and C. Richardson, *Proc. Natl. Acad. Sci. U.S.A.* **82,** 1074 (1985).
[21] C. P. Wolk, J. Elhai, T. Kuritz, and D. Holland, *Mol. Microbiol.* **7,** 441 (1993).

With this approach, the expression of the *hepA* gene (formerly *hetA*)[22] within single cells in a filament could be visualized clearly hours earlier than without amplification, and the time course of induction of the gene under nitrogen deprivation conditions could be studied. Similar systems based on cloned genes for the DNA-dependent RNA polymerases of coliphages SP6 and T3, and corresponding promoters, were much less effective in *Anabaena* sp. strain PCC 7120.[23]

lux Genes as Reporters of Circadian Rhythms

As discussed extensively elsewhere in this volume,[24] the small genome and the diversity of techniques available for the genetic manipulation of *Synechococcus* sp. strain PCC 7942 made it the prokaryote of choice for studying prokaryotic circadian rhythms[25] (see Fig. 1). Interestingly, *Synechococcus* sp. sustains a ca. 24-hr cycle even under conditions in which the cells divide three or more times per day.[26,27] Promoterless *luxAB* genes proved to be a very powerful tool for studying circadian rhythms and permitted the continuous monitoring of such rhythms over many cycles.[28] Light emission from thousands of colonies was acquired by use of a cooled CCD camera, automated to sample signals from a large number of petri dishes.[29] Almost all luminescent colonies manifested circadian rhythms with a variety of phase relationships, i.e., the circadian clock appeared to affect the entire metabolism of the cell.[29] Other measures of circadian cycling were much less sensitive than *luxAB*-based luminescence.[30]

Biosensing with lux Genes in Cyanobacteria

Cyanobacteria occupy a wide diversity of ecological niches, from desert to aquatic environments. Because of the paucity of naturally luminescent organisms outside of marine ecosystems, *lux*-tagged cyanobacteria could

[22] D. Holland and C. P. Wolk, *J. Bacteriol.* **172,** 3131 (1990).

[23] C. P. Wolk, unpublished observations.

[24] C. Andersson and S. S. Golden, *Methods Enzymol.* **305** [37] 2000 (this volume).

[25] S. S. Golden, M. Ishiura, C. H. Johnson, and T. Kondo, *Annu. Rev. Plant Physiol. Plant Mol. Biol.* **48,** 327 (1997).

[26] T. Mori, B. Binber, and C. H. Johnson, *Proc. Natl. Acad. Sci. U.S.A.* **93,** 10183 (1996).

[27] T. Kondo, T. Mori, N. V. Lebedeva, S. Aoki, M. Ishiura, and S. S. Golden, *Science* **275,** 224 (1997).

[28] T. Kondo, C. A. Strayer, R. D. Kulkarni, W. Taylor, M. Ishiura, S. S. Golden, and C. H. Johnson, *Proc. Natl. Acad. Sci. USA* **90,** 5672 (1993).

[29] Y. Liu, N. F. Tsinoremas, C. H. Johnson, N. V. Lebedeva, S. S. Golden, M. Ishiura, and T. Kondo, *Gene Dev.* **9,** 1469 (1995).

[30] Y. Liu, S. S. Golden, T. Kondo, M. Ishiura, and C. H. Johnson, *J. Bacteriol.* **177,** 2080 (1995).

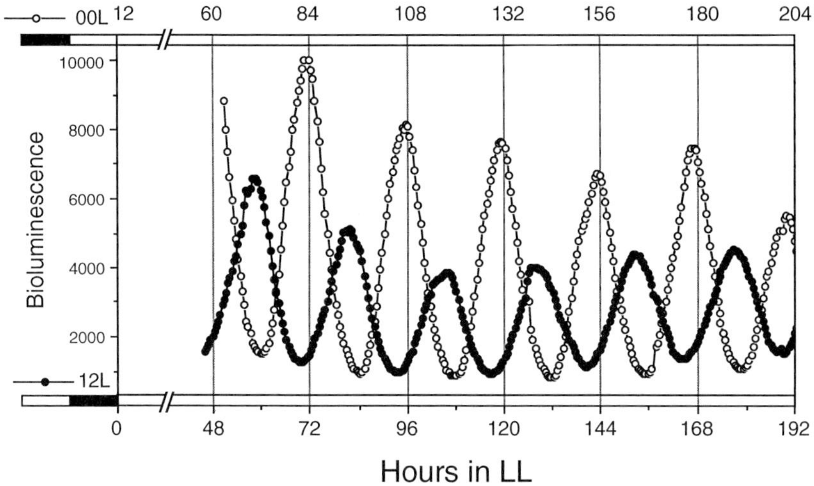

FIG. 1. Circadian rhythm of bioluminescence in continuous light. The transformed derivative AMC149 of *Synechococcus* sp. strain PCC 7942 was cultured at 30° under a 12-hr/12-hr light–dark (LD) cycle (46 mol/m^2 sec) with continuous shaking (100 rpm) and was then transferred to vials for the measurement of bioluminescence. The two traces are from cultures that were entrained previously to LD cycles that were 12 hr out of phase. The last LD cycles preceding continuous light (LL) are illustrated on the abscissa (open bar, light period; filled bar, dark period). Reproduced from Kondo *et al., Proc. Natl. Acad. Sci. U.S.A.* **90,** 5672 (1993). Copyright (1993) National Academy of Sciences, with permission from the publisher and authors.

be conspicuous in many environments. To construct such cyanobacterial strains, *luxAB* could be transcribed from a constitutive promoter placed at a "neutral site" in the genome.[31] Sensitive light-measuring devices would enable the detection of a small number of luminescent organisms, permitting the monitoring of population dynamics in natural habitats.[8,32] *luxAB* could be detected independently by the polymerase chain reaction (PCR).

luxAB can also be put under the control of promoters inducible by such toxic compounds as cyclic hydrocarbons[8] and biologically available ions of heavy metals.[33] Such chimeric[8] constructions have been introduced into various bacteria, including cyanobacteria,[33] to permit the rapid, sensitive monitoring of particular pollutants.

[31] S. A. Bustos and S. S. Golden, *Mol. Gen. Genet.* **232,** 221 (1992).

[32] J. Elhai, *J. Appl. Phycol.* **6,** 177 (1994).

[33] J. L. Erbe, A. C. Adams, K. B. Taylor, and L. M. Hall, *J. Ind. Microbiol.* **17,** 80 (1996).

lux Constructs

From available clones of luciferase genes from *V. fischeri, V. harveyi, P. leiognathi,* and *X. luminescens,* conveniently manipulable cassettes were generated that bear *luxCDABE* free of their native promoters or from which most of *luxA* and *luxB* were deleted, leaving *luxC, luxD,* and *luxE* intact for the generation of fatty aldehydes (Table I).[4] In addition, plasmids pRL1145 and pRL1461 bear cassettes of, respectively, *X. luminescens lux CDABE* and *luxC, luxD,* and *luxE* that carry their original promoters. Cassettes of *luxAB* from *V. fischeri* and *V. harveyi* are available with diverse restriction site ends (Table I). Tn5-1058 derivatives that bear *V. fischeri luxAB, V. harveyi luxAB,* or *luxCDABE* from the four different luminescent bacteria as transcriptional reporters have been constructed and shown to transpose in *Anabaena* sp. strain PCC 7120 (Table I). Integrating vectors (pRL561, pRL579)[9] and a replicating vector (pRL488)[9] with *V. fischeri luxAB* as a reporter are also available for PCC 7120.

Procedures

Exogenous Supply of Aldehyde

Cells expressing *luxAB* produce light only upon the addition of aliphatic aldehydes of appropriate chain lengths. Such an addition should not be necessary if cells are adequately expressing, in addition to *luxAB,* genes *luxC, luxD,* and *luxE* for aldehyde production. The natural aldehydic substrate for *luxA* and *luxB* is believed to be tetradecanal,[34] although for exogenous additions, *n*-decanal (Sigma, St. Louis, MO) is superior for screening due to its greater volatility and possibly a greater ability to cross the bacterial membrane.[35] Erbe *et al.*[33] have used *n*-dodecanal in the cyanobacterium *Synechococcus* sp. strain PCC 7942.

n-Decanal has been provided in a variety of ways: (i) It may be dissolved either in ethanol or in nonionic detergents.[1,33] (ii) A 0.1% (v/v) emulsion has been prepared in a 20-mg/ml solution of bovine serum albumin in water (Sigma Chemical Co.).[5,9] The emulsion is prepared by cavitating the mixture thoroughly in an acid-washed glass tube until a faint blue translucence is seen. Prior to addition, 10% (v/v), to a suspension of *Anabaena* sp., the emulsion is allowed to age overnight, a procedure that increases the duration of luciferase activity.[9] (iii) To provide *n*-decanal in the gas phase, several microliters of pure *n*-decanal can be spread inside the top of a glass petri dish whose bottom holds cyanobacterial colonies (discussed further later).

Prolonged exposure to exogenous aldehyde leads to the lysis of many

[34] S. Ulitzur and J. W. Hastings, *Proc. Natl. Acad. Sci. U.S.A.* **75,** 266 (1978).

[35] C. Miyamoto, M. Boylan, A. Graham, and E. Meighen, *Methods Enzymol.* **133,** 70 (1986).

TABLE I

PLASMIDS AND TRANSPOSONS BEARING *lux* CASSETTES[a]

Plasmid	Drug resistance	*lux* genes	Sites for excision of *lux* cassette	Native promoter	Transposon, or plasmid based on pRL11, bearing *lux* cassette
V. fischeri					
pJHD7744[b]	Ap	*CDABE*	—	Yes	
pRL1114	Km	*CDABE*	*Xba*I, *Sal*I, *Pst*I, *Nru*I, *Eco*RI, *Bam*HI	No	Tn*5*-1113a
pRL1116	Ap	*CDABE*	*Kpn*I, *Bam*HI, *Eco*RI	No	
pRL1022a[c]	Km	*AB*	*Bam*HI, *Xba*I, *Sal*I (and *Acc*I), *Pst*I	No	Tn*5*-1063a
pRL1031a[d]	Ap	*AB*	*Bam*HI, *Xba*I, *Kpn*I, *Sst*I, *Eco*RI	No	
pRL1118	Km	*CD-E*	*Xba*I, *Sal*I, *Pst*I, *Sph*I, *Nru*I, *Xho*I, *Eco*RI, *Bam*HI	No	pRL1456
pRL1143	Ap	*CD-E*	*Kpn*I, *Xba*I, *Sal*I, *Pst*I, *Sph*I, *Nru*I, *Xho*I, *Bam*HI, *Eco*RI	No	
V. harveyi					
pT7VhSacH1[e]	Ap	*CDABEGH*	—	Yes	
pRL208	Km	*CDABE*	*Sph*I, *Kpn*I	No	Tn*5*-1407
pRL92[f]	Ap	*AB*	*Bam*HI, *Xba*I, *Sal*I, *Sph*I, *Hin*dIII	No	Tn*5*-1062
pRL1410	Km	*CD-E*	*Sph*I, *Kpn*I	No	pRL1460
P. leiognathi					
pP125521[g]	Ap	*CDABEG* (+*ribBAH*)	*Xba*I	Yes	
pRL1458	Km	*CDABE*	*Pst*I, *Sal*I (*Acc*I), *Xba*I, *Sma*I, *Eco*RI	No	Tn*5*-1406
pRL1440	Km	*CD-E*	*Pst*I, *Sal*I (*Acc*I), *Xba*I, *Sma*I, *Kpn*I, *Eco*RI	No	pRL1447
X. luminescens					
pXI[h]	Ap	*CDABE*	*Eco*RI	Yes	
pRL1145	Km	*CDABE*	*Pst*I, *Sal*I, *Bam*HI, *Sma*I, *Kpn*I, *Sst*I, *Eco*RI	Yes	Tn*5*-1419
pRL1461	Km	*CD-E*	*Pst*I, *Sal*I, *Bam*HI, *Sma*I, *Kpn*I, *Sst*I, *Eco*RI	Yes	pRL1472a
pRL1425	Km	*CDABE*	*Pst*I, *Sal*I, *Bam*HI, *Sma*I, *Kpn*I	No	
pRL1433	Km	*CD-E*	*Pst*I, *Sal*I, *Bam*HI, *Sma*I, *Kpn*I	No	pRL1465

[a] Adapted from F. Fernández-Piñas and C. P. Wolk, *Gene* **150,** 169 (1994), with permission from Elsevier Science-NL, Sara Burgerhartstraat 25, 1055 KV Amsterdam, The Netherlands. References and constructions, unless otherwise specified, are presented in this reference.

[b] Source of the *Hin*dIII–*Nco*I sequence presented by T. O. Baldwin, J. H. Devine, R. C. Heckel, J. W. Lin, and G. S. Shadel, *J. Biolum. Chemilum.* **4,** 326 (1989).

[c] pRL1022a,b: *luxAB*-bearing *Bam*HI fragment of pRL488 [J. Elhai and C. P. Wolk, *EMBO J.* **9,** 3379 (1990)] cloned in two orientations between the *Bam*HI sites of pRL498 [J. Elhai and C. P. Wolk, *Gene* **68,** 119 (1988)].

[d] pRL1031a: *luxAB*-bearing *Xba*I fragment of pRL1022b[c] cloned between the *Xba*I sites of pRL500 (idem).

[e] C. M. Miyamoto, M. Boylan, A. F. Graham, and E. A. Meighen, *J. Biol. Chem.* **263,** 13393 (1988); E. Swartzman, C. Miyamoto, A. Graham, and E. Meighen, *J. Biol. Chem.* **265,** 3513 (1990).

[f] pRL92: *Bst*XI (trimmed with S1 nuclease)–*Pvu*II fragment of pTB7 [T. O. Baldwin, T. Berends, T. A. Bunch, T. F. Holzman, S. K. Rausch, L. Shamansky, M. L. Treat, and M. M. Ziegler, *Biochemistry* **23,** 3663 (1984)] cloned between the *Sma*I sites of pRL171 [J. Elhai and C. P. Wolk, op. cit. (1988)].

[g] C. Y. Lee, R. B. Szittner, and E. Meighen, *Eur. J. Biochem.* **201,** 161 (1991).

[h] Received courtesy of E. A. Meighen.

cells of *Anabaena* sp. strain PCC 7120, and likely other cyanobacteria. Perhaps to circumvent the toxicity of *n*-decanal to *Synechococcus* sp. strain PCC 7942, Kondo and Ishiura[36] and others[25–30] supplied the aldehyde in the gas phase from solutions [0.5 to 10% (v/v)] of *n*-decanal in vacuum pump oil, soybean oil, or dimethyl sulfoxide.

Endogenous Supply of Aldehyde

Toxicity may possibly result from the accumulation of long-chain fatty acids that are generated by the luciferase-catalyzed oxidation of fatty aldehydes. The exogenous addition of aldehyde can be avoided by the provision of aldehyde-biosynthetic genes (see later); even if such a provision is nonsaturating for the reaction, the products of those genes may dissipate the acids produced from, and thus may reduce the toxicity resulting from, the exogenously supplied aldehyde.

Tn*5* derivatives that contain *luxCDABE* from four different luminescent bacteria (Table I) were used as transcriptional reporters in *Anabaena* sp. strain PCC 7120. The intensity of the luminescence of colonies resulting from transposition was measured in the absence and presence of exogenous aldehyde. When *luxCDABE* was fused to a strong promoter as a result of transposition, endogenously produced aldehyde was not limiting for luminescence; however, in the absence of added aldehyde, *luxCDABE* undermeasured the strength of weak promoters. Similarly, luminescence resulting from a fusion of *V. fischeri luxCDABE* to a metal-responsive *smt* operator/promoter region of *Synechococcus* sp. strain PCC 7942 could be increased by the addition of aldehyde.[33] Thus, it is essential to test whether endogenously generated aldehyde limits luminescence in any particular application.

The expression of *X. luminescens luxC, luxD,* and *luxE* with its native promoter in pDU1 derivative pRL1472a largely circumvented the aldehyde limitation for light emission, yet did not lead to toxicity, in *Anabaena* sp. strain PCC 7120. Neither growth nor differentiation was affected significantly. Similar constructions using *luxC, luxD,* and *luxE* from *V. fischeri, V. harveyi,* and *P. leiognathi* were not as satisfactory, nor were constructions using *X. luminescens* DNA in other plasmids.[4,23] Plasmid pRL1472a was subsequently used successfully to provide aldehyde in derivatives of PCC 7120 and of *Nostoc ellipsosporum* in which *luxAB* from either *V. fischeri* or *V. harveyi* (Fig. 2) was fused to different promoters of interest[12,16]; *X. luminescens luxC, luxD,* and *luxE* genes fused to the very strong *psbA1* promoter of *Synechococcus* sp. strain PCC 7942 and integrated at a "neutral site" of the genome of that cyanobacterium[31] permitted monitoring of circadian rhythms.[25]

[36] T. Kondo and M. Ishiura, *J. Bacteriol.* **176,** 1881 (1994).

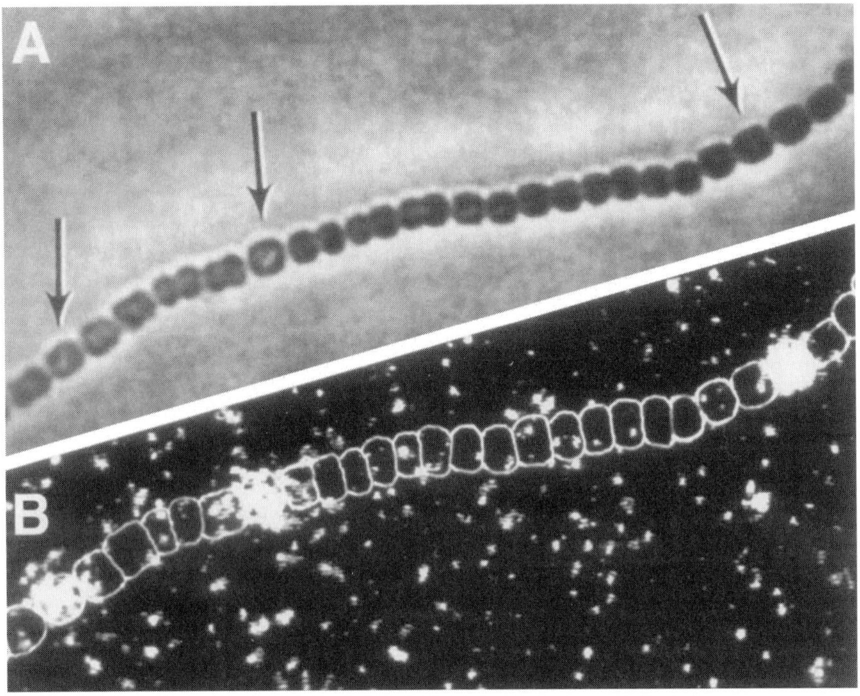

FIG. 2. Light emission from self-luminescent filaments of *hepA* mutant *Anabaena* sp. DR1069a [pRL 1472a] after 17 hr of nitrogen deprivation. This mutant bears a fusion of *Vibrio harveyi luxAB* to *hepA*. (A) A bright-field video image and (B) the corresponding image of luminescence recorded during a 20-min exposure. In A, arrows point to developing heterocysts. Adapted from F. Fernández-Piñas and C. P. Wolk, *Gene* **150,** 169 (1994), with permission from Elsevier Science-NL, Sara Burgerhartstraat 25, 1055 KV Amsterdam, The Netherlands.

Luciferase Assays *in Vivo:* Instrumentation

Measurements of Cells in Suspension

Because the products of transcription and translation of *luxAB* turn over and the light produced by luciferase does not accumulate (although it can be integrated over time), *lux* permits continuous *in vivo* monitoring of gene expression in cyanobacteria as well as in other organisms. Luminescence of cells can be measured by means of a luminometer or, more expensive but more generally available, a scintillation counter, such as those produced by Beckman that are sensitive to photons of visible light. A scintillation counter should be operated in chemiluminescent mode, i.e., coincidence counting should be disengaged. One or 2 ml of a dilute suspension of cells (e.g., 2×10^5 cells/ml) is placed in a suitable vial and, if

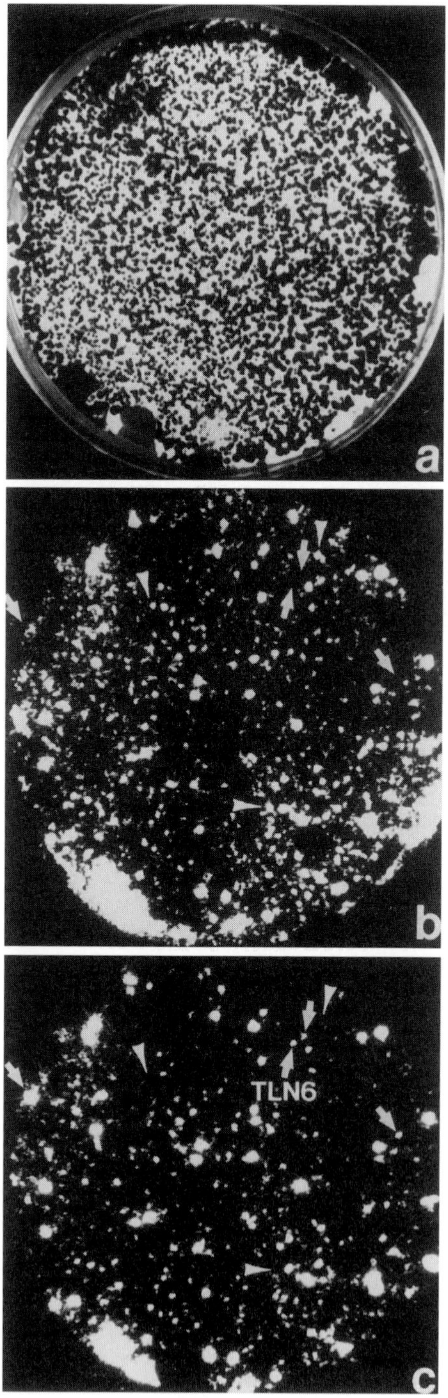
a
b
TLN6
c

necessary, is supplemented with a solution or emulsion of *n*-decanal.[5,18,19,33] Methodology for the continuous, automated measurement of luminescence for studying circadian rhythms[25–30,36] is described elsewhere in this volume.[24]

It is convenient to express luminescence in units of absolute radiant flux, i.e., quanta (photons) emitted per second, upon calibration by the method of Hastings and Weber.[37] To capture very weak luminescence signals, Kondo *et al.*[27] processed the output from a low-noise photomultiplier tube (Hamamatsu R2693P) by a photon-counting unit (Hamamatsu C3866) that removes thermal noise. The results may be normalized to units of cyanobacterial biomass, e.g., chlorophyll *a* content.[18,19,38]

Measurements of Luminescence of Colonies on a Petri Dish

For measurements of short duration, the most sensitive method, but also the most expensive, for screening many colonies on a petri dish employs a macro-lens coupled optically to a Hamamatsu photonic system. In such a system (ours is Model C1966-20 from Photonics Microscopy Inc., Oakbrook, IL), a photon-counting tube consisting of a photocathode, a two-stage microchannel plate, and a phosphor screen is in turn coupled optically to a high-performance, low-lag vidicon, Peltier-cooled to reduce electronic background.[11] For more prolonged measurements, a cooled CCD camera serves well.[29] Least expensive but also least sensitive is photographic detection,[39] for which prolonged exposures and highly sensitive film are normally required. When *lux*-tagged cyanobacteria are grown on sterile filters (e.g., detergent-free, presterilized Nucleopore REC-85 filters or Millipore HATF filters[40]) on an agar surface, their medium can be changed by transfer of a filter from one petri dish to another (Fig. 3). Image analysis then permits visualization of changes of colonial luminescence in response to a change

[37] J. W. Hastings and G. Weber, *J. Opt. Soc. Am.* **53,** 1410 (1963).
[38] T. A. Black, Y. Cai, and C. P. Wolk, *Mol. Microbiol.* **9,** 77 (1993).
[39] L. J. Kricka and G. H. G. Thorpe, *Methods Enzymol.* **133,** 404 (1986).
[40] J. Elhai and C. P. Wolk, *Methods Enzymol.* **167,** 747 (1988).

FIG. 3. (a) White filter bearing thousands of green colonies of *Anabaena* sp. strain PCC 7120 derived from transposition of Tn*5*-1063, a transposon that generates transcriptional fusions to *V. fischeri luxAB* on transposition. Luminescent images of the colonies prior to nitrogen deprivation (b) and after 6.5 hr of deprivation of fixed nitrogen (c). Colonies indicated with arrows show increased luminescence in response to nitrogen deprivation. Colonies indicated by arrowheads show reduced luminescence. Reproduced from C. P. Wolk, Y. Cai, and J.-M. Panoff, *Proc. Natl. Acad. Sci. U.S.A.* **88,** 5355 (1991). Copyright (1991) National Academy of Sciences, with permission.

in environmental conditions (Fig. 3) or a developmental change intrinsic to the cells.

As an alternative to monitoring the luciferase activity of colonies, whose interior and exterior cells are apt to be exposed to different conditions, the activity of homogeneous spots of cells from liquid cultures can be monitored and quantitated. Portions of 2–20 μl are spotted on cut pieces of sterile, detergent-free filters atop an agar-solidified medium. At different times before the luminescence is measured, pieces of filter that bear spots are transferred to a second petri dish that incorporates a different medium (for a response to altered temperature or certain other variables, no such transfer is needed). Imaging of the second petri dish shows the differences of luminescent intensity resulting from different periods of residence on the second medium, i.e., a time course.[11–13] An advantage of this procedure is that it provides automatic compensation of relative luminescence for possible small changes in ambient temperature, instrumentation voltage, or other variables that might affect individual measurements at different time points.

If not generated endogenously, an aldehyde substrate of luciferase, usually *n*-decanal, is added. Several microliters of pure aldehyde can be spread with a cotton swab on the inner surface of the glass cover of a petri dish. To avoid toxicity, luminescence is monitored during or after brief exposure of colonies to aldehyde. To eliminate the remaining aldehyde, sterile air is blown across the petri dish, and the glass cover with *n*-decanal is replaced by an aldehyde-free petri dish cover.[11] The method of Kondo and Ishiura[36] permits the monitoring of luminescence for periods of time in excess of 1 week.[24]

Measurements at the Level of Single Cells

As shown long ago by Eckert,[41–43] bioluminescence can be visualized and quantitated at and below the level of single living cells of a eukaryotic alga. Such imaging is now possible with commercially available equipment. For example, the Hamamatsu Model C1966-20 system mounted on a research microscope in a light-free room has localized expression to individual cells (vegetative cell, heterocyst or akinete)[4,9,12,16,21,38] in filaments of *Anabaena* sp. strain PCC 7120 and *N. ellipsosporum* (Fig. 2).

A drop of a suspension of cyanobacterial cells is placed on a slide, and a cover glass is placed over it and sealed with liquified wax. If cells are not bioluminescent autonomously, 1 μl of an emulsion of aldehyde is added

[41] R. Eckert, *Science* **147,** 1140 (1965).
[42] R. Eckert, *Science* **151,** 349 (1966).
[43] R. Eckert and G. T. Reynolds, *J. Gen. Physiol.* **50,** 1429 (1967).

before the sample is sealed. The slide is placed on a microscope stage that should be kept at a desired, constant temperature. Depending on the brightness of the strain, light can be collected from anywhere from seconds to 10–20 min. For strains that generate their own aldehyde so that toxicity of the substrate is not a problem, the collection time can be extended to 1 hr, although electronic background increases continuously. Another concern is that cells move neither vertically (out of focus) nor longitudinally during imaging. As a control against this possibility, it is advisable to capture a bright-field image of the organism both before and after photon imaging and to check that the bright-field images are superimposable.

Other Considerations

Measurements in (Microaerobic) Heterocysts

Although $FMNH_2$ must be present for bacterial luciferase to function, the possibility remains that the provision of $FMNH_2$ can be rate limiting for luminescence in cyanobacteria. The possible effect of the energy drain of luciferase activity has also not been assessed. Because the light-producing reaction requires O_2, the activity of luciferase is O_2 limited in the microaerobic interior of heterocysts, a fact that has permitted a rough estimation of the pO_2 within the heterocyst.[44] One consequence is that the use of luciferase as a reporter of gene expression during heterocyst differentiation has been limited, to an extensive degree, to the period before the internal pO_2 of the heterocyst decreases greatly. However, because O_2 is required for formation of the glycolipid layer of the heterocyst envelope,[45] and proper formation of that layer is required if respiration is to diminish the internal pO_2 under aerobic conditions,[46,47] differentiation under microaerobic conditions allows O_2 to enter the heterocyst readily upon return of *Anabaena* sp. to aerobic conditions. This strategy has permitted visualization of the expression of *nifHDK* and *hetR* in mature heterocysts.[4,9]

No concerted advantage has yet been taken, in cyanobacteria, of the differing thermal stabilities of luciferase from different sources. Thus, the rapid inactivation of *V. fischeri* LuxAB at 37° (it is stable at 30°) could be used to "erase" a prior record of transcriptional reporting, whereas the

[44] J. Elhai and C. P. Wolk, Abstr. VII Internat. Symp. Photosynthetic Prokaryotes, Amherst, abstr. 114B (1991).
[45] R. Rippka and R. Y. Stanier, *J. Gen. Microbiol.* **105,** 83 (1978).
[46] A. E. Walsby, *Proc. R. Soc. Lond. B* **226,** 345 (1985).
[47] M. A. Murry and C. P. Wolk, *Arch. Microbiol.* **151,** 469 (1989).

stability of *X. luminescens* LuxAB at 42°[48] would commend its use as a reporter in a strain that favors such higher temperatures.

Green Fluorescent Protein (GFP) and LacZ as Alternative Reporters of Gene Expression

A notable achievement of luciferase in cyanobacteria was transcriptional reporting at the level of single living prokaryotic cells. We shall consider the advantages and disadvantages of luciferase relative to two more recent systems for single-cell reporting: fluorescence microscopy with the green fluorescent protein and with fluorogenic substrates of β-galactosidase (LacZ).

The natural as well as modified forms of GFP from the jellyfish *Aequorea victoria* are now used extensively as reporters of gene expression. The fluorescence microscope needed is much less expensive than the photon-counting imaging device needed for luciferase and, unlike measurements with luciferase, there is no difficulty in focusing in such a way as to gain information from numerous filaments simultaneously. Exogenous substrates or cofactors are not needed, assays can be continuous and *in vivo,* and the reporter is highly sensitive.[49] The principal limitation is low background fluorescence around 510 nm. Buikema has used this reporter to localize the expression of particular genes in wild-type *Anabaena* sp. strain PCC 7120 and in several mutants of that strain.[50] There is a delay of fluorescence of ca. 1.5 hr while the protein undergoes a spontaneous cyclization/oxidation reaction,[51] but in a brighter variant of the protein[52] that time interval has been reduced to 20–30 min. However, photobleaching may be a problem. Upon treatment of GFP with sodium dithionite or ferrous sulfate, the fluorescence of the GFP is bleached and remains so under argon but is restored upon the addition of air,[53] suggesting that microaerobiosis in the heterocyst may influence reporting.[50] Because little reduction of fluorescence is seen within several days after treatment with chloramphenicol, the technique may be limited to viewing induction but not turnover of GFP.[54]

[48] R. Szittner and E. A. Meighen, *J. Biol. Chem.* **265,** 16581 (1990).

[49] M. Chalfie, Y. Tu, G. Euskirchen, W. W. Ward, and D. C. Prasher, *Science* **263,** 802 (1994).

[50] R. Haselkorn, *in* Nitrogen Fixation: Fundamentals and Applications (I. A. Tikhonovich, N. A. Provorov, V. I. Romanov, and W. E. Newton, eds.), p. 29. Kluwer, Dordrecht, 1995.

[51] A. Crameri, E. A. Whitehorn, E. Tate, and W. P. C. Semmer, *Nature Biotechnol.* **14,** 315 (1996).

[52] R. Heim, A. B. Cubitt, and R. Y. Tsien, *Nature* **373,** 663 (1995).

[53] S. Inouye and F. I. Tsuji, *FEBS Lett.* **351,** 211 (1994).

[54] W. J. Buikema, personal communication.

Transcription of the *nif* genes of *A. variabilis* has been localized with a LacZ reporter after 1 to 2 hr of incubation with the fluorogenic β-galactoside, C12-FDG (Molecular Probes, Inc., Eugene, OR).[55] Unlike when *luxAB* is used, cells must be permeabilized (killed) and fixed in 0.01% glutaraldehyde prior to the reaction with the fluorogenic substrate. The substrate is expensive, but very little is needed per reaction.[56] Chromogenic substrates of LacZ have not been identified that work well for *in situ* localization. To avoid photobleaching, cells are suspended in one drop of Vecta shield mounting medium (H-1000) (Vector Laboratories, Inc., Burlingame, CA), an antibleaching agent. Self-fluorescence was very dim as compared to the expression of LacZ (unless transcriptional amplification[21] can be used), but may limit the technique to use with strong promoters.[56] Unlike Lux and (indirectly) GFP, the reaction is not influenced by the pO_2.

Acknowledgments

We thank W. J. Buikema and T. Thiel for information concerning the use of GFP and LacZ as alternative reporters. This work was supported by the U.S. Department of Energy under Grant DE-FG02-90ER20021 and by D.G.I.C.Y.T. No. PB 93-0274.

[55] T. Thiel, E. M. Lyons, J. C. Erker, and E. Ernst, *Proc. Natl. Acad. Sci. U.S.A.* **92,** 9358 (1995).

[56] T. Thiel, personal communication.

[37] Application of Bioluminescence to the Study of Circadian Rhythms in Cyanobacteria

By CAROL R. ANDERSSON, NICHOLAS F. TSINOREMAS, JEFFREY SHELTON, NADYA V. LEBEDEVA, JUSTIN YARROW, HONGTAO MIN, and SUSAN S. GOLDEN

Introduction

Cyanobacteria are the simplest organisms known to display circadian rhythms—rhythms of gene expression and physiology that serve to anticipate changes in the day/night cycle and which persist under constant conditions. *Synechococcus* sp. strain PCC 7942, a unicellular obligate photoautotroph, offers many advantages for elucidating the molecular mechanism of

0076-6879/00 $30.00

the circadian clock, because of its small genome size (~2.7 Mb[1] and M. Sugiura, personal communication), efficient systems of homologous recombination,[2,3] and the feasibility of saturation mutant screens.[4]

Luciferase is an ideal reporter for studying circadian rhythms by virtue of its short half-life, real-time reporting, and nondestructive assay. The endogenous luciferase of the marine dinoflagellate *Gonyaulax* displays circadian rhythms of both flash and glow (see paper by Morse and Fritz, this volume).[5] Firefly luciferase is being used to monitor circadian rhythms in higher plants[6] and in *Drosophila*.[7] We are utilizing bacterial luciferase technology together with the genetic malleability of *Synechococcus* sp. strain PCC 7942 to study circadian rhythmicity in this cyanobacterium. The bioluminescent reporter has allowed the automation of monitoring circadian regulated gene expression from thousands of colonies simultaneously.[4]

This article describes the luciferase constructs and genetic tools we have developed to identify and characterize components of the circadian oscillator itself, as well as the signal transduction pathways leading to and from the clock. These methods and plasmids are applicable to the study of other biological questions as well.

luxAB and *luxCD-E* Vectors

Our original reporter construct comprises the *Vibrio harveyi luxAB* structural genes (encoding the luciferase enzyme) fused transcriptionally downstream of the promoter of the *Synechococcus* sp. strain PCC 7942 *psbAI* gene (P_{psbAI}), which encodes a photosystem II component.[8] This transcriptional fusion is linked to a spectinomycin (Sp) resistance selectable marker and is flanked by sequences from a "neutral site" of the cyanobacterial genome: a locus that is known to tolerate the introduction of transgenes without any apparent phenotype (neutral site I, NSI; GenBank accession number U30252). Because the *Escherichia coli* plasmid is not capable of

[1] T. Kaneko, M. Matsubayashi, M. Sugita, and M. Sugiura, *Plant Mol. Biol.* **31,** 193 (1996).
[2] S. S. Golden, J. Brusslan, and R. Haselkorn, *Methods Enzymol.* **153,** 215 (1987).
[3] N. F. Tsinoremas, A. K. Kutach, C. A. Strayer, and S. S. Golden, *J. Bacteriol.* **176,** 6764 (1994).
[4] T. Kondo, N. F. Tsinoremas, S. S. Golden, C. H. Johnson, S. Kutsuna, and M. Ishiura, *Science* **266,** 1233 (1994).
[5] D. S. Morse, L. Fritz, and J. W. Hastings, *TIBS* **15,** 262 (1990).
[6] A. J. Millar, S. R. Short, N. H. Chua, and S. A. Kay, *Plant Cell* **4,** 1075 (1992).
[7] C. Brandes, J. D. Plautz, R. Stanewsky, C. F. Jamison, M. Straume, K. V. Wood, S. A. Kay, and J. C. Hall, *Neuron* **16,** 687 (1996).
[8] T. Kondo, C. A. Strayer, R. D. Kulkarni, W. Taylor, M. Ishiura, S. S. Golden, and C. H. Johnson, *Proc. Natl. Acad. Sci. U.S.A.* **90,** 5672 (1993).

replicating in *Synechococcus*, transformation results from homologous recombination at the neutral site, which causes insertion of the transgene and marker and loss of the flanking and vector sequences.[2] The resulting transformants are bioluminescent when supplied with an exogenous substrate, a long chain aldehyde. *n*-Decanal (Sigma), dissolved to 3% (v/v) in vacuum pump oil or vegetable oil and placed in an upturned microcentrifuge tube lid, is positioned within an agar plate with cyanobacterial colonies to allow a constant supply of substrate in the vapor phase.[9] The bioluminescence is rhythmic in an LD cycle (alternating light/dark cycle of 12/12 hr) and after transfer to LL (conditions of constant light), and the circadian rhythm reflects the rhythm of endogenous *psbAI* expression.[10]

An exogenous supply of substrate can be problematic because higher concentrations of decanal are lethal, while gradual lowering of the concentration of available substrate leads to reduced amplitudes of bioluminescence rhythms. Additionally, adopting a microplate measurement device (Packard TopCount), as developed by the laboratory of Steve Kay for monitoring rhythms of firefly luciferase bioluminescence in *Arabidopsis* and *Drosophila*,[6,7] prompted us to engineer strains that express endogenous substrate to eliminate the need to supply decanal to each individual well.

We have developed autonomously bioluminescent strains by expressing the *Xenorhabdus luminescens luxC-DE* genes encoding enzymes of the fatty acid reductase complex, which assembles aldehyde for the luciferase reaction.[11] High-level expression of *luxC-DE* genes in a strain rhythmically expressing *luxAB* produces autonomous bioluminescence. The rhythm reflects expression of the luciferase enzyme encoded by *luxAB*, suggesting that the reaction substrates (long chain fatty aldehyde, $FMNH_2$, and O_2) are all available in excess.

We have constructed a series of vectors allowing recombination into a second neutral site, NSII (GenBank accession number U44761). These vectors carry either kanamycin (Km) or chloramphenicol (Cm) resistance markers, either promoterless or with P_{psbAI} fusions to either *V. harveyi luxAB* or *X. luminescens luxC-DE* genes. The backbone vectors and the promoterless *luxAB* and *luxCD-E* vectors have a number of unique restriction sites for cloning upstream. The *luxAB* vector should prove ideal for testing various promoters/enhancers for both qualitative and quantitative expression patterns. The promoterless *luxAB* vector pAM1580 (Fig. 1) has a number of compatible restriction sites (*Sal*I/*Xho*I, *Xba*I/*Nhe*I) as well as

[9] T. Kondo, and M. Ishiura, *J. Bacteriol* **176,** 1881 (1994).
[10] Y. Liu, S. S. Golden, T. Kondo, M. Ishiura, and C. H. Johnson, *J. Bacteriol.* **177,** 2080 (1995).
[11] F. Fernandez-Pinas and C. P. Wolk, *Gene* **150,** 169 (1994).

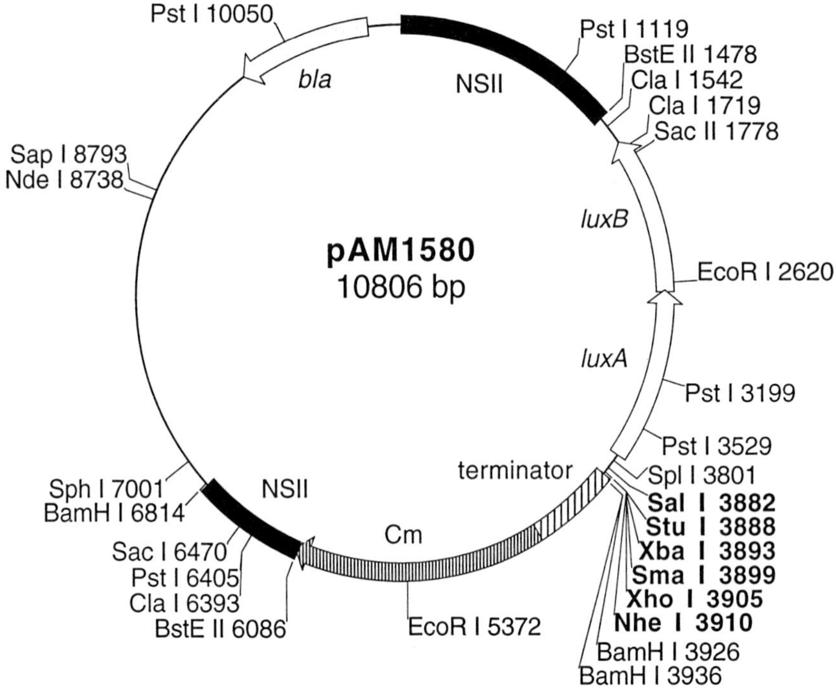

FIG. 1. Map of the promoterless *luxAB* vector pAM1580. This vector allows cloning of a promoter of interest upstream of the *Vibrio harveyi luxAB* genes and subsequent transfer into NSII of the *Synechococcus* sp. strain PCC 7942 genome. The vector backbone is pBR322, allowing for both transformation of and conjugal transfer to *Synechococcus* sp. strain PCC 7942 (an advantage as this strain occasionally loses its transformation ability) and conferring Ap resistance in *E. coli.* The *V. harveyi luxAB* genes (GenBank accession number M10961) are preceded by a number of unique restriction sites (shown in bold type) for insertion of the promoter of interest. In addition, the vector is designed so that when a promoter is cloned into the *Stu*I site and an upstream element into the *Sma*I site, the orientation of the upstream element can be flipped by digestion with *Xba*I and *Nhe*I followed by religation of the compatible restriction sites, and the orientation of the promoter plus upstream element can be flipped by cutting with the compatible *Xho*I and *Sal*I and religation. A strong transcriptional terminator, *rrnB,* minimizes the possibility of flanking genomic or vector sequences affecting transcriptional activity of the promoter of interest. A gene conferring Cm resistance is the selectable marker for *Synechococcus* and *E. coli* transformation. The *luxAB*-marker cassette is flanked by NSII sequences (GenBank accession number U44761), allowing an apparent double recombination event that transfers the reporter cassette into a neutral site of the *Synechococcus* sp. strain PCC 7942 genome.

blunt cloning sites (*Sma*I, *Stu*I), making it possible to flip potential enhancers to test their effect in alternate orientations. The presence of a transcriptional terminator upstream of the promoter cloning sites (or P_{psbAI} promoter) reduces the possibility of transcriptional interference on the promoter of interest.

This variety of vectors allows construction of several reporter strains with different antibiotic markers, allowing subsequent genetic manipulations such as transposon mutagenesis and hit-and-run allele replacement (described later). Maps and full sequences of NSII vectors are available

TABLE I
PLASMIDS FOR ENGINEERING BIOLUMINESCENT REPORTER GENES IN *Synechococcus* sp. Strain PCC 7942

Plasmid[a]	Promoter	*lux* genes	Marker[b]	Description
pAM1303[c]	None	None	Sp^r	NSI vector; cloning sites *Not*I, *Bam*HI, and *Sma*I
pAM1414	None	*luxAB*	Sp^r	Derivative of pAM1303; cloning sites *Not*I and *Bam*HI
pAM1504	None	*luxCD-E*	Sp^r	Derivative of pAM1303; cloning sites *Not*I and *Bam*HI
pAM1518	P_{psbAI}	*luxCD-E*	Sp^r	Derivative of pAM1504; directs luciferase substrate synthesis in *Synechococcus* sp. strain PCC 7942
pAM1573	None	None	Cm^r (Ap^r)	NSII vector; cloning sites *Nhe*I, *Xho*I, *Sma*I, *Xba*I, *Stu*I, *Sal*I, and *Eco*RV
pAM1579	None	None	Km^r (Ap^r)	NSII vector; cloning sites *Nhe*I, *Xba*I, *Stu*I, *Sal*I, and *Eco*RV
pAM1580	None	*luxAB*	Cm^r (Ap^r)	Derivative of pAM1573; cloning sites *Nhe*I, *Xho*I, *Sma*I, *Xba*I, *Stu*I, and *Sal*I
pAM1607	None	*luxCD-E*	Km^r (Ap^r)	Derivative of pAM1579; cloning sites *Nhe*I, *Stu*I, and *Sal*I
pAM1619	P_{psbAI}	*luxCD-E*	Km^r (Ap^r)	Derivative of pAM1607; directs luciferase substrate synthesis in *Synechococcus* sp. strain PCC 7942
pAM1667	None	*luxCD-E*	Cm^r (Ap^r)	Derivative of pAM1573; cloning sites *Nhe*I, *Xho*I, *Sma*I, *Stu*I, and *Sal*I
pAM1706	P_{psbAI}	*luxCD-E*	Cm^r (Ap^r)	Derivative of pAM1667; directs luciferase substrate synthesis in *Synechococcus* sp. strain PCC 7942

[a] All are based on the pBR322 replicon and can be mobilized for conjugation from *E. coli.*

[b] Used for the selection of cyanobacterial transformants (homologous recombination at indicated neutral site); if an additional marker is available for selection in *E. coli,* it is shown in parentheses.

[c] *E. coli* hosts that carry the *rpsL* allele (e.g., DH10B and HB101) have high intrinsic resistance to Sp, which is the only selectable marker on this plasmid and its derivatives.

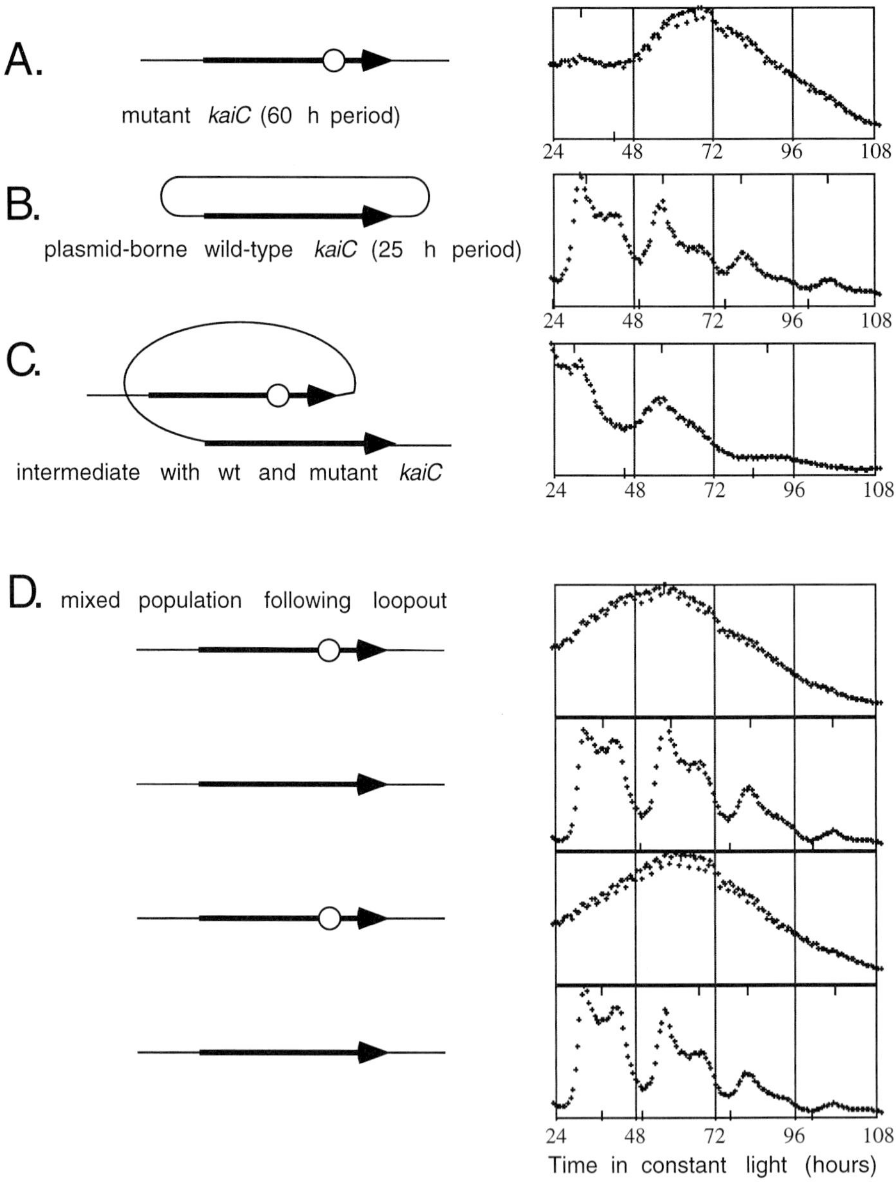

FIG. 2. "Hit and run" can introduce specific point mutations. We used a two-step allele replacement method to introduce specific circadian clock alleles into *Synechococcus* sp. strain PCC 7942. A 3.1-kb *Bam*HI genomic fragment containing a wild-type copy of the clock gene *kaiC* was inserted into pRL278 (GenBank accession number L05083), which carries Km resistance and the *sacB* gene, to create pAM1736. A bioluminescent circadian clock mutant of *Synechococcus* sp. strain PCC 7942, lp60,[4] which carries a single missense mutation in the *kaiC* gene and which has a long free-running period under conditions of constant light (A), was transformed with pAM1736 (B), resulting in insertion of the entire plasmid by a single

on-line at http://ACS.TAMU.EDU/~ssg7231/index.html. Characteristics of the *luxAB* and *luxCD-E* plasmids available are summarized in Table I.

Hit-and-Run Allele Replacement

An important tool for analyzing protein function is the ability to examine the effect of single amino acid changes in a protein of interest, as well as introducing in-frame deletions or insertions. Gene disruption is accomplished easily in *Synechococcus* sp. strain PCC 7942,[2] but insertion of a selectable marker within an open reading frame introduces the possibility of polar effects on upstream or downstream gene expression. In-frame, unmarked deletions are more desirable for unambiguous assignment of a mutant phenotype to a particular gene. A procedure known as hit-and-run allele replacement is established in yeast,[12] mice,[13] and some prokaryotes.[14] The *Bacillus subtilis sacB* gene has been used to provide negative selection for an integrated plasmid as an intermediate step to introduce gene inactivations, deletions, and insertions in a range of bacteria such as *E. coli*[14] and *Anabaena*.[15] The *sacB* gene, encoding a secretory levansucrase, is conditionally lethal in many Gram-negative bacteria in the presence of 5% sucrose.

We have extended the use of a gene inactivation vector, pRL278 (GenBank accession number L05083), designed to create gene knockouts in *Anabaena* sp. strain PCC 7120[15] to introduce defined single nucleotide changes in the *Synechococcus* sp. strain PCC 7942 genome without a residual marker. Vector pRL278 carries both positive and negative selectable markers. A Km resistance gene allows selection for plasmid integration into the genome, whereas the *sacB* gene allows subsequent selection for excision of the plasmid when the medium is supplemented with sucrose. The gene of interest, or a portion thereof, is cloned into pRL278 (Fig. 2B).

[12] R. Rothstein, *Methods Enzymol.* **194,** 281 (1991).
[13] P. Hasty, R. Ramirez-Solis, R. Krumlauf, and A. Bradley, *Nature* **350,** 243 (1991).
[14] A. J. Link, D. Phillips, and G. M. Church, *J. Bacteriol.* **179,** 6228 (1997).
[15] Y. Cai and C. P. Wolk, *J. Bacteriol.* **172,** 3138 (1990).

crossover event at the homologous site in the genome following selection for Km resistance. Colonies carrying the lp60 mutant and wild-type *kaiC* alleles show an intermediate circadian phenotype of bioluminescence (C). Individual Km-resistant colonies were grown in liquid medium lacking Km and plated on medium containing 5% (w/v) sucrose and lacking Km to select for loss of the integrated plasmid. Individual Km-sensitive colonies were screened for their circadian phenotype by monitoring bioluminescence. As expected, the population contained a mixture of colonies expressing either the original lp60 phenotype or the newly acquired wild-type phenotype (D).

Transformation and selection on Km result in transformants in which the entire plasmid has integrated into the homologous site in the genome by a single crossover event, resulting in duplication of the gene of interest (Fig. 2C).[2] After culturing in the absence of Km, cells are plated on medium lacking Km and are supplemented with 5% sucrose to select for plasmid excision, the reverse of the integration event (Fig. 2D).

Depending on the position of the recombination event, the gene of interest remaining in the genome may be either the original allele or the new, plasmid-introduced allele. Subsequent screening (by phenotype, if known, or sequence) allows determination of the allele carried. *In vitro* site-directed mutagenesis of cloned genes allows limitless alleles to be swapped into the cyanobacterial genome.

We have demonstrated the efficacy of this protocol in introducing known circadian clock alleles into different reporter backgrounds. A mutant strain with a 60-hr free-running period could be converted to a wild-type reporter strain and vice versa. This protocol can also be used for moving a mutation identified in a mutagenesis screen into a "clean" background, to eliminate the possibility of secondary mutations influencing subsequent phenotype characterization, or for mutant rescue, to verify the involvement of a mutant allele in conferring a particular phenotype.[16] This two-step procedure can also be used for analyzing alleles of essential genes, as it does not require viability of a null mutant if the plasmid is designed such that the intermediate carries at least one functional copy of the gene.

Protocol for "Hit and Run"

1. Transform the cyanobacterial[2] strain with a pRL278 vector containing the genomic DNA of interest. Select for individual transformants on modified BG11[17] (N+ in all cases for this organism) that contains Km. [Other *sacB* vectors carry different antibiotic markers, giving a greater degree of flexibility, e.g., pRL277 (streptomycin/Sp resistance, accession L05082) and pRL271 (Cm resistance, accession L05081); however, selection for Km is more reliable than for Cm in *Synechococcus* sp. strain PCC 7942, and we have tested this protocol only with pRL278.] These colonies will be the result of a single crossover event, with the entire plasmid inserted into the genome.[2]
2. Pick at least two individual transformants and culture in liquid BG11 lacking Km. Because the *sacB* gene readily picks up spontaneous

[16] Y. Ouyang, C. R. Andersson, T. Kondo, S. S. Golden, and C. H. Johnson, *Proc. Natl. Acad. Sci.* U.S.A. **95,** 8660 (1998).

[17] S. A. Bustos and S. S. Golden, *J. Bacteriol.* **173,** 7525 (1991).

mutations that inactivate the gene, it is important to check more than one independent transformant. (It may be advantageous to maintain the cultures under Km selection for a few days before withdrawing the selection to ensure complete segregation of the multiple chromosome copies so that all copies carry the Km resistance marker.)

3. Plate 100 μl serial dilutions (10^{-1}, 10^{-2}, 10^{-3}, 10^{-4}) of these strains on BG11 supplemented with 5% sucrose and lacking Km to select for loop out of the integrated plasmid. Incubate plates at 30° in constant light. Colonies should be visible in 5–6 days. There will probably be a lawn or patch of lawn on some plates.
4. Choose approximately 100 colonies from the sucrose plates to verify the loop out event by confirming Km sensitivity. Pick colonies that come up early, as background colonies will also grow up with more time. Streak in a grid pattern first on plates containing Km and then on plates lacking Km. (Colonies that are Km resistant probably have a mutated *sacB* gene that allowed escape from the sucrose selection. In most experiments we see no Km-resistant strains, but occasionally half or more of the colonies are Km resistant, indicating that the initial Km-resistant merodiploid chosen probably already had a mutated *sacB* gene.)
5. Screen the Km-sensitive loop out strains for your chosen phenotype. The colonies should contain a mix of the original phenotype (due to retention of the original allele) and the phenotype conferred by the newly introduced allele. The ratio of the two phenotypes will vary (and this variation does not appear to simply correlate with the position of the mutation within the fragment of genomic DNA inserted). As controls in this screen, we always include strains carrying the original allele, a strain carrying the allele we are trying to convert to (if available), and a Km-resistant strain still carrying the pRL278-based vector and, thus, the gene duplication which, for circadian alleles, often shows a phenotype intermediate between that of the individual alleles.

This allele replacement protocol should also be useful in suppressor screens to test whether a suppressor mutation is intragenic or extragenic. Transform the suppressed strain with a pRL278-based plasmid containing the initial mutant allele plus flanking DNA covering the entire gene. Loop out as described earlier. If the mutation is extragenic, all loop out strains should retain the suppressed phenotype, as the allele replacement will never replace the suppressing mutation. However, if the mutation is intragenic, the loop out strains should contain a mix of the suppressed mutant and the original mutant phenotype, as the allele replacement will sometimes replace the suppressing mutation with wild-type DNA, but will always leave the original mutation.

Transposon Mutagenesis of *Synechococcus* sp. Strain PCC 7942 Using Conjugation from *E. coli*

Transposon mutagenesis is a powerful method of introducing mutations as the mutant gene is tagged by the inserted foreign DNA. In addition to null mutants, transposons can also produce more subtle mutations due to over- or underexpression of adjacent genes.

We have developed a conjugation procedure for *Synechococcus* sp. strain PCC 7942,[3] which is a modification of Elhai and Wolk's method for *Anabaena* sp. strain PCC 7120, using Tn*5* transposons developed by Wolk *et al.*[18] We use a Tn*5*-based transposon plasmid, pAM1037, which is a derivative of pRL1058 that has been modified by the addition of a strong outward-reading promoter. This promoter, designated $P_{rbcL\text{-}glnA}$, is a tandem arrangement of the *rbcL* and *glnA* promoters from *Anabaena* sp. strain PCC 7120 (T. S. Ramasubramanian and J. Golden, unpublished results). We have shown that this tandem promoter directs strong transcription in *Synechococcus* sp. strain PCC 7942 with a class I circadian phase[19] like that of P_{psbAI} and at about half the level of P_{psbAI} (Fig. 3). Insertion of this promoter should allow isolation of Tn*5* mutants that result from over- or underexpression of adjacent genes as well as insertional nulls.

Protocol for Tn5 Mutagenesis

Preparation of E. coli cells

1. Grow 10–20 ml overnight cultures of *E. coli* strains AM1037 (Km resistant) and AM1460 (Km resistant) in TB (Terrific Broth[20]) supplemented with the appropriate antibiotics. AM1037 carries the Tn*5*-based transposon plasmid pAM1037 (a derivative of pRL1058 from Peter Wolk). AM1460 contains the conjugal plasmid pRK2013 with the conjugal transfer function of RK2.[21,22]

[18] C. P. Wolk, Y. Cai, and J. M. Panoff, *Proc. Natl. Acad. Sci. U.S.A.* **88,** 5355 (1991).

[19] Y. Liu, N. F. Tsinoremas, C. H. Johnson, N. V. Lebedeva, S. S. Golden, M. Ishiura, and T. Kondo, *Genes Dev.* **9,** 1469 (1995).

[20] K. D. Tartof and C. A. Hobbs, *Focus* **9,** 12 (1987).

[21] D. H. Figurski and D. R. Helinski, *Proc. Natl. Acad. Sci. U.S.A.* **76,** 1648 (1979).

[22] M. F. Cohen, J. G. Wallis, E. L. Campbell, and J. C. Meeks, *Microbiology* **140,** 3233 (1994).

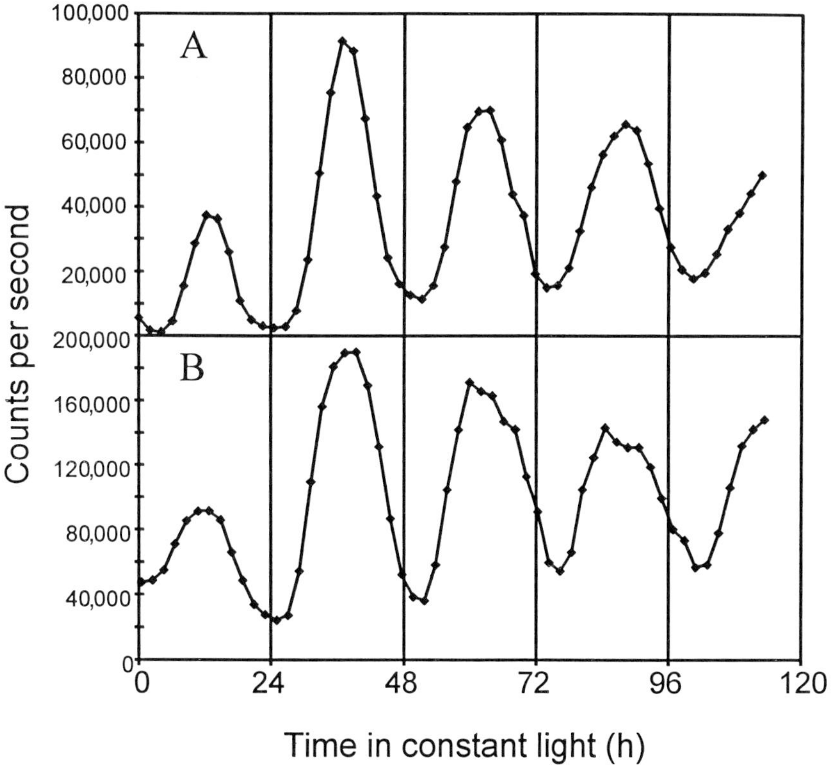

FIG. 3. The *Anabaena* PCC 7120 tandem promoter $P_{rbcL\text{-}glnA}$ drives strong expression in *Synechococcus* sp. strain PCC 7942. The $P_{rbcL\text{-}glnA}$ promoter was inserted into pRL1058,[18] creating pAM1037, to act as a strong outward reading promoter after insertion of the Tn*5* transposon into the cyanobacterial genome. Fusion of the $P_{rbcL\text{-}glnA}$ promoter to *luxAB* confirmed that this promoter drives strong expression in *Synechococcus* sp. strain PCC 7942 with the same circadian phase of expression as the *psbAI* promoter (class I). (A) Bioluminescence from the $P_{rbcL\text{-}glnA}$::*luxAB* fusion and (B) bioluminescence from the P_{psbAI}::*luxAB* fusion. Strains were entrained to a 12-hr light : 12-hr dark cycle and then released into constant white light at time 0.

2. Pellet 10 ml of each *E. coli* strain at 3000 rpm for 5 min.

3. Wash cells twice with 10 ml of sterile water (resuspend pellets gently, do not vortex). Resuspend strains gently in 10 ml of sterile water.

Preparation of Synechococcus sp. Strain PCC 7942 Cells

1. Pellet 100 ml of an appropriate reporter strain of (Km sensitive) *Synechococcus* sp. strain PCC 7942, grown to an OD_{750} of 0.5, at 3000 rpm for 10 min.

2. Gently resuspend cells in 10 ml of fresh BG11.

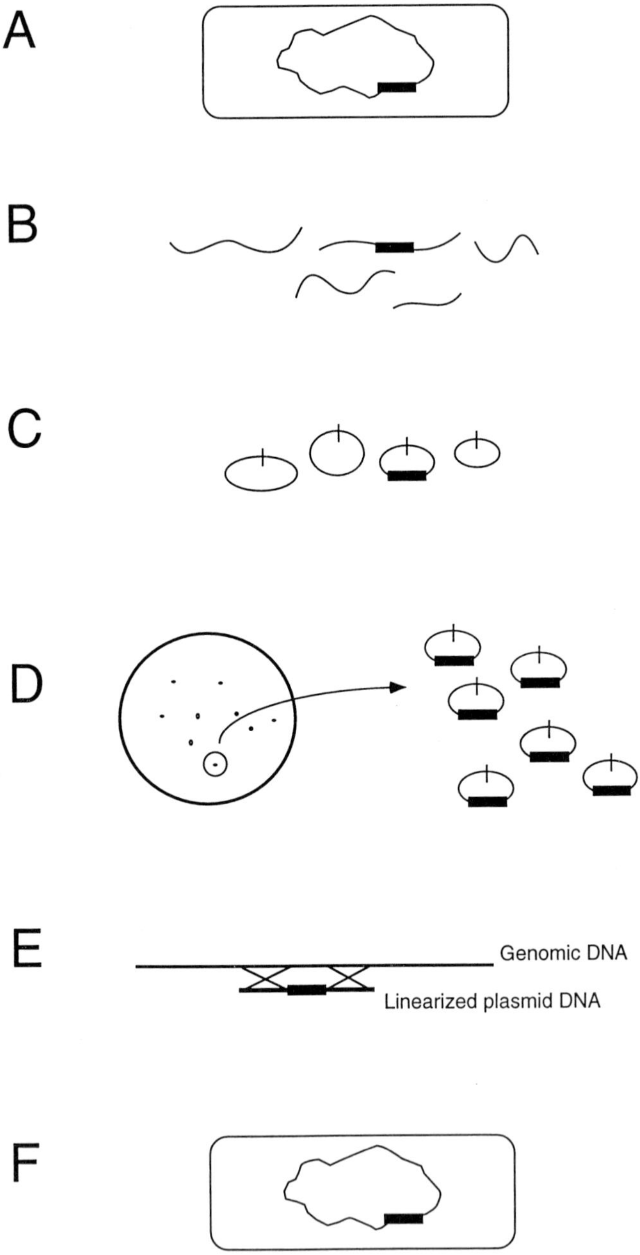
A
B
C
D
E
Genomic DNA
Linearized plasmid DNA
F

Conjugation Protocol

1. Mix 10 ml of *Synechococcus* cells and 5 ml of each *E. coli* strain. Vortex very gently or invert tube to mix cells.

2. Plate 400 μl of the mix on BG11 plates supplemented with 5% (v/v) LB medium (use 40–50 plates when ready to screen for mutants).

3. Grow cells in very low light intensity (30 $\mu E \cdot m^{-2} \cdot sec^{-1}$ with a spherical radiometer = approximately 10 $\mu E \cdot m^{-2} \cdot sec^{-1}$ with a flat meter) for 2–3 days at 30°, without antibiotic.

4. Underlay[2] with 50 μg/ml Km (final concentration after diffusion through agar) and transfer plates to higher light intensities (100–300 $\mu E \cdot m^{-2} \cdot sec^{-1}$). Small Km-resistant colonies will appear 3–5 weeks after the lawn dies, about 50–200 per plate.

The transposon can be recovered by digestion of total DNA from the clones with the restriction enzyme *Kpn*I or *Stu*I, neither of which cuts within the transposon. The genomic DNA is diluted, ligated, and used to transform *E. coli.* The resulting plasmid will carry *Synechococcus* sp. strain PCC 7942 DNA flanking the transposon. Digestion with the same enzyme to linearize the plasmid creates DNA that will transform PCC 7942 or a derivative reporter strain at the site of the original transposition to check linkage of the insertion event with the mutant phenotype, as outlined in Fig. 4.

We have provided a general protocol, although individuals have adopted their "preferred" protocol. The following parameters may influence the efficiency of the protocol in different situations.

FIG. 4. Flow chart for verification of linkage of the Tn*5* insertion with the mutant phenotype. Following conjugation, the transposon carried on pAM1037 is inserted into the *Synechococcus* sp. strain PCC 7942 genome (A). Genomic DNA from a selected mutant of interest is extracted and digested with *Kpn*I or *Stu*I, which do not cut within the inserted transposon sequence, so the transposon will be flanked by genomic DNA sequences (B). The digested DNA is diluted (to promote intramolecular over intermolecular ligation), circularized by ligation, and used to transform *E. coli* (C). Only circularized DNA carrying the transposon sequence will possess an origin of replication for *E. coli,* so *E. coli* colonies selected for Km resistance will carry the rescued transposon plus flanking genomic DNA. Plasmid DNA isolated from individual Km-resistant *E. coli* colonies (D) is linearized by digestion with the original restriction enzyme (either *Kpn*I or *Stu*I) and used to transform a wild-type *Synechococcus* sp. strain PCC 7942 reporter strain. Transformation with linear rather than circular plasmid DNA ensures that the subsequent genomic recombination event is a double crossover targeted to the homologous site of the flanking genomic DNA, rather than a single crossover event resulting in insertion of the entire plasmid (E). Km-resistant *Synechococcus* sp. strain PCC 7942 colonies will carry the transposon insertion at the same locus as in the original mutant (F). Confirmation of the mutant phenotype of the reconstructed mutant verifies linkage of the phenotype with the interrupted locus.

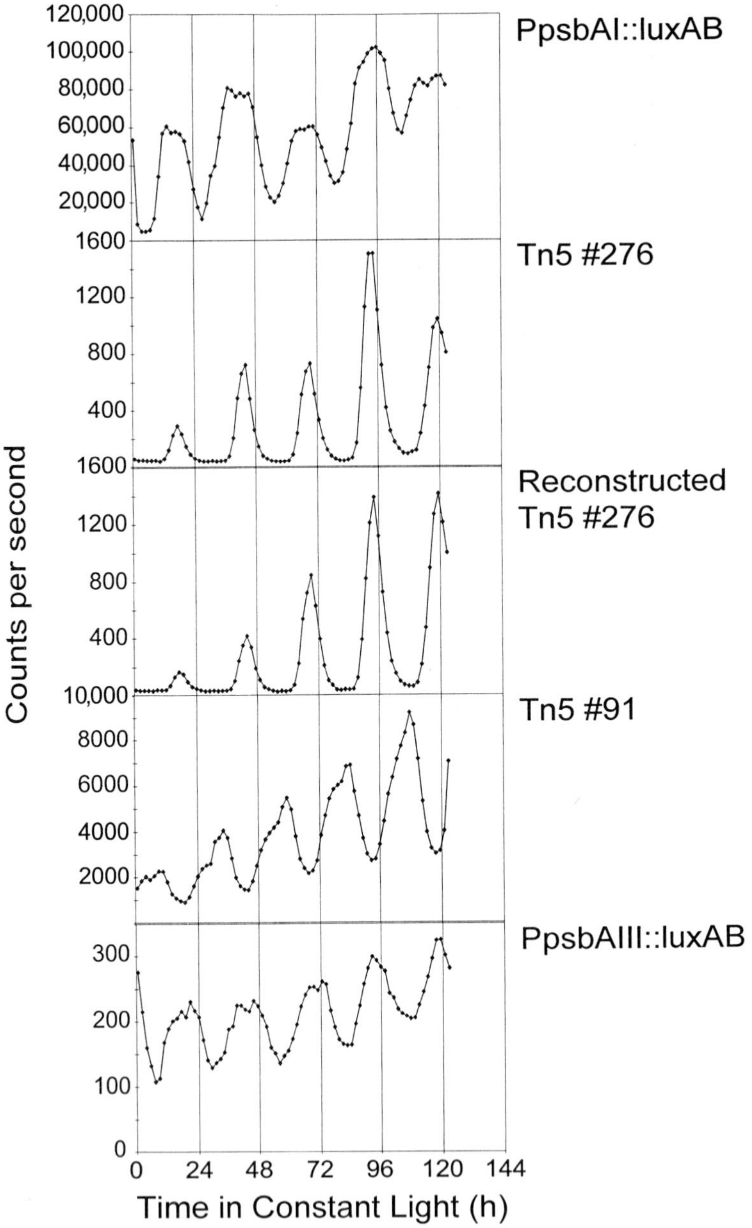
PpsbAI::luxAB
Tn5 #276
Reconstructed Tn5 #276
Tn5 #91
PpsbAIII::luxAB
Counts per second
Time in Constant Light (h)
120,000
100,000
80,000
60,000
40,000
20,000
1600
1200
800
400
10,000
8000
6000
4000
2000
300
200
100
0
24
48
72
96
120
144

- A variation in steps 3 and 4 is to incubate only 1 day in low light before adding the antibiotic and then incubate another day in low light after drug addition before increasing the fluence. With either protocol, the aim is to allow maximal time for transposition before strong selection without encouraging development of a lawn.
- The time required for colony formation (related to die back of the lawn) varies from 10–14 days to 3–6 weeks, as does the frequency of false positives after long incubations. In general, it is advisable to pick colonies that appear earliest to avoid the possibility of working with Km escapes.

Random Insertion of *luxAB*

Promoter trapping is an elegant means of identifying novel genes with interesting developmental, spatial, or temporal expression patterns. Additionally, the tag facilitates subsequent cloning and analysis of genes of interest.

A previous technique for random insertion of a promoterless *luxAB* cassette utilized integration of a plasmid library carrying random genomic fragments linked to *luxAB*.[19] After integration of the plasmid at the homologous genomic site, the *luxAB* cassette will be linked to genomic DNA and there will be a duplication of the genomic region. Plasmid integration appears to be inherently unstable in *Synechococcus* sp. strain PCC 7942, necessitating constant selection for the presence of the integrated plasmid. Also, the use of an intermediate plasmid library created in *E. coli* introduces the possibility of bias in the resulting pool of "random" transformants.

FIG. 5. Random transposon insertion of a promoterless *luxAB* identifies various patterns of circadian gene expression. Bioluminescence of a strain entrained to a 12-hr light : 12-hr dark cycle and then released into constant white light at time 0. (Top) A trace for our standard reporter strain carrying P_{psbAI}::*luxAB* for comparison of bioluminescence levels and phase. Tn*5* #276 was obtained by insertion of the promoterless *luxAB* transposon carried on pRL1063a. This strain displays a very high amplitude rhythm of bioluminescence with sharp peaks in a class 1 phase (peaks in late subjective day). Subsequent recovery of the transposon and flanking genomic DNA (as outlined in Fig. 4) revealed the *luxAB* cassette had inserted within an ORF encoding a predicted helix-turn-helix motif, as found in numerous transcriptional repressors. Recombination of the rescued transposon into a reporter strain to reconstruct the original mutation recreated the phenotype, confirming linkage. Tn*5* #91 is another strain from the *luxAB* transposition pool, but with a less common class 2 phase (peaks in late subjective night). The $P_{psbAIII}$::*luxAB* reporter demonstrates that, even with very low levels of bioluminescence, robust circadian rhythmicity is apparent.

We have used a Tn*5*-*luxAB* transposon[18] to randomly integrate a promoterless *luxAB* cassette throughout the *Synechococcus* sp. strain PCC 7942 genome. Direct conjugation of the transposon from *E. coli* should eliminate some of the potential bias of the library method. Transfer of the *luxAB* cassette into a cyanobacterial strain stably expressing the *luxCD-E* genes (transformed by pAM1518 or pAM1706, Table I) will generate autonomously bioluminescent strains. We have utilized this method to identify tagged strains with various phases and waveforms of circadianly expressed bioluminescence (Fig. 5). Of the Km-resistant ex-conjugants obtained, approximately 73% had detectable levels of bioluminescence. Of these bioluminescent strains, 92% displayed clearly rhythmic bioluminescence (J. L. Shelton and S. S. Golden, unpublished results).

Protocol for Conjugation of Tn5-luxAB

1. Grow *E. coli* strain AM1460 (pRK2013 in HB101, conjugal transfer function of RK2, Km resistant) and a strain containing pRL1063a, the promoterless *luxAB* Tn*5* (Km resistant), overnight. Harvest and wash as for the conjugation protocol described earlier.
2. Grow *Synechococcus* strain AMC395, which carries the P_{psbAI}::*luxCD-E* cassette linked to Sp resistance in NSI (pAM1518). Harvest and wash as for the conjugation protocol described earlier.
3. Mix 10 ml cyanobacterial cells with 5 ml of each resuspended *E. coli* strain. Plate 400 μl on BG plates supplemented with 5% (v/v) LB medium. Incubate at low light, for example, in a transluscent Rubbermaid container for 1 day.
4. The following day, underlay agar with Km to a final concentration of 50 mg/ml. Continue incubation at low light for 1 day.
5. Transfer plates to higher light intensity ($\sim$300 $\mu E \cdot m^{-2} \cdot sec^{-1}$).
6. Colonies can be picked after about 10 days and continue to appear for about 4 weeks. After this, the frequency of false positives that have escaped Km selection increases.

Acknowledgments

We thank Dr. C. P. Wolk for *lux* plasmids and Tn*5* derivatives and advice regarding reporter plasmid constructions, Dr. C. H. Johnson for instigating our switch to the Packard TopCount microarray scintillation counter, C. A. Strayer for assistance in converting to the Packard TopCount system, Drs. S. A. Kay and J. D. Plautz for Microsoft Excel macros for analysis of TopCount data, and members of the S. and J. Golden laboratories for helpful discussions, especially U. Nair for Fig. 4. This work was supported by grants from the National Science Foundation (MCB-9513367), the National Institutes of Health (GM37040), and the International Human Frontier Science Program (RG0385/1996-M with co-PIs C. H. Johnson, M. Ishiura, and T. Kondo).

[38] Construction of *lux* Bacteriophages and the Determination of Specific Bacteria and Their Antibiotic Sensitivities

By S. ULITZUR and J. KUHN

General Principles of Detecting Bacteria with Introduced DNA

One of the main goals of medical and industrial microbiology is the identification of specific types of bacteria. In addition, the antibiotic susceptibility of pathogens is of great interest in clinical microbiology. A wide variety of techniques have been developed for these purposes and include, among others, those of classical microbiology, immunological techniques, DNA probes for hybridization, and the polymerase chain reaction. In all cases, speed and accuracy are paramount. We have developed a rapid method of detecting specific bacterial types and determining their sensitivity or resistance to antibiotics in raw samples.[1]

Our method is based on the introduction of foreign DNA into target bacteria. Conjugation, transformation, and transduction can be used to this end. The transferred segment carries a reporter gene(s). In most cases, we have used *luxA* and *luxB,* specifying the two subunits of bacterial luciferase. The tester reagent, whether it be naked DNA, bacteriophage containing these genes, or a conjugatively competent strain, must emit little or no light. After their entry into the target bacteria, the *lux* genes are expressed and luminescence results. Both bacteriophage and naked DNA are metabolically inactive and dark as are bacteria of medical and industrial importance. The emission of light indicates the presence of a targeted bacterium. In addition, when a sample is found to be positive, the antibiotic sensitivity of the organism can be determined by running the assay in the presence of physiological levels of an antibiotic. Sensitivity is reflected by a reduction in light output.

Media

Luria broth (10 g tryptone, 5 g yeast extract, 5 g NaCl per liter, adjusted to pH 7.0 with 1 *N* NaOH). FT (fortified tryptone for phage λ) contains

[1] S. Ulitzur and J. Kuhn, U.S. Patent 4,861,709 (1989).
[2] A. Janik, E. Juni, and G. A. Heym, *J. Clin. Microbiol.* **4,** 71 (1976).
[3] R. E. Bawdon, E. Juni, and E. M. Britt, *J. Clin. Microbiol.* **5,** 108 (1977).
[4] J. C. Sanford, F. D. Smith, and J. A. Russell, *Methods Enzymol.* **217,** 483.

0076-6879/00 $30.00

10 g tryptone, 5 g NaCl, 2 g maltose per liter, and $MgSO_4$ at a final concentration of 10 m*M*. Plates for bacteria contain 15 g agar per liter; for phage, 10 g. The top layer agar for phage has 7 g agar per liter. Antibiotics are added as necessary: ampicillin (50 μg/ml), kanamycin (30 μg/ml), and streptomycin (2.5 μg/ml).

Bacterial Strains

Several collections of *Escherichia coli* and *Salmonella* were used as test organisms. The *E. coli* strains were: The Selander collection of 72 strains,[5] which is believed to be a representative sample of this organism; 25 Kaufman strains which were the kind gift of Y. Schecter (Ministry of Health, Jerusalem); strains isolated from human urine were obtained from Central Health Services, Haifa; strains isolated from the water supply were provided by Mekorot Water Company of Israel. Collections of *Salmonella* were obtained from Hans Fey and were mostly his strains that are resistant to phage Felix 01; a large collection of strains from Central Health Services, Haifa, many of which have been typed immunologically.

A number of strains for genetic manipulations were received from various sources or were common laboratory strains. *Salmonella typhimurium* and its mutants were obtained from the Salmonella Stock Center. *Escherichia coli* strains for genetic work were all K12 derivatives and appropriate mutant strains are available from the E. coli Stock Center. The relevant genotypes of these strains are given in the text.

Phage Strains

λ, *M13, Mu, P1, P22, N4, Φ80, T5, T7,* and *PRD1* are common laboratory phages.[6] The genotypes of various mutants are indicated in the text. *Felix 01* and G_{47} (referred to here as *OE1*) were the kind gift of Fey.[7] Among the many phages isolated from sewage or chicken farms, three are mentioned here: *4.1, 7.1,* and *LOL.*

Plasmids

Many different plasmids for transposition were constructed and these are based on pNK861 and pNK862,[8] which contain the target sequences

[5] H. Ochman and R. K. Selander, *J. Bacteriol.* **157,** 690 (1984).

[6] R. Calendar, "The Bacteriophages." Plenum Press, New York, 1988.

[7] H. Gudel and H. Fey, *Zbl. Bakt. Hyg. I. Abt. Orig. A* **249,** 220 (1981).

[8] D. Morisato and N. Kleckner, *Cell* **39,** 181 (1984).

for the Tn10 transposase and a gene for ampicillin resistance. The gene for this transposase is present on these plasmids under *plac* control, but lies outside the region that is transposed. pBTK5 contains the entire *lux* operon cloned as a *Sal*I fragment directly from *Vibrio fischeri*.[9] Segments containing the *lux* genes from different organisms have been described.[10] The plasmid containing *sup7* (*supU* = *trpT* for UAG) was obtained from Michael Yarus[11]; those with *supE* (*glnV* for UAG) and *supF* (*tyrT* for UAG) were made by subcloning from phages NM762 and NM781.[12] Preparation of small amounts of DNA was done by the boiling method[13]; larger amounts were prepared by CsCl–ethidium bromide density gradients or Qiagen Midi Prep kits.

Phage Stocks

Small quantities of phage were made by the plate lysate method. Large quantities were prepared as liquid lysates. Liquid lysates can be concentrated by precipitation of the phage by polyethylene glycol 6000[14] or by centrifugation at 47000*g*. For DNA analysis, concentrated phages were purified further on CsCl equilibrium density gradients (0.8 g/ml of cesium chloride, 24,000 rpm using an SW30 rotor in an ultracentrifuge). After dialysis to remove CsCl, the extraction of phage DNA was performed with water-saturated phenol.

Measurement of Light Emission

Scintillation counters or equivalent luminometers can be used to measure low levels of light. For maximum sensitivity, scintillation counters should be operated at the ^3H setting with the coincidence control in the off position. The optimal temperature for measurement is 28° for the *V. harveyi* and *V. fischeri* luciferases. Strong light should be avoided during reading. Under conditions of low humidity, the use of an antistatic spray is recommended to avoid sparks due to friction created by the motion of the vials in the counter. In addition, the first vial usually gives elevated luminescence and should be a blank. Sterile glass vials (prepared with a

[9] J. Engebrecht, K. Nealson, and M. Silverman, *Arch. Microbiol.* **146,** 35 (1983).

[10] S. Almashanu, A. Tuby, R. Hadar, R. Einy, and J. Kuhn, *J. Biolumin. Chemilumin.* **10,** 157 (1995).

[11] L. A. Raftery, J. B. Egan, S. W. Cline, and M. Yarus, *J. Bacteriol.* **158,** 849 (1984).

[12] N. E. Murray, W. J. Brammar, and K. Murray, *Mol. Gen. Genet.* **150,** 53 (1977).

[13] D. S. Holmes and M. Quigley, *Anal. Biochem.* **114,** 193 (1981).

[14] T. Maniatis, E. F. Fritsch, and J. Sambrook, "Molecular Cloning." Cold Spring Harbor Laboratory Press, Cold Spring Harbor, NY, 1982.

lignin stopper, which is removed before use) should preferably be made of high-quality glass as some types made with inferior glass have high levels of fluorescence. Aliquots of 1 to 10 ml can be assayed. Integrative counting is usually done for 6 or 12 sec but 1 to 2 min can be used when more sensitivity is required. Controls are the assay system with bacteria but without bacteriophage and the bacteriophage without its host.

Addition of Aldehyde: The Nazid Mix

In most of the phages carrying *lux,* only the *luxAB* genes were introduced and the aldehyde-synthesizing genes, *luxCDE,* were missing. A long chain aliphatic aldehyde must therefore be added to the assay system shortly before (2 min is sufficient) the determination of luminescence. Although tetradecanal shows the highest luminescence in cell-free extracts of *V. fischeri,*[15] aldehyde permeability becomes a factor in *in vivo* experiments. In most gram-negative bacteria, the presence of lipopolysaccharide in the cell wall decreases the permeability of aldehydes. This phenomenon is more pronounced with C_{14} and C_{12} than with C_9 and C_{10} aldehydes. At certain concentrations, polymyxin B, ethanol, and Triton X-100 promote the entrance of aldehydes into the cell. In addition, NaN_3 has been shown to enhance luciferase activity *in vivo.* All of these have been incorporated into an aldehyde-containing mix named "Nazid." Nazid contains 1 ml of decyl aldehyde, 5 ml of ethanol, 1 ml of Triton X-100, 10 mg of polymyxin B, and 2 g of NaN_3, which is made up to 100 ml with H_2O. The solution is stable for at least several weeks at 4°. Shortly before determining luminescence, the Nazid mix is shaken vigorously and diluted 1:100 into the assay system (10 μl/ml of assay mix).

Reporter Genes

Detection of microorganisms by introduced DNA requires that this DNA carry information encoding some easily measurable function. Reporter genes should express a function that is lacking in the target bacterium, in any other bacterial species that might be present in the sample, and in the sample material itself. A number of reporter genes fit this criterion and among these are the green fluorescent protein, luciferase, and enzymatic activities present in extreme thermophiles. Our studies have focused mainly on bacterial luciferase encoded by the *luxA* and *luxB* genes, which are found in a few marine species. The activity of bacterial luciferase depends on the availablity of oxygen, $FMNH_2$, and a long chain aldehyde.

[15] S. Ulitzur and J. W. Hastings, *Proc. Natl. Acad. Sci. U.S.A.* **76,** 265 (1979).

Conjugation

Certain bacteria (donors) have the capacity to transfer their DNA to other cells (recipients). This process, termed conjugation, is mediated by certain plasmids (fertility plasmids). These plasmids may mediate their own transfer or cause part or all of the bacterial chromosome to be transferred when the plasmid is integrated in the bacterial chromosome. Some fertility plasmids such as *RP4* mediate transfer between distantly related bacteria, whereas others such as F^+ have a limited range in this respect.

Strain AT2446, which is *HfrH* (F^+ is integrated into the bacterial chromosome), was made doubly lysogenic for $\lambda L28$ (containing the entire *lux* operon; its construction is described later) and λcI^+. The latter strain provides an λatt site and encodes functions necessary for integration and stable lysogenization, all of which are lacking in $\lambda L28$. After infection, nonlysogenic cells were eliminated by challenging the cells with $\lambda cI^- b2$. Survivors were isolated that emit light and produce phage after exposure to UV.

HfrH transfers *thr* and *leu* (0 min) early and the site of λ integration 17 min later. When the lysogenized *HfrH* is crossed with a nonlysogen, the transfer of λ leads to zygotic induction because the recipient is lacking the λ repressor. During the lytic cycle of λ, the *lux* genes are expressed from the powerful p_L promoter of the phage. Although there is some background light from the *Hfr* donor, it was found that this can be greatly reduced by growth prior to mating in a sublethal concentration of streptomycin (2.5 μg/ml), which selectively inhibits, the induction of luminescence. Figure 1 shows the development of luminescence for different concentrations of recipient (W3110, F^-) in the presence of a fixed concentration of donor (10^7 cells/ml). The donor had been propagated in LB broth containing streptomycin for 2 hr and then the culture was centrifuged twice and washed with the same medium without the antibiotic. After mixing the donor with the recipient, the mating mixture was incubated without shaking at 37° for 30 min and then placed in a scintillation counter at 24° for repeated light determination. No addition of aldehyde was necessary because $\lambda L28$ contains the entire *lux* operon.

Transformation

Transformation takes place naturally in a number of organisms, whereas in others special treatments such as calcium shock or electroporation are necessary. Because naked DNA is the test reagent, the problem of gene expression before introduction of the DNA is nonexistent.

When the introduced genetic material contains a functional origin of replication, such as those occurring in plasmids, host specificity may be lost. This drawback may be overcome by using fragments of DNA lacking

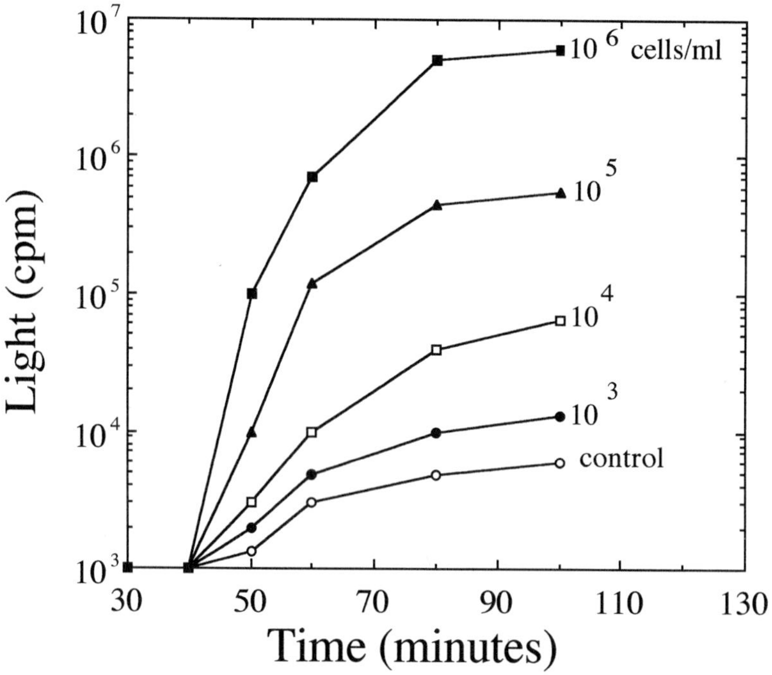

FIG. 1. Time course of luminescence in conjugational crosses. A cross between HfrH (λL28) and W3110 is shown.

such an origin. Such an introduced fragment can become integrated into the chromosome by recombination. When homologous recombination occurs via host sequences flanking the foreign gene or genes, excellent specificity would be expected because closely related bacterial species usually exhibit great divergence in sequence at the DNA level or differ in certain regions that may be absent in one of them while present in the other. However, a loss of specificity could still transpire due to transitory gene expression, similar to that found in abortive transduction. Chromosomal integration could perhaps be made obligatory by using reporter gene DNA fragments that only become linked to a promoter via integration.

Transformation might find use in those rare species that have naturally high levels of transformation but which lack bacteriophages such as *Neisseria gonorrhoeae*.[2,3] Advances in introducing DNA with high efficiency via electroporation and biolistic projectiles[4] have made transformation more attractive. However, in practice the efficacy of such a method is in doubt due to the problems accompanying sample preparation and genera such as *Salmonella* that are relatively refractory to all these techniques.

Transduction

The aforementioned problems would seem to be overcome by using phage-mediated transduction to introduce the foreign DNA containing a reporter gene. Phage infection usually leads to the rapid and efficient injection of the genetic material enclosed within the phage particle. While some phages can transfer sections of bacterial chromosomes from a donor cell to a recipient cell (classical transduction), it is obviously much more efficient to include the reporter gene(s) within the DNA (or RNA, for a few) of the phage genome itself. Large quantities of purified phage lacking bacterial DNA and RNA are easy to prepare. The main problems that need to be solved are, therefore, the isolation of a phage or phages that are specific for the bacterial species to be tested and introduction of the reporter gene(s) into the phage genome.

Phage Specificity

By far the most difficult problem inherent to the use of phage is the isolation of phage specific for a particular species or type and that also infects all the different natural strains of that species. Many phage are species specific but only infect some of the strains therein whereas others spill over and infect some strains from other species. In addition, there may exist problems such as the presence of restriction systems or relatively poor adsorption in specific strains. Good coverage may require a cocktail of several phage types. The specificity problem may not be relevant in cases where a sample is normally sterile such as blood or central nervous system fluid or where related bacteria are unlikely to be found as is the case for *Mycobacterium tuberculosis* infections of the lung. When appropriate phage are available, the construction of *lux* derivatives may be laborious but the chances of success are great.

Synthesis of Recombinant Phage

There are a number of ways in which reporter gene DNA can be inserted into a bacteriophage genome. We have used direct cloning, cloning and recombination, and transposition to incorporate the *lux* genes into various bacteriophages and examples of these methods are detailed later. In most cases, direct cloning is not possible because many phage genomes are large, transformation of phage genomes is relatively poor, many restriction endonucleases used for cloning have multiple sites within the phage genome, and the function(s) of the substituted region is unknown. Transposition is relatively efficient and easily performed but the detection of the transposed phage is often difficult. A third approach is a combination of subcloning phage DNA, insertion of *lux* genes, and homologous recombination. When this is coupled with selective techniques, *lux* phages can be isolated and these can be made in such a way that they are environmentally safe.

Direct Cloning of *lux* Genes into Phage Genomes

The entire *lux* operon of *V. fischeri* is contained within a 9-kb *Sal*I fragment. This was cloned directly into λ *Charon 30*,[16] a cloning vector with *Sal*I sites flanking a region that is dispensable for the lytic growth of λ. Both possible orientations were obtained. In λ *L1* the *luxR* gene is in the same orientation as the leftward operon (p_L) of λ, whereas *luxI, C, D, A, B,* and *E* genes are in the opposite orientation. Light emission begins late in the infective cycle because the accumulation of the *lux* inducer, synthesized by the product of the *luxI* gene, is relatively slow. However, the external addition of the *lux* inducer leads to rapid light production. In λ *L28,* the orientation is reversed and the *luxI-E* genes are transcribed from the p_L promoter, which results in the early onset of luminescence. Luminescence as a function of phage and bacterial concentrations is shown in Fig. 2. It is apparent that at high phage concentrations the light output decreases. This is probably due to lysis from without.

M13 is a male-specific phage of *E. coli* and infects cells that carry the *F* or *colVB* plasmid, regardless of whether these plasmids are free or integrated into the bacterial genome. Elegant derivatives for cloning have been developed by Messing[17] and the *luxA* and *luxB* genes have been cloned into *M13mp18* using fragments with appropriate flanking restriction endonuclease cleavage sites.

Transposition of *lux* Genes into Various Bacteriophages

Many, if not most, phages can incorporate a limited amount of extraneous DNA without losing their ability to package the resultant genome. In addition, from the limited number of phage types for which sufficient information is available, it is clear that most phage genomes contain genes that are "nonessential" for growth under laboratory conditions. Thus, for many phage, direct transposition is possible.

In other phage, the strictures of packaging may prevent transposition to the wild-type phage and deleted derivatives will have to be prepared.

[16] D. Rimm, D. Horness, J. Kucera, and F. R. Blattner, *Gene* **12,** 301 (1980).

[17] J. Norrander, T. Kempe, and J. Messing, *Gene* **26,** 101 (1983).

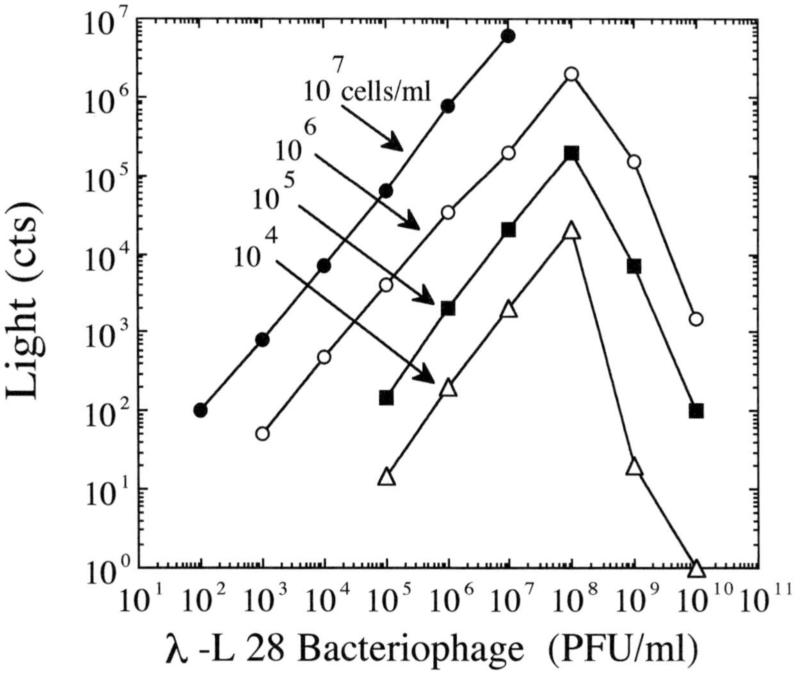

FIG. 2. Light formation in cells infected by λL28. *E. coli* strain W3110 growing logarithmically in FT medium at 37° was diluted to give different cell concentrations as determined by viable count and *λL28* was then added. Different concentrations of the infecting phage were used. The tubes were incubated at 28°. Light output was determined after 45 min as described in the text.

The deletion of dispensable genes will then permit transposition of reporter genes. Deleted derivatives can often be enriched for by exposure to chelating agents such as EDTA or citrate[18] coupled with, in some cases, a mild heat treatment. When enough is known about a specific phage genome, deletions may be isolated by subcloning a segment, removing some of that segment with appropriate restriction enzymes, and recombining the deletion into the complete genome (described later).

A large number of synthetic transposons were prepared from the plasmids pNK861 and pNK862. Among the more than 75 such constructs (an example of which is shown in Fig. 3) are those without a promoter within the region to be transposed, those with the *E. coli plac* promoter, those with the powerful *p*R promoter of λ, and those with the promoter for the

[18] J. S. Parkinson and R. J. Huskey, *J. Mol. Biol.* **56,** 369 (1971).

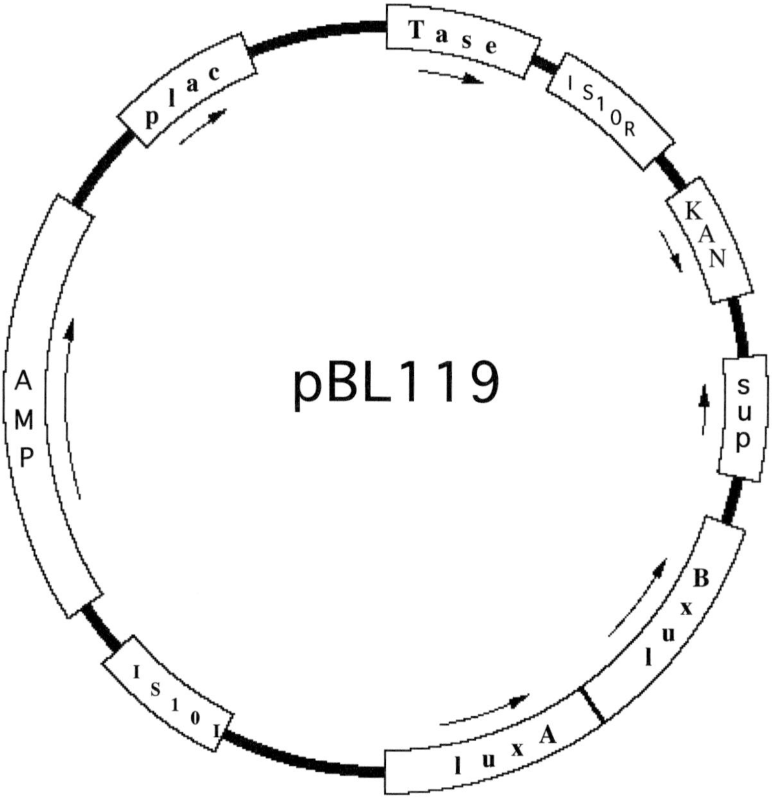

FIG. 3. An example of a *lux* transposition plasmid for isolating *lux* phages. pBL119 was constructed from pNK862, which contains to *bla* gene for ampicillin resistance; *ptac,* a promoter that is inducible by isopropyl thiogalactoside (IPTG); a gene for the *Tn10* transposase (*Tase*) that is transcribed from *ptac*; target sites for the transposase (represented as squares); the kanamycin resistance gene from IS903 (*kan*); the amber suppressor, sup7 (*sup*); and the promoterless *luxA* and *luxB* from *V. fischeri.* When cells containing this plasmid are infected by a phage in the presence of IPTG, transposition of the *kan-sup-lux* region can occur.

kanamycin resistance gene of *IS903.* Transposons lacking a promoter are expected to express the transposed genes from phage promoters when they are inserted into an appropriate region in a functional orientation. The transposable region contains the *lux* genes, *sup7*, and *Kan*^{R} in various combinations.

The isolation of transposed phage depends on a number of factors. When the target phage is temperate, selection for lysogens that have gained resistance to kanamycin is straightforward and many such events have been characterized for λ.[19] For lytic phage, kanamycin resistance or comparable traits might be used to enrich transposed phage carrying antibiotic resistance by adding the antibiotic some minutes after infection. However, in our hands, this has not been found to be effective. There are two more tedious ways that have usually been found to lead to the isolation of luminescent phage. One of these involves the use of an amber suppressor and amber mutations. The other is the direct screening of the phage lysate for *lux* phages.

Direct screening of phage subjected to transposition has been done both by the examination of individual plaques and pooling of phage particles. In the latter, groups of 10, 100, or 1000 phage are added to bacteria in early logarithmic phase and these are amplified by subsequent growth. Prior to lysis of the culture, aldehyde is added and the luminescence is measured. Alternatively, fresh bacteria can be added after lysis and light monitored in the presence of aldehyde. A second method is to plate the transposed lysate such that individual plaques are formed. The plaques are examined directly for luminescence by spraying them with Nazid or subjecting them to aldehyde vapors and then checking them with a special luminometer designed for this purpose. This luminometer is constucted from a circular photomultiplier tube encased in a metal cannister with a movable but tightly fitting top. The petri plate can be secured to the top by means of adjustable screws. The phototube can be partially covered so that only a small area is screened. The plate is rotated slowly and a luminescent plaque causes a sudden rise in light emission. The relevant section is again subdivided until the luminescent plaque is in a small area of the plate that is then removed, sterilely, eluted, and replated. This enrichment is continued until a well-isolated single luminescent plaque is found.

Many luminescent phages have been isolated by transposition: *4.1, 7.1,* λ, *Φ80, LOL, N4, OE1, P1, P22, PRD1,* and *T5.* This technique failed to yield the desired *lux* phage with some other phages. In the latter case, it is worthwhile to employ some selection scheme for transposition events. Amber suppressors should have widespread applicability in this regard.

lux Phages Made by Cloning, Insertion, and Recombination

The use of recombinant DNA techniques provides a precise way to introduce *lux* genes into bacteriophage. Sections of DNA from a bacteriophage of interest are cloned by standard procedures. If nothing is known about the physical map of the phage DNA, a restriction enzyme cleavage map or (preferably) the DNA sequence of the cloned section must be determined. The *lux* genes are then introduced into specific sites through

[19] J. Kuhn, "Molecular Genetics of Bacteria and Phages," p. 77. Cold Spring Harbor Laboratory, Cold Spring Harbor, NY, 1993.

restriction enzyme cleavage and ligation. To enable subsequent transfer of the *lux* insert into the phage genome via homologous recombination, the sites chosen must be flanked with phage DNA and it is recommended that the flanking regions be at least 1 kb in length to promote recombination. When the site chosen is in an essential gene, no viable recombinants will result. Therefore, it is desirable to have some information about which genes are nonessential. Where possible, the isolation of deletions mentioned earlier is an excellent approach. In some cases the function of a gene product can be surmised from the homology of the amino acid sequence of its product to proteins of known function. When these ways are nonproductive, it is probably quickest to insert the *lux* genes at multiple sites and hope that some inserts are in nonessential genes.

Three luminescent types of phage have been made by the technique of cloning and recombination: *T7*, λ, and *Felix 01* (often simply termed *01*).

λ lux

The construction of λ with promoterless *luxA* and *luxB* from *V. fischeri* and *V. harveyi* has been described.[10] A section of λ containing the *N* gene was cloned into a plasmid, and the *lux* genes were inserted in an *Eco*RV site that is within the nearby *ssb* gene of λ. The recombinant phage could be selected for by infecting a permissive strain (*supE* strain C600) carrying the plasmid with an *N* gene double amber (N_7N_{53}) and plating the progeny on a sup^- host (W3110), which selects for N^+ phage. However, even without selection there are not less than one recombinant per 300 progeny.

Felix 01 lux

Felix 01[20] is a bacteriophage that infects almost all *Salmonella* strains and thus is potentially valuable for the testing of foodstuffs, patients with severe diarrhea, and carriers. After the initial report of this phage's specificity for *Salmonella*[21] and its efficacy in testing for this genus, many groups have subsequently employed it as a diagnostic test. Using a simple spot test, Fey *et al.*[22] found that 96% of all strains from clinical and food samples were susceptible to this phage. Their samples (22,000 strains) were collected in Switzerland over an extensive period. A different study in Sweden[23] found that 99.5% of the strains examined were sensitive to *Felix 01*. The phage does, however, attack a few *E. coli* strains.

The genetic material of this bacteriophage is linear, double-stranded

[20] A. Felix and B. R. Callow, *Br. Med. J.* **2,** 127 (1943).
[21] W. B. Cherry, B. R. Davis, P. R. Edwards, and R. B. Hogan, *J. Lab. Clin. Med.* **44,** 51 (1954).
[22] H. Fey, E. Burgi, A. Margadant, and E. Boller, *Zbl. Bakt. Hyg. I. Abt. Orig. A* **240,** 7 (1978).
[23] E. Thal and L. O. Kallings, *Nord. Vet. Med.* **7,** 1063 (1955).

DNA with a length of about 80 kb. Biological information has aided us in the construction of a *Felix 01* containing the *lux* genes. λ*red* mutants fail to grow on *E. coli* that is *recA*.[24] We discovered that one of the *Felix 01* clones permits λ*red* mutants to propagate on *recA* strains. The *Felix 01* gene responsible has also been named *red* and it is non-essential. There are several open reading frames that lie immediately downstream of the *Felix 01 red* gene. Fortunately, at least one of them is essential. *red* and the downstream genes of *Felix 01* were deleted from an appropriate *Felix 01* subclone by cleavage with restriction enzymes, and a promoterless segment with *sup7, luxA,* and *luxB* was substituted in the correct transcriptional orientation. The orientation of the genes in this segment of *Felix 01* has been inferred from DNA sequence data.

A number of amber nonsense mutants were isolated from the wild-type phage by nitrous acid mutagenesis and subsequent testing of plaques on sup^+ and sup^- *Salmonella* strains. Several of these ambers were crossed to each other, and a phage with two amber mutations was isolated. In lysates of the double amber, reversion to the wild type is negligible.

The plasmid with the inserted *lux* genes was transferred to a sup^+ strain of *Salmonella* and then infected with the double amber phage. The amber mutations lie outside the cloned segment used. Recombinants that pick up the sup^+ of the plasmid will be able to grow on sup^- strains. The sup^+ recombinants will also, by necessity, contain the *lux* genes. However, the recombinants should not be able to grow because they now lack an essential gene. This can be overcome by constructing a plasmid expressing the essential gene(s). To prevent the loss of the *lux* insert from the recombinant *Felix 01* phage, the complementing plasmid should only contain the essential gene but not *red* or sequences outside the deletion. Due to a lack of homology, recombination is prevented. These *lux* phages are thus "locked," i.e., unable to grow in nature unless they recombine with a wild-type phage, which leads to the loss of the insert. Such phage are environmentally safe.

Felix 01 containing *lux* detects most of the *Salmonella* strains tested but only a very few of the many *E. coli* strains. A number of other enteric bacteria were examined but were negative.

Use of Phage-Infected Cells to Detect Antibiotics and Antibiotic Sensitivity

Several sensitive tests for antibiotics using the *lux* system have been described.[25,26] We have developed two tests that use either *lux* bearing

[24] G. R. Smith, *in* "Lambda II" (R. Hendrix, J. Roberts, F. Stahl, and R. Weisberg, eds.), p. 175. Cold Spring Harbor Laboratory Press, Cold Spring Harbor, NY, 1983.

[25] A. Naveh, I. Potasman, H. Bassan, and S. Ulitzur, *J. Appl. Bact.* **56,** 457 (1983).

[26] S. Ulitzur, *Methods Enzymol.* **133,** 275.

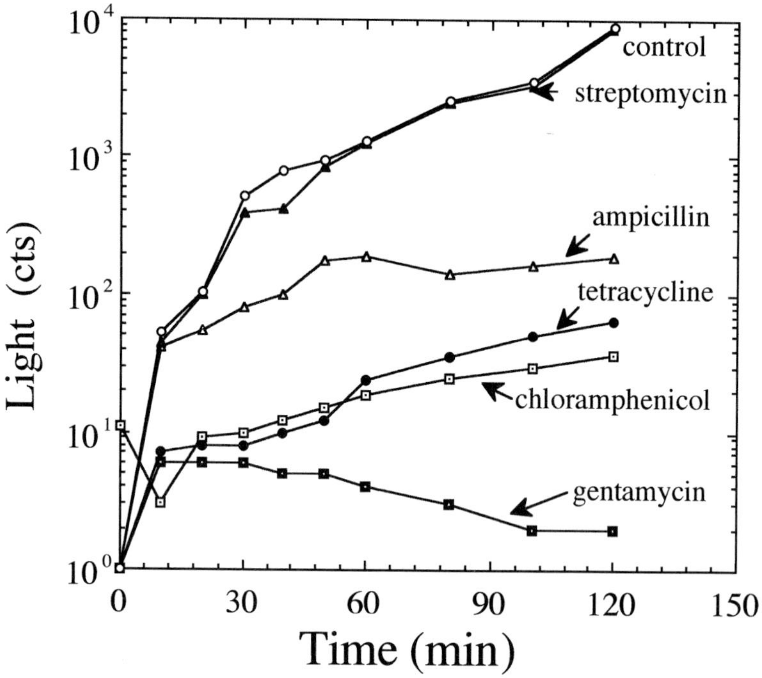

FIG. 4. Determination of antibiotic sensitivity in urine. A fresh urine sample was contaminated with 10^6/ml *E. coli* strain W3110 *strA* (streptomycin resistance). The urine was diluted fourfold into FT medium with and without different antibiotics (all at a concentration of 35 μg/ml). $\lambda L28$ was then added to a final concentration of 10^8/ml. The vials were incubated at 28°, and luminescence of the cultures was determined at various times with a luminometer.

phage or plasmids containing the *lux* genes under phage control. The latter of these is composed of the *luxA* and *luxB* genes transcribed from the powerful *p*R promoter of λ. The promoter is under the control of the λ repressor expressed from the *cI* gene in pHG276.[27] The tester cells are heated in buffer to 50°, which inactivates any previously synthesized luciferase and also the λ repressor. The cells can be used directly or after lyophilization. In the latter case the cells are rehydrated when convenient. Light is emitted very shortly after the bacteria are added to an antibiotic-free sample, but the amount of light is reduced greatly by antibiotics that affect transcription, translation, or the electron transport chain. Antibiotics that affect cell wall synthesis, such as penicillins, are not detected.

A second type of test is to add an *E. coli* strain to the sample and infect

[27] G. S. A. B. Stewart, S. Lubinsky-Mink, and J. Kuhn, *Plasmid* **15,** 182 (1986).

the cells at various times with a *lux*-bearing phage such as $\lambda L28$. A reduction or absence of light after infection again indicates the presence of an inhibitory compound. Penicillins can be detected by waiting an appropriate interval before adding the phage reagent. An example of the kind of results that are obtained with such tests is presented in Fig. 4. The host cells were *strA*, which confers a streptomycin-resistant phenotype. These tests are very applicable for the food industry where the presence of antibiotics or other inhibitory substances is undesirable.

In an analogous manner, the sensitivity of pathogens to antibiotics can be determined. When samples tested by the *lux* phage show that a pathogen is present, the test can be repeated adding a physiologically relevant amount of an antibiotic. A similar test employing our technology has been developed for *Mycobacterium* using phage and firefly luciferase.[28]

[28] W. R. Jacobs, R. G. Barletta, R. Udani, J. Chan, G. Kalkut, G. Sosne, T. Kieser, G. J. Sarkis, G. F. Hatfull, and B. R. Bloom, *Science* **260,** 819 (1993).

[39] Luciferase Gene as Reporter: Comparison with the CAT Gene and Use in Transfection and Microinjection of Mammalian Cells

By S. Gelmini, P. Pinzani, and M. Pazzagli

Introduction

Comparison of the relative merits of different luciferases with nonluminescent reporters is complicated by the wide variety of biological systems and experimental conditions with which they have been used. Evaluation of the performance of luciferases as reporter genes can also be influenced by the fact that measurements of light emission are often expressed in arbitrary units, which can vary between instruments and assay conditions. However, luciferase genes have several advantages over traditional reporter genes, including the possibility of making measurements in living cells. This article reports on data related to several aspects, including (i) comparison of the performances of various reporter genes; (ii) data about the stability of the bacterial and firefly luciferases and how we can solve them; (iii) problems of reproducibility in transfection studies and how we can improve precision in experimental results; and (iv) microinjection of reporter genes.

METHODS IN ENZYMOLOGY, VOL. 305
0076-6879/00 $30.00

Performance of Luminescent Reporter Genes and Comparison with CAT

In order to evaluate the various luciferase systems as reporter genes, we chose to express them from the glucocorticoid-inducible promoter of mouse mammary tumor virus (MMTV). Transfection was performed in murine Ltk$^-$ cells by a DEAE/dextran-mediated procedure. The chloramphenicol acetyltransferase (CAT) gene from *Escherichia coli* was included for comparison in this study. The luciferase genes investigated were the *luc* gene from firefly *Photinus pyralis* and the *luxAB5* gene, a translational fusion of the α and β subunits of bacterial luciferase from *Vibrio harveyi* (VH). The *luc* gene of *P. pyralis* is the most commonly used bioluminescent reporter and has been expressed successfully in numerous animal and plant cells.[1-5]

The luciferase of *V. harveyi* is a well-characterized,[6] heterodimeric enzyme (see Baldwin and Ziegler[7]); the *lux* genes have been used as reporters mainly in prokaryotic cells. Because the expression of two genes (*luxA* and *luxB*) would make the *lux* system less suitable for use in eukaryotes, we have used a fused *luxAB* construction[8] for the expression of bacterial luciferase in mammalian cells. Successful application of fused bacterial luciferase in yeast has been reported.[9,10] In order to overcome the limitations that result from the use of arbitrary light units in the luminescent assays, we have employed reference materials and calibration curves that allow us to express the results in terms of molar amounts of active enzyme within samples. We have also investigated the possibility of quantitating luciferase in transfected mammalian cells using a procedure that takes advantage of its high detection sensitivity and assay facility.

Materials and Methods

A detailed list of reagents, the procedures for the various enzyme assay methods, and information on the construction of the plasmids used in this

[1] J. R. deWet, K. V. Wood, M. DeLuca, D. R. Helinski, and S. Subramani, *Mol. Cell. Biol.* **7,** 725 (1987).
[2] M. Schneider, D. W. Ow, and S. H. Howell, *Plant Mol. Biol.* **14,** 935 (1990).
[3] A. J. Palomares, M. DeLuca, and D. R. Helinski, *Gene* **81,** 55 (1989).
[4] A. R. Brasier, J. E. Tate, and J. F. Habener, *BioTechniques* **7,** 1116 (1989).
[5] I. H. Maxwell and F. Maxwell, *DNA* **7,** 557 (1988).
[6] C. H. Roelant, D. A. Burns, and W. Scheirer, *BioTechniques* **20,** 914 (1996).
[7] T. O. Baldwin and M. M. Ziegler, *in* "Chemistry and Biochemistry of Flavoenzymes" (F. Müller, ed.), Vol. III, p. 467. CRC Press, Boca Raton, FL, 1992.
[8] L. J. Chlumsky, Ph.D. Thesis, Department of Biochemistry and Biophysics, Texas A&M University, College Station, Texas (1991).
[9] M. Boylan, J. Pelletier, and E. A. Meighen, *J. Biol. Chem.* **264,** 1915 (1989).
[10] D. Manen, M. Pougeon, P. Damay, and J. Geiselmann, *Gene* **186,** 197 (1997).

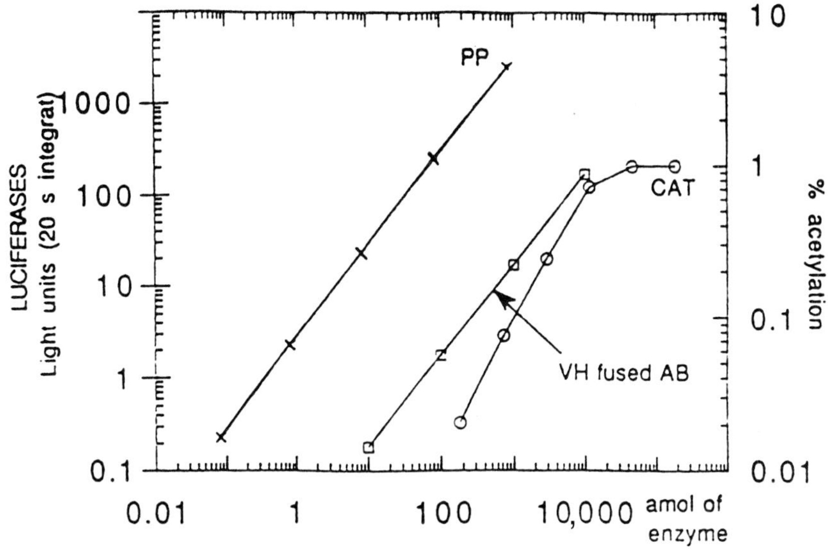

FIG. 1. Calibration curves of firefly and bacterial luciferases and CAT.

study have been reported by Pazzagli *et al.*[11] Transfections of Ltk$^-$ cells in standard 100-mm plates by the DEAE-dextrandimethyl sulfoxide shock procedure were performed as described previously.[12]

Results and Discussion

The calibration curves established for firefly luciferase, bacterial luciferase, and CAT are presented in Fig. 1. The performance of each enzyme assay with various tests is summarized in Table I. The signal derived from the various luciferases was linear over the entire range tested; furthermore, the bioluminescent assays were 20- to 2000-fold more sensitive than the CAT assay. The firefly luciferase assay was about 100-fold more sensitive than the bacterial luciferase assay.

Calibration curves for the various enzymes investigated clearly show that firefly luciferase can be assayed with very high detection sensitivity (0.05 amol/sample). This sensitivity is about one order of magnitude higher than that reported by Subramani and DeLuca[13] and may be due to both the addition of CoA in the reaction and the use of small test tubes rather than the vials that earlier investigators used for their assays. Characteristics

[11] M. Pazzagli, J. H. Devine, D. O. Peterson, and T. O. Baldwin, *Anal. Biochem.* **204,** 315 (1992).

[12] M. G. Toohey, K. L. Morley, and D. O. Peterson, *Mol. Cell. Biol.* **6,** 4526 (1986).

[13] S. Subramani and M. DeLuca, *Genet. Eng.* **10,** 75 (1988).

TABLE I
ANALYTICAL PERFORMANCES OF ENZYME ASSAYS RELATED TO VARIOUS REPORTER GENES[a]

	Firefly luciferase PP	Bacterial Luciferase LuxAB5 (VH)	CAT
Source of standard	BMB	TOB laboratory	BMB
Sensitivity	0.05 amol	5 amol	100 amol
Intraassay precision			
Low range	CV = 1.8%	CV = 5.0%	
High range	CV = 3.1%	CV = 6.2%	ND
Interassay precision	CV = 10.6%	CV = 12.3%	
Recovery test			
Low range	102.1 ± 4.8%; *n* = 6	93.7 ± 4.1%; *n* = 4	
High range	96.1 ± 3.0%; *n* = 6	93.0 ± 5.1%; *n* = 4	ND
Effect of Triton X-100	No effect to 4%	Reduction to one-third activity at 1%	

[a] ND, not determined; BMB, Boehringer Mannheim; TOB, T. O. Baldwin.

of the firefly luciferase appear to be entirely satisfactory in terms of linearity, precision, and absence of interference from biological sources for use in assays of the type described here. The CAT assay appears to be significantly less sensitive than any of the bioluminescent assays.

Expression of Bacterial Luciferase. Plasmid pLSVH was used for the transfection of Ltk^- cells. A very low light signal, close to the background of the assay, was detected in cell lysates when the cells were grown without dexamethasone (DXM) stimulation (7.34 ± 3.7 amol/100 μg of total protein; $n = 5$). In dishes where DXM was added, the expression of VH luciferase increased to 57.9 ± 18.5 amol/100 μg of total protein ($n = 8$). We postulated that the low level of expression of bacterial luciferase in mammalian cells could be due to the temperature at which the mammalian cells were incubated (37°). To test this theory, transfected cells with the same amount of plasmid were shifted to 30° at the time of the addition of DXM. After 15 hr, harvested cells yielded more than 10 times the expression of bacterial luciferase at 37° (without DXM = 72.9 ± 26.3 amol/100 μg of total protein; with DXM = 965.0 ± 97.6 amol/100 μg of total protein).

Expression of CAT and Firefly Luciferase. Results of the expressed enzyme quantity, using different amounts of the various plasmids (10 and 5 μg/dish, respectively) and normalized to 100 μg of total protein, are summarized in Figs. 2A and 2B for CAT and in Figs. 2C and 2D for *P. pyralis* firefly luciferase.

In Figs. 3A and 3B the time courses of expression of the enzyme activities are also reported. Increments due to the addition of DXM (final concentra-

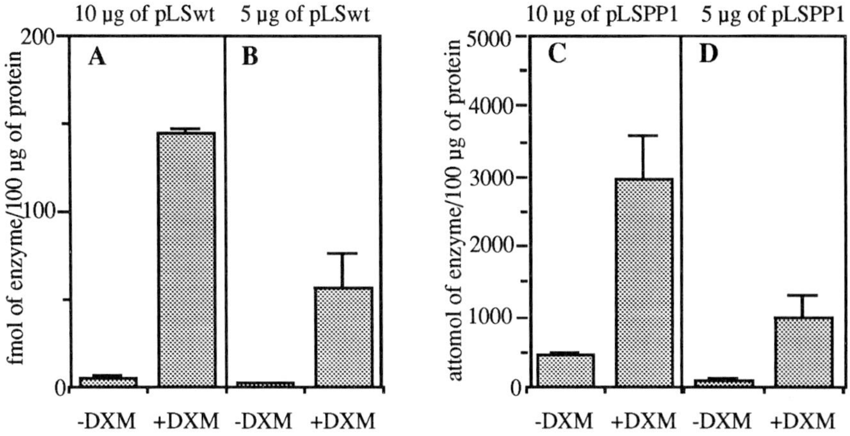

FIG. 2. Expression of CAT (A and B) and firefly luciferase activities in Ltk⁻ cells (C and D).

tion 10^{-7} *M*) to transfected cells were about 28 for CAT activity and about 8 for *P. pyralis* luciferase. The increment values were not affected by the amount of plasmid used for the transfection.

Results from transfection experiments show a different behavior of the various reporter genes under the experimental conditions described in this article. The expression of the bacterial luciferase in mammalian cells was very low (less than 100 amol/100 mg of total protein after stimulation with DXM), with an increase of more than 10-fold when cells were grown at a lower temperature (30°). These data are consistent with the results of Escher

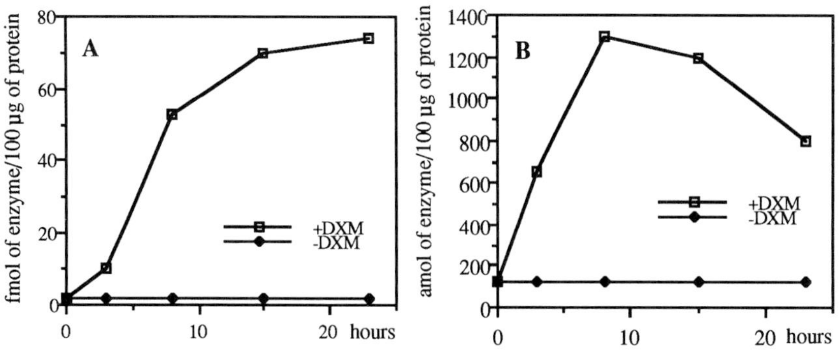

FIG. 3. Time course of the expression of CAT (A) and firefly luciferase (B) activities in transfected cells.

et al.[14] obtained in expression studies of a bacterial luciferase fusion protein in *E. coli.* In fact, the authors reported that the fusion of the α and β subunits (expressed in *E. coli* in the monomer or the dimer form) does not significantly affect the light-emitting properties when compared to the wild-type luciferase; however, in contrast to the heterodimeric wild-type enzyme, the enzymatic activity of the fusion is extremely sensitive to elevated growth temperature due not to heat lability, but presumably to temperature-sensitive aggregation or degradation of the polypeptide. Our results, together with a relatively low sensitivity of detection of the enzyme and a more complicated injection procedure in the assay method in comparison to the procedure for the firefly luciferase, suggest that the bacterial *luxAB* fusion gene is less suitable as a reporter gene in mammalian cells.

The expression of *P. pyralis* firefly luciferase shows a significant difference when compared to the expression of CAT. A major difference is in the absolute amount of enzyme produced, both with and without DXM stimulation. Cells transfected in the absence of DXM had a luciferase content about 10-fold less than CAT, whereas after 15 hr of DXM treatment, the content of luciferase was about 50- to 70-fold less than CAT. Although several papers have reported a comparison of CAT and *P. pyralis* firefly luciferase as reporter genes,[1,13,15–17] these studies did not employ calibration curves to quantitate the results, and consequently these data cannot be directly compared to ours. However, de Wet *et al.*[1] reported an overall 30-fold higher sensitivity for transfection with the luciferase reporter compared to CAT. Because we have shown here that the firefly luciferase assay is about 1000- to 2000-fold more sensitive than the CAT assay, an increased expression of CAT of 10- to 70-fold relative to luciferase would result in an overall increased sensitivity near the 30-fold reported by de Wet *et al.*[1]

Other striking differences observed in the expression of CAT and luciferase are in the time course of DXM induction and in the DXM induction ratio (the ratio of enzyme activity expressed in the presence and absence of hormone). While CAT expression is induced about 30-fold to a relatively constant steady-state level within 15 hr after the addition of DXM, the expression of firefly luciferase is induced maximally only 6- to 8-fold and reaches a peak concentration 10 hr after DXM stimulation and then decreases slowly. This time course for luciferase expression has been observed

[14] A. Escher, D. J. O'Kane, J. Lee, and A. A. Szalay, *Proc. Natl. Acad. Sci. U.S.A.* **86,** 6528 (1989).

[15] S. J. Gould and S. Subramani, *Anal. Biochem.* **175,** 5 (1988).

[16] O. Schwartz, J. L. Virelizier, L. Montagnier, and U. Hazan, *Gene* **88,** 197 (1990).

[17] T. M. Williams, J. E. Burlein, S. Ogden, L. J. Kricka, and J. A. Kant, *Anal. Biochem.* **176,** 28 (1989).

in several studies.[5,16,17] One possible explanation for these observations is a differential stability of the CAT and luciferase gene products *in vivo.*

Thermal Stability of Various Luminescent Reporter Genes

Hill *et al.*[18] have demonstrated that separate α and β subunits of the VH bacterial luciferase maintain a certain thermal stability in contrast to the various monomeric luciferases based on fused subunits in a unic plasmid. They have attempted to increase the thermal stability of the fused protein using functional chimeric fusions and, in some cases, they obtained higher light emissions even at 37°. O'Kane and Gibson[19] have studied native bioluminescent bacteria isolated from seawater that occurs at an elevated temperature. They found several thermal-labile and -resistant forms of VH bacterial luciferases and some of them were able to emit light even at 45°. These findings should permit the preparation of plasmids suitable for transfection experiments in mammalian cells based on bacterial luciferases. In any case, the overall lower sensitivity of the bacterial luciferase reporter system in comparison to the PP firefly luciferase, together with a more complicated luminescent reaction, seems to limit the use of the VH luciferase as a reporter gene in mammalian cells.

Thompson *et al.*[20] have reported that CAT has a half-life of about 50 hr compared with a half-life of only 3 hr for the PP firefly luciferase. Dement'eva *et al.*[21] have tried to influence enzyme stability by mutations of Cys residues in the polypeptide chain and, in some cases, they obtained higher thermal stability. Using random mutagenesis, White *et al.*[22] obtained mutant clones producing thermostable PP firefly luciferase at 43°. Thompson *et al.*[20] observed that PP firefly luciferase instability is due to its sensitivity to proteolysis and that the degradation of the encoded enzyme can be prevented by the addition of luciferin analogs into the culture medium.

All of these findings seem to suggest the possibility of having available plasmids encoding thermostable luciferases in the near future. Such en-

[18] P. J. Hill, J. P. Throup, and G. S. A. B. Stewart, *in* "Bioluminescence and Chemiluminescence, Status Report" (A. A. Szalay, L. J. Kricka, and P. E. Stanley, eds.), p. 122. Wiley, Chichester, 1993.

[19] D. J. O'Kane and B. G. Gibson, *in* "Bioluminescence and Chemiluminescence, Status Report" (A. A. Szalay, L. J. Kricka, and P. E. Stanley, eds.), p. 153. Wiley, Chichester, 1993.

[20] J. F. Thompson, L. S. Hayes, and D. B. Lloyd, *Gene* **103,** 171 (1991).

[21] E. L. Dement'eva, G. Kutuzova, and N. N. Ugarova, *in* "Bioluminescence and Chemiluminescence, Fundamentals and Applied Aspects" (A. K. Campbell, L. J. Kricka, and P. E. Stanley, eds.), p. 415. Wiley, Chichester, 1994.

[22] P. J. White, C. R. Lowe, D. J. Squirrel, and J. A. H. Murray, *in* "Bioluminescence and Chemiluminescence, Fundamentals and Applied Aspects" (A. K. Campbell, L. J. Kricka, and P. E. Stanley, eds.), p. 419. Wiley, Chichester, 1994.

zymes would improve the overall utility of luciferases as reporter genes. However, one of the main drawbacks in the actual use of luciferases as reporter genes is that, because of their relative instability *in vivo* and *in vitro,* they produce more variable results than CAT.

Improvements in Reproducibility Using Luminescent Reporter Genes in Expression Experiments

Several problems (i.e., PP firefly relative instability, variability in efficiency in transient transfection experiments, critical timing for harvesting of cells) can affect reproducibility within and between assays in gene expression studies based on luminescent reporter genes. We have investigated two possible alternatives in order to overcome this problem, i.e., the use of stable transfected cell lines or the splitting of transient-transfected cells.

Use of Stable Transfected Cell Lines

The PP firefly luciferase gene has been used as a reporter gene to construct chimeric cellular models in which luciferase expression mimics a natural steroid hormonal response. Once bound to a hormone, the steroid receptor becomes activated and can interact specifically with short specific DNA sequences called hormone responsive elements (HRE). Cells containing the endogenous receptor of interest are stably transfected with a plasmid carrying the firefly luciferase structural gene downstream from a hormonal regulatory sequence. The use of firefly luciferase as the reporter gene is particularly helpful as luciferase activity can be detected in intact cells without cell lysis. This is especially useful in the cloning and subcloning of stable-transfected cells.[23]

In some cases it is also possible to transfect cells with plasmids encoding native receptors or a chimeric receptor. Chimeric receptors in particular have been developed in order to specifically evaluate the biological activity of steroids; they are constructed by the fusion of sequences of the DNA-binding domain of yeast Gal4 protein with the hormonal-binding domain (HBD) of a given steroid receptor (see Fig. 4). Thus only after binding with the steroid hormone are these chimeric receptors able to interact with a specific exogenous DNA sequence composed of 17M(×5) bases placed upstream of the firefly luciferase gene.[24]

[23] D. Gagne, P. Balaguer, E. Demirpence, C. Chabret, F. Trousse, J. C. Nicolas, and M. Pons, *J. Biolumin. Chemilumin.* **9,** 201 (1994).

[24] N. Jausons-Loffreda, S. Gelmini, M. Pazzagli, P. Balaguer, S. Roux, M. Fuentes, M. Pons, and J. C. Nicolas, *J. Biolum. Chemilum.* **9,** 217 (1994).

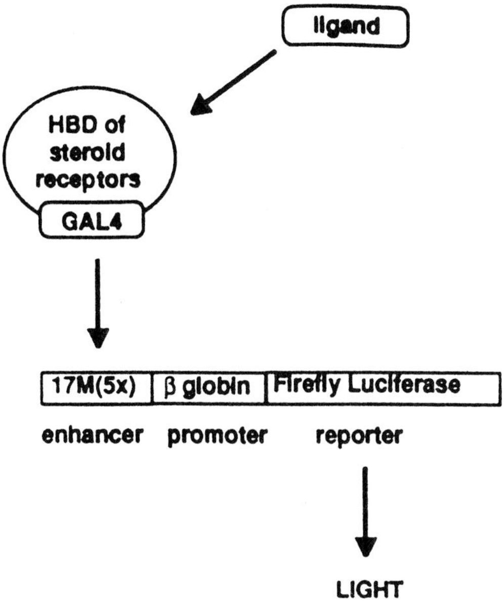

FIG. 4. Schematic representation of the model based on the use of chimeric Gal receptors.

Selection Procedure. As an example of the preparation of stable transfected cells,[24] HeLa cells are cotransfected with plasmid pAG-60 for resistance to neomycin or geneticin, the reporter plasmid p(17M)5-bGlob-Luc, and the chimeric receptor using the calcium phosphate precipitation technique.[25] Transfected cells are then selected according to their geneticin resistance: geneticin is usually introduced 1 or 2 days after transfection (1 mg/ml) in the routine medium and cells are cultured for 3 weeks. Geneticin-resistant clones are then allowed to grow in the routine medium alone until they can be detected visibly. A second selection is then carried out by measuring the inducible luciferase activity in these clones; cells are then incubated for 24 hr with medium containing or lacking the specific steroids (controls received only 0.2% ethanol). Luciferase activity is assayed in intact cells using a single photon-counting camera luminometer. Once localized, the luminescent clones are harvested, spread out, and allowed to grow in culture flasks. The selection of positive stable-transfected cells is performed using a photon-counting camera luminometer coupled with an imaging analysis system. After incubation of the clones with the hormones, the

[25] J. Sambrook, E. F. Fritsch, and T. Maniatis, *in* "Molecular Cloning: A Laboratory Manual" (C. Nolan, ed.). Cold Spring Harbor Laboratory Press, Cold Spring Harbor, NY, 1989.

medium is replaced by a medium containing 0.33 mmol/liter of the substrate luciferin. Luminescence values are measured for 5 to 10 min integration time, allowing localization of the luminescent clones. Only the clones for which induction factors are high (about 10 or more) are selected.

Results. An example of the application of stable-transfected cell lines is reported in Fig. 5. Using HeLa cells cotransfected with plasmid pAG-60 for resistance to neomycin or geneticin, the reporter plasmid p(17M)5-bGlob-Luc, and the chimeric receptor, as described by Jausons-Loffreda *et al.*,[24] it is possible to modulate the light emission of the transfected cells by addition into the medium of different concentrations of specific ligands.

Discussion. The preparation of stable cell lines is an attractive and already established technology that theoretically represents an unlimited source of homogeneous transfected cells suitable for extensive studies of regulation of gene expression. This approach, however, requires a com-

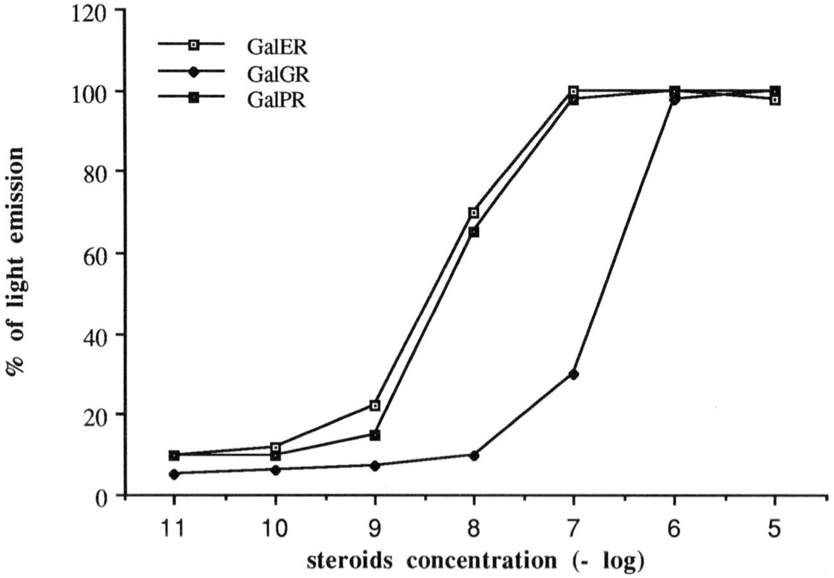

FIG. 5. Dose–response curves of chimeric receptors in stable cell lines. Cells were incubated for about 20 hr with various concentrations of their ligand. Agonistic effects were expressed as a percentage of maximum induction obtained for each chimeric receptor with its specific ligand. The GalER-expressed cell line was incubated with estradiol, the GalGR-expressed cell line was incubated with the glucocorticoid analog dexamethasone, and the GalPR cell line was incubated with the progesterone analog R5020. Modified from N. Jausons-Loffreda *et al., J. Biolum. Chemilum.* **9,** 217 (1994).

plicated and time-consuming methodology. Moreover, expensive instrumentation (such as a low-light imaging system) is usually necessary for the selection of suitable clones. In some cases, differences in assay results have been observed due to the use of different cell lines. Finally, for unknown reasons, the luminescence of some of the stable cell lines becomes hormone independent after a few passages.

Splitting of Transient-Transfected Cells

This second procedure takes full advantage of the rapidity of the bioluminescent method and the high detection sensitivity of the PP luciferase together with the low interfering effects of the biological matrix and of detergents on its luminescence activity.

With this procedure, transfection is limited to a single or a few dishes, but 40 hr after the shock, at the time of the addition of the DXM, cells are just washed and fresh medium is added for an additional 48 hr. The plate is then trypsinized, the cells are counted, and $1–2 \times 10^4$ cells are split into sterile assay tubes (12×75-mm glass tubes) with 0.5 ml of medium. After 4 hr, DXM can be added to the medium (final concentration 10^{-7} M) for the induction of gene expression. Usually between 15 and 30 tubes can be prepared from a single dish for expression experiments in which the variability of transfection is avoided. In an experiment of reproducibility within assays, transient transfected cells were divided among 15 tubes and then DXM (10^{-7} M) was added. Measurements of light emission were very reproducible (mean $\pm$ SD = 1076 $\pm$ 35, n = 15, CV 3.2%).

Variability between assays was also satisfactory when results were expressed as a percentage of the signal obtained from the DXM 10^{-7} M point (see Fig. 6). The absolute level of luminescence intensity was variable mainly due to differences in the efficiency of transfection, but the use of transfected cells divided from a single plate avoids this source of variability.

The bioluminescent reaction for the measurement of the expressed luciferase in this case can be performed in lysed cells using the following procedure: aspirate the medium, wash three times with 0.5 ml of phosphate-buffered saline, add 350 μl of the bioluminescence assay mixture[11] and 50 μl of Triton X-100 (1:100), mix the tube gently for about 30 sec, insert the tube into the luminometer, and inject 100 μl of 1 mM D-luciferin. As usual, light emission is measured at 20 sec integration. If the addition of Triton X-100 (and thus the consequent lysis of the cells) is avoided, it is still possible to measure some light emission from transfected cells. In a similar procedure described by de Wet *et al.*,[1] CV-1 cells, stably expressing luciferase, were grown on coverslips and assayed without performing the lysis in

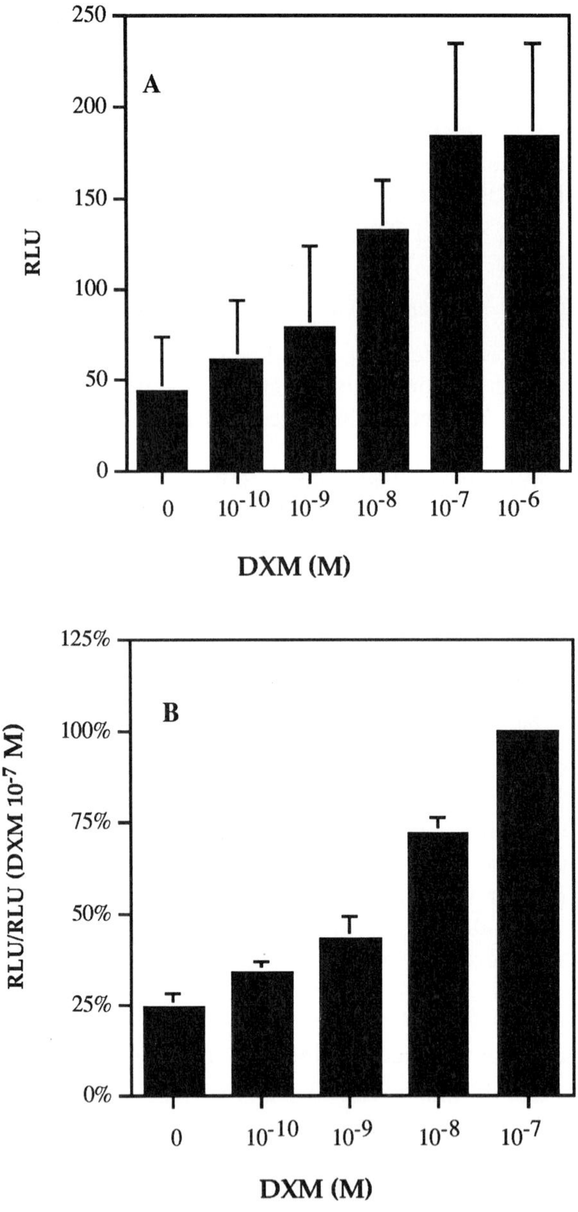

FIG. 6. Reproducibility of curves obtained in transient transfection experiments of LTK$^-$ cells after division of the transfected cells into tubes and the subsequent addition of various amounts of DXM into the medium. Data are expressed both as relative light units (RLU) obtained at each DXM concentration (A) and as a percentage of the signal obtained from the point DXM 10^{-7} *M* (B).

sterile scintillation vials. They observed about sevenfold lower light emission from intact cells compared to the light emission after lysis with Triton X-100; this lower level of light emission was attributed to incomplete permeability of the cell membrane to D-luciferin. However, they did not report that the light emission in intact cells is always correlated to the total amount of luciferase expressed in the cells after lysis. Figure 7 shows results of an experiment in which, from a single dish, 36 test tubes were prepared and then half of them were stimulated with DXM. At different times after DXM stimulation, tubes were assayed for firefly luciferase using both of the procedures described earlier (with and without lysis), and the results were evaluated with a linear regression test. In Fig. 7, arbitrary light units instead of the molar concentration of enzymes are reported because the "*in vivo*" assay cannot be standardized.

Results show that under our assay conditions the procedure without lysis produces about one-third of the light observed from lysed cells; the kinetics of light emission was constant for several minutes. After lysis, however, the higher light intensity obtained showed different kinetics, with the light emission slowly decaying after the peak. These kinetics results are in agreement with those described by deWet *et al.*[1] The correlation between the results obtained with or without lysis of the transfected cells was highly significant ($P > 0.001$; $y = 5.252 + 2.708x$; $R = 0.9384$).

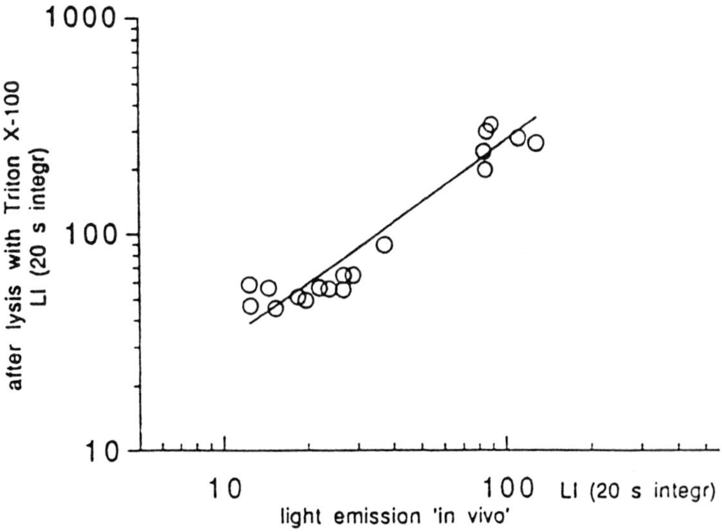

FIG. 7. Correlation between light emission with or without lysis by Triton X-100 ($y = 5.252 + 2.708x$; $R = 0.9384$).

The major advantages of the use of luciferases as reporter genes are related mainly to the high detection sensitivity and to the facility of the assay method. The possibility of reducing the number of transfected cells for the assay of the reporter gene makes the firefly luciferase suitable for more extensive studies on the characteristics of the various plasmids and for more precise studies of gene expression and regulation. The use of detergent lysis of transfected cells directly on dishes with Triton X-100 in order to avoid the scraping and the freeze/thaw cycles has been reported.[4,16,17] These procedures facilitate the assay of the reporter gene and make it more reproducible. This article reports on an additional procedure characterized by high sensitivity and assay facility that should prove useful for molecular biological application. Using the splitting of transfected cells it is possible to monitor gene expression without introducing in the results the variability due to the transfection procedure. With this method we have evaluated the correlation between the light signal emitted by transfected cells with or without lysis with Triton X-100. The light emission obtained

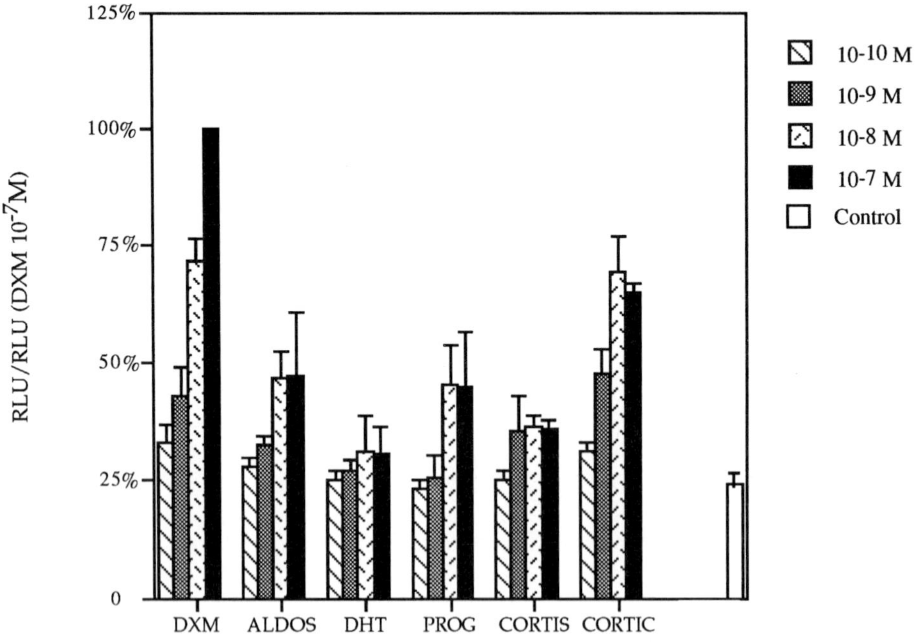

FIG. 8. Transfection of pLSPP plasmid into the LTK$^-$ cell line. Cells were treated with different steroid molecules (DXM, dexamethasone; ALDOS, aldosterone; DHT, dihydrotestosterone; PROG, progesterone; CORTIS, cortisol; CORTIC, corticosterone), each at four different concentrations, all in the same experiment. Results are expressed as a percentage of the firefly luciferase enzyme activity obtained with 10^{-7} *M* DXM.

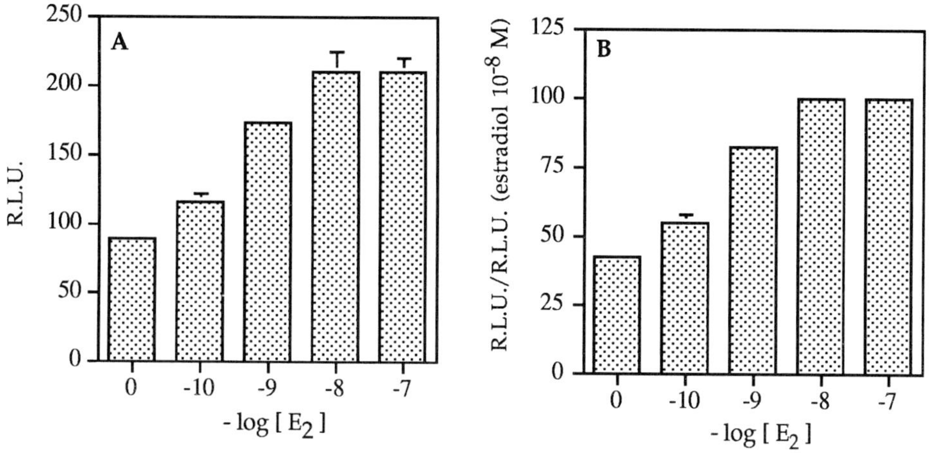

FIG. 9. Firefly luciferase expression in the MCF7 cell line transfected with pERELuc and stimulated with estradiol at concentrations varying from 10^{-10} to 10^{-7} *M*. Results are expressed both as relative light units obtained at each estradiol (E_2) concentration (A) and as a percentage of the highest enzyme activity obtained with 10^{-8} *M* E_2 (B). Results are mean values of three different experiments. Each set of data was obtained in triplicate.

from intact cells as described by de Wet *et al.*[1] could be affected by parameters other than the amount of firefly luciferase, such as intracellular concentration of ATP, Mg^{2+}, permeability of D(−)luciferin, or compartmentalization of the luciferase as observed in the expression of firefly luciferase in transgenic plants.[2] Our results demonstrate that, even in the absence of cell lysis with Triton X-100, although the light signal is reduced in amplitude, it is correlated to the amount of intracellular luciferase. Yang and Thomason[26] have synthesized a new photolyzable luciferase substrate [1-(4,5-dimethoxy-2-nitrophenyl)diazoethane (DMNPE) caged luciferin] that crosses cell membranes into intact cells readily.

Applications. Reporter genes regulated by hormone-inducible promoters have been widely used for evaluation of the biological activity of natural and newly synthesized steroid substances. Time-consuming and complex procedures are often needed for such studies. In order to underline the sensitivity and practicability of our method in the screening of biological activities of steroid hormones, Fig. 8 shows results obtained in a single transfection experiment that allows evaluation of the biological activities of six different steroids on the same cell line at four different concentrations. Figure 9 shows results from another experiment demonstrating the applica-

[26] J. Yang and D. B. Thomason, *BioTechniques* **15,** 848 (1993).

tion of the same system, transfection of MCF-7 cells with a different plasmid (pERELuc) whose hormone-responding element is stimulated selectively by estrogens. After transfection and the splitting procedure, luminescence of the cells was stimulated with estradiol (E_2) ranging from 10^{-7} to 10^{-10} *M* concentration.

Microinjection of Reporter Genes

Principles

The introduction of reporter genes into cells is used commonly to study the regulation of gene expression. Indeed, the traditional methods of transfection (transfection with DEAE-dextran, calcium phosphate precipitation, electroporation, etc.) are very useful but possess a number of limitations (time-consuming procedures, need of standardization for each type of cell, risk of cell damage, lack of specificity for a given cell type within a mixed population of cells, need of relatively high number of cells and amount of plasmid DNA). Microinjection helps to overcome some of these problems, even if it requires highly sensitive and specific methods to monitor gene expression in the limited number of cells that can be microinjected. We have investigated the use of a plasmid (pLSPP) containing the MMTV-LTR sequence as the promoter (a glucocorticoid-inducible region) and the firefly luciferase gene from *Photinus pyralis* (PP) as the reporter in microinjection experiments of mammalian cells. Two different cell lines were tested: murine Ltk$^-$ fibroblasts[11] and human Saos 2 osteosarcoma cells. For comparison, the plasmid pLSPP was also used in transfection experiments of Ltk$^-$ cells by the DEAE-dextran procedure.

Materials and Methods

Ltk$^-$ and Saos 2 cells are grown at 37° in 5% CO_2, respectively, in Dulbecco's modified Eagle's medium (DMEM; GIBCO, Paisley, UK) and in Coon's modified Ham's F-12 medium (Imperial, UK) supplemented with 10% fetal calf serum (FCS, GIBCO). Dexamethasone and D-luciferin are from Sigma Chemical (St. Louis, MO). Coenzyme A (CoA; lithium salt) and DEAE-dextran are from Pharmacia (Turku, Finland).

Plasmid pLSwt has been used as a parental vector for the construction of the pLSPP plasmid. This plasmid bears the wild-type sequence of the MMTV-LTR linked to the gene of firefly luciferase and SV40 sequences containing the small t intron and early poly(A) addition signals. To construct the plasmid pLSPP, the CAT gene of pLSwt is replaced with the luciferase gene using unique *Nhe*I and *Hpa*I restriction sites that flank the CAT gene. The construction of this plasmid has been described in detail.[11] Plasmid

DNA is amplified in *E. coli* HB101 cells and purified by alkaline-SDS lysis and CsCl gradient ultracentrifugation.

The automatic microinjection system used in this study is an AIS (Automated Injection System, Zeiss, Oberkochen, Germany), consisting of a phase-contrast inverted microscope (Axiovert 35 M), a charge-coupled device (CCD) video camera, and a microinjector (Eppendorf, Hamburg, Germany). This system allows axial injections into as many as 1000–1500 cells per hour. The hardware and software of the system control the positioning of the scanning stage, the movement of the micromanipulator, and the CCD camera for the transmission of the microscopic image to the TV monitor.[27] The volume injected into the cells is controlled by the time during which the capillary is inside the cell. The injection pressure chosen is 150 hPa with an injection time of 0.5 sec. The cells are seeded on glass coverslips (20 × 20 mm) in 35-mm dishes and cultured for 24 hr in serum-free medium before microinjection. At least two coverslips are microinjected during the same session. The plasmid is injected at a concentration of about 1 mg/ml. After injection the cells are incubated for 24 hr in complete medium; those adhering to one coverslip are stimulated with DXM (1×10^{-7} *M*) in the complete medium for 16–18 hr.

Transfection of Ltk^- cells (500,000 cells per dish) by the DEAE-dextran-dimethyl sulfoxide shock procedure is performed as described by Pazzagli *et al.*[11] using the same time intervals and the concentration of DXM for stimulation reported in the microinjection procedure. One-tenth of the cell lysate is used for the luciferase assay.

Cells are washed with 2 × 2.5 ml of cold luminescence buffer (15 *M* potassium phosphate, 8 m*M* $MgCl_2$, pH 7.4) and then scraped in 1 ml of the same buffer containing 2 m*M* ATP.[24] Cells are lysed by three cycles of freezing in liquid nitrogen and thawing at 37°. Luciferase activity is measured from an aliquot (300 μl plus 50 μl of a solution of 3% Triton X-100 in the luminescence buffer). Luminescence is measured, for an integration time of 10 sec, after the injection of 100 μl of luminescence buffer containing 6×10^{-4} *M* D-luciferin and 6×10^{-4} *M* coenzyme A. Luciferase activity is assayed using a Biolumat LB 9500 luminometer (Berthold, Wildbad, Germany).

Comparison between Microinjection and DEAE-Dextran Transfection in Ltk^- Cells

We have compared the results of microinjection and DEAE-dextran transfection experiments with the pLSPP plasmid using the same cell line. The results, expressed as relative light units, are shown in Table II.

[27] R. Pepperkok, M. Zanetti, R. King, D. Delia, W. Ansorge, and L. Philipson, *Proc. Natl. Acad. Sci. U.S.A.* **85,** 6748 (1988).

TABLE II

COMPARISON BETWEEN RESULTS OF MICROINJECTION AND TRANSFECTION EXPERIMENTS USING PLASMID pLSPP AND Ltk^- CELLS[a]

	n	DEAE-dextran transfections in Ltk^- cells of pLSPP (mean ± SD)	*n*	Microinjection in Ltk^- cells of pLSPP plasmid (mean ± SD)
−DMX	4	1452 ± 550 CV = 37.8%	4	52.6 ± 30.9 CV = 58.8%
+DMX	4	7540 ± 3825 CV = 50.7%	4	353.0 ± 139.5 CV = 39.4%
Ratio	4	6.3 ± 1.7 CV = 26.2%	4	7.5 ± 2.0 CV = 26.6%

[a] Data are expressed as relative light units; *n* is the number of separate experiments. *Ratio* is the increment in light emission induced by DXM calculated for each experiment. Data are reported after subtraction of the background.

A low but significant light signal (about three times the background of the assay performed in lysates of cells that were neither microinjected nor transfected) was detected when the cells were grown without DXM stimulation. This low but significant level of expression of firefly luciferase seemed due to both the presence of serum in the medium (and thus to the presence of endogenous steroids) and the MMTV promoter elements located within about 80 bp of the transcription initiation site that mediate basal levels of transcription and that are distinct from the hormone responsive elements (HRE) located from about 200–80 bp upstream of the transcription initiation site as discussed previously.[11] When DXM was added, an increase of about sevenfold was observed in both the microinjection and the DEAE-dextran transfection method. However, both these techniques showed high variability with respect to the absolute values of light emission (−DXM and/or +DXM).

The variability seemed to reflect a number of factors that influence the relative amount of the plasmid that enters into the cells with both DEAE-dextran transfection and microinjection. However, when the results were evaluated in terms of induction ratio due to DXM stimulation (the ratio between light emission from cells cultured with/without DXM), the CV value was reduced significantly. This observation held, within the same experiment of induction of the luciferase due to hormone activity, for both the methods investigated. In conclusion, these data suggest that variables affecting the reproducibility of DEAE-dextran transfection or of microinjection are probably very different, but they are relevant and can limit the interpretation of results significantly when

TABLE III
COMPARISON BETWEEN RESULTS OF MICROINJECTION OF pLSPP PLASMID IN Ltk⁻ AND Saos 2 CELLS[a]

	n	Microinjection in Saos 2 cells of pLSPP plasmid (mean ± SD)	*n*	Microinjection in Ltk⁻ cells of pLSPP plasmid (mean ± SD)
−DMX	4	33.3 ± 14.4 CV = 43.2%	4	52.6 ± 30.9 CV = 58.8%
+DMX	4	253 ± 109 CV = 43.0%	4	353.0 ± 139.5 CV = 39.4%
Ratio	4	6.9 ± 1.0 CV = 14.4%	4	7.5 ± 2.0 CV = 26.6%

[a] For abbreviations, see Table II.

expressed in terms of absolute amounts of products of gene expression. This is particularly true when data from experiments performed in different sessions are considered. Only the use of an internal control in the methodology, such as in our case the use of the induction ratio by DXM, allows evaluation of the entity of the plasmid incorporation and its functionality within the cells.

Comparison of Microinjection with Ltk⁻ Cells and with Saos 2 Cells

We have microinjected the plasmid pLSPP into Saos 2[28] cells and compared the results obtained with the microinjection of the same plasmid into Ltk⁻ cells. Results are shown in Table III.

Both basal and stimulated results in human Saos 2 osteosarcoma cells were similar to those obtained with Ltk⁻ cells; the ratio values were also comparable. The high variability of the absolute values of light emission (−DXM and/or +DXM) was also comparable in both cell lines, whereas the CV value of the ratio was reduced significantly. Finally, we tested the possibility of monitoring gene expression with the pLSPP plasmid in microinjection experiments with a limited number of cells. Five hundred cells (instead of 1000) were used, but extremely low levels of light emission were measured (close to the background of the assay) even after the addition of DXM.

These data suggest that variables affecting the reproducibility of microinjection are probably dependent on the technology itself and are not

[28] S. B. Rodan, I. Imay, M. A. Thiede, G. Wesolowsky, D. Thompson, and Z. Bar Shavit, *Cancer Res.* **47,** 4961 (1987).

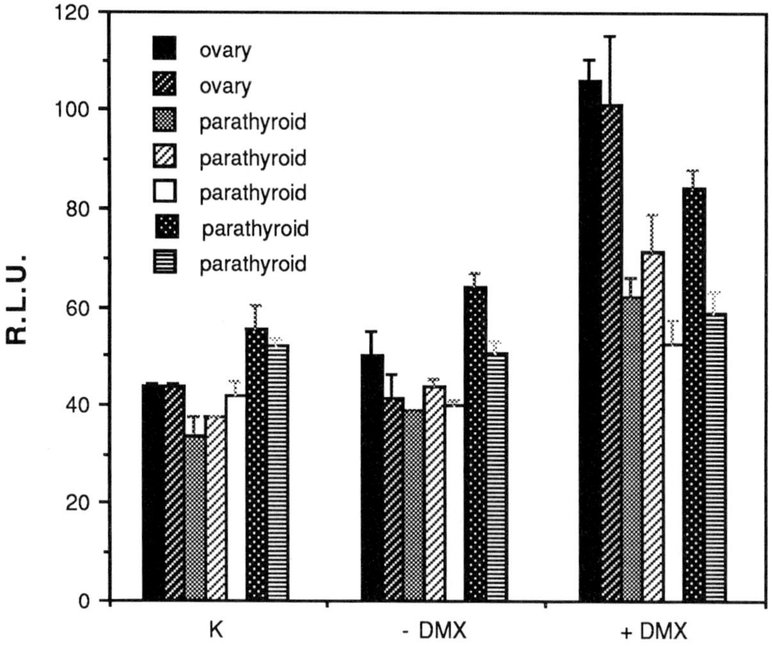

FIG. 10. Microinjection of the plasmid pLSPP into primary cultures of human tumor cells. K, light emission in nontransfected cells; −DXM, light emission in transfected cells without the addition of DXM; +DXM, light emission in transfected cells after the addition of DXM (10^{-7} *M*).

correlated to cell variability. However, they confirm that microinjection is a suitable method for the introduction of reporter gene constructs into different cell types without the need of optimization required for traditional transfection methods. We have microinjected the plasmid pLSPP into primary cultures of human tumor cells using the procedure reported previously. The results in terms of light emission following the addition of DXM (10^{-7} *M*) are illustrated in Fig. 10.

Even if data reported in Fig. 8 confirm the feasibility of using microinjection in primary cultures of human tumor cells, the firefly luciferase method is not satisfactory in terms of detection sensitivity. The monitoring of gene expression in a limited number of microinjected cells sometimes requires the study of time courses or dose-dependent studies and the current level of sensitivity is not sufficient for such studies.[29,30]

[29] D. O. Peterson, K. K. Beyfuss, and K. L. Morley, *Mol. Cell. Biol.* **7,** 1563 (1987).
[30] A. S. Alberts, J. A. Frost, and A. M. Thorburn, *DNA Cell Biol.* **12,** 935 (1993).

[40] *In Situ* Hybridization and Immunohistochemistry with Enzyme-Triggered Chemiluminescent Probes

By ALDO RODA, MONICA MUSIANI, PATRIZIA PASINI, MARIO BARALDINI, and JEAN E. CRABTREE

In situ hybridization (ISH) and immunohistochemistry (IHC) techniques provide important tools for the localization of specific nucleic acids and antigens within individual cells. The preservation of cellular and tissue morphology allows determination of the spatial interaction between cells or between pathogens and cells and correlation of the molecular information with the morphological structure.[1]

Nowadays, ISH and IHC techniques are used extensively in research and routine laboratories and have been applied in different biomedical fields, including oncology, microbiology, virology, transplantation, and inherited disorders.[2–6]

In situ hybridization is based on the complementary binding of a nucleotide probe to a specific target nucleic acid sequence, which can be endogenous DNA, mRNA, viral nucleic acid, or bacterial sequence. Labeled nucleic acid probes are required, and indirect labels are usually preferred for their increased sensitivity compared to direct labels. Different molecules, including biotin, fluorescein, and digoxigenin, are widely used to label the specific genetic probe.[7] Bound probes are revealed indirectly by streptavidin, antifluorescein antibody, and antidigoxigenin antibody, respectively, which have been coupled covalently with a signaling moiety that is usually an enzyme such as alkaline phosphatase (AP) or horseradish peroxidase (HRP).

In IHC the probes used are highly specific antibodies that bind to antigens such as proteins, enzymes, and viral or bacterial products. The bound specific antibody is revealed indirectly by species-specific or class-specific secondary antibodies conjugated to enzymes such as AP and HRP.

[1] J. H. Wilcox, *J. Histochem. Cytochem.* **41,** 1725 (1993).
[2] J. A. Matthews and L. J. Kricka, *Anal. Biochem.* **169,** 1 (1988).
[3] J. J. Leary and J. L. Ruth, *in* "Nucleic Acid and Monoclonal Antibody Probes" (B. Swaminathan and G. Prakash, eds.), p. 33. Dekker, New York, 1989.
[4] E. P. Diamandis, *Clin. Chim. Acta* **194,** 19 (1988).
[5] D. V. Pollard-Knight, *in* "Bioluminescence and Chemiluminescence: Current Status" (P. E. Stanley and L. J. Kricka, eds.), p. 83. Wiley, Chichester, 1991.
[6] P. R. Wenham, *Ann. Clin. Biochem.* **29,** 598 (1992).
[7] K. Misiura, I. Durrant, M. E. Evans, and M. J. Gait, *Nucleic Acids Res.* **18,** 4345 (1990).

0076-6879/00 $30.00

In both techniques these enzymes can be revealed using different detection methods, including colorimetry, fluorescence, and chemiluminescence (CL). In the colorimetric system the enzyme substrate is converted to a colored product that precipitates at the site of the biospecific recognition. This system has been validated and is the most widely used, but the results are only qualitative and a higher sensitivity is often required. The use of fluorescent substrates provides higher potential sensitivity, but fluorescent-based detection systems have the drawback of nonspecific background fluorescence, which is present in all biological samples and reduces the specificity and sensitivity of the system. Moreover, quantitative analysis is still a problem as the emitted fluorescence is a function of the potency of the exciting light. Chemiluminescence detection, i.e., the use of an enzyme substrate that is converted to a light-emitting product, has potentially the highest detectability. Theoretically, down to 10^{-18}–10^{-21} mol of enzyme can be measured[8] if the CL reaction has a high efficiency, and the detection device can measure light down to a few photons level. Novel chemiluminescent substrates for AP[9–12] and enhanced chemiluminescent reagents for HRP[13–15] have been developed, providing increased detectability of these enzymes over colorimetric or fluorescent substrates and reaching similar sensitivity to that obtained with radiolabeled probes. Moreover, chemiluminescent systems can give precise and accurate quantitative results as the luminescent signal intensity is proportional to the enzyme activity present in the reaction.

Chemiluminescent Reactions

Different CL substrates for AP, based on adamantyl 1,2-dioxetane phenyl phosphate derivatives, are now available commercially. CSPD (PE-Tropix, Bedford, MA) and Lumi-Phos Plus (Lumigen, Southfield, MI) have been used in our studies. These substrates allow the detection of a few zeptomoles of AP using a photomultiplier tube[8] or attomoles of enzyme

[8] I. Bronstein, B. Edwards, and J. C. Voyta, *J. Biolumin. Chemilumin.* **4,** 99 (1990).

[9] I. Bronstein and L. J. Kricka, *J. Clin. Lab. Anal.* **3,** 316 (1989).

[10] I. Bronstein, J. C. Voyta, and B. Edwards, *Anal. Biochem.* **180,** 95 (1989).

[11] A. P. Schaap, H. Akhavan-Tafti, and L. J. Romano, *Clin. Chem.* **35,** 1863 (1989).

[12] S. Beck and H. Köster, *Anal. Chem.* **62,** 2258 (1990).

[13] J. A. Matthews, A. Batki, C. Hynds, and L. J. Kricka, *Anal. Biochem.* **151,** 205 (1985).

[14] G. H. G. Thorpe and L. J. Kricka, *Methods Enzymol.* **133,** 331 (1986).

[15] G. H. G. Thorpe and L. J. Kricka, *in* "Bioluminescence and Chemiluminescence: New Perspectives" (J. Schölmerich, R. Andreesen, A. Kapp, M. Ernst, and W. G. Woods, eds.), p. 199. Wiley, Chichester, 1987.

using a video camera[16] and have a glowing kinetics with a steady-state light emission lasting several minutes, which permits easy handling and analysis of the samples.

For HRP detection, two luminol-based reagents have been used: the enhanced chemiluminescent reagent ECL (luminol/enhancer/H_2O_2) (Amersham, Amersham, UK) and SuperSignal Ultra (luminol/enhancer/stable peroxide) (Pierce, Rockford, IL). Both reagents contain substances, such as *p*-iodophenol, which enhance the light production deriving from the HRP-catalyzed oxidation of luminol by hydrogen peroxide, thus permitting detection of HRP with very high efficiency. The kinetics is of a glow type with a steady-state light emission maintained for at least 10–15 min.

The half-life and the diffusion of the light-emitting species in the CL cocktail are critical points to adapt these CL systems, primarily developed for light measurement and not for imaging. In order to achieve precise and accurate localization of the enzyme-labeled probe on a target surface, the light emission should occur at the site of the primary biospecific recognition. Previous studies[16,17] demonstrated that the CL substrates used in the present work are suitable for this purpose, as the half-life of excited species is very short and thus diffusion in the cocktail solution plays a negligible role, allowing sharp localization of the enzyme-labeled probe on the target surface. On the contrary, when excited species have a long half-life, they are more likely to decay to the ground state and emit light after diffusing into the solution, resulting in poor spatial resolution.

Low-Light Imaging Device

The chemiluminescent signal from the samples has to be detected using a photon imaging instrumentation, based on a charge-coupled device (CCD) video camera or a Vidicon pick up tube, connected to a light microscope. These instruments not only allow quantification of the emitted light at a single photon level, but also permit localization of the signal on a target surface. The microscope should have a lens coupling system as simple as possible and objectives with the highest numerical aperture compatible with focal aberration and depth of field in order to minimize the potential loss of light deriving from the CL reaction.

We used the Luminograph LB 980 (EG&G Berthold, Bad Wildbad, Germany), which is a high-performance, low-light imaging system able to

[16] A. Roda, P. Pasini, M. Musiani, S. Girotti, M. Baraldini, G. Carrea, and A. Suozzi, *Anal. Chem.* **68,** 1073 (1996).

[17] A. Roda, P. Pasini, M. Baraldini, M. Musiani, G. Gentilomi, and C. Robert, *Anal. Biochem.* **257,** 53 (1998).

detect any type of luminescent emission (400–700 nm) over a wide range of intensities (sensitivity range 50 plx–10 lx at 490 nm). The video system consists of a high dynamic range video camera (1″, Saticon), which is a Vidicon-type tube with Se-As-Tl light target photoconductor (Siemens, Karlsruhe, Germany) linked to an image intensifier by high-transmission lenses. The video camera is connected to a Model BH-2 optical microscope (Olympus Optical, Tokyo, Japan) using a C-mount adapter. The microscope is provided with objectives of different magnifying power (SPlan 10, 20, 40, Olympus Optical, Tokyo, Japan) and is enclosed in a dark box to prevent contact with external light. The system is controlled by a PC provided with the software Lum (Siemens, Karlsruhe, Germany) for image processing and quantitative analysis.

A new instrumentation (Night Owl, EG&G Berthold) has been developed and used preliminarily. The system is based on a slow-scan, back-illuminated, cooled CCD camera that provides very high detection sensitivity due to high quantum efficiency (40% at 650 nm). In addition, with this instrument the static chemiluminescent image does not require an intensification step, which could affect the image quality and the signal-to-noise ratio negatively by increasing instrumental noise.

The detection system for chemiluminescence ISH and IHC operates in three steps: cell and tissue structure images are recorded in transmitted light, then the luminescent signal is acquired, following the addition of the CL cocktail, with an optimized photon accumulation usually lasting 1 min; computer elaboration of the luminescent signal with pseudocolors corresponding to the light intensity and overlay of the pseudocolor and live images allow the spatial distribution of the target analyte to be localized and evaluated.

Slides are examined without coverslips, which could impede chemiluminescent detection, reducing the amount of light reaching the objective by internal reflectance phenomena.

Quantitative Analysis

The system used, i.e., video camera-optical microscope, has been optimized and validated previously in terms of magnification factor and pixel size achieved using different objectives.[16]

The light emission from each sample can be quantified by defining a fixed area and summing the total number of photon fluxes from within that area, which corresponds to a known number of pixels of known dimension. The light emission is then expressed as photons/second/surface area.

For example, in ISH for cytomegalovirus detection, control uninfected cells and control negative biopsy specimens provide the threshold background levels for nonspecific binding of the probe. Threshold values are calculated analyzing a mean of 50 cells from control negative smear samples and 10 microscopic fields of sections from reference negative biopsies. The average values ±SD values of the background light emission (expressed as photons/second/surface area) are then calculated. The average value of the background signal plus fivefold its standard deviation is considered the threshold value above which the chemiluminescent signal resulting from the hybridized nucleic acid can be determined as positive. The net light signal from the sample is obtained automatically on the screen after subtraction of the threshold values using appropriate software.

Chemiluminescence *in Situ* Hybridization

Specimen Preparation

Centrifuge cellular specimens from body fluids at 1000 rpm for 10 min at room temperature, resuspend the cells in 0.15 *M* phosphate-buffered saline (PBS), pH 7.4, and smear or cytospin cells onto pretreated or silanated glass slides. For blood nucleated cells, a previous separation from peripheral blood by Ficoll gradients is recommended. Fix the samples with 4% paraformaldehyde in PBS for 30 min. After fixation, wash the slides three times in PBS for 10 min each and then dehydrate the samples with ethanol washes (30, 60, 80, 95, and 100%) for 5 min each. Air dry and store the slides at 4°. Samples can be stored for several months with no decrease in their interaction with the hybridization probes.

Tissue samples fixed in formalin and embedded in paraffin or tissue samples frozen in liquid nitrogen and stored at −80° until use can also be suitable for chemiluminescence *in situ* hybridization.

Cut 3- to 5-μm-thin sections from either paraffin-embedded or frozen tissue blocks and place them on pretreated or silanated slides. Dewax paraffin-embedded tissue sections by two 5-min incubations in xylene and then wash in absolute ethanol. Fix and dehydrate frozen sections as described earlier for cellular samples.

Probes

Probes used for chemiluminescence ISH generally consist of double-stranded DNA probes obtained from cloned DNA fragments, but they can also consist of synthetic oligonucleotides, single-stranded DNA probes generated by polymerase chain reaction, or single-stranded antisense RNA probes. Because of their small size, synthetic oligonucleotide probes have good penetration properties, and the fact that they are single stranded rules out the possibility of annealing with complementary strands in the hybridization solution. Cloned probes, which are usually longer, have the advantage of having more sites available for labeling and of recognizing more of the target DNA, resulting in a stronger signal. Probes constructed in our laboratory and commercially available probes have been used (Table I).

Specificity Controls for Chemiluminescence ISH

The following control experiments are necessary to prove that the chemiluminescence hybridization reaction is detecting target sequences specifically.

No signal has to be revealed when mock-infected samples are treated with the labeled probe and the CL detection system.

A CL signal has to be detected when reference target-positive samples are hybridized with labeled probe and treated with the CL detection system.

No signal has to be observed when target-positive samples are treated with the vector control DNA-labeled probe and the CL detection system.

No signal has to be revealed in target-positive samples after hybridization with the unlabeled probe, followed by treatment with the CL detection system.

Target-positive samples have to be completely negative after hybridization with the fluorescein- or digoxigenin-labeled probe when incubation with the enzyme-labeled antibody is either omitted or replaced by incubation with nonimmune serum.

TABLE I
LABELED PROBES USED TO DETECT VIRAL DNAS BY CHEMILUMINESCENCE *in Situ* HYBRIDIZATION

Viral DNA	Labeled probe
Cytomegalovirus	Digoxigenin-labeled probe[a]
Parvovirus B19	Digoxigenin-labeled probe[b]
Human papillomavirus	Fluorescein-labeled probe (Dako, Glostrup, Denmark)
	Digoxigenin-labeled probe (Kreatech, Amsterdam, The Netherlands)
	Biotinylated probe (Biohit, Helsinki, Finland)
Herpes simplex virus	Biotinylated probe (Enzo Biochem, New York NY)

[a] G. Gentilomi, M. Musiani, M. Zerbini, G. Gallinella, D. Gibellini, and M. La Placa, *J. Immunol. Methods* **125,** 177 (1989).

[b] A. Azzi, K. Zakrzewska, G. Gentilomi, M. Musiani, and M. Zerbini, *J. Virol. Methods* **27,** 125 (1990).

Hybridization Reaction

Several protocols for hybridization reactions have been described with minor modifications, generally involving a sample pretreatment followed by the hybridization reaction with the labeled probe.

Hydrate cells/sections briefly in PBS and then place them in 0.02 *N* HCl for 10 min. After three washes with PBS, treat samples with 0.01% Triton X-100 in PBS for 2 min. After a further three washes with PBS, treat samples with pronase (0.5 mg/ml) in 0.05 *M* Tris–HCl buffer, pH 7.6, containing 5 m*M* EDTA for 5 min. After pronase treatment, wash the slides twice with PBS containing 2 mg/ml of glycine. After these treatments, postfix cell monolayers and tissue sections with 4% paraformaldehyde in PBS for 5 min and wash twice with PBS containing 2 mg/ml of glycine. Then dehydrate samples by ethanol washes (30, 60, 80, 95, and 100%).

Incubate samples with 10 μl of the hybridization mixture, which consists of 50% deionized formamide, 10% dextran sulfate, 250 μg/ml of carrier calf thymus DNA, and 2 μg/ml of the labeled probe (Table I) in 2× SSC buffer (0.3 *M* NaCl, 0.03 *M* sodium citrate, pH 7.0). Denature together the samples and the hybridization mixture containing the labeled probe by heating in an 85° water bath for 6 min and then hybridize at 37° for 3 hr. After hybridization, wash samples with stringent conditions in order to minimize nonspecific hybridization: three washes at 34° with 50% formamide in 2× SSC for 3 min each, followed by two 3-min washes in 2× SSC at 34°, and finally one 3-min wash in 2× SSC at room temperature.

Chemiluminescent Detection

Wash the slides briefly in a 150 m*M* Tris–HCl buffer, pH 7.5, containing 150 m*M* NaCl and incubate with the appropriate HRP- or AP-labeled detection systems. In detail, (a) incubate the samples hybridized with fluorescein-labeled probes for 30 min at 37° with HRP- or AP-conjugated Fab fragments to fluorescein isothiocyanate (Dako, Glostrup, Denmark) following the manufacturer's instructions; (b) incubate the samples hybridized with the digoxigenin-labeled probe for 30 min at 37° with HRP- or AP- conjugated antidigoxigenin Fab fragments (Kreatech, Amsterdam, The Netherlands) following the manufacturer's instructions; and (c) incubate the samples hybridized with the biotinylated DNA probe for 30 min at 37° with HRP- or AP-conjugated streptavidin (Biohit, Helsinki, Finland) following the manufacturer's instructions.

To detect HRP-labeled hybridized probes, add 20–40 μl of luminol-based chemiluminescent reagent (ECL or SuperSignal Ultra), incubate 2 min in the dark at room temperature, and acquire the light signal from the located specimen.

To detect AP-labeled hybridized probes, wash samples for 2 min with equilibration buffer (100 m*M* Tris–HCl, 100 m*M* NaCl, 50 m*M* $MgCl_2$, pH 9.5), add 20–40 μl of dioxetane phosphate-based chemiluminescent substrate (CSPD or Lumi-Phos Plus), and acquire the light emission after 15–20 min incubation in the dark at room temperature.

The CL signal is collected for 1 min. The image of the same field is acquired under transmitted light to allow localization of the specific signal. The CL signal could be recorded several times within 5–10 min.

Double Hybridization Reaction

Chemiluminescence ISH can also be used for the simultaneous detection of different DNA sequences in the same sample utilizing both HRP and AP as reporter enzymes. In the double hybridization reaction, probes to different targets have to be labeled and detected using different enzymatic systems.

Probes with different labels can be cohybridized with samples following the protocol described previously and then detected separately. The chemiluminescent detections of the two probes are performed sequentially.

After cohybridization, wash samples three times (5 min each) with 0.05% Triton X-100 in 100 m*M* Tris–HCl buffer, pH 7.6, containing 150 m*M* NaCl, add 20–40 μl of the chemiluminescent substrate for HRP (ECL or SuperSignal Ultra) and measure the light output after a 2-min incubation in the dark at room temperature. After acquisition of the signal, wash samples three times for 2 min each in equilibration buffer (100 m*M* Tris–HCl, 100 m*M* NaCl, 50 m*M* $MgCl_2$, pH 9.5), incubate for 15–20 min in the dark at room temperature with 20–40 μl of the chemiluminescent substrate for AP (CSPD or Lumi-Phos Plus), and measure the light output from the second substrate. The detection and analysis of the signals are performed as described previously.

In the double chemiluminescence ISH, the enzymatic reaction of HRP is usually performed before that of AP. This is done because the HRP/luminol reaction reaches a steady-state light output rapidly, which is maintained for a relatively short time, and then the signal decreases while the AP/dioxetane reaction slowly reaches the steady-state light emission, which then lasts for a relatively long time. Short washes to remove the HRP substrate are thus sufficient to avoid any interfering photon emission during the AP reaction. Even if for reasons of rapidity of the assay the enzymatic reaction of HRP is usually performed before that of AP, consistent results are obtained when the substrate reaction sequence is reversed and longer washes are performed.

Results and Discussion

Chemiluminescence ISH has proved to be very sensitive, being able to detect as few as 10 to 50 viral genome copies in infected cells using both biotinylated probes and digoxigenin-labeled probes with HRP and AP chemiluminescent detections, respectively. In fact, positive chemiluminescent signals were obtained in HeLa cells, which are known to contain about 10 to 50 integrated genome copies of human papillomavirus (HPV) 18.[18,19] Chemiluminescence ISH has proved more sensitive than *in situ* hybridization followed by colorimetric detection and almost as sensitive as *in situ* hybridization followed by ^{35}S autoradiography; in fact, the detection of HPV DNA in HeLa cells, which was observed by chemiluminescence, was not obtained with colorimetric methods,[19,20] and the chemiluminescent signal resolution was comparable to that obtained with ^{35}S autoradiography.[18] Moreover, with chemiluminescence ISH to detect B19 parvovirus DNA in bone marrow cells, all specimens from patients with a diagnosed B19 infection proved positive with a higher number of positive cells/specimen compared to colorimetric detection, thus permitting an easier evaluation of the sample.[21]

Chemiluminescence ISH allows quantification of the results; in fact, results on smears of Caski and HeLa cells (which are known to contain 500–600 copies of HPV 16 DNA and 10–50 copies of HPV 18 DNA, respectively) demonstrated that the luminescent signal changed in proportion to the known numbers of viral genome copies per cell.[19] In addition, increasing values of emitted photons/cell corresponding to the presence of hybridized cytomegalovirus DNA in infected cells could be found in cells fixed at 48, 60, 72, and 96 hr postinfection following the cytomegalovirus replication cycle (Fig. 1).[22]

In the double chemiluminescence ISH, positive signals for the presence of both herpes simplex virus DNA and cytomegalovirus DNA were local-

[18] P. Lorimier, L. Lamarcq, A. Negoescu, C. Robert, F. Labat-Moleur, F. Gras-Chappuis, I. Durrant, and E. Brambilla, *J. Histochem. Cytochem.* **44,** 665 (1996).

[19] M. Musiani, M. Zerbini, S. Venturoli, G. Gentilomi, G. Gallinella, E. Manaresi, M. La Placa Jr., A. D'Antuono, A. Roda, and P. Pasini, *J. Histochem. Cytochem.* **45,** 1 (1997).

[20] P. Lorimier, L. Lamarcq, F. Labat-Moleur, C. Guillermet, R. Bethier, and P. Stoebner, *J. Histochem. Cytochem.* **41,** 1591 (1993).

[21] M. Musiani, A. Roda, M. Zerbini, G. Gentilomi, P. Pasini, G. Gallinella, and S. Venturoli, *J. Clin. Microbiol.* **34,** 1313 (1996).

[22] M. Musiani, A. Roda, M. Zerbini, P. Pasini, G. Gentilomi, G. Gallinella, and S. Venturoli, *Am. J. Pathol.* **148,** 1105 (1996).

ized sharply in infected cells with no cross-reactions and low background (Fig. 2). The developed double chemiluminescence ISH can thus be a useful tool for sensitive and specific diagnosis of dual viral infections[23] and for the detection of multiple genetic sequences inside cells. However, in comparison with other existing ISH methods to detect two targets, the chemiluminescence ISH has the limitation of requiring sequential staining and signal visualization.

With chemiluminescence ISH, because the positive signal is considered the one above threshold values, an objective evaluation of the results could be achieved without any microscopic training, thus minimizing doubts about positive or negative results. Moreover, chemiluminescence ISH offers a permanent record of the reactions as all the images of the samples are stored in the computer and these images can be printed or sent for an evaluation in other laboratories using floppy disks or other computer networks.

In conclusion, the chemiluminescent ISH assay can be a useful tool for the sensitive detection of viral infection and the study of specific genetic sequences inside cells. The use of the chemiluminescent ISH assay, which can be applied to detect one or two target nucleic acids in cell smears, archival frozen, and paraffin-embedded tissue samples, may also be promising for estimating and quantifying nucleic acids present in tissue samples or cellular smears and for imaging gene expression in cells, provided a strong standardization of the methods, reagents, and samples.

Chemiluminescence Immunohistochemistry

Specimen Preparation

Place biopsy specimens directly into Histocon (Cellpath PLC, Hemel Hempstead, UK), a proprietary product that optimally preserves tissues prior to freezing. To freeze, place the biopsy into a methyl cellulose-based mounting fluid, e.g., OCT (Tissue Tek II, Miles Inc, Elkart, IN), on a cork disk and orientate so that sectioning will result in transverse sections and freeze in isopentane cooled over liquid nitrogen. Store frozen biopsies at $-70°$ until sectioned.

Cut 5-μm-thin sections in a cryostat microtome and collect onto a Multispot slide (CA Hendly, Laughton, UK). Fix dried slides in acetone for 5 min, redry, wrap in cling film, e.g., Glad wrap or Saran wrap, and

[23] G. Gentilomi, M. Musiani, A. Roda, P. Pasini, M. Zerbini, G. Gallinella, M. Baraldini, S. Venturoli, and E. Manaresi, *BioTechniques* **23,** 1076 (1997).

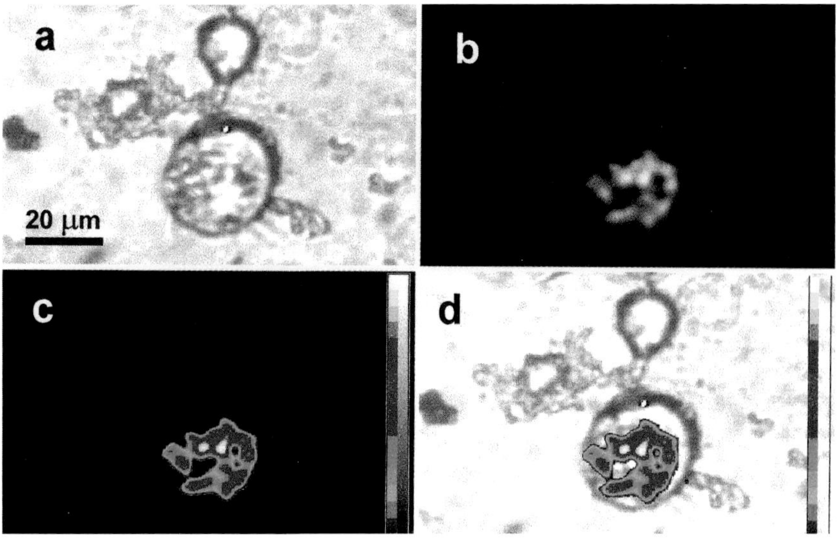

FIG. 1. Chemiluminescence *in situ* hybridization revealing cytomegalovirus DNA in human fibroblasts fixed at 96 hr postinfection. Live image (a), luminescent signal (b), pseudocolor processed luminescent image (c), and overlay of luminescent and live images (d). Reproduced with permission from M. Musiani, A. Roda, M. Zerbini, P. Pasini, G. Gallinella, and S. Venturoli, *Am. J. Pathol.* **148,** 1105 (1996).

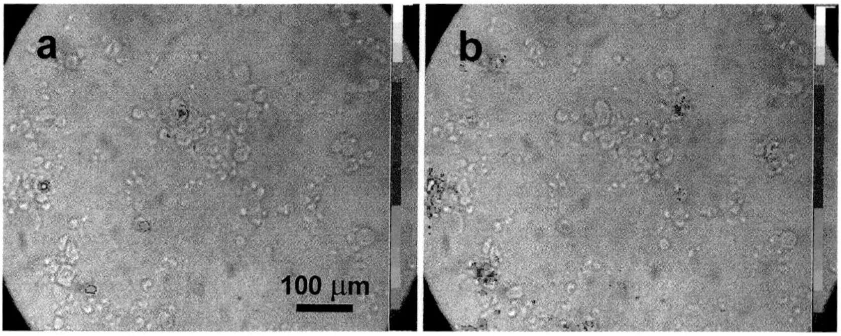

FIG. 2. Chemiluminescence *in situ* hybridization revealing herpes simplex virus DNA (a) and cytomegalovirus DNA (b) in infected cells in the same specimen. Reproduced with permission from G. Gentilomi, M. Musiani, A. Roda, P. Pasini, M. Zerbini, G. Gallinella, M. Baraldini, S. Venturola, and E. Manaresi, *BioTechniques* **23,** 1076 (1997).

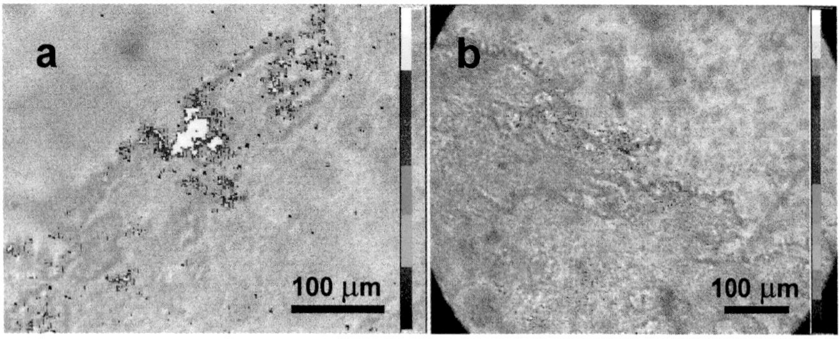

FIG. 3. Chemiluminescence imaging of *H. pylori* on gastric mucosal sections. Detection with antiurease antibody (a) and detection with anti-VacA antibody (b). This patient had a positive antibody titer to VacA. Light microscopy of toluidine blue-stained sections confirmed foci of bacteria corresponding to the chemiluminescent signal.

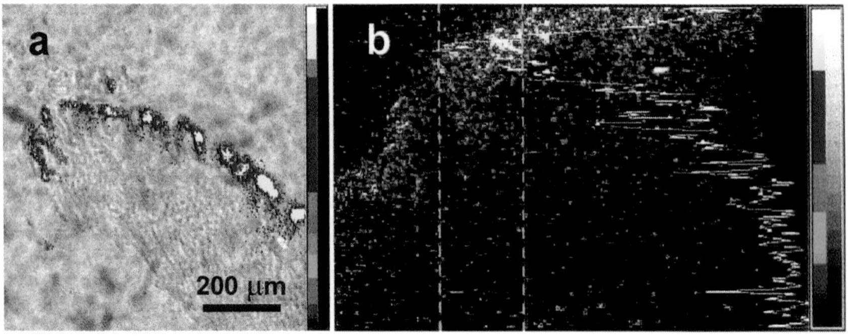

FIG. 4. Sections of gastric mucosa of *H. pylori*-infected patients showing high immunolabeling (IL-8) in epithelial cells (a) and quantitative analysis of variations in mucosal immunoreactivity (b). Within the lamina propria, the highest concentrations of IL-8 immunoreactivity are in the subepithelial region and decrease vertically through the lamina propria.

store at $-20°$. For quantitative studies, the cryosection thickness should be uniform.

Determination of Working Dilutions of Antibody Preparations

For every first and second layer antibody used, the optimal working dilutions should be determined to give good immunochemiluminescence and avoid nonspecific background chemiluminescence. First-layer antibodies should be diluted in PBS containing 1% bovine serum albumin and 0.1% azide and tested at 10-fold dilutions. Many hybridoma supernatants, however, can be used neat. Second-layer antibodies (goat antimouse IgG or goat antirabbit Ig, etc.) should be diluted in PBS containing 0.1% azide and 5% goat serum (or appropriate species-specific sera). Each new batch of enzyme-conjugated antisera should be tested to determine the optimal working dilution.

Specificity Controls for Indirect Immunochemiluminescence

There are a number of situations in which false-positive immunochemiluminescence may occur. This can arise from the use of nonspecific or cross-reacting antibodies or background endogenous enzymatic activity. To ensure that chemiluminescence is a result of specific immunolabeling, it is essential to include appropriate controls. In general, where polyclonal antibodies are used it is important to include nonimmune sera in the primary layer. Immunochemiluminescence with monoclonal antibodies should include an irrelevant monoclonal antibody of the same isotype. The specificity of immunolabeling should be confirmed by absorption controls to demonstrate that immunoreactivity is abolished by prior absorption with an excess of a specific protein but not by absorption with irrelevant substances. Additional controls should include omission of the first-layer antibodies to check for nonspecific binding of the enzyme-conjugated second-layer antibodies to components in the tissue section. Endogenous phosphatase activity in the luminal membrane of intestinal epithelial cells precludes the use of alkaline phosphatase as a label for the chemiluminescence detection of antigens. In the gastric mucosa, where endogenous alkaline phosphatase activity in 5-μm sections is not detectable except in regions of intestinal metaplasia, *Helicobacter pylori* bacterial and host antigens have been detected successfully by chemiluminescence using phosphatase-labeled probes.

Immunolabeling of Human Gastric Mucosa

Allow stored wrapped slides to warm to room temperature before unwrapping. Incubate sections with first-layer antibodies (Table II) and appro-

TABLE II
SPECIFIC ANTIBODIES USED TO DETECT BACTERIAL AND MUCOSAL ANTIGENS BY CHEMILUMINESCENCE IMMUNOHISTOCHEMISTRY

Antigen	Antibody
Helicobacter pylori urease enzyme	Polyclonal rabbit antiurease[a]
H. pylori VacA cytotoxin	Polyclonal rabbit anti-VacA[b]
Interleukin-8 (IL-8)	Monoclonal mouse anti-IL-8[c]
Inducible nitric oxide synthase (iNOS)	Polyclonal rabbit anti-iNOS (Affinity Bioreagents, Cat. No. PA3-030, Cambridge Biosciences, Cambridge, UK)

[a] From Dr. C. Clayton (Glaxo Wellcome Medical Research Center, UK).
[b] From Dr. J. Telford (Chiron Biocine, Siena).
[c] From Dr. I. J. D. Lindley (Novartis Research Institute, Vienna).

priate controls for 1 hr at room temperature. There is no need to cover the whole area of the spot on the slide, just the sections. Ten microliters per spot will suffice. Sections should be labeled in a humidified chamber and should not be allowed to dry during the immunolabeling. Wash in Tris-buffered saline (TBS) for 30 min with stirring.

Remove excess TBS but leave the area of section moist. Incubate with appropriate second-layer antibodies, e.g., phosphatase-labeled goat antirabbit (Dako, Glostrup, Denmark) or phosphatase-labeled goat antimouse (Sigma, St. Louis, MO), for 1 hr at room temperature in a humidified chamber. Wash in TBS for 30 min with stirring. Keep slides in TBS at room temperature until addition of the chemiluminescent substrate.

Chemiluminescent Detection

Incubate sections with 20–40 μl of the AP chemiluminescent substrate (CSPD or Lumi-Phos Plus) for 15–20 min in the dark at room temperature and acquire the chemiluminescent signal.

The chemiluminescent signal is collected for 1 min. The image of the same field is acquired under transmitted light to allow localization of the specific signal. The CL signal could be recorded several times within 5–10 min.

Results and Discussion

Although we have used unstained sections for optimal chemiluminescent detection, staining following CL collection allows the determination of specific structure or sites of *H. pylori* infection for comparative analysis (Fig. 3).

The amount of bound antibody can be quantified by determining the chemiluminescence signal from a defined area of the tissue sections. This approach has been adopted to determine iNOS immunoreactive protein in gastric epithelial cells in relation to *H. pylori* infection. For secreted products of epithelial cells, for example, IL-8, image processing using software allows variations in chemokine distribution within epithelial cells and mucosal tissues to be examined. Image processing allows focal areas of high IL-8 immunoreactivity to be determined in epithelial cells, allowing the potential analysis of host epithelial mediator changes in relation to infection with pathogenic agents such as *H. pylori* (Fig. 4).

The main advantage of using the immunochemiluminescence technique over the colorimetric technique is that it allows the quantification of tissue antigens. Furthermore, antigen detection can be more sensitive[20] and improves the localization of infectious pathogens markedly (Fig. 3). Discrete foci of pathogens, e.g., *H. pylori,* are determined using antibodies against conserved bacterial proteins, e.g., urease.

Alternatively, the immunochemiluminescent approach could be used to identify a bacterial phenotype using antisera against known strain-specific virulence determinants, e.g., VacA.[24] Such an approach could allow identification of a bacterial phenotype *in situ* without microbial culture.

Another advantage of the technique is that it allows examination of the immunopathogenic aspects of infections. It can be used to investigate host–pathogen interactions, with particular respect to inflammatory mediators in epithelial cells such as IL-8 and iNOS, which are upregulated in response to infectious agents.

Future approaches could be double immunochemiluminescence for quantifying the epithelial expression of mediators in relation to luminal pathogenic agents.

Concluding Remarks

The overall results obtained in both ISH and IHC demonstrate that chemiluminescence detection can be used successfully for such techniques.

The chemiluminescent detection of AP- or HRP-labeled probes allows improvement in their detectability compared to colorimetric detection. Furthermore, chemiluminescence uses relatively simple instru-

[24] J. L. Telford, P. Ghiara, M. Dell'Orco, M. Comanducci, D. Burroni, M. Bugnoli, M. F. Tecce, S. Censini, and A. Covacci, *J. Exp. Med.* **179,** 1653 (1994).

mentation and avoids the use of radioactive probes to achieve similar performance.

The detection of AP- and HRP-labeled antibodies, required for the recognition of primary antibody–antigen or cDNA–target genetic sequence reactions, is carried out using commercially available CL substrates with optimized kinetics and suitable for imaging purposes. Their steady-state kinetics further simplifies the instrumentation required for chemiluminescence ISH and IHC imaging.

Commercially available cooled CCD or similar video cameras with or without image intensification can be connected easily to a common optical microscope by a C-mount adapter. The only variation in respect to conventional light microscopy is that the system must be kept in the dark. CL detection can be applied to different kinds of samples such as single cells, frozen, and paraffin-embedded tissue sections, and the specimens are prepared following the same protocols developed for ISH and IHC techniques with other detection systems.

Furthermore, once standardized, CL imaging is suitable for the quantitative detection of AP- or HRP-labeled probes, and this represents an important improvement over conventional colorimetric or fluorometric methods.

The CL mapping of the analytes in the specimens requires three steps: (1) the transmitted light image collection; (2) the addition of the CL cocktail and the acquisition of the emitted light; and (3) the superimposition of the CL image with the live image to localize the analyte properly. The multistep procedure would be facilitated by the use of an optical microscope equipped with a computer-assisted device able to locate the same sample spots precisely. With an automated system, the detection of two or more ISH or IHC reactions in the same specimen would also be easier, thus increasing the diagnostic power of such techniques.

A further advancement must be the improvement of the transmitted light image quality by staining the samples with proper dyes able to better define the morphological structure. This could be done either after the CL measurement or before it, provided that the staining does not interfere with the CL reaction.

Acknowledgments

The authors thank Dr. I.J.D. Lindley (Novartis Research Institute, Vienna), Dr. J. Telford (Chiron Biocine, Siena), and Dr. C. Clayton (Glaxo Wellcome Medical Research Center, UK), respectively, for antibodies to IL-8, VacA, and recombinant urease. This work was supported by grants from The British Council, the European Commission (contract number IC 18CT950024), the Yorkshire Cancer Research Campaign, the University of Bologna (funds for selected research topics), the MURST, and the CNR (target project on "Biotechnology"). We thank Sarah Perry for her skillful technical assistance.

[41] Blood Phagocyte Luminescence: Gauging Systemic Immune Activation

By ROBERT C. ALLEN, DAVID C. DALE, and FLETCHER B. TAYLOR, JR.

Introduction*

Acute defense against infecting microbes is achieved through a complex information-effector mechanism. Neutrophil leukocytes, monocytes, and eosinophil leukocytes serve as direct phagocyte effectors of microbe killing action, but to properly execute their microbicidal function, these blood phagocytes must respond to information in the form of molecular signals generated by the humoral and cellular components of the acute inflammatory response. These signals direct phagocyte–endothelial contact, diapedesis, and locomotion to the site of inflammation. Ultimately phagocyte–microbe contact results in recognition, degranulation, phagocytosis, and activation of redox metabolism.[1–3]

The molecular signals that direct phagocyte function are varied and

* Abbreviations used: rhC5a, recombinant human complement C5a anaphylatoxin; C5a, complement C5a anaphylatoxin; IL-8, interleukin 8; rhG-CSF, recombinant human granulocyte colony-stimulating factor; G-CSF, granulocyte colony-stimulating factor; PAF, platelet-activating factor (1-*O*-hexadecyl-2-*O*-acetyl-sn-glycero-3-phosphorylcholine); LTB_4, leukotriene B4; C3b, complement C3b opsonin; C3bi, complement C3bi opsonin; HMP, hexose monophosphate; *S*, net spin quantum number; $NADP^+$, nicotinamide adenine dinucleotide phosphate, oxidized; NADPH, nicotinamide adenine dinucleotide phosphate, reduced; HO_2, hydrodioxylic acid; O_2^-, superoxide anion; $^1O_2^*$, singlet molecular oxygen; MPO, myeloperoxidase; EPO, eosinophil peroxidase; HRP, horseradish peroxidase; *E*, potential in volts; ΔE, change in potential in volts; ΔG, change in free energy; CLS, chemiluminigenic substrates; dL/dt, luminescence intensity or velocity; Ox, oxygenating agent; Φ_{CL}, chemiluminescence quantum yield; Φ_{PMT}, quantum efficiency of a photomultiplier tube; C, substrate; C_{native}, native substrate; DBA^{++}, dimethylbiacridinium (bis-*N*-methylacridinium nitrate), lucigenin; POX, peroxidase; NMA, *N*-methyacridone; π^*, pi antibonding orbital; $OONO^-$, peroxynitrite anion; DBSS, dimethylbiacridinium (lucigenin) balanced salt solution; LBSS, luminol balanced salt solution; hC-OpZ, complement-opsonized zymosan, human; PMA, phorbol 12-myristate 13-acetate; FMLP, *N*-formyl-methionyl-leucyl-phenylalanine; LPS, lipopolysaccharide (endotoxin); R, opsonin receptor; P, phagocyte; R/P, opsonin receptor expression per phagocyte; COR, circulating opsonin receptor; MOR, maximum opsonin receptor; DF, discriminant function; Eff_{Lum}, luminometer efficiency; F_{Geo}, geometry factor; Eff_{Net}, net efficiency; RLU, relative light unit; PM_{HMP}, phagocyte HMP shunt metabolism.

[1] S. J. Klebanoff and R. C. Clark, "The Neutrophil: Function and Clinical Disorders." North-Holland, Amsterdam, 1978.

[2] R. C. Allen and B. A. Pruitt, Jr., *Arch. Surg.* (*Chicago*) **117,** 133 (1981).

[3] R. C. Allen and D. L. Stevens, *Curr. Opin. Infect. Dis.* **5,** 389 (1992).

0076-6879/00 $30.00

diverse. Blood phagocytes respond to the complement anaphylatoxin C5a, cytokines such as interleukin(IL)-8, colony-stimulating factors such as G-CSF and GM-CSF, and tumor necrosis factor (TNF). Phagocytes are also highly sensitive to lipid-derived metabolic products, such as platelet-activating factors (PAF) and leukotrienes (LTB4). In addition, phagocytes can respond directly to microbial products such as *N*-formylmethionyl peptides and endotoxin. These small molecular weight products diffuse from their site of generation and establish a concentration gradient that directs phagocyte chemotaxis to the site of inflammation.

The immune system recognizes and labels target microbes by binding specific antibodies directly, i.e., immunoglobulins such as IgM, IgG, and IgA. Activation of the alternative and classical pathways of complement produces anaphylatoxin C5a and also covalently fix C3b and C3bi to the target microbe.[4] The immunoglobulins and complement molecules that bind to the target microbe are the opsonins that mediate phagocyte–microbe contact recognition by way of opsonin-specific phagocyte receptors. Ligation triggers phagocytosis (i.e., invagination of the phagocyte membrane allowing the cell to engulf the microbe), phagosome formation (i.e., the membrane-enclosed microbe within the space of the cell), and ultimately phagolysosome formation (i.e., the fusion of azurophilic lysosomal granule with the phagosome).[1]

The morphologic events of phagocytosis are associated with extreme changes in phagocyte metabolism. Phagocytosis is associated with a large increase in glucose metabolism via the dehydrogenases of the hexose monophosphate (HMP) shunt and with a large increase in oxygen consumption that is unrelated to mitochrondrial metabolism.[5,6] This metabolic activation, commonly referred to as respiratory burst, generates the oxidants and oxygenating agents required for effective microbe killing.[7–9]

Phagocyte Oxygenation Activity and Native Luminescence

The oxidants and oxygenating agents generated by activated phagocytes accomplish microbicidal action by a type of limited combustion. The spin symmetry restrictions that limit the direct reaction of oxygen (O_2) with

[4] G. D. Ross, "Immunobiology of the Complement System." Academic Press, New York, 1986.

[5] A. J. Sbarra and M. L. Karnovsky, *J. Biol. Chem.* **234,** 1355 (1959).

[6] F. Rossi, D. Romeo, and P. Patriarca, *J. Reticulendothel. Soc.* **12,** 127 (1972).

[7] G. Y. N. Iyer, D. M. F. Islam, and J. H. Quastel, *Nature* **192,** 535 (1961).

[8] R. C. Allen, R. L. Stjernholm, and R. H. Steele, *Biochem. Biophys. Res. Commun.* **47,** 679 (1972).

[9] R. C. Allen, S. J. Yevich, R. W. Orth, and R. H. Steele, *Biochem. Biophys. Res. Commun.* **60,** 909 (1974).

microbes are overcome by the phagocyte through a sequence of electron exchange reactions.[10–12] In chemical combustion (i.e., burning), an organic substrate is radicalized, i.e., homolytic cleavage increasing the net spin quantum number (S) of the substrate. By this process, an $S = 0$ molecule is cleaved thermally to yield one $S = +1/2$ molecule and one $S = -1/2$ molecule, i.e., a single nonradical molecule is cleaved to yield two radical products. The O_2 we breathe is a paramagnetic diradical molecule with an S value of 1. Radicalization increases the S value of a substrate and facilitates spin allowed reaction with O_2. The reaction of O_2 ($S = 1$) with a radical substrate ($S = 1/2$) yields a radical product ($S = 1/2$). This $S = 1/2$ product can in turn react with an additional O_2 ($S = 1$) in the radical propagation process of burning.

Conservation of spin symmetry limits the direct reactivity of O_2 with biological molecules. Phagocytes employ a converse approach to overcoming the spin symmetry limitations that restrict direct O_2 reactivity. Instead of increasing the S value of substrate, phagocytes decrease the S value of oxygen. Phagocyte redox metabolism provides the energy and mechanisms that decrease the S value of O_2 from $S = 1$ to $S = +1/2$ (or $-1/2$), and finally to $S = 0$. By lowering the molecular spin quantum number of O_2, phagocytes eliminate the spin symmetry barrier that prevents direct O_2 reactivity and, by doing so, phagocytes realize the enormous reactive potential of O_2 as a microbicidal agent. Phagocyte combustion is controlled and effective against the broad spectrum of microbes.

NADPH Oxidase and Respiratory Burst Metabolism

The phagocyte metabolic pathways responsible for O_2-dependent microbicidal action and luminescence are depicted schematically in Fig. 1.[8,9,13–15] Glucose is dehydrogenated via the HMP shunt (also known as the pentose pathway and phosphogluconate pathway). Glucose-6-phosphate : $NADP^+$ oxidoreductase and 6-phosphogluconate : $NADP^+$ oxidoreductase generate the NADPH required for the univalent reduction of O_2.

NADPH oxidase (NADPH : O_2 oxidoreductase) catalyzes the one equivalent, i.e., 1 electron (e^-) plus 1 proton (H^+), reduction of O_2. In doing so the S of O_2 is changed from 1 to 1/2. The spin multiplicity of an atom or molecule is equal to the absolute value its S multiplied by two

[10] R. C. Allen, *Stud. Org. Chem.* **33,** 425 (1988).
[11] R. C. Allen, *Quimica Nova* **16,** 354 (1993).
[12] R. C. Allen, *Environ. Health Perspect.* **102**(Suppl. 10), 201 (1994).
[13] R. C. Allen, *Front. Biol.* **48,** 197 (1979).
[14] R. C. Allen, *Biochem. Biophys. Res. Commun.* **63,** 675 (1975).
[15] R. C. Allen, *Biochem. Biophys. Res. Commun.* **63,** 683 (1975).

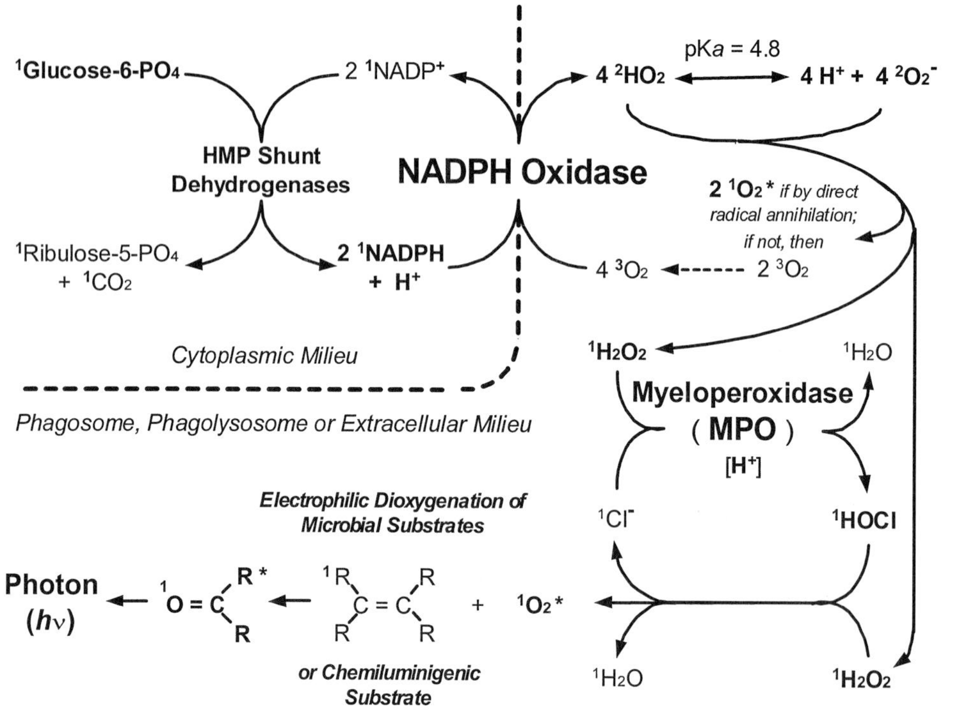

FIG. 1. Diagram depicting the major enzymatic systems responsible for phagocyte microbicidal metabolism and oxygenation activities and the relationship of these activities to photon emission. In the schema the activities of the cytoplasmic milieu are separated from those of the phagosome–phagolysosome–extracellular milieu. The superscripted number that precedes each molecular symbol (i.e., ^{1}for singlet, ^{2}for doublet, and ^{3}for triplet multiplicity) depicts the multiplicity of each reactant. Multiplicity is related to the spin quantum number (S) by the equation: $|2S| + 1$ = multiplicity ($''$).

plus one, i.e., $|2S| + 1$. One equivalent reduction changes the multiplicity of diradical O_2 from triplet [i.e., $|2(1)| + 1 = 3$], indicated by the superscript 3 (i.e., $^3O_2$) as shown in Fig. 1, to doublet [i.e., $|2(1/2)| + 1 = 2$] multiplicity hydrodioxylic acid, indicated by the superscript 2 (i.e., $^2HO_2$). Because the activities of the HMP shunt dehydrogenases are directly dependent on the availability of $NADP^+$, NADPH oxidase activity exerts pivotal control over the rate of respiratory burst metabolism.[6]

Hydrodioxylic acid has a pK_a of 4.8.[16] As such, the one equivalent reduction of O_2 to HO_2 by NADPH oxidase provides a mechanism for acidifying the phagolysosome.[13] Oxidase activity drives the pH of the phago-

[16] B. H. J. Bielski and A. O. Allen, *J. Phys. Chem.* **81,** 1048 (1977).

lysosome in the direction of the pK_a of HO_2. The acid dissociation of HO_2 yields superoxide anion ($^2O_2^-$). As the pH approaches the pK_a, the ratio of HO_2 to O_2^- approaches one, and the anionic repulsion that prevents direct superoxide radical–radical annihilation is eliminated.[16] Under acidic to neutral conditions, the direct disproportionation reaction

$$^2HO_2 + {}^2O_2^- \xrightarrow{\text{(acid milieu)}} {}^1H_2O_2 + {}^1O_2^* \quad (1)$$

can proceed via a singlet surface, i.e., by doublet–doublet annihilation, yielding the singlet multiplicity products H_2O_2 and singlet molecular oxygen, $^1O_2^*$.[13,17,18] Antithetically, if the reaction is sequential and involves high multiplicity metal-catalyzed intermediates, the products can be of mixed multiplicity, i.e.,

$$^2O_2^- + \text{metal}^{+n} \rightarrow \text{metal}^{+(n-1)} + {}^3O_2 \quad (2a)$$
$$^2O_2^- + \text{metal}^{+(n-1)} \rightarrow \text{metal}^{+n} + {}^1H_2O_2 \quad (2b)$$

Equations (2a) and (b) describe the type of reaction catalyzed by superoxide dismutase.[19] Note that singlet multiplicity hydrogen peroxide ($^1H_2O_2$) is produced by either mechanism. H_2O_2 is the substrate for phagocyte haloperoxidase activity.

Haloperoxidase-Catalyzed Oxidations and Oxygenations

Myeloperoxidase (MPO), i.e., $Cl^-:H_2O_2$ oxidoreductase, is an enzyme common to neutrophil leukocytes and monocytes. A similar haloperoxidase with a bromide (Br^-) halide cofactor requirement, i.e., eosinophil peroxidase (EPO), is found in eosinophil leukocytes. MPO is reported to constitute approximately 5% of the dry weight of human neutrophils.[20] In the unstimulated neutrophil, MPO is located in azurophilic lysosomal granules.[1] Following phagocytosis, azurophilic granules fuse with the phagosome to create the phagolysosome. This fusion brings MPO into direct contact with the phagocytized microbe.

As described in Fig. 1, MPO catalyzes the acid-optimum, peroxide-dependent oxidation of chloride (Cl^-) to hypochlorous acid (HOCl).[14,15,21] Once generated, HOCl can serve as a direct dehydrogenating agent. It can

[17] A. U. Khan, *Science* **168,** 476 (1970).
[18] W. H. Koppenol, *Nature* **262,** 420 (1976).
[19] J. M. McCord and I. Fridovich, *J. Biol. Chem.* **244,** 6049 (1969).
[20] J. Schultz and K. Kaminker, *Arch. Biochem.* **96,** 465 (1962).
[21] K. Agner, *Proc. 4th Intl. Cong. Biochem. Vienna* **15,** 64 (1958).

also react with amines to produce chloramines and other products.[22,23] Chloride-dependent MPO activity is bactericidal and fungicidal.[24,25] HOCl can also react with the second H_2O_2 to yield singlet oxygen $^1O_2*$.[26,27]

This type of oxidative–reductive coupling can be defined quantitatively by the Nernst relationship.[11–13,28] The net potential, expressed as change in E (ΔE) in volts, for the primary reaction catalyzed by MPO:

$$^1H_2O_2 + {}^1Cl^- + H^+ \xrightarrow{\text{(MPO acid milieu)}} {}^1H_2O + {}^1HOCl \quad (3)$$

is expressed

$$\Delta E_1 = (RT/nF)\ln[([H_2O][HOCl])/([H_2O_2][Cl^-][H^+])] \quad (4)$$

where R is the gas constant, T is the absolute temperature, n is the number of electrons per gram equivalent transferred, F is a Faraday, and ln $[([H_2O][HOCl])/([H_2O_2][Cl^-][H^+])]$ is the natural log of the products divided by the reactants. At pH values of 5, 6, and 7, the ΔE_1 values are 0.134, 0.104, and 0.075 V, respectively. These values correspond to free energies (ΔG's) of −6.16, −4.80, and −3.44 kcal/mol, respectively.

Once generated, HOCl can react directly with a second H_2O_2 molecule:

$$^1H_2O_2 + {}^1HOCl \rightarrow {}^1H_2O + H^+ + {}^1Cl^- + {}^1O_2* \quad (5)$$

This secondary reaction is described by the Nernst equation

$$\Delta E_2 = (RT/nF)\ln[([H_2O][H^+][Cl^-][^1O_2*])/([H_2O_2][HOCl])] \quad (6)$$

At pH values of 5, 6, and 7, the ΔE_2 values are 0.960, 0.990, and 1.019 V, and the corresponding ΔG's are −21.63, −22.99, and −24.35 kcal/mol, respectively. Even with the generation of singlet oxygen, the net reaction

$$2\,{}^1H_2O_2 \rightarrow 2\,{}^1H_2O + {}^1O_2* \quad (7)$$

has a ΔE_{net} of 1.09 V and a ΔG of −27.79 kcal/mol.

Note in Fig. 1 that from the point of H_2O_2 generation forward, the $S = 0$ for all reactants and products, i.e., the reactions are nonradical. The reduction of O_2 to HO_2 and its further reduction to H_2O_2 decreased the

[22] J. M. Zgliczynski, T. Stelmaszynska, W. Ostrowski, J. Naskalski, and J. Sznajd, *Eur. J. Biochem.* **4,** 540 (1968).

[23] J. M. Zgliczynski, T. Stelmaszynska, J. Domanski, and W. Ostrowski, *Biochim. Biophys. Acta* **235,** 419 (1971).

[24] S. J. Klebanoff, *J. Bacteriol.* **95,** 2131 (1968).

[25] R. I. Lehrer, *J. Bacteriol.* **99,** 361 (1969).

[26] H. Taube, *J. Gen. Physiol.* **49,** 29 (1965).

[27] M. Kasha and A. U. Khan, *Ann. N.Y. Acad. Sci.* **171,** 5 (1970).

[28] M. Pourbaix, "Atlas of Electrochemical Equilibria in Aqueous Solutions." Pergamon, Oxford, 1966.

spin multiplicity of oxygen from $S = 1$ to $S = 1/2$ and on to $S = 0$. In compliance with Wigner spin conservation rules, the $S = 0$ product H_2O_2 can participate directly in spin allowed reactions with other $S = 0$ organic molecules, but in its fully protonated form, H_2O_2 is a relatively poor direct oxidant. The pK_a of H_2O_2 is 11.65. As such, the optimum hydrogen ion (H^+) concentration for direct peroxide reactivity is four logs or ten thousandfold lower than that of neutral physiologic medium, and about six logs or a million fold lower than that of the acid phagolysosomal space.

The $S = 0$ products of MPO catalysis, such as HOCl, chloramine, and $^1O_2^*$ can react directly with $S = 0$ biologic substrates at neutral physiologic pH and in the acid phagolysosomal milieu, and these reactants play a direct role in the phagocyte microbicidal action. Dioxygenation reactions yield intermediate dioxetanes and endoperoxides that disintegrate, yielding singlet multiplicity electronically excited carbonyl products. Relaxation of these excited carbonyls by photon emission produces the native chemiluminescence associated with phagocyte microbicidal function.[8,9,13]

Native versus Chemiluminigenic Substrate-Dependent Luminescence

Native phagocyte luminescence is direct proof that phagocyte microbicidal activity is combustive. The high exergonicities required for electronic excitation and photon emission are not achieved by most biochemical reactions. High exergonicity is characteristic of combustive dioxygenation reactions. As illustrated in Fig. 1, phagocyte luminescence is dependent on and proportional to respiratory burst metabolism and its resulting oxygenation activity.[8,29,30] Native phagocyte luminescence provides a window from which to view the temporal kinetics of microbicidal combustion.

Native phagocyte luminescence has two major limitations as a diagnostic technique. First, the yield of excited products from dioxygenations of native substrates is relatively low in comparison to bioluminescence reactions. Second, the substrates presented are numerous and vary with the nature of the microbe or the particulate material phagocytized. Introducing chemiluminigenic substrates (CLS) overcomes both of these limitations.[29,31–33] A

[29] R. C. Allen, *in* "Chemical and Biological Generation of Excited States" (G. Cilento and W. Adam, eds.), p. 309. Academic Press, New York, 1982.

[30] R. C. Allen and T. F. Lint, *in* "Analytical Applications of Bioluminescence and Chemiluminescence" (E. Schram and P. Stanley, eds.), p. 589. State Printing & Publishing, Westlake Village, CA, 1979.

[31] R. C. Allen and L. D. Loose, *Biochem. Biophys. Res. Commun.* **69,** 245 (1976).

[32] R. C. Allen, *in* "Bioluminescence and Chemiluminescence, Basic Chemistry and Analytical Applications" (M. A. DeLuca and W. D. McElroy, eds.), p. 63. Academic Press, New York, 1981.

[33] R. C. Allen, *Methods Enzymol.* **133,** 449 (1986).

CLS is an exogenous molecule that is susceptible to oxygenation by one or more phagocyte-derived agents and ultimately yields an electronically excited product. The exogenous CLS serves as a high quantum yield substitute for the native microbial substrates described in Fig. 1.

The intensity of light emission (dL/dt) is dependent on the phagocyte generation of oxygenating agents (Ox) and on the nature and chemiluminescence quantum yield (Φ_{CL}) of the substrate (C) oxygenated. This relationship is expressed by the rate equation

$$dL/dt = k[Ox][C] \tag{8}$$

where k is the proportionality constant and the brackets indicate concentrations. The sensitivity for detecting an oxygenating agent is dependent on the substrate oxygenated. Under native conditions, C represents numerous different molecules of relatively low Φ_{CL}. The relationship of dL/dt to phagocyte-generated oxygenating agents (Ox) is uncertain due to the variability and low efficency of native substrate oxygenation reactions.

Introducing a CLS, such as the cyclic hydrazide luminol or biacridimium compound lucigenin, into the phagocyte measurement milieu overcomes the limitations imposed by native substrate variability and low Φ_{CL}. Cyclic hydrazides and biacridinium compounds are organic substrates with Φ_{CL}'s many orders of magnitude greater than native microbial substrates. When presented at a nonlimiting concentration, a CLS increases luminescence intensity in proportion to the ratio of its Φ_{CL} relative to the mean Φ_{CL} of the native substrates it replaces. Furthermore, introducing a CLS defines the substrate and restricts the type of oxygenation activity measured. When present in excess, a CLS can compete successfully with native substrate (C_{native}) for available Ox. When the concentration of CLS far exceeds available oxygenating agent, i.e., $[C] \ggg [Ox]$, CLS oxygenation is effectively the sole source of luminescence and is nonlimiting. As such, the reaction simplifies to

$$dL/dt = k[Ox] \tag{9}$$

Under such conditions, luminescence measurements provide a quantitative gauge of phagocyte oxygenation of the particular CLS in a manner analogous to measuring substrate-dependent product generation in an enzyme reaction.[33]

NADPH Oxidase Activity Measured as Acridinium Chemiluminescence

Phagocyte NADPH oxidase catalyzes the univalent reduction of O_2 to HO_2, and in an acid milieu, these products disproportionate yielding H_2O_2.

The acridinium salt lucigenin [i.e., dimethylbiacridinium (DBA^{++}) also know as bis-*N*-methylacridinium] has a long history of use as a luminescence substrate[34,35] and has been applied to measure the products of phagocyte oxidase activity.[29,32,33] Under acid to neutral pH conditions, lucigenin ($^1DBA^{++}$) can undergo one electron reduction:

$$^1DBA^{++} + e^- \rightarrow {}^2DBA^+ \quad (10)$$

Any appropriate radical-reducing agent can serve as the one electron source, including metals and possibly O_2^-. Reduction changes the *S* value of lucigenin from 0 to 1/2, and as such, the reduced product is a doublet multiplicity radical ($^2DBA^+$). This radical can react directly with doublet superoxide ($^2O_2^-$) in the spin-allowed doublet–doublet, anion–cation annihilation reaction[32,33]:

$$^2O_2^- + {}^2DBA^+ \rightarrow {}^1DBA\text{-}O_2 \quad (11)$$

The moloxide intermediate ($^1DBA\text{-}O_2$) produced is an unstable singlet multiplicity dioxetane that disintegrates yielding one electronically excited *N*-methyacridone ($^1NMA^*$) and one ground state *N*-methyacridone (^1NMA):

$$^1DBA\text{-}O_2 \rightarrow {}^1NMA + {}^1NMA^* \quad (12)$$

Singlet excited NMA* relaxes by electronic transition from the pi antibonding (π^*) orbital of the carbonyl to the nonbonding (n) orbital of carbonyl oxygen with energy released as a photon:

$$^1NMA^* \rightarrow {}^1NMA + \text{Photon(hv)} \quad (13)$$

At physiological pH, the overall reaction can be described by the rate relationships

$$dL/dt = k[e^-]^1[\text{Ox } {}^2O_2^-]^1[\text{C } {}^1DBA^{++}]^1 \text{ or } k[\text{Ox } {}^2O_2^-]^1[\text{C } {}^2DBA^+]^1 \quad (14)$$

The reaction is first order with respect to $^1DBA^{++}$. However, the reaction order can be complex regarding superoxide. Depending on the source of the radical electron, the reaction may be first order or possibly second order relative to O_2^-. The requirement for initial radical reduction to produce the DBA^+ intermediate may be responsible for the delayed and unusual temporal kinetic pattern of the lucigenin-dependent phagocyte luminescence response.[33]

Liochev and Fridovich[36] reported that radically reduced lucigenin can

[34] K. Gleu and W. Petsch, *Angew. Chem.* **48,** 57 (1935).
[35] J. R. Totter, *Photochem. Photobiol.* **3,** 231 (1964).
[36] S. I. Liochev and I. Fridovich, *Arch. Biochem. Biophys.* **337,** 115 (1997).

undergo air oxidation to yield O_2^- and lucigenin. They state that lucigenin "can thus mediate O_2^- production in systems, such as glucose plus glucose oxidase, in which there is ordinarily no O_2^- production. Luc^{2+} luminescence can thus be used as the basis for assaying superoxide dismutase activity but should not be used for measuring, or even detecting, O_2^-."[36] Regarding the use of lucigenin in the study of phagocyte oxygenation activity, it is important to appreciate that both O_2^- and H_2O_2 are products of NADPH oxidase. If an investigator is interested in specifically measuring the generation of O_2^-, then the advice of Liochev and Fridovich applies.[36] However, if an investigator is interested in measuring the general oxidase activity of phagocytes, then the lucigenin luminescence methodology can provide a highly sensitive and relatively reproducible approach to analysis.[12,33]

That lucigenin luminescence is complex and that the source of the electron for initial radical reduction can be other than O_2^- were points considered in the original description of lucigenin luminescence as a measure of phagocyte oxidase activity.[32] "The univalent reduction of lucigenin can be affected by flavin or metal catalysis. Superoxide anion may directly serve as the univalent reducing agent. Radical–radical annihilation reaction between univalently reduced lucigenin and an additional superoxide anion can yield a dioxetane product."[32] Another interesting feature of lucigenin is that its reduction appears to require an electron and not an equivalent (i.e., an electron plus a proton). Lucigenin has chemical characteristics in common with the viologens (i.e., bipyridyl componds).[33] These molecules are unusual in that their reduction potential is influenced only minimally by pH, i.e., proton transfer plays a minor role in reduction.[37]

Essentially no luminescence is observed when lucigenin is exposed to physiologically appropriate concentrations of H_2O_2 under mildly acid conditions, and only a very weak signal is observed at physiologic pH. However, alkaline conditions favor the direct reaction of DBA^{++} with the conjugate base of peroxide[34,35]:

$$^1DBA^{++} + {}^1HO_2^- + {}^1OH^- \rightarrow {}^1DBA\text{-}O_2 + {}^1H_2O \qquad (15)$$

At alkaline pH the overall reaction can be described by the rate relationship

$$dL/dt = k[\text{Ox}\,{}^1HO_2^-]^1[\text{C}\,{}^1DBA^{++}]^1 \qquad (16)$$

The reactants and products are all of singlet multiplicity molecules, and the reaction is first order with respect to peroxide and first order with respect to lucigenin. At near neutral physiologic pH the reaction of Eq. (16) may be responsible for a very small portion of lucigenin-dependent

[37] W. M. Clark, "Oxidation-Reduction Potentials of Organic Systems." Williams & Wilkins, Baltimore, 1960.

phagocyte luminescence. Because O_2^- and H_2O_2 are both products of NADPH oxidase, the link between lucigenin luminescence and phagocyte oxidase activity is maintained.

The observation of McCapra and Hann[38] is important with respect to the role of spin conservation in lucigenin reactivity. The authors reported that the two electron reduction of lucigenin yields a product, i.e., ^{1}DBA, that reacts directly with $^1O_2^*$, i.e.,

$$^1\text{DBA} + {}^1O_2 \rightarrow {}^1\text{DBA-}O_2 \tag{17}$$

Note that the reaction yields the same moloxide intermediate. Luminescence results when this intermediate disintegrates according to Eqs. (12) and (13).

Measurement of Phagocyte Oxygenation Activity as Luminol Luminescence

In 1928 Albrecht reported that luminol produced intense luminescence in alkaline solutions of H_2O_2.[39] This chemical phenomenon stimulated interest and research into the mechanisms of chemiluminescence and in the use of luminol for the luminescence detection of oxidants.[40–42] In 1964 Dure and Cormier reported that luminol could substitute for luciferin in the bioluminescence reaction catalyzed by *Balanoglossus biminiensis* luciferase and that horseradish peroxidase (HRP) catalyzed a luminol luminescence reaction kinetically similar to that responsible for *B. biminiensis* luciferin–luciferase bioluminescence.[43]

Luminol was the first CLS applied to the study of phagocyte oxygenation activity. The high Φ_{CL} of luminol suggested that it might be applied as a CLS for measuring the oxidative activity of phagocytes.[31] With respect to Fig. 1, luminol serves as the high-Φ_{CL} substitute for the native microbial substrates.[31–33] In 1976 Allen and Loose reported that luminol increased the luminescence of neutrophil leukocytes by several orders of magnitude relative to the native luminescence response and that macrophages, which produce a very low and difficult to detect native luminescence, could be measured easily by introducing luminol as a CLS.[31] As a substrate, luminol had the major advantages of relatively high Φ_{CL} and relatively low toxicity.[29]

[38] F. McCapra and R. A. Hann, *J. Chem. Soc. Chem. Commun.* P, 442 (1969).
[39] H. O. Albrecht, *Z. Phys. Chem.* **136,** 321 (1928).
[40] E. H. White and D. F. Roswell, *Acc. Chem. Res.* **3,** 54 (1970).
[41] K. D. Gundermann, *Top. Curr. Chem.* **46,** 61 (1974).
[42] J. Lee and H. H. Seliger, *Photochem, Photobiol.* **4,** 1015 (1965).
[43] L. S. Dure and M. J. Cormier, *J. Biol. Chem.* **239,** 2351 (1964).

Haloperoxidases, such as MPO and EPO, can catalyze both classical peroxidation and haloperoxidation reactions. Both catalytic activities can ultimately produce luminol dioxygenation, yielding aminophalate and luminescence. However, these catalytic activities are distinctly different with respect to pH and cofactor requirement and proceed by different reaction pathways.

Myeloperoxidase-Catalyzed Luminol Oxygenation

Haloperoxidase-catalyzed luminol luminescence proceeds by a nonradical mechanistic pathway.[12,33,44] H_2O_2 does not react directly with luminol to yield luminescence under neutral to acid pH conditions. However, HOCl, a product of MPO-catalyzed H_2O_2-dependent oxidation of Cl^- as described in Eq. (3), can directly dehydrogenate luminol under such conditions. The reaction is by a nonradical mechanism and yields diazaquinone as the product:

$$^1\text{HOCl} + {}^1\text{luminol} \rightarrow {}^1\text{H}_2\text{O} + \text{Cl}^- + {}^1\text{luminol(diazaquinone)} \quad (18)$$

This oxidation is a two-equivalent dehydrogenation of luminol. It is not a dioxygenation, and the energy liberated is not sufficient to produce electronic excitation or luminescence. However, the product diazaquinone can react with an additional molecule of peroxide to yield excited aminophthalate and, ultimately, luminescence[44,45]:

$$^1\text{H}_2\text{O}_2 + {}^1\text{luminol(diazaquinone)} \rightarrow {}^1\text{N}_2 + {}^1\text{aminophthalate}^* \quad (19a)$$
$$^1\text{aminophthalate}^* \rightarrow {}^1\text{aminophthalate} + \text{Photon}(h\nu) \quad (19b)$$

This second reaction is a true dioxygenation yielding electronically excited aminophthalate with the excited product relaxing to ground state by photon emission. In the context of the previously considered reactions, the rate relationship can be described as

$$dC\,{}^1\text{luminol}_{(\text{diazaquinone})}/dt = k[\text{Ox}\,{}^1\text{HOCl}]^1[\text{C}\,{}^1\text{luminol}]^1 \quad (20a)$$
$$dL/dt = k[\text{Ox}\,{}^1\text{H}_2\text{O}_2]^1[\text{C}\,{}^1\text{luminol}_{(\text{diazaquinone})}]^1 \quad (20b)$$

Additional pathways are also possible. HOCl and, to a lesser extent, chloramine can react directly with an additional H_2O_2 to yield $^1O_2^*$ as described in Eq. (5).[27] The $^1O_2^*$ generated can react directly to dioxygenate luminol:

$$^1\text{luminol} + {}^1\text{O}_2 \rightarrow {}^1\text{N}_2 + {}^1\text{aminophthalate} + \text{Photon}(h\nu) \quad (21)$$

[44] R. C. Allen, "Haloperoxidase Acid Optimum Chemiluminescence Assay System." U.S. Patent 5,556,758, (1996).

[45] S. Mueller, H. D. Riedel, and W. Stremmel, *Anal. Biochem.* **245,** 55 (1997).

Both pathways start with the same primary MPO-catalyzed oxidation of halide as described in Eq. (3). By either pathway the net stoichiometry of the MPO-catalyzed luminol luminescence reaction is the same.[44] An initial H_2O_2 is consumed in generating the HOCl that is itself consumed in luminol dehydrogenation. A second H_2O_2 dioxygenates the luminol diazaquinone producing excited aminophthalate or it can react with HOCl to yield $^1O_2$. For this alternative pathway the rate relationship would be

$$dL/dt = k[\text{Ox}\ ^1O_2{}^*]^1[\text{C}\ ^1\text{luminol}]^1 \quad (22)$$

The reaction pathways described by Eqs. (18), (19a), (19b), and (21) are exclusively singlet multiplicity ($S = 0$) reactions and yield identical products.

Classical Peroxidase by Radical Mechanism

Haloperoxidases can also catalyze the classical halide-independent peroxidase reaction. In the initial reaction of this pathway, H_2O_2 reacts with the peroxidase (POX) oxidizing the enzyme by two eqivalent[46]:

$$^1H_2O_2 + \text{POX} \rightarrow 2\,^1H_2O + \text{POX Cpd I} \quad (23)$$

The resulting oxidized POX Cpd I can oxidize luminol by one equivalent yielding luminol radical (^{2}luminol) and generating POX Cpd II[47]:

$$\text{POX Cpd I} + {}^1\text{luminol} \rightarrow {}^2\text{luminol} + \text{POX Cpd II} \quad (24)$$

The doublet multiplicity luminol radical produced can participate directly in the spin allowed reaction with appropriate radicals such as O_2^-,

$$^2\text{luminol} + {}^2O_2^- \rightarrow {}^1N_2 + {}^1\text{aminophthalate}^- + \text{Photon}(h\nu) \quad (25)$$

Reaction by this type may be responsible for a small portion of the extraphagolysosomal luminol-dependent luminescence of phagocytes.[29]

As an additional consideration, the peroxide-dependent radical oxidation of luminol could also proceed by nonperoxidase mechanisms. If generated by phagocyte metabolism, the hydroxyl radical (OH) could also oxidize luminol by one electron:

$$^1\text{luminol} + {}^2\text{OH} \rightarrow {}^1H_2O + {}^2\text{luminol} \quad (26)$$

This type of radical activity may also be involved in the low-intensity, luminol-dependent luminescence of phagocytes lacking MPO.[29,33,48]

The partially reduced POX Cpd II can be reduced further by reaction

[46] B. Chance, *Adv. Enzymol.* **12,** 153 (1951).

[47] M. J. Cormier and P. M. Prichard, *J. Biol Chem.* **243,** 4706 (1968).

[48] G. A. Merrill, R. Bretthauer, J. Wright-Hicks, and R. C. Allen, *Lab. Anim. Sci.* **46,** 530 (1996).

with a second luminol molecule, completing the catalytic cycle by regenerating POX:

$$\text{POX Cpd II} + {}^{1}\text{luminol} \rightarrow {}^{2}\text{luminol} + \text{POX} \tag{27}$$

However, this reaction is relativley slow and tends to exert a rate-limiting effect. As an alternative approach, a molecule with reducing capacity greater than luminol can be added to enhance POX regeneration:

$$\text{POX Cpd II} + \text{enhancer} \rightarrow \text{enhancer}_{(\text{oxidized})} + \text{POX} \tag{28}$$

Enhancer effects are described in the work of Thorpe and Kricka.[49]

As described previously, luminol can substitute for the luciferin of the *B. biminiensis* bioluminescence reaction. Likewise, horseradish peroxidase can substitute for *B. biminiensis* luciferase in this bioluminescence reaction.[43] In these reactions, luminescence intensity is first order with respect to H_2O_2 and second order with respect to luminol or luciferin.[50,51] The kinetic work by Cormier and Prichard[47] suggests that the light-emitting step in the HRP-catalyzed reaction is complicated and may involve two luminol radicals plus H_2O_2.

With respect to reaction order, the previously described acid-optimum haloperoxidase-catalyzed luminol luminescence mechanism presents the opposite relationship. The haloperoxidase luminescence reaction is first order with respect to luminol, but second order with respect to H_2O_2.[44] These kinetic observations are consistent with the reaction mechanisms described in Eqs. (3), (5), (18), (19), and (21) and with the schematic of Fig. 1.

Peroxynitrite Mechanism

Macrophage and chicken heterophil leukocytes possess oxidase activity but are deficient in MPO. Investigating the luminescence activities of these phagocytes can be used to elucidate the role of haloperoxidase in CLS oxygenation. The lucigenin luminescence response of MPO-deficient chicken phagocytes is consistent with that observed from MPO-positive human leukocytes. However, the luminol luminescence responses of these MPO-deficient leukocytes are diminished greatly relative to those of MPO-positive leukocytes.[48] An alternative approach can be taken by studying the activity of MPO-deficient macrophages. The microbicidal capacity and the luminol luminescence activity of MPO-deficient macrophages are increased greatly by preexposing these macrophages to haloperoxidase.[52]

[49] G. H. G. Thorpe and L. J. Kricka, *Methods Enzymol.* **133,** 331 (1986).

[50] J. Lee and H. H. Seliger, *Photochem, Photobiol.* **15,** 227 (1972).

[51] W. R. Seitz, *Methods Enzymol.* **57,** 445 (1978).

[52] D. L. Lefkowitz, J. A. Lincoln, K. R. Howard, R. Stuart, S. S. Lefkowitz, and R. C. Allen, *Inflammation* **21,** 159 (1997).

The observation that even a small luminol-dependent luminescence is observed from MPO-deficient phagocytes raises questions as to the nature of the responsible oxygenating agent(s). Phagocytes are reported to possess nitric oxide (NO) synthase activity.[53,54] Like O_2^-, NO is a doublet multiplicity radical. NO can react readily with O_2^- in a doublet–doublet annihilation reaction similar to that described in Eq. (1) to yield singlet multiplicity peroxynitrite anion ($^1OONO^-$) as the product.[55,56] Peroxynitrite has been reported to react with luminol to yield luminescence.[57] Such a reaction might proceed according to

$$^1\text{luminol} + {}^1OONO^- \rightarrow {}^1NO^- + {}^1N_2 + {}^1\text{aminophthalate} + \text{Photon}(h\nu) \quad (29)$$

This nonradical mechanism of luminol luminescence is a spin-allowed oxygenation reaction similar to that described in Eq. (21)

Based on detection of the 1268-nm photon ($h\nu_{1268\text{ nm}}$) emission associated with $^1O_2^*$ relaxation:

$$^1O_2^* \rightarrow {}^3O_2 + \text{Photon}(h\nu_{1268\text{nm}}) \quad (30)$$

Khan reported that in neutral to mildly acid conditions, peroxynitrite disintegrates, yielding singlet oxygen[58]:

$$^1OONO^- + H^+ \rightarrow {}^1HNO + {}^1O_2^* \quad (31)$$

As such, a peroxynitrite-dependent luminol luminescence reaction might proceed by several different pathways. However, an attempt to demonstrate the participation of peroxynitrite in the blood phagocyte luminol-dependent luminescence reaction using a conventional NO synthase inhibitor and superoxide dismutase was unsuccessful.[59]

Materials and Methods for Whole Blood Luminescence Analysis

Reagents

The luminescence measurements of phagocyte oxygenation activities described herein were performed using reagents specifically designed for

[53] H. H. W. Schmidt, R. Seifert, and E. Bohme, *FEBS Lett.* **244,** 357 (1989).

[54] C. D. Wright, A. Mulsch, R. Busse, and H. Osswald, *Biochem. Biophys. Res. Commun.* **160,** 813 (1989).

[55] N. V. Blough and O. C. Zafirou, *Inorg. Chem.* **24,** 3502 (1985).

[56] M. Saran, C. Michel, and W. Bors, *Free Radic. Res. Commun.* **10,** 221 (1990).

[57] J. F. Wang, P. Komarov, H. Sies, and H. DeGroot, *Biochem. J.* **279,** 311 (1991).

[58] A. U. Khan, *J. Biolumin, Chemilumin.* **10,** 329 (1996).

[59] R. C. Allen, P. R. Stevens, T. H. Price, G. S. Chatta, and D. C. Dale, *J. Infect. Dis.* **175,** 1184 (1997).

the differential measurement of whole blood oxidase and MPO-dependent activities and blood phagocyte opsonin receptor expression (EOE Inc., Little Rock, AR; presently available from DCS Innovative Diagnostik-Systeme, Hamburg, Germany).[3,12,60–70] Chemical stimuli and priming agents are coated to the inner surface of polystyrene test tubes that serve as the raction vessel for luminescence measurement. Both lucigenin and luminol media were used for testing. For measurements of circulating (COR) and maximum opsonin-receptor (MOR)-dependent luminescence activities, the blood phagocytes were stimulated with complement-opsonized zymosan in the absence and in the presence of an optimum quantity of priming agent to ensure maximal opsonin receptor expression.

The reagents employed include the following:

i. Blood diluting medium (BDM): a 5 m*M* 2-(*N*-morpholine)ethanesulfonate (MES)-buffered balanced salt solution deficient in divalent cations [pH 7.2, 280–295 mOsmol/kg; <0.06 endotoxin units (EU)/ml].
ii. Dimethylbiacridinium (DBA^{++}, i.e., lucigenin) balanced salt solution (DBSS): a 5 m*M* 3-(*N*-morpholine)propanosulfonate (MOPS)-buffered balanced salt solution containing 139 mEq/liter Na^+, 5.0 mEq/liter K^+, 1.3 m*M* Ca^{2+}, 0.9 m*M* Mg^{2+}, 142 mEq/liter Cl^-, 0.8 m*M* H_nPO_4, plus 0.2 m*M* 10, 10′-dimethyl-9,9′ biacridinium dinitrate (bis-*N*-methylacridinium nitrate), 5.5 m*M* D-glucose and 0.05% (w/v) human albumin; pH 7.2, 280–295 mOsmol/kg, and <0.06 EU/ml.

[60] R. C. Allen, "Chemiluminescence Assay of in Vivo Inflammation." U.S. Patent 5,108,899 (1992).

[61] V. Witko-Sarsat, L. Halbwachs-Mecarelli, I. Sermet-Gaudelus, G. Bessou, G. Lenoir, R. C. Allen, and B. Descamps-Latscha, *J. Infect. Dis.* **179,** 151 (1999).

[62] G. E. Brown, G. M. Silver, J. Reiff, R. C. Allen, and M. P. Fink, *J. Trauma* **46,** 297 (1999)

[63] D. L. Pitrak, K. M. Mullane, M. L. Bilek, P. Stevens, and R. C. Allen, *J. Lab. Clin. Med.* **132,** 284 (1998).

[64] P. S. Wollert, M. J. Menconi, H. Wang, B. P. O'Sullivan, V. Larkin, R. C. Allen, and M. P. Fink, *Shock* **2,** 362 (1994).

[65] T. J. VanderMeer, M. J. Menconi, B. P. O'Sullivan, V. A. Larkin, H. Wang, R. L. Kradin, and M. P. Fink, *J. Appl. Physiol.* **76,** 2006 (1994).

[66] G. S. Chatta, T. H. Price, R. C. Allen, and D. C. Dale, *Blood* **84,** 2923 (1994).

[67] V. Witko-Sarsat, R. C. Allen, M. Paulais, A. T. Nguyen, G. Bessou, and B. Descamps-Latscha, *J. Immunol.* **157,** 2728 (1996).

[68] D. L. Stevens, A. E. Bryant, J. Huffman, K. Thompson, and R. C. Allen, *J. Infect. Dis.* **170,** 1463 (1994).

[69] F. B. Taylor, Jr., P. A. Haddad, G. T. Kinasewitz, A. K. C. Chang, G. Peer, A. Li, and R. C. Allen, *Exp. Biol. Abs.* 694 (1997).

[70] G. E. Brown, J. Reiff, R. C. Allen, G. M. Silver, and M. P. Fink, *J. Leukocyte Biol.* **62,** 837 (1997).

iii. Luminol balanced salt solution (LBSS): a 5 m*M* MES-buffered salt solution containing 139 mEq/liter Na^+, 5.0 mEq/liter K^+, 1.3 m*M* Ca^{2+}, 0.9 m*M* Mg^{2+}, 142 mEq/liter Cl^-, 0.8 m*M* H_nPO_4, plus 0.15 m*M* luminol (5-amino-2,3-dihydro-1,4-phthalazinedione) and 5.5 m*M* D-glucose; pH 7.2, 280–295 mOsmol/kg and <0.06 EU/ml.
iv. Complement-opsonified zymosan, human (hC-OpZ): 2×10^9 yeast cell wall units/ml of normal saline.
v. Precoated stimulus tubes (12×50-mm polystyrene tubes) for direct opsonin-independent activation of blood phagocytes:

   a. High-dose phorbol 12-myristate 13-acetate (PMAa) tubes: prefabricated tubes coated with 5 nmol of PMA.
   b. Low-dose PMA (PMAb) tubes: prefabricated tubes coated with 10 pmol of PMA.
   c. High-dose *N*-formyl-Met-Leu-Phe (FMLPa) tubes: prefabricated tubes coated with excess (1 μmol) FMLP.

vi. Precoated opsonin receptor primer tubes with sufficient immunomodulator to produce the maximum expression of phagocyte opsonin receptors:

   a. PAF tubes: prefabricated tubes coated with 10 pmol 1-*O*-hexadecyl-2-*O*-acetyl-sn-glycero-3-phosphorylcholine.
   b. rhC5a (C5a) tubes: prefabricated tubes coated with 20 pmol of recombinant human C5a (rhC5a).
   c. Leukotriene B_4(LTB_4) tubes: prefabricated tubes coated with 10 pmol of LTB_4.
   d. Low-dose *N*-formyl-Met-Leu-Phe (FMLPb) tubes: prefabricated tubes coated with 100 pmol of FMLP.
   e. Uncoated (blank) tubes.

Alternative media preparations, primers, and stimuli can be used. For best results, media should be maintained sterile and essentially free of endotoxin. The pH, ionic strength, osmolality, final divalent cation concentrations, and CLS concentration should be formulated to ensure optimum conditions for phagocyte function and measurement.[33]

Method

Blood is collected into sterile evacuated tubes containing 50 μl 15% K_3EDTA (7.5 mg/5 ml blood) as anticoagulant (Becton-Dickinson, New Jersey). The anticoagulated whole blood (0.1 ml) is diluted in BDM (9.9 ml) and loaded into the luminometer (Berthold EG&G LB953 Axis-modified, Wildbad, Germany). The luminometer is heat controlled and set to 37° throughout testing.

Basal and PMA-stimulated (opsonin receptor-independent) NADPH oxidase activities are measured after the luminometer injection of 0.1 ml (100 μl) of diluted whole blood (a 1 μl equivalent of the original blood) into uncoated and PMAb-coated (10 pmol/tube) tubes containing 0.6 ml DBSS (120 nmol DBS^{++}/tube), respectively.

Basal and PMA-stimulated (opsonin receptor-independent) oxidase-driven MPO oxygenation activities were measured by the injection of 0.1 ml of diluted blood into uncoated and PMAa-coated (5 nmol/tube) tubes containing 0.6 ml LBSS (90 nmol luminol/tube), respectively.

Circulating (COR) and maximum opsonin receptor (MOR)-dependent oxygenation activities were measured by injecting 0.1 ml of diluted blood into uncoated and PAF-coated (10 pmol/tube) tubes containing 0.1 ml hC-OpZ ($\sim 2 \times 10^8$ complement-opsonized zymosan particles/tube) and 0.6 ml LBSS, respectively.

Luminescence measurements were taken in triplicate over a 20-min interval. The results are expressed as either activity per volume of blood tested (i.e., counts/20 min/μl blood) or specific activity per phagocyte (i.e., counts/20 min/phagocyte). The number of phagocytes per microliter of blood was determined from the routine hematology blood count. The total leukocyte count minus the total lymphocyte count equals the total phagocyte count. Monocytes and eosinophil leukocytes yield luminescence activities similar to neutrophil leukocytes, but in normal healthy donors, these phagocytes are present in relatively low numbers, and in the absence of neutropenia, eosinophilia, or monocytosis, the contribution of these phagocytes to the total blood luminescence activity is minimal.

Subjects and Conditions of Testing

The human G-CSF studies described were conducted at the University of Washington. The subjects received included young (20–30 years) and elderly (70–80 years) volunteers of both sexes in good health. Seven young and five elderly subjects no drug, and seven young and six elderly subjects received a subcutaneous dose of 300 μg recombinant human G-CSF (rhG-CSF) daily (Amgen, Thousand Oaks, CA). The blood specimens for luminescence analysis were obtained in the morning approximately 24 hr after the previous injection and just prior to the daily injection of rhG-CSF as described previously.[59,66]

The human endotoxin, i.e., lipopolysaccharide (LPS), studies were conducted at the University of Oklahoma Health Science Center. The subjects included four young males in good health. The four subjects received 4 ng of *Escheichia coli* endotoxin (Lot F, Bureau of Biologics, Rockville, MD) per kilogram body weight. After the intravenous lines were established and the baseline preinfusion specimen was drawn, LPS was infused intravenously. Blood specimens were taken during the early inflammatory period at 1, 2, and 4 hr post-LPS infusion, during the mid inflammatory period at 8 and 12 hr post-LPS infusion, and during the late inflammatory period 24 hr post-LPS infusion.[69]

Data Collection and Analysis

A four-letter code (*XXYZ*) was used for ease in identifying and differentiating the various luminescence measurements taken.[59] As shown in Table I, the first two letters of the code (*XX*) describe the priming agent or chemical stimulus coating the reaction tube. BK, PA, PB, C5, PF, LT FA, and FB represent uncoated (blank), high-dose PMA, low-dose PMA, C5a, PAF, LTB4, high-dose FMLP, and low-dose FMLP-coated tubes, respectively. The third letter (*Y*) describes the type of opsonin stimulus employed, e.g., N indicates no opsonin, C indicates complement-opsonized zymosan, and I indicates immunoglobulin-opsonized zymosan. The forth letter (*Z*) describes the chemiluminigenic substrate employed with L indicating luminol and D indicating DBA^{++} (dimethylbiacridinium; lucigenin). For example, PBND represents the activity of blood in response to low-dose PMA (PB) with no opsonin stimulus (N) using lucigenin (D) as the chemiluminigenic substrate. BKCL represents activity in response to uncoated blank tubes (BK) with complement-opsonized zymosan (C) as the

TABLE I

STIMULUS–CHEMILUMINIGENEIC SUBSTRATE TEST CODE

XXYZ code	Immune primer agent (*XX*)	Chemical stimulant (*XX*)	Opsonin stimulant (*Y*)	Chemiluminigenic substrate (*Z*)
BKNN	None (BK)	None (BK)	None (N)	None (N)
BKND	None (BK)	None (BK)	None (N)	Lucigenin (D)
PBND	None	PMA, 10 pmol (PB)	None (N)	Lucigenin (D)
BKNL	None (BK)	None (BK)	None (N)	Luminol (L)
PANL	None	PMA, 5 nmol (PA)	None (N)	Luminol (L)
FANL	None	FMLP, 1 μmol (FA)	None (N)	Luminol (L)
BKNL	None (BK)	None (BK)	None (N)	Luminol (L)
PFNL	PAF, 10 pmol (PF)	None	None (N)	Luminol (L)
LTNL	LTB_4, 10 pmol (LT)	None	None (N)	Luminol (L)
C5NL	rhC5a, 20 pmol (C5)	None	None (N)	Luminol (L)
FBNL	FMLP, 1 μmol (FB)	None	None (N)	Luminol (L)
BKCL	None (BK)	None (BK)	Complement opsonized zymosan (C)	Luminol (L)
PFCL	PAF, 10 pmol (PF)	None	Complement opsonized zymosan (C)	Luminol (L)
LTCL	LTB_4, 10 pmol (LT)	None	Complement opsonized zymosan (C)	Luminol (L)
C5CL	rhC5a, 20 pmol (C5)	None	Complement opsonized zymosan (C)	Luminol (L)
FBCL	FMLP, 1 μmol (FB)	None	Complement opsonized zymosan (C)	Luminol (L)

opsonin stimulus and using luminol (L) as the chemiluminigenic substrate.

All data were collected and stored initially using Berthold LB953 luminometer software. Data were then entered into a relational database, Paradox for Windows (Borland, version 5). Statistical analyses, including multiple discriminant function analysis, were performed using the SPSS for Windows 8.0 software (SPSS, Chicago, IL).

Luminescence Measurement of Blood Phagocyte Function

Basal and PMA-Stimulated Oxidase Activities

NADPH oxidase activity is common to all circulating blood phagocytes and can be measured as the luminescence product of lucigenin reductive dioxygenation.[29,33] PMA-stimulated oxidase activity per volume of blood (PBND) is directly proportional to the blood phagocyte count and can be evaluated by regression analysis.[59] This proportionality remains relatively constant even under conditions associated with inflammation and stimulated myelopoiesis.

PBND activity and the phagocyte count of healthy untreated controls and healthy subjects tested in the pretreatment and early (first 2 days) G-CSF treatment periods were compared with the PBND activity and phagocyte count measured after 5 to 6 days of G-CSF treatment.[59] The calculated regression coefficient B, the standard error of B, and the Student's t value for B are presented in Table II. Values for the Pearson correlation coefficient (r), the r^2, the ANOVA F value, and the significance (probability) of F are also presented. PMA-stimulated lucigenin-dependent (PBND) luminescence shows a linear relationship to the phagocyte count

TABLE II
UNSTIMULATED AND PMA-STIMULATED BLOOD OXIDASE ACTIVITY UNDER DIFFERENT CONDITIONS OF MARROW ACTIVATION

Code	Untreated and G-CSF treated	Unstd. coeff. B	Unstd. coeff. Std. err.	t	Pearson corr. coeff. (r)	r^2	ANOVA F	Sig.
BKNN	Days 0–3	0	2	−0.3	0.012	0.000	0	0.795
BKNN	Days 5–6	1	1	1.4	0.108	0.012	2	0.161
BKND	Days 0–3	59	4	13.6	0.525	0.276	184	0.000
BKND	Days 5–6	99	9	10.8	0.638	0.408	116	0.000
PBND	Days 0–3	820	21	38.6	0.869	0.755	1494	0.000
PBND	Days 5–6	883	32	27.6	0.905	0.819	763	0.000

per volume blood tested. As such, the value of B is a measure of the specific activity per phagocytes:

$$B = \frac{(\text{counts/20 min/}\mu\text{l blood})}{(\text{phagocytes/}\mu\text{l blood})} = \text{counts/20 min/phagocyte} \qquad (32)$$

Essentially the same relationship holds when data are expressed relative to the neutrophil count per volume of blood tested. When PBND data for Day 0–3 measurements are expressed as counts/20 min/neutrophil, the r value is 0.865; when expressed as count/20 min/phagocyte, the r value is 0.869. For activity comparison, the luminometer background noise (BKNN) is also presented. Note that BKNN activity is relatively low and does not correlate with the phagocyte count.

Phagocyte counts increase following the initiation of G-CSF treatment, and the counts remain elevated throughout the treatment period. As illustrated by data of Table II, the increase in basal (BKND) activity following G-CSF treatment was also small. Although about 10-fold higher than BKND activity, PMA-stimulated oxidase (PBND) activity per phagocyte during the pretreatment and initial treatment period (820 $\pm$ 21 counts/20 min/phagocyte) is essentially unchanged following 5 to 6 days of G-CSF treatment (883 $\pm$ 32 counts/20 min/phagocyte). Relative to luminol-dependent activities, the lucigenin-dependent luminescence measurements of oxidase activity per phagocyte are stable and typically show only minor changes even in patients undergoing immune activation or in subjects treated with G-CSF to stimulate bone marrow production. The stability of PBND activity is such that this measure of oxidase capacity per volume of blood is functionally equivalent to the phagocyte count per volume of blood tested.

Basal and PMA-Stimulated Oxidase-Driven Myeloperoxidase Activities

The H_2O_2 produced by NADPH oxidase is the substrate for the MPO generation of microbicidal oxidants and oxygenating agents. This oxidase-driven MPO activity can be functionally measured as the luminescence product of luminol dioxygenation. The statistics relating basal (BKNL), high-dose PMA-stimulated (PANL), and high-dose FMLP-stimulated (FANL) luminol luminescence activities to the phagocyte count per microliter of blood are presented in Table III. The blood specimens were from the same healthy subjects described for Table II.

As with the previously considered oxidase activity, both PANL activity and FANL activity correlate with the number of phagocyte in the 1 μl of blood tested. However, unlike PBND activity, the regression coefficients (B values) relating PANL and FANL activities to the phagocytes count per specimen show large increases following 5 to 6 days of G-CSF treat-

TABLE III
UNSTIMULATED AND PMA-STIMULATED BLOOD MYELOPEROXIDASE ACTIVITY UNDER DIFFERENT CONDITIONS OF MARROW ACTIVATION

Code	Untreated and G-CSF treated	Unstd. coeff. *B*	Unstd. coeff. Std. err.	*t*	Pearson corr. coeff. (*r*)	r^2	ANOVA *F*
BKNL	Days 0–3	24	3	8.0	0.340	0.116	63
BKNL	Days 5–6	68	4	16.1	0.778	0.605	259
PANL	Days 0–3	4,152	103	40.3	0.878	0.771	1628
PANL	Days 5–6	11,346	260	43.6	0.958	0.918	1903
FANL	Days 0–3	784	22	35.7	0.852	0.726	1273
FANL	Days 5–6	3,594	106	33.8	0.933	0.871	1143

ment.[59] The graphics of Fig. 2 illustrate the temporal kinetic pattern and magnitude of the luminol luminescence responses of a representative healthy control subject on the second day of G-CSF treatment (Fig. 2a) and the same subject following 5 days of G-CSF treatment (Fig. 2b). In each graphic the bottom curve represents basal measurements and the top curve represents PMA-stimulated measurements. The phagocyte count

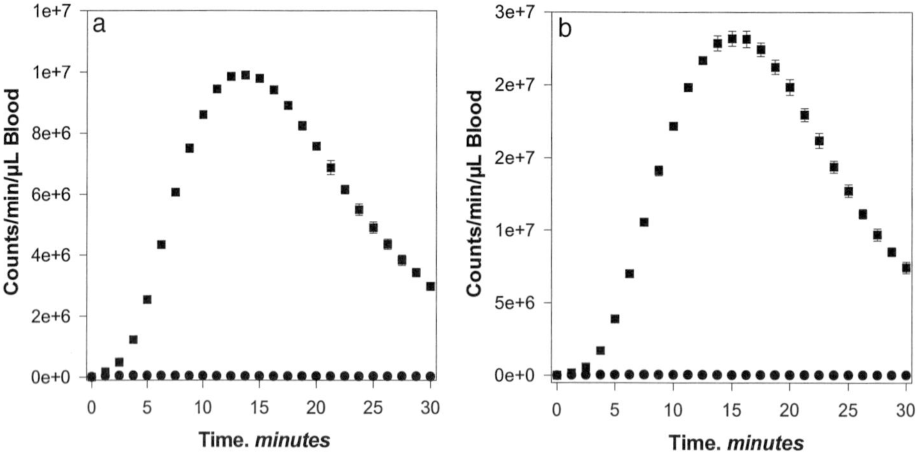

FIG. 2. Basal (BKNL) and PMA-stimulated (PANL) luminol luminescence responses of blood (1 μl equivalent) from a healthy human subject plotted as luminescence intensities against time. The temporal kinetics of the luminol luminescence responses were studied on the second day (a) and on the fifth day (b) of G-CSF (300 μg rhG-CSF/day) treatment. Bottom curves represent BKNL measurements (●) and top curves represent PANL measurements (■). The phagocyte count was 18,090/μl blood on the second post-G-CSF treatment day (a) and 26,880/μl blood on the fifth posttreatment day (b).

increased from a pretreatment count of 4240 phagocytes/μl blood to 18,090 on the second day of G-CSF treatment (activity of Fig. 2a) and 26,880 on the sixth day of treatment (activity of Fig. 2b). Note that MPO-dependent specific activities per phagocyte increase following marrow stimulation, i.e., myelopoiesis. For this subject the PANL activity per phagocyte changed from 1702 counts/20 min/phagocyte pretreatment to 6930 on the second day and onto 10,210 on the sixth day of G-CSF treatment. Note in Table III that the B value for PANL increases almost threefold by Days 5–6. When high-dose FMLP was used as the stimulus (FANL), the B value increases by greater than fourfold over this same period of marrow stimulation. Treatment with G-CSF appears to exert a functional hypertrophic as well as a hyperplastic effect on neutrophil leukocytes.[59]

As described earlier, luminol-dependent luminescence measures MPO and, to a minor extent, non-MPO oxygenation activities. The available evidence, including data presented herein, supports the conclusion that these phagocyte activities are not constant but changes with the state of inflammation and cytokine stimulation. The luminol luminescence activity per phagocyte is subject to the state of *in vivo* immune activation and myelopoietic activity.

Opsonin Receptor-Dependent Oxygenation Activities

Although the chemical stimulation of blood phagocyte metabolism, i.e., direct activation with PMA and FMLP, can provide useful information regarding oxidase and MPO functional capacities, this approach of activation is artificial. It is independent of phagocyte opsonin receptor expression and, as such, does not involve the physiologic sequence of opsonin–opsonin receptor ligation, phagocytosis, phagosome formation, and ultimately azurophilic granule fusion and phagolysosomal formation as required for microbicidal action. Using opsonized particles as stimulants more closely approximates the physiologic condition. Opsonin-based analysis requires the functional expression and participation of phagocyte opsonin receptors in the metabolic activation process. In the presence of a relative excess of opsonin, phagocyte opsonin receptor expression exerts the rate-limiting effect on opsonin ligation and the metabolic activation that follows. Under such conditions, the initial luminol luminescence response is proportional to the degree of phagocyte integrin and opsonin receptor expression.[3,12,59–62]

Phagocyte opsonin receptor expression increases in proportion to the degree of *in vivo* or *in vitro* exposure to immunomodulator agents such as PAF, LTB_4, C5a, TNF-α, and IL-8. Oxidase activity is a membrane and specific (secondary) granule-associated enzyme function whereas MPO is a component of azurophilic (primary) granules. MPO functional activity is ultimately dependent on the peroxide product of oxidase activity and, as such, both specific and azurophilic degranulation are required for optimum MPO-dependent activity. Specific degranulation is required for optimal phagocytic activity. Phagocyte contact with an opsonized microbe results in opsonin–opsonin receptor ligation, phagocytosis, and phagosome formation. The fusion of azurophilic lysosomal granules with the phagosome forms the phagolysosome and activates MPO. These morphologic events are linked to metabolic activation, phagolysosomal acidification, and the generation of oxidants and oxygenating agents. The resulting oxidase-driven MPO dioxygenation activity can be quantified as luminol-dependent luminescence.

The BKCL activity reflects the opsonin receptor-dependent activity of the circulating phagocytes in blood. The circulating opsonin receptor (COR)-dependent activity reflects the *in vivo* state of phagocyte opsonin receptor expression.[3,60–62] The maximum opsonin receptor (MOR) activity is measured on the same blood specimen using the same opsonin stimulus plus a quantity of exogenous immunomodulator sufficient to prime maximum opsonin receptor expression per phagocyte. Table IV describes the COR and various MOR activities (i.e., PFCL, C5CL, LTCL, and FBCL) of blood from healthy subjects during the pre and initial G-CSF period and during the 5–6 days post-G-CSF treatment period. Figure 3 presents BKCL and PFCL activities of the same pretreatment subject described in Fig. 2. Note that during the initial 10-min interval the PFCL activity is greater than eightfold higher than the BKCL activity. Table V presents

TABLE IV

UNPRIMED AND PRIMED COMPLEMENT-OPSONIZED ZYMOSAN-STIMULATED BLOOD MYELOPEROXIDASE ACTIVITY UNDER DIFFERENT CONDITIONS OF MARROW STIMULATION

Code	Untreated and G-CSF treated	Unstd. coeff.		t	Pearson corr. coeff. (r)	r^2	ANOVA F
		B	Std. err.				
BKCL	Days 0–3	5,396	143	37.7	0.864	0.746	1422
BKCL	Days 5–6	8,120	272	29.9	0.917	0.841	892
PFCL	Days 0–3	12,885	160	80.8	0.965	0.931	6524
PFCL	Days 5–6	19,039	290	65.6	0.981	0.962	4305
C5CL	Days 0–3	12,159	172	70.5	0.955	0.911	4972
C5CL	Days 5–6	18,005	305	59.0	0.977	0.954	3480
LTCL	Days 0–3	11,718	141	83.0	0.967	0.934	6892
LTCL	Days 5–6	14,498	289	50.2	0.968	0.937	2518
FBCL	Days 0–3	10,572	144	73.5	0.958	0.918	5407
FBCL	Days 5–6	16,954	287	59.0	0.977	0.954	3481

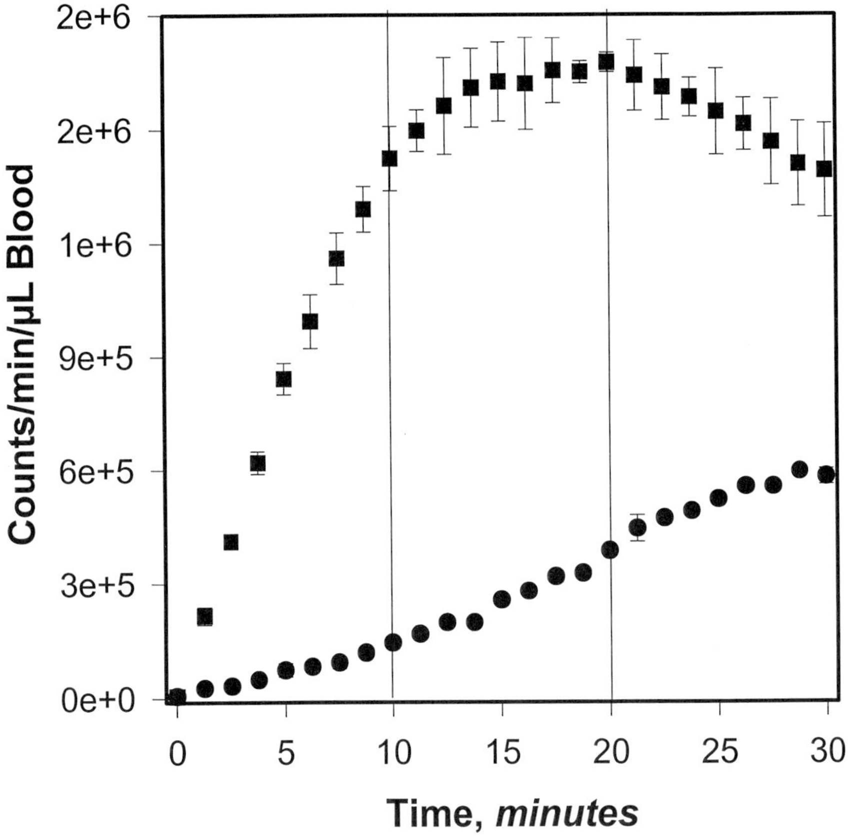

FIG. 3. Circulating (COR) and maximum opsonin receptor (MOR)-dependent luminol luminescence responses of a blood specimen (1 μl equivalent) from a healthy control subject plotted against time. The temporal kinetics of the BKCL (COR) and PFCL (MOR) activities are depicted by the bottom curve (●) and the top curve (■), respectively. The specimen was from the same pretreatment control subject described in Fig. 2. The phagocyte count was 4,240/μl blood.

the inflammatory activation ratios for the pre and initial treatment (0–3 days) period and for the 5–6 days post-G-CSF treatment period. Table V presents ratio data calculated from both 10- and 20-min integral luminol luminescence data. The standard ratios are calculated for each individual subject, and for comparison, the ratios are also calculated for the entire group taken as a whole.

As shown by the velocity measurements of Fig. 3, BKCL and PFCL activities do not reach maximum intensity during the initial 10-min interval, but PFCL does reach maximum luminescence intensity by 20 min. The

TABLE V
IMMUNE OR INFLAMMATORY ACTIVATION RATIOS OF CIRCULATING (COR) TO PAF, C5a, LTB4, AND FMLP MAXIMALLY PRIMED (MOR) BLOOD PHAGOCYTE OXYGENATION ACTIVITIES FOLLOWING CONTACT WITH COMPLEMENT-OPSONIZED ZYMOSAN AND MEASURED AS MYELOPEROXIDASE-DEPENDENT LUMINOL LUMINESCENCE[a]

		Ratio values			
		10-min interval		20-min interval	
Ratio conditions	G-CSF treatment	Individual	Group	Individual	Group
BKCL/PFCL	Pre and initial (Days 0–3)	0.197	0.210	0.264	0.419
BKCL/PFCL	Post (Days 5–6)	0.264	0.275	0.346	0.426
BKCL/C5CL	Pre and initial (Days 0–3)	0.204	0.216	0.259	0.444
BKCL/C5CL	Post (Days 5–6)	0.261	0.279	0.344	0.451
BKCL/LTCL	Pre and initial (Days 0–3)	0.247	0.265	0.323	0.460
BKCL/LTCL	Post (Days 5–6)	0.344	0.364	0.430	0.560
BKCL/FBCL	Pre and initial (Days 0–3)	0.296	0.314	0.316	0.510
BKCL/FBCL	Post (Days 5–6)	0.354	0.363	0.397	0.479

[a] Blood was obtained from healthy donors in the pre-treatment and early G-CSF treatment (Days 0–3) phase and during the late G-CSF treatment (Days 5–6) phase.

attainment of maximum luminescence intensity correlates with maximum phagocyte metabolic activity, i.e., phagocyte metabolism has reached some rate-limiting condition. Luminescence measurements of opsonin receptor-dependent function should focus on the initial time interval. During the initial poststimulation period, phagocyte integrin expression exerts rate-limiting control over the luminescence response. In the initial period, luminescence is not limited by phagocyte metabolic response capacity.

The relationship of the COR to MOR activities changes with the state of *in vivo* inflammatory activation. Figure 4 depicts BKCL and PFCL activities of a healthy human subject prior to and 8 hr after receiving 4 ng/kg body weight *E. coli* endotoxin (LPS). The pre-LPS treatment curves of Fig. 4a show that initial PFCL activity is approximately fourfold greater than BKCL activity. Injection of a small quantity of LPS produced a relatively strong *in vivo* inflammatory response. This systemic inflammation exerts a large effect on the state of priming of the circulating blood phagocyte. As shown in Fig. 4b, 8 hr after LPS-induced activation there is a large alteration in the COR to MOR relationship. During LPS-activated systemic inflammation, BKCL activity approaches PFCL activity. Exogenous priming agents are unable to elicit a response from the *in vivo*-activated circulating phagocytes. The increased luminescence per microliter blood reflects

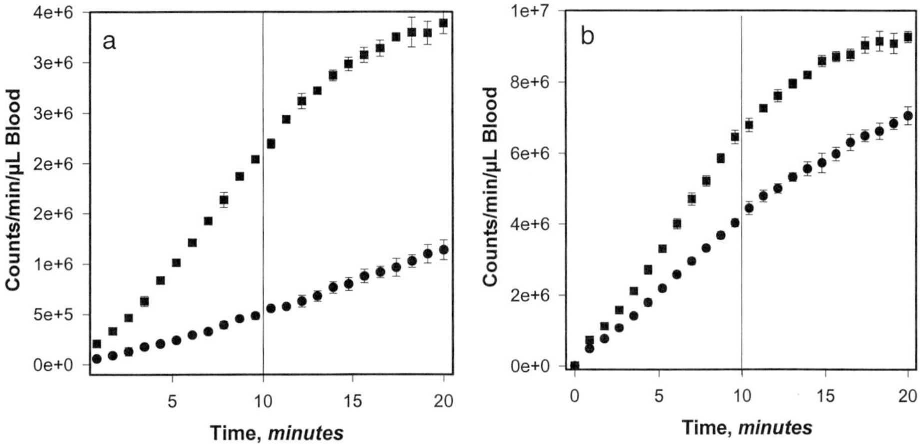

FIG. 4. BKCL and PFCL activities of a 1-μl blood specimen from a healthy human control plotted against time. Activities are shown before (a) and 8 hr (b) after the subject was injected with LPS [4 ng *Escherichia coli* LPS (endotoxin) per kg body weight]. The bottom curve (●) and the top curve (■) depict BKCL and PFCL responses, respectively. The phagocyte count was 4,360/μl for the pre-treatment blood (a), and 9,360/μl for the 8 hr post-LPS blood (b).

the increase in the neutrophil leukocyte count (neutrophilia) observed during this period.

As described in Table VI, basal oxidase (BKND) and basal MPO (BKNL) activities also show a relatively large increase during this early inflammatory period, but these activities return to the pre-LPS range by 8 to 12 hr post-LPS infusion. These findings are consistent with direct *in vivo* activation of the circulating phagocyte during the early phase of LPS-activated systemic inflammation.[69]

Ratio of Circulating of Maximum Opsonin Receptor-Dependent Activities

Opsonin stimulation of phagocyte oxygenation activity is dependent on opsonin receptor and especially integrin (R) expression per phagocyte (P), i.e., the R/P.[3,61,62] When integrins are expressed maximally and opsonin stimulus is not limiting, the phagocyte generation of oxygenating agents (Ox/P) reflects the maximum rate of opsonin-dependent metabolic activation per phagocyte, i.e., the MOR activity. As described previously, phagocyte capacity changes with the state of immune activation and myelopoiesis. As such, the rate of oxygenation activity (dOx/dt) per volume of blood is dependent on the number of phagocytes (P), the opsonin receptor expression per phagocyte (R/P), the capacity of the phagocyte to generate oxy-

TABLE VI
RATIOS OF CIRCULATING (COR) TO PAF, C5a, AND FMLP MAXIMALLY PRIMED (MOR) BLOOD PHAGOCYTE OXYGENATING ACTIVITIES FOLLOWING CONTACT WITH COMPLEMENT-OPSONIZED ZYMOSAN AND (hC-OpZ) AND MEASURED AS MYELOPEROXIDASE-DEPENDENT LUMINOL LUMINESCENCE[a]

Condition	Temporal relationship to LPS infusion	Ratio (10-min interval)	
		Mean	SD
BKCL/PFCL	Pre-LPS	0.250	0.036
BKCL/PFCL	Early (1 to 4 hr post)	0.680	0.113
BKCL/PFCL	Mid (8 to 12 hr post)	0.671	0.177
BKCL/PFCL	Late (24 hr post)	0.419	0.091
BKCL/C5Cl	Pre-LPS	0.336	0.047
BKCL/C5Cl	Early (1 to 4 hr post)	0.770	0.080
BKCL/C5Cl	Mid (8 to 12 hr post)	0.759	0.189
BKCL/C5Cl	Late (24 hr post)	0.553	0.158
BKCL/FBCL	Pre LPS	0.450	0.041
BKCL/FBCL	Early (1 to 4 hr post)	0.832	0.052
BKCL/FBCL	Mid (8 to 12 hr post)	0.857	0.162
BKCL/FBCL	Late (24 hr post)	0.660	0.136
		Luminescence ± SD (counts/20 min/ phagocyte)	
PFCL (MOR MPO)	Pre-LPS	9,132	2262
PFCL (MOR MPO)	Early (1 to 4 hr post)	12,035	2934
PFCL (MOR MPO)	Mid (8 to 12 hr post)	11,705	2008
PFCL (MOR MPO)	Late (24 hr post)	10,577	2363
BKNL (Basal MPO)	Pre-LPS	30	6
BKNL (Basal MPO)	Early (1 to 4 hr post)	152	94
BKNL (Basal MPO)	Mid (8 to 12 hr post)	53	29
BKNL (Basal MPO)	Late (24 hr post)	48	16
BKND (Basal oxidase)	Pre-LPS	28	13
BKND (Basal oxidase)	Early (1 to 4 hr post)	111	104
BKND (Basal oxidase)	Mid (8 to 12 hr post)	31	17
BKND (Basal oxidase)	Late (24 hr post)	55	30

[a] Opsonin receptor-dependent PFCL activity is also presented. Basal oxidase (BKND) and basal oxidase-driven MPO (BKNL) activities are also presented. The blood was obtained from healthy donors pre- and postinfusion of LPS (4 ng/kg body weight). The inflammatory phases are divided into the early period (1 to 4 hr post-LPS), the midperiod (8–12 hr post-LPS), and the late period (24 hr post-LPS).

genating agents (Ox/P), and the availability of opsonin stimulus (O) as described by

$$d\mathrm{Ox}/dt = k[\mathrm{R/P}][\mathrm{Ox/P}][\mathrm{P}][\mathrm{O}] \quad (33)$$

When [O] is not rate limiting, i.e., when [O] $\ggg$ [R], the relationship simplifies to

$$d\mathrm{Ox}/dt = k[\mathrm{R/P}][\mathrm{Ox/P}][\mathrm{P}] \quad (34)$$

The R/P is dependent on phagocyte (P) exposure to immunomodulator agents *in vivo* and/or *in vitro.* BKCL activity provides a rough measure of the circulating R/P, whereas PFCL activity reflects the maximum R/P, i.e., R_{max}/P. Expressing the activity as the ratio of these measurements, i.e., an immune or inflammatory activation index, obviates variable differences such as the concentration of phagocytes [P] or oxygenation capacity per phagocyte (Ox/P). The [P] and the Ox/P are the same for both the numerator and the denominator. The ratio is proportional to the degree of integrin expression per circulating phagocyte.[3,61,62] Phagocyte opsonin receptor expression is linked to opsonin-stimulated oxygenation activity:

$$\mathrm{Ratio} = \frac{d\mathrm{Ox}/dt}{d\mathrm{Ox}_{max}/dt} = \frac{[\mathrm{R}]}{[\mathrm{R}_{max}]} \quad (35)$$

When luminol is not rate limiting, oxygenation activity (Ox) is proportional to luminescence activity (L). Therefore, if both luminol (C) and the opsonin stimulus (O) are not rate limiting, the relationship of circulating to maximum opsonin receptor can be expressed as

$$\mathrm{Ratio} = \frac{d\mathrm{L}/dt}{d\mathrm{L}_{max}/dt} = \frac{[\mathrm{R}]}{[\mathrm{R}_{max}]} \quad (36)$$

This ratio gauges the *in vivo* state of immune activation or inflammation.[3,60,67]

The ratios measured in the G-CSF and LPS studies described are presented in Tables V and VI, respectively. Table V presents the calculated individual and group ratios for both 10- and 20-min luminescence measurement intervals. Comparison of the results suggests that the individually calculated 10-min luminescence ratio provides the best measure of opsonin receptor functional expression, and as such, only this form of the ratio is presented in Table VI. The initial 10-min interval BKCL/PFCL ratios were 0.20 for the pre-G-CSF treatment controls and 0.25 for the pre-LPS treatment controls. These values are consistent with those of previously published results.[59,65] The BKCL/PFCL ratio showed a small increase to 0.26 following 5 to 6 days of G-CSF treatment, but this value still falls within the normal range. In contrast, relatively large increases in the inflammation ratio are observed following the infusion of LPS. As described in Table VI, the BKCL/PFCL ratio increased to 0.68 in the early post-LPS treatment period, remained at 0.67 during the midtreatment period, and was still relatively high at 0.42 for the 24-hr post-LPS treatment measurement. These observations support the position that the opsonin receptor-dependent activity of circulating phagocytes is proportional to the *in vivo* state of immune activation or inflammation and that the luminescence ratio provides a functional gauge of the state of systemic immune activation.[3,61,62] This approach has been applied successfully to the study of infectious disease patients.[61,68]

Multiple Discriminant Function Analysis of Blood Luminescence

Basal and opsonin receptor-independent stimulated luminescence measurements also provide useful diagnostic information about the *in vivo* state of immune activation. Basal oxidase (BKND) activity and basal MPO (BKNL) activity provide a direct measure of circulating phagocyte activation. Increased BKND and BKNL activities are observed in conditions of systemic immune activation and inflammation. BKND appears to be especially sensitive in this regard.

Opsonin receptor-independent stimulation of blood phagocytes is based on the use of exogenous chemical stimuli to assess some limit of functional capacity. As described, previously PBND oxidase activity is relatively constant and is proportional to the circulating phagocyte count under a broad variety of test conditions. Conversely, PANL activity increases in conditions that stimulate myelopoiesis.[59]

Combined analysis of these different blood luminescence-based measurements can be applied to diagnostic assessment of *in vivo* immune activation. Discriminate analysis is the statistical technique of choice when the dependent variable is categorical (i.e., nominal or nonmetric) and when the independent variables are metric. When three or more classifications are involved the technique is referred to as multiple discriminate analysis (MDA).[68,71,72] In MDA the linear combination of the two or more independent variables is obtained so as to maximally discriminate between the defined categories. In the present case the goal is to differentiate healthy controls from subjects with altered states of immune activation or myelo-

[71] J. F. Hair, Jr., R. E. Anderson, and R. L. Tatham, "Multivariate Data Analysis with Readings." MacMillan, New York, 1987.

[72] J. D. Jobson, "Applied Multivariate Data Analysis," Vol. II. Springer-Verlag, New York, 1991.

poiesis as observed following treatment with LPS or G-CSF and to differentiate G-CSF activation from LPS activation and, if possible, to further differentiate the time period postactivation with either stimulant. The statistical decision rule is to maximize the between-group variance relative to the within-group variance.

For the present analysis the times pre-and post-G-CSF and LPS treatment will be considered as conditions of the categorical dependent variable, and the several luminescence measurements will be considered as metric-independent variables. MDA involves deriving the linear combination of independent variables so as to maximize between-group variance relative to the within-group variance:

$$Z = W_1X_1 + W_2X_2 = W_3X_3 + \cdots + W_nX_n \tag{37}$$

where Z is the discriminate score, W is the discriminate weight, and X is the independent variable. A separate Z value is generated for each discriminate equation employed in analysis. For the present analysis the independent luminescence variables are: X_1 is basal oxidase (BKND/20 min/μl blood) activity, X_2 is stimulated oxidase (PBND/20 min/μl blood) activity, X_3 is basal MPO (BKNL/20min/μl blood) activity, X_4 is PMA-stimulated MPO (PANL/20 min/μl blood) activity, X_5 is circulating opsonin receptor (BKNL/20 min/μl blood) activity, X_6 is maximum opsonin receptor (PFCL/20 min/μl blood) activity, and X_7 is the BKCL/PFCL ratio at 10 min. Each variable is multiplied by its derived discriminate weight and the products are added to generate the composite discriminate score for each blood specimen analyzed.

Differences in the units and in the ranges of the independent variables make unstandardized coefficients poor indicators of the relative contributions of the variables to overall discrimination. However, the standardized canonical discriminant function (DF) coefficients presented in Table VII provide an accurate estimate of the relative contribution of the variables to the overall discrimination. Table VII also presents eigenvalues, canonical correlations, Wilk's λ and χ^2 values for DF1 and DF2.[71,72] The eigenvalue is the ratio of the between-group sum of squares to the within-group sum of squares, and as such, its size is a useful measure of the spread of the group centroids in the corresponding dimension of the multivariant space. The group centroid value is analogous to the group mean. Wilk's λ is used to test the null hypothesis; it is the proportion of the total variance in the discriminant score not explained by differences among the groups. Wilk's λ can be transformed to a variable with an approximate χ^2 distribution for the evaluation of differences between group centroids.

LPS infusion stimulated systemic inflammatory activation and increased phagocyte integrin–opsonin receptor expression as indicated by large in-

TABLE VII
MULTIPLE DISCRIMINANT (DF) STATISTICS

Independent variables	Standardized canonical DF coefficients: Function 1	Function 2
BKNL (20 min)	−0.219	0.296
PANL (20 min)	0.449	−0.711
BKCL (20 min)	−0.681	1.059
PFCL (20 min)	0.678	0.562
BKCL/PFCL (10 min)	0.151	0.397
BKND (20 min)	−0.007	−0.648
PBND (20 min)	0.671	−0.582
Eigenvalue	6.700	2.123
Canonical correlation	0.933	0.824
Wilk's λ	0.015	0.118
χ^2	2529	1293
Conditions of the dependent variable	**Group centroids: Function 1**	**Function 2**
Controls, untreated	−0.997	−0.554
Early (1–4 hr post-LPS)	−1.455	1.994
Mid (8–12 hr post-LPS)	−0.887	6.066
Late (24 hr post-LPS)	−0.740	3.141
1–2 days post-rhG-CSF	3.686	−0.355
5–6 days post rhG-CSF	7.367	0.123

creases in the BKCL/PFCL ratio. In addition, LPS infusion increased basal oxidase and basal MPO activities. Conversely, prolonged G-CSF treatment produced severalfold increases in opsonin receptor-independent activities such as PANL and FANL, but did not cause large increases in the basal activities or the BKCL/PFCL ratio. Each luminescence parameter tested has its own unique pattern of activity with respect to immune activation and inflammation. For the present DF analysis, the luminescence parameters BKND, PBND, BKNL, PANL, BKCL, PFCL, and BKCL/PFCL at 10 min were the independent variables used to measure the circulating blood phagocytes. The nature and temporal conditions of *in vivo* stimulation with G-CSF and LPS [i.e., untreated controls, Days 1–2 post-G-CSF, Days 5–6 post-G-CSF, early (1–4 hr) post-LPS, mid (48–12 hr) post-LPS, and late (24 hr) post-LPS groups] served as the grouping or dependent variable.

Every independent variable was measured on each blood specimen, and each sample was run in triplicate. The G-CSF study group contained

146 blood specimens (438 measurements) from 25 healthy subjects. Twelve healthy subjects were treated with 300 μg G-CSF per day (87 measurements). Ten blood specimens (30 measurements) were tested on Days 0–1 and 19 blood specimens (57 measurements) were tested on Days 5–6.

The LPS study group contained blood specimens from four healthy untreated controls (12 measurements), four early LPS-treated subjects measured at 1, 2, and 4 hr (36 measurements), four mid LPS-treated subjects measured at 8 and 12 hr (24 measurements), and four late LPS-treated subjects measured at 24 hr (12 measurements). The standardized canonical coefficients for discriminant function 1 and 2 are presented in Table VII. Eigenvalues for DF1 and DF2 are 6.7 and 2.1, respectively, and the canonical correlations are 0.93 and 0.82, respectively. Wilk's λ for DF1 and DF2 are 0.015 and 0.118, respectively.

Figure 5 presents the composite luminescent activities plotted in a two-dimensional field defined by the canonical discriminant functions. The abscissa DF1 is labeled "myelopoietic stimulation." This DF has good discriminatory power for assessing the state of bone marrow stimulation or myelopoietic activity. Note that DF1 is capable of discriminating untreated controls from short-term G-CSF-treated subjects and, in addition, is capable of discriminating short term G-CSF-treated subjects from long-term G-CSF-treated subjects. The ordinate DF2 is labeled "acute inflammation" and has good discriminatory power for assessing the state of inflammation. Note that DF2 is not only capable of differentiating untreated from LPS-treated subjects, but also differentiates the early from the later phases of inflammation. Numerical values for the axes are analogous to the standard deviations values of traditional statistical analysis, and the group centroid values are analogous to the group means.

As described in Table VII, the group centroid for untreated healthy controls has the coordinates −1.00, −0.55 (DF1, DF2). Initial *in vivo* exposure to G-CSF produced a large excursion to the right, yielding a group centroid with the coordinates 3.69, −0.36. Prolonged exposure produced an even larger excursion in the same direction, yielding a group centroid with the coordinates of 7.37, 0.12.

In vivo exposure to LPS produced an excursion in the upward direction and is defined relative to the DF2 or "acute inflammation" axis. Early (1–4 hr) post-LPS changes show a large upward movement with a minor movement to the left, yielding a group centroid with the coordinates −1.46, 1.99. By midperiod (8–12 hr) the post-LPS group centroid is much higher and is shifted slightly to the right with the coordinates −0.89, 6.07. The late post-LPS (24 hr) group centroid is directed downward and slightly more to the right with the coordinates −0.74, 3.14. This movement to the right may reflect early secondary myelopoietic stimulation induced by LPS

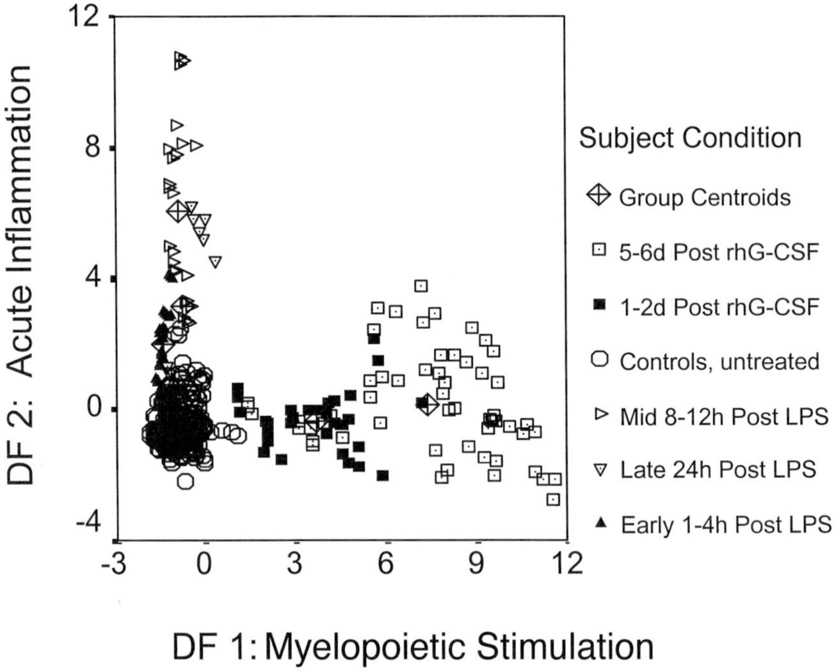

FIG. 5. Multiple discriminant analysis of the composite phagocyte luminescent activities plotted in a two-dimensional field defined by the canonical discriminant functions DF1 and DF2. The abscissa DF1 labeled "myelopoietic stimulation" discriminates for bone marrow stimulation (myelopoietic activity), and the ordinate DF2 labeled "acute inflammation" discriminates for systemic inflammatory activity. The independent variables used are (1) BKND, (2) PBND, (3) BKNL, (4) PANL, (5) BKCL, and (6) PFCL activities expressed as counts/20 min/μl blood and (7) the BKCL/PFCL ratio based on the 10-min integral luminescence response. The dependent or grouping variable was the stimulus-temporal condition. The dependent variable is subcategorized as (1) untreated healthy human controls, (2) initial (1–2 days) post-G-CSF-treated, (3) late (5–6 days) post-G-CSF-treated, (4) early (1–4 hr) post-LPS-treated, (5) mid (8–12 hr) post-LPS-treated, and (6) late (24 hr) post-LPS-treated human subjects. The related statistics are presented in Table VII.

treatment. By 48 hr the post-LPS values have essentially returned to the control group.

The graphic depiction of the MDA presented in Fig. 5 illustrates the power of the composite blood luminescence measurements as applied to the measurement and differentiation of systemic inflammatory activation and myelopoietic stimulation. Previous work has also demonstrated the applicability of this approach to monitoring infectious disease patients.[68]

Relationship of Phagocyte Metabolism to Luminescence

Does whole blood luminescence analysis provide a quantitative measure of phagocyte respiratory burst metabolism? The relationship linking phagocyte metabolism and oxygenation activity to luminescence has been considered earlier in this presentation. When a nonrate-limiting concentration of luminol is employed as C, the rate of light emission is directly dependent on the rate and quantity of dioxygenation agents (Ox) generated by the phagocyte

$$dL/dt = k[\mathrm{Ox}] \tag{38}$$

where Ox represents H_2O_2, HOCl, and other phagocyte-derived oxidants and oxygenating agents. If the reaction milieu presents nonrate-limiting conditions with respect to the reaction components, if the quantity of phagocytes tested is known, and if the detection sensitivity of the instrument is known, then it should be possible to directly relate the measured luminescence to the metabolic activity of phagocytes tested. What follows is a rough attempt to quantify these relationships.

Luminometer Efficiency

Photomultiplier tube (PMT)-based luminometers detect photons by amplifying the electrons emitted by photoelectric events at the photocathode. Although a multitude of factors influence luminometer performance, the major factors determining luminescence detection efficiency ($\mathrm{Eff}_{\mathrm{Lum}}$) are (1) the quantum efficiency (Φ_{PMT}) of the PMT or detector device relative to the wavelength of light detected and (2) the geometry factor (F_{Geo}), i.e.,

$$(\mathrm{Eff}_{\mathrm{Lum}}) = (\Phi_{\mathrm{PMT}})(F_{\mathrm{Geo}}) \tag{39}$$

Φ_{PMT} is the ratio of the number of electronic counts registered relative to the number of photons of the specified wavelength striking the photocathode. F_{Geo} is defined as the ratio of emitted light striking the photocathode relative to the total light emitted by the specimen.

Many commercially available luminometers use a bialkali PMT with a Φ_{PMT} of about 0.2 for blue photons such as those emitted by excited aminophthalate. The F_{Geo} is more variable and can be greater than 0.6 for instruments approximating a 4π collection geometry. Instruments with multiple specimen capacity typically have a lower F_{Geo} value. The F_{Geo} of the instrument used for the present studies is estimated to be about 0.2. The $\mathrm{Eff}_{\mathrm{Lum}}$ for such an instrument is thus estimated to be

$$\mathrm{Eff}_{\mathrm{Lum}} = (0.2)(0.2) = 0.04 \tag{40}$$

For every 100 blue photons emitted by the specimen, about 4 are registered as counts or relative light units (RLU).

Luminol Luminescence Quantum Yield

The luminescence quantum yield (Φ_{CL}) of luminol is the ratio of the number of photons emitted relative to the number of luminol dioxygenations under specific conditions of testing. The Φ_{CL} is influenced by the reaction milieu and by quantum mechanical factors related to the spin state of the reaction product and the fluorescence quantum yield (Φ_{Fluor}).[50]

The Φ_{CL} for the aqueous OCl^--H_2O_2 luminol reaction at pH 11.5 and for the methylene blue photosensitized luminol reaction, both reactions are known to generate $^1O_2{}^*$, is relatively high at about 1×10^{-2} (0.01); i.e., 1 photon is emitted per 100 reactions.[50] However, the Φ_{CL} decreases with increased acidity. Based on the work of Lee and Seliger[50] and the unpublished observations of Allen, the Φ_{CL} is estimated to be about 3×10^{-4} at pH 7, 3×10^{-5} at pH 6, and 6×10^{-6} at pH 5.

Overall Luminescence Measurement Efficiency

The overall or net luminescence efficiency (Eff_{Net}) is the product of the luminometer efficiency and the chemical efficiency as defined by the relationship:

$$Eff_{Net} = (Eff_{Lum})(\Phi_{CL}) \tag{41}$$

If the luminol reaction is measured at a pH of 6, the Eff_{Net} is a product of the Eff_{Lum} (0.04) multiplied by the Φ_{CL} (3×10^{-5} at pH 6); i.e., $Eff_{Net} = (0.04)(3 \times 10^{-5}) = 1.2 \times 10^{-6}$. Therefore, if luminol is the CLS and the pH of the reaction milieu is 6, about 1 count (RLU) will be detected for every million (10^6) luminol dioxygenation events. At pH 5 the Φ_{CL} is decreased to 6×10^{-6}, and the value of Eff_{Net} is decreased to 2.4×10^{-7}.

Phagocyte Metabolism Measured as Luminescence

Phagocytosis activates respiratory burst metabolism and generates the oxidants and dioxygenating agents that react with a CLS such as luminol to yield luminescence. When the conditions are set properly and the value of the Eff_{Net} can be estimated, the luminescence response can be related directly to phagocyte metabolic activity. As such, the Eff_{Net} can be used to relate the measured luminescence of a blood specimen to the phagocyte metabolic activity of that blood specimen. This approach is analogous to using an extinction coefficient to relate light absorbance to chemical concentration in absorbance spectroscopy.

Phagocyte oxygenation activity measured as luminescence intensity (i.e., dL/dt) can be expressed as the integral or sum of luminescence counts measured over a stated time interval, e.g., the total RLU measured over a 20-min period (counts/20 min). Phagocyte luminescence activity can also be expressed as the activity per volume of blood (counts/20 min/μl blood) or as the specific activity per phagocyte (counts/20 min/phagocyte).

The product of luminescence activity divided by the Eff_{Net} is an estimate of the metabolic activity of the phagocytes in the specimen of blood measured. As illustrated in Fig. 1, phagocyte hexose monophosphate shunt metabolism (PM_{HMP}) generates the reducing equivalents required for O_2 reduction to O_2H (i.e., H^+ and O_2^-) and ultimately to H_2O_2. Therefore,

$$dH_2O_2/dt = k\, PM_{HMP} \tag{42}$$

where PM_{HMP} indicates phagocyte HMP shunt metabolism in response to the activation of phagocyte NADPH oxidase.

Each glucose-6-PO_4 metabolized via the HMP shunt generates the four reducing equivalents required to reduce 2 $NADP^+$ to 2 NADPH. In turn, the oxidase uses 2 NADPH to catalyze the reduction of 2 O_2, ultimately generating 2 H_2O_2. One H_2O_2 can serve as the substrate for the MPO-catalyzed oxidation of Cl^- to HOCl. HOCl can dehydrogenate luminol directly to its diazaquinone as described in Eq. (18). The second H_2O_2 can react with HOCl to yield singlet oxygen or react with the diazaquinone. Both reactions yield the same dioxygenated product and luminescence. By the pathway described in Eqs. (18) and (19) or the pathway of Eq. (21), the net dioxygenation of luminol by phagocyte MPO is proportional to the phagocyte generation of H_2O_2. As such, the phagocyte luminescence (L) is directly proportional to the quantity of glucose metabolized multiplied by Eff_{Net}:

$$L_{(counts/time/phagocyte)} = (PM_{HMP})(Eff_{Net}) \tag{43}$$

As described in Table IV, the maximally primed, opsonized zymosan-stimulated luminol-dependent luminescence (e.g., C5CL, PFCL) activity per blood neutrophil is about 10,000 counts/20 min/neutrophil.[12,59] Greater than 80% of blood phagocyte luminol-dependent luminescence is MPO dependent.[33,59] The majority of MPO activity is assumed to be phagolysosomal, and as such, the mean pH is expected to be in the range between 7 and 5. At a pH of 6 the value of the Eff_{Net} will be approximately 1.2×10^{-6}. Under such conditions the luminescence-based estimation of phagocyte HMP shunt metabolism (PM_{HMP}) is

$$\begin{aligned} PM_{HMP} &= (10{,}000 \text{ counts/20 min/neutrophil})/(1.2 \times 10^{-6}) \\ &= 0.8 \times 10^{10} \text{ glucose-6-PO}_4 \text{ metabolized/20 min/neutrophil} \\ &= 1.6 \times 10^{10}\ H_2O_2 \text{ generated/20 min/neutrophil} \end{aligned} \tag{44}$$

TABLE VIII

REPORTED VALUES FOR NEUTROPHIL METABOLIC ACTIVITIES[a]

	Ref.
Glycolytic metabolism	
$4.7 \pm 0.5 \times 10^{10}$ molecules glucose consumed/60 min/neutrophil (n = 5)	8, 73
Hexose monophosphate shunt (phosphogluconate pathway)	
$1.5 \pm 0.6 \times 10^{10}$ molecules glucose consumed/60 min/neutrophil (n = 5)	8. 73
Oxygen (O_2) consumed	
$0.68 \pm 0.09 \times 10^{10}$ molecules O_2 consumed/4 min/neutrophil (n = 6)	74
$1.64 \pm 0.09 \times 10^{10}$ molecules O_2 consumed/20 min/neutrophil (n = 3)	75
$3.13 \pm 0.18 \times 10^{10}$ molecules O_2 consumed/10 min/neutrophil (n = 12)	76
$8.11 \pm 1.55 \times 10^{10}$ molecules O_2 consumed/60 min/neutrophil (n = 24)	77
Superoxide (O_2^-) generated	
$0.28 \pm 0.06 \times 10^{10}$ molecules O_2^- generated/4 min/neutrophil (n = 6)	74
$0.57 \pm 0.03 \times 10^{10}$ molecules O_2^- generated/4 min/neutrophil (n = 3)	78
$5.40 \pm 0.60 \times 10^{10}$ molecules O_2^- generated/10 min/neutrophil (n = ?)	79
Hydrogen peroxide (H_2O_2) generated	
$0.23 \pm 0.03 \times 10^{10}$ molecules H_2O_2 generated/4 min/neutrophil (n = 6)	74
$5.27 \pm 0.81 \times 10^{10}$ molecules H_2O_2 generated/10 min/neutrophil (n = 7)	76
$5.82 \pm 1.02 \times 10^{10}$ molecules H_2O_2 generated/60 min/neutrophil (n = 10)	80
$7.44 \pm 0.42 \times 10^{10}$ molecules H_2O_2 generated/60 min/neutrophil (n = 20)	81
$0.16 \pm 0.01 \times 10^{10}$ molecules H_2O_2 generated/1 min/neutrophil (n = 20)	81
$7.12 \pm 0.24 \times 10^{10}$ molecules H_2O_2 generated/90–120 min/neutrophil (n = 24)	82

[a] Expressed as molecules of glucose or oxygen consumed or molecules of products generated per measurement interval per neutrophil.

If the phagolysosome pH is assumed to be 5.0, the $\mathrm{Eff}_{\mathrm{Net}}$ equals about 2.4×10^{-7} and the relationship is

$$\begin{aligned} \mathrm{PM_{HMP}} &= (10{,}000 \text{ counts/20 min/neutrophil})/(2.4 \times 10^{-7}) \\ &= 4 \times 10^{10} \text{ glucose-6-PO}_4\text{/20 min/neutrophil} \\ &= 8 \times 10^{10}\ \mathrm{H_2O_2} \text{ generated/20 min/neutrophil} \end{aligned} \tag{45}$$

These values are in reasonably good agreement with the values for HMP shunt, O_2 consumption, O_2^- generation, and H_2O_2 generation reported in the literature when these activities are normalized and expressed per neutrophil as described in Table VIII.[8,73–82]

[73] R. C. Allen, R. L. Stjernholm, M. A. Reed, T. B. Harper, S. Gupta, R. H. Steele, and W. W. Waring, *J. Infect. Dis.* **136,** 510 (1977).

[74] F. Rossi, P. Patriarca, and D. Romeo, *in* "The Reticuloendothelial System, a Comprehensive Treatise" (A. J. Sbarra and R. R. Strauss, eds.) Vol. 2, p. 153. Plenum Press, New York, 1980.

[75] R. K. Root, J. A. Metcalf, N. Oshino, and B. Chance, *J. Clin. Invest.* **55,** 945 (1975).

[76] R. K. Root and J. A. Metcalf, *J. Clin. Invest.* **60,** 1266 (1977).

[77] R. S. Weening, D. Roos, and J. A. Loos, *J. Lab. Clin. Med.* **83,** 570 (1974).

In closing, luminescence analysis provides unsurpassed sensitivity for measuring the temporal kinetics of phagocyte metabolic activation and oxygenation activity. The information–effector relationship linking the state of the circulating phagocyte to the state of systemic immune activation opens broad opportunities for employing blood phagocyte luminescence as a gauge of the *in vivo* state of immune activation.[3,12,61–63,68]

[78] J. T. Curnutte and B. M. Babior, *Blood* **45,** 851 (1975).
[79] R. S. Weening, R. Wever, and D. Roos, *J. Lab. Clin. Med.* **85,** 245 (1975).
[80] J. W. T. Homan-Muller, R. S. Weening, and D. Roos, *J. Lab. Clin. Med.* **83,** 198 (1975).
[81] C. F. Nathan, *J. Clin. Invest.* **80,** 1550 (1987).
[82] C. F. Nathan, S. Srimal, C. Farber, E. Sanchez, L. Kabbash, A. Asch, J. Gailit, and S. D. Wright, *J. Cell Biol.* **109,** 1341 (1989).

Section VII

Bioluminescence as an Educational Tool

[42] Demonstrations of Chemiluminescence

By FRANK MCCAPRA

Introduction

Although the chemist may envy the biologist with his descriptions of the dramatic displays of bioluminescence in nature, exciting a sense of wonder in the beholder, the chemist can provide complementary displays, much more easily summoned up, by reactions emitting bright chemiluminescence. The value of such displays, of course, lies not only in the pleasure they give, but in the way in which the underlying chemistry can be linked to the chemistry of bioluminescence. In addition, there are a number of scientific principles such as entropy, energy conversion, the nature of light, the electromagnetic spectrum, and so on inherent in the reactions so demonstrated, providing an excellent basis for a variety of pedagogic purposes.

This article does not cover all possible demonstrations, but should provide a means of devising those appropriate to the reader's purpose, whether it is merely to excite some admiration or to enlarge upon the principles already mentioned. Each description will include a suggestion for mounting the display, a summary of the chemical reactions involved, and, where applicable, a link to the relevant bioluminescence. It is written for interested scientists in general (from a chemical or biochemical background) and specialists must look elsewhere for more definitive treatments. It is intended to be largely self-contained, and there is plenty of literature elsewhere for those digging more deeply. The chemistry of chemiluminescent and bioluminescent reactions is discussed in an earlier article in this volume.[1]

Some demonstrations can be mounted with readily available materials from the laboratory and chemical supply companies, but others require some effort in synthesis. Even where the materials are available commercially, the scale of a demonstration required to seduce a large audience makes it an expensive affair. Synthetic procedures for the key reagents are therefore included. If the reader is not skilled in practical organic chemistry, a useful stratagem is to persuade the organizer of an undergraduate organic chemistry practical class to include the synthesis (and of course the light reaction) in the syllabus for the course. Many of the examples here contain lessons in synthesis (most procedures are well within the skills of a second- or third-year undergraduate student, with supervision) and they will provide

[1] F. McCapra, *Methods Enzymol.* **305** [1] 2000 (this volume).

0076-6879/00 $30.00

useful pegs on which to hang instruction in concepts in photochemistry, thermodynamics, and so on. The descriptions are given with a variety of demonstrators in mind, ranging from those in a high school, who would presumably confine themselves to commercially available materials, to those with access to an organic chemistry laboratory. Safety is of course paramount, especially if the audience is young, and although in over 200 demonstration lectures the author has harmed neither himself nor the watchers, the only suitable demonstrator is one who has had practical scientific experience, and can be expected to interpret the warnings given here, as befits his or her own circumstances. That being said, there are few chemical demonstrations, with the exceptions to be indicated, that are as safe. The few hazardous reagents are noted.

Although some of the reactions are bright enough to be seen in low-level ambient light, it has to be said at the outset that all will benefit from being shown in as near total darkness as can be arranged. I have often been invited to lecture in schools with quite inadequate means of blackout of the windows, on bright June days, with disappointment evident among all concerned. Each chemiluminescent compound is suited to a particular form of demonstration, but except where noted, it would be possible, with a little experimentation, to substitute another if availability is a problem. The demonstrations are grouped according to the chemiluminescent compound, rather than by type of display. It is to be expected that in assembling a demonstration lecture, a variety of effects will be drawn from reactions of several compounds. Finally, these demonstrations carry with them the age old warning attached to any public display—rehearse, rehearse, rehearse. The experiments are all certain to work in the proper circumstances, but the exact effect and the maximum light yield given the user's own conditions of apparatus, solvents, and so on can only be determined by a degree of trial and error. The quantities and proportions are not immutable, and experimentation around the parameters given here is recommended.

Luminol

This compound, whose chemiluminescence was discovered in 1928, has been featured in many demonstrations.[1a] Although it is synthesized easily,[2] it is inexpensive enough to purchase and use in quantity if desired.[3]

[1a] B. Shakhashiri, "Chemical Demonstrations: A Handbook for Teachers of Chemistry," Vol. 1, p. 125. University of Wisconsin Press, Madison, WI, 1981.

[2] L. F. Fieser, "Experiments in Organic Chemistry," p. 199. Heath, Boston, 1957.

[3] Aldrich Chemical Co., 3-aminophthalhydrazide 12,307-2.

Synthesis[2]

This is a typical second-year undergraduate laboratory experiment. First put a flask containing 15 ml of water on the steam bath to get hot. Then heat a mixture of 1 g of 3-nitrophthalic acid and 2 ml of an 8% aqueous solution of hydrazine in a test tube until the solid is dissolved, add 3 ml of triethylene glycol, and clamp the tube in a vertical position about 2 in above a Bunsen burner or hot plate with heat transfer. Insert a thermometer and an aspirator connected to a suction pump and boil the solution vigorously to distill off the excess water (110–130°). Let the temperature rise rapidly until (3–4 min) it reaches 215°. Remove the burner, or from the hot plate, note the time, and, by intermittent gentle heating, maintain a temperature of 215–220° for 2 min. Remove the tube, cool to about 100° (crystals of the product often appear), add the 15 ml of hot water, cool under the tap, and collect the light yellow granular nitro compound (dry weight, 0.7 g). The nitro compound need not be dried and can be transferred at once, for reduction, to the uncleaned test tube in which it was prepared. Add 5 ml of 10% sodium hydroxide solution, stir with a rod, and, to the resulting deep brown-red solution, add 3 g of sodium hydrosulfite dihydrate (molecular weight 210.15). Wash the solid down the walls with a little water. Heat to the boiling point, stir, and keep the mixture hot for 5 min, during which time some of the reduction product may separate. Then add 2 ml of acetic acid, cool under the tap and stir, and collect the resulting precipitate of light yellow luminol. The filtrate on standing overnight usually deposits a further crop of luminol (0.1–0.2 g). The luminol is sufficiently pure for the demonstrations. The synthesis can be scaled up very easily.

Demonstrations

The easiest of all the demonstrations is that produced by dissolving luminol in a dipolar aprotic solvent such as dimethyl formamide (DMF), dimethyl sulfoxide, or *N*-methyl acetamide (hexamethylphosphoramide is in this class but there have been adverse reports on its toxicity) and adding a very strong base. The more usual solvent is dimethyl sulfoxide[4] (DMSO). While it is true that the DMSO should be relatively dry and free from its major impurity, dimethyl sulfide, the solvent obtained from freshly opened containers as sold, is best for our purposes. The base used is most conveniently potassium hydroxide pellets, with sodium hydroxide being less effective. These should be fresh and relatively free from surface carbonate, although this is only a problem when the deposit is excessive. Sodium hydroxide can be satisfactory, but in general it is inferior, probably as a

[4] Aldrich Chemical Co., methyl sulfoxide M8, 108-2.

result of a lower aqueous solubility and the varying amounts of water in the "dry" DMSO. Another base that can be used is potassium *tert*-butoxide,[5] producing the brightest light, but not actually the best demonstration. The luminol need only be reasonably pure, and there is no need to recrystallize it (a rather difficult operation). Commercial material used "as is" is usually excellent. The most convenient container is a 1 liter Erlenmeyer (conical) flask with a fairly wide neck and secure stopper, either a rubber bung or ground glass. If the latter is used, beware the seizure of the stopper, which can occur in the presence of strong base, if the flask is put aside for any length of time. Make sure there is no leakage, as the flask will be shaken vigorously.

Safety

Solid KOH, NaOH, and potassium *tert*-butoxide are powerful, corrosive alkalis. Handle only with gloves. DMSO and DMF are absorbed through the skin, but are not particularly toxic. Luminol has no known hazards, but should obviously not be ingested.

Materials. KOH pellets; potassium *tert*-butoxide; DMSO; luminol; and a conical flask with good stopper.

Procedure

1. KOH pellets (100 g or sufficient to cover the bottom of the flask) are placed in the flask and barely covered with fresh DMSO (about 60–100 ml). Luminol [200 mg or enough to cover (heaped) a 25 cent coin] is added and the flask is stoppered. Vigorous shaking in the dark will result in the DMSO becoming luminescent with a blue-green light. The time to maximum brightness is variable (depending on water content, purity, etc.), but provided not too much luminol has been used and the other materials are fresh, the interval is never much more than about 1 min. An opportunity is provided for comments, serious or otherwise, while shaking. No heat is generated (except in the demonstrator!) and, provided a hand is kept over the stopper, it is completely safe.
2. Dissolve/suspend potassium *tert*-butoxide (2.0 g) in DMSO (200 ml) in a suitable container (e.g., the 1-liter flask mentioned earlier) and add luminol (100 mg). Shake vigorously. The light is brighter and appears more quickly, but the drama is less!

[5] Aldrich Chemical Co., potassium *tert*-butoxide 15,667-1.

Comments

Using twice the amount of DMSO increases the time to maximum brightness, but the light is more visible. The requirement for oxygen is demonstrated by leaving the flask with the smaller amount of DMSO to settle. A dark solution with a luminescent meniscus (of a depth depending on the amount of luminol used) shows the diffusion of oxygen. Filling the flask with pure oxygen before stoppering gives an enhanced light emission. Increasing the amount of luminol (say twofold) requires more shaking and may give quenching of the light, but the light will last for about 24 hr, provided the shaking is seldom and brief.

Mechanism

Luminol(I) forms the dipotassium salt on the surface of the KOH and probably reacts with oxygen to form superoxide ion and the luminol radical anion. Cage recombination gives a peroxide that rearranges to form the peroxide precursor to the excited state. An outline reaction scheme is shown, but the last word on the mechanism of this, the first of the strongly chemiluminescent compounds, has not been said. An excellent analysis of the possibilities has been given by White and Roswell.[6] Transition metals such as copper(II) and iron(III) are required in aqueous systems to achieve the same oxidation. The diazaquinone(II) is an intermediate in both media.

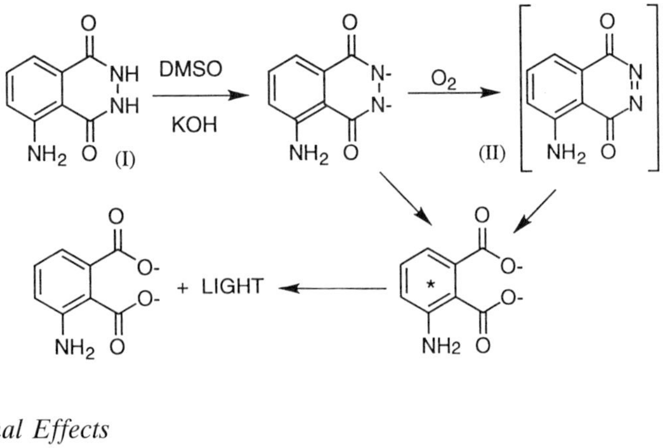

Additional Effects

Fluorescent energy acceptors can be added to change the greenish blue light to the color of the fluorescence of the additive. Some of these are

[6] E. H. White and D. F. Roswell, *in* "Chemi- and Bioluminescence" (J. G. Burr, ed.), p. 215. Dekker, New York, 1985.

expensive (although synthesized easily[7]) and it may be as well to reduce the scale to fit the budget. Predissolve the fluorescer (some are not easily soluble) in the DMSO and, if reducing the scale, add the base as a saturated solution of KOH in water, as described later.

The fluorescer (see later,[7] 1 mg) is dissolved in DMSO (10 ml), and luminol (50 mg) is added and dissolved. Several drops of a saturated KOH solution [about 40% (w/v)] are then added and the well-stoppered container is shaken as before to provide the color of light indicated:

Green: 9,10-bis(phenethynyl)anthracene
Yellow: 1,8-dichloro-9,10-bis(phenethynyl)anthracene
Salmon: Rubrene
Red: 5,12-bis(phenethynyl)naphthacene

Other Experiments

Safety

Sodium hypochlorite (domestic bleach) is caustic. The oldest demonstration involving luminol is also one of the easiest to prepare.[8] Luminol (100 mg, a large spatulaful) is dissolved in NaOH (0.1 *M,* 200 ml), and the solution is added to 200 ml of commercial chlorine bleach (about 5% sodium hypochlorite). Blue light is produced. It is not the most efficient reaction, but it is very easy to do. Be careful with the bleach, which in this undiluted form is very caustic to the eyes and can damage clothing.

Another old experiment[9] reflects one of the uses of luminol: the detection and analysis of transition metals. Potassium ferricyanide was used in the early examples, but a more sustained light is produced using the cuprammonium ion.[1]

Materials

Sodium carbonate; ammonium carbonate; copper sulfate; sodium bicarbonate; hydrogen peroxide (30% i.e. 100 vol); water; luminol. The water need not be distilled if the supply is reasonably clean; the salts can be any of the hydrated forms available to hand. The hydrogen peroxide can be 3%, requiring a proportionate change in the amount used, but the higher

[7] Obtainable from the Aldrich Chemical Co. or synthesized in P. Hanhela and D. Paul, *Aust. J. Chem.* **34,** 1669 (1981) II. Violet and blue emitters, *Aust. J. Chem.* **34,** 1687 (1981). III. Yellow and red fluorescent emitters, *Aust. J. Chem.* **34,** 1701 (1981)

[8] H. O. Albrecht, *Z. Physik. Chem.* **136,** 321 (1928); W. H. Fuchsman and W. G. Young, *J. Chem. Educ.* **53,** 548 (1976).

[9] E. H. Huntress, L. N. Stanley, and A. S. Parker, *J. Chem. Educ.* **11,** 142 (1934).

concentration is more usually found in the chemistry laboratory and is in fact more stable, almost indefinitely so, in a refrigerator at 4°. It is in any case best for several of the later experiments described here.

Safety

Copper sulfate is toxic if ingested, and sodium carbonate is moderately caustic and dangerous to the eyes. Thirty percent hydrogen peroxide causes skin burns: a white, painful burn, which on small areas is transient. Absorption onto combustible surfaces with high surface areas such as paper towels can later result in spontaneous combustion. Always soak any spills with water before disposal.

Procedure

To make solution A: Dissolve Na_2CO_3 (5.0 g) in water (500 ml) and add luminol (0.25 g) with stirring, until dissolved. Add $NaHCO_3$ (25.0 g), $CuSO_4$ (0.5 g), and $(NH_4)_2CO_3$ and stir until all is dissolved. Make up to a final volume of 1 liter with water.

To make solution B: Add hydrogen peroxide [5.0 ml of 30% (v/v)] to water (1 liter)

Demonstration. The volumes in this demonstration have been made deliberately high so that a variety of mixing effects can be obtained. These effects are limited only by the ingenuity of the demonstrator, and only pointers are given here. The simplest involves pouring both A and B from a height so that the streams mix before reaching a very large beaker or other glass vessel. Alternatively, two large (500-ml) dropping funnels can be arranged to drip solutions A and B into a large glass filter funnel to which is attached a spiral of translucent Tygon or polyethylene tubing 1 cm in diameter before exiting into a collecting vessel. A glass spiral produces a marginally better effect, but is of course somewhat more difficult to make.

Additional Effects

Make up two separate solutions of fluorescein and rhodamine (any version) by adding 0.5 g of each to 100 ml of 0.1 *M* NaOH. Small amounts of these solutions (to be determined by experiment) are added *alternately* to the flowing luminol solution via the filter funnel, giving a gorgeous three-color effect in the Tygon spiral. The quantum yield is enhanced by the transfer of energy to the more efficient fluorescers, and the increase in light intensity adds to the overall effect. The unused fluorescer solution can be kept for several weeks.

Clock Reactions and Other Curiosities

Chemists have been fascinated by periodic, usually color, reactions for a very long time. Various clock and oscillating reactions[10] have challenged theorists to describe the sometimes bizarre effects in kinetic terms. Perhaps the most well known of these is the Balousov-Zhabotinski oscillating reaction. An extra dimension is added by coupling these reactions to light emission. Not all are suitable for demonstration to a large audience, but they are certainly sufficiently intriguing to include on the right occasions. E. H. White, probably the most influential of chemists with a major interest in the mechanism of chemiluminescence, described[11] the first clock reaction using luminol.

A Luminol Clock Reaction

An aqueous solution of luminol, hydrogen peroxide, ammonia, and a complex cyanide starts colorless and dark, and after a short and variable delay, in the dark, produces a flash of light. The delay in producing the light—the clock effect—is mirrored by a change in the solution from colorless to blue when the reaction is viewed in daylight.

Safety

Potassium cupric cyanide is as toxic as potassium cyanide, and acid must *never* come into contact with the solutions. If complex cyanide is unavailable, it can be made by adding the stoichiometric amount of cupric chloride and potassium cyanide.

Procedure

Use distilled or deionized water for all solutions. Dissolve luminol (2 g) in 100 ml concentrated ammonia (15 *M*) in a 1-liter volumetric flask and make up to the mark. Make up 50 ml of a 0.1 *M* solution of potassium cupric cyanide [$K_2Cu(CN)_3$] by dissolving the salt (1.1 g) in exactly 50 ml of distilled water. Make up 500 ml of a 3% hydrogen peroxide solution from 50 ml of 30% hydrogen peroxide.

For each of the clock reactions, place 100 ml of the luminol solution in a suitable flask and add 5.0 ml of the complex cyanide solution, followed by 30 ml of the 3% hydrogen peroxide solution. Mix by stirring while adding the hydrogen peroxide to avoid high local concentrations, which can cause a premature start to the reaction. Light will be emitted in a 2-sec flash after

[10] M. Orban, *J. Am. Chem. Soc.* **108,** 6983, (1986).
[11] E. H. White, *J. Chem. Educ.* **34,** 275 (1957).

a delay of 10 sec; by varying the hydrogen peroxide concentration,[1] the delay can be increased roughly as shown.

Delay (seconds)	Volume of H_2O_2 (ml of 3%)
10	30
15	20
25	10
60	5

With precision, speed, and a fair amount of practice, it is possible to have a pattern of light both in space and in time on the demonstration bench. It is, of course, not necessary to use all of the different delays, as the "clocks" are started in sequence by the addition of the hydrogen peroxide, which obviously provides a pattern in time on its own.

The effect is enhanced if, for example, five different individually shaped glass containers are used (*not* wine glasses, Coke bottles, etc.) and if the audience is prevented from seeing the preparations by a simple screen. The presenter should distance himself from the bench when the display starts, meanwhile describing the chemistry in simple terms. Because the induction period is essentially determined by the time it takes to oxidize the complexing cyanide ion, to give free catalytic cuprammonium ion (see the description of the mechanism), the delay can be manipulated by changing the concentration of the complex cyanide or adding KCN (there should be no free copper ion). There may be some value in looking at these alterations if this demonstration is featured. Experimentation can be carried out in the light, without luminol, as the solutions turn bright blue in coincidence with light emission.

Mechanism[11]

$$[Cu(CN)_3]^{2-} + H_2O_2 \xrightarrow[NH3]{slow} [Cu(NH_3)]^{2+} + CO_3^{2-} + CNO + HO^- \quad (1)$$

$$[Cu(NH_3)]^{2+} + CN^- \xrightarrow{fastest} \text{copper(I) complex} + (CN)_2 \quad (2)$$

$$\text{copper(I) complex} + H_2O_2 \xrightarrow[NH_3]{fast} [Cu(NH_3)]^{2+} + HO^- \quad (3)$$

$$[Cu(NH_3)]^{2+} + H_2O_2 + \text{Luminol} \longrightarrow \text{light} + \text{products} \quad (4)$$

Luminol is acting as an indicator for the catalytic-free cuprammonium ion; reactions (1), (2), and (3) account for the "clock," and an increase in cyanide concentration "ties up" this ion. Conversely, an increase in H_2O_2 concentration destroys cyanide ion and decreases the time to the appearance of the blue color of the free cuprammonium ion and the accompanying flash of light.

Other Clock Reactions

Shakhashiri[1] has described variations on the reaction, and if the use of the very toxic cyanides is a concern, an alternative method requiring the synthesis of a deuteroferriheme has been described.[12] More elaborate versions,[13] such as the oscillating reactions of the Belousov-Zhabotinski type, may be of interest to kineticists looking for an unusual and arresting undergraduate laboratory experiment. On the whole, they are less suitable for demonstration to larger audiences due to their smaller scale.

A Chemiluminescent Ammonia Fountain

Although most chemiluminescent reactions are attractive when demonstrated in simple flasks on the bench, extending the display with movement and the use of "classical" chemistry apparatus such as glass tubing provides a sense of occasion. One such demonstration is obtained by combining the ammonia fountain with luminol chemiluminescence.[1,14] It is based on the partial vacuum created by the introduction of water into a very large flask of ammonia vapor, the vacuum so formed then sucking in the two solutions, A and B, of the classical luminol reaction described earlier.

The apparatus consists of a round flask, preferably of 2.0-liter size, although a 1-liter flask can be used, with reduction of the volumes of solutions A and B to 500 ml (see Fig. 1; it is assumed that proper support and clamping of the flask are provided). This is fitted with a rubber stopper (this must fit well), bored to accept 6-mm glass tubing, and a length of glass tubing pushed through to within about 5 cm of the wall of the flask. A long syringe needle is inserted alongside the tubing (best done first). A short length of Tygon tubing is placed over the end of the glass, and the flask is filled with ammonia gas from a lecture bottle.[15] Keep the flow rate low so as not to overload the venting from the needle. Quickly attach the flask to the rest of the apparatus (see Fig. 1) and connect a syringe containing 10

[12] P. Jones, J. E. Frew, and N. Scowen, *J. Chem. Educ.* **64,** 70 (1987).

[13] H. Brandl, S. Albrecht, and T. Zimmermann, *in* "Chemiluminescence and Bioluminescence: Molecular Reporting with Photons" (J. W. Hastings, L. J. Kricka, and P. E. Stanley, eds.), p. 196. J. Wiley, Chichester, 1996.

[14] N. C. Thomas and J. H. Dreisbach, *J. Chem. Educ.* **67,** 339 (1990).

[15] The ammonia can also be produced by heating 10 g of ammonium chloride with 10 g of calcium hydroxide in 10 ml of water according to B. Z. Shakhashiri, "Chemical Demonstrations: A Handbook for Teachers of Chemistry," Vol. 2, p. 202.

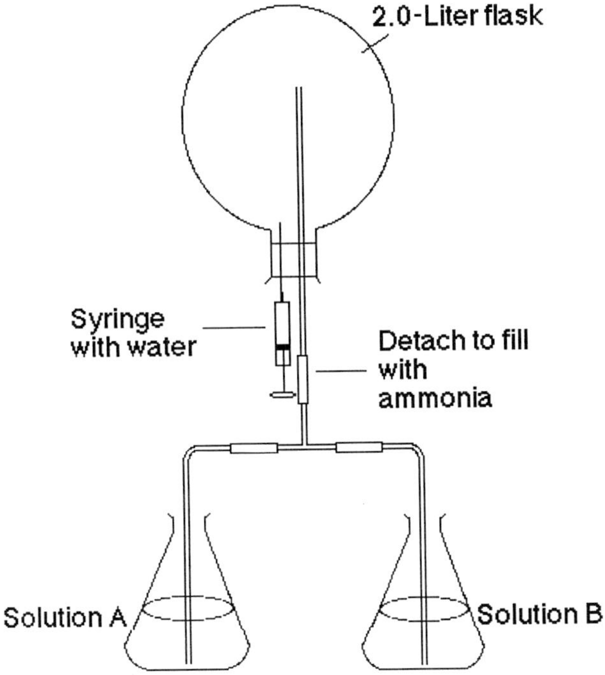

FIG. 1. Apparatus for the chemiluminescent ammonia fountain.

ml water to the needle. Leave an air bubble to avoid premature injection of the water during connection. It is a good idea to have a few trial runs to see how long the setup remains viable as the ammonia may slowly dissolve or leak. Wrapping the mouth of the flask in Parafilm will increase security. Inject the water in total darkness. A fountain of blue light starts up and should run for 1–2 min. This certainly requires practice. Be sure to dry the flask between runs.

Additional Effects

Add 10 mg of dry fluorescein or rhodamine to the large empty flask before setting up for a multicolor effect.

Lucigenin

Like luminol, the chemiluminescence of this compound (10,10′-dimethyl-9,9′-biacridinium dinitrate) was examined in the 1930s, and it is equally available from commercial sources.[16] Oxidation by hydrogen peroxide in alkaline aqueous ethanol results in a blue-green light of moderate intensity, usually lasting about 15 min depending on conditions. The reaction is not a clean one, and the green light derives from energy transfer to fluorescent by-products. This confuses the investigation of the mechanism,[17] but is of no consequence for the demonstration. The complexity of the reaction is actually an advantage in this context in that changes in color by the addition of fluorescein (yellow) and rhodamine (red) are intensified by the efficient energy transfer from the by-products.

The intensity of light emission is lower than that from the aqueous luminol reaction, but it is longer lasting, and the yellow and red light from energy transfer is more striking. These attributes make it particularly suitable for chemiluminescent "sign writing," although one can mount an effective static demonstration if desired.

Mechanism

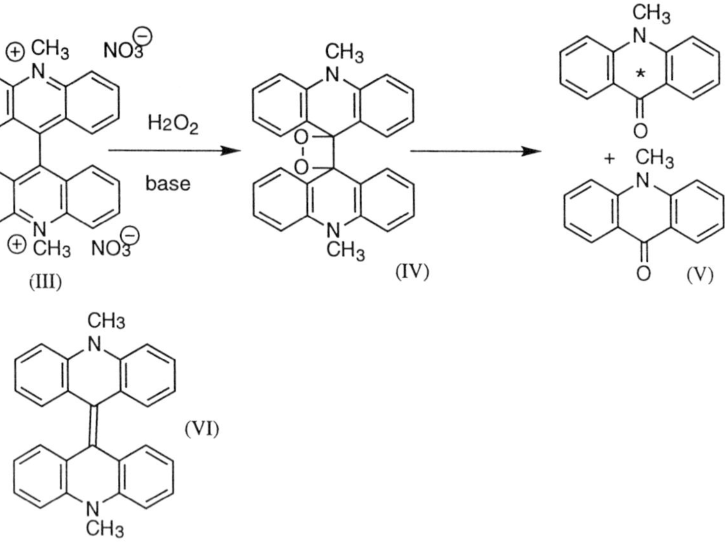

The mechanism is considerably more complex than that shown here, and the literature[17] should be consulted by the specialist. To explain the demonstration, the outline is sufficient. The addition of hydrogen peroxide to the ring system gives a dioxetane [which incidentally can be synthesized

[16] Aldrich Chemical Co., bis-*N*-methylacridinium nitrate B4920-3.

[17] R. Maskiewicz, D. Sogah, and T. C. Bruice, *J. Am. Chem. Soc.* **101,** 5347 (1979); (b) R. Maskiewicz, D. Sogah, and T. C. Bruice, *J. Am. Chem. Soc.* **101,** 5355 (1979).

by the addition of singlet oxygen to biacridylidene(VI)]. This decomposes to give one *N*-methylacridone(V) molecule in the excited state and one in the ground state. There is insufficient energy to populate both. The biacridylidene(VI), among other by-products of the lucigenin reaction, accepts energy from the excited *N*-methylacridone (NMA)(V) to give the overall greenish light observed. Compare this color with the clean blue emission from NMA itself in the acridinium ester experiment.

"Cold Light"

For over 20 years, the author has started his lecture demonstrations, especially those for lay audiences, with what amounts to a chemical neon sign, with the words "cold light" fashioned from 1-cm glass tubing. This older term for chemiluminescence emphasizes the intriguing central feature of the phenomenon.

Materials

Although lucigenin chemiluminescence can be catalyzed by solutions of sodium hydroxide, the by-products generate a dark solution and milky precipitate. A much cleaner result is obtained as follows.

Solution A. Make up a *saturated* solution of lucigenin by shaking an excess (about 1 gr) in 1 liter of ethanol (any grade). The excess solution can be stored for many months.

Solution B. 100 ml of ammonia (0.880, 15 *M*) add ethanol to make 1 liter. Solutions A and B are used in equal proportions. To the selected portion of solution A is added 30% hydrogen peroxide in the ratio of 2 ml per liter of A. This is stable enough to be made up several hours before the demonstration. The same volume of solution B is then added to give light. Lesser volumes of B can also be used. Two further solutions are prepared. Fluorescein (500 mg) is dissolved in 2 *M* NaOH (100 ml), and rhodamine (200 mg) is dissolved in ethanol (100 ml).

Apparatus

Many alternatives to the apparatus to be described can be devised, but the basic design used by the author can serve as a model. To a sheet of blockboard or plywood about 1 cm thick and about 60 cm square are screwed two "wings" 60 × 30 cm. Two pieces of laboratory scaffolding long enough to bridge the approximately 58-cm gap between the wings are each furnished with two right angle bosses, with each one placed 15 cm from each end. The two lengths of scaffolding are attached across the rear of these wings about 15 cm from the top and bottom of the wings, acting as braces. Two long pieces of the scaffolding are threaded through the bosses, and rings supporting two 500-ml dropping funnels are placed about 30 cm above the top of the wood. The ends of the dropping funnels just enter a large filter funnel clamped between and below them. The author was fortunate in having the services of an expert glass blower, Mr. Kenneth Pyke of Sussex University, who constructed the words "cold light" in continuous script in 1-cm-wide glass tubing. The entry to the "c" emerged from the back of the 60-cm square front, and the exit from the "t" went into the back of the apparatus to take a plastic or rubber tube, leading to a container for the effluent. A glass stop cock placed here allows one to stop the display at the best view of the colors, if desired, but remember to close off the supply from the two funnels. The lettering is supported at intervals by twists of copper wire pushed through two small holes on either side of the tubing. A plastic tube leads from the filter funnel to the entry to the "c" in the back.

The dropping funnels are almost filled with the solutions A and B (including the hydrogen peroxide addition) and the flow is adjusted to fill the lettering. The front of the apparatus should be covered in a black cloth, which is removed as soon as the taps are opened, all in total darkness. (Hint: use an earlier display such as the luminol-DMSO flask as a "torch" to see what you are doing, meanwhile complaining about the power cut or outage). Small quantities (ca. 5 ml) of the fluorescer solutions are added, one after the other to the filter funnel, from time to time, creating a triple color chemiluminescent sign. If a good glass blower cannot be found, it is possible to make the sign from Tygon tubing. In this case the wire twists have to be fairly frequent, and a better effect is obtained by blacking out the interconnecting nonscript portions of the text by black adhesive tape.

A similar device using luminol has been described,[18] but the author has found that lucigenin (equally available) gives a better result. Of course, a simpler, but still effective display can be constructed using a long spiral of Tygon tubing, running from the filter funnel to a large beaker, as before. The addition of the concentrated fluorescer solutions during the experiment is also effective in this method of presentation.

TMAE [tetrakis(dimethylamino)ethylene]

The abbreviation for tetrakis(dimethylamino)ethylene[19] is usually pronounced "Tammy" and is regarded with affection by demonstrators of chemiluminescence, as one only needs to take the top off the readily purchased bottle to have this spontaneously chemiluminescent material react

[18] D. B. Phillips, *J. Chem. Educ.* **70,** 773 (1993).

[19] Aldrich Chemical Co., tetrakis(dimethylamino)ethylene 23,423-0

with oxygen to give an unearthly green glow! A moderately interesting demonstration is obtained by pouring about 10 ml into a large spherical flask and rotating the flask so that the liquid coating glows. (Keep the flask stoppered. There will be enough oxygen, and TMAE smells powerfully of dimethylamine.) TMAE is an extremely electron-rich olefin with a very low ionization potential, causing it to react spontaneously with oxygen. The involvement of the dioxetane shown has not been fully substantiated, but is very likely. The strong electron-donating properties of TMAE allow the formation of a complex between it and the dioxetane. It is presumed that the charge donation catalyzes the decomposition of the dioxetane, with excitation of the fluorescent TMAE. Thus the light you see is from the fluorescence of the starting material.

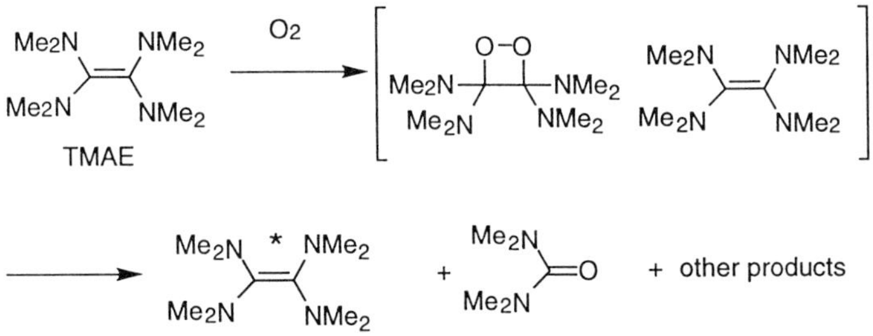

However, there really is only one way to demonstrate TMAE chemiluminescence, and that is to write a luminescent message on paper, on a firm backing, facing the audience, in the total darkness of the lecture room. Among well-educated audiences, "Fiat Lux" goes down well. A truly magic marker can be made easily by taking a 5- or 10-ml glass vial with a screw cap and narrow neck and *very tightly* wedging a rolled wick in the neck. The wick should project about 1 mm. Everything depends on the skill with which this is done and on the selection of the paper or preferably cloth material for the wick. Several materials are suitable, but varying local availability precludes a specific recommendation. Some TMAE is then injected with a syringe through the wick and the cap is screwed on to provide a most effective chemiluminescent pen. Keep the wick end lowest while writing. If the cap fitting is good, and the wick well made, the pen can be stored and reused dozens of times. Note that only glass vials and polypropylene or phenol-formaldehyde closures should be used, as TMAE will dissolve plastics such as polystyrene.

Acridinium Esters

These are available commercially, but are extremely and unreasonably expensive for demonstrations. Their synthesis is easy, and because they are used extensively in clinical diagnostics, being detectable down to the attomole range with the simplest chemistry, their demonstration helps to answer[20] a frequently asked question: What use is chemiluminescence? The light is an attractive clear royal blue and can be made to flash on a very short time scale, unlike many other demonstrations. The effect can be startlingly bright.

Synthesis

Safety

Acridine, as its name implies, can irritate the skin and mucous membranes. Potassium cyanide is highly toxic. Dimethyl sulfate, like all alkylating agents, is potentially carcinogenic.

The least expensive route starts with acridine, but a reasonable compromise is to purchase acridine-9-carboxylic acid. The last step, methylation, has been considered a problem in that the methylating agents are toxic. However, the involatile dimethyl sulfate is handled easily and is effective.

Acridine-9-carboxylic acid. Acridine (25 g, 0.14 mol) is dissolved in ethanol (110 ml) and the solution is added to glacial acetic acid (8 ml). A solution of KCN (12 g, 0.31 mol) in 20 ml water is added and the mixture is stirred under reflux for 1 hr. After cooling, the precipitate is filtered off nd washed with 2 *M* NaOH solution and water. It is dissolved in chloroform and dried over anhydrous Mg_2SO_4. Evaporation of the filtrate to dryness and recrystallization from *n*-propanol yields acridine-9-nitrile (19.6 g, 70%) (mp 185–186°, ir 2222 cm^{-1}, CN).

The acridine-9-nitrile (15 g, 70 mmol) is added to concentrated sulfuric acid (120 ml) and heated on a steam bath for 2.5 hr. It is cooled to 0° in an iced water bath and sodium nitrite (55 g, 0.8 mol) is added slowly. It is important to allow time for each portion of sodium nitrite to react as otherwise there will be excessive frothing during the next stage. Brown fumes are evolved. The mixture is then heated carefully until no more gas is given off. It is then stirred at 100° for 2 hr, cooled, and poured slowly into iced water to precipitate the yellow product. This is filtered off, washed with water, and sucked dry. The product is then dissolved in the minimum of 2 *M* NaOH solution and filtered through a sintered funnel. The deep red filtrate solution is treated with concentrated HCl until the

[20] A. K. Campbell, "Chemiluminescence, Principles and Application in Biology and Medicine." Horwood/VCH, Chichester, 1988.

yellow precipitate has permanently reformed. This is filtered off, washed with water, and sucked dry. It is dried further at 55° under reduced pressure for 24 hr (yield 14.7 g, 89%) (ir 1650 cm^{-1}, acid C=O).

Acridine-9-carbonyl chloride. Acridine-9-carboxylic acid (10 g, 41 mmol) is refluxed in a large excess of redistilled thionyl chloride under dry conditions until solution is complete (ca. 3 hr). The reagent is evaporated off (any residue of thionyl chloride being removed by coevaporation with dry benzene) to give a yellow powder that is used without further purification (ir 1780 cm^{-1}, Cl—C=O).

Phenyl acridine-9-carboxylate. Acridine-9-carbonyl chloride (about 10 g, from the previous experiment) is stirred in dry pyridine, forming a brown suspension after a few minutes. Phenol (4 g, about 10% molar excess) is dissolved in a few milliliters of pyridine and added to the suspension. The reaction mixture is stirred for 8–12 hr at room temperature and is then poured into a stirred mixture of ice and concentrated hydrochloric acid to precipitate the product. This is filtered off, washed with water, and sucked dry. Recrystallization gives the pure product in 85% yield. Recrystallization from toluene gave beige needles, mp 186–188°. Nmr δ 7.3–8.42 complex aromatic pattern, ir 1740 cm^{-1}.

Phenyl 10-methylacridinium-9-carboxylate methosulfate. Phenylacridine-9-carboxylate (500 mg, 1.7 mmol) is dissolved in 40 ml dry toluene. Dimethyl sulfate (4 ml, 42 mmol) is added and, after stirring at 100° for 5 hr, the cooled precipitated salt is filtered and washed with toluene to dissolve excess dimethyl sulfate. The solid is then washed with diethyl ether and allowed to suck dry. The product is a bright yellow powder (635 mg, 89%). Recrystallization from ethanol affords a pure sample.

A reproducible melting point could not be obtained. Acridinium salts often show this behavior. Nmr (d_6-DMSO) δ 4.97 (3H, s; N—CH_3). δ 7.42–9.07 (12H, m; aromatic H's) ir 1608 cm^1 (aromatic C=C), 1760 cm^{-1} ester C=O ms 315 (M^+ + H), 299 (M^+ + H—CH_3), 178 (206-CO), 206 (299-C_6H5).

Demonstrations

1. Advantage can be taken of the rapid rate of reaction by dissolving 50–100 mg of acridinium phenyl ester(VII) in 100 ml of 95% ethanol in a 250-ml conical flask and adding 100 μl of 30% hydrogen peroxide or its more dilute equivalent. The slow, regular addition of drops of 5% NaOH from a burette while lecturing or performing other tasks in near darkness creates a worthwhile effect. The reaction is almost independent of the quantities given and it is an extremely easy demonstration to set up. The burette can be replaced by a pipette with a teat for a brief demonstration.

2. This is also an easy experiment and like the previous one, can be arranged very quickly with guaranteed success. The author uses a 1000-ml measuring cylinder, but any long glass tube container will do. Dissolve about 1 g of NaOH in 100 ml distilled or deionized water in the cylinder by swirling and fill to the top mark with 95% ethanol. Add about 0.5 ml of 30% hydrogen peroxide (or its equivalent). Mix thoroughly by inversion. The cylinder can be prepared ahead of time. In the dark, add, by gentle tapping from a vial, a stream of fine crystals of acridinium 9-phenyl ester (20–50 mg is plenty) to create a drift of scintillations. After a suitable wait for appreciation, invert the whole cylinder, covered by your hand, to give a remarkably bright display. This experiment is very tolerant of quantities, and estimates are acceptable. The crystals should be as fine as possible, but should *not* be powdered, as they must descend by gravity, yet be small enough to dissolve readily. A fast recrystallization with cooling and shaking usually does the job. Do, however, try it out first. For reasons not yet examined, energy transfer to other fluorescers is poor.

Mechanism

Hydrogen peroxide is more nucleophilic than hydroxyl ion and the peroxide(VIII) is formed in about 90% yield. The dioxetanone(IX) is only a transient and decomposes in a fast step to excited *N*-methylacridone. The fluorescence of this is blue (442 nm). Its formation is made possible by the acidic phenol acting as a good leaving group. Many other leaving groups with similar properties have been used. It is interesting (and almost certainly significant) that the same "rules" apply to the oxalate esters.

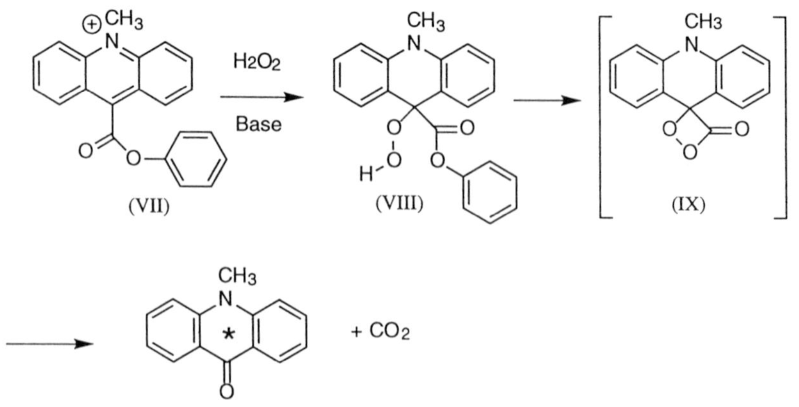

Acridan 9-phenyl Carboxylate

The light from the reaction of this compounds has the same color as that from the acridinium ester, but the mechanism is identical in principle to that for the reaction of firefly luciferin and coelenterazine. These two compounds give beautiful chemiluminescent reactions, but are very difficult to synthesize, so that the acridan becomes a good substitute. The simplicity and power of the demonstration, and the exact correspondence with the light-emitting step of the biological reactions, make it a good, easily transported, single demonstration for any lecture on bioluminescence, where a full-scale demonstration lecture is inappropriate.

Demonstration

Extremely small quantities give a very bright light and demonstrate the simplicity underlying the biochemical reactions in that removal of a proton by a strong base results in a spontaneous reaction with atmospheric oxygen to give the light. Cover the bottom of a 100-ml conical flask with a single layer of KOH pellets and add about 20 ml of DMSO. This mixture is stable indefinitely if stoppered tightly. A few specks of the acridan ester added from a spatula, followed by shaking, give an instantaneous bright light. The experiment can of course be scaled up to resemble the luminol reaction in DMSO described already, but with a much brighter and bluer light.

Synthesis

Phenyl 10-methylacridan-9-carboxylate. Phenyl acrididinium-9-carboxylate (1 g) is refluxed with zinc dust (0.3 g) in an acetic acid solution (10 ml) until the intermediate purple color vanishes. Any unreacted zinc dust is removed by filtration, and the product is precipitated by the addition of water. If the precipitate is difficult to filter, extraction with ether followed by drying over Mg_2SO_4 followed by evaporation of the solvent gives good recovery. Recrystallization from ethanol gives white needles, mp 116–118°, in 60% yield. Nmr δ 3.42 (3H, s; NCH_3); 5.17(1H, s; C9-H). 6.84–7.40 (13H, m; aromatic H's).

Connection between Chemiluminescence and Bioluminescence

The *acridan* phenyl ester reacts to form the dioxetanone, but by an autoxidation mechanism, as it is reduced compared to an acridinium ester. This is a very general mechanism and, in terms of the light-generating step, identical in principle to the mechanism pertaining in the firefly and those marine organisms using coelenterazine as their luciferins. The repetition of the following general reaction scheme can be seen in the reactions of both firefly luciferin(X) and the acridan(XI).

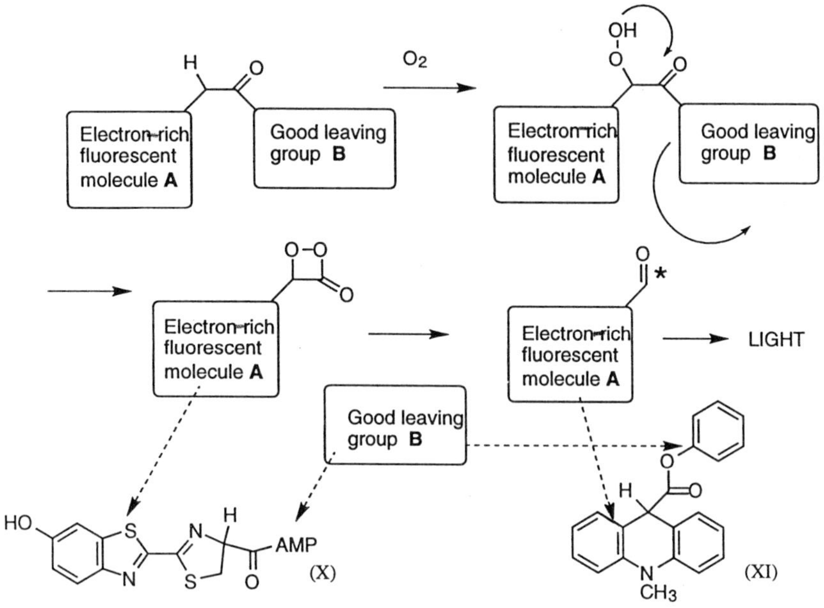

Active Oxalate Esters[21]

These compounds were developed by a team at American Cyanamid, led by Dr. Michael Rauhut,[22] as the most effective exploitation of chemiluminescence as a source of illumination. He had calculated that with a fully efficient reaction (quantum yield = 1.0) and a highly soluble compound to give a high power density and using 1 liter of a 5.0 M solution, a light source equivalent to a 40-W incandescent bulb burning for up to 2 weeks could be devised. Devices based on this system are now familiar worldwide, mainly in the form of "light sticks" of various sorts, justifying this suggestion, and standing as a tribute to the detailed development work carried out by his team. The remarkably high efficiency (ϕ_{es} = 60%) of the basic reaction provides really effective illumination with a much longer lifetime than

[21] A. G. Mohan, *in* "Chemi- and Bioluminescence" (J. G. Burr, ed.), p. 245. Dekker, New York, 1985.

[22] M. M. Rauhut, *Acc. Chem. Res.* **2,** 80 (1969).

any other reaction. Although the refinements introduced by American Cyanamid are of necessity missing from the more convenient approach described here, this is still an impressive display. The principal reactant, bis(2,4,6-trichlorophenyl)oxalate, can be purchased, but its synthesis is very straightforward and desirable if it is to be used on any significant scale.

Preparation of Chemiluminescent Oxalate Esters

Safety

Oxalyl chloride is volatile and toxic. Use a fume hood until it is all added to the flask. Nitrobenzene is an unpleasant toxic solvent, absorbable through the skin. Avoid using (see below for an alternative) unless the operators are skilled. The phenols are corrosive, and dinitrophenol stains the skin. Use gloves. Benzene is potentially carcinogenic.

Synthesis[23]

Bis(2,4-dinitrophenyl)oxalate (DNPO). 2,4-Dinitrophenol (9 g) is placed in a three-necked flask with toluene[24] (150 ml) and the phenol is dried by azeotropic distillation (usually requires the removal of 30–50 ml). The solution is cooled in an ice bath to 10° after adding a dropping funnel with drying tube to the flask. Dry triethylamine (6.5 ml) is added to the mixture in the flask, and oxalyl chloride (4.5 ml) in dry toluene (50 ml) is added via the dropping funnel over 20 min. The flask is stirred by a magnetic stirrer during this operation. Remove the ice bath and stir at room temperature for 2 hr. A thick suspension forms during these operations. Concentrate (Buchi apparatus or other means of removal of solvent under reduced pressure) this suspension to a still filterable mass, and filter under vacuum. Chloroform (see next step) can be used to effect the transfer of residues to the funnel. Wash, on the filter, with a total of 100 ml of chloroform, in portions, to remove triethylamine hydrochloride. This material is suitable for demonstrations, but residual triethylamine hydrochloride is an inhibitor of the reaction if present in quantity. For the very best material, recrystallization from ethyl acetate (the ester is not very soluble) or nitrobenzene can be

[23] M. M. Rauhut, J. J. Bollyky, G. B. Roberts, M. Loy, H. Whitman, and A. V. Iannotta, *J. Am. Chem. Soc.* **89,** 6515, (1967). A. G. Mohan and N. J. Turro, *J. Chem. Educ.* **51,** 528 (1974).

[24] Benzene is much more convenient, but there are restrictions on its use in undergraduate laboratories.

carried out. The dinitrophenyloxalate is tedious to recrystallize, for the average demonstration, use crude. The yield is about 5 g (50%); mp 191–195°. Store in a desiccator.

Bis-2,4,6-trichlorophenyloxalate (TCPO). Use the stoichiometric amount of trichlorophenol in the experiment just described, but this time, make sure that the transfer to the funnel can be made without chloroform, as the ester is soluble in this solvent. Wash rapidly with water and suck dry. Transfer to a freshly charged vacuum desiccator and dry thoroughly. Surprisingly, this easily hydrolyzed compound can be used directly for the demonstration after the water wash, but for storage it must be very dry. The yield is about 7 g (60%).

The trichlorophenyloxalate is recrystallized easily from toluene, with excellent results. This is the ester of choice for demonstrations, and recrystallization is worthwhile in this case. The mp is 192–194°.

Demonstrations

Small amounts of light are detectable in the absence of an added fluorescer, probably as a result of energy transfer to fluorescent impurities. However, on the addition of one of a large number of mainly aromatic hydrocarbon-based fluorescers, bright light of any desired color is obtained. The fluorescers can even be mixed to give white light! Make up three chemiluminescing solutions blue, yellow, and green (see later). These can be mixed to achieve a perfect white light without difficulty, simply judging by eye.

Several requirements of the very efficient commercial system have been abandoned in the interests of safety and convenience, but the difference is barely noticeable on the time scale of a lecture demonstration. The changes mainly affect lifetime, and there is clearly no virtue in a display lasting 13 hr during a lecture! The recommended solvents are dialkyl phthalates, and although other solvents do work, the increase in efficiency in the phthalates is so great as to make them essential. In addition, they are among the safest, although it is essential to dispose of the spent solvent in properly designated waste disposal containers as they should not be allowed to contaminate water supplies.

Solution A. Bis(2,4,6-trichlorophenyl) oxalate (500 mg) and 9,10-diphenylanthracene (50 mg) are dissolved in diethyl (or dialkyl) phthalate (100 ml).

Solution B. Add 20 ml of *tert*-butanol to 80 ml diethyl phthalate and add 1 ml of 30% hydrogen peroxide. Add sodium salicylate (10 mg) and mix.

Mix solutions A and B. Both separate solutions are stable for about

12 hr and even longer if moisture is excluded from A. Blue light appears and reaches a maximum intensity in about 5 min. The light should last for at least 30 min at a useful intensity, but if this is not attained, simply increase the concentration of the oxalate (undissolved oxalate in the flask will act as a reservoir, with shaking from time to time). Blue is the weakest color for two reasons. The human eye is more sensitive in the green and yellow regions of the spectrum, and the catalytic effect of fluorescers with lower ionization potentials is less. However, the light is still impressive.

Note that water is a competitor with hydrogen peroxide for the oxalate, and the hydrolysis of the oxalate destroys the light. The commercial devices use 98% peroxide, a very dangerous and corrosive material in its undiluted state and difficult to obtain. This, and the rigorous exclusion of water generally, together with a much more soluble ester, explains the much better results from the light sticks. Professional chemists can handle it safely, however, and demonstrations have been described using it.[23] The original literature[22,23] can also be consulted for procedures and explanation. Note that it is present in a completely safe, highly diluted form in *tert*-butanol in the inner glass tubes of the light sticks.

Alternative Procedure

A simpler procedure for use when time is short, or when portability is required, involves adding 1 ml of 30% hydrogen peroxide to 100 ml of the phthalate solvent in a glass container and adding 100 mg of the oxalate which has been premixed in a mortar with 10 mg sodium salicylate and 20 mg of the fluorescer. Vigorous shaking gives strong light that is a little slow to develop to its maximum intensity. Vials of the solid mixture can be prepared ahead of time, and although the mixture does deteriorate, it can be kept for several months in a desiccator.

Other Fluorescers (Brighter Light)

Most of the classical fluorescers such as fluorescein and rhodamine are poor sensitizers of this reaction, with best results undoubtedly being obtained with the polynuclear aromatic hydrocarbons. The brightest (green) light is obtained using 9,10-bis(phenylethynyl)anthracene. Rubrene gives a very bright yellow orange light (it is not very soluble, and best results are obtained by predissolving as much as possible in the phthalate solvent). It is not used for very long-lived displays, as it photobleaches, and is degraded by the hydrogen peroxide. However, it is remarkably effective for lecture demonstrations, where this disadvantage can be ignored. The very stable 1,8-dichloro-9,10-bis(phenylethynyl)anthracene gives a clear yellow light. The efficient emission of red light is much harder to obtain, and in some of the commerical light sticks, the plastic outer tube is itself red fluorescent and acts as a filter for the yellow-green light actually produced by the reaction. Rhodamine will give a relatively short-lived, but strong, red light, but if you are going to use this, the oxalate should be bis(2,4-dinitrophenyl) oxalate and sodium salicylate or any other basic catalyst should not be used. Violanthrone gives weak red light, with solubility being a problem, and 5,12-bis(phenethynyl)naphthacene (see the discussion on luminol effects) is preferable even if the light does not have the blood red character of rhodamine.

Magic Light

All of the descriptions of the demonstrations of chemiluminescence are open to improvement and innovation by the user, and as an indication of the possibilities, the author suggests the following little bit of magic. In a mortar and pestle grind three *separate* mixtures, each containing 50 mg of bis(2,4,6-trichlorophenyl) oxalate(XII), 10 mg of sodium salicylate, and 10 mg of, respectively diphenylanthracene, bis(phenylethynyl)anthracene, and rubrene. You can use any more convenient multiple of these quantities. Transfer the material, using portions of about 50 mg of each, if more was prepared, to three separate 10-ml conical flasks. Add about 5 ml of *dry* acetone to each (ignore the undissolved material). Take a good-quality paper napkin or serviette and, using a Pasteur pipette, cover the serviette in random splotches or spot a pattern, using all three colors, and covering all the paper. Fold very tightly, when dry, and place in the bottom of a small opaque glass or ceramic container. During the demonstration, display a clean serviette and a flask of water-white solution B (the phthalate/butanol/hydrogen peroxide mixture; see general oxalate demonstration described earlier). Claim the total absence of ingredients. By sleight of hand, "lose" the clean serviette in the container (or up your sleeve if you really are a magician), pour on enough solution B to just wet the paper, and unfold the Jacob's coat of many colors, switching off the house lights at the same time. The last might require an assistant.

Mechanism

Details of the mechanism are still under investigation, and several peroxy esters are implicated as the energy-rich intermediate. No direct evidence for the dioxetanone(XIII) exists. However, provided that one is aware of the complexity, the simple scheme shown will satisfy most of the audience. The dioxetanone and other peroxy esters may form a charge transfer complex with the hydrocarbons and then decompose to give the excited state of the hydrocarbon directly within the complex. The reaction is the most efficient synthetic chemiluminescent example discovered to date, having a quantum yield of 60%, much greater than any bioluminescent reaction, except that of the firefly.

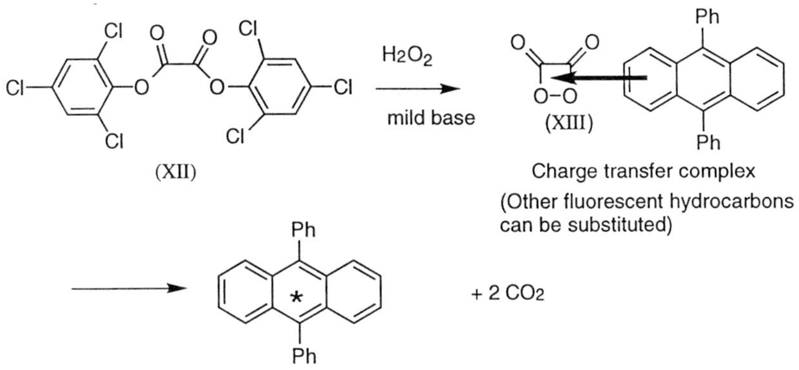

Why Do These Reactions Produce Light?

In the expectation that these demonstrations will be presented by the general chemist or biochemist, the details of the mechanisms have been kept to a minimum. The question of light emission has not yet been answered to everyone's satisfaction. Sadly, the reactions that comprise the most convenient demonstrations are among the hardest to explain. This gives an unnecessarily gloomy picture, and it is certainly possible to provide some solid ground on which to stand. The acridan esters and, by extension, the acridinium esters react by the same mechanism as do firefly luciferin and coelenterazine. In its broad outline, this is a proven mechanism. The scheme given earlier for these compounds does not include the final step: that of the population of the excited state. This is more complicated, but can be addressed simply by saying that the four-membered ring has electronic properties that lead to an avoidance of the ground state. Provided there is enough energy in the decomposition of this ring, then an electronically excited state will be generated, and the electron-rich, aromatic portion of the molecule will ensure fluorescence from this state. The other luciferins (including the bacterial system) cannot be described in the same terms, but progress is being made.

Other Light-Emitting Reactions

Because one of the attractions of chemiluminescence is the unusual nature of the source of the light, it is interesting to compare it with other uncommon light sources. Easily visible examples include fluorescence, phosphorescence, electroluminescence, and triboluminescence. Most of these phenomena can be used to interrelate many basic photophysical and chemical concepts, as pointed out in the introduction. Each lecturer can be expected to bring their own preferred examples to illustrate, and the following is only an indication of the possibilities.

Fluorescence can be excited after the chemiluminescent reaction is over by shining a "black lamp"—a source of ultraviolet light—on the flasks. Phosphorescence can be demonstrated using the many adhesive stars and planets sold in most toy shops for use on the ceilings of children's rooms, with the same lamp. Keep them in the dark in a container until the demonstration. The "instantaneous" nature of the prompt fluorescence (excited state lifetime about 10^{-8} secs) can be contrasted with the very much longer lived phosphorescence, lasting for several minutes, if not hours. This long life exposes the excited triplet state to quenching in fluid solution, whereas the short-lived fluorescent singlet state emits before it can be quenched. It is worth pointing out that triplet states are indeed formed in chemiluminescent reactions, but that the foregoing circumstances explain why emission from them is rearely visible. These considerations can lead to discussions of singlet and triplet states, kinetic (collision) theory, spin conservation, the spectral sensitivity of the human eye versus electronic detection, and many other topics. An additional demonstration of the influence of temperature on the kinetics of reactions is provided by the strong quenching effect of cooling on the intensity of the longer lived chemiluminescence demonstrations. A commonly asked question (even by mature scientists!) is whether the spent reaction can be reactivated. Comparison with the ashes of a fire and a reminder of the second law of thermodynamics and the concept of entropy are timely at this point!

Triboluminescence[25] (the emission of light from stressed solids) can also be demonstrated, but only from relatively weak point sources, so that it is not a suitable topic for large lecture rooms. However, questions are often

[25] R. Angelos, J. I. Zink, and G. E. Hardy, *J. Chem. Educ.* **56,** 413 (1979) and references cited; A. J. Walton, *Adv. Phys.* **26,** 887 (1977).

raised about it because it is much more common than is usually realized. A related example is the blue light observed when a piece of Scotch tape (Sellotape) is stripped from a surface in the dark, and although not triboluminescence by the strictest definition, the light is produced by a gas discharge in air (the emission is from excited N_2 molecules), excited by the recombination of electrostatically produced positive and negative charges. A similar, but not identical, mechanism operates in triboluminescence proper. Apparently, almost half of all compounds are triboluminescent, but only a few have been investigated.

The earliest report is that by Sir Francis Bacon, who noticed the light produced on scraping loaf sugar. A modern, brighter form is that of the sweet or candy known as Lifesavers. If either a sugar cube or a (opaque, white) Lifesaver is crushed, a flash of light is seen. It is blue in the case of sugar and green from the candy. In the latter case, there is energy transfer from the excited nitrogen molecule to the fluorescent flavoring agent, methyl salicylate, or oil of wintergreen. Uranyl nitrate also shows quite bright green flashes when crystals are crushed, and again energy transfer to the very fluorescent uranyl ion in the solid is responsible. This may well be the most effective demonstration of triboluminescence and is relatively inexpensive, especially if the powder produced is recrystallized. *N*-Acetylanthranilic acid crystals[26] also show the phenomenon, with a similar mechanism. The last two compounds have crystals soft enough to be crushed between two substantial watch glasses or glass plates, making them easier to demonstrate to a small audience. There is more than one mechanism, but the commonest and best understood is that, under stress, the crystal acquires two opposite charges on opposite sides of an imminent fracture and, on separation of the surfaces, an electrical gas discharge occurs, exciting the nitrogen molecules of the air, with emission in the blue and ultraviolet regions of the spectrum. If the medium is fluorescent, energy transfer results in light of the expected color.

In electroluminescence, electrons are promoted to the conduction band of a phosphor, such as zinc sulfide, by alternating current, which them ionizes a fluorescent atom. Recombination of an electron with the atom forms the excited state, and the emission of light ensues. These phosphor screens are commonly used to backlight liquid crystal displays, and examples are easy to obtain for demonstration purposes. Note that the active displays used in more recent lap-top computers use p-n transistor or semi-conducting junction materials and are considerably more sophisticated. The principles are, however, the same.

[26] J. Erikson, *J. Chem. Educ.* **55,** 688 (1972).

[43] Bioluminescence as a Classroom Tool for Scientist Volunteers

By MARA HAMMER and JOSEPH D. ANDRADE

"One of the best tools any teacher can have is a person who knows and understands science and technology."[1] A practicing scientist.

Background

A major shift in science education is now being implemented. There is a growing interest in hands-on, discovery-based science education. There is also interest in treating science as a coherent, integrated field of inquiry that can deal with complex multidisciplinary subjects and problems.[2]

Bioluminescence is an ideal subject with which to experience the scientific process and critical science concepts and themes, particularly in upper elementary and junior high school.[3] We have developed bioluminescent dinoflagellate cultures that enable teachers and students to experience bioluminescence, as well as closed ecosystems, circadian rhythms, and a variety of principles related to protozoa, optics, and chemistry. Much of the experience is conducted in the dark. Science in the dark has been an effective way to reduce science anxiety and fears and to encourage teachers to develop fresh, positive, and instructive attitudes toward hands-on science in their classrooms.

The intrigue of bioluminescence reaches far beyond the science classroom. For example, the beautiful blue light can be an inspiration to students interested in creative writing. Experiences with the dinoflagellate, Pyrocystis lunula (little fire moons), can begin with a poem:

Little fire moons dance in the sea
Making the waves glow mysteriously
Little fire moons come play with me.
And we will dance together in the sea

[1] Sharing Science With Children: A Survival Guide for Scientists and Engineers, North Carolina Museum of Life and Science, P.O. Box 15190, Durham, North Carolina 27704.

[2] F. J. Rutherford and A. Ahlgren, "Science for All Americans (The Project 2061 Report)." Oxford Univ. Press, 1990.

[3] M. G. Jones, Bioluminescence: Activities that will receive glowing reviews. *The Science Teacher,* Jan. 19 (1993).

0076-6879/00 $30.00

Tiny artists can be captivated by their mystery and beauty. Aldo Leopold, the noted conservationst, relates an account of when he first became a conservationist. He saw a gray wolf die and noticed the "fierce green fire" in her eyes. It was a moment he never forgot.[4] Perhaps seeing the "cold blue fire" of nature's living lights can inspire others in the same way.

There is a growing awareness of the need for scientists and other professionals as role models in school classrooms. Students need to learn science skills in elementary school that can be built upon in the upper grade levels. Unfortunately, many elementary school teachers do not have any science background and are, therefore, uncomfortable teaching science. In many cases they do not have the resources to conduct meaningful science experiments. Bioluminescence is a wonderful tool to help students and teachers start thinking and acting like scientists. Practicing scientists are needed as regular, serious volunteers in the classroom. By volunteering your time at a local elementary, middle, or high school, being a guest speaker at an after school science club, being a friendly resource for local teachers, or allowing field trips through your laboratory, you can sow the seeds of science and make a lasting impact on impressionable young minds and on their teachers.

The Need for Scientists in Science Education

The structure of education in the United States is based on the concept that what students learn in elementary school is built upon in middle and high school. "Failure to educate and excite students in science in the early years is a primary reason for the inadequacy of the learning of science in later years."[5,6] Unfortunately, many students do not receive a background in science. What they do receive is often book learning, rote answers and memorization.[7] They are not taught to think like scientists. They are not taught to find a problem and to test a hypothesis. Scientists in the classroom provide a vital link in encouraging today's students to become tomorrow's responsible, informed citizens and scientists. In most cases, when the student reaches college, it is much harder to capture their interest, and it is a daunting task to teach them basics they should have learned in middle and high school.

The way science is commonly taught leads many students to believe it

[4] A. Leopold, "A Sand County Almanac." Ballantine Books, New York, 1959.

[5] National Research Council, "Fulfilling the Promise: Biology Education in the Nation's Schools," p. 6. National Academy Press, 1990.

[6] Academic Preparation in Science: "Teaching for the Transition from High School to College," 2nd ed. The College Entrance Examination Board, New York, 1990.

[7] Foundations: The Challenge and Promise of K-8 Science Education Reform, National Science Foundation, NSF 97-76, p. 68.

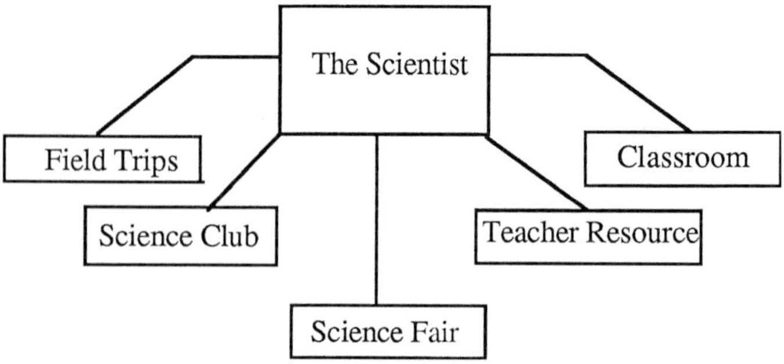

FIG. 1. Ways in which practicing scientists can participate in school-based science education.

is a "hard subject." A possible reason for this attitude is the difficulty of relating abstract concepts to concrete, real-life situations. The National Research Council has recommended that "the major objective of elementary school science courses should be to foster positive attitudes about and respect for the natural world, rather than to acquire detail. Children in grades 3–6 might be especially sensitive, because it is in those grades that "likes" and "dislikes" become established. . . . Exciting experiences in the early grades can lead to science-related career decisions."[5] This is why it is vital that scientists become involved in precollege education.

Involvement

There are many ways to become involved in K–12th grade science education and many resources are available to help you have a productive, enjoyable visit and to establish a relationship with the students and teacher. Some of these resources include "Chemists in the Classroom," a brochure produced by the American Chemical Society,[8] and "Sharing Science with Children: A Survival Guide for Scientists and Engineers"[1] (Fig. 1).

Becoming involved is the first and sometimes the most difficult step. A good way to begin is to participate in one of the national "Weeks." National Science and Technology Week has been designated by the National Science Foundation as the last week in April, National Chemistry Week is usually in early November, and National Engineering Week is the third week in February. Other professional societies also support community outreach programs through schools, museums, and other institutions. Contact your

[8] "Chemists in the Classroom" Prehigh School Science Office, American Chemical Society, 1155 Sixteenth St., NW, Washington, DC 20036, phone: (202)452-2113.

national professional organization to find out what programs and resources are available. If you are a parent, talk to your child's teacher or school principal about volunteering in the classroom.

Planning your visit: Find out at what level the teacher and the students feel comfortable. Ask for suggestions. Consult with the teacher about the level and depth of your presentation. Find out if the topic, demonstrations, and experiments are appropriate for the age and attention span of your audience. Teachers are not experts in teaching science, but they are experts in teaching children. Remember, you are there to be a role model, to present a lesson and experience on a specific topic, and to offer career information. One of the most interesting topics is *you* and what *you* do.[8]

Presenting the topic: Involve the students in the activities. Doing is a lot more interesting than lecturing. Involve the teacher as well as the students. Do not let the teacher leave. Make it clear to the teacher that your participation is a partnership with him/her. One of your goals is to educate and empower the teacher, as well as the students. Ask questions that encourage students to make predictions, offer explanations, state an opinion, or offer conclusions. Be aware of your vocabulary. Define words that students may not know and relate those terms to common everyday experiences. Use analogies and metaphors. It may be helpful to give the teacher a vocabulary list before your visit. You must connect and relate your topics to the students and to the world around them. It is important that the students leave with the connection between experimental science and practical applications. Remember to emphasize the importance of safety. Demonstrate to students the proper way to handle and dispose of chemicals. If the students are required to wear safety gear, set the example. Before you leave, help set up an experiment that the class can continue. Give the students an assignment that they can complete on their own. Invite them to write, call, or e-mail if they have any questions.[1,8]

Finally, follow up with the teacher and the class. This is essential to reinforce the concepts that you presented. Ways to follow up can include a return visit to build on what you taught previously or offering suggestions to the teacher on activities that she/he can do with the class to reinforce what was taught. These could include an extension of the science concept, a related science topic, the history of that aspect of science, an experiment that can be done as a class, or ways to connect the science topic with art, literature, or other subjects in the curriculum.

Other ways to become involved are to participate as a judge at a science fair, act as a consultant/advisor to students on their science fair projects, organize or be a guest speaker at an after school science club, provide summer research opportunities for high school students and teachers in

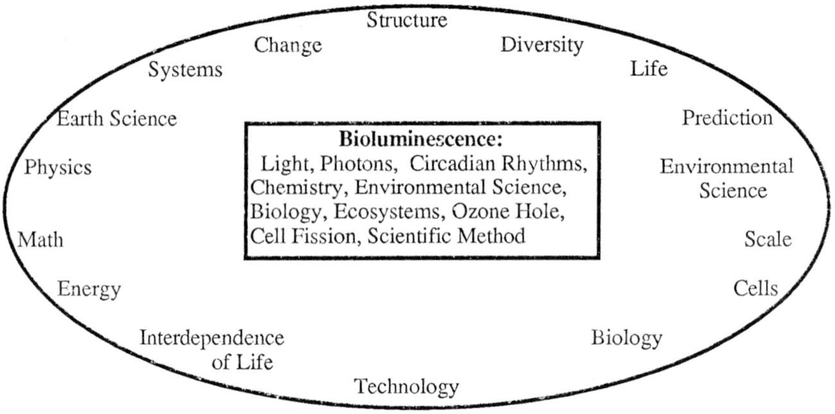

FIG. 2. Some of the Project 2061-Benchmarks for Science Literacy concepts and themes that can be experienced using bioluminescence.

your laboratory or institution, and organize field trips through your research facilities or to a local spot of scientific interest.[9]

What You Can Do with Bioluminescent Dinoflagellates

Benchmarks for Science Literacy–Project 2061[10] is a set of guidelines established by the American Association for the Advancement of Science. The benchmarks are guidelines for what all students should know and be able to do in science at given grade levels (Fig. 2).

Bioluminescent dinoflagellates can be an effective way to teach and reinforce these basic science concepts. Bioluminescent dinoflagellates are an ideal teaching tool because they are unique. Few students or teachers have ever experienced them, so they do not have any preconceived notions about what to expect. It is a subject they have never failed. Much of the teaching and discovery is done in the dark where people are inherently less intimidated.

Bioluminescent dinoflagellates are available from many sources.[11] The

[9] A. K. Campbell Field Trip, 2–3 September, 1994, Marine Biological Association, Plymouth, *in* "Bioluminescence and Chemiluminescence: Fundamentals and Applied Aspects" (A. K. Campbell, L. J. Kricka, and P. E. Stanley, eds.), p. 232. Wiley, New York, 1994.

[10] American Association for the Advancement of Science, "Benchmarks for Science Literacy (Project 2061)." Oxford Univ. Press, New York, 1993.

[11] Source for bioluminescent dinoflagellates: Provasoli-Guillard National Center for the Culture of Marine Phytoplankton (CCMP), Bigelow Laboratory for Ocean Science, P.O. Box 475, McKnown Point, West Boothday Harbor, Maine 04575-0475. http://ccmp.bigelow.org, phone: (207)633-2173, fax: (207)633-6584 Carolina Biological, e-mail http://living@carolina.com, phone: 1-800-334-5551, fax: 1-800-222-7112.

University of Utah's Center for Science Education and Outreach[12] provides Pyrocystis lunula sealed in small polyethylene bags, essentially a "miniecosystem." Pyrocystis fusiformis is available through Carolina Biological. We also use dried "sea fireflies" (Vargula hilgendorfi) in some of our demonstrations (available from Sigma).

By involving students and teachers in the exploration of topics and phenomena they have never seen before, they experience the scientific process in action and learn basic science skills and processes. Dinoflagellates motivate the asking of questions; the students become detectives solving the mysteries of the world around them. Students learn to understand that hypotheses are to be tested and retested. A simple question such as "why is the light blue" can become a whole research project involving light, photons, optics, evolution, adaptation, and the optical properties of water.

It is very interesting to involve the students and the teacher in the nature of luminescence itself—what is light, where does it come from, how does it originate?—and to observe, compare, and contrast such phenomena as thermal luminescence, fluorescence, phosphorescence, chemiluminescence, and bioluminescence. Chemiluminescent demonstrations are readily available and are exciting for the students.[13,14] An explanation of the difference between warm and cold light can lead into a discussion of energy efficiency. For example, in a conventional light bulb, 90% of the total energy disappears as heat. The bioluminescent enzyme reaction produces light with almost no heat, which makes it nearly 100% efficient.[15,16] Various fluorescent minerals and a small UV black light provide a great deal of material for observations, hypotheses, experiments, and discussions. The fluorescent whitener used in most laundry detergents and present in significant quantity on the students' own clothes is a very popular topic. Chemiluminescent light sticks, available in most hardware stores, as well as in science-related gift shops, have been used and experienced by many of the children and teachers, but with little scientific thought or consideration. Once the concept of chemically generated light, so called cool light, is firmly established, then the connection can be made that bioluminescence is basically chemical luminescence produced by living organisms.

[12] Center for Science Education and Outreach, University of Utah, Bioengineering, 50 S Central Campus DR. RM 2480, Salt Lake City, Utah 84112-9202, phone (801)581-4171. e-mail:http://www.utah.edu/cise/bioluminescence.html

[13] H. Brandl, S. Albrecht, and T. Zimmermann, *in* "Bioluminescence and Chemiluminescence: Molecular Reporting with Photons" (J. W. Hastings, L. J. Kricka, and P. E. Stanley, eds.), p. 196. Wiley, New York, 1997.

[14] F. McCapra, *Methods Enzymol.* **305** [42] 2000 (this volume).

[15] S. Brolin and G. Wettermark, "Bioluminescence Analysis," p. 1. VCH, New York, 1992.

[16] M. Toner, *Int. Wildlife* May/June, p. 33 (1994).

The firefly analogy is always useful and popular. Fireflies have a short season and are unavailable in some parts of the country at any time. A short film clip of fireflies in action is incredibly popular. A particularly good video of bioluminescent organisms, including a good section on fireflies, is David Attenborough's "Talking to Strangers," a program in the *Trials of Life* series.[17] From fireflies one can get into a discussion of marine bioluminescence, including the more spectacular jelly fish and giant squid type species well illustrated in the Attenborough video. We usually use the excuse that it is hard to carry the organisms around, so we prefer to work with microscopic ones, the dinoflagellates. It is useful to begin this discussion with marine bioluminescence as observed by sailors and fishermen throughout history: the blue glow of the sea and of the waves.

Dinoflagellates are unicellular, flagellated organisms. They are found throughout the world in both marine and freshwater environments and are responsible for many of the algae blooms known as red tides. Bioluminescent dinoflagellates are found only in marine habitats. There are no known freshwater bioluminescent dinoflagellates. Most, but not all, of the bioluminescent dinoflagellates are photosynthetic. The best known bioluminescent dinoflagellates are *Gonyaulax polyedra, Noctiluca scintillans, Pyrodinium bahamense,* and several species of Pyrocystis.[18] These dinoflagellates can be cultured in the laboratory or classroom. They require a light/dark cycle with 6–8 hr of darkness in a 24-hr period. They also require moderate temperatures, generally between 11.5 and 21.5°. Bioluminescent dinoflagellates are regulated by a circadian rhythm. They produce light most intensely during their night phase. For classroom presentations the dinoflagellates should receive light at "night" and dark by "day," reversing their normal cycle, to maximize light production during normal classroom hours. In dinoflagellates, bioluminescence is produced in response to mechanical stimulation. Light is also produced when weak acid is added to a sample of dinoflagellates.

Closed Ecosystems

We seal our dinoflagellates in low-density polyethylene bags, using a heat sealer. Each bag becomes its own colony. When we visit classrooms we often get questions about how living things can live in a sealed environment, which leads to a wonderful discussion on closed ecosystems. We talk about

[17] David Attenborough, "Talking to Strangers." *The Trials of Life* series, available from Ambrose Video, 1290 Avenue of the Americas, Suite 2245, New York, NY 10104. The last 15 min of the video are on bioluminescence.

[18] F. J. R. Taylor, "Biology of Dinoflagellates." Blackwell Scientific Publications, Palo Alto, CA, 1987.

what living things need to survive and the different requirements for plants and animals. One of the requirements of *P. lunula* is light. Dinoflagellates are considered "masters of photosynthesis."[19,20] They have the ability to harvest light in wavelengths that green plants miss and are able to convert light into stored energy very efficiently.[21] After we discuss the importance of light, we talk about nutrient requirements and how the dinoflagellates use vitamins and minerals from the fortified sea water. The bags we use are semipermeable, allowing the exchange of gases. We also discuss Biosphere2[22] and the closed systems that will be required for long duration space travel.

Division Rates

Cell division can be studied with the bioluminescent dinoflagellates. Students can set up microscopes focusing on a single cell and can watch daily to see the changes as the cell divides. *P. lunula* divides approximately every 7 days. However, at any point in time a colony of dinoflagellates will have cells in all stages of cell division. A fun activity for students is a cell division scavenger hunt, where the students look for, and draw, as many stages of the dinoflagellates' division process as they can find. For students who have never used a microscope, the dinoflagellates make an interesting subject to learn microscope skills and to teach concepts such as scale. Many elementary classrooms do not have microscopes, so you may need to borrow and bring one or more.

Circadian Rhythms

In humans, a biological clock, or circadian rhythm, is responsible for when we get tired, hungry, and why we feel jet lag. In bioluminescent dinoflagellates, the circadian rhythm regulates when the cell should bioluminesce and photosynthesize.[23,23a,24] In most cases, the dinoflagellates will only produce light during the "night." The intensity of continuous light

[19] R. Lewis, Photosynthesis the Plankton Way. *Photonics Spectra,* Sept., p. 46 (1996).

[20] E. Hoffman, P. M. Wrench, F. P. Sharples, R. G. Hiller, W. Welte, and K. Diederichs, "Structural Basis of Light Harvesting by Carotenoids: Peridinin-Chlorophyll-Protein from *Amphidinium carterea,*" *Science* **272,** 1788 (1996).

[21] A. S. Moffat, Form Follows Function When Plants Harvest Light, *Science* **272,** 1743 (1996).

[22] D. A. Sieloff, "Biosphere 2: A World in Our Hands." Bisophere Press, Oracle, Arizona, 1995.

[23] D. Morse, L. Fritz, and J. W. Hastings, "What is the Clock? Transitional Regulation of Circadian Bioluminescence," *Biochem. Physiol. 94 C* 129 (1989).

[23a] T. Roenneberg and W. Taylor, *Methods Enzymol.* **305** [7] 2000 (this volume).

[24] P. Colepicolo, T. Ronneberg, D. Morse, W. Taylor, and J. W. Hastings, Circadian Regulation of Bioluminescence in the Dinoflagellate *Pyrocystis lunula, J. Phycol.* **29,** 173 (1993).

has a large effect on how the natural clock runs. Different colors of light have different effects on dinoflagellates. For example, bright red light makes the clock "tick" slower, lengthening its day. Continuous blue light shortens the phytoplankton's day.[25] You can reset the biological clock in your dinoflagellates so that day is night, manipulating them to glow during class time. To do this, give the dinoflagellates light at night and keep them in the dark during the day. For *P. lunula,* 3 to 4 days are needed for complete resetting of the clock.

Several classroom experiments can focus on circadian rhythms. For example, the physiological changes that occur during the dinoflagellates' night and day phases can be seen by putting one dinoflagellate colony on a regular light–dark cycle where it gets light during the day and darkness at night and a second colony on an alternate cycle where it gets light at night and is in the dark during the day. The class can compare both colonies under the microscope at the same time. Another experiment could be to deprive the dinoflagellates of light completely, give them light continuously, give them varying degrees (1, 2, 5 hr) of light a day or give them different intensities of light and observe the luminescence and physiology of the dinoflagellates over time.

Toxicity/Environment

Bioluminescent dinoflagellates have been used to test sea water for the toxicity of pollutants. Classroom projects involving bioluminescent bacteria can be used as well.[26,27] To perform this experiment, several different household chemicals can be used. Students can bring the chemicals from home or you can provide them. Chemicals can include laundry soap, bleach, vinegar, milk, baking soda, and oil. Before anything is added to the sample, students should be able to observe and roughly quantitate the dinoflagellates' bioluminescence. A small amount of each chemical is then added to the sample of bioluminescent dinoflagellates. It is important to deal with control experiments and basic chemical parameters such as pH. There are, of course, many measures of toxicity. One might be an acute toxicity, in which the presence of the chemicals directly impacts the chemical/biological nature of the dinoflagellates, directly impairing their ability to produce light. Another form of toxicity might be more "chronic" in nature, perhaps

[25] W. J. Cromie, Bioluminescence: Evolution's Novelty. *Harvard Gazette,* Feb. 16, p. 13 (1990).

[26] V. A. Kratasyuk, A. M. Kuznov, and J. I. Gitelson, Bacterial Bioluminescence in Ecological Education, *in* "Bioluminescence and Chemiluminescence: Molecular Reporting with Photons" (J. W. Hastings, L. J. Kricka, and P. E. Stanley, eds.) p. 177. Wiley, New York, 1997.

[27] D. Lapota, D. E. Rosenberger, and D. Duckworth, A Bioluminescent Dinoflagellate Assay for Determining Toxicity of Coastal Waters, "Bioluminescence and Chemiluminescence: Fundamentals and Applied Aspects" (A. K. Campbell, L. J. Kricka, and P. E. Stanley, eds.), p. 156. Wiley, New York, 1994.

impairing the organism's health and well-being over a period of several days. Still another may have to do with the ability of the organism to continue to divide and expand the population. The degree of "toxicity" could be measured by how long the dinoflagellates continue to glow.

The conclusions from such experiments can be related to the effects of pollution on the environment. Photosynthetic dinoflagellates and other marine phytoplankton utilize carbon dioxide from the atmosphere. On a global scale, annual phytoplankton photosynthesis is roughly equivalent to that of plants and trees on land. Organic matter, including phytoplankton, settles slowly to the deep ocean in the form of dead particulate matter. This organic carbon is colonized by bacteria and other microorganisms and slowly degrades to carbon dioxide as it descends to the ocean floor. This whole process creates a carbon sink that serves to concentrate carbon dioxide in the deep sea. If phytoplankton cease to exist, the deep sea reservoir of carbon dioxide would eventually equilibrate with the atmosphere, tripling the atmospheric carbon dioxide and altering the earth's climate through the greenhouse effect.[28] This discussion can lead to the topics of photosynthesis, the ozone hole,[29] the greenhouse effect, and the food web. Photosynthetic dinoflagellates are primary producers.[30] Much of the life in the ocean is dependent on phytoplankton, without which there would be very little life in the world's oceans.[28,31]

By integrating these different concepts students can see that although science is normally separated into small parts in order to be investigated and studied, all the parts are interrelated and interdependent. Bioluminescent dinoflagellates are a valuable tool for teaching integrated science themes and concepts.

Our Experiences[32,33]

When we introduce students to bioluminescence using dinoflagellates, we make sure the room is completely dark. A room with no windows is

[28] S. W. Chisholm, What Limits Phytoplankton Growth? *Oceanus* **35,** 36 (1992).

[29] R. C. Smith, B. B. Prezelin, K. S. Baker, R. R. Bidigare, N. P. Boucher, T. Coley, D. Karentz, S. MacIntyre, H. A. Matlick, D. Menzies, M. Ondrusek, Z. Wan, and K. J. Waters, "Ozone Depletion: Ultraviolet Radiation and Phytoplankton Biology in Antarctic Waters," *Science* **225,** 925 (1992).

[30] K. Steidger, Phytoplankton Ecology: A Conceptual Review Based on Eastern Gulf of Mexico Research. *CRC Crit. Rev. Microbiol.,* Sept., p. 49 (1973).

[31] R. Baum, Phytoplankton Seen as an Atmospheric CO_2 Control. *C & EN,* May 28, p. 7 (1990).

[32] J. D. Andrade, J. Tobler, M. Lisonbee, and D. Min, *in* "Bioluminescence and Chemiluminescence: Status Report" (A. A. Szalay, L. J. Kricka, and P. E. Stanley, eds.), p. 69. Wiley, New York, 1993.

[33] J. D. Andrade, M. Lisonbee, and D. Min, *in* "Bioluminescence and Chemiluminescence: Fundamentals and Applied Aspects" (A. K. Campbell, L. J. Kricka, and P. E. Stanley, eds.), p. 371. Wiley, New York, 1994.

ideal but black paper or aluminum foil over windows and in cracks works well. We allow the students to become dark adapted. While their eyes are adjusting, we talk about how the eyes dark adapt. In some cases, we excite a light stick and pass it around. We explain chemical reactions and how the chemicals in the light stick make cold light. We also use wintergreen lifesavers to introduce the students to triboluminescence.[34] When the students' eyes have adapted and the room is completely black, we gently and unexpectedly toss out the dinoflagellates in sealed bags. Ideally every student catches their own bag. This is when the fun really starts. The questions start to pour out. What is it? Why is the light blue? When we tell them that the contents of the bags are alive, the students want to know what is in the bags, how do they get air? Why do they only make light when they are shaken? Will shaking kill them? We encourage them to formulate hypotheses and to discard hypotheses. After many of the questions are addressed we turn on the lights. The students are usually a bit disappointed when the bags appear to contain only murky water. We usually use a microscope so the students can see the dinoflagellates. This introduction is only the first step. Now the students can do experiments using *their* dinoflagellates. They take the bags home and are encouraged to continue their observations and experiments.

What Others Are Doing

Dr. Mariam Polne-Fuller, at the University of California, Santa Barbara, uses bioluminescent dinoflagellates to enhance lessons for K–12 grades. She has found bioluminescent dinoflagellates to be a wonderful tool for sharing the excitement of research and the fun of learning and teaching. Some of these lessons are

K–3	Light, Invisible Partners, the Scientific Method, and Let Us See the Lights in the Dark Sea
3–6	The Living Cell, Plankton, and the Surprises in Science
7–12	Water Column Migrators, Intracellular Motion, and Do Plants See Light.[35]

Dr. Polne-Fuller runs the Young Marine Scientists[36] program as part of the Marine Science Institute at the University of California, Santa Barbara.

[34] W. L. Dills, Jr., "The Great Wintergreen Candy Experiment," *Science Scope,* Nov./Dec., p. 24 (1992).

[35] M. Polne-Fuller, Personal Correspondence, e-mail: polne@lifesci.ucsb.edu.

[36] Young Marine Scientists -http://research.ucsb.edu/msi/msitexts/YMSP/ymsp.htm

This program provides research opportunities for undergraduates and K–12 teachers and their students. The participants are involved in hands-on activities using practical materials. Through this program, teachers and students are offered direct exposure to working scientists.

Real Scientist is a commercially available kit that utilizes bioluminescent bacteria and CD-ROM technology to provide hands-on science experiences.[37]

Green fluorescent protein (GFP)[38] from the bioluminescent jellyfish *Aequorea victoria* has been used in cell and developmental biology as a fluorescent marker for gene expression.[39,40] These techniques are being developed for high school laboratory use. Kratasyuk *et al.*,[26] at the Russian Academy of Science have used bioluminescence as a tool in ecological education by using bioluminescent bacteria to test toxic substances and media. There is a new coloring book on bioluminescence by Edith Widder (published by the Harbor Branch Oceanographic Institute, 1998), which is an ideal way to reach very young children. There are also several excellent web pages featuring bioluminescence[41] and articles on bioluminescence for science education.[42]

Summary

There is a great need for practicing scientists to volunteer their time and expertise in the K–12th grade science classroom. We have found that bioluminescence is a fun and exciting way to teach basic science concepts and is an excellent tool for the volunteering scientist. We have had very positive reactions from both teachers and students. The excitement of the

[37] P. E. Andreotti, T. Berthold, P. E. Stanley, and F. Berthold, *in* "Bioluminescence and Chemiluminescence: Molecular Reporting with Photons" (J. W. Hastings, L. J. Kricka, and P. E. Stanley, eds.), p. 181. Wiley, New York, 1997. Microlab Systems Corporation, Boca Raton, FL 33428.

[38] Green Fluorescent Protein, http//util.ucsf.edu/sedat/marsh/gfp_gateway.html http://www.wam.umd.edu/~cha/research.htm

[39] M. W. Cutler, D. F. Davis, and W. W. Ward, *in* "Bioluminescence and Chemiluminescence: Fundamentals and Applied Aspects" (A. K. Campbell, L. J. Kricka, and, P. E. Stanley, eds.), p. 383. Wiley, New York, 1994.

[40] D. A. Yernoon, P. V. Reddy, D. F. Davis, and W. W. Ward, *in* "Bioluminescence and Chemiluminescence: Molecular Reporting with Photons" (J. W. Hastings, L. J. Kricka, and P. E. Stanley, eds.), p. 192. Wiley, New York, 1997.

[41] The Bioluminescence Web Page http://lifesci.ucsb.edu/~biolum/http://cdlib.phys.msu.su/GenPhys/Optica/HM/Lec26/biolumin.htm

[42] http://biotech.biology.arizona.edu/primer/readings/bioluminescence.html

students when they first see bioluminescence is contagious. Bioluminescent dinoflagellates are one of the easiest ways to introduce students to this fascinating topic. Many activities and experiments can be done using the bioluminescent dinoflagellates and many students and teachers could benefit from your knowledge and expertise. See you in the classroom![43]

[43] Please share your bioluminescent classroom experiences with Mara, e-mail: pid@msn.com, or Joe, e-mail: joe.andrade@m.ccutah.edu.

[44] Green Fluorescent Protein in Biotechnology Education

By William W. Ward, Gavin C. Swiatek, and Daniel G. Gonzalez

General Introduction

People of all ages seem to have a natural fascination with bioluminescence. Small children delight in catching fireflies on hot summer evenings and older children and adults marvel at the displays of marine bioluminescence ("phosphorescence") seen in many locations throughout the world's oceans. The association of bioluminescence with night time and with remote ocean depths and the relative isolation, from bioluminescence habitats, of industrialized, urbanized people adds to the mystery of biological light. For these reasons, educational tools that involve bioluminescence are capable of motivating and captivating students of all age levels.

Until fairly recently it was difficult for most educators to take full advantage of bioluminescence tools in the classroom. Benthic and pelagic luminescent fish and invertebrates are very difficult to catch and nearly impossible to maintain in aquaria. Fireflies are easy to collect (but only appear in midsummer when many schools are in recess) and the complex life cycle of the insect makes year-round culturing exceedingly difficult. Culturable bioluminescent bacteria and dinoflagellates have provided some access to biological light in the classroom, but only for the more dedicated and industrious teachers. In the last decade or so, access to bioluminescent materials for classroom use has improved dramatically. This access is the result of recombinant DNA technology, which places exotic bioluminescence materials within the reach of all teachers of biotechnology. One such material now in fairly common use in education is the green fluorescent protein (GFP) of *Aequorea victoria*.

Green fluorescent proteins occur naturally in a variety of bioluminescent

0076-6879/00 $30.00

cnidarians (coelenterates) where they act as secondary emitters in calcium-triggered bioluminescence flashes that appear to function in nature by startling predators. The green fluorescent proteins were first discovered as accessory bioluminescence proteins in the jellyfish *A. victoria,*[1] the colonial hydroid *Obelia geniculata,*[2] and the sea pansy *Renilla reniformis.*[3] Their biophysical function is to convert would-be blue bioluminescence (from oxyluciferin) into green light emission[4,5] ostensibly to enhance the transmittance of emitted light through chlorophyll-laden estuarine waters. The GFPs trap excitation energy from excited-state oxyluciferin (λ max = 480 nm) and, by a radiationless mechanism,[5] emit light spectrally identical to their own fluorescence emissions (λ max = 509 nm).

Classroom Applications

The first documented educational application for GFP[6] was our $5\frac{1}{2}$-day-long intensive short course "Protein Purification: Isolation, Analysis, and Characterization of GFP," offered for the first time in 1989 by the Cook College Office of Continuing Professional Education (OCPE) and now presented six times annually by the Center for Research and Education in Bioluminescence and Biotechnology (CREBB) at Rutgers University. Experiments created by CREBB are also used in the Cook College undergraduate courses "Introductory Biochemistry Laboratory" and "Problem Solving in Biochemistry." During the initial 6-year history of "Protein Purification," source material was the excised marginal rings (50–100 per three student group) of the jellyfish *A. victoria,* animals that had been individually collected at the University of Washington's marine facility, Friday Harbor Laboratories, at Friday Harbor, Washington. In those years, the relative inaccessibility of the starting material restricted potential teaching applications to those persons doing primary research on GFP (probably six or eight laboratories worldwide) or those willing to make jellyfish-collecting trips to Washington State. Thus, in those years, only James Slock and colleagues at Kings College, Pennsylvania, developed their own GFP-based undergraduate laboratory courses and exercises.[7]

[1] O. Shimomura, F. H. Johnson, and Y. Saiga, *J. Cell. Comp. Physiol.* **59,** 223 (1962).

[2] J. G. Morin and J. W. Hastings, *J. Cell. Physiol.* **77,** 313 (1971).

[3] J. E. Wampler, K. Hori, J. Lee, and M. J. Cormier, *Biochemistry* **10,** 2903 (1971).

[4] J. G. Morin, *in* "Coelenterate Biology" (L. Muscatine and H. Lenhoff, eds.), p. 397. Acaemic Press, New York, 1974.

[5] W. W. Ward, *Photochem. Photobiol. Rev.* **4,** 1 (1979).

[6] M. W. Cutler, D. F. Davis, and W. W. Ward, *in* "Bioluminescence and Chemiluminescence: Fundamentals and Applied Aspects" (A. K. Campbell, L. J. Kricka, and P. E. Stanley, eds.), p. 383. Wiley, New York, 1994.

[7] J. A. Slock, "International Symposium on GFP," p. 13. Abst., New Brunswick, NJ, 1997.

The situation changed dramatically in 1992 when the gene for *A. victoria* GFP was sequenced and cloned, [8] in 1993 when the chromogenic hexapeptide was characterized,[9] and in 1994 when a functional, fluorescent gene product was expressed in *Escherichia coli* and *Caenorhabditis elegans.*[10] The success with cloning has made GFP widely available for laboratory exercises in biotechnology education. Thus, in the mid-1990s, an existing Rutgers continuing education course in "Recombinant DNA Technology" was redesigned to incorporate GFP as the experimental centerpiece.[11] At the same time the "Protein Purification" course offered by CREBB switched over its source material from jellyfish GFP to recombinant GFP. Other college-level GFP-based courses and biotechnology exercises have now sprung up at New Jersey's Middlesex County College,[12] Wilkes University, Pennsylvania,[13] and the Medical College of Wisconsin.[14] Bio-Rad Laboratories, in cooperation with many teachers and scientists, has created a series of self-contained biotechnology laboratory kits, the "Biotechnology Explorer" series, for high schools and postsecondary institutions. Several of these kits utilize an arabinose-induced *E. coli* plasmid (Bio-Rad's pGLO system) to illustrate molecular biology techniques with GFP or to generate sufficient levels of GFP for *in vitro* protein chemistry experiments. The kits may be used as designed or they may serve as the material basis for a variety of custom-designed exercises.

In 1998, CREBB offered for the first time two new hands-on short courses, which, like the "Protein Purification" course, utilize recombinant GFP throughout. The first of these, "Affinity Chromatography of Recombinant Proteins," offered in cooperation with QIAGEN, introduces three hands-on laboratory exercises: purification of histidine-tagged GFP from an *E. coli* crude extract by Ni-NTA-immobilized metal ion affinity chromatography, purification of anti-GFP polyclonal antibodies on a column of immobilized GFP, and substrate affinity chromatography of an enzyme attached covalently to GFP by gene fusion. A second hands-on laboratory

[8] D. C. Prasher, V. K. Eckenrode, W. W. Ward, F. G. Prendergast, and M. J. Cormier, *Gene* **III,** 229 (1992).

[9] C. W. Cody, D. C. Prasher, W. M. Westler, F. G. Prendergast, and W. W. Ward, *Biochemistry* **32,** 1212 (1993).

[10] M. Chalfie, Y. Tu, G. Euskirchen, W. W. Ward, and D. C. Prasher, *Science* **263,** 802 (1994).

[11] D. A. Yernool, P. V. Reddy, D. F. Davis, and W. W. Ward, *in* "Bioluminescence and Chemiluminescence: Molecular Reporting with Photons" (J. W. Hastings, L. J. Kricka, and P. E. Stanley, eds.), p. 192. Wiley, New York, 1997.

[12] C. S. Mintz, H. E. Fisher, and W. W. Ward, "International Symposium on GFP," p. 13. Abst., New Brunswick, NJ, 1997.

[13] M. Donahue, B. Weidlich, G. Milevich, and W. B. Terzaghi, "International Symposium on GFP," p. 13. Abst., New Brunswick, NJ, 1997.

[14] T. Herman, "International Symposium on GFP," p. 14. Abst., New Brunswick, NJ, 1997.

course, "HPLC-MS Workshop: A Problem Solving Approach," taught in partnership with ThermoQuest Institute, teaches basic and advanced topics of high-performance liquid chromatography and mass spectrometry using GFP, throughout, in a problem-solving setting.

CREBB courses are taken by industrial laboratory scientists and by community college and high school teachers and advanced students. The dominant feature of each of these courses is the central role played by GFP, which serves as a motivating tool and as a source of stimulation and curiosity for all who see its brilliant fluorescence. Unlike most bioluminescence exercises that rely on a brief luciferin-luciferase flash reaction and that may require near total darkness for proper visualization, GFP can be detected continuously by eye in a fully lighted room with a simple handheld mineral light (long wave ultraviolet light with λ max = 365 nm). The only criticism of GFP as an educational tool seems to be that it is simply too perfect; course participants bemoan the fact that when a course ends they return to their "ordinary," invisible proteins.

Immobilized Metal Affinity Chromatography of GFP

The GFP-based CREBB course in "Protein Purification," an overview of which has been published,[6] has been modified to include an exercise in immobilized metal ion affinity chromatography (IMAC). The NSS1 variant of wild-type recombinant GFP[15] in the pBAD vector under arabinose control was modified genetically by the C-terminal addition of six histidine residues following a four amino acid flexible linker. This variant (F64C), with a phenylalanine to cysteine substitution at position 64, has the unique property, among described variants of GFP, of "greening" (chromophore maturation) in the near absence of molecular oxygen. Thus, *E. coli* cultures carrying the NSS1 plasmid may be sealed (for safety and convenience) without losing their ability to form mature, fluorescent GFP molecules.

Escherichia coli cells carrying the NSS1 vector are grown at 37° in 100 ml of LB medium to late log phase ($A_{660} \geq 1.0$) under ampicillin (0.1 mg/ml) selection in the presence of 0.2% L-arabinose. Cells are harvested by centrifugation (10,000*g*, 10 min) and lysed by the addition of lysozyme (1 mg/ml) at pH 6.0 with repeated cycles of freezing and thawing.[16] The crude lysate is clarified by centrifugation (20,000*g*, 20 min) and then taken up in a pH 8.0 buffer consisting of 50 m*M* sodium phosphate and 100 m*M* sodium

[15] H. E. Fisher, A. Sawyer, and W. W. Ward, "International Symposium on GFP," p. 39. Abst., New Brunswick, NJ, 1997.

[16] B. H. Johnson and M. H. Hecht, *Bio/Technology* **12,** 1357 (1994).

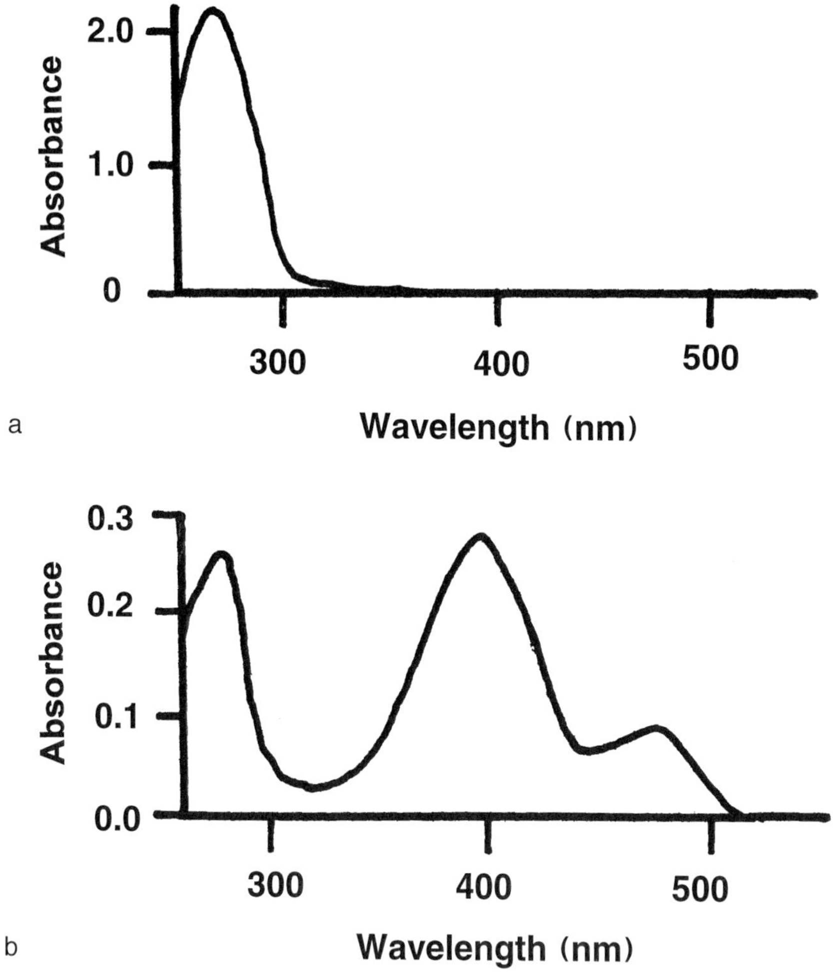

FIG. 1. (a) Absorption spectrum of an 85-ml sample of *E. coli* crude extract prior to IMAC. Spectrum recorded from 250 to 550 nm. (b) Absorption spectrum of a 0.6-ml sample of an IMAC-purified *E. coli* extract recorded from 260 to 550 nm. A calculated absorption ratio (A_{398}/A_{279}) of 1.06 for this sample is indicative of purity in excess of 90%.

chloride with 0.02% sodium azide (loading buffer). This extract, totaling 85 ml, contains 165 mg of total protein, as estimated by absorbance at 280 nm, and 0.3 mg of GFP, as determined by calibrated fluorometry using a GFP standard solution as described by González and Ward.[17] An absorption

[17] D. G. González and W. W. Ward, *Methods Enzymol.* **305** [16] 2000 (this volume).

spectrum of the centrifuged crude extract is shown in Fig. 1a. At a flow rate of 2 ml/min, the sample was applied to a 7 × 45-mm column (2.0 ml) of Clontech Talon resin with immobilized cobalt ions, preequilibrated with loading buffer. Virtually the entire 165 mg of total protein was accounted for in the first 105 ml of column wash through. A 10 m*M* imidazole wash in the same loading buffer released an additional 0.2 mg of bound protein. Further washing with 100 m*M* imidazole in loading buffer released highly purified GFP with no detectable contaminants. In this single purification step, which consumed about 2 hr of running time, the sample was purified 540-fold and concentrated 47-fold. The column elution profile is shown in Fig. 2 and an absorption spectrum of the pure GFP appears in Fig. 1b.

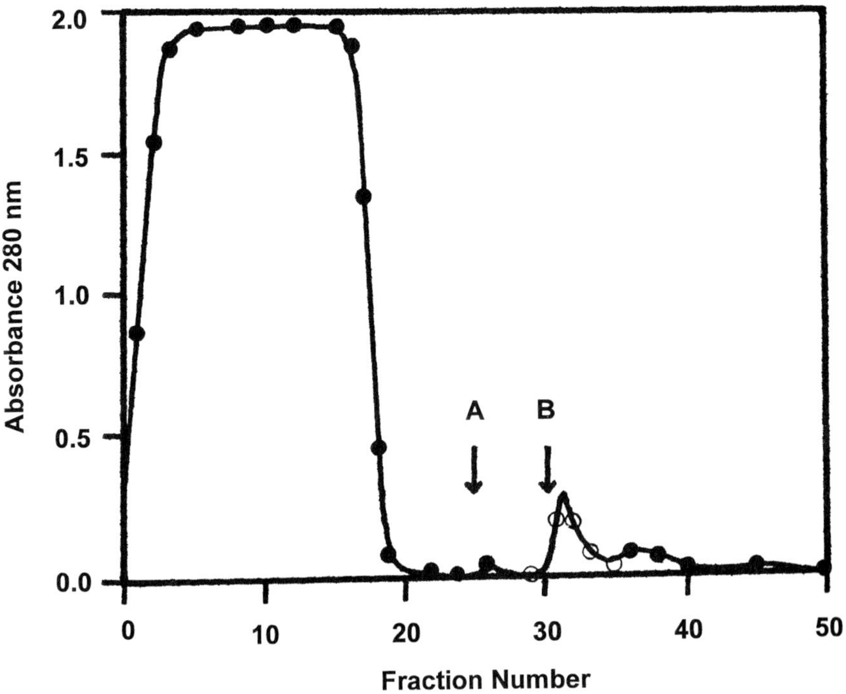

FIG. 2. Elution profile of an 85-ml sample of *E. coli* crude extract on a Clontech Talon IMAC column. Column dimensions: 7 × 45 mm (2.0 ml). Sample loaded and eluted in 50 m*M*, pH 8.0, sodium phosphate buffer containing 100 m*M* sodium chloride and 0.02% sodium azide. At the arrow labeled A, a pulse of 10 m*M* imidazole in loading buffer was initiated. At the arrow labeled B, a second pulse of 100 m*M* imidazole in loading buffer was initiated. Also, at arrow B, the fraction size was reduced from 5.25 ml (fractions 1 through 29) to 0.6 ml (fractions 30 through 50). Solid circles represent absorbance at 280. Open circles represent fluorescence measurements with excitation at 365 nm and emission at 510 nm.

Qualitatively similar IMAC purification of His_6-tagged GFP from crude bacterial lysates on Clontech Talon or QIAGEN Ni-NTA resins has been achieved under a variety of chromatographic conditions. Results similar to those reported previously have been observed at pH 7.0, 8.0, and 8.8. We have used phosphate or Tris–HCl buffers with or without added sodium chloride and we have eluted the columns with linear gradients of imidazole or with stepwise additions as shown in Fig. 2. Linear flow rates as high as 5 cm/min have been utilized with no apparent loss of column resolution.

Conclusion

In many years of college level teaching, we have not seen a more effective teaching tool than GFP. Now that GFP is so well characterized and so readily available, its use in college and precollege instruction is bound to expand. We can envision, for example, a variety of GFP-based laboratory courses at the college level in such areas as molecular biology, biochemistry, genetics, cell biology, and biophysics. Appendix I describes some of the properties of GFP that can be utilized in laboratory instruction. Appendix II lists some of the major companies that market GFP-based products. For further information, consult the interesting websites listed in Appendix III.

Acknowledgments

The authors acknowledge past instructors in "Protein Purification," especially Dr. Mark Cutler who wrote the laboratory manual and Dr. Diane Davis who has been involved in Rutgers continuing education courses for more than a decade. This work is supported by the Advanced Technological Education program of NSF (DUE #96-02356).

Appendix I: Properties of GFP

Visualized easily with a hand-held UV lamp (λ max = 365 nm)
Stable in *E. coli* cells for months and in solution for years
Cultured and maintained easily; tolerated well by *E. coli* and other organisms
Available in multiple color mutants: blue, green, and yellow emitters
Low molecular weight (27,000 monomer)
Dimerizes reversibly; dissociation constant varies with mutant form
Cloned into many organisms, including yeast, *Arabidopsis,* and *Drosophila*
High culture yields: up to 5 mg GFP per 25-ml *E. coli* culture

C-term and N-term both accept "histidine tags"
His-tagged GFP can be purified to >95% by IMAC
Pure protein, polyclonal antibodies, and monoclonals available commercially
Plasmids, genes, cultures, and kits available commercially
Rate of chromophore maturation varies with mutant strain
Oxygen required for chromophore maturation ("greening")
A nearly oxygen-independent variant is available (NSS1 variant)
C-term truncation mutants fluoresce (up to 10 amino acid deletion)
Denatured protein (nonfluorescent) retains a yellow chromophore
Denatured protein renatures easily and rapidly
Fluorescence unaffected by formaldehyde treatment
Fluorescence is stable over wide pH range (pH 5 to pH 12)
Molecular structure is known; computer exercises available[14]
Easily related to "real-world" uses: marker for gene induction, metastasis indicator, drug discovery marker

Appendix II: Companies That Provide GFP Clones, Vectors, Fusion Protein Kits, Purified Proteins, and Antibodies

Aurora Biosciences
BABCO (Berkeley Antibody Co.)
Bio-Rad Laboratories
Chemicon International
Clontech Laboratories
Invitrogen
Life Technologies
Molecular Probes
Novus Molecular
Packard Instruments
Panvera
Pharmingen
Quantum Biotechnologies, Inc
Shimadzu (fluorometers)
Turner Designs (fluorometers)

Appendix III: Interesting GFP Websites

http://www.rci.rutgers.edu/~meton/protein.html
http://www.wam.umd.edu/~cha/gfp.htm
http://www.mc.vanderbilt.edu/vumcdept/mpb/piston/files/gfp.html
http://www.maxgen.com/nf_gfppaper.html

http://www.rochester.edu/College/BIO/olmstedlab/olmstedhp.html
http://www.uncg.edu/%7Ecnstewar/gfp.htm
http://www.image1.com/products/metagfp/index.html
http://www.bioc.rice.edu/Bioch/Phillips/Papers/gfpbio.html
http://www.qbi.com/Products/autofluorescent.asp
http://www.biorad.com/889168.html
http://www.clontech.com/clontech/Manuals/GFP/TOC.html

Author Index

Numbers in parentheses are footnote reference numbers and indicate that an author's work is referred to although the name is not cited in the text.

A

C

D

E

F

G

H

I

J

K

L

M

N

O

P

Q

R

S

T

U

V

W

X

Y

Z

Subject Index

A

B

D

E

F

G

M

N

O

Q

U

V

W

X

Y

ISBN 0-12-182206-0
90038
9 780121 822064